SMART AND FLEXIBLE ENERGY DEVICES

SMART AND FLEXIBLE ENERGY DEVICES

Edited by
Ram K. Gupta and Tuan Anh Nguyen

CRC Press is an imprint of the
Taylor & Francis Group, an **informa** business

First edition published 2022
by CRC Press
6000 Broken Sound Parkway NW, Suite 300, Boca Raton, FL 33487-2742

and by CRC Press
4 Park Square, Milton Park, Abingdon, Oxon, OX14 4RN

CRC Press is an imprint of Taylor & Francis Group, LLC

Library of Congress Cataloging-in-Publication Data
Names: Gupta, Ram K., editor. | Nguyen, Tuan Anh (Chemist), editor.
Title: Smart and flexible energy devices / edited by Ram K. Gupta, Tuan Anh Nguyen.
Description: First edition. | Boca Raton, FL : CRC Press, 2022. | Includes bibliographical references and index.
Identifiers: LCCN 2021043801 (print) | LCCN 2021043802 (ebook) | ISBN 9781032033242 (hbk) | ISBN 9781032033266 (pbk) | ISBN 9781003186755 (ebk)
Subjects: LCSH: Electric batteries. | Solar cells. | Fuel cells. | Flexible electronics.
Classification: LCC TK2896 .S585 2022 (print) | LCC TK2896 (ebook) | DDC 621.31/242--dc23/eng/20211110
LC record available at https://lccn.loc.gov/2021043801
LC ebook record available at https://lccn.loc.gov/2021043802

ISBN: 978-1-032-03324-2 (hbk)
ISBN: 978-1-032-03326-6 (pbk)
ISBN: 978-1-003-18675-5 (ebk)

DOI: 10.1201/9781003186755

Typeset in Times
by MPS Limited, Dehradun

Contents

Preface ix
Editors xi
Contributors xiii

Chapter 1 Smart and Flexible Energy Devices: Principles, Advances, and Opportunities 1

   Tenzin Ingsel and Ram K. Gupta

Chapter 2 Innovation in Materials and Design for Flexible Energy Devices 23

   Ayushi Katariya and Jyoti Rani

Chapter 3 Basics and Architectural Aspects of Flexible Energy Devices 43

   T. Bendib, M. A. Abdi, H. Bencherif, and F. Meddour

Chapter 4 Characterization Techniques of Flexible Energy Devices 59

   Endersh Soni, Ujjwal K. Prajapati, Ayushi Katariya, and Jyoti Rani

Chapter 5 Micro- and Nanofibers-Based Flexible Energy Devices 81

   Arunima Reghunadhan, Jiji Abraham, and P. S. Sari

Chapter 6 3D Printed Flexible Energy Devices 99

   Dongxu He and Zheling Li

Chapter 7 Environmental Impact of Flexible Energy Devices 119

   H. Bencherif, A. Meddour, T. Bendib, and M. A. Abdi

Chapter 8 Metal Oxide-Based Catalysts for Flexible and Portable Fuel Cells: Current Status and Future Prospects 133

   Bincy George Abraham, S. Vinod Selvaganesh, Raghuram Chetty, and P. Dhanasekaran

Chapter 9 Flexible Fuel Cells Based on Microbes 157

   Hamide Ehtesabi

Chapter 10 Flexible Silicon Photovoltaic Solar Cells 171

   Pratik Deorao Shende, Krishna Nama Manjunatha, Iulia Salarou, and Shashi Paul

Chapter 11 Flexible Solar Cells Based on Metal Oxides 197

   Soner Çakar, Mahmut Özacar, and Fehim Findik

Chapter 12 Inorganic Materials for Flexible Solar Cells.. 211

    Mozhgan Hosseinnezhad and Zahra Ranjbar

Chapter 13 Efficient Metal Oxide-Based Flexible Perovskite Solar Cells.. 227

    Subhash Chander and Surya Kant Tripathi

Chapter 14 Flexible Solar Cells Based on Chalcogenides... 241

    Kulwinder Kaur, Nisika, and Mukesh Kumar

Chapter 15 Perovskite-based Flexible Solar Cells 261

    Rushi Jani and Kshitij Bhargava

Chapter 16 Quantum Dots Based Flexible Solar Cells.. 285

    Sandeep Kumar, Prashant Kumar, Arup Mahapatra, and Basudev Pradhan

Chapter 17 A Method of Strategic Evaluation for Perovskite-Based Flexible Solar Cells.............. 307

    Figen Balo and Lutfu S. Sua

Chapter 18 Flexible Batteries Based on Li-ion.. 323

    Solen Kinayyigit and Emre Bicer

Chapter 19 Flexible Batteries Based on Na-ion... 345

    Jun Mei

Chapter 20 Flexible Batteries Based on K-ion.. 359

    Yu Liu, Yuzhen Sun, Zhiyuan Zhao, Xiaowei Yang, and Rong Xing

Chapter 21 Flexible Batteries Based on Zn-ion 375

    Haobo Dong, Ivan P. Parkin, and Guanjie He

Chapter 22 Fabrication Techniques for Wearable Batteries 397

    Ifra Marriam, Hiran Chathuranga, Cheng Yan, Mike Tebyetekerwa, and Shengyuan Yang

Chapter 23 Carbon-Based Advanced Flexible Supercapacitors.. 417

    Anuj Kumar and Ram K. Gupta

Chapter 24 2D Materials for Flexible Supercapacitors.. 441

    Wei Ni and Lingying Shi

Chapter 25 Flexible Supercapacitors Based on Metal Oxides.. 461

    Haradhan Kolya, Chun-Won Kang, Sarbaranjan Paria, Subhadip Mondal, and Changwoon Nah

Chapter 26 Recent Advances in Transition Metal Chalcogenides for Flexible Supercapacitors 481

Somoprova Halder, Souhardya Bera, and Subhasis Roy

Chapter 27 MOFs-Derived Metal Oxides-Based Compounds for Flexible Supercapacitors 501

Charu Goyal, Anuj Kumar, Ghulam Yasin, and Ram K. Gupta

Chapter 28 Textile-Based Flexible Supercapacitors 519

Yasin Altin and Ayse Bedeloglu

Chapter 29 Current Development and Challenges in Textile-Based Flexible Supercapacitors 539

Sunny R. Gurav, Rajendra G. Sonkawade, Maqsood R. Waikar, Akash S. Rasal, and Rakesh K. Sonker

Chapter 30 Flexible Supercapacitors Based on Nanocomposites 551

Fang Cheng, Xiaoping Yang, Wen Lu, and Liming Dai

Chapter 31 Textile-Based Flexible Nanogenerators 575

Ünsal Ömer Faruk Çelik Bedeloğlu Ayşe, and Borazan İsmail

Index 589

Preface

Sustainable energy can be produced via fuel cells and solar cells, while produced energy can be successfully stored in batteries and supercapacitors. The novelty of a flexible device is the incorporation of a flexible electrode or substrate material to combine structural flexibility without harming the energy generation or storage capacity. The formation of such substrates is challenging due to morphologies' inhomogeneity and dimensional stability on bending. Other challenges on the materials side are electrical, electronic, optical, and mechanical properties which should be transparent (in the case of solar cells), lightweight, and electrochemically stable under various conditions. The architectural aspects of the devices play a crucial role in industrialization. For example, proton-exchange membrane fuel cells (PEMFCs) provide high-energy density; however, traditional PEMFCs are usually too heavy, rigid, and bulky to be used in flexible devices.

Electronics for flexible device applications require reliable manufacturing to develop suitable materials, innovative technology, and tools to carry advanced research. Efficient, flexible, and mechanical robust energy devices presently pose as a bottleneck to the fast-growing wearable flexible technologies. The main purpose of this book is to provide current state-of-the-art development in the field of smart and flexible devices that will be attractive to students, researchers, engineers, and scientists working in the area of materials development, technology, and advanced energy devices. This book provides detailed information about smart and flexible devices, their working principle, materials, and challenges for further advancement. Basic and architectural aspects of flexible devices and various methods used to characterize and test are covered. Recent developments in smart devices using additive manufacturing are discussed. Flexible fuel cells, as well as photovoltaic devices using nanomaterials, are covered to provide insight into the current development and future aspects of these devices. The flexible batteries and supercapacitor market is expected to increase exponentially due to their wide applications in wearable and smart electronics. Detail about materials, methodology, and emerging technology in the development of flexible metal-ion batteries and supercapacitors are covered in this book.

Ram K. Gupta, Ph.D.
Associate Professor
Department of Chemistry
Kansas Polymer Research Center
Pittsburg State University
Pittsburg, Kansas, United States

Tuan Anh Nguyen, Ph.D.
Institute for Tropical Technology
Vietnam Academy of Science and
Technology
Hanoi, Vietnam

Editors

Dr. Ram Gupta is an associate professor at Pittsburg State University. Dr. Gupta's research focuses on conducting polymers and composites, green energy production and storage using biowastes and nanomaterials, optoelectronics and photovoltaics devices, organic-inorganic hetero-junctions for sensors, bio-based polymers, flame-retardant polymers, bio-compatible nanofibers for tissue regeneration, scaffold and antibacterial applications, corrosion inhibiting coatings, and bio-degradable metallic implants. Dr. Gupta has published over 230 peer-reviewed articles, made over 300 national, international, and regional presentations, chaired many sessions at national/international meetings, edited many books, and written several book chapters. He has received over two and a half million dollars for research and educational activities from many funding agencies. He is serving as editor-in-chief, associate editor, and editorial board member of numerous journals.

Dr. Tuan Anh Nguyen has completed his BSc in Physics from Hanoi University in 1992, and his Ph.D. in Chemistry from Paris Diderot University (France) in 2003. He was a visiting scientist at Seoul National University (South Korea, 2004) and the University of Wollongong (Australia, 2005). He then worked as a postdoctoral research associate and research scientist at Montana State University (USA), 2006-2009. In 2012, he was appointed as head of Microanalysis Department at the Institute for Tropical Technology (Vietnam Academy of Science and Technology). He has managed 4 Ph.D. theses as thesis director and 3 are in progress; He is editor-in-chief of "Kenkyu Journal of Nanotechnology & Nanoscience" and founding co-editor-in-chief of "Current Nanotoxicity & Prevention". He is the author of 4 Vietnamese books and editor of 32 Elsevier books in the Micro & Nano Technologies Series.

Contributors

M. A. Abdi
Higher National School of Renewable Energy,
    Environment & Sustainable Development
Batna, Algeria
and
LEPCM
University of Batna 1
Batna, Algeria
and
LEPCM laboratory
Department of Physics
University of Batna 1
Batna, Algeria

Bincy George Abraham
Department of Chemical Engineering
Indian Institute of Technology Madras
Chennai, Tamil Nadu, India

Jiji Abraham
Department of Chemistry
Vimala College
Thrissur, Kerala, India

Yasin Altin
Department of Polymer Materials Engineering
Bursa Technical University
Bursa, Turkey
and
Department of Chemistry
Ordu University
Ordu, Turkey

Figen Balo
Department of Industrial Engineering
Firat, University
Elâzığ, Turkey

Ayse Bedeloglu
Department of Polymer Materials Engineering
Bursa Technical University
Bursa, Turkey

Çelik Bedeloğlu
Polymer Materials Engineering Department
Bursa Technical University
Bursa, Turkey

H. Bencherif
University of Mostefa Benboulaid Batna 2
Batna, Algeria

T. Bendib
Higher National School of Renewable Energy,
    Environment & Sustainable Development
Batna, Algeria
and
LEPCM
University of Batna 1
Batna, Algeria
and
LEPCM laboratory
Department of Physics
University of Batna 1
Batna, Algeria

Souhardya Bera
Department of Chemical Engineering
University of Calcutta
Kolkata, India

Kshitij Bhargava
Department of Electrical and Computer Science
    Engineering
Institute of Infrastructure, Technology,
    Research and Management
Ahmedabad, Gujarat, India

Emre Bicer
Battery Research Laboratory
Faculty of Engineering and Natural Sciences
Sivas University of Science and Technology
Sivas, Turkey

Soner Çakar
Zonguldak Bulent Ecevit University
Department of Chemistry
Zonguldak, Turkey
and
Biomaterials, Energy, Photocatalysis, Enzyme
    Technology, Nano & Advanced Materials, Additive
    Manufacturing, Environmental Applications and
    Sustainability Research & Development Group
Sakarya University
Sakarya, Turkey

Subhash Chander
Centre for Advanced Study in Physics
Department of Physics
Punjab University Chandigarh
Chandigarh, India
and
Department of Chemical Sciences
Indian Institute of Science Education and
    Research Mohali
Mohali, India

Hiran Chathuranga
School of Mechanical, Medical and Process
    Engineering, Faculty of Engineering
Queensland University of Technology
Brisbane, Queensland, Australia

Fang Cheng
College of Chemical Science and Engineering
Yunnan University
Kunming, China
and
Institute of Energy Storage Technologies
Yunnan University
Kunming, China

Raghuram Chetty
Department of Chemical Engineering
Indian Institute of Technology Madras
Chennai, Tamil Nadu, India

Liming Dai
School of Chemical Engineering
University of New South Wales
Sydney, NSW, Australia

P. Dhanasekaran
CSIR-Central Electrochemical Research
    Institute-Madras Unit
CSIR Madras Complex
Chennai, Tamil Nadu, India

Haobo Dong
Christopher Ingold Laboratory
Department of Chemistry
University College London
London, United Kingdom

Hamide Ehtesabi
Faculty of Life Sciences and Biotechnology
Shahid Beheshti University
Tehran, Iran

Ömer Faruk
Polymer Materials Engineering Department
Bursa Technical University
Bursa, Turkey

Fehim Findik
Biomaterials, Energy, Photocatalysis, Enzyme
    Technology, Nano & Advanced Materials,
    Additive Manufacturing, Environmental
    Applications and Sustainability Research &
    Development Group
Sakarya University
Sakarya, Turkey
and
Faculty of Technology, Department of
    Metallurgical and Materials Engineering
Sakarya University of Applied Science
Sakarya, Turkey

Charu Goyal
Department of Chemistry
GLA University
Mathura, Uttar Pradesh, India

Ram K. Gupta
Department of Chemistry
Kansas Polymer Research Center
Pittsburg State University
Pittsburg, Kansas, USA

Sunny R. Gurav
Department of Physics
Shivaji University
Kolhapur Maharashtra, India

Somoprova Halder
Department of Chemical Engineering
University of Calcutta
Kolkata, India

Guanjie He
Christopher Ingold Laboratory
Department of Chemistry
University College London
London, United Kingdom
and
School of Chemistry
University of Lincoln
Brayford Pool, Lincoln, United Kingdom

Dongxu He
College of Materials and Chemistry
    & Chemical Engineering
Chengdu University of Technology
Chengdu, China

Mozhgan Hosseinnezhad
Department of Organic Chemistry
Institute for Color Science and Technology
Tehran, Iran

Tenzin Ingsel
Department of Chemistry
Kansas Polymer Research Center
Pittsburg State University
Pittsburg, Kansas, USA

Borazan, İsmail
Bartın University
Bartin, Turkey
and
Polymer Materials Engineering Department
Bursa Technical University
Bursa, Turkey

Rushi Jani
Department of Electrical and Computer Science
    Engineering
Institute of Infrastructure, Technology, Research
    and Management
Ahmedabad, Gujarat, India

Chun-Won Kang
Department of Housing Environmental Design,
    and Research Institute of Human Ecology
College of Human Ecology
Jeonbuk National University
Jeonju, Jeonbuk, Republic of Korea

Ayushi Katariya
Department of Physics
Maulana Azad National Institute of Technology
Bhopal, Madhya Pradesh, India

Kulwinder Kaur
Functional and Renewable Energy Materials
    Laboratory
Indian Institute of Technology Ropar
Punjab, India

Solen Kinayyigit
Nanocatalysis and Clean Energy Technologies
    Laboratory
Institute of Nanotechnology
Gebze Technical University
Gebze, Kocaeli, Turkey

Haradhan Kolya
Department of Housing Environmental Design,
    and Research Institute of Human Ecology
College of Human Ecology
Jeonbuk National University
Jeonju, Jeonbuk, Republic of Korea

Anuj Kumar
Nano-Technology Research Laboratory
Department of Chemistry
GLA University
Mathura, Uttar Pradesh, India

Mukesh Kumar
Functional and Renewable Energy Materials
    Laboratory
Indian Institute of Technology Ropar
Punjab, India

Prashant Kumar
Department of Energy Engineering
Central University of Jharkhand
Brambe, Ranchi, Jharkhand, India

Sandeep Kumar
Department of Energy Engineering
Central University of Jharkhand
Brambe, Ranchi, Jharkhand, India

Yu Liu
Institute of New Energy on Chemical Storage and
    Power Sciences
Yancheng Teachers University
Yancheng, Jiangsu, China
and
Department of Chemical Engineering and Waterloo
    Institute for Nanotechnology
University of Waterloo
Waterloo, Ontario, Canada

Wen Lu
College of Chemical Science and Engineering
Yunnan University
Kunming, China
and
Institute of Energy Storage Technologies
Yunnan University
Kunming, China

Arup Mahapatra
Department of Energy Engineering
Central University of Jharkhand
Brambe, Ranchi, Jharkhand, India
and
Centre of Excellence in Green and Efficient
    Energy Technology
Central University of Jharkhand
Brambe, Ranchi, Jharkhand, India

Krishna Nama Manjunatha
Emerging Technologies Research Center
De Montfort University
The Gateway Leicester, United Kingdom

Ifra Marriam
School of Mechanical
Medical and Process Engineering, Faculty of
    Engineering
Queensland University of Technology
Brisbane, Queensland, Australia

F. Meddour
Higher National School of Renewable Energy
    Environment & Sustainable Development
Batna, Algeria
and
LAAAS
Department of Electronics
University of Batna 2
Batna, Algeria

A. Meddour
University Mostefa Benboulaid Batna
Batna, Algeria

Jun Mei
School of Chemistry and Physics
Queensland University of Technology
Brisbane, Queensland, Australia
and
Centre for Materials Science
Queensland University of Technology
Brisbane, Queensland, Australia

Subhadip Mondal
Department of Polymer-Nano Science and
    Technology
Jeonbuk National University
Jeonju, Jeonbuk, Republic of Korea

Changwoon Nah
Department of Polymer-Nano Science and
    Technology
Jeonbuk National University
Jeonju, Jeonbuk, Republic of Korea

Wei Ni
State Key Laboratory of Vanadium and Titanium
  Resources Comprehensive Utilization
ANSTEEL Research Institute of Vanadium &
  Titanium (Iron & Steel)
Chengdu, China
and
Material Corrosion and Protection Key Laboratory
  of Sichuan Province
Sichuan University of Science and Engineering
Zigong, China
and
Vanadium and Titanium Resource Comprehensive
  Utilization Key Laboratory of Sichuan Province
Panzhihua University
Panzhihua, China

Nisika
Functional and Renewable Energy Materials
  Laboratory
Indian Institute of Technology Ropar
Punjab, India

Mahmut Özacar
Faculty of Science & Arts, Department of Chemistry
Sakarya University
Sakarya, Turkey
and
Biomaterials, Energy, Photocatalysis, Enzyme
  Technology, Nano & Advanced Materials,
  Additive Manufacturing, Environmental
  Applications and Sustainability Research &
  Development Group
Sakarya University
Sakarya, Turkey

Sarbaranjan Paria
Department of Polymer-Nano Science and
  Technology
Jeonbuk National University
Jeonju, Jeonbuk, Republic of Korea

Ivan P. Parkin
Christopher Ingold Laboratory
Department of Chemistry
University College London
London, United Kingdom

Shashi Paul
Emerging Technologies Research Center
De Montfort University
The Gateway Leicester, United Kingdom

Basudev Pradhan
Department of Energy Engineering
Central University of Jharkhand
Brambe, Ranchi, Jharkhand, India
and
Centre of Excellence in Green and Efficient
  Energy Technology
Central University of Jharkhand
Brambe Ranchi, Jharkhand, India

Ujjwal K. Prajapati
Department of Physics
Maulana Azad National Institute of Technology
Bhopal, M.P., India

Sari P.S.
Department of Polymer Science and Rubber
  Technology
CUSAT, Kochi

Jyoti Rani
Department of Physics
Maulana Azad National Institute of Technology
Bhopal, M.P., India

Zahra Ranjbar
Department of Surface Coatings and Novel
  Technologies-Institute for Color Science and
  Technology
Tehran, Iran
and
Center of Excellence for Color Science and
  Technology
Tehran, Iran

Akash S. Rasal
Department of Chemical Engineering
National Taiwan University of
  Science and Technology
Taipei, Taiwan

Arunima Reghunadhan
Department of Chemistry
Milad-E-Sherif Memorial College Kayamkulam
Alappuzha, Kerala, India
and
School of Energy Materials
Mahatma Gandhi University
Kottayam, Kerala, India

Subhasis Roy
Department of Chemical Engineering
University of Calcutta
Kolkata, India

Iulia Salarou
Emerging Technologies Research Center
De Montfort University
The Gateway Leicester, United Kingdom

S. Vinod Selvaganesh
Department of Chemical Engineering
Indian Institute of Technology Madras
Chennai, Tamil Nadu, India

Ling-Ying Shi
College of Polymer Science and Engineering
State Key Laboratory of Polymer Materials
Engineering Sichuan University
Chengdu, China

Endersh Soni
Department of Physics
Maulana Azad National Institute of Technology
Bhopal, Madhya Pradesh, India

Rajendra G. Sonkawade
Department of Physics
Shivaji University Kolhapur
Maharashtra, India

Rakesh K. Sonker
Department of Physics
Acharya Narendra Dev College
University of Delhi
Delhi, India

Lutfu S. Sua
School of Entrepreneurhsip and Business
    Administration, AUCA
Bishkek, Kyrgyz Republic

Yuzhen Sun
Institute of New Energy on Chemical Storage and
    Power Sources
Yancheng Teachers University
Yancheng, Jiangsu, China

Mike Tebyetekerwa
School of Engineering
The Australian National University
Canberra, Australian Capital Territory, Australia

Surya Kant Tripathi
Centre for Advanced Study in Physics
Department of Physics
Punjab University Chandigarh
Chandigarh, India

Ünsal
Polymer Materials Engineering Department
Bursa Technical University
Bursa, Turkey

Maqsood R. Waikar
Department of Physics
Shivaji University
Kolhapur Maharashtra, India
and
Department of Engineering Physics
Padmabhooshan Vasantraodada Patil Institute of
    Technology
Sangli, Maharashtra, India

Rong Xing
Institute of New Energy on Chemical Storage and
    Power Sources
Yancheng Teachers University
Yancheng, Jiangsu, China

Cheng Yan
School of Mechanical, Medical and Process Engineering
Faculty of Engineering
Queensland University of Technology
Brisbane, Queensland, Australia

Shengyuan Yang
State Key Laboratory for Modification of
    Chemical Fibers and Polymer Materials
College of Materials Science and Engineering
Donghua University
Shanghai, China

Xiaowei Yang
Institute of New Energy on Chemical Storage and
    Power Sources
Yancheng Teachers University
Yancheng, Jiangsu, China

Xiaoping Yang
College of Chemical Science and Engineering
Yunnan University
Kunming, China
and
Institute of Energy Storage Technologies
Yunnan University
Kunming, China

Ghulam Yasin
Institute for Advanced Study
Shenzhen University
Shenzhen, China

Zhiyuan Zhao
Institute of New Energy on Chemical Storage
    and Power Sources
Yancheng Teachers University
Yancheng, Jiangsu, China

1 Smart and Flexible Energy Devices: Principles, Advances, and Opportunities

Tenzin Ingsel and Ram K. Gupta

Department of Chemistry, Kansas Polymer Research Center,
Pittsburg State University, Pittsburg, Kansas, USA

CONTENTS

1.1   Introduction.. 1
1.2   Flexible supercapacitors 2
      1.2.1   Flexible supercapacitors based on carbon 2
      1.2.2   Flexible supercapacitors based on metal oxides and sulfides............................ 3
      1.2.3   Flexible supercapacitors based on nanocomposites 5
1.3   Flexible batteries 5
      1.3.1   Flexible Li-ion and Li-sulfur batteries 7
      1.3.2   Flexible metal–air batteries... 12
1.4   Flexible proton exchange membrane fuel cells.. 14
1.5   Flexible solar cells 14
      1.5.1   Dye-sensitized flexible solar cells 14
      1.5.2   Perovskite-based flexible solar cells 16
1.6   Conclusion 19
References... 19

1.1 INTRODUCTION

Flexible devices are an essential group of electronics with versatile and innovative applications. Flexible electronics can function normally, even when subjected to various deformations such as stretching, bending, and twisting. Their potential applications range from wearable devices, renewable energy storage and production, medical care, consumer electronics, and many more. Electronics for flexible applications require reliable manufacturing to develop suitable materials, innovative technology, and tools to carry advanced research [1]. From a scientific perspective, materials that can power flexible devices remain a bottleneck for their development. For the past few years, multiple studies have been carried out to actualize power source materials that are mechanically flexible, strong, and suitable for versatile applications. Despite all the technological advances in the flexible device area, development in flexible energy conversion and storage systems is still in its infancy [2]. Ideal flexible energy conversion and storage device or power source materials for electronics exhibit suitable deformation properties; these include being bendable, stretchable, foldable and having operational safety and secure electrical functioning. Figure 1.1 illustrates an example of a futuristic, smart wearable glove that can sense, convert, and store various energy forms [3].

The following subsections will explore principles of different flexible energy storage and conversion devices, their recent advances, and growth opportunities. The first section will cover flexible supercapacitors based on carbon material, metal oxides, sulfides, and nanocomposites. A few concepts

DOI: 10.1201/9781003186755-1

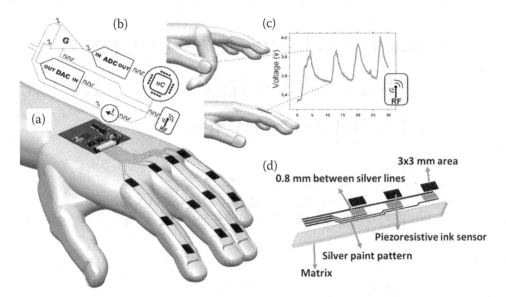

FIGURE 1.1 (a) Sensors and readout circuit schematics. (b) Piezoresistive multisensory. (c) Example of sensor response. (d) Example of patterned sensor configuration employed for application in each finger for characterization. Adapted with permission [3]. Copyright (2018) American Chemical Society.

to look forward to in the flexible supercapacitors include two common types of supercapacitors based on their energy storage mechanism and wearable textile-based supercapacitors; in addition, comparisions are made in energy and power density of pristine carbon materials and nanocomposites comprised of carbon and metal oxides. The flexible battery section discusses flexible lithium-ion, lithium-sulfur, and metal-air. Some recent progress in flexible battery technologies is explained, where unique electrode design prototypes pave the way for their large-scale production and practical applications. The flexible fuel cell part presents motivation for exploring their advantages. For the flexible solar cell part, third-generation low-cost, flexible dye-sensitized solar cells and perovskite solar cells are reviewed.

1.2 FLEXIBLE SUPERCAPACITORS

1.2.1 FLEXIBLE SUPERCAPACITORS BASED ON CARBON

Supercapacitors (SC) are energy storage devices known for their relatively simple assembly, high power density, quick charge-discharge rates, and high cyclic performance. Such properties make supercapacitors suitable for their use in flexible electronics. Supercapacitors can be divided into two categories based on how the charge is stored in the device. In electrical double-layer capacitors (EDLCs), electrical energy is stored via ion adsorption and desorption at the electrode/electrolyte interface. In pseudocapacitors, power is mainly generated and accumulated by redox reactions occurring at the surface [4,5]. Different types of materials are suitable for supercapacitors functioning via these two distinct charge-storage principles. Carbon-based materials are widely known to be applied in EDLCs, while pseudocapacitors use metal sulfides, metal oxides, and conducting polymers. EDLCs and pseudocapacitors have their advantages and disadvantages. For instance, EDLCs have high cycling stability, but only moderate capacitance. On the other hand, pseudocapacitors have poor cycling stability but high capacitance. This brings us to hybrid supercapacitors, which are made up of selectively chosen and synthesized nanomaterials for nanocomposites that show enhanced energy density without compromising their power density and cyclic life. This is a perfect segue into the subsections on discussions about flexible su-percapacitors based on carbon, metal oxides, metal sulfides, and nanocomposite materials [2].

When it comes to carbon-based materials for flexible supercapacitors, carbon materials of different morphology and structure, such as graphene, carbon nanotubes (CNTs), and carbon fiber, are widely investigated [6]. Due to their high conductivity, large surface area, and desirable mechanical prop-erties, such carbon materials are fit for applications in flexible devices. Graphene is a two-dimensional structure with monolayers of carbon atoms where the carbon bonds are sp^2 hybridized. Graphene represents a distinguished group of carbon materials because of its outstanding electrical and thermal conductivity, mechanical flexibility, and relatively low production cost [7]. Supercapacitors based on graphene can deliver a decent specific capacitance of about 550 F/g, theoretically. However, due to the restacking of graphene sheets during the fabrication process and the cutback in their specific surface area, the specific capacitance delivery is lower than the theoretical limit. Laser reduction of graphene oxide generated graphene with heavily reduced restacking of graphene sheets [8]. The supercapacitor device was constructed by employing two identical laser-reduced graphene film (LSG) electrodes sandwiched with a polymer gel electrolyte. An extremely thin device was obtained with a thickness of less than 100 μm. The LSG electrode showed a high specific surface area and high specific capacitance of 204 F/g, with high energy and power density. The LSG electrode displayed outstanding cycle stability. The device proved highly flexible, with its bending having no adverse effect on the electrochemical performances [9].

Carbon-based materials have also been used as capacitive sensors, a subset of strain sensors where mechanical strain gets converted to electrical output [10]. In fact, in an ideal capacitive strain sensor, a strain is proportional to the capacitance. Equation 1.1 represents the strain-capacitance relationship.

$$C = \varepsilon_0\varepsilon_r S/d \quad \dots \tag{1.1}$$

Where ε_0 is the vacuum dielectric constant (F/m), ε_r denotes the dielectric constant of the electrolyte, S is the size of the electrode accessible to ions (m^2), and d is the thickness of the dielectric layer [10]. A carbon-based capacitive strain sensor was fabricated as knittable fiber. The core was made up of rubber and played a role as a dielectric layer, and carbon nanotube served as the active material [11]. The fiber supercapacitor device effectively absorbed shear and tensile stresses and converted them to capacitance during deformation. The device showed high capacitance, strain linearity, and high sensitivity to me-chanical force. Figure 1.2 shows a schematic image of the twisted and stretchable fiber, as well as the actual fiber sewn into a glove, where it shows high structural reversibility (stretched up to 60%) [11]. The fiber was also helically wrapped around a glass tube to demonstrate no fiber damage or delami-nation. The scanning electron microscopes (SEM) further displayed the surface microstructure of the fiber-based sensor. Such design can potentially be applied in wearable energy devices, such as flexible solar cells, batteries, and supercapacitors, among many others.

A textile-based EDLC was fabricated where an activated carbon was screen printed onto a carbon fiber cloth. The carbon cloth was knitted with patterns and acted as a current collector, as shown in Figure 1.3 [12]. The textile EDLC showed suitable capacitance, performance, and mechanical stability to be stretched and bent to 180°. Such knitted electrodes can find ample opportunities for applications in wearable smart energy harvesting and storage system where the system can capture external energy from the surroundings, such as body heat, radio frequencies, sunlight, and body heat.

1.2.2 FLEXIBLE SUPERCAPACITORS BASED ON METAL OXIDES AND SULFIDES

As discussed earlier, metal oxides and sulfides find their role in being applied as pseudocapacitance materials for pseudocapacitors. Materials like Fe_2O_3, MnO_2, and CoO_x are suitable low-cost materials for applications in flexible pseudocapacitors [2]. In a flexible symmetrical supercapacitor device fabricated by utilizing $NiCo_2O_4$ nanoparticles as the active material, the specific capacitance was

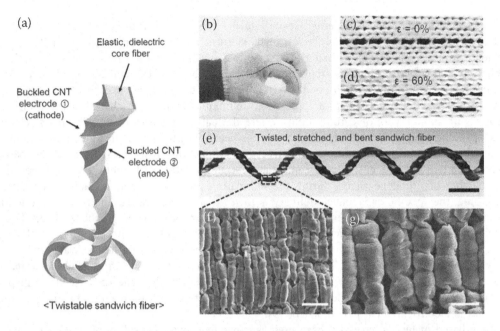

FIGURE 1.2 (a) Schematic image of a twisted sandwich-structured fiber device composed of two CNT electrodes with silicon rubber as the core. Digital images of (b) the fiber sewn into a glove and enlarged images of (c) before and (d) after subjecting the test sample to 60% tensile strain with a scale bar (8 mm). (e) Photograph of helically wrapped fiber on a glass tube with scale bar (50 mm). Microstructural mapping of fiber during its relaxation after an applied strain at low magnification (f) and high magnification (g). Adapted with permission [11]. Copyright (2016) American Chemical Society.

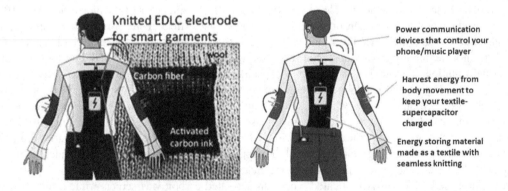

FIGURE 1.3 Schematic image demonstrating textile-based supercapacitor's design with piezoelectric materials and antennas for energy harvesting. Adapted with permission [12]. Copyright (2013) Royal Society of Chemistry.

recorded as high as 117 F/g at a current density of 0.625 A/g with an operating voltage of 3 V [13]. Nanoparticles of $NiCo_2O_4$ were synthesized by a hydrothermal process to obtain monodispersed particles of 50–60 nm size with octahedron crystal structures. The device delivered similar capacitance values at different bending angles ranging from 0 to 180°. The $NiCo_2O_4$ nanoparticles in its three-electrode configuration provided an outstanding specific capacitance of 1260 F/g and power density of 2.8 kW/kg. Such foldable supercapacitor is highly anticipated to be applied in lightweight, portable devices owing to its promising charge storage abilities and reversible deformation and recovery, without affecting their electrochemical performances.

A cobalt oxide and perovskite oxide $PrBaMn_{1.7}Co_{0.3}O_{5+\delta}$ (PBMCO)-based hybrid supercapacitor (Co_3O_4-PBMCO) was fabricated on three-dimensional nickel wire [14]. The supercapacitor delivered the highest specific capacitance of 1571 F/g with a 1.48 kW/kg power and 215.80 Wh/kg energy density. Figure 1.4 demonstrates the mechanism of charge transfer in Co_3O_4-PBMCO, cyclic voltammetry (CV), charge-discharge profiles, specific capacitance, and cyclic stability up to 30,000 charge-discharge cycles. Three main advantages were observed in the perovskite oxide-based flexible ultracapacitor: (1) quick charge transfer was facilitated by Co_3O_4, which in turn helped deliver high power and energy density, (2) the presence of defect sites in Co_3O_4-PBMCO enabled oxygen anions to intercalate inside the perovskite, and (3) the three-dimensional nickel wire acted as a suitable metal current collector for flexible supercapacitor, and dip coating of the active material on the nickel wire provided a cost-effective device fabrication.

In general, transition metal sulfides have higher electrochemical activity than transition metal oxides owing to their narrow bandgap, and replacing oxygen with sulfur provides higher ionic conductivity. Metal sulfides have gained much attention due to their high theoretical capacities, low cost, and interesting electrochemical properties [15]. Facile hydrothermal synthesis formulated different metal sulfides: FeS, NiS, CoS, and CuS deposited on nickel foam (NF) for ultracapacitor applications [16]. The NiS/NF and FeS/NF showed a hierarchical structure where the FeS/NF delivered the highest specific capacitance of about 2000 F/g at 2 A/g of current density. Such high-performance devices can be used in electric vehicles, flexible electronics, and solar cells. Metal oxides and sulfides provide pseudocapacitive charge-storage abilities, and carbon-based materials provide EDLC properties. Therefore, their nanocomposites generally find their niche by compositing EDLC and pseudocapacitive materials in a device. The next section describes nanocomposites for flexible devices.

1.2.3 FLEXIBLE SUPERCAPACITORS BASED ON NANOCOMPOSITES

In previous discussions, we have established the importance of fabricating flexible devices with high mechanical stability and energy density without compromising their cycle life and power density. A widespread technique is to coalesce EDLCs and pseudocapacitors' advantages to achieve higher power and energy densities. Figure 1.5 demonstrates the incentives in compositing materials like Co_3O_4 that possess high-density faradic actions and carbon material like CNT, graphene, and reduced graphene oxides (rGO) that can offer a high surface area and electrical conductivity [17]. In that work, Co_3O_4@ NiO was electrodeposited on the CNT yarn, and then graphene was coated on a flexible nanocomposite supercapacitor. Such a supercapacitor not only provided high volumetric capacitance, energy, and power densities, but its yarn design can be weaved into textiles for wearable electronic devices. Among all the materials used for pseudocapacitors, conducting polymers are intrinsically flexible and suitable for applications in flexible energy devices. A nanocomposite of poly(3,4-ethylene-dioxythiophene) (PEDOT) and cellulose paper (CP) were synthesized, as shown in Figure 1.6(a), by a multistepped vapor-phase polymerization method [18]. Figure 1.6(b) illustrates the assembled supercapacitor's configuration with the conducting polymer and cellulose paper (PEDOT/CP) composite possessing the dual role of being both the current collector and electrode. The cellulose paper participates in the device as a separator and polyvinyl alcohol/sulfuric acid (PVA/H_2SO_4) as an electrolyte. Minor changes in the capacitance occurred when the fabricated supercapacitor was subjected to 90° bending and 45° rotating. A flexible 3D graphene was electrodeposited with MnO_2 to obtain a flexible supercapacitor [19]. The graphene/MnO_2 composite was used as electrodes with a polymer separator, and the whole device was encapsulated by polyethylene terephthalate (PET), a flexible polymer to construct a flexible freestanding supercapacitor. The fabricated symmetrical supercapacitor was of thickness ~800 µm and weighed less than 10 mg.

1.3 FLEXIBLE BATTERIES

For the successful commercialization of wearable electronics, it is necessary to develop materials with high electrochemical performance and fabricate devices having lightweight batteries that can be flexible.

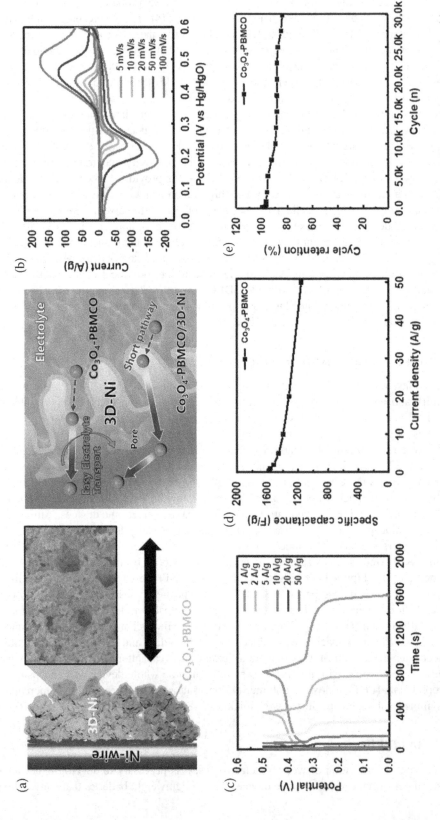

FIGURE 1.4　(a) Illustration of 3D-PBMCO/Ni's energy storage mechanisms. (b) CV profiles at various scan rates. (c) GCD profiles at various current densities. (d) Specific capacitance at different current densities of the cobalt oxide-perovskite oxide/nickel wire-based electrode. (e) Cyclic stability of the electrode. Adapted with permission [14]. Copyright (2020) American Chemical Society.

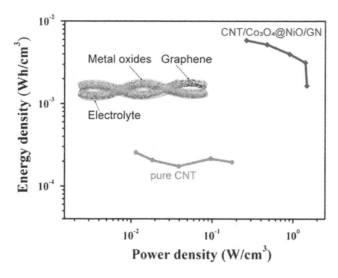

FIGURE 1.5 Volumetric energy density versus power density of pure CNT and CNT/Co$_3$O$_4$@NiO/GN composites for flexible devices. Adapted with permission [17]. Copyright (2020) American Chemical Society.

This section covers different types of flexible batteries, such as Lithium-ion (Li-ion) Li-sulfur, and metal-air batteries.

1.3.1 FLEXIBLE LI-ION AND LI-SULFUR BATTERIES

Li-ion batteries (LIB) are highly sought after due to their competitive power and energy density. These unique features make them desirable for applications in electric vehicles, portable electronics, and electrical grids. Here is the role of cathode and anode in LIB during the charging and discharging. During the discharge cycle, the anode undergoes lithium deintercalation and oxidization while the cathode undergoes reduction. The opposite occurs during the charge cycle [20]. In this subsection, intercalation and conversion-type cathodes and anodes will be discussed with their advances and challenges in flexible devices. Figure 1.7 depicts some relevant electrode material groups with their specific capacity versus electrode potential [21]. Some of the electrode groups are experimentally available, and some Li-ion materials have been theoretically studied. Intercalation cathodes can store guess ions where these ions can be introduced and removed reversibly from the cathode. Intercalation cathodes include compounds, such as transition metal oxides, metal chalcogenides, and polyanion compounds. On the other hand, conversion-type cathodes experience a solid-state reduction and oxidation reaction during the charge and discharge process. Conversion-type cathodes include metal halides and lithium sulfides, selenium and tellurium, and iodine compounds [21].

For a Li-ion battery to be flexible, intuitively, all the components, including the anode, cathode, separator, and plastic outer package, require the ability to withstand bending and stretching without losing their electrochemical performances. Standard binders and metal current collectors used in LIBs are impractical for flexible LIBs. In general, carbon materials used in LIBs face capacity limitations. They possess safety concerns as well as low performance due to the formation of Li dendrites between a separator and the electrode. On the other hand, electrode materials like Ge, LiFePO$_4$, Si, LiMO (M = Ni, Mn, Co) show poor conductivity but superior capacitance. On top of that, LIB's electrodes need to withstand volume change upon Li insertion and removal. Therefore, to combat shortcomings associated with these materials, few strategies to obtain flexible LIBs include altering materials' nanostructures and incorporating active electrode materials into adjustable carbon support [2]. Li-ion batteries applied in flexible applications are heavily hindered because of a lack of inherently

(a)

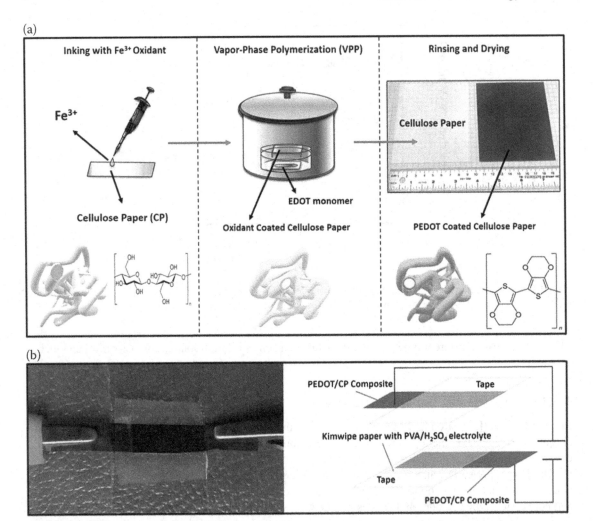

FIGURE 1.6 (a) Vapor-phase polymerization method to obtain PEDOT coated CP. (b) Flexible solid-state supercapacitor fabricated on the left and its scheme of design on the right. Adapted with permission [18]. Copyright (2020) American Chemical Society.

flexible functional materials. Here are some examples of the strategies used in fabricating flexible electrodes with the opportunities and challenges for flexible and wearable devices.

Thin-film electrode design provides an intuitive strategy for flexible devices and has been embraced by flexible commercial batteries [22]. However much improved the bendability is, of thin-film, its low capacity poses a critical disadvantage, coupled with expensive and sophisticated fabrication methods undesirable for scalability purposes. Therefore, investigators have delved into other electrode designs, such as island-bridge structures, wavy-structured electrodes, porous electrodes, Miura fold, percolating network, and 1-D and textile [22]. Note that the materials used range from metal oxides to carbon and carbon-metal oxide composites. A battery with an island-bridge electrode design, as shown in Figure 1.8a, used low modulus thin Si elastomer as a substrate. This provides an opportunity for the active material to transfer stress to the elastomers. $LiCoO_2$ was utilized as a cathode and $Li_4Ti_5O_{12}$ as an anode. This segmented, interconnected structure achieved reversible stretchability up to 300% while maintaining a decent capacity [23].

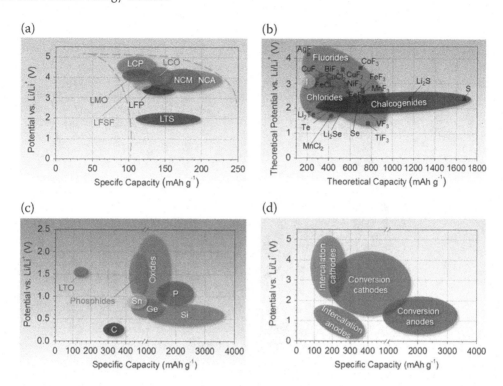

FIGURE 1.7 Average range of specific capacity vs. discharge potentials of (a) experimentally possible intercalation-type cathodes, (b) theoretically known conversion-type cathodes, (c) experimentally studied conversion-type anodes, and (d) all the different types of conversion and intercalation-type electrodes. LCO: lithium cobalt oxide, LCP: lithium cobalt phosphate, LFP: lithium iron phosphate, LFSF: lithium iron fluorosulfate, LMO: lithium manganese oxide, LTO: lithium titanium oxide, LTS: lithium titanium sulfide, NCA: lithium nickel cobalt aluminum oxide, NCM: lithium nickel cobalt manganese oxide. Adapted with permission [21]. Copyright (2015) Elsevier.

A wavy electrode design was implemented where the battery was able to withstand a strain of more than 400% with a minor change in the specific capacity from an initial specific capacity of 115 mA h/g to 112 mA h/g after 500 stretching cycles [24]. The schematic of the wavy electrode design is illustrated in Figure 1.8(b), where CNT/lithium manganese oxide (LMO) material was used as an arched cathode and CNT/LTO-CNT as arched anode with a gel electrolyte. CNT acts as the current collector and a structural framework, whereas LMO and lithium titanium oxide (LTO) nanoparticles act as electrochemically active materials. Porous structured LIB electrodes fabricated using a nanocomposite of germanium quantum dot (Ge-QD), nitrogen-doped graphene (NG), graphene foam (NGF), and poly (dimethyl) siloxane (PDMS), [Ge-QD@NG/NGF/PDMS], displayed relatively high capacity due to their versatile design (Figure 1.9). However, porous electrode structures lack uniformity and dependable characterizations [25]. In Miura fold-type battery designs, i.e., folding and cutting patterning, kirigami and origami designs-based batteries can be fabricated [26]. This process yields highly flexible batteries (over 150%); however, their complex fabrication limits their scale-up potentials. Metal or carbon conductive fillers can be embedded in a percolation network that could be another strategic technique for obtaining stretchable current collectors. Gold nanoparticles (AuNPs) and layers of gradient-assembled polyurethane composite for LIB exploited AuNPs' conductive properties and the polymer's stretchability. Incorporating a polymer-percolating network in flexible electrode design does have positive implications on self-healing properties and providing highly elastic material. However, the fabrication is rather sophisticated with relatively low conductivity [27]. In a study, anode and cathode made up of carbon nanotube composites were

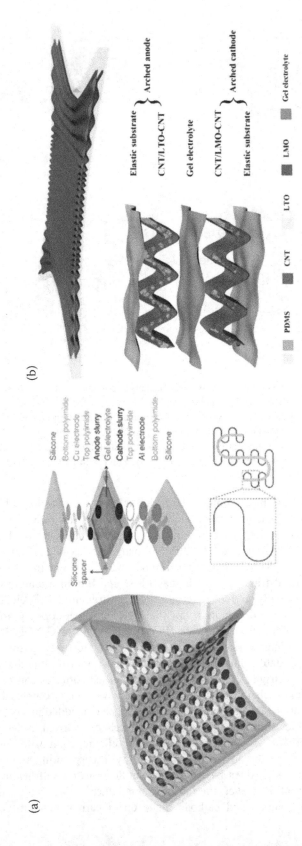

FIGURE 1.8 (a) Stretchable battery design with serpentine geometries. Adapted with permission [23]. Copyright (2013) Springer Nature. (b) Schematics of a stretchable battery with wavy electrode design. Adapted with permission [24]. Copyright (2015) John Wiley and Sons.

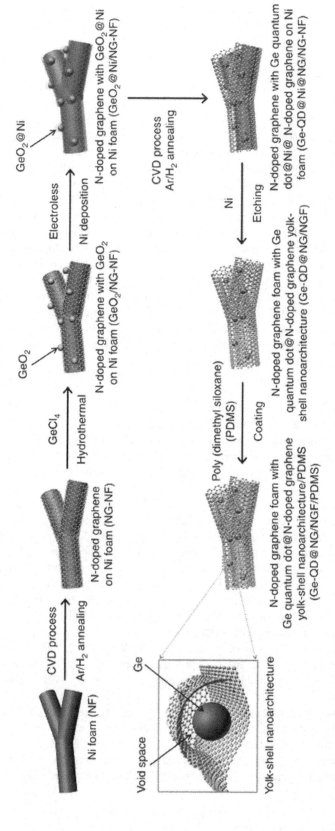

FIGURE 1.9 Fabrication process of Ge-QD@NG/NGF/PDMS. Adapted with permission [25]. Copyright (2017), Springer Nature (Published under Creative Commons CC BY license).

accumulated onto a cotton fiber to obtain a high-performance, wearable energy storage device [28]. Textile or one-dimensional electrode designs are highly flexible and suitable for applications in smart clothing; however, they generally have low cell capacity.

Lithium-sulfur (Li-S) batteries are highly anticipated energy storage systems with high theoretical capacities, i.e., Li: 3860 mA h/g and S: 1675 mA h/g [29]. However, there are many drawbacks, such as deficient mechanical flexibility, Li dendrite formation, and low energy density due to the heavyweight of battery components. Investigations are being carried out in obtaining Li anode and S cathode with high coulombic efficiency and stability. Li-S are advantageous from the point of view of sustainability because they prevent the usage of a disproportionate amount of Li; this outcome is because Li's production is costly and environmentally unfriendly. Numerous strategies have been applied in obtaining Li anode by developing high surface area Li host, utilizing unusual new designs, nanostructuring, and compositing, among many others. The current collectors, such as Cu and Al, are not ideal for folding in batteries as they undergo plastic deformation. Another common issue is the electrochemically active material gets detached from the current collector when folded. Therefore, to address such problems, multiple works point toward utilizing flexible carbon; carbon nanotube and carbon fiber, among many others, function as current collectors. In a study, different carbon interlayers participating in Li-S batteries were compared. Carbon materials included graphene, CNT, and a mixture of CNT-graphene, among which CNT was able to replace the current collector and provide excellent battery performances [30]. Different active materials patterning is being applied to address the delamination problem associated with the active material. For instance, the investigators used CNT as a current collector and deposited their active material (Li) in checkerboard patterning [31]. All in all, even though Li-ion batteries are commercial batteries, Li-S batteries could help develop more low-cost battery technologies for flexible applications.

1.3.2 FLEXIBLE METAL–AIR BATTERIES

Metal-air batteries use metals as anode materials and ambient air as a cathode in aqueous or aprotic electrolytes. Metal-air batteries are very attractive due to their high specific capacity and energy density compared to metal-ion batteries. For commercialization of metal-air batteries, many challenges, such as passivation of the metal anode, efficiency of catalysts, and cost-effective electrolytes, have to be overcome.

Among many metals as anodes, Li metal offers a high theoretical specific capacity. Lithium-air battery system surpasses other metal-air batteries, such as Fe-air, Zn-air, Mg-air, Na-air, and Al-air in terms of theoretical and practical specific energy delivery capabilities [32]. However, metal-air batteries using zinc, aluminum, and magnesium are explored due to a safe operation of the device, material accessibility, and benign nature. Zinc-air batteries (ZABs), for instance, have attracted huge attention because of their environmental friendliness, high theoretical energy density (<1084 Wh/kg), and low cost [33]. Therefore, zinc-ion batteries are an attractive choice for applications in wearable devices. The working principles of metal-air batteries are different than metal-ion batteries. Metal-air batteries require an efficient electrocatalyst showcasing bifunctional abilities to carry out oxygen evolution (OER) and oxygen reduction reaction (ORR), as shown in Figure 1.10. Enhanced active site exposure is aspired in these electrocatalysts for their metal-air battery applications by implementing strategies such as constructing heterogeneous interfaces, obtaining nanoparticles, hybridizing with carbon materials, and elemental doping [34].

Numerous efforts have been invested in obtaining zinc-air batteries that could electrochemically function under stable conditions while subjected to deformations such as twisting, folding, bending, and stretching. Standard flexible zinc-air batteries utilize hydrogels as an electrolyte, air cathode, and zinc anode, where all the components meet the flexibility criteria. However, the problem remains that the metallic current collectors on which zinc is electroplated or the zinc plates (foils or films) used are

(a)

(b)

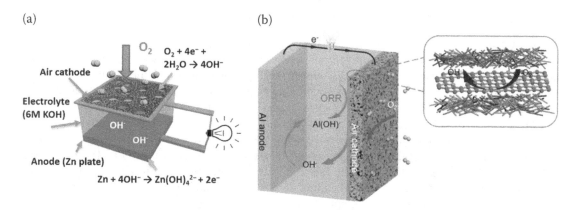

FIGURE 1.10 (a) Zinc-air battery system setup. Adapted with permission [40]. Copyright (2019) American Chemical Society. (b) Schematics of aluminum-air battery. Adapted with permission [39]. Copyright (2020) American Chemical Society.

inherently rigid and display irreversible shape change under deformation. On the other hand, nickel foam is attractive material of choice for the fabrication of air cathode and current collector because of its high conductivity properties and air permeability, beneficial for electrochemical properties of metal-air batteries. However, nickel foam is unfortunately not mechanically robust enough to withstand constant bending deformation. The kirigami design-inspired flexible zinc-air batteries of zinc and nickel-coated polyurethane sponges were developed [33]. The kirigami configured device performed stably under twisting and rolling deformations.

Carbon and metal oxide-based electrocatalysts studies for OER and ORR have dominated the quest for electrochemically active materials applied in zinc-air batteries. Bifunctional electrocatalysts, for instance, transition metal alloys, CoFe, and Fe/Fe_3C embedded in carbon showed activity toward ORR and OER, where CoFe was found out to be the active site for OER, and Fe/Fe_3C encapsulated in carbon for the ORR [35]. Core-shell configuration with Co_3O_4 nanowire as core and NiFe-layered double hydroxide as shell proved active toward ORR and OER. Metal oxides in a nanowire (NW) array configuration like Co_3O_4 NW are popular choices for ORR electrocatalysts, owing to their excellent structural stability, large surface area, high conductivity, and tunable electronic structure [36]. Metal-organic-framework-derived cobalt-based carbon nanocomposite showed trifunctional (OER, ORR, HER) properties with self-powered overall water splitting [37]. The fabricated electrode utilized in the zinc-air batteries exhibited stable performance under stressed conditions, and hence, is paving the way for next-generation flexible and wearable self-powered energy storage systems. Many investigators have explored carbon materials for flexible metal-air batteries. CNT fiber bifunctional air catalyst, for instance, provided not only high flexibility but also prominent energy density (<830 Wh/kg) and capacity (<690 mAh/g) [38].

Aluminum-air batteries for flexible energy storage systems are gaining traction because of their high theoretical energy density (~2790 Wh/kg), low cost, and environmentally friendly nature [39]. Other metal-air batteries have been strategically designed, where polymer or carbon-based materials are utilized as substrates and active materials are integrated via printing, vapor deposition, and coating. A honeycomb-inspired aluminum-air battery exhibited an energy density of 1620 mWh/g and a specific capacity of 1203 mAh/g. The battery showed stable electrochemical functioning even with 2500% stretching [39]. Although flexible metal-air batteries are an upcoming technology with multiple challenges hindering their commercial use, unusual effective designs like honeycomb and kirigami configurations could provide more insight into enhancing metal-air batteries' flexibility and performances.

1.4   FLEXIBLE PROTON EXCHANGE MEMBRANE FUEL CELLS

Proton-exchange membrane fuel cells (PEMFC) could theoretically outdo battery technologies in energy-density delivery because hydrogen has an energy density of 39.7 kWh/kg. Fuel cells' high energy density and decent power density could be highly beneficial when applied in flexible electronics. However, a flexible PEMFC system is not commercially available in today's technology because it isn't easy to obtain a flexible hydrogen generator and the fuel cell [41]. Initiatives have been undertaken to make fuel cell components flexible and high performing. For instance, a study looked at the fuel-cell clamp plate fabricated with stiffness-controlled polymeric material, polydimethylsiloxane (PDMS) [42]. It was observed that the stiffness-controlled PDMS could forgo additional fuel cell fasteners to add to the fuel cell's weight and stiffness. Intriguingly, the faradaic and ohmic impedances decreased under bending conditions. In another lead on the flexible fuel cell for potential energy applications, an adaptable and flexible hydrogen generator made of distinctive aerogel catalyst was synthesized [41]. The silica aerogel catalyst demonstrated capabilities in storing formic acid in its pores and catalyzing the formic acid to hydrogen, as shown in Figure 1.11. When the aerogel catalyst was incorporated into a PEMFC stack, the device delivered an energy density of about 135 Wh/kg.

Figure 1.12 delineates the differences between traditional PEMFCs and a prototype flexible PEMFC. In the prototype flexible PEMFC, the cathode side has forgone the heavy plate, and the anode has been replaced by cheap plastic that is light and flexible [43]. The electrode used in the prototype device was made up of carbon paper and CNT composite. In available flexible devices, graphene, Ag or Si nanowire, conducting polymer, CNTs, and carbon cloth are utilized because they possess high electrical conductivity and surface area. Traditionally, 80% of the conventional PEMFC's weight was due to the bipolar plates, responsible for 30% of the production cost and 90% of the volume occupation [43]. Therefore, it is intuitive to utilize cost-effective, lightweight, and suitable carbon-based materials like CNT for flexible PEMFC. Flexible fuel cell research is still in its infancy; however, incentives in studying flexible fuel cells' characteristics include their high theoretical energy density and the zero charging time necessary for batteries.

1.5   FLEXIBLE SOLAR CELLS

The energy world has witnessed three generations of solar cells introducing solar cells made up of different materials and configurations to provide unique power conversion efficiencies. Early solar cells employ silicon wafers to convert sunlight energy to electricity, whereas modern photovoltaic (PV) cells use semiconducting materials, such as silicon (amorphous, single crystal, and multi-crystalline silicon); CdTe thin film; copper indium gallium selenide (CIGS) thin-film; nanocrystal-based; and dye-sensitized solar cells, among many others [44]. The bandgap of PV semiconductors largely determines the wavelengths of light that can be absorbed to convert to electrical energy. This section will discuss third-generation type flexible PV cells.

1.5.1   DYE-SENSITIZED FLEXIBLE SOLAR CELLS

Standard older PV cells were not built for flexible energy storage applications because of rigid substrates and components used. Second-generation PV cells, copper indium gallium selenide (CIGS) thin-film, gained traction because they provided an edge over conventional Si-based solar cells by providing a more straightforward manufacturing process and cost. However, because of selenium, indium, and gallium's finite reserve, their development has been highly hindered. Dye-sensitized solar cells (DSSCs), on the other hand, are known for their versatility, straightforward production, flexibility, and benign nature. In general, the DSSCs consist of four components: a dye sensitizer, a counter electrode (C or Pt), a semiconductor (n-type TiO_2 and p-type NiO), and a redox

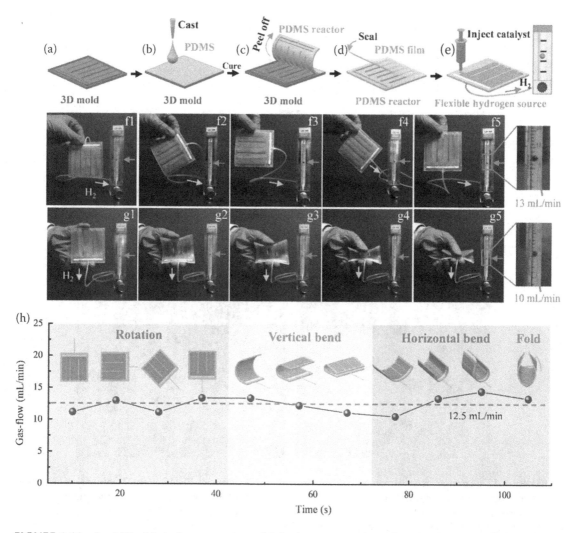

FIGURE 1.11 (a–e) Flexible hydrogen generator fabrication process schematics. (f1–f5) Adaptability test where the hydrogen source is rotated at different angles during gas production. (g1–g5) Flexibility test where the hydrogen generator is bent at different angles and gas flow rate is measured. (h) Gas-flow vs. time during rotation, vertical bend, horizontal bend, and fold. Adapted with permission [41]. Copyright (2020) American Chemical Society.

mediator. Multiple studies have reported indium tin oxide (ITO) and titania-based DSSCs to reach a photoelectric conversion efficiency (PCE) of ~11%. Compared to expensive, standard monocrystalline Si-based solar cells with ~18% PCE, ITO-based cost-efficient DSSCs are a particular type of solar cell. One factor that limits their applications in flexible devices is the brittleness of ITO films [2].

Therefore, multiple efforts have been put into developing transparent conducting oxides (TCO)-free configured DSSCs. TCO/glass substrates include ITO and fluorine-doped indium oxide (FTO), which takes up approximately 20–30% of the total materials cost [45]. One of the first studies on Pt and TCO-free DSSCs demonstrated conductive polymer use in replacement of the conventional TCO substrate used initially as a counter electrode. Such relief provided the DSSC with favorable characteristics, such as fast charge transport, good flexibility, and high catalytic ability [46]. Graphene/conducting polymer (PEDOT) composite film of high flexibility employed in a DSSC as a

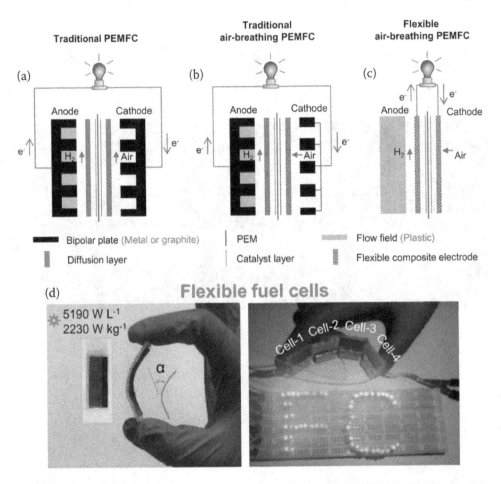

FIGURE 1.12 Schematics of structure and workings of (a) Traditional PEMFC. (b) Traditional air-breathing PEMFC where air flows into the PEMFC. (c) Prototype novel PEMFC devoid of heavy, costly, and bipolar rigid plates (graphite or metal) as a current collector. (d) Flexibility and action of a fuel cell. Adapted with permission [43]. Copyright (2017) American Chemical Society.

counter electrode delivered a performance that could rival conventional DSSCs requiring Pt or TCO substrate [47].

Fiber-structured DSSC could be an attractive choice for application in wearable energy-conversion devices. As illustrated in Figure 1.13, $Co_{0.85}Se$ nanosheets and polyaniline (PANI) were grown on carbon fibers' surfaces [48]. The synthesized material played the role of the counter electrode, and the titania nanotubes cultivated on Ti wire as the photoanode. The PANI provided an enormous surface area to lower the charge transfer resistance and enhance the material's catalytic activity. In addition, the $Co_{0.85}Se$ was able to catalyze the redox reaction of the I^-/I_3^--based electrolyte. The cell's photovoltaic performance indicates a novel design strategy for flexible next-generation fiber-shaped DSSCs.

1.5.2 Perovskite-Based Flexible Solar Cells

Perovskite solar cells (PSCs) are third-generation of PV cells of recent discovery. They possess multiple pros over standard first- and second-generation silicon and thin film-based solar technologies. Perovskites are a group of chemical compounds distinguished by the formula ABX_3, where X represents halogen elements like Cl^-, I^-, Br^-, whereas A and B represent cations with different sizes.

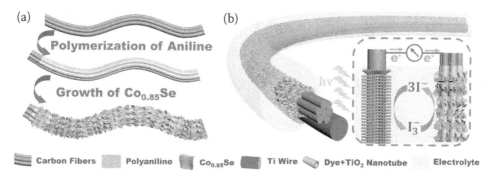

FIGURE 1.13 (a) Fiber-shaped DSSC based on PANI and cobalt selenide counter electrode and (b) schematic of Fiber-based DSSCs. Adapted with permission [48]. Copyright (2019) American Chemical Society.

PSCs possess high PCE, low cost, and solution processability. They are compatible with flexible substrates and therefore can potentially help realize flexible PSC (F-PSC) technologies. In the past few years, the F-PSCs witnessed accelerated development where their PCE improved from ~2% to ~19%, indicating high potentials in being applied in bendable and portable devices [49]. However, the flexible versions of PSCs cannot surpass the performance of standard PSCs containing rigid substrates, which is a challenge for researchers in the coming years. Different flexible PSCs have been prepared using vacuum deposition, spray deposition, roll-to-roll construction, and spin-coating, among many others [49].

An emerging subgroup of perovskites, organic-inorganic halide perovskites, is a highly anticipated group of materials for photovoltaics applications. These organic-inorganic halide perovskites have favorable properties, such as low exciton binding energy and high absorption coefficient. Unlike common oxide-based perovskites with big band-gap, hybrid perovskites have an optimal band gap for sunlight's visible spectrum (1.2–3.0 eV) [50]. A novel laser crystallization technique facilitated the control of hybrid perovskite morphology and crystal structure. The reliable laser crystallization technique employed a near-infrared laser type (λ = 1064 nm) with a 1 m/min production scan rate. The solar cell obtained was of inverted-type organic-inorganic hybrid perovskites on flexible polymer substrates with a PCE of 8% [50]. In another initiative on actualizing flexible PSCs, a customized electrospinning process produced methylammonium triiodoplumbate perovskite ($MeNH_3PbI_3$) fibers with a polymeric material polyvinylpyrrolidone (PVP) as the support matrix [51]. Figure 1.14(a) sheds light on the electrospinning technique used in obtaining the hybrid perovskite fiber with the morphological image of the electro-spun fiber. The finite element simulation in Figure 1.14(b) speculates hybrid perovskite in its fiber geometry possessing a higher potential in absorbing power from the sun than its layered thin film geometry. Such nanostructuring technique provides a promising route to obtaining perovskite-based optoelectronic devices with enhanced light absorption efficiencies [51]. An advantage of the electrospinning technique is that it can be potentially employed to upscale the production process from a lab-scale application, unlike spin-coating used for thin-film solar cells.

Titania is a commonly used n-type material for flexible and rigid PSCs. Different titania deposition techniques, such as nanoparticle spin coating, atomic layer deposition (ALD), e-beam evaporation, and sputtering, are employed in obtaining PSCs. For instance, a titania layer synthesized in a flexible PSC with the aid of a radio frequency magnetron sputtering achieved a PCE higher than 15% [52]. Flexible perovskite solar cells require more research because they are known to have faced durability and stability issues where the power conversion efficiency drops over time when material degrades. However, they possess many advanced properties for applications in flexible optoelectronic devices, and cutting-edge nanostructuring and hybridizing with stable materials could help realize their efficiencies and help actualize their large-scale use.

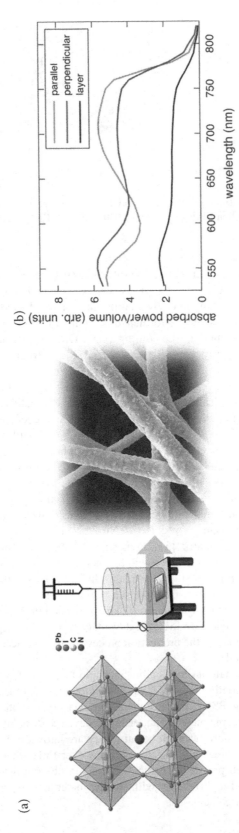

FIGURE 1.14 (a) Schematics showcasing the concept of the one-stepped electrospinning process of perovskite fibers. (b) Simulation comparing 300 nm thick perovskite layer vs. perovskite fiber with 300 nm diameter and length. The absorbed power is nearly three times larger in the fiber geometry. Adapted with permission [51]. Copyright (2019) American Chemical Society.

1.6 CONCLUSION

The development in flexible power sources with excellent mechanical and electrochemical performance is still in its infancy. Realization of cost-effective, reliable wearable devices, implantable electronics, foldable and flexible laptops, and smartphones largely pivot on various types of flexible power sources. Therefore, flexible power sources' development should not solely be limited to obtaining durable and robust LIBs for flexible applications. Other flexible power sources, such as supercapacitors, solar cells, fuel cells, and metal-air batteries, could provide suitable power and storage fit for versatile applications. In this chapter, potential and existing flexible power sources were discussed, and hopefully, the readers will realize the intricacies and complexities involved in this area of research.

REFERENCES

1. Shivakumar S (2015) The flexxible electronics opportunity. In: *The Flexible Electronics Opportunity*. National Academies Press, pp 41–53.
2. Li L, Wu Z, Yuan S, Zhang XB (2014) Advances and challenges for flexible energy storage and conversion devices and systems. *Energy Environ Sci* 7:2101–2122.
3. Costa P, Carvalho MF, Correia V, Viana JC, Lanceros-Mendez S (2018) Polymer nanocomposite-based strain sensors with tailored processability and improved device integration. *ACS Appl Nano Mater* 1: 3015–3025.
4. Bhoyate S, Kahol PK, Gupta RK (2019) Nanostructured materials for supercapacitor applications. In: Thomas PJ, Revaprasadu N (eds) *Nanoscience: Volume 5*. The Royal Society of Chemistry, pp 1–29.
5. Bhoyate S, Kahol PK, Gupta RK (2020) Broadening the horizon for supercapacitor research via 2D material systems. In: Revaprasadu N (ed) *Nanoscience: Volume 6*. The Royal Society of Chemistry, pp 120–149.
6. Niu Z, Liu L, Sherrell P, Chen J, Chen X (2013) Flexible supercapacitors – Development of bendable carbon architectures. In: *Nanotechnology for Sustainable Energy*. American Chemical Society, pp 101–141, SE–5.
7. Zequine C, Bhoyate S, de Souza F, Arukula R, Kahol PK, Gupta RK (2020) Recent advancements and key challenges of graphene for flexible supercapacitors. In: Singh L, Mahapatra DM (eds) *Adapting 2D Nanomaterials for Advanced Applications*. American Chemical Society, pp 3–49.
8. Chee WK, Lim HN, Zainal Z, Huang NM, Harrison I, Andou Y (2016) Flexible graphene-based supercapacitors: A review. *J Phys Chem C* 120:4153–4172.
9. El-Kady MF, Strong V, Dubin S, Kaner RB (2012) Laser scribing of high-performance and flexible graphene-based electrochemical capacitors. *Science* 335:1326–1330.
10. Li S, Xiao X, Hu J, Dong M, Zhang Y, Xu R, Wang X, Islam J (2020) Recent advances of carbon-based flexible strain sensors in physiological signal monitoring. *ACS Appl Electron Mater* 2:2282–2300.
11. Choi C, Lee JM, Kim SH, Kim SJ, Di J, Baughman RH (2016) Twistable and stretchable sandwich structured fiber for wearable sensors and supercapacitors. *Nano Lett* 16:7677–7684.
12. Jost K, Stenger D, Perez CR, McDonough JK, Lian K, Gogotsi Y, Dion G (2013) Knitted and screen printed carbon-fiber supercapacitors for applications in wearable electronics. *Energy Environ Sci* 6: 2698–2705.
13. Kumar L, Boruah PK, Das MR, Deka S (2019) Superbending (0–180°) and high-voltage operating metal-oxide-based flexible supercapacitor. *ACS Appl Mater Interfaces* 11:37665–37674.
14. Kang KN, Lee H, Kim J, Kwak MJ, Jeong HY, Kim G, Jang JH (2020) Co_3O_4 exsolved defective layered perovskite oxide for energy storage systems. *ACS Energy Lett* 5:3828–3836.
15. Khan MD, Aamir M, Sohail M, Bhoyate S, Hyatt M, Gupta RK, Sher M, Revaprasadu N (2019) Electrochemical investigation of uncapped $AgBiS_2$ (schapbachite) synthesized using in situ melts of xanthate precursors. *Dalt Trans* 48:3714–3722.
16. Ikkurthi KD, Srinivasa Rao S, Jagadeesh M, Reddy AE, Anitha T, Kim H-J (2018) Synthesis of nanostructured metal sulfides via a hydrothermal method and their use as an electrode material for supercapacitors. *New J Chem* 42:19183–19192.
17. Zhou Q, Chen X, Su F, Lyu X, Miao M (2020) Sandwich-structured transition metal oxide/graphene/carbon nanotube composite yarn electrodes for flexible two-ply yarn supercapacitors. *Ind Eng Chem Res* 59: 5752–5759.

18. Li B, Lopez-Beltran H, Siu C, Skorenko KH, Zhou H, Bernier WE, Whittingham MS, Jones WE (2020) Vaper phase polymerized PEDOT/Cellulose paper composite for flexible solid-state supercapacitor. *ACS Appl Energy Mater* 3:1559–1568.

19. He Y, Chen W, Li X, Zhang Z, Fu J, Zhao C, Xie E (2013) Freestanding three-dimensional graphene/Mno$_2$ composite networks as ultralight and flexible supercapacitor electrodes. *ACS Nano* 7:174–182.

20. Franco Gonzalez A, Yang NH, Liu RS (2017) Silicon anode design for lithium-ion batteries: Progress and perspectives. *J Phys Chem C* 121:27775–27787.

21. Nitta N, Wu F, Lee JT, Yushin G (2015) Li-ion battery materials: Present and future. *Mater Today* 18: 252–264.

22. Zhao Y, Guo J (2020) Development of flexible Li-ion batteries for flexible electronics. *InfoMat* 2:866–878.

23. Xu S, Zhang Y, Cho J, Lee J, Huang X, Jia L, Fan JA, Su Y, Su J, Zhang H, Cheng H, Lu B, Yu C, Chuang C, Kim TIL, Song T, Shigeta K, Kang S, Dagdeviren C, Petrov I, Braun PV, Huang Y, Paik U, Rogers JA (2013) Stretchable batteries with self-similar serpentine interconnects and integrated wireless recharging systems. *Nat Commun* 4. doi:10.1038/ncomms2553

24. Weng W, Sun Q, Zhang Y, He S, Wu Q, Deng J, Fang X, Guan G, Ren J, Peng H (2015) A gum-like lithium-ion battery based on a novel arched structure. *Adv Mater* 27:1363–1369.

25. Mo R, Rooney D, Sun K, Yang HY (2017) 3D nitrogen-doped graphene foam with encapsulated germanium/ nitrogen-doped graphene yolk-shell nanoarchitecture for high-performance flexible Li-ion battery. *Nat Commun* 8:13949.

26. Song Z, Wang X, Lv C, An Y, Liang M, Ma T, He D, Zheng YJ, Huang SQ, Yu H, Jiang H (2015) Kirigami-based stretchable lithium-ion batteries. *Sci Rep* 5:1–9.

27. Gu M, Song W-J, Hong J, Kim SY, Shin TJ, Kotov NA, Park S, Kim B-S (2019) Stretchable batteries with gradient multilayer conductors. *Sci Adv* 5:eaaw1879.

28. Weng W, Sun Q, Zhang Y, Lin H, Ren J, Lu X, Wang M, Peng H (2014) Winding aligned carbon nanotube composite yarns into coaxial fiber full batteries with high performances. *Nano Lett* 14:3432–3438.

29. Chang J, Shang J, Sun Y, Ono LK, Wang D, Ma Z, Huang Q, Chen D, Liu G, Cui Y, Qi Y, Zheng Z (2018) Flexible and stable high-energy lithium-sulfur full batteries with only 100% oversized lithium. *Nat Commun* 9:1–11.

30. Li M, Wahyudi W, Kumar P, Wu F, Yang X, Li H, Li L-J, Ming J (2017) Scalable approach to construct free-standing and flexible carbon networks for lithium–sulfur battery. *ACS Appl Mater Interfaces* 9:8047–8054.

31. Li L, Wu ZP, Sun H, Chen D, Gao J, Suresh S, Chow P, Singh CV, Koratkar N (2015) A foldable lithium–sulfur battery. *ACS Nano* 9:11342–11350.

32. Gilligan GE, Qu D (2015) Chapter 12 – Zinc-air and other types of metal-air batteries. In: Menictas C, Skyllas-Kazacos M, Lim Tmbt-A (eds) *B for M and L-SES. Woodhead publishing series in energy.* Woodhead Publishing, pp 441–461.

33. Qu S, Liu B, Wu J, Zhao Z, Liu J, Ding J, Han X, Deng Y, Zhong C, Hu W (2020) Kirigami-inspired flexible and stretchable zinc–air battery based on metal-coated sponge electrodes. *ACS Appl Mater Interfaces* 12:54833–54841.

34. Zheng X, Cao Y, Zheng X, Cai M, Zhang J, Wang J, Hu W (2019) Engineering interface and oxygen vacancies of NixCo1–xSe2 to boost oxygen catalysis for flexible Zn–air batteries. *ACS Appl Mater Interfaces* 11:27964–27972.

35. Peng Z, Wang H, Xia X, Zhang X, Dong Z (2020) Integration of CoFe alloys and Fe/Fe3C nanoparticles into N-doped carbon nanosheets as dual catalytic Active sites to promote the oxygen electrocatalysis of Zn–air batteries. *ACS Sustain Chem Eng* 8:9009–9016.

36. Guo X, Hu X, Wu D, Jing C, Liu W, Ren Z, Zhao Q, Jiang X, Xu C, Zhang Y, Hu N (2019) Tuning the bifunctional oxygen electrocatalytic properties of core–shell Co3O4@NiFe LDH catalysts for Zn–air batteries: Effects of interfacial cation valences. *ACS Appl Mater Interfaces* 11:21506–21514.

37. Singh T, Das C, Bothra N, Sikdar N, Das S, Pati SK, Maji TK (2020) MOF derived Co3O4@Co/NCNT nanocomposite for electrochemical hydrogen evolution, flexible zinc-air batteries, and overall water splitting. *Inorg Chem* 59:3160–3170.

38. Pendashteh A, Palma J, Anderson M, Vilatela JJ, Marcilla R (2018) Doping of self-standing CNT fibers: Promising flexible air-cathodes for high-energy-density structural Zn–air batteries. *ACS Appl Energy Mater* 1:2434–2439.

39. Shui Z, Liao X, Lei Y, Ni J, Liu Y, Dan Y, Zhao W, Chen X (2020) MnO$_2$ synergized with N/S codoped graphene as a flexible cathode efficient electrocatalyst for advanced honeycomb-shaped stretchable aluminum–air batteries. *Langmuir* 36:12954–12962.

40. Cai X, Lai L, Zhou L, Shen Z (2019) Durable freestanding hierarchical porous electrode for rechargeable zinc–air batteries. *ACS Appl Energy Mater* 2:1505–1516.
41. Wang H, Bai C, Zhang T, Wei J, Li Y, Ning F, Shen Y, Wang J, Zhang X, Yang H, Li Q, Zhou X (2020) Flexible and adaptable fuel cell pack with high energy density realized by a bifunctional catalyst. *ACS Appl Mater Interfaces* 12:4473–4481.
42. Chang I, Park T, Lee J, Lee HB, Ko SH, Cha SW (2016) Flexible fuel cell using stiffness-controlled endplate. *Int J Hydrogen Energy* 41:6013–6019.
43. Ning F, He X, Shen Y, Jin H, Li Q, Li D, Li S, Zhan Y, Du Y, Jiang J, Yang H, Zhou X (2017) Flexible and lightweight fuel cell with high specific power density. *ACS Nano* 11:5982–5991.
44. Sharma S, Jain KK, Sharma A (2015) Solar cells: In research and applications—A review. *Mater Sci Appl* 06:1145–1155.
45. Yoo K, Kim J-Y, Lee JA, Kim JS, Lee D-K, Kim K, Kim JY, Kim B, Kim H, Kim WM, Kim JH, Ko MJ (2015) Completely transparent conducting oxide-free and flexible dye-sensitized solar cells fabricated on plastic substrates. *ACS Nano* 9:3760–3771.
46. Lee KS, Lee HK, Wang DH, Park N-G, Lee JY, Park OO, Park JH (2010) Dye-sensitized solar cells with Pt- and TCO-free counter electrodes. *Chem Commun* 46:4505–4507.
47. Lee KS, Lee Y, Lee JY, Ahn J-H, Park JH (2012) Flexible and platinum-free dye-sensitized solar cells with conducting-polymer-coated graphene counter electrodes. *ChemSusChem* 5:379–382.
48. Zhang J, Wang Z, Li X, Yang J, Song C, Li Y, Cheng J, Guan Q, Wang B (2019) Flexible platinum-free fiber-shaped dye sensitized solar cell with 10.28% efficiency. *ACS Appl Energy Mater* 2:2870–2877.
49. Guo J, Jiang Y, Chen C, Wu X, Kong X, Li Z, Gao X, Wang Q, Lu X, Zhou G, Chen Y, Liu JM, Kempa K, Gao J (2019) Nondestructive transfer strategy for high-efficiency flexible perovskite solar cells. *ACS Appl Mater Interfaces* 11:47003–47007.
50. Jeon T, Jin HM, Lee SH, Lee JM, Park H Il, Kim MK, Lee KJ, Shin B, Kim SO (2016) Laser crystallization of organic-inorganic hybrid perovskite solar cells. *ACS Nano* 10:7907–7914.
51. Bohr C, Pfeiffer M, Öz S, Von Toperczer F, Lepcha A, Fischer T, Schütz M, Lindfors K, Mathur S (2019) Electrospun hybrid perovskite fibers: Flexible networks of one-dimensional semiconductors for light-harvesting applications. *ACS Appl Mater Interfaces* 11:25163–25169.
52. Li J, Han G, Vergeer K, Dewi HA, Wang H, Mhaisalkar S, Bruno A, Mathews N (2020) Interlayer engineering for flexible large-area planar perovskite solar cells. *ACS Appl Energy Mater* 3:777–784.

2 Innovation in Materials and Design for Flexible Energy Devices

Ayushi Katariya and Jyoti Rani
Department of Physics, Maulana Azad National Institute of Technology, Bhopal, India

CONTENTS

2.1   Introduction...24
2.2   Materials 25
    2.2.1   Inorganic nanomaterials 25
        2.2.1.1   1D materials...25
        2.2.1.2   2D materials...26
    2.2.2   Organic materials 26
        2.2.2.1   Polymers 26
        2.2.2.2   Other organic materials 27
2.3   Structural requirements 27
    2.3.1   Flexible substrates and membranes 27
    2.3.2   Thickness of compound/active layer 27
2.4   Wearability assessments 28
    2.4.1   Softness...28
    2.4.2   Stretchability: The residual strain 29
2.5   Self-healing mechanism 29
    2.5.1   Intrinsic self-healing polymers with reversible bonds 29
    2.5.2   Self-healing through exhaustion of healing agents 30
2.6   Design of flexible energy devices 30
2.7   Flexible energy storage and conversion devices...30
    2.7.1   Energy conversion devices..31
        2.7.1.1   Nanogenerator (NGs) 31
        2.7.1.2   Photovoltaic 31
        2.7.1.3   Other flexible generators 32
    2.7.2   Energy storage devices (ESDs) 33
        2.7.2.1   Flexible batteries (FBs) 33
        2.7.2.2   Supercapacitors (SCs)...35
2.8   Configuration designs for flexible ESDs 37
    2.8.1   1D configuration of ESDs..37
        2.8.1.1   Fiber-type..37
        2.8.1.2   Spring types 38
        2.8.1.3   Spine type 38

DOI: 10.1201/9781003186755-2

2.8.2   2D configuration of ESDs... 38
       2.8.2.1   Layered sandwich configuration 39
       2.8.2.2   Planar interdigital configuration.. 39
       2.8.2.3   Other 2D novel configurations... 39
2.8.3   3D configuration ESDs 39
       2.8.3.1   Origami/Kirigami/honeycomb-based structures........................ 39
2.9   Summary.. 40
References... 40

2.1 INTRODUCTION

Research on flexible/portable energy devices has attracted tremendous interest and has guided the evolution of new power sources for numerous types of wearable/flexible electronics. The establishment of flexible electronics, namely smart mobile devices, roll-up devices, wearable devices, energy devices, bio-sensor, etc., has attracted great interest. Compared to traditional electronics, wearable electronics are cheaper, thin, wearable, bendable, portable, and even implantable. Among all the flexible/portable devices, portable EDs relevant to energy storage and conversion are among the most remarkable fields due to the universal energy crisis and the necessity of portability. Recently, accessible energy devices have been too bulky and rigid for the future of wearable electronics. Therefore, the advancement of flexible EDs that acquire both high power/energy density and excellent durability and flexibility to power numerous systems and demands to be thin, light, and flexible becomes crucial. Additionally, these energy devices want to be operational under numerous mechanical distortions, such as twisting, stretching, and even bending.

The flexible energy devices, such as batteries, photodetectors, solar cells, supercapacitors (SCs), nanogenerators (NGs), RFIDs, display devices, fuel cells, etc., have attracted significant attention from researchers. These flexible energy devices could show steady performance after repeated twisting and stretching. Every component of a portable/flexible energy device, such as current collectors of the batteries and active layer of optoelectronic devices, must be flexible and mechanically stable. Moreover, the active material has also acquired high theoretical specific capacity, good electrical conductivities for energy storage applications or great quantum yield, decent bandgap for optoelectronic devices relevant to energy conversion applications. The main technical issues with the current flexible energy devices lie in the intrinsic inflexibility of the materials used to fabricate the energy devices, resulting in weak contact between the device components. To overcome these issues, the innovation in design and synthesis of novel materials is required. For illustration, flexible components that are mechanically robust need to be invented, i.e., flexible cathode, anode, current collectors, active layer, separator, and electrolytes. For the materials used for flexible devices, such as inorganic nanomaterials such as 1D materials, 2D materials, 3D materials, and organic materials such as small molecules, polymers generally show outstanding mechanical properties. For the production of flexible EDs, 2D materials are satisfactory materials. These 2D materials have numerous outstanding optical, electrical, mechanical, and thermal properties. In these materials, flexibility can be significantly increased by decreasing the thickness. Also, 1D materials play an important role in flexible devices; 1D materials such as nanoribbon, nanotubes, and nanowires (NWs) have numerous distinct electrical, mechanical, and thermal properties from bulk materials. Carbon nanotubes (CNTs) are a favorable material for wearable/portable energy devices due to their outstanding mechanical flexibility, intrinsic carrier mobility, and conductivity. Organic materials such as polymers are the most favorable material for flexible substrate due to its excellent intrinsic stretchability. Polymer plays a key role in the flexible devices field because the majority of flexible/portable devices need a bendable, flexible, and stretchable substrate. Flexible devices that used polymer substrate are easy to fabricate since polymer shows outstanding stretchability; we also discussed the structural requirement and wearability assessment of flexible EDs, and the most important configurational designs of 1D, 2D, and 3D ESDs.

2.2 MATERIALS

In traditional electronic industries, silicon is a widely used material, while flexible electronics have limited applications because of the intrinsic stiffness of silicon wafer and the complex manufacturing producers of its microstructures. Thus, the innovation of novel and highly flexible/wearable materials is of significant importance. Various materials possibly used to manufacture flexible energy devices are categories into two parts (i) inorganic materials and (ii) organic materials. Organic materials include carbon nanotube [1], nanowires, nanoribbons, graphene [2], liquid metals, GaAs, and conductive polymer [3], and small organic molecules, organometallics complexes. Conventional inorganic materials show excellent thermal and electric conductivity, but they also have obvious demerits, such as terrible stretchability, portability, and fragile and terrible bendability. To overcome those difficulties, scientists created a series of architecture for flexible energy devices. However, to solve this issue completely, new inorganic materials, mainly nanomaterials, attracted great attention. The low-dimension nanomaterials have excellent chemical and physical properties and also outstanding intrinsic flexibility. Hence, inorganic nanomaterials are suitable for the fabrication of flexible energy devices. Organic materials have excellent intrinsic flexibility, which makes them highly acceptable for flexible EDs. The fabrication processes of organic materials are easier than those for inorganic materials.

2.2.1 INORGANIC NANOMATERIALS

2.2.1.1   1D materials

1D materials have attracted the great attention of researchers due to their distinctly superior mechanical, thermal, electrical, and physical properties from bulk materials.

CNT is a favorable candidate for energy storage and conversion devices thanks to its conductivity, mechanical flexibility, and intrinsic carrier mobility. Yakobson *et al.* built a structural molecular dynamic approach to examine the distortion of CNTs. When CNTs were extended by appealed tension, Yakobson anticipated that the maximum tension of CNTs could be an outreach to 15%, revealing the great flexibility/portability of CNTs [4]. The apparent anticipation was recognized by experimental outcomes. Numerous CNTs materials acquire various traps, so the pure CNT thread in the earlier report can tolerate only 7–8% strains because the cracking will develop when the tension enhances to its maximum limit. In this instance, several methods further strengthen the stretchability and flexibility of CNT threads. One method embraced several polymer materials as an assisted channel to manufacture carbon-polymer composite ribbons, and another to strengthen the flexibility of original CNTs threads by originating several portable/flexible structures. For example, Shang *et al.* identified a yarn-derived spring such as CNTs cable. In this instance, CNTs yarn could enlarge many times its ideal length, and the maximum strain was 285% [5]. With these procedures, the stretchability of CNTs materials will be significantly enhanced, making these materials highly appropriate for flexible energy devices. Rather than CNTs, several materials play key roles in flexible energy devices.

Silicon is a conventional material mostly used in semiconductor fields. It has several drawbacks, such as silicon cannot be stretched and bendable. To overcome these issues, silicon nanoribbon material was invented. Roger *et al.* have done numerous researches in this field to enhance the flexibility/stretchability of silicon materials. In addition, III–V group compounds were also analyzed because materials have a broad direct bandgap, high electron mobility, broad range of working temperature, and high saturated drift velocity. Contrasted with metal materials, CNTs and polymer materials have poor conductivity, due to which they have bounded applications and performance. In this instance, metallic NWs have attracted significant attention from the scientific community. Lee *et al.* have published that a novel highly flexible, stretchable, highly conductive metallic electrode was manufactured with the help of long Ag NWs. The long Ag NWs should maintain excellent performance with stretching. The resistance of NWs is rarely changed. According to the latest reviews, mainly the metallic nanowires researches focused on inquiring the substitution of indium tin oxide (ITO) and investigating the flexible energy device applications of metallic nanowires [6].

2.2.1.2 2D materials

When the material's vertical and lateral size decreases to the nanometer range, the anti-bending and stretching properties will be enhanced appreciably. The category includes 2D materials; it comprises independent, freestanding layer materials having a thickness of one atom, or even a few atoms. These 2D materials have outstanding optical, mechanical, electrical, and thermal properties. The stretchability of 2D material can be improved significantly with the decrement of material thickness. These materials are highly acceptable for the manufacturing of flexible energy devices.

Graphene is a classical 2D material that acquires excellent mechanical, chemical, and electrical properties [7]. These outstanding properties have currently made this material a quickly growing star on the view of material science. Flexible electronics have a broad scope of research fields. The applications in this field using graphene contain biological sensors, energy conversion and storage devices, and high-performance electrical and optical devices. The flexible and transparent electrodes play a major role in the manufacturing of flexible EDs. The regular economical transparent electrode materials are ITOs, while the stiff properties obstruct the manufacturing of transparent flexible electrodes. Due to the ultrathin property, graphene acquires outstanding flexibility and light transmittance. Simultaneously, graphene also has superior electrical conductivity, making it highly appropriate for the fabrication of flexible transparent electrodes. A review of the considerable quantity of literature based on graphene explains that graphene is well suited for the fabrication of ESDs thanks to its superior mechanical properties and high coplanar electrical conductivity. Because of the zero-band gap of graphene material, there are plenty of leakage currents circumstances in the applications with graphene as active material. So, the scientist uses several chemical approaches to open the bandgap and improve graphene. Graphene oxide (GO) is fabricated to functionalize graphene utilizing a powerful oxidizer like $KMnO_4$ to oxidize graphene. GOs have outstanding chemical and physical characteristics, and they have wider applications in the area of flexible electronics like energy storage device [8].

Apart from graphene materials, various other 2D materials are also available; these can be utilized to fabricate flexible energy devices. Transition metal dichalcogenides are a category of 2D materials having high photoconductivity, carrier mobility, environmental sensitivity, and thickness-dependent electronic band structure.

2.2.2 ORGANIC MATERIALS

2.2.2.1 Polymers

The collection of polymer materials plays a significant role in the portable/flexible electronic research area because the maximum of flexible electronics demands bendable, flexibility, and stretching substrates. Polymer is the most appreciable material for flexible energy devices thanks to its outstanding intrinsic flexibility [9]. Polydimethylsiloxane (PDMS) is the highly used substrate material for the fabrication of flexible/portable electronic devices. Apart from PDMS, various other polymer materials are used for flexible energy devices. The main characteristic of polymer materials is their intrinsic stretchability/flexibility. Semiconductor polymers are cheaper polymers than conventional inorganic semiconductors. CNTs have commonly been combined into polymers to develop composite materials having high conductivity, flexibility, and stretchability. Elastic-conductive polymers acquire higher stretchability than conventional metal wires, and their electrical features sustained their earlier class when considering the high strain. Apart from these qualities, the flexible EDs utilize polymer materials that are simple to manufacture with printing technology. Electrically conducting polymers have been effectively analyzed as electrode material thanks to their mechanical flexibility, fast kinetic charge/discharge, low cost, and high electrical conductivity. The conducting polymer, mechanical properties permit simple processing and can be fabricated as self-healing/self-supporting flexible thin film. These features of conducting polymers make them appreciably advanced electrode materials for the fabrication of flexible energy devices.

2.2.2.2 Other organic materials

Beyond polymer materials, there are various other materials, namely organometallic complexes and small organic molecules. Organometallic complexes are the type of materials containing a huge amount of fascinating mechanical, magnetic, electrical, and optical properties. The remarkable application of these materials in the field of optoelectronics and energy devices makes these materials appropriate.

2.3 STRUCTURAL REQUIREMENTS

The distortion restrictions of devices commonly depend on the counter amid the acquired strain and strength of active materials. Except for the advancement of novel active materials that can naturally flex [10], configurational design, which is derived from the concept of minimizing the acquired strain, plays a crucial part in flexible EDs technology. The mechanical distortion method is complex in realistic applications. Hence, the evaluation of distorted mechanics and electrical properties alteration can lead to the flexible configurational design of EDs.

2.3.1 Flexible substrates and membranes

Electronics films are generally placed on a rigid substrate. Hence, maximum stress is assembled on the active material; the resulting electronic device cannot undergo considerable bending distortion [11]. Elastic substrate takes maximum bending strain, and the higher strain limits of devices are mostly based upon the stretchability/flexibility of the substrate. By enhancing the structural design of active electrode materials, their constraint can be conquered. Essentially, smooth organic electrode active materials [12] endure low charge transport; therefore, a significant structural design unlocks the opportunity of fabricating flexible devices by utilizing the wide range of rigid inorganic materials with outstanding stability and high theoretical performance [13].

In EDs, the approach of the flexible substrate has been expanded to other components. Thus, the advancement of flexible energy devices is launched by using the flexible electrode. For example, a wide range of flexible carbon-based materials like graphene, CNT films, graphene foam, graphene oxides, carbon nanofiber paper are broadly used as a current collector for flexible energy devices. Compared to conventional metal foils current collector, flexible carbon-based materials show strong adhesion, high contact area, and outstanding flexibility to electrode materials.

The approach to using flexibillily integrates the flexibility of electrolytes and separators for highly flexible energy devices. Some carbon textile derivatives were found from natural fabrics, such as paper, silk, and cellulose; copper textile was also selected as a membrane by integrating its fibers with the active layer. Simultaneously, flexible gel polymer electrolytes and solid-state electrolytes are excepted as being necessary for flexible EDs to combine buckle/twisted package [14].

In the process of enhancing mechanical distortion, the flexible approach can be appealed to each and every individual component of an encapsulated EDs. Numerous flexible conductive substrates have been applied to replace the conventional stiff substrate. To fabricate high-performance flexible energy devices, flexible separators are combined. However, other favorable methods must be utilized to strengthen the connection between the flexible substrate and active layer, thus provide the electronic devices under distortion.

2.3.2 Thickness of compound/active layer

The development of nanomaterials with small thickness of a few hundred nanometers can enhance the flexibility and twistability of flexible EDs. However, the thickness of the electrically active layer decreases the extent of the power source and affects the final performance of the energy device due to the quantity decrement of active materials. The thickness of the active layer should be less than the thickness of the flexible device substrate. Therefore, to control the substrate thickness, important attention has been given to integrated flexible devices.

2.4 WEARABILITY ASSESSMENTS

2.4.1 SOFTNESS

One of the essential aims of the flexible/portable EDs is comfortability, which means making it wearable with textile-level texture [15]. The convenience of flexible energy devices has never been mentioned in the researches. Concerning the convenience analysis of textile, we accept that a wearable energy device should be both soft and flexible. Considering the truth that flexible energy devices generally acquire a leather-like texture with a covered polyelectrolyte [16], we suggest a softness variable to estimate the softness of the energy devices. The softness of flexible energy devices can be analyzed by commercially available fabric and leather softness tester, as displayed in Figure 2.1. The extension of the flexible energy devices produced is inscribed as softness (unit = mm). The huge extension height means superior softness.

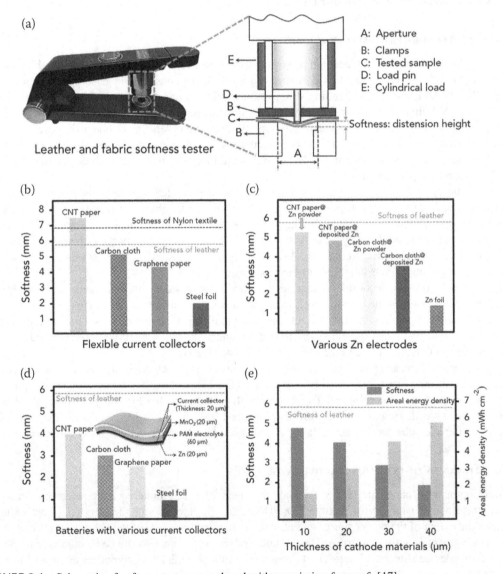

FIGURE 2.1 Schematic of softness tester, reproduced with permission from ref. [17].

A flexible current collector is a crucial component in flexible energy storage devices. We calculated the softness of numerous generally used flexible current collectors, as well as the economically available nylon and leather textile. Among them, CNTs display the best softness, which is nearly three times greater than the steel foil and even greater than nylon textile and commercial leather. Other carbon-based substrates, such as carbon cloth and graphene paper, also display satisfactory softness. That said, the coated structure, flexible/bendable carbon-based materials carbon cloth and CNT paper are superior to serve as conductive substrates for flexible EDs.

2.4.2 STRETCHABILITY: THE RESIDUAL STRAIN

Wearable and flexible EDs occasionally need stretchability. In contrast to flexible devices, stretchable energy storage devices handle huge strain and shape distortion, and therefore declare excessive demands for structural designs and materials. For both devices and materials, variable stretchability is difficult to achieve but the main character because any slight irreversible mechanical distortion will ultimately disgrace device performance [18]. For practical applications, portable/stretchable EDs don't yet work accurately when being stretched, and they retrieve both their mechanical strain and electrochemical performance upon elimination of external force. Until now, several scientists have devoted considerable efforts to creating energy devices significantly more stretchable; however, they have paid limited attention to their mechanical recoverability.

2.5 SELF-HEALING MECHANISM

For stretchable and flexible EDs, the unavoidable mechanical damage and structural fracture that happens during the repeated bending and stretching contribute to electrochemical performance distortion and even safety problems [19]. Particularly when the stretchable/flexible EDs are combined with implanted biological electronics, the life span of the power source determines the service duration, due to the complex and risky replacement of any supercapacitor or damaged batteries. Self-healing materials are motivated by natural skin; they can self-repair the electrochemical properties and mechanical strength after the unacceptable mechanical destructions [20]. Triggered by external motivation and damages, self-healing polymeric materials have been efficiently used to attain the improvement of the damaged electrodes. In this case, after introducing the self-healing potential, the durability and reliability of EDs could be extended effectively.

To establish the high-performance stretchable/flexible electrode for energy devices, a series of self-healing polymers has been investigated according to the regeneration method of ligand metal bonding, supramolecular network, reversible crosslinking among polymer chain, hydrogen bonding, etc. The self-healing mechanism can be classified into two types: (i) intrinsic self-healing polymers with reversible bonds and (ii) self-healing through exhaustion of healing agents.

2.5.1 INTRINSIC SELF-HEALING POLYMERS WITH REVERSIBLE BONDS

These healing mechanisms depend on the reversible generation of poor chemical bonding of polymer chains containing dynamic covalent bonds [21] and noncovalent bonds [22]. Upon damage, poor chemical bonds like hydrogen bonds and S-S bonds were broken. These broken bonds could restore at the cracked interfaces due to the efficient properties of the polymer chain.

In the prior case, the healing method is a classical supramolecular assembly with effective a noncovalent bond containing metal coordination bonding, hydrogen bonding, electrostatic cross-linking, and others. Depending on the mobility of the polymer chains and the kinetic constants of supramolecular assembly, the healing procedure classically happens in series, ranging from minutes to hours. To boost the healing procedure, external stimuli like moisture and heat can be applied to enhance the assembly activities and the chain dynamics. For the former, various types of effective covalent bonds containing imine, disulfide, olefin, ester, acylhydrazone, diels-alder reaction are appropriate for the growth of such intrinsic self-healing materials. Because of the poor kinetic ability of covalent generation, as compared to the noncovalent case, exterior stimulus like pH change, heat, catalyst, and light is always required to trigger the healing process.

Note that the self-healing procedure can be designed with the consolidation of practical guest materials inside or on top of the polymer matrix in a composite system [23]. The reformation of bonding can happen either between polymeric chains and other guest materials or nanostructure or two polymeric chains, or two or more different polymers.

2.5.2 SELF-HEALING THROUGH EXHAUSTION OF HEALING AGENTS

Different from the above-mentioned category, self-healing under this healing mechanism needs the consumption of materials that are pre-encapsulated in the fabrication. Thus, this self-healing mechanism is also known as extrinsic self-healing. The highly reported technique for self-healing via the exhaustion of the healing agent depends on microcapsule filler introduced by White *et al.* Functional liquid-based agents including catalyst, metal alloy, and monomers are embedded in microcapsules, which are encapsulated into a polymer matrix that supplies the healing sites. These microcapsules are created to be shattered by mechanical fracture. While microcapsules damage, the healing agents are released from these microcapsules at the affected portions to repair the cracks by chemical reactions such as physically filling the gaps or polymerization [24]. The healing procedure is introduced once the cracks are created/ formed and does not need any external impetus.

This approach can also be appropriate to numerous polymer matrix and distinct kinds of practical liquid healing agents, permitting the self-healing composites to acquire various specific properties. For example, to develop self-healing conductors, the liquid metal is required as a healing agent. In these conductors, the transport and release of liquid agents can restore the damaged conductive pathway to revive conductivity. However, numerous healing cycles are hard to attain at the same location because of the limited amount of healing agents accessible in the healed region. To overcome this problem, the microvascular self-healing system has been developed. This system contains a refilling channel consisting of self-healing agents.

2.6 DESIGN OF FLEXIBLE ENERGY DEVICES

Stretchable and flexible electronics that can contribute outstanding adaptable properties for daily wear, and even be attached to the human skin and inserted into the human body as biodegradable electrical devices, can normally be accomplished by rational usage of architectural and mechanical design techniques [25] that depend on rational design rules. Recently, several novels and unique ideas on architectural design have been carried to flexible electronics, and noticeable advancement has been made in the stretchability/flexibility of distortable batteries and supercapacitors (SCs) with innovative circuit configurations, such as origami design, bridge island design, and cable/wire patterns with the usage of complexed mechanical design approaches.

Fiber-based clothes, which offer outstanding flexibility and stretchability in all dimensions, use cable/ wire-like samples of EDs; this clothing has been broadly analyzed by decreasing the size of 2D perpendicular to the twisting direction utilizing gel and solid-state electrolytes and wire-shaped current collector/electrodes. These unique fiber-shaped devices offer huge development in future stretchable devices because of their wearability in numerous wearable and portable electronic textiles.

2.7 FLEXIBLE ENERGY STORAGE AND CONVERSION DEVICES

This segment will examine distinct types of energy conversion and storage devices. The devices to be discussed contain batteries (Bs), supercapacitors (SCs), nanogenerators (NGs), solar cells, etc. Note that each type of device includes more than one form of material.

2.7.1 ENERGY CONVERSION DEVICES

2.7.1.1 Nanogenerator (NGs)

Energy conversion devices that can convert numerous energy sources (such as wind, solar, thermal, mechanical energy) into electrical energy have gained value currently [26]. Mechanical energy is comprehensive and sufficient in the environment. Triboelectric NGs with superior functioning for mechanical energy conversion are evolving impressively. They also are proving exceptionally interesting for wearable/portable applications. The traditional triboelectric NGs, designed triboelectric layers with nano and microstructure, were employed to enhance the output power by expanding the contact surface area. However, the nano and microstructure of the designed triboelectric layers can be easily disrupted after a long operational time, dominating the distortion of overall performance. Figure 2.2 shows the schematics of some of the NGs. The most favorable solution would be to provide the triboelectric NGs with self-healing potential, which can revive the destructed nano and microstructure of the triboelectric layer and eventually recover the overall performance of NGs. Kim *et al.* reported self-healing shape memory polymer-based triboelectric NGs, which illustrated the self-healing potential to revive the overall performance of NGs [29]. The lifetime and tolerance of triboelectric NGs could be enhanced appreciably through the self-healing strategy of shape memory polymer. Tang *et al.* reported the paper-based triboelectric NGs based on coincidence of contact electrostatic induction, and electrification has been created into numerous configurations for the conversion of low-energy mechanical energy [30]. Paper has all the features that a substrate material has, such as having a low cost, being lightweight, biocompatibility, and flexibility. The piezoelectric and conductive materials can be finely covered to the paper substrate and even stick to the substrate while stretching, twisting, and bending.

2.7.1.2 Photovoltaic

Perovskite solar cells (PSCs), used in photovoltaic (PV) devices with low cost and excellent power-conversion efficiency (PCE) in third-generation PV technologies, have experienced rapid development over the past decade due to their various advantages, such as low-cost production, low density, and flexibility. Figure 2.3 shows simple schematics of an SC. The scope of electrical characteristics from dielectric to superconductivity revealed by perovskites can be recognized as the broader scopes of physical characteristics revealed by a single material. Currently, organic-inorganic halide perovskites have displayed outstanding performance as semiconductors appealed in thin-film transistors and light-emitting diodes, as first recorded by Schmid, Gebauer, and Mitzi, respectively. With regard to solar cells, an inorganic-organic halide perovskite, particularly $CH_3NH_3PbI_3$, was initially involved in an SC as a sensitizer in a liquid-electrolyte based DSSC structure by Kojima *et al.* in 2009; it had a low efficiency of 3.8% [27]. Lead-based perovskite has various limitations, and they are toxic to the environment. Hui-jing Du *et al.* has reported a lead-free perovskite solar cell with PCE of 23.36%; it has the advantages of low cost and simple processing techniques [31]. Flexible OSCs have attracted rapid interest for use in wearable, portable electronics and roll type applications thanks to their mechanical flexibility and

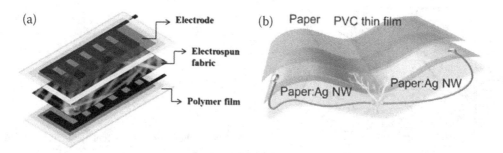

FIGURE 2.2   (a) Nanogenerator (NG), (b) Paper based NG, reproduced with permission from refs. [27,28].

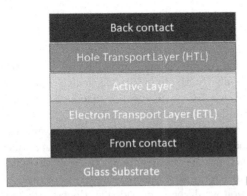

FIGURE 2.3 A schematic of a solar cell.

lightweight are feasible with the necessity of soft materials and curved surface. Because of their quick development, the performance efficiency of organic solar cells (OSCs) has attained as high as 17% for tandem OSC and 14% for single-layer OSC with rigid glass substrate [32,33]. To date, the PCE of flexible OSCs is lower than the cell with the rigid glass substrate, which is nearly 16% on ITO. In the beginning, the flexible OSCs with PCE of 1.5% were reported by Kaltenbrunner in 2012 [34]. The efficiency of flexible OSCs has since enhanced rapidly to 16%, as reported in 2020 for non-fullerene OSCs [35]. To improve the efficiency of flexible OSCs, numerous flexible substrates must be examined to use in place of ITO, such as hybrid composite electrodes, metallic nanostructures, and carbon-based materials.

2.7.1.3   Other flexible generators

Fuel cells can work with tremendously high electrical efficiencies of 60–70%, which are appraised as ideal and perfect energy conversion devices [36]. A stretchable on-chip fuel cell was manufactured on a flexible/bendable cycloolefin polymer film rather than a rigid and breakable silicon film, as shown in Figure 2.4. The performance efficiency of the stretchable fuel cell was similar to that of the breakable and rigid silicon cell. This design gives an efficient solution that fixes the high cost and brittleness of conventional silicon-based on-chip fuel cells, as shown in Figure 2.4. Harvesting energy from the mechanical motion of the human body and converting it into electricity is a powerful approach for developing self-powered, environmentally friendly, and low-cost flexible and portable/stretchable devices.

Zhong et al. has reported a paper-based NG, which converts external mechanical energy into electricity. The film of Ag is deposited by thermal evaporation on the paper available to form 'Ag/paper,' and then the composite paper is spin-coated with PTFE to produce 'PTFE/Ag/paper,' and finally assembled

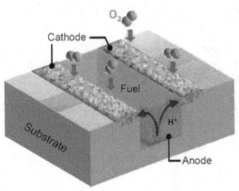

FIGURE 2.4 Schematic of a fuel cell, reproduced with permission from ref. [37].

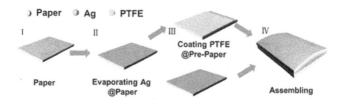

FIGURE 2.5 Process for manufacturing NGs, reproduced with permission from ref. [38].

with 'Ag/paper' to build NG, as shown in Figure 2.5. This device shows the highest output power density of 90.6 µW/cm². It can be encapsulated with energy storage units to store the pulse energy.

2.7.2 Energy storage devices (ESDs)

2.7.2.1 Flexible batteries (FBs)

The developing technology and demand progress push the batteries market to constantly inquire for new applications. While scientists are striving for emerging low-cost and high-energy-density batteries to inscribe the huge-sized electric vehicles and grid applications, batteries are evolved and designed in a unique technique, becoming ultrathin, rollable, stretchable, and flexible. The batteries manage huge opportunities to promote roll-up display, wearable electronics, IOT, and implantable medical robots.

2.7.2.1.1 Li-ion flexible batteries (LiBs)

LiBs are recognized as the main sources of power for flexible smartphones and wearable electronics, thanks to their long life span, high-energy density, lack of memory effect, and high voltage (>3.5 V); these features enhance the design advancement of flexible devices. The recent task falls into reducing the gap amid lab-scale technologies and industrial device requirements. All these characteristics of Li-ion batteries are appropriate for portable electronics. Thus, there is no uncertainty that enabling the flexibility of LiBs will have appreciable meaning. A scientist evolved a J. flex battery to be fast-charging, solid-state, and highly flexible. J. flex can completely twist without abdicating power capability, which can assist a huge scope of companies using new technologies to append novel and unique design, efficiency, and comfort to their devices, as shown in Figure 2.6(a, b). Unexpectedly, belt-type batteries can acquire approx. 80% retention capacity after 1000 cycles at 1C.

In flexible/wearable batteries, while integrating the wireless charge setup, it excludes the universal series bus port; it is excluded because it shows negative performance on the manufacturing of flexible and thin batteries. Panasonic displays the flexible battery with capacities ranges of 18, 42, 67 mAh, with a thickness of 0.55 nm, as shown in Figure 2.6(c). To the battery setup, the wireless charging ability is added to wearable/flexible batteries. These batteries have the capability to twist up to ±25°/100 mm and bend up to a radius of 25 mm. FBs with small bending curvature radius are attracting appreciable interest. Samsung showcases a model of flexible and thin batteries with a curve-shaped pattern at the inter Battery exhibition in 2015, as shown in Figure 2.6(d). LG Chem evolved the flexible wire batteries, which can bend into a 15 mm radius and display excellent tolerance to mechanical distortion, as shown in Figure 2.6(e). One of the important factors used to evaluate flexible batteries is safety. Extraordinary flexible batteries with high capacity are discussed by Prologium Technology Co., Ltd.; these batteries have flexible circuit board packaging and printing technologies, and they display outstanding tolerance to destructive testing and high temperatures. The cells can be as huge as $182 \times 230 \times 0.4 \, \text{mm}^3$, with a capacity of 1000–1400 mAh, as shown in Figure 2.6(f). These batteries can tolerate high temperatures and will not explode. These batteries stay strong in safety tests for tears, punctures, gunshots, and even impact, as shown in Figure 2.6(g).

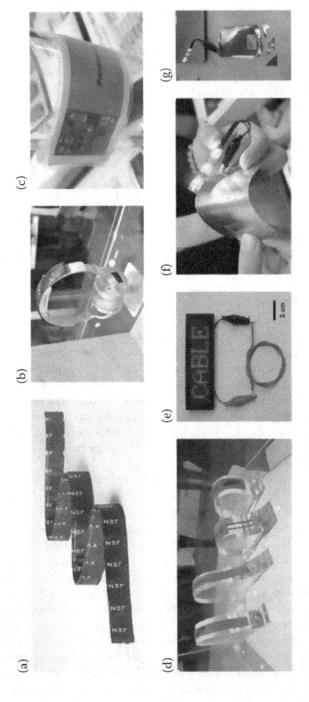

FIGURE 2.6 Flexible batteries, reproduced with permission from refs. [39,40].

2.7.2.1.2 *Other flexible batteries*

Rather than LiBs, various other chemistry-based FBs have also significantly evolved superior characteristics; batteries include silver-zinc (Ag-Zn), lithium-manganese oxide (Li-MnO$_2$), and zinc-manganese oxide (Zn-MnO$_2$), respectively. Silver-zinc (Ag-Zn) batteries are composed of zinc anode and a silver-oxide cathode, which display high working stability and high power/energy density. Ag-Zn batteries are used in military equipment, where cost is not the major feature, such as rockets, satellites, and spacecraft. Imprint energy displays the modeled Ag-Zn batteries with polymer electrolytes, which were asserted to deliver a variety of features encapsulated into the flexible, wrist-worn product that joined wireless communication, as shown in Figure 2.7.

Lithium-manganese oxide (Li-MnO$_2$) batteries consist of Li as an anode and MnO$_2$ as a cathode, with an actual output voltage of ~3.0 volts. In these batteries, high power density is an outstanding application. Brightvolt developed these batteries with the capacity of 10–48 mAh and 0.45 mm thickness, as shown in Figure 2.7(b). These flexible batteries contained a polymer electrolyte; this nontoxic and printed Li-metal paste confirmed the safety requirement. Zn-MnO$_2$ batteries consist of Zn as anode and MnO$_2$ as cathode, and an aqueous NH$_4$Cl solution electrolyte has promoted the main battery marketplace thanks to easy manufacturing, high safety, and low cost. The Zn–MnO2 batteries show an actual voltage of 1.5 V and display their excellent performances with 15°C–30°C temperature. These types of wearable/flexible batteries, as shown in Figure 2.7(a), are broadly used in music cards, RFID electronics tags, and smart cards.

2.7.2.2   Supercapacitors (SCs)

Flexible SCs are tremendously interesting for a significant amount of developing portable and wearable thin consumer products. The innovation of the flexible/portable supercapacitor is the actualization of flexible/wearable substrate and electrode materials to integrate configurational flexibility with essentially high-power density of a flexible supercapacitor (Figure 2.8). Flexible supercapacitors mainly are of two types: (a) pseudo capacitors (PCs) and (b) electric double-layer capacitors (EDLCs). EDLCs are based on the non-faradaic, while PCs are based on the faradaic energy storage process. Flexible SCs are a tremendously interesting and encouraging field compared to batteries, thanks to their high operation time, fast charging and discharging rate, mechanical flexibility, and high power density. In flexible SCs, the flexible and highly conducting carbon chain is used as both the current collector and electrodes.

Currently, a broad set of novel pseudocapacitive electrode materials have been evaluated, with the target of enhancing the energy/power density of flexible SCs. PCs have a great charge-storing capability in comparison to EDLCs. But PCs are bounded because of their low cyclic stability and high cost. A huge range of transition metal oxides is popular for their encouraging pseudocapacitive nature. Traditionally, various binary metal oxides such as RuO$_2$, Mo$_2$O$_3$, NiO, V$_2$O$_5$, Co$_3$O$_4$, iron oxide, etc. have displayed excellent power and energy density. Currently, quasi-solid state highly flexible SC devices are manufactured by sandwiching electrodes of NiCo$_2$O$_4$ nanocrystal on GO (graphene oxide) substrate [42]. The metal oxide frameworks (MOFs) for flexible SCs can be adapted in three different ways: a composite electrode, electrode materials, and a flexible substrate. The developing way in a flexible substrate is wood and biomass-derived substrates. Currently, a novel low crystalline nanoparticle of FeOOH anode materials covered on carbon-fiber cloth has acquired both great power and energy density [41]. Carbon fabric is an appropriate carbon chain for flexible/portable supercapacitors with regard to flexibility, stiffness, and strength, and it is mainly fabricated twill, plain, and satin-weaving techniques. Numerous manufacturing techniques, such as printing, weaving, filtration, chemical vapor deposition, dip-drying, or evaporation, are used to build a carbon chain by hydrogen bonding and Vander Waals interaction of carbon particles. The other scientific efforts are to evolve robust manufacturing routes and solution-based procedures for large-scale fabrication of flexible supercapacitors. New and unique electrodes and electrolyte materials are being examined to allow the high-temperature functioning of flexible SCs. Thus, supercapacitors are environmental-friendly and rapidly accelerating technology for energy storage devices.

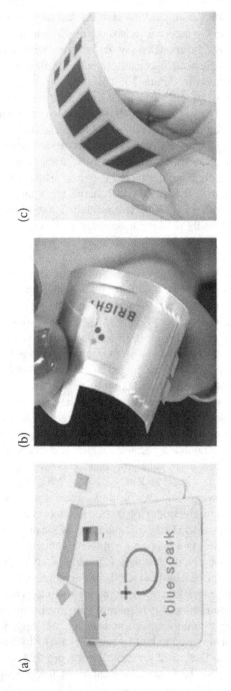

FIGURE 2.7 Other types of flexible batteries (FBs), reproduced with permission from ref. [39].

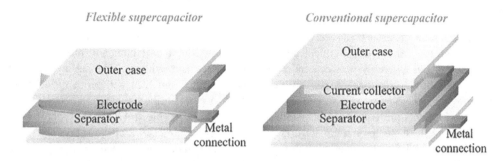

FIGURE 2.8 Schematic diagram of flexible SC and conventional SC, reproduced with permission from ref. [41].

2.8 CONFIGURATION DESIGNS FOR FLEXIBLE ESDS

The configurational designs play a major role in defining the stretchability/flexibility of ESDs. To encounter the creative demands and determining applications, designs from 1D to 3D structures, as shown in Figure 2.9, have been developed.

2.8.1 1D CONFIGURATION OF ESDs

2.8.1.1 Fiber-type

This type of structure was established to incorporate batteries and supercapacitors (SCs) into portable electronic applications of ultimate stretchability/flexibility. Fiber-type configurations can be classified into three types: coaxial fiber, helix fiber, and parallel fiber. Fiber-shaped device configuration ESDs illustrate the chances of instantly combining them with portable electronics to power general-purpose smart fabrics. All three structures are developed from the smooth sandwiched system used in conventional 2D ESDs.

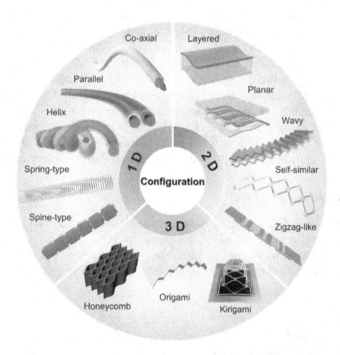

FIGURE 2.9 Configurational designs of 1D, 2D, 3D ESDs, reproduced with permission from ref. [43].

The coaxial fiber-type configuration is fabricated by affixing one fiber electrode to the other fiber electrode of opposite electrode material using gel electrolyte. This can allocate a large interfacial area amid two electrodes to improve the electrochemical performance. However, it is technically more demanding to recognize the coaxial assembly between inner and outer electrodes without inducing any short circuit. Because of their omni-directional flexibility, these fiber batteries and SCs will probably observe wide applications when integrated into smart clothes and textiles. A parallel fiber structure indicates that the two distinct fiber electrodes are connected in parallell, confined by a gel electrolyte. This type of fiber configuration is simple to scale up on a flat substrate and also easy to fabricate. It also displays high flexibility but bounded stretchability, while the helix fiber configuration includes two twisted electrodes without any need for a flat and smooth substrate. It generally shows similar flexibility/wearability and improved stretchability, in comparison with parallel configuration devices. These twisted-type, fiber-shaped batteries and supercapacitors are simpler to interlink into fabrics to deliver as portable ESDs.

2.8.1.2 Spring types

Spring-type structures can be established from threadlike devices. The coiled/twisted spring-type structure could provide the design with excellent elasticity and stretchability. This ability makes it self-reliant from the intrinsic characteristics of electrode materials. Because of the excellent elasticity and stretchability, spring-type batteries and SCs could deliver as power sources to build appreciable flexible electronic devices. However, the comparative low loading weight of active materials for fiber-type and spring-type devices limits their applications.

2.8.1.3 Spine type

A spine-type configuration includes various interconnected stiff batteries and supercapacitor divisions combined by soft connectors. The soft connector may also provide the spine configuration devices with average mechanical stretchability/wearability and a specific amount of stretchability, managing the high electrochemical performance subject to mechanical distortion. The batteries and the supercapacitor divisions with huge loading mass of active materials and traditional sandwiched structures can offer a remarkably enhanced overall energy supply in comparison to spring and fiber type configuration. Because of their growing size, the spine-type configuration devices generally get encapsulated into niche applications and into smart textiles to supply as fiber-type configurational devices.

Altogether, the 1D configurational ESDs display a variety of leading properties, such as significant omnidirectional flexibility, diminutive size, and being lightweight according to their linear structural designs. However, 1D configurational energy devices exhibit relatively lower energy density because of the bounded active materials loading; the overall device's energy could improve to a certain extent by increasing the loading of the active material. The application required high energy density, 1D configurational devices, which are not likely to fulfill the demands. Therefore, 2D and even 3D stretchable/flexible ESDs must be fabricated to fulfill the demand of high energy density energy storage devices.

2.8.2 2D CONFIGURATION OF ESDs

2D configuration batteries and supercapacitors can be manufactured with remarkably enlarged electrode size and increased active materials loading, compared to 1D configuration ESDs [44]. Moreover, depending on the layered electrode configuration, electrode laminations can remarkably enhance the loading of active materials and overall energy of the devices. 2D configurational devices can be divided into three types:

    i. Layered sandwich,
   ii. Planar interdigital, and
  iii. Other new 2D device configurations.

2.8.2.1 Layered sandwich configuration

Layered sandwich-type configuration is the generally utilized structure for materialistic batteries and supercapacitors, namely cylindrical batteries and pouch cells. However, stretchability and flexibility are tough to recognize with traditional layered sandwich structural design in which numerous separator and electrode layers are tightly connected and closed into a cylindrical case or rigid box. Moreover, the stiff microstructure of the active materials, such as activated carbon bulk powder and lithium cobalt oxide, controls the stretchability and flexibility of the electrodes. Even slightly twisting, bending, stretching, and flexing could initiate cracks in the active materials, resulting in critical performance degradation or safety problems. Therefore, to acquire stretchability and flexibility, CNT and graphene have been used as active materials. According to the large size of the electrode and dense packing, the layered sandwich structural devices can acquire great total energy supply compared to 2D flat configuration devices. The layered sandwich configurational devices are highly suitable for large-scale development. The layered sandwich configurational devices have a limitation of their stretching tolerance and bendability because flexible batteries and supercapacitors are exposed to critical stretching and bending conditions, and the electrode layers endure extrusion force and tension in different directions.

2.8.2.2 Planar interdigital configuration

To resolve extrusion force and inner tension that occurred in sandwich configuration devices, the planar interdigital type structural was introduced with two electrodes modeled in one plane. The co-planar model provides the same extrusion stress and tension to both the electrodes under mechanical distortion, significantly stifling the movement amid different layers. The liquid/gel electrolyte or gap between negative and positive electrodes can separate two electrodes and encourage effective ion transport between the two electrodes. Currently, the interdigital structure has become the influencing design because the open channel between the electrodes improves the performance rate. Moreover, the planar devices with an interdigital configuration can simply be encapsulated into planar chips, connected with the function of information processing, harvesting, and sensing. However, these configuration devices, such as supercapacitors and batteries, still endure serious challenges of relatively low loading of active materials and limited stretchability.

2.8.2.3 Other 2D novel configurations

However, the planar interdigital configuration energy storage devices provide high flexibility, but the features of stretchability are still not satisfactory. Currently, significant research activity has been devoted to introducing new 2D configurations to acquire improved flexibility and stretchability, such as zig-zag like structure, self-similar geometries, and wavy structures.

In conclusion, 2D configuration devices generally provide a bunch of remarkable properties, such as being highly flexible, ultrathin, and lightweight. These properties allow 2D stretchable/flexible ESDs to be encapsulated into a huge range of portable and wearable electronics.

2.8.3 3D CONFIGURATION ESDs

2.8.3.1 Origami/Kirigami/honeycomb-based structures

To fulfill the need for 3D stretchable functional devices and electronics, 3D configuration devices with 3D origami/kirigami or honeycomb architecture have been introduced and provide outstanding uniaxial stretchability and tunable flexibility, along the direction of stretchability. The architecture can be encapsulated with foldable smart sunshields, wavy roofs, and solar panels. The a la carte design can fulfill the requirements and delight customers [45].

2.9 SUMMARY

Recent developments in structural design and characterization of novel materials have facilitated numerous applications in the area of flexible/portable energy devices to grow as reality. The applications such as flexible solar cells, nanogenerators, fuel cells, batteries, supercapacitors, etc., will transform the culture of today's world. This chapter covers flexible energy devices consisting of structural requirement, flexible material used, wearability assessment, self-healing mechanism, and configurational designs of ESDs. Numerous new materials produce various possibilities and options for scientists to evolve a new era. Numerous nanomaterials and organic materials have outstanding characteristics and are attracting lots of interest. In particular, conjugated polymer is attracting significant interest thanks to its intrinsic stretchability, wearability, and flexibility. The device's stretchability and flexibility have been enhanced considerably via these structural and configurational designs. Thus, the flexible energy devices' configurational designs are becoming more and more attractive. All configurational designs have illustrated the huge potential for the near future. That said, enhancing mechanical features, acquiring excellent flexibility, and maintaining excellent capacity simultaneously are still demanding for energy devices.

REFERENCES

1. Xia M, Cheng Z, Han J, et al. (2014) Extremely stretchable all-carbon-nanotube transistor on flexible and transparent substrates. *Appl Phy Lett*, 105(14): 143504.
2. Bae S, Kim H, Lee Y, et al. (2010) Roll-to-roll production of 30-inch graphene films for transparent electrodes. *Nat Nanotechnol*, 5(8): 574.
3. Chen J, Liu Y, Minett AI, et al. (2007) Flexible, aligned carbon nanotube/conducting polymer electrodes for a lithium-ion battery. *Chem Mater*, 19(15): 3595.
4. Yakobson BI, Brabec CJ, Bernholc J (1996) Nanomechanics of carbon tubes: instabilities beyond linear response. *Phys Rev Lett*, 76(14): 2511.
5. Shang Y, He X, Li Y, et al. (2012) Super-stretchable spring-like carbon nanotube ropes. *Adv Mater*, 24(21): 2896.
6. Hecht DS, Hu L, Irvin G (2011) Emerging transparent electrodes based on thin films of carbon nanotubes, graphene, and metallic nanostructures. *Adv Mater*, 23(13): 1482.
7. Geim AK, Novoselov KS (2007) The rise of graphene. *Nat Mater*, 6(3): 183.
8. Wu X, Sprinkle M, Li X, et al. (2008) Epitaxial-graphene/graphene-oxide junction: an essential step towards epitaxial graphene electronics. *Phys Rev Lett*, 101(2): 026801.
9. Onorato J, Pakhnyuk V, Luscombe CK (2017) Structure and design of polymers for durable, stretchable organic electronics. *Polym J*, 49(1): 41.
10. Oh JY, Rondeau-Gagne S, Chiu YC, Chortos A, Lissel F, Wang GN, Schroeder BC, Kurosawa T, Lopez J, Katsumata T, Xu J, Zhu C, Gu X, Bae WG, Kim Y, Jin L, Chung JW, Tok JB, Bao Z (2016) Intrinsically stretchable and healable semiconducting polymer for organic transistors. *Nature*, 539: 411.
11. Gleskova, H, Wagner, S (1999) Failure resistance of amorphous silicon transistors under extreme in-plane strain, *Appl Phys Lett*, 75: 3011.
12. Nishide H, Oyaizu K (2008) Toward flexible batteries. *Science*, 319(5864): 737–738.
13. Sun Y, Rogers JA, (2007) Inorganic semiconductors for flexible electronics. *Adv Mater*, 19(15): 1897–1916.
14. Sun P, Lin R, Wang Z, Qiu M, Chai Z, Zhang B, Meng H, Tan S, Zhao C, Mai W (2017) Rational design of carbon shell endows TiN@C nanotube-based fiber supercapacitors with significantly enhanced mechanical stability and electrochemical performance. *Nano Energy*, 31: 432.
15. Zhai S, Karahan HE, Wei L, Qian Q, Harris AT, Minett AI, Ramakrishna S, Ng AK, Chen Y (2016) Textile energy storage: Structural design concepts, material selection and future perspectives. *Energy Stor Mater*, 3: 123–139.
16. Huang Y, Tang Z, Liu Z, Wei J, Hu H, Zhi C (2018) Toward enhancing wearability and fashion of wearable supercapacitor with modified polyurethane artificial leather electrolyte. *Nanomicro Lett*, 10: 38.
17. Li H, Tang Z, Liu Z, Zhi C (2019) Evaluating flexibility and wearability of flexible energy storage devices. *Joule*, 3: 613–619.
18. Liu W, Song MS, Kong B, Cui Y (2017) Flexible and stretchable energy storage: Recent advances and future perspectives. *Adv Mater*, 29: 1603436.

19. Chen D, Wang D, Yang Y, Huang Q, Zhu S, Zheng Z (2017) Self-healing materials for next-generation energy harvesting and storage devices. *Special Issue: Renew Energ Convers Storage*, 7(23): 1700890.

20. Kang J, Tok BJ, Bao Z (2019) Self-healing soft electronics. *Nat Elect* 2: 144–150.

21. Voorhaar L, Hoogenboom R (2016) Supramolecular polymer networks: hydrogels and bulk materials. *Chem Soc Rev*, 45: 4013.

22. Taynton P, Ni H, Zhu C, Yu K, Loob S, Jin Y, Qi HJ, Zhang W (2016) Repairable woven carbon fiber composites with full recyclability enabled by malleable polyimine networks. *Adv Mater*, 28: 2904.

23. Wu Q, Wei J, Xu B, Liu X, Wang H, Wang W, Wang Q, Liu W (2017) A robust, highly stretchable supramolecular polymer conductive hydrogel with self-healability and thermo-processability. *Sci Rep*, 7: 41566.

24. Zhang X, Tang Z, Tian D, Liu K, Wu W (2017) A self-healing flexible transparent conductor made of copper nanowires and polyurethane. *Mater Res Bull*, 90: 175.

25. Zhang Y, Huang Y, Rogers JA (2015) Mechanics of stretchable batteries and supercapacitors. *Curr Opin Solid State Mater Sci*, 19: 190.

26. Zhao Z, Yan C, Liu Z, Fu X, Peng LM, Hu Y, Zheng Z (2016) Machine-washable textile triboelectric nanogenerators for effective human respiratory monitoring through loom weaving of metallic yarns. *Adv Mater*, 28: 10267.

27. Wu C, Kima TW, Sung S, Park JH, Li F (2018) Ultrasoft and cuttable paper-based triboelectric nanogenerators for mechanical energy harvesting. *Nano Energy*, 44: 279–287.

28. Shetty S, Mahendran A, Anandhan S (2020) Development of a new flexible nanogenerator from electrospun nanofabric based on PVDF/talc nanosheets composites. *Soft Matter*, 16: 5679–5688.

29. Lee JH, Hinchet R, Kim SK, Kim S, Kim SW (2015) Shape memory polymer-based self-healing triboelectric nanogenerator. *Energy Environ Sci*, 8: 3605.

30. Tang Q, Guo H, Yan P, Hu C (2020) Recent progresses on paper-based triboelectric nanogenerator for portable self-powered sensing systems. *EcoMat*, 2: e12060.

31. Akihiro Kojima, Kenjiro Teshima, Yasuo Shirai, Tsutomu Miyasaka (2009, May 6) Organometal halide perovskites as visible-light sensitizers for photovoltaic cells. *J Am Chem Soc*, 131(17): 6050–6051.

32. Du HJ, Wang WC, and Zhu JZ (2016) Device simulation of lead-free perovskite solar cells with high efficiency. *Chin Phys B*, 25: 108802.

33. Meng LX, Zhang YM, Wan XJ, Li CX, Zhang X, Wang YB, Ke X, Xiao Z, Ding LM, Xia RX, Yip HL, Cao Y, Chen YS (2018) Organic and solution-processed tandem solar cells with 17.3% efficiency. *Science*, 361 (6407): 1094–1098.

34. Li H, Xiao Z, Ding LM, Wang JZ (2018) Thermostable single-junction organic solar cells with a power conversion efficiency of 14.62%. *Sci Bull*, 63(6): 340–342.

35. Andreani LC, Bozzola A, Kowalczewski P, Liscidini M, Redorici L (2019) Silicon solar cells: Toward the efficiency limits. *Adv Phys X*, 4(1): 1–25.

36. Winter, M, Brodd, RJ (2004) What are batteries, fuel cells, and supercapacitors. *Chem Rev*, 104: 4245.

37. Li L, Zhong W, Yuan S, Zhang XB (2014) Advances and challenges for flexible energy storage and conversion devices and systems. *Energy Environ Sci*, 7: 2101.

38. Zhong QZ, Zhong JW, Hu B, Hu QY, Zhou J, Wang ZL (2013) A paper-based nanogenerator as a power source and active sensor. *Energy Environ Sci*, 6: 1779.

39. Kong L, Tang C, Peng HJ, Huang JQ, Zhang Q (2020) Advanced energy materials for flexible batteries in energy storage. *A Review, SmartMat*, 1: 1–35.

40. Gupta RK, Candler J, Palchoudhury S, Ramasamy K, Gupta BK (2015) Flexible and high performance supercapacitors based on NiCo(2)O(4)for wide temperature range applications. *Sci Rep*, 5: 15265.

41. Palchoudhury S, Ramasamy K, Gupta RK, Gupta A (2019) Flexible supercapacitors: A materials perspective. *Front Mater* 5: 83.

42. Owusu KA, Qu L, Li J, Wang Z, Zhao K, Yang C, et al. (2017) Low-crystalline iron oxide hydroxide nanoparticle anode for high-performance supercapacitors. *Nat Commun*, 8: 14264.

43. Tong X, Tian Z, Sun J, Tung V, Kaner RB, Shao Y (2021) Self-healing flexible/stretchable energy storage devices. *Materials Today*, 44: 78–104.

44. Muzaffara A, Basheer AM, Deshmukha K, Thirumalaib J (2019) A review on advances in hybrid supercapacitors: Design, fabrication and applications. *Renew Sust Energ Rev*, 101: 123–145.

45. Ning X, Wang X, Zhang Y, Yu X, Choi D, Zheng N, Kim DS, Huang Y, Zhang Y, Rogers JA (2018) Assembly of advanced materials into 3D functional structures by methods inspired by Origami and Kirigami. *A Review*, 5(13): 1800284.

3 Basics and Architectural Aspects of Flexible Energy Devices

T. Bendib, and M. A. Abdi

Higher National School of Renewable Energy, Environment & Sustainable Development, HNS-RE2SD, Batna, Algeria

LEPCM, University of Batna 1, Batna, Algeria

H. Bencherif

University of Mostefa Benboulaid Batna 2, Batna, Algeria

F. Meddour

Higher National School of Renewable Energy, Environment & Sustainable Development, HNS-RE2SD, Batna, Algeria

LAAAS, Department of Electronics, University of Batna 2, Algeria

CONTENTS

3.1 Introduction 44
3.2 Nanotechnology for flexible energy devices 44
3.3 Architectural concepts, structures, and materials for flexible solar cells 45
    3.3.1 Flexible dye-sensitized solar cells FDSSCs 45
        3.3.1.1 Structure design and basic concept 45
        3.3.1.2 Flexible materials and fabrication process for FDSSCs 46
    3.3.2 Quantum dot synthesized solar cell (QDSSCs) 47
        3.3.2.1 Structure design and basic concept 47
        3.3.2.2 Flexible materials and fabrication process for QDSSCs 48
    3.3.3 Toward other flexible photovoltaic technologies 48
        3.3.3.1 Inorganic materials based flexible solar cells 49
        3.3.3.2 Organic materials based flexible solar cells 50
3.4 Architectural concepts, structures, and materials for flexible batteries 50
    3.4.1 Lithium-ion batteries (LIBs) 51
        3.4.1.1 Structure design and basic concept 51
        3.4.1.2 Flexible materials for LIB structures 51
    3.4.2 Zinc-ion batteries (ZIBs) 52
        3.4.2.1 Structure design and basic concept 52
        3.4.2.2 Flexible materials for ZIB structures 53
    3.4.3 Flexible batteries advancement 53
3.5 Architectural concepts, structures, and materials for flexible supercapacitors 53
    3.5.1 Structure design and basic concept 54
    3.5.2 Flexible materials for SCs structures 54

DOI: 10.1201/9781003186755-3

3.6   Conclusion.. 56
References.. 56

3.1   INTRODUCTION

During the past decades, the improvement and implementation of flexible energy devices into flexible electronics has been a field of interest for many researchers due to the devices' exceptional properties, such as flexibility, being lightweight, seamless integration, shape controllability with high mechanical resistance, and reliability [1,2]. Moreover, the design and the structure of integrated energy devices attracted great attention because of the significant potential applications in bendable, wearable, stretchable, and portable electronics [3]. Therefore, it is necessary to develop flexible energy conversion and storage devices, besides the considerable progress in the field of solid-state flexible devices.

In recent years, a wide range of flexible energy devices, such as solar cells, batteries, and supercapacitors, have been thoroughly investigated for powering future wearable and portable electronics. Hence, achieving high performance in terms of energy storage and conversion, and long and reliable cycling with secure operation, becomes a key challenge. As a result, it's essential to obtain effective devices to be applied to the energy conversion and storage systems with continuous and stable functioning. Innovatively designed materials are prerequisites to providing desired architectures and electrochemical performance for flexible energy devices in all aspects. Many research works have been devoted to exploring suitable materials for applications in flexible batteries, supercapacitors, and solar cells [4–6]. To date, several materials and structures have been introduced to shift from conventional energy devices to new flexible forms; different examples of desired flexible energy devices are recently reported in the literature [7]. Despite the technological progress achieved so far, this is still an emerging field, which will lead to challenging enhancements and new opportunities for scientific researchers to overcome many fundamental and industrial issues that currently obstruct the technological aspects of flexible energy conversion and storage in practical applications.

In this chapter, we present fundamental architectures of flexible energy devices, covering the basic materials design and fabrication process. The main goal of this chapter is to offer a comprehensive view of knowledge of flexible energy devices configuration and structural designs to better understand the fundamental principles, including device architecture and designed materials related to the associated technology. In the photovoltaic section, we especially highlight the importance of third-generation solar cells, owing to their performance, and discuss the possibility to emerge such devices for flexible photovoltaic applications through appropriate designs; then we look outward to flexible photovoltaic technologies, including inorganic and organic materials. In the energy storage section, we offer for consideration fundamental aspects in the design of energy storage materials for developing flexible storage devices, in particular, batteries and SCs.

3.2   NANOTECHNOLOGY FOR FLEXIBLE ENERGY DEVICES

Flexible energy devices are becoming more prevalent in our daily lives, including wearable, bendable, and flexible electronic equipment such as medical devices, displays, mobile phones, etc. Recently, the most developed flexible electronics still use rigid-structured batteries as a power source; these are referred to as pseudo-flexible devices. Therefore, developing flexible and portable energy devices is more vital than ever [8,9]. Significant research has focused on the creation of new flexible energy devices that meet requirements for rapid advances in solid-state flexible devices based on polymer and 2D materials. On the one hand, the advancement of these devices gives rise to promising new energy applications. Flexible current collectors, electrodes, conductive polymers, and innovative electrochemical mechanisms, on the other hand, open up the way for flexible energy devices. Thus, the field of energy conversion and storage is being improved even more by highly developed nanotechnology. Nanotechnology has accelerated the development of traditional energy conversion and storage applications, making them more efficient and environmentally friendly.

Some flexible devices, including flexible displays and smartphones and other prototypes, have attracted tremendous interest. The integration of flexible energy storage devices into these smart devices as a power source is required to make people's lives more comfortable. To sustain the strain created under extreme environmental conditions, these flexible devices should have high mechanical robustness. Furthermore, for difficult applications like electric cars, space satellites, and buildings with curved surfaces, flexibility presents a vital parameter. In this context, new materials are required to obtain the superb performance of flexible devices. Thus, higher quality materials can have the ability to overcome the performance constraints in specific applications.

Recent advancements in nanomaterials and ultrathin film-fabrication processes have paved the way for the creation of extremely flexible energy sources, such as supercapacitors, batteries, and solar cells [10,11]. Recent research has taken advantage of conducting nanostructure materials based on thin films such as 0D-nanoparticles, 1D-nanotubes, and 3D-nanoflakes and sheets, which have immunity to mechanical deformation. Additionally, thin films also ensure scalability, multilayer integration, and make compatibility more feasible. Furthermore, chemical and physical vapor deposition, preparation of the solution, sputtering, and sophisticated lithographic patterning have all been used to create mechanically robust flexible power sources [10,11]. For flexible energy applications, nanomaterials such as electrodes with appropriate shape, electroactive nanomaterials, porous electrocatalysts, and stretchable structural designs have been shown to play a crucial role in manufacturing the energy conversion and storage devices, and they may also overcome some issues related to flexible applications.

3.3 ARCHITECTURAL CONCEPTS, STRUCTURES, AND MATERIALS FOR FLEXIBLE SOLAR CELLS

Along with the enormous technological growth in the industry, we are increasingly facing environmental problems like global warming and pollution due to the high demand for fossil fuels. Today, it is necessary to find effective routes for energy conversion to gradually limit fossil fuel consumption. Even though the earth receives 440 EJ/h per year, the amount of the total solar energy is still higher than the annual energy consumption [12]. As a result, photovoltaics opens relevant challenges for active research and development in energy conversion, such as solar cells. Crystalline silicon-based cells are considered as widely used devices in actual photovoltaic applications and occupy more than 80% of industrial PV cells [13]. Nevertheless, energy consumption is increasing, requiring new development of PV technology to overcome serious limitations, e.g., cost, power conversion efficiency (PCE), stability, weight, fabrication process, as well as using rare materials and purity grade.

In recent few years, other kinds of solar cells classified as third-generation cells, such as dye-sensitized and quantum dots-sensitized solar cells (DSSCs and QDSSCs), solar cells (QDSSCs), and organic solar cells have been proposed due to their versatile properties, e.g., esthetics, transparency, and flexibility that meet the demands of commercialization.

3.3.1 FLEXIBLE DYE-SENSITIZED SOLAR CELLS FDSSCs

3.3.1.1 Structure design and basic concept

In 1991, Grätzel and O'Regan reported the prototype of DSSCs using Ru-complex-sensitized colloidal nanostructures with an initial PCE of 7.9% [14]. This prototype has now been enhanced to reach a value of 13% under standard solar conditions AM1.5 [15].

Aside from the conventional solar cells, DSSCs were reported as interesting devices owing to their low manufacturing cost and ease of fabrication. A typical DSSC is shown in Figure 3.1. It is structured

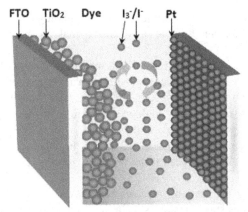

FTO TiO$_2$ Dye I$_3^-$/I$^-$ Pt

FIGURE 3.1 Typical architecture of a dye-sensitized solar cell (DSSC).

with a mesoporous dye-sensitized semiconductor (dye-modified TiO$_2$) as photo-anode, platinum (Pt) as a counter electrode, and iodide (I$^-$/I3$^-$) redox couples as an electrolyte. Figure 3.1 shows the basic operation of DSSC. The light photon induces excitation of electrons, in the sensitizers, which will be injected, via (HOMO to LUMO), to the conduction band of TiO$_2$. These electrons are responsible for the oxidation of the dye-sensitizer, as well as the redox couples dissolved in the electrolyte and hence reduced via the counter electrode. In other words, electrons acquired from the reduced redox couples are responsible for the dye-sensitizers regeneration at the cathode.

3.3.1.2 Flexible materials and fabrication process for FDSSCs

In general, DSSCs use heavy and rigid substrate, like glass, which controls the practical use of DSSCs. To obtain flexible DSSCs, we need initially to engineer flexible substrate, which is considered as an important component to enhance the DSSC proprieties. So, an appropriate substrate is required to replace a glass substrate for flexibility, lightness, impact resistance, and transparency. Transparent flexible substrates like indium tin oxide (ITO) substrate on polyethylene naphthalate (PEN) or polyethylene terephthalate (PET) are the alternative substrates that realize the device's flexibility. However, these materials present disadvantages from the technological process, which needs to apply high-temperature (>150 °C) to fabricate the electrode. Hence, high temperature leads to poor crystallinity; by the way, it also limits the performance of flexible solar cells. To overcome these constraints, several investigations have been considered to involve new materials, instead of Pt, as the electrode to enhance the device performance for commercial use. Due to the excellent catalytic property, high conductivity flexibility, and mechanical stability, polymer composites and graphene and CNTs based materials can substitute the traditional electrode materials [16]. Lee et al. spun coated graphene/PEDOT film on a plastic substrate to fabricate a FDSSC. The PCE of the FDSSCs based on coated graphene/PEDOT film deposited on PET substrate achieves 6.3%, which is competitive to that using sputtered Pt/ITO [17]. On the other hand, Sun et al. realized a counter-electrode based on polymer composite called polyaniline PANI, which is deposited on flexible graphite via chemical polymerization. The resulting PANI-based counter-electrode in FDSSCs, performs PCE of 7.4% [18]. Table 3.1 lists a comparison of FDSSCs performance in terms of PCE and the preparation process of the counter electrode. The low costs and simple fabrication process of graphene-based counter electrodes enable them to be promising candidate materials for new flexible DSSCs development.

Photoanode is another important part of the FDSSCs. TiO$_2$ is considered a promising candidate to prepare the photoanode due to its excellent mechanical and electrical properties. TiO$_2$ pastes typically contain TiO$_2$ nanoparticles, wherein organic solvents and different additives should be heated to around 500 °C to ensure nanoparticles interconnection, film adherence, and high absorbance structure [20].

TABLE 3.1

Comparison of PCE of FDSSCs based on diverse counter-electrode

Counter-electrode/substrate	Material	Preparation method	η (%)	References
Pt/ITO/PET	Metallic	Sputtering	6.7	[19]
PEDOT/Graphene/PET	Carbon	Spin coating	6.3	[17]
PANI/ flexible graphite	Polymer composites	Chemical polymerization	7.4	[18]

On the other hand, plastic-conducting electrodes like ITO/PET and ITO/PEN cannot support such conditions. However, some efforts have focused on this issue, using submicrometer-sized TiO_2 building blocks to create ITO/PEN-based FDSCs; these have been assembled on the substrates after the sinteration at around 650 °C [21].

3.3.2 QUANTUM DOT SYNTHESIZED SOLAR CELL (QDSSCs)

3.3.2.1 Structure design and basic concept

In nanoscale, the chemical and physical properties of materials become unstable. QDs could be an ideal solution that gives special properties, especially for photoelectronic devices. As an example, QDSSCs can be considered as one of the most photo-electronic candidates using quantum confinement. Hence, the absorber materials-based QDs is a prerequisite in QDSSCs due to their significant absorption efficiency and wide-bandgap synthesis; this increases the electron-hole pairs photo-excited and obtains the best energy band alignment. In addition, a huge number of materials that mainly include QDs (Pb-, Cu-, and Cd-based QDs can be exploited as sensitizers for QDSSCs, to offer a wider range of high-quality elaboration of PV devices.

Figure 3.2 depicts the standard QDSSCs structure. The QDSSCs consist of mesoporous film-based metal oxide compact materials such as TiO_2, SnO_2 and ZnO deposited over transparent conducting oxide (TCO) layer to form a substrate. These kinds of nanomaterials are usually used as photo-anode to re-absorb the incident light via scattering and accelerate the electron transport in QDSSCs; hence, an important quantity of QDs can be easily absorbed. The S^{2-}/S_x^{2-} redox couple is injected inside QDSSCs

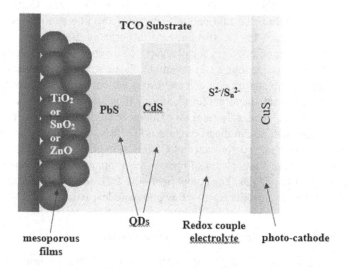

FIGURE 3.2 The standard architecture of typical QDSSCs.

as an electrolyte, which is catalyzed by the counter electrode to fabricate a photo-cathode electrode (CuS). For silicon solar cells, a space-charge zone is created at the interface of PN junction, producing a built-in electric field that induces the separation of electron-hole pairs photoexcited. However, the separation behavior is qualitatively different in QDSSCs due to the absence of the built-in electric field. Therefore, the QDs absorb from the light photons to create electron-hole pairs wherein the electrons move from the conduction band of QDs to that of the photoanode (TiO_2, SnO_2, ZnO, etc.). The adequate variance in the conduction bands of these metal oxide materials leads to a sufficient driving force of the electron transfer, which is preferred to a shift from the high energy level to a low one. In the same manner, the holes in the valence band of QDs shift to the catalyzed electrolyte, which can reduce QDs oxidation. The photoexcited electrons in the PbS/CdS QDs shift toward the external circuit via TCO. Therefore, the decay of these electrons can be observed due to the recombination mechanisms, such as trapping processes.

3.3.2.2   Flexible materials and fabrication process for QDSSCs

Recently, the main goal of most researchers was to characterize and synthesize QDs for possible application in the field of QDSSCs. A lot of materials that use metals, such as Cd^-, Pb and Cu^- and chalcogenide semiconductors such as S, Se, and Te, have been blended as QDs. For example, Pb- and Cd-based QDs are mostly used as sensitizers for QDSSCs owing to their high-quality material in terms of suitable band gaps and simple fabrication processes. Moreover, the choice of the QDs should satisfy the bandgap alignment with the metal oxide semiconductor's conduction band inside the QDSSCs and hence improve the performance of QDSSCs. For example, the QDSSCs devices which use only PbS as QDs have exhibited poor performance; however, employing PbS and CdS based QDs simultaneously for preparing QDSSCs leads to an enhancement of their PCE [22]. In addition, the photoanode is another important component that should be engineered. It is usually made of metal oxides, which are inserted inside the device to load QDs and then transfer electrons via the metal oxides-based photoanode.

The photoexcited charges can be influenced by the structure of the composite metal oxides, which affect their transfer and recombination as well as the deposition of QDs. As a result, metal oxide research has become one of the most important areas of investigation for improving photovoltaic performance in QDSSCs. Furthermore, the charge separation, transfer, and collection are all influenced by the varied architectures of oxide semiconductors. The greater density and surface ratio of nanoparticle P-type and N-type metal oxides can emphasize that QDs have a large absorption capability, which is advantageous for electrolyte penetration. Metal oxides deposited on flexible plastic sheets or thin metal as photoanode are potential materials to meet the requirements of flexibility in QDSSCs. Conventional QDSSCs are made on TCO substrates, wherein the fabrication process of the photoanode-layered films undergoes relatively high temperatures near 450 °C to eliminate the organic additives and ensure the chemical connectivity between the designed particles for electrical connection. [23]. Despite flexible metal substrate supporting high-temperature levels, it can substitute plastic substrate but leads to optical absorption losses when QDSSCs are illuminated from the front. From the backside, the transparency of the plastic conductive substrates leads to satisfactory illumination, which successfully meets the requirement for enhanced solar cells absorption, such as indium tin oxide/polyethylene terephthalate (ITO/PET), indium tin oxide/polyethylene naphthalene-2,6-dicarboxylate (ITO/PEN), polymers, etc. The main factors in achieving an improved photovoltaic performance of QDSSCs; in other words, fast charge transport, great optical absorption, and effective suppression of charge recombination are the shape of the photoanode films and the alloyed structure of QDs. For comparison, the PCE of QDSSCs with different structures-based nanomaterials, which are introduced to act as photoanodes, is given in Table 3.2.

3.3.3   Toward other flexible photovoltaic technologies

Over the past decade, solar cells of the first generation have fulfilled the initial fantasy of generating electricity directly from sunlight. People began to consider ways for this technology to be more portable,

TABLE 3.2

Comparison of QDSSC efficiency with different QDs and nanomaterials based photo-anode

Photo-anode	Sensitized quantum dot	η (%)	Reference
TiCl$_4$-modified SnO$_2$	PbS/CdS	1.60	[24]
SnO$_2$/HMS	CdS/CdSe	2.50	[25]
SnO$_2$/ZnO	CdS	3.41	[26]
SnO$_2$	CdS/CdSe	4.37	[27]

wearable, and flexible, which has been largely facilitated by the development of thin-film PV. This advanced technology is considered more beneficial than traditional crystalline Si-based PV, which is heavier and more fragile but not more expensive. Besides, organic photovoltaic is considered as another promising PV candidate in the field of flexible energy conversion and may be compared to inorganic thin-film devices. One of the most attractive features of this organic device is that it is constituted of organic components that are generally flexible, and it can be made over a multilayer thin-film process with low cost, such as roll-to-roll, spin-coating, and ink-jet printing, etc. In this regard, a summary will be provided in the field of flexible solar cells responding to new opportunities and challenges revealed by these devices.

3.3.3.1 Inorganic materials based flexible solar cells

Crystalline silicon is the most favored material used to manufacture solar cells due to its extremely advantagous material redundancy, high carrier mobility, widely ranging spectral absorption, and advanced technology that have a leading market share [28]. Large-scale silicon-based bulk wafer technology has fulfilled the prerequisites for PV applications. However, downscaling in thickness to a few micrometers can significantly reduce material usage, as well as the cost. In comparison with silicon-based bulk wafers, thin-film silicon can help high-performance flexible solar cells since it limits the recombination process in the device. Meanwhile, the method followed when slicing wafers with a wire saw to prepare thin films creates a huge volume of micron-sized silicon powder, or "kerf" [29]. To limit this drawback, new methods have been investigated to manufacture thin films for low-cost flexible PVs. Hydrogenated amorphous and microcrystalline silicon thin films are highlighted as alternative materials for the fabrication of flexible substrates due to their high flexibility, which is appropriate for low-cost PV applications.

Besides silicon-based materials discussed above, semiconductor composite materials, such as CIGS, CdTe, and III–V semiconductors have also been considered as promising candidates for low-cost PV on flexible substrates with high efficiency and good stability. The thin-film fabrication processes of these materials are usually thermal evaporation or sputtering, which is a convenient multilayer thin-film microscale method. Besides, the roll-to-roll deposition method can also be implemented to grow these active materials on flexible substrates, such as flexible polyimide film, to realize flexible thin-film CIGS solar cells with an achieved PCE of more than 18% compared to those made on rigid substrates [30]. Comparing to the inorganic solar cells that use CIGS, solar cells that use CdS/CdTe thin-film as substrate exhibit relatively low PCE, more than 11%, achieved on flexible substrates molybdenum foil substrate [31]. However, solar cells based on III–V composite materials, such as thin-film GaAs multi-junction, become competitive candidates in the flexible solar cells field due to their ability to be stretched and their mechanical robustness. Their property of mechanical flexibility allows them to be appropriate for better incorporation into daily products with PCE over 44% [32]. For specific applications, the toxicity of GaAs and CdTe needs to be considered. In this context, environmentally friendly alternative materials, such as organic materials, are desired as feasible candidates for new-generation flexible PV applications.

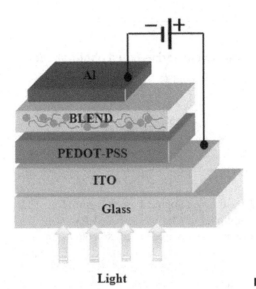

Light   **FIGURE 3.3**   General configuration of OSCs.

3.3.3.2   Organic materials based flexible solar cells

The advantages of low-cost fabrication and rich natural sources, together with flexibility, mean the organic solar cells have attracted significant attention from academic and industrial researchers to realize practical applications. A typical organic solar cell (OSC) is generally constituted of ITO deposited on a rigid glass substrate as a transparent electrode, as shown in Figure 3.3. It is manufactured by a rigid fabrication process that avoids OSCs to be flexible. Therefore, considerable efforts have been focused on developing new innovative transparent conducting oxide (TCO) to fabricate OSCs on flexible or stretchable substrates. Carbon-based (graphene, CNT), metal grid-based (Ag and Cu) materials, and conducting polymers (PEDOT:PSS) can be the alternatives for increasing the conductivity, as well as the transparency, to obtain high-quality flexible electrodes. Overall, the fabrication process of OSCs is fundamentally based on nontoxic organic materials deposited layer by layer onto plastic rolls via multilayer thin-film techniques such as spray-coating, roll-to-roll printing, and ink injecting.

More knowledge about how processing parameters affect polymer composition structure intrinsic morphology, stability, and PCE will be crucial in the near future. The unique properties of these features, compared to other conventional solar cell technologies, make them competitive candidates in the future PV market. The highest OSCs efficiency of over 14% is recorded in academia and over 13% in industrial markets [33,34].

3.4   ARCHITECTURAL CONCEPTS, STRUCTURES, AND MATERIALS FOR FLEXIBLE BATTERIES

The working principle of batteries is mainly based on the electrochemical process, which stores energy in chemical form and then converts it into electrical energy to serve as power sources for implantable/portable electronic devices. Power sources toward miniaturization are in progress to evaluate bendability, porosity, and adaptability to the full wearable device for possible development of flexible/wearable electronics and as a solution for sustainable and equitable operation. The creation of extremely efficient wearable and flexible power sources necessitates material science considerations. This part summarizes and analyzes the scientific progress on wearable and flexible batteries, such as lithium-ion batteries (Li-ion), lithium-sulfur batteries (Li–S), and zinc ion battery Zn–ion [35].

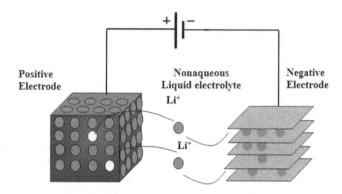

FIGURE 3.4 Structure design and working basics of Li-ion batteries in the discharge process.

3.4.1 LITHIUM-ION BATTERIES (LIBS)

3.4.1.1 Structure design and basic concept

Because of their nontoxic nature and advantages to other power sources in terms of high negative standard redox potential, favorable capacity, long cycle life, and high safety and low density, LIBs are currently used much more in wearable devices. Several reports in the scientific literature have highlighted materials features of electrodes and electrolytes, engineered to produce flexible batteries [36]. The structure design of a typical LIB is shown in Figure 3.4. It is composed of a negative electrode that typically uses carbon materials as an anode, electrolyte (e.g., nonaqueous liquid solution), and a positive electrode, typically layered $LiCoO_2$ as a cathode.

The de-intercalation mechanism of Li-ions enables their transition from the layered $LiCoO_2$ and their transportation via the electrolyte; afterward, they will be intercalated within layers of anode materials during the charging process. This phenomenon is reversed through the discharge process. During these two mechanisms, namely charge and discharge processes, the energy can be delivered and stored by delivering lithium ions back and forth between both electrodes anode and cathode.

3.4.1.2 Flexible materials for LIB structures

Fabrication of the whole flexible LIB needs more investigation to know the capability and the limitation of each battery component. Flexible anode-based materials have attracted a lot of interest for advanced flexible devices. The anode of a LIB was formerly made of lithium metal. This metal was quickly abandoned due to the risk of producing lithium dendrite, which might break the separator, causing an internal short circuit and potentially dangerous explosion. In this regard, finding a safe anode as an alternative component becomes a critical issue.

Dey reported in 1971 that lithium alloys containing metals may be produced in organic electrolytes using an electrochemical process, and hence, sustain the study of these alloys as alternative anode materials [37]. The internal stress induced by the large volume changes in lithium alloys causes pulverization and cracking of anodes and leads to poor cyclic stabilities of lithium alloys, as well as loss of electrical conduction. The use of lithiated carbonaceous (graphite) anode instead of lithium metal reduces the formation of irregularly shaped lithium and improves the cyclic lifetime of the battery with safety stability. However, due to the special reaction process, graphite's theoretical capacity is relatively limited. To overcome this problem, some promising examples in which CNTs and fiber electrodes are generating great interest are to fabricate electrode materials with outstanding mechanical resistance, flexibility, and conductivity. In addition, the combination of CNT/lithium manganate (LMO) composite and CNT/Si deposited on cotton fiber substrate provides the hybrid layered structure for both electrodes anode and cathode, respectively, allowing fabrication of coaxial fiber-shaped LIB with higher performance [38]. Another structure designed as a sandwich can be implemented using CNT sheets deposited on polymers (PDMS), providing highly stretchable and conductive electrodes [39]. Moreover, graphene has also been considered the most important component in wearable and flexible LIBs owing to their equitable mechanical stress resistance and conductivity. Its presence in LIBs improves the reversible capacity and the anode-cycling stability. Besides, other materials like 3D nickel nitride nanosheets $ZnCo_2O_4$ arrays were successfully integrated into clothes using carbon fiber and thin LIB devices with higher flexibility and electrochemical performance. These kinds of materials cause researchers to pay attention and consider using alternative battery materials for the fabrication of LIB anodes with higher mechanical strength conductivity and specific surface area. Electrolytes and separators are also promising components of LIBs in stretchable device applications. Solid-state gel and polymer electrolytes, such as succinonitrile gel polymer, crystal-polyethylene oxide, and crosslinked gel poly (vinylidene fluoride-co-hexafluoropropylene), electrolyte are typically used in LIB due to their excellent tensile strength, elasticity, superior electrochemical stability, cycling performance, and ionic conductivity [40].

3.4.2 ZINC-ION BATTERIES (ZIBs)

3.4.2.1 Structure design and basic concept

Owing to the intrinsic features of ZIB, such as diversity of potential electrolytes, the higher redox potential of zinc, improved safety, reduced toxicity, and non-flammable, low-cost, and long cycle life, Zn-ion batteries are quickly gaining the attention of many scientific researchers. As shown in Figure 3.5, flexible ZIBs consist of typical components: a polymer electrolyte-based polyethylene oxide holding TiO_2 nanoparticles, a current collector consisting of carbon nanofibers, an anode consisting of Zn foil, and a cathode consisting of MnO_2/SWCNTs for hosting Zn ions. Similar to LIBs, the ZIBs use an intercalation mechanism where zinc ions react at both electrodes, anode and cathode, and travel from the anode to the cathode across a polymer-based electrolyte. In other words, zinc metal at the anode is dissolved into the electrolyte as zinc ions and simultaneously absorbed into the cathode via the electrolyte during discharge. During charge, this mechanism is reversed.

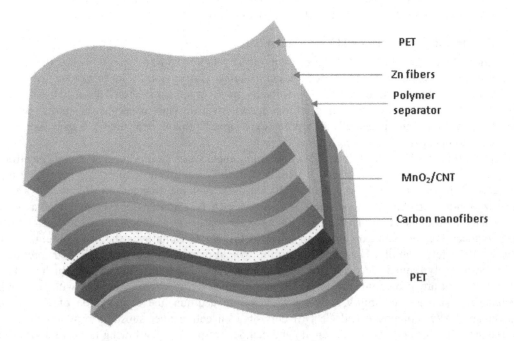

PET

Zn fibers

Polymer separator

MnO_2/CNT

Carbon nanofibers

PET

FIGURE 3.5 Schematic illustration of typical components of a flexible ZIB.

3.4.2.2 Flexible materials for ZIB structures

Several composite materials have been used to fabricate and enhance ZIB performance in terms of flexibility and long cycle life capability. As an example, a ZIB with aqueous electrolyte, Zn/carbon anode, and Co_3O_4/Ni foam cathode has been fabricated with an energy-density reach of 241 Whk/g [41]. Another kind of designed materials: C-fiber papers with P-doped graphitic carbon nitride, Zn/ coating MnO_2 nanoparticles, and 3D carbon nanotubes/a Ni foam cathode, are widely used for flexible ZIB as promising materials to power wearable and flexible devices [42,43]. Despite multiple research that has identified the best composite materials and designs, creating environmentally acceptable and safe wearable rechargeable batteries is still a challenge. Besides, a coaxial-fiber ZIB is another competitive battery to the conventional one due to its coaxial form, which offers, under long-term bending, enhanced performance stability and mechanical flexibility as well as modest charge conductivity and ohmic contact overpotentials. On one hand, the utilization of organic solvent to prepare aqueous electrolytes provides the conventional batteries a solution related to safety problems. On the other hand, it also guarantees high flexibility for the device. As an example, zinc triflate combined with electrolytes based on polymers and crosslinked porous gelatin can improve conductivity, and degradation issues under several environmental conditions, making it more convenient and more applicable for commercial wearable electronics.

3.4.3 FLEXIBLE BATTERIES ADVANCEMENT

Currently, various research works are working on how to make each component–electrolyte, anodes, and cathodes–independently flexible. Meanwhile, some research works are focused on how to properly realize assembly to create a completely flexible device. All practical components of such a battery may be well engineered to achieve free-standing and bending features, as mentioned in the previous sections. The requirement for a simple way to properly enable them to ensure high flexibility and areal capacity of devices becomes ever more important. Electronic devices that can be printed are becoming more common and give rise to new research ideas for future studies related to flexible nano-energy devices, which include flexible ion batteries. It would have a huge impact on the integration, architecture, and implementation of energy storage devices if all their components: separator, both electrodes (anode and cathode), charge collector, and electrolyte, have to be well chosen to effectively design them as dyes and assembled all together to form a full battery pack. Shortly, the application of such energy devices will become more flexible and portable. Flexible ion batteries are capable of powering our everyday lives in a variety of ways, thanks to the numerous advantages mentioned previously.

Consequently, adaptable ion batteries deserve special consideration. However, there are other issues with this subject. For example, the high cost of LIBs, including the cost of constructing flexible ones, is still a matter to be considered. For a flexible application, consistent performance cannot always be assured, especially during bending or folding. Some industrial problems are tremendously complicated and require teams of experts from various domains to address them before commercialization. As an example, the flexible LIB's instability will result in a reduced lifetime and even safety issues. Designing appropriate nanostructures for all of the constituent parts of ion batteries has been shown to have a considerable, favorable impact on their performance, paving the way for the fabrication of high-quality, flexible, and portable batteries in the near future by overcoming the actual issues.

3.5 ARCHITECTURAL CONCEPTS, STRUCTURES, AND MATERIALS FOR FLEXIBLE SUPERCAPACITORS

Supercapacitors (SCs), also named ultracapacitors or electrochemical capacitors, have recently gained attention owing to their enormous power density, high energy density, and long cycling life (>100,000 cycles). They are linking the gap between traditional capacitors and batteries. SCs have a higher energy density than traditional capacitors by several orders of magnitude. In comparison to conventional capacitors, SCs store a huge amount of charge and quickly deliver more power than batteries. They are commonly utilized in different industrial power and energy management applications. Recently, flexible SCs are considered a promising kind of energy-storage device because of the outstanding advantages, including reduced their design, high flexibility, being lightweight, and high operating temperature range. In this regard, several research works have been focused to design flexible SC devices to meet the different demands of the modern market, such as flexible and wearable electronic devices [44].

3.5.1 STRUCTURE DESIGN AND BASIC CONCEPT

As shown in Figure 3.6(a), a typical SC has metal-based current collectors, positive and negative electrodes, electrolyte, and a separator, which are all packaged like a sandwich. The main differences between the flexible SCs and the conventional ones are that the flexible SCs don't need to use current collectors and binders but require that the electrode, as well as the electrolyte, must be flexible Figure 3.6(b). Thus, an electrolyte with high conductivity and a mechanical and thermal stability can work as a full separator without needing an additional separator. Moreover, the advantage of the lightweight and bendable plastic package makes the flexible SCs more suitable in some applications in comparison with the conventional ones.

3.5.2 FLEXIBLE MATERIALS FOR SCS STRUCTURES

Pure carbon materials have provided significant opportunities for developing advanced flexible electrodes over the last few decades. This is due to their good performance and characteristics that include being lightweight and having high conductivity, controllable porous structure, high-temperature stability, high surface area, compatibility in composite materials, and low cost. CNTs and graphene, as carbon-based materials, have attracted many investigations for the development of flexible SCs due to their particular chemical and physical characteristics. As an example of mechanical flexibility, Pushparaj's group [45] reports an electrode designed with an aligned CNT using embedded nanoporous cellulose papers emerged in ionic liquid electrolyte. Another interesting structure, with a specific power density of 200,000 W k/g, specific energy density between 30 and 47 Whk/g, and cycling life of more than 40,000 cycles, is based on single-walled CNTs (SWNTs) coating to create high conductive electrodes and current collectors [46]. A comparative review of the applications of SWNT-treated papers and textiles to fabricate electrodes for SCs is demonstrated by Cui's group [47]. In addition to the paper-based method, sponges made up of many small polyester or cellulose fibers with a hierarchical macroporous nature have been used to substitute the traditional substrate to fabricate flexible SCs electrodes. A simple method called 'dipping and drying' was used to create nanostructured MnO_2/CNT/sponge hybrid electrodes. The sponge's macroporous nature, combined with the porous nature of the electrodeposited MnO_2, resulted in a double-porous electrode structure, significantly improved the performance of MnO_2/CNT/sponge SCs due to its good conductivity, and completed electrolyte accessibility to MnO_2. As a result, this structure delivers an energy density of 31 Wh/kg and specific power of 63 kW/kg. Laser-scribed graphene films, reported by Kaner's group [48], are employed as SCs electrodes, unless they are current collectors.

It is advantageous to improve energy density as well as power density as well as cycle stability exhibiting high charge conductivity and areal surface area of 1738 S/m and 1520 m^2/g, respectively. Nano-fibrous felts of carbide-derived carbon (CDC) is another type of material, based on electrospun titanium carbide nanobelts (TiC); it is widely used as a new electrode material for SCs due to its high mechanical flexibility, robustness, and ability to be used without any additive binders [49]. A carbon materials (CNT) based screen-printing method is used to infiltrate the porous structures into cloth and polyester microfibers, reaching higher mass loading and surface-area capacitance in comparison with the above-discussed methods [50]. In addition, H_2SO_4/polyvinyl alcohol gel electrolyte was combined with polyaniline to fabricate electrodes for polymer SCs. However, the SCs fabricated through this method exhibit outstanding electrochemical performance, such as specific capacitance of 350 F/g for electrode

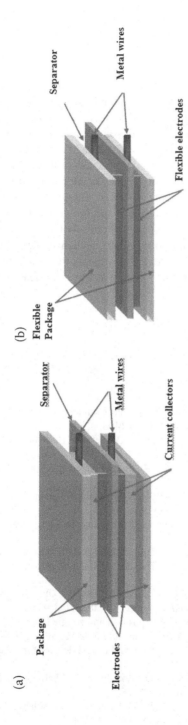

FIGURE 3.6 Schematic representation of supercapacitor (a) conventional and (b) flexible.

materials, good cycling performance after 1000 cycles, and a leakage current of 17.2 μA [51]. Flexible fiber supercapacitors were discovered; they feature two fiber electrodes, an electrolyte, and a helical spacer wire. The inclusion of this spacer wire allows for optimal separation of the two fiber electrodes, as well as scalability of the preparation process.

3.6 CONCLUSION

In the field of energy conversion and storage, flexible energy devices have been widely explored for commercial electronic products. The fabrication of flexible energy devices requires flexible materials, which is the main challenge in meeting the need for advanced bendable, portable, and stretchable devices. Although solid-state technology has advanced, some limitations of fabrication still exist and need to be remedied. In this chapter, we have provided fundamental designs and architectures related to flexible energy conversion and storage devices. Different kinds of flexible solar cells, batteries as well as SCs, are presented to understand the requirement needed for advanced applications. Materials, which are mainly introduced to fabricate basic components of flexible energy devices, are the fundamental key that should be well-designed to improve such devices in a specific application. Designing nanostructures is the most important aspect in the development of high chemical and physical nanomaterial properties for future applications in energy conversion and storage area, and hence, they can overcome many key issues related to flexible applications. Lastly, we believe that research in the flexible energy field will motivate innovations of materials with high-performance, as well as novel fabrication processes for future energy devices with desired properties.

REFERENCES

1. Khan Y., Thielens A., Muin S., Ting J., Baumbauer C., Arias A. C. (2020). A new frontier of printed electronics: flexible hybrid electronics. *Advanced Materials*, 32(15), 1905279.
2. Li L., Wu Z., Yuan S., Zhang X. B. (2014) Advances and challenges for flexible energy storage and conversion devices and systems. *Energy & Environmental Science*, 7(7), 2101–2122.
3. Wang B., Facchetti A. (2019) Mechanically flexible conductors for stretchable and wearable e-skin and e-textile devices. *Advanced Materials*, 31(28), 1901408.
4. Zhu Y. H., Yang X. Y., Liu T., Zhang X. B. (2020) Flexible 1D batteries: recent progress and prospects. *Advanced Materials*, 32(5), 1901961.
5. Sharma K., Arora A., Tripathi S. K. (2019). Review of supercapacitors: Materials and devices. *Journal of Energy Storage*, 21, 801–825.
6. Das, S., Sudhagar, P., Kang, Y. S., & Choi, W. (2014). Graphene synthesis and application for solar cells. *Journal of Materials Research*, 29(3), 299–319.
7. Wang X., Shi G. (2015) Flexible graphene devices related to energy conversion and storage. *Energy & Environmental Science*, 8(3), 790–823.
8. Nathan A., Ahnood A., Cole M. T., Lee S., Suzuki Y., Hiralal P., Bonaccorso F., Hasan T., Garcia-Gancedo L., Dyadyusha A. (2012) Flexible electronics: The next ubiquitous platform. *Proceedings of the IEEE*, 100, 1486–1517.
9. Wang X., Lu X., Liu B., Chen D., Tong Y., Shen G. (2014) Flexible energy-storage devices: Design consideration and recent progress. *Advanced Materials*, 26(28), 4763–4782.
10. Wang X., Yushin G. (2015) Chemical vapor deposition and atomic layer deposition for advanced lithium ion batteries and supercapacitors. *Energy & Environmental Science*, 8(7), 1889–1904.
11. Lipomi D. J., Bao Z. (2011) Stretchable, elastic materials and devices for solar energy conversion. *Energy & Environmental Science*, 4(9), 3314–3328.
12. Reddy A. K., Annecke W., Blok K., Bloom D., Boardman B., Eberhard A., Ramakrishna J. (2000) Energy and social issues. *World Energy Assessment*, 39–60.
13. Avrutin V., Izyumskaya N., & Morkoç, H. (2011). Semiconductor solar cells: Recent progress in terrestrial applications. *Superlattices and Microstructures*, 49(4), 337–364.
14. O'Regan B., Grätzel M. (1991) A low-cost, high-efficiency solar cell based on dye-sensitized colloidal TiO_2 films. *Nature*, 353(6346), 737–740.

15. Mathew S., Yella A., Gao P., Humphry-Baker R., Curchod B. F., Ashari-Astani N., ... & Grätzel M. (2014) Dye-sensitized solar cells with 13% efficiency achieved through the molecular engineering of porphyrin sensitizers. *Nature Chemistry*, 6(3), 242–247.

16. Weerasinghe H. C., Sirimanne P. M., Simon G. P., Cheng Y. B. (2012) Cold isostatic pressing technique for producing highly efficient flexible dye-sensitised solar cells on plastic substrates. *Progress in Photovoltaics: Research and Applications*, 20(3), 321–332.

17. Lee K. S., Lee Y., Lee J. Y., Ahn J. H., Park J. H. (2012) Flexible and platinum-free dye-sensitized solar cells with conducting-polymer-coated graphene counter electrodes. *ChemSusChem*, 5(2), 379–382.

18. Sun H., Luo Y., Zhang Y., Li D., Yu Z., Li K., Meng Q. (2010) In situ preparation of a flexible polyaniline/carbon composite counter electrode and its application in dye-sensitized solar cells. *The Journal of Physical Chemistry C*, 114(26), 11673–11679.

19. Fu N., Xiao X., Zhou X., Zhang J., Lin Y. (2012) Electrodeposition of platinum on plastic substrates as counter electrodes for flexible dye-sensitized solar cells. *The Journal of Physical Chemistry C*, 116(4), 2850–2857.

20. Ito S., Chen P., Comte P., Nazeeruddin M. K., Liska P., Péchy P., Grätzel M. (2007) Fabrication of screen-printing pastes from TiO_2 powders for dye-sensitised solar cells. *Progress in Photovoltaics: Research and Applications*, 15(7), 603–612.

21. Huang F., Chen D., Li Q., Caruso R. A., Cheng Y. B. (2012) Construction of nanostructured electrodes on flexible substrates using pre-treated building blocks. *Applied Physics Letters*, 100(12), 123102.

22. Chen Y., Tao Q., Fu W., Yang H., Zhou X., Su S., ... & Li M. (2014) Enhanced photoelectric performance of PbS/CdS quantum dot co-sensitized solar cells via hydrogenated TiO_2 nanorod arrays. *Chemical Communications*, 50(67), 9509–9512.

23. Fu F., Feurer T., Weiss T. P., Pisoni S., Avancini E., Andres C., ... & Tiwari A. N. (2016) High-efficiency inverted semi-transparent planar perovskite solar cells in substrate configuration. *Nature Energy*, 2(1), 1–9.

24. Huang Q., Li F., Gong Y., Luo J., Yang S., Luo Y., ... & Meng Q. (2013) Recombination in SnO_2-based quantum dots sensitized solar cells: the role of surface states. *The Journal of Physical Chemistry C*, 117(21), 10965–10973.

25. Ganapathy V., Kong E. H., Park Y. C., Jang H. M., Rhee S. W. (2014) Cauliflower-like SnO_2 hollow microspheres as anode and carbon fiber as cathode for high performance quantum dot and dye-sensitized solar cells. *Nanoscale*, 6(6), 3296–3301.

26. Patil S. A., Shinde D. V., Bhande S. S., Jadhav V. V., Huan T. N., Mane R. S., Han S. H. (2013) Current density enhancement in ZnO/CdSe photoelectrochemical cells in the presence of a charge separating SnO_2 nanoparticles interfacing-layer. *Dalton Transactions*, 42(36), 13065–13070.

27. Xiao J., Huang Q., Xu J., Li C., Chen G., Luo Y., ... & Meng Q. (2014) CdS/CdSe co-sensitized solar cells based on a new SnO_2 photoanode with a three-dimensionally interconnected ordered porous structure. *The Journal of Physical Chemistry C*, 118(8), 4007–4015.

28. Swanson R. M. (2006) Photovoltaics Res. *In Appl*, 14, 443–453.

29. Qingfeng L., Hongtao H., Yan J., Huiying F., Paichun C., Dongdong L., Yan Y., Zhiyong F. (2014) Flexible photovoltaic technologies. *The Journal of Physical Chemistry C*, 2(7), 1233–1247.

30. Chirilă A., Buecheler S., Pianezzi F., Bloesch P., Gretener C., Uhl A. R., ... & Tiwari A. N. (2011) Highly efficient Cu (In, Ga) Se$_2$ solar cells grown on flexible polymer films. *Nature Materials*, 10(11), 857–861.

31. Kranz L., Gretener C., Perrenoud J., Schmitt R., Pianezzi F., La Mattina F., ... & Tiwari A. N. (2013) Doping of polycrystalline CdTe for high-efficiency solar cells on flexible metal foil. *Nature Communications*, 4(1), 1–7.

32. Lee J., Wu J., Ryu J. H., Liu Z., Meitl M., Zhang Y. W., ... & Rogers J. A. (2012) Stretchable semiconductor technologies with high areal coverages and strain-limiting behavior: Demonstration in high-efficiency dual-junction GaInP/GaAs photovoltaics. *Small*, 8(12), 1851–1856.

33. Li S., Ye L., Zhao W., Yan H., Yang B., Liu D., ... & Hou J. (2018) A wide band gap polymer with a deep highest occupied molecular orbital level enables 14.2% efficiency in polymer solar cells. *Journal of the American Chemical Society*, 140(23), 7159–7167.

34. Organic Photovoltaic World Record Efficiency of 13.2%. News accessed on Nov 12, 2021 (https://eepower.com/news/organic-photovoltaic-world-record-efficiency-of-13-2/#)

35. Zhu Y. H., Yang X. Y., Liu T., Zhang X. B. (2020) Flexible 1D batteries: Recent progress and prospects. *Advanced Materials*, 32(5), 1901961.

36. Song W. J., Yoo S., Song G., Lee S., Kong M., Rim J., ... & Park S. (2019) Recent progress in stretchable batteries for wearable electronics. *Batteries & Supercaps*, 2(3), 181–199.

37. Dey A. N. (1971). Electrochemical alloying of lithium in organic electrolytes. *Journal of the Electrochemical Society*, 118(10), 1547.
38. Guan G., Deng J., Ren J., Pan Z., Zhuang W., He S., ... & Peng H. (2017) Tailorable coaxial carbon nanocables with high storage capabilities. *Journal of Materials Chemistry A*, 5(42), 22125–22130.
39. Ren J., Zhang Y., Bai W., Chen X., Zhang Z., Fang X., ... & Peng H. (2014) Elastic and wearable wire-shaped lithium-ion battery with high electrochemical performance. *Angewandte Chemie*, 126(30), 7998–8003.
40. Kim I., Kim B. S., Nam S., Lee H. J., Chung H. K., Cho S. M., ... & Kang C. (2018) Cross-linked poly (vinylidene fluoride-co-hexafluoropropene)(PVDF-co-HFP) gel polymer electrolyte for flexible Li-ion battery integrated with organic light emitting diode (OLED). *Materials*, 11(4), 543.
41. Wang X., Wang F., Wang L., Li M., Wang Y., Chen B., ... & Huang W. (2016) An aqueous rechargeable Zn//Co$_3$O$_4$ battery with high energy density and good cycling behavior. *Advanced Materials*, 28(24), 4904–4911.
42. Zeng Y., Zhang X., Meng Y., Yu M., Yi J., Wu Y., ... & Tong Y. (2017) Achieving ultrahigh energy density and long durability in a flexible rechargeable quasi-solid-state Zn–MnO$_2$ battery. *Advanced Materials*, 29(26), 1700274.
43. Ma T. Y., Ran J., Dai S., Jaroniec M., Qiao S. Z. (2015) Phosphorus-doped graphitic carbon nitrides grown in situ on carbon-fiber paper: Flexible and reversible oxygen electrodes. *Angewandte Chemie*, 127(15), 4729–4733.
44. Ghouri A. S., Aslam R., Siddiqui M. S., Sami S. K. (2020) Recent progress in textile-based flexible supercapacitor. *Frontiers in Materials*, 7, 58.
45. Pushparaj V. L., Shaijumon M. M., Kumar A., Murugesan S., Ci L., Vajtai R., ... & Ajayan P. M. (2007) Flexible energy storage devices based on nanocomposite paper. *Proceedings of the National Academy of Sciences*, 104(34), 13574–13577.
46. Peng C., Zhang S., Jewell D., Chen G. Z. (2008) Carbon nanotube conducting polymer composites for supercapacitors. *Progress in Natural science*, 18(7), 777–788.
47. Hu L., Cui Y. (2012) Energy and environmental nanotechnology in conductive paper and textiles. *Energy & Environmental Science*, 5(4), 6423–6435.
48. El-Kady M. F., Strong V., Dubin S., Kaner R. B. (2012) Laser scribing of high-performance and flexible graphene-based electrochemical capacitors. *Science*, 335(6074), 1326–1330.
49. Presser V., Zhang L., Niu J. J., McDonough J., Perez C., Fong H., Gogotsi Y. (2011) Flexible nano-felts of carbide-derived carbon with ultra-high power handling capability. *Advanced Energy Materials*, 1(3), 423–430.
50. Jost K., Pere, C. R., McDonough J. K., Presser V., Heon M., Dion G., Gogotsi Y. (2011) Carbon coated textiles for flexible energy storage. *Energy & Environmental Science*, 4(12), 5060–5067.
51. Meng C., Liu C., Chen L., Hu C., Fan S. (2010) Highly flexible and all-solid-state paperlike polymer supercapacitors. *Nano Letters*, 10(10), 4025–4031.

4 Characterization Techniques of Flexible Energy Devices

Endersh Soni, Ujjwal K. Prajapati, Ayushi Katariya, and Jyoti Rani
Department of Physics, Maulana Azad National Institute of Technology, Bhopal, M.P, India

CONTENTS

4.1 Introduction.. 59
4.2 Characterization techniques for flexible energy devices.. 61
    4.2.1 Scanning electron microscopy 61
    4.2.2 Transmission electron microscopy.. 65
    4.2.3 X-ray diffraction.. 68
    4.2.4 Cyclic voltammetry 69
    4.2.5 Galvanostatic charge-discharge test... 70
    4.2.6 Electrochemical impedance spectroscopy 72
    4.2.7 Atomic force microscopy.. 73
    4.2.8 Secondary ion mass spectroscopy 73
    4.2.9 Inductively coupled plasma-mass spectroscopy.. 76
    4.2.10 Fourier transform infrared spectroscopy 77
4.3 Summary.. 79
References.. 79

4.1 INTRODUCTION

Electronic devices have become an important part of the human lifestyle. Humans have become very dependent on electronic gadgets like smartphones, smartwatches, smart screens, healthcare electronics, etc. All of these electronics require power to operate. Many batteries are incompatible with flexible electronic devices like smart cards, electronic papers, and devices that require high current pulses like wireless communication and light-emitting devices. That's why developments of flexible Energy device (ED) are important to fulfill the demand for the quickly evolving flexible electronic devices. Considerable efforts are being made in the last few years to design high-performance flexible EDs with high-energy storing capacity, high power density, long life, stable cycling, and safe operation. Various researchers are interested in discovering appropriate and effective materials for electrodes/electrolytes, preferable cell configuration, and structural design.

The concept of flexibility is related to rigidity. Flexibility emphasizes on deformation of material with the elastic, inelastic, or plastic stress-strain relationship. Ideally, a flexible device should possess characteristics like foldability, bendability, twistability, stretchability with stable electrical performance, and safe operation. To impart these characteristics to the devices, in current years, scientists have made significant efforts to recognize the mechanical flexibility of all EDs by building each component to be flexible. Flexible EDs are not bound by bulky design and configurational limitations, unlike conventional energy devices, but they should be small, lightweight, stable, and highly efficient in different mechanical distortion. Materials used in traditional energy devices are brittle and not compatible with flexible electronics devices. For example, in traditional lithium-ion batteries (LIBs), electrode active materials

DOI: 10.1201/9781003186755-4

are covered on metals like Al and Cu, which easily detach from the contact surface on bending and do not restore to their original shape. To provide flexibility to the flexible energy devices, each component should be made of materials that are shape comfortable, highly efficient, nontoxic, cheaper, non-flammable, and scalable. However, all flexible energy devices demonstrated to date do not meet these requirements. The huge problem developers of flexible EDs face is to fabricate reliable, efficient, and shape comfortable components. Another problem is to find a suitable packaging material that can prevent damage to integral components of the device and ensure the safe operation of the device under different conditions. Significant scientific efforts are devoted to developing suitable materials. Nanostructured materials have shown growth in flexible power sources to boost the flexible electronics field.

Flexible energy devices may be of two kinds: energy storing devices (ESDs) and energy conversing devices. Supercapacitors (SCs) and batteries are the examples of ESDs, while solar cells and fuel cells are types of energy conversion devices. Supercapacitors have gained considerable attention due to its advantages. Supercapacitors have high charging-discharging rates, high energy density, long operational life, and simple structures concerning LIBs. These qualities make supercapacitors a favorite candidate of power-storing devices in the area of flexible electronics. Supercapacitors can be classified into three categories based on the charge storing mechanism. Three types are electrical double-layer capacitors (EDLCs), pseudo-capacitors, and hybrid supercapacitors, which use both charge storing mechanisms of EDLCs and pseudo-capacitors. Table 4.1 displays a comparison of EDLCs, pseudo-capacitors, and hybrid SCs.

LIBs have been used traditionally as a main power source in flexible electronics because of their low self-discharge rate, long-term cyclability, high operating voltages, and high energy density. But these are too bulky and rigid to be used as a power source for flexible devices. A flexible LIB's needs are made up of flexible, bendable, and stretchable components and outer packaging. Traditionally used materials (generally, metals and brittle materials) for LIBs are not recommended here. Materials with high capacity, mechanical flexibility, cycling stability, and superior conductivity are highly desired for flexible LIBs. Nano-engineered materials like carbon materials, including carbon nanofibers, CNTs remove and graphene, and metal oxide nanowires, have shown their capabilities to be used as an electrode in flexible LIBs. For electrolytes, liquids are preferred conventionally just because of their good physical contact and excellent conductivity with electrodes. But liquids cannot be used in flexible LIBs because they have many drawbacks, such as safety problems, the requirement for separators, and poor performance under mechanical distortion. Solid-state electrolytes like gel polymer electrolytes, which have mechanical flexibility, excellent conductivity, low flammability, and low rates of electrolyte leakage, are strong candidates for the electrolyte of flexible LIBs.

Now, concerning flexible energy conversion devices, these are the energy generators that harvest ambient energy; for instance, the convert to electricity sunlight, heat, vibration, or human body displacements. This method is an effective and impressive way to generate cheaper, eco-friendly electrical

TABLE 4.1

Comparison of EDLCs, pseudo-capacitors, and hybrid SCs. Reproduced with permission from ref. [6]

Types of capacitors	Materials	Capacitors	Cycling stability
EDLCs	Carbon	Moderate	Excellent
Pseado-capacitors	MO_x^a, Conducting Polymer	High	Poor
Hybrid capacitors	Carbon-MO_x/conducting polymer	High	Moderate

*where MO_x represent transition metal.

power that can be utilized to power flexible electronics. A flexible solar cell is one of the flexible energy conversion devices that have gained tremendous attention from the research community due to their capacity to convert sunlight (easily available in ambient) into electricity. A traditional solar cell is built with planner sandwiched configurations and rigid substrates that limit its application in flexible electronics. A wire-shaped solar cell is gaining attention due to its three-dimensional flexibility. Fu *et al.* reported a wire-shaped dye-sensitized solar cell with excellent flexibility and PCE of 7.02% [1]. This PCE was achieved with Pt wire as cathode and TiO_2 as an anode. It has also been reported that Pt wire can be replaced with CNTs thin films to improve flexibility, but PCE is relatively low.

For fabricating high-performance flexible energy devices, suitable materials for each component are required [2]. To date, there is a lack of materials that possess all required properties at the same time. Researchers are trying to modify the properties of these materials to make them suitable for flexible energy devices. To modify properties, it is necessary to examine the property at each step. Hence, characterization is a step-by-step process by which the structure and properties of the material can be examined and measured. Characterization develops the scientific understanding of the materials. Some techniques are used to measure overall performance, while some are used to evaluate the performance of individual components. Many techniques are used for the characterization of flexible energy devices like SEM, TEM, XRD, NMR, GCD test, etc. This chapter briefly explains some of the most common characterization techniques used for flexible devices characterization. It also gives an idea of the use of these characterization techniques.

4.2 CHARACTERIZATION TECHNIQUES FOR FLEXIBLE ENERGY DEVICES

4.2.1 SCANNING ELECTRON MICROSCOPY

Scanning electron microscopy (SEM) is a characterizing technique used to determine the chemical composition, size of the constituent particle, and the topography of the material surface. It is a microscope that uses a beam of an electron instead of light. It scans the material surface and generates a high-resolution image. SEM uses an accelerated beam of electrons, which scans the sample surface in a roaster-like fashion (Figure 4.1). Initially, these electron beams have some specific amount of energy, but after an interaction with the sample surface, that energy dissipates. Due to the electron matter interaction, two types of the electron produce: secondary electrons (SE) and backscattered electrons (BSE). Both types of electrons carry different information about the sample surface. Secondary electrons are produced due to inelastic interaction between the accelerated electron beam and sample surface, while backscattered electrons are generated due to elastic interaction. These SE and BSE are collected using some collector or detector to generate the image. BSE originated from the deep insight of the sample thus can give details about the composition, crystallography, and magnetic field of the solid sample surface. SE originated from the outermost surface of the specimen thus gives information about the surface texture and topography. There are several operating modes of the SEM, each of them corresponding to the method of collection of different types of signals that are arising after interaction with sample material.

1. **Emission mode:** The most common mode of SEM operation is emission mode in which secondary electrons emitted from the sample are collected by the detector.
2. **Reflective mode:** This mode is preferable for the cauterization of bulk materials. In this mode, the detector collects the primary electrons backscattered from the sample.
3. **Absorptive mode:** This mode is complimentary to the above two modes. An electrical lead is connected to the specimen. Current flowing from the lead to earth is utilized as a signal.
4. **Transmission mode:** In this mode, the detector collects the transmitted electrons from the sample. This mode enables SEM to examine thicker specimens with high contrast.

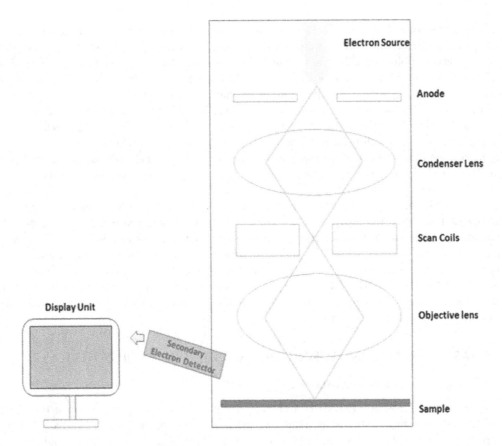

FIGURE 4.1   Schematic diagram of SEM.

5. **X-ray mode:** For the X-ray mode, the emitted x-rays are collected and used as the signal. Either the X-rays are used as they arise (nondispersive), or particular wavelengths are selected (dispersive) with the aid of a crystal spectrometer or pulse-height analyzer. This mode is the basis of the X-ray probe microanalyzer.

The SEM consists of the following parts: electron beam source (electron gun), lenses (condenser and objective), sample holder, detectors for preferred mode, and display devices along with other infrastructures, such as power source, vibration-free floor, vacuum system, room free of ambient electric and magnetic field, and cooling system.

SEM is similar to the optical stereo-binocular microscope to observe the morphology and shape of the specimen. Electrons are produced by electron guns, generally through thermionic heating. Generated electrons are accelerated to a voltage between 1-40kV, and then they are allowed to pass through a combination of different electron lenses and aperture to make a narrow and extremely focused beam of electrons. The sample is mounted at the sample stage. Through an objective lens, a focused beam is made to fall on the specimen. The position of the beam on the specimen is controlled by a scanning coil placed above the objective lens. These coils allow the beam to scan over the whole sample surface. As a result of electron matter interaction, secondary and backscattered electrons are produced, and they carry the information about the sample. Suitable detectors are used to detect produced secondary electrons, which helps to generate the high-resolution image on the display unit.

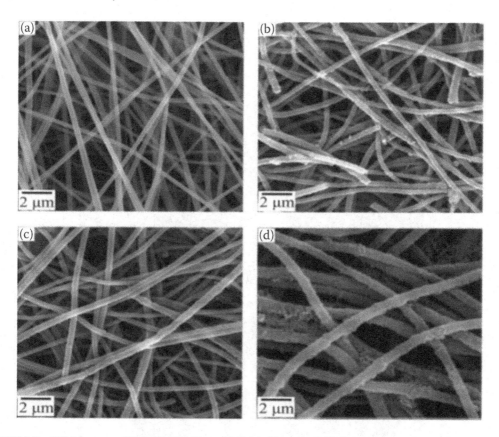

FIGURE 4.2 SEM images of PAA/PANI fibers (a) PAA fiber, (b) PAA/PANI 1, (c) PAA/PANI 2, (d) PAA/ PANI 3. Reproduced with permission from ref. [5] copy right (2013) ACS Publication.

A morphological study was done using SEM by Simotwo et al. [3] of high-purity polyaniline (PANI) nanofibers (Figure 4.2). PANI is one of the oldest organic semiconducters and attracts interest due to its conducting nature, flexibility, and mechanical properties. Due to easy fabrication techniques and cheaper, stable, and good electrical activity, PANI becomes an important material to be utilized in flexible energy devices. Electrospinning the PANI93 solution provides results in a nonwoven and continuous nanofiber structure with accurate interfiber porosity. Adding CNT to PANI does not bring any negative difference in electron-transporting. Here, we discuss the morphological SEM images of PANI with distinct concentrations of aniline and polymerization solution on the surface of poly (amic acid) (PAA) fibers. The PAA forerunner polyimide is soluble in an organic solvent and makes easily uniform fibers with constant diameter [4]. The avarage diameter of PAA before polymerization is near 300 nm, as shown in Figure 4.2(a). Due to the situ polymerization of aniline, PANI accumulates onto the surface of PAA fibers, making a thin film co-axially. Figure 4.2(b, c, d) represents the PAA/PANI 1, 2, 3, which is basically PAA fiber after accumulation of PANI onto its surface. With different concentration of aniline solution, the accumulation of PANI on PAA fiber is in different amount. Due to this difference, there is variation in the diameter of PAA/PANI nanofibers. The average diameter of PANI 1, 2 and 3 was found near 400 nm, 450 nm, and 550 nm, respectively.

Lin *et al.* [6] discussed the use of SEM images of various materials that can be used in the fabrication of flexible energy devices for their morphological study. Flexible lithium-ion batteries have synthetic graphite and lithium cobalt oxide (LCO) as the active material for electrodes, respectively, because of their steady electrochemical romance.

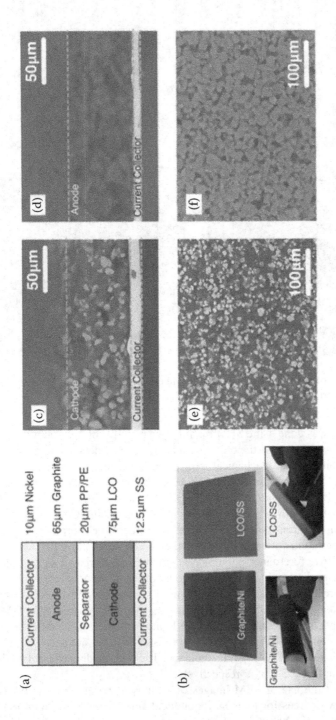

FIGURE 4.3 (a) Cross-sectional schematic of the lithium-ion battery. (b) Optical image of graphite electrode on nickel foil and LCO electrode on stainless steel. Cross-sectional SEM (c) LCO, and (d) graphite electrodes, respectively. Topographical SEM (e) LCO, and (f) graphite electrodes, respectively. Reproduced with permission from ref. [8].

Figure 4.3(a) shows the systematic layer separation of anode, cathode, and current collector layers. Here the active material is LCO and graphite at the cathode and anode ends. In the battery, active particles generate at the anode side mixed with conductive additives at the cathode side and so increase the electronic conductivity. Both are separately non-flexible and rigid, but after the mixing with polymer binder, a porous structure is formed. This structure absorba the tensile stress generated at the electrode during the flexing. Figure 4.3(b) shows the optical image of the active material LCO (lithium cobalt oxide) and graphite flat and the pen structure with the diameter is 10 mm. Figure 4.3(c, d) shows the cross-sectional and topographical SEM micrographs of the LCO and graphite electrodes [7]. Similarly, in Figure 4.3(e, f), the SEM images display that active material and conductive additives are well dispersed. Theoretically, the thickness of the Li-ion batteries was 182.5 µm, but after the integration within aluminum-laminated pouches, battery thickness enhanced to 420 µm [8].

4.2.2 Transmission electron microscopy

Transmission electron microscopy (TEM) is a microscopy technique that provides details about the inner structure of the specimen. It works on the principle of the interaction between the extremely focused and uniformly dense current of electron beam and thin-film sample. In the TEM method, an electron beam is transmitted via the thin sample, which interacts with the sample's inner structure. Some of the electrons are absorbed while rest are either reflected or transmitted through the sample, and then an image is created on the imaging device; the image shows the sample inner structure. That TEM image provides the information about sample inner structure, including molecule arrangement, size, depth morphology, and explains the new series of physical properties, such as catalytic, magnetic, electronic, and optical, etc. [9–13]. The TEM image can give an idea that how nano bridges were formed between the nanoparticles as soon as sunlight is exposed. These morphological changes, along with surface sulfidation that changes the dissolution rate of nanoparticles, result in decreases in toxicity. Furthermore, TEM can be used to examine the process of biodegradation of the nanostructured polymeric materials covered by bacteria. The biodegradation of the particle results in colloidal accumulation, which affects their cytotoxicity and mobility [14,15]. The transmission electron microscope has three main constitutive systems [16].

1. The first is an electron gun that produces a high-voltage electron beam.
2. The second one is the system consisting of the objective lens, sample holder (movable), intermediate lens, prior lenses, and objective lens. This whole arrangement helps TEM to generate the image. Different types of lenses are used to focus the beam on the sample. A scan coil is placed near the objective lens, which controls the position of an electron beam on the specimen. The image formed will be real and highly magnified.
3. The next one is the image-producing system, which is perceptible to human eyes. This system converts an electron image into a real image. This system comprises a fluorescent screen for focusing and viewing the image and a digital charge-coupled devices (CCD) camera for everlasting records. Other than this, a vacuum system is there; it encloses the whole arrangement. The vacuum system consists of pumps and their connected valves and gauges, and power supplies are necessary. The condenser lens is placed between the electron gun and sample, which restricts the angular aperture and intensity of the electron beam.

TEM involves the emission of a high-voltage electron beam from a tungsten filament (cathode) by electrical heating; the shaft of the electron beam is drawn toward an anode (magnetic lenses) and passes through an aperture (Figure 4.4). The beam traverses the aperture and next moves through an electromagnetic condenser, objective, intermediate, and projector lens. The focused electron beam is transmitted through a sample thickness of 50 nm (which is semitransparent for negative charge carrier and imports details about the configuration of the sample); the beam is loaded on a grid inserted in the path and manipulated by a goniometer. The part of the beam absorbed is scattered and transmitted through an

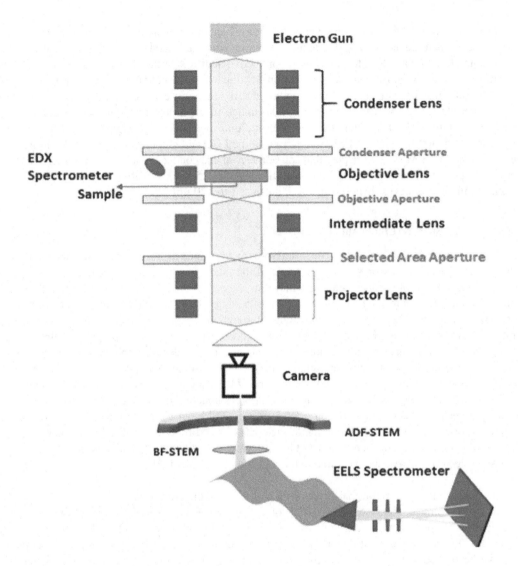

FIGURE 4.4 Set-up for a TEM instrument.

objective aperture and projected by a projector lens after being corrected by intermediate lenses on the fluorescence screen. A set of magnetic lenses is then used to magnify and vary the spatial resolution of the image, and then it gets recorded by hitting a light-sensitive sensor or fluorescent screen such as a CCD camera fitted in either the side or the bottom of the photographic plate. TEM electron scattering, rather than differences in absorbance, produce variation in the image. TEMs produce 2D white and black images [17,18].

The TEM characterization technique is used in a solid and liquid cell to study the lithiation process for Li-ion batteries. TEM is capable of providing valuable and real-time details and observation on the linked effects of the electrode electrochemical kinetics, structural evolution, mechanical degradation [19].

Figure 4.5(a) shows the schematic picture of the static experimental system for all solid-electrochemical cell: using the "liquid cell" with class liquid electrolyte and LiCoO2 counter electrode, while the "solid cell" with Li metal electrode for Si lithiation test. Figure 4.5(b–d) shows the pristine Si nanowires classis nanostructure the study used. The nanowire grows along the plane (112) direction,

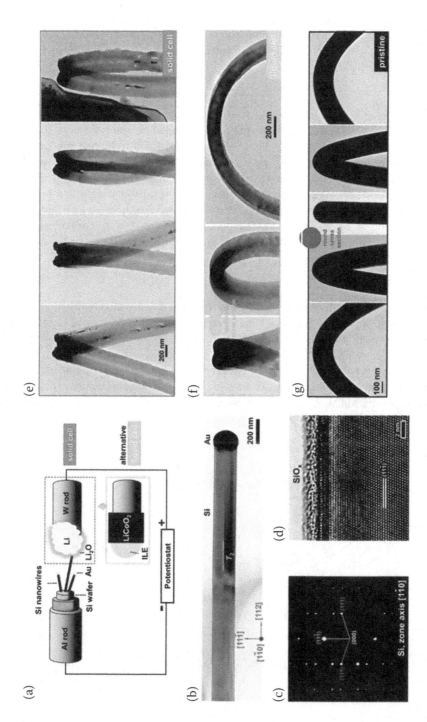

FIGURE 4.5 TEM study for the lithiation process for Si anode. Schematic illustrations of the open solid cell and liquid cell setup inside a TEM. b–d, Microstructure of a typical pristine Si nanowire, e.g. Reproduced with permission from ref. [20].

typically with 180° twin boundary parallel to plane (111) (Figure 4.5(b, c)) and native oxide layer of 2 nm thick on the surface (Figure 4.5(d)). Impressively, after lithiation, the volumetric expansion of nanowire in the solid cell was extremely anisotropic, revealing the dumbbell-shaped cross section (Figure 4.5(e)). Similar phase transformation and anisotropic expansion occurred in solid cell was also noticed in liquid cell (Figure 4.5(f)), and both are completely different to the round shape of pristine Si nanowire. Figure 4.5(g) reveals that the phase deformation and transformation were intrinsic to the process insertion of Li into Si.

4.2.3 X-RAY DIFFRACTION

X-ray diffraction (XRD) is a powerful technique that gives detail concerning the crystalline grain size, crystalline structure, lattice parameters, and nature of the phase. XRD technique can be also used on samples in powder form, generally, after drying their analogous colloidal solutions. This technique is poor for amorphous materials because generated peaks are too broad [21]. In XRD, a monochromatic beam of x-ray incident on the specimen (Figure 4.6). Interaction results in scattering x-rays from atoms within the target material. Scattered x-rays undergo destructive and constructive interference, known as diffraction. The diffraction of x-ray is defined by Bragg's Law (Figure 4.7). According to Bragg's Law, an x-ray beam is incident on the crystal face at certain angles θ (theta).

$$n\lambda = 2d \sin\theta \quad \ldots\ldots\ldots\ldots\ldots \tag{4.1}$$

where, n = integer; λ = wavelength of incident x-ray beams, and d = distance between atomic layers in the crystal. Bragg's law is often used to describe the interference pattern of x-rays in crystals; diffraction is used to produce information about the configuration of all matter's state with any beam, whose wavelength is near to the value of interplanar spacing, e.g., beams of ions, electrons, neutrons, and protons [22]. For example, XRD can be used for the characterization of polycrystalline material to obtain crystal parameter from diffraction images. Figure 4.8 shows the pure polyacrylonitrile (PAN) based CNF (PAN-CNF) and also 40wt% MnACAC (40 Mn@CNF). The Mn@CNF are used in the supercapacitor as

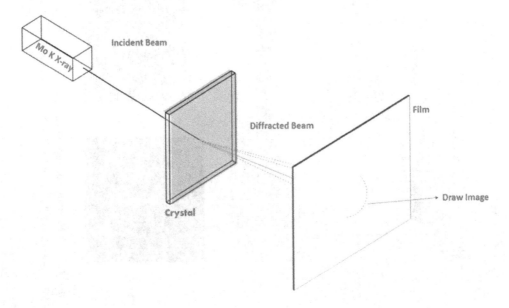

FIGURE 4.6 Basic schematic of XRD.

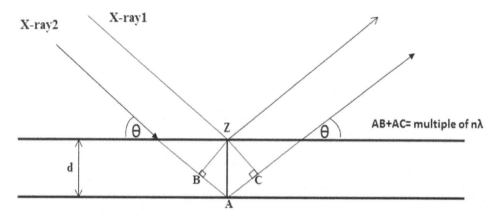

FIGURE 4.7   Bragg's Law wave diagram.

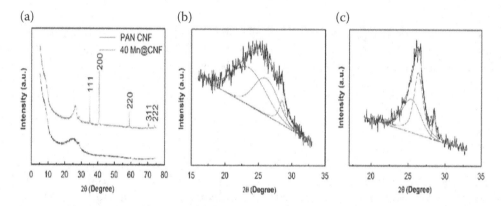

FIGURE 4.8   (a) XRD spectra of pure PAN-CNF and 40 Mn@CNF. Peak fittings of the d002 peak of the XRD pattern, (b) PAN-CNF and (c) 40 Mn@CNF. Reproduced with permission from ref. [23].

the fiber electrode. XRD patterns analyzed broadened (002) diffraction peaks specifying the distinct crystalline regions in PAN-CNF and 40 Mn@CNF [23].

4.2.4   Cyclic voltammetry

Cyclic voltammetry (CV) is a technique used in electrochemistry to measure the current produced when a voltage is applied. The CV technique is used to know the electrical performance of ESDs. It works on the principle of three electrodes (working, reference, and counter electrode). The potential between working and reference electrodes is measured. This applied voltage generates an excitation signal, as shown in Figure 4.9. The slope of the excitation signal is the measure of the scan rate. This technique is capable of generating forward scan, as well as a reverse scan within a minute. Hence, it can be used to determine the mechanism of energy-storing devices [24].

Electrochemical performance can be evaluated based on current response, according to the applied voltage. The following mathematical expression can be used for calculating EDLC capacitance.

$$C = \frac{Q_{total}}{2\Delta V} \ldots \ldots \ldots \ldots \ldots \quad (4.2)$$

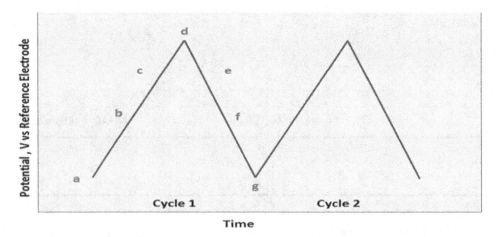

FIGURE 4.9 Cyclic voltammogram excitation signal.

Where, Q_{total} is the total charge in coulombs, it can be evaluated by calculating the integral area of the CV curve. ΔV is the difference in voltage between the two terminals of the device.

Specific capacitance can be express by the equation below

$$C_s = \frac{C}{m} \dots\dots\dots\dots\dots \tag{4.3}$$

Where C is the capacitance that can be determined using the above equation, and m is the active mass of the associated material of both electrodes. Tanwilaisiri *et al.* [25] worked with a 3D-printed supercapacitor. They applied potential in the range of 0–0.8 V at different scan rates of 0.02, 0.06, and 0.10 V/s. They found that with the increase in scan rates, capacitance (C) decreases from 182 mF to 32 mF. The decrement in capacitance that results from increasing the scan rate is due to a reduction in the efficiency of the ion diffusion process. Generally, at lower scan rates, ions get enough time to diffuse through pores of the electrode, resulting in high capacitance. Specific capacitance calculated using expression (4.2) was found to be 193.20 mF/g for a capacitance of 182 mF. Wang et al. [26] also used the CV method for the characterization of supercapacitors. They found specific capacitance of supercapacitors as 441, 399, 347, 265, and 182 F/g at the different scan rates of 5, 10, 20, 50, and 100 mV/s, respectively. Based on the information provided by the CV technique, it is possible to conclude that, what type of electrochemical reaction (such as capacitive or faradic) is occurring within the cell during charging and discharging. Generally, the voltammogram can be modified by changing the rate of multielectron transfer; thus, the electrons transfer rate can be a basis to distinguish specific types of reaction.

Figure 4.10(a) indicates the cyclic voltammetry (CV) curves with different scanning rates of 10 Mn@CNF based on every type of fiber supercapacitor. Most surprisingly, to congregate flexible pouch cell, the flexible free-standing electrode film can be used. Figure 4.10(b) shows the dynamic CV curve of pouch cell with bending angle from 0° to 180°. During the bending process, the dynamic CV curves are remarkably stable and refer the pontential of CNF electrode as flexible/bendable device.

4.2.5 GALVANOSTATIC CHARGE-DISCHARGE TEST

Performing parameters of supercapacitors like capacitance, energy density, power density, and cyclic stability can be evaluated using the galvanostatic charge-discharge (GCD) method. GCD test works on the time-dependent responsive potential. This technique can provide information about the electrochemical reaction taking place at SC. Generally, GCD measurement consists of two steps: first, charging

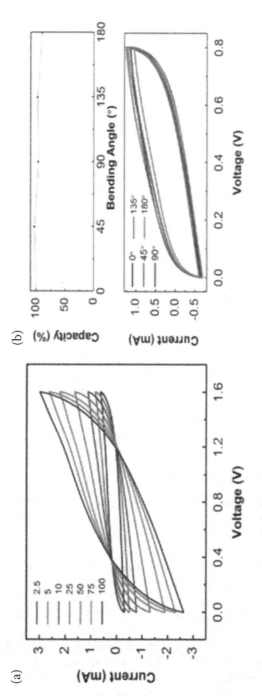

FIGURE 4.10 The CV curves of the Mn@CNF based all type fiber SCs with distinct scan rates, showing great capacitive performance and the figure displays the CV curves of the pouch cell with different bending angles, reproduced with permission from ref. [23].

a SC by a constant current (charging test) and second, discharging a SC in a specific time and within a given voltage limit (discharging test). The capacitance can be calculated by using Eq. (4.4)

$$C = \frac{i\Delta V}{\Delta t} \dots\dots\dots\dots\dots\dots\dots\dots\dots \quad (4.4)$$

Where i is the discharging current, ΔV denotes given discharge voltage, and Δt is the time taken by SC to get fully discharged. Specific capacitance can be calculated simply by dividing the Eq. (4.4) by m, where m is the active mass of both electrodes. The equation for specific capacitance can be expressed as Eq. (4.5)

$$C_s = \frac{i\Delta V}{m\Delta t} \dots\dots\dots\dots\dots\dots\dots\dots\dots \quad (4.5)$$

Energy and power densities can be calculated using the following mathematical expressions:

$$E = \frac{C_s \Delta v}{2} \dots\dots\dots\dots\dots\dots\dots\dots\dots \quad (4.6)$$

$$P = \frac{E}{\Delta t} \dots\dots\dots\dots\dots\dots\dots\dots\dots \quad (4.7)$$

Where, Cs is the specific capacitance and Δt denotes time taken by SC to get fully discharge and ΔV represents the given voltage at which SC has been discharged. One major use of GCD is to measure the stability performance of SCs. Cao et al. [27] learned that the specific capacitance of capacitors is as high as 520 F/g at 0.5 A/g, and energy and power densities were found to be 58.5 Wh/kg and 22.5 W/kg, respectively. Only 82% of the initial value of capacitance was retained after use 800 cycles at a current density of 3 A/g. Hou et al. [28] also used GCD to check the performance of their SCs. They found specific capacitance and energy density as high as 286 F/g and 39.7 Wh/kg, respectively, at 0.5 A/g. They observed that 88.6% of the initial capacitance was retained after 3,000 cycles.

4.2.6 ELECTROCHEMICAL IMPEDANCE SPECTROSCOPY

Electrochemical impedance spectroscopy (EIS) is an important and complex characterization tool in electrochemical research to assess the electrochemical performance of electrode and electrolyte materials. EIS is capable of differentiating the influence of different components of a device on its performance. An individual contribution of each component can be estimated, which helps in designing the devices with improved performance. The sensitivity of EIS is very high for the changes occurring at the surface of device components and therefore can provide a wealth of information that other instruments can't. For example, due to bending, twisting, swelling, or after removal of corrosion protective coating, some changes may occur at the surface of active materials, which can be observed by EIS. EIS can be analyzed by both the Nyquist diagram and Bode plot.

Generally, in EIS measurements, impedance data are recorded at open potential by applying the alternating potential of small amplitude over a wide range of frequencies (from 0.01 to 100 kHz). The capacitance can be calculated from EIS as C = 1/ (2πf |Z|) (using a linear portion of a log |Z| vs. log f curve: Bode plot), where |Z| is the imaginary part of impedance and f is the frequency. Impedance is analogous to resistance but it describes a complex circuit (which exhibits capacitance, inductance, or mass diffusion) showing a nonlinear current-voltage relationship. EIS relates theoretical circuit elements to the actual electrochemical process, which takes place in materials of energy-storing devices. EIS

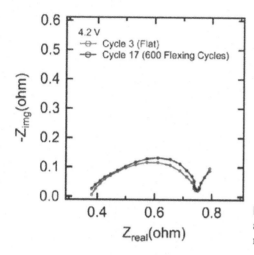

FIGURE 4.11 EIS curves of the battery under a flat state and after bending 600 times. Reproduced with permission from ref. [8].

technique was used for flexible Li-ion battery. It was performed before and after the flexing to understand the effect of flexing on the performance and stability of the device. Figure 4.11 shows comparative data of impedance measurement by EIS before and after flexing [8].

4.2.7 Atomic force microscopy

Atomic force microscopy (AFM) creates a three-dimensional image of the sample surface with high magnification. Similar to SEM, AFM is used to study of size, shape, and aggregation of inner particles. The basic principle of AFM is measuring the interacting force that may be repulsive or attractive amid the scanning probe and the specimen surface. The measurement of force is done through the probe tip, which connects with the cantilever, brought in the proximity of the specimen surface. AFM operates in the three modes, which are based on the physical interaction between the sample and scanning probe: (1) contact mode, (2) non-contact mode, and (3) tapping mode [29]. The non-contact mode has no physical contact between probe and specimen. This mode is used for soft material since it does not damage the specimen surface. The contact mode has the physical contact between probe and sample so it provides high-resolution images and better characteristics. Contact mode has constant repulsive forces onto the probe tip [30]. The tapping mode has a short-time physical interaction between sample and probe so it provides attractive and repulsive forces. The tip contact at the specimen surface appears invalid during a short period that it provides a short-time high-resolution image [31]. AFM consists of equipment setup, which includes controller, position-sensitive photodetector, AFM probe or cantilever, laser source, and piezoelectric ceramic tube; another one is a computer where we get the 3D image (Figure 4.12).

4.2.8 Secondary ion mass spectroscopy

A secondary ion mass spectroscopy (SIMS) is a surface analysis technique that is used to analyze the molecular and chemical composition of solid material's surface (Figure 4.13). In SIMS, sputtering of the specimen is done using a focused primary ion beam. A secondary ion ejected from the sample surface is collected and analyzed with the help of some mass analyzer. The emission of secondary ions from the sample surface gives information about molecular and chemical composition. SIMS is a destructive technique. The formation of secondary ions is a result of the interaction of primary ions with the outmost surface of the specimen [32]. SIMS has two modes: (1) static and (2) dynamic. In static SIMS, a small dose of a focused and controlled primary ion beam is used to sputter the sample. An ultra-high vacuum is required in static SIMS because of a low dose of primary ion. Sometimes it is considered as a non-destructive technique due to the low dose of primary ion beam because damage to the surface is

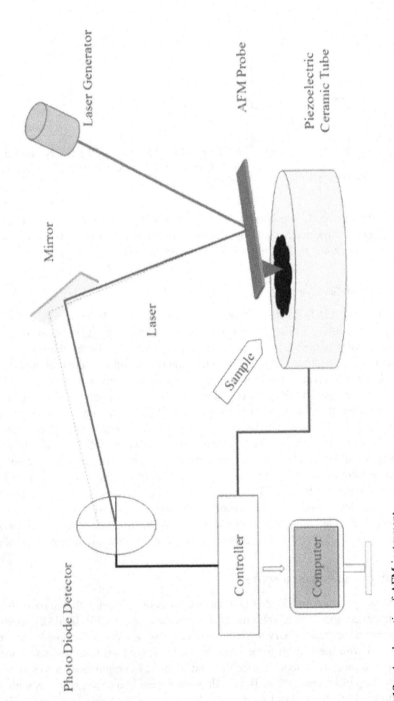

FIGURE 4.12 A schematic of AFM instrument.

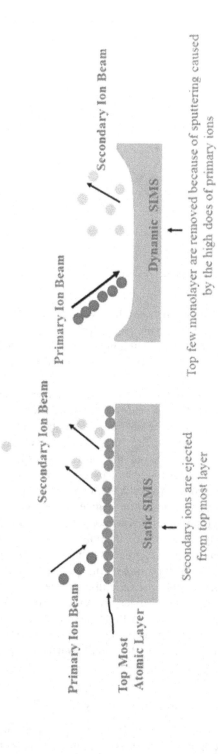

FIGURE 4.13 Basic principles of secondary ion emission: Static versus dynamic mode.

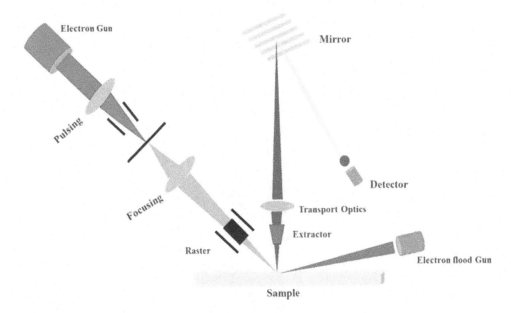

FIGURE 4.14   TOF-SIMS instrumentations setup.

negligible. Static mode SIMS is useful to extract the information about the outmost nano-surface of the sample. In dynamic SIMS, a high dose (a large number of ions per unit area) of primary ion is used, and therefore, a moderate vacuum is required. Due to the high dose of primary ions, information about the sub-surface, up to few nanometers, can be obtained. This mode of SIMS is preferred to detect the trace of impurities in the sample [33].

Three types of mass analyzers are used for analyzing secondary ions, known as time of flight (Tof), magnetic sector, and quadrupole. In Tof mass analyzer, the time taken by the secondary ion to reach the detector is measured; this is called the time of flight. Since the velocity of the ions depends upon the mass-to-charge ratio of ions, by measuring the time taken to travel a known distance, velocity can be calculated. Hence, the mass-to-charge ratio of the ion can be estimated. Figure 4.14 shows the setup of the Tof mass analyzer. In a magnetic sector mass analyzer, two types of sectors are used: electrostatic and magnetic sector. The electrostatic sector selects the ions of fixed kinetic energy, while the magnetic sector allow to the pass ions of specifice mass-to-charge ratio. A quadrupole mass analyzer consists of four parallel cylindrical rods. Each opposite pair of rods is connected electrically. These quadrupole arrangements only allow ions of a certain mass-to-charge ratio. Since each element of the periodic table has a different mass-to-charge ratio, the element present in the sample can be predicted based on the evaluated value of the mass-to-charge ratio. Time of flight SIMS (Tof-SIMS) is a surface-sensitive analytical method that uses a pulsed ion beam to remove molecules from the very outermost surface of the sample. The particles are removed from the atomic monolayer on the surface. These particles are then accelerated in a flight tube, and their mass is determined by measuring the exact time at which they reach the detector. Tof-SIMS referred to the static SIMS because the low current primary ion beam is used for the sample surface ions, molecules, and molecules clusters analysis.

4.2.9   INDUCTIVELY COUPLED PLASMA-MASS SPECTROSCOPY

An inductively coupled plasma-mass spectroscopy (ICP-MS) technique is used for elemental analysis of the samples. It is characterized by high sensitivity, selectivity, robustness, wide dynamic range, and

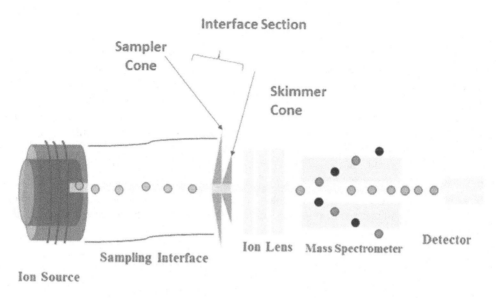

FIGURE 4.15 Basic ICP-MS setup.

virtual matrix independence [34]. It permits the reliable elemental and quantification composition characterization of materials. ICP-MS can also evaluate the size distribution and concentration of materials [35]. ICP-MS has six major steps for the analysis: plasma discharge devices, sample introduction, interface, ion-focusing lenses, mass spectrometer, and detector [36]. The main role of plasma in the ICP-MS is to ionize the sample. RF supply (800–1,600 W with an operating frequency of 27.5MHz or 40.2 MHz) is mostly used in plasma generation. To form inductively coupled plasma, ions are transferred to gas through an induction coil wound around an ion source assembly, as display in Figure 4.15. Plasma location depends on the wounded induction coil. Solid samples are analyzed through electrothermal vaporization or laser ablation. After vaporization, the sample goes into the ICP-mass spectrometer. An interface section consists of a group of the two interface cones (sampler and skimmer cone). The sampler cone radius is higher than the skimmer cone. The sampler cone collects the ionized ion from the ion source and is focused by a skimmer cone [37]. The series of cones work to focus the ion on the lens. The mass spectrometer is located amid the ion lens and the electron detector. The mass analyzer separates ions from the multi-ion beam, which depends on the m/q ratio. The separated beam goes directly to the detector here, measuring the individual ion's current. Three main types of mass analyzers are used: (1) quadrupole, (2) magnetic sector, and (3) time of flight (TOF). The ions are detected using an electron multiplier, which is a continuous or discrete dynode electron multiplier.

4.2.10 FOURIER TRANSFORM INFRARED SPECTROSCOPY

In Fourier transform infrared (FTIR) spectroscopy, infrared light is passed over the specimen. A small amount of the IR radiation is transmitted through the specimen, and the remainder of the IR radiation is absorbed by the sample. Then the resultant spectrum displays the molecular absorption and molecular transmission, generating a molecular image of the specimen. FTIR provides information about a sample-like identification of the material, determines the quality of the sample, determines the amount of content in any mixture, determines the structure, detects impurity in the sample, and studies the chemical reaction. FTIR spectroscopy is based on Lambert-Beer´s law; according to this law, the amount of energy transmitted and absorbed by a sample is proportional to the absorptivity and the concentration of the specimen.

$$i.e., \quad A = \varepsilon bC \quad \ldots \quad (4.8)$$

The FTIR instrument includes a combination of fixed and movable mirrors, one IR source, a beam splitter, and a detector. First, a light beam that has IR wavelength is passed over a beam splitter, and the beam is split into two parts; the first half reaches a fixed mirror, while the second half goes on the moveable mirror moving at a constant velocity. These split beams are reflected and recombined to create an interference pattern reflecting the destructive and constructive of the recombination due to path differences between the beams. This interference pattern is sent to samples, and the transmitted portion is detected through a detector (Figure 4.16).

In the perovskite solar cell $FA_x Cs_{1-x} PbX_3$ (X = bromine, iodine) FTIR provides interaction between the sulfur (S) donor group and Pb^{+2} due to Lewis's acid-base. Through the spectrum, the redshift of the C=S vibration from 729 cm^{-1} in pure SN interacts with PbI_2. Figure 4.17 shows the formation of intermediate $SN - PbI_2$ adduct weakening the C=S bond strength caused by the interaction with the PbI_2 as a Lewis acid [38].

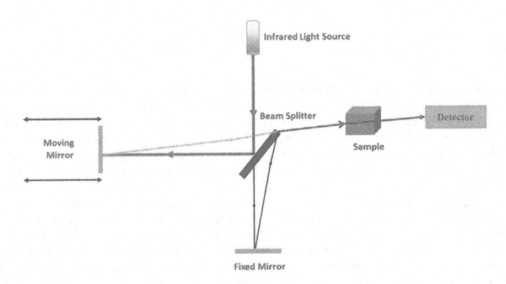

FIGURE 4.16 Instrument setup for FTIR spectroscopy.

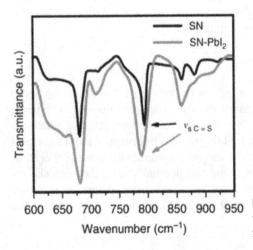

FIGURE 4.17 FTIR spectra the arrows indicate the stretching vibration peak of C=S in the films. Reproduced with permission from ref. [38].

4.3 SUMMARY

Researchers synthesize and analyze different materials for different properties so that new materials can be successfully used in flexible devices. For the examination of properties and characteristics, many tools and techniques are needed. These techniques are used to characterize the materials to make sure the material possesses the required properties. The characterization can be defined as "the step-by-step process by which material's structure and properties can be probed and measured." It is a basic and crucial process to advance material science. This chapter discusses many characterization techniques that are particularly useful for energy devices.

REFERENCES

1. Fu YP, Lv ZB, Hou SC, Wu HW, Wang D, Zhang C, Chu ZZ, Cai X, Fan X, Wang ZL and Zou DC, (2011) *Energy Environ. Sci.* 4: 3379.
2. Tominaga S, Nishizeko H, Mizuno J, and Osaka T, (2009) *Energy Environ. Sci.* 2: 1074.
3. Simotwo SK, DelRe C, and Kalra V, (2016) Supercapacitor electrodes based on high-purity electrospun polyaniline and polyaniline–carbon nanotube nanofibers, *ACS Appl. Mater. Interfaces* 8: 21261–21269.
4. Hung C, Wang S, Zhang H, Li T, Chen S, Lai C, and Hou H, (2006) *Eur. Polym. J.* 42: 1099–1104.
5. Miao YE, Fan W, Chen D, and Liu T, (2013) *ACS Appl. Mater. Interfaces* 5: 4423–4428
6. Li L, Wu Z, Yuan S, and Zhang XB, (2014) *Energy Environ. Sci.* 7: 2101.
7. Jansen AN, Amine K, Newman AE, Vissers DR, and Henriksen GL, (2002) Low-cost, flexible battery packaging materials, *JOM* 54: 29–32.
8. Ostfeld AE, Gaikwad AM, Khan Y, and Arias AC, (2016) High-performance flexible energy storage and harvesting system for wearable electronics. *Sci. Rep.* 6: 26122
9. Reimer L, and Kohl H, (2009) *Transmission electron microscopy physics of image formation*, Springer, New York, 51: 1–15.
10. Daniel MC, and Astruc D, (2004) *Chem. Rev.* 104: 293.
11. Pankhurst Q, Connolly J, Jones SK, and Dobson J, (2003) *J. Phys. D: Appl. Phys.* 36: R167–R181.
12. Nurmi JT, Tratnyek PG, Sarathy V, Baer DR, Amonette JE, Pecher K, Wang C, Linehan JC, Matson DW, Penn RL, and Driessen MD, (2005) *Environ. Sci. Technol.* 39: 1221.
13. Astruc D, Lu F, and Aranzaes JR, (2005) *Angew. Chem. Int. Ed.* 44: 7852.
14. Cheng Y, Yin L, Lin S, Wiesner M, Bernhardt E, and Liu J, (2011) *J. Phys. Chem. C* 115: 4425.
15. Kirschling TL, Golas PL, Unrine JM, Matyjaszewski K, Gregory KB, Lowry GV, and Tilton RD, (2011) *Environ. Sci. Technol.* 45: 5253.
16. Fultz B, and Howe JM, (2001) *Transmission electron microscopy and diffractometry of materials*, Springer-Verlag, Berlin, Heidelberg, 101.
17. Goldstein JI, Yakowitz H, Newbury DE, Lifshin E, Colby JW, and Coleman JR, (1975) *Practical scanning electron microscopy: Electron and ion microprobe analysis*, Springer, New York.
18. Thomas J, Goringe MJ, (1979) *Transmission electron microscopy of materials*, A Wiley-Interscience Publication, Hoboken, NJ.
19. Jun L, Tianpin W, and Khalil A, (2017) State-of-the-art characterization techniques for advanced lithium-ion batteries. *Nat. Energy* 2: 17011.
20. Liu XH, Zheng H, Zhong L, Huang S, Karki K, Zhang LQ, Liu Y, Kushima A, Liang WT, Wang JW, Cho J-H, Epstein E, Dayeh SA, Picraux ST, Zhu T, Li J, Sullivan JP, Cumings J, Wang C, Mao SX, Ye ZZ, Zhang S, and Huang JY, (2011) *AnisotropicSwelling and Fracture of Silicon Nanowires during Lithiation* 11: 3312–3318.
21. Upadhyay S, Parekh K, and Pandey B, (2016) *J. Alloys Compd.* 678: 478.
22. Li W, Zamani R, Rivera P Gil, Pelaz B, Ibanez M, Cadavid D, Shavel A, Alvarez-Puebla RA, Parak WJ, Arbiol J, and Cabot A, (2013) *J. Am. Chem. Soc.* 135: 7098.
23. Liu X, Marlow MN, Cooper SJ, Song B, Chen X, and Brandonb NP, (2018) Flexible all-fiber electrospun supercapacitor. *J Power Sources* 384: 264–269.
24. Kissinger PT, and Heineman WR, (1983) Cyclic voltammetry. *J. Chem. Edu.* 60: 702.
25. Tanwilaisiri A, Xu Y, Zhang R, Harrison D, Fyson J, and Areir M, (2018) Design and fabrication of modular supercapacitors using 3D printing. *J Energy Stor.* 16: 1–7.

26. Wang K, Wang Z, Wang X, Zhou X, Tao Y, and Wu H, (2018) Flexible long-chain-linker constructed Ni-based metal-organic frameworks with 1D helical channel and their pseudo-capacitor behavior studies. *J. Power Sources* 377: 44–51.

27. Cao P, Fan Y, Yu J, Wang R, Song P, and Xiong Y, (2018) Polypyrrole nanocomposites doped with functional ionic liquids for high-performance supercapacitors. *New J. Chem.* 42: 3909–3916.

28. Hou Z, Lu H, Yang Q, Zhao Q, and Liu J, (2018) Micromorphology-controlled synthesis of polypyrrole films by using binary surfactant of Span80/OP10 via interfacial polymerization and their enhanced electrochemical capacitance. *Electrochim. Acta* 265: 601–608.

29. Jagtap RN, and Ambre AH, (2006) Overview literature on atomic force microscopy (AFM): Basics and its important applications for polymer characterization. *Ind. J. Eng. Mater. Sci.* 13: 368–384.

30. Jalili N, and Laxminarayana K, (2004) A review of atomic force microscopy imaging systems: Application to molecular metrology and biological sciences. *Machatronics* 14: 907–945.

31. Rajagopal SR Achary, (2010) *Characterization of individual NPs and applications of NPs in mass spectrometry*, Ph.D. thesis, A&M University, Texas.

32. Harkness KM, Cliffel DE, and Mclean JA, (2010) *Analyst* 135: 868.

33. Allabashi R, Stach W, Escosura-Muniz AD, Liste-Calleja L, and Merkoci A, (2009) *J. Nanopart. Res.* 11: 2003.

34. Nageswaran G, Choudhary YS, and Jagannathan S, (2017) Inductively coupled plasma mass spectrometry, Chapter 8.

35. Wilschefski SC, and Baxter MR, (2019) Inductively coupled plasma mass spectrometry: Introduction to analytical aspects. *Clin Biochem Rev* 40(3): 115–133.

36. Case CP, Ellis L, Turner JC, and Fairman B, (2001) Development of a routine method for the determination of trace metals in whole blood by magnetic sector inductively coupled plasma mass spectrometry with particular relevance to patients with total hip and knee arthroplasty. *Clin Chem* 47: 275–280.

37. Hu J, Deng D, Liu R, and Lv Y, (2018) Single nanoparticle analysis by ICPMS: A potential tool for bioassay. *J Anal at Spectrom* 33: 57–67.

38. Dongqin B, Xiong L, Jovana VM, Dominik JK, Norman P, Jingshan L, Thomas L, Pierre M, Lyndon E, Shaik MZ, and Michael G, (2018) Multifunctional molecular modulators for perovskite solar cells with over 20% efficiency and high operational stability. *Nat. Commun* 9: 4482.

5 Micro- and Nanofibers-Based Flexible Energy Devices

Arunima Reghunadhan

Department of Chemistry, Milad-E-Sherif Memorial College Kayamkulam, Alappuzha, Kerala, India

School Of Energy Materials, Mahatma Gandhi University, Kottayam, Kerala, India

Jiji Abraham

Department of Chemistry, Vimala College, Thrissur, Kerala, India

P. S. Sari

Department of Polymer Science and Rubber Technology, Cochin University of Science and Technology, Kochi, Kerela, India

CONTENTS

5.1 Introduction of nanofibers and microfibers 81
    5.1.1 Carbon fibers 82
    5.1.2 Biopolymer fibers 82
    5.1.3 Aramid fibers 82
    5.1.4 Ceramic fibers 82
5.2 Flexible energy devices based on nanofibers 84
    5.2.1 Inorganic fibers for flexible energy devices 86
    5.2.2 Metallic fibers for energy devices 86
    5.2.3 Carbon-based fibers for energy devices 87
    5.2.4 Biobased fibers for energy devices 89
        5.2.4.1 Cellulose-based fibers 90
        5.2.4.2 Keratin and chitin fiber composites 93
5.3 Electrospun fibers for flexible energy devices 94
5.4 Conclusions 95
References 95

5.1 INTRODUCTION OF NANOFIBERS AND MICROFIBERS

"Nanofiber" is the generic term describing fibers with the dimensions in the nanometer, while "microfibers" cover the fibers with micrometer size. Different types of materials, such as carbon, natural and biodegradable synthetic polymers, semiconductors, and composites, can be used to fabricate nano- and microfibers [1]. Current ways for nano/microfiber fabrication embrace carbon dioxide optical maser supersonic drawing, solution blow spinning, plasma-induced synthesis, centrifugal jet spinning, electrohydrodynamic direct writing, etc. These materials can be used in many advanced applications because of their ability to form networks of extremely porous mesh with significant interconnectivity between their pores. Emergent requests of nanofibers in energy generation and storage comprise batteries, fuel cells, supercapacitors, solar cells, hydrogen storage and generation, piezoelectricity, etc. Incipient submissions of nanofibers in water treatment and conservational redressal include water treatment using the principles of ultrafiltration, photocatalysis, and chemical and gas sensing, etc. The very demanding role in healthcare and biomedical engineering includes tissue engineering (TE) and regenerative medicine, wound dressing/healing drug and therapeutic agent delivery, biological sensing, etc. [2]. Figure 5.1 shows various synthetic approaches used and applications of nanofibers. The following sections discuss various types of fibers.

5.1.1 CARBON FIBERS

Carbon nanofibers (CNFs) obligate diameters of 50–200 nm. Carbon nanofibers can be prepared via catalytic methods, thermal methods, chemical vapor deposition, and electrospinning. In the vapor deposition method, the gas-phase molecules are disintegrated at high temperatures, and carbon is dumped in the presence of a transition metal compound. A consequent growth of the fiber around the catalyst particles is detected [3]. Catalytic-thermal-chemical vapor deposition can be used to make cup-slanted carbon nanofibers and platelet carbon nanofibers (Figure 5.2) [5]. Electrospinning polyacrylonitrile trailed by stabilization and carbonization was developed as a direct and suitable method to create incessant carbon nanofibers [6]. CNF composites are the ideal candidate in numerous fields, for instance, electrical devices, electrode materials for batteries and supercapacitors, and sensors.

5.1.2 BIOPOLYMER FIBERS

Nanofibers encompassing biopolymers, for instance, chitosan, alginate, cellulose/chitin, alginate/carboxymethyl (CM) chitosan, collagen/poly(lactide-co-glycolide) (PLGA), and alginate/soy are extensively cast off in many sectors. Nano effects of these biopolymer fibers contain augmented quantum efficiency, remarkably in high surface energy, elevated surface reactivity, high thermal and electrical conductivity, and high strength-to-weight ratios [7]. These nanofibers are of curiosity for diverse submissions fluctuating from filtration, antibacterial coatings, drug-release formulations, tissue engineering, wound healing, sensors, and so on [4]. Biological claims of biopolymer-based nanofibers have attracted tremendous consideration due to their inherent biocompatibility, biodegradability, and low immunogenicity [8].

5.1.3 ARAMID FIBERS

Aramid nanofibers (ANFs) have outstanding properties like high surface area, high aspect ratio, upright mechanical properties, and high-temperature resistance. ANFs have extensive applications in the fields of polymer reinforcement, battery separators, adsorption filtration, electrical insulation, and supercapacitor electrodes. Four methods are used for the preparation of aramid fibers, including mechanical disintegration, electrospinning deprotonation, and bottom-up polymerization-induced self-assembly [9].

5.1.4 CERAMIC FIBERS

Ceramic nanofibers include nanofibers synthesized from TiO_2, silica, iron oxide (Fe_2O_3), nickel titanate, nickel oxide, and zinc oxide (ZnO). Ceramic nanofibers can be prepared via chemical vapor deposition and electrospinning methods. Ceramic nanofibers can be used in various areas, including sensors, catalysts, batteries, environmental science, energy technology, filters, separators, modifications for lithium-ion batteries, and dye-sensitized solar cell electrodes [10].

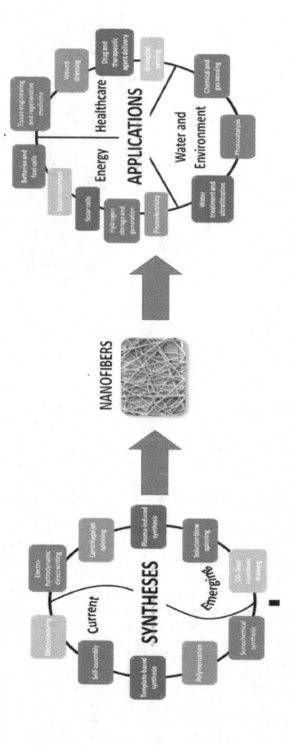

(a)   (b)   (d)

(c)

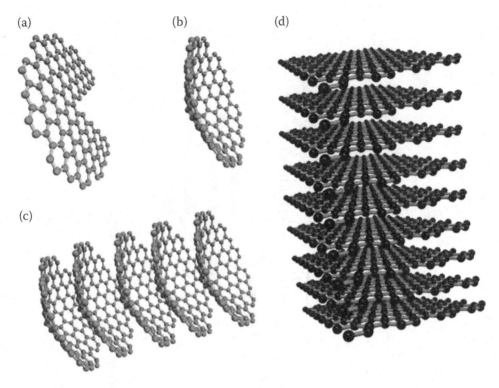

FIGURE 5.2 Schematic demo of the establishment of cup-stacked carbon nanofiber (CNF) (a–c) structure; and the rest is the platelet CNF structure (Adapted with permission from ref. [4], Copyright (2014), American Chemical Society).

5.2 FLEXIBLE ENERGY DEVICES BASED ON NANOFIBERS

Polymer nanofibers can be used to construct flexible energy devices in various fields, including photonics, electronics, energy generation, and micromechanics. Because of their unique characteristics, like flexibility, chemical composition, greater mechanical performance (e.g., stiffness and tensile strength), and achievable functionality, these can be identified as the building blocks for photodetectors, photovoltaic devices, piezoelectric and thermoelectric generators, actuators, field-effect transistors, light-emitting devices, and lasers [11]. One-dimensional (1D) nanofibers systematized via electrospinning procedures have found widespread use in different energy-related applications [12]. Electrospun nanofibers are also used in flexible nonvolatile transistor-memory devices. Highly supple freestanding porous PVDF nanofibers were fabricated using an electrospun process and used as electrode material in triboelectric nanogenerator (TENG) by Huang et al. [13]. Figure 5.3 represents the fabrication method and the resultant sole materials. The flexible matreials were placed inside the footwear sole, and the layers are represented schematically in the same figure. The TENG-based liners were lenient, stretchy, and lightweight; thus, they warrant maximum coziness for the person who is wearing them and were cast off to transform mechanical energy to light. Inorganic nanofibers can also be used for the fabrication of flexible piezoelectric nanogenerators. Manganese-doped nanofibers organized via an electrospinning technique displayed an effective piezoelectric coefficient, which was five times higher than that of Mn free nanofibers [15]. Nanofibers of zinc oxide, lead zirconate titanate (PZT), cadmium sulfide (CdS), $BaTiO_3$, etc., were similarly deciphered mechanical energy into electrical energy [16]. Piezoelectric nanofibers, for instance, PVDF and PZT, are exceedingly flexible and laid back to fabricate for probable incorporation in implantable and flexible devices, along with textile applications, such as electric clothing [14].

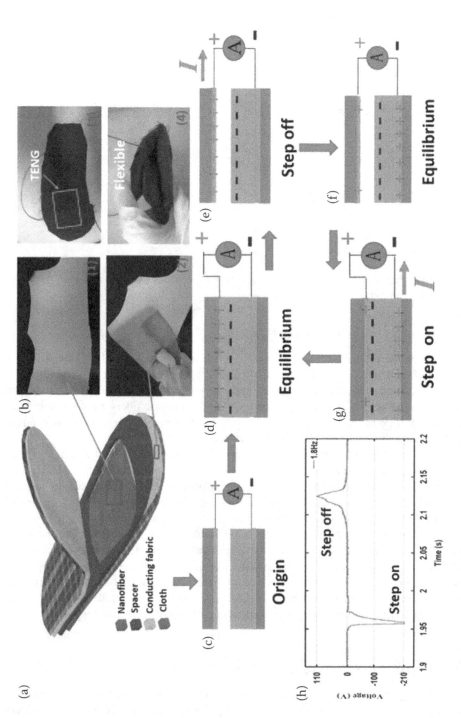

FIGURE 5.3 Schematic illustration of the assembly of the fabricated insole (b) digital snaps of the as-spun PVDF nanofibers on the conducting material. Working mechanism(c-g) of the TENG based device and (h) the conforming V-time curve (Adapted with permission from ref. [14], Copyright (2014), Elsevier).

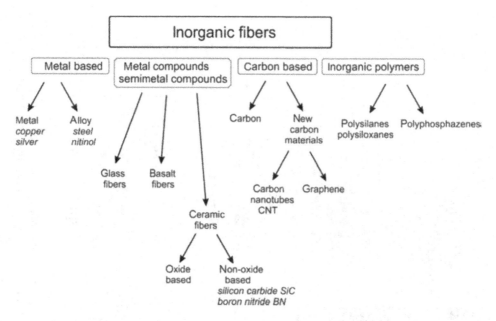

FIGURE 5.4 Different types of inorganic fibers used for energy applications (Adapted with permission from ref. [19], Copyright (2005), Springer).

5.2.1 Inorganic Fibers for Flexible Energy Devices

Inorganic fibers are a special class of fibers obtained from inorganic materials. They possess extraordinary thermal and mechanical properties. Their superior properties, including the ease of processing, thermal resistance, resistance to chemical agents, and specific strength, make them suitable for engineering applications rather than textile applications [17,18]. Inorganic fibers are accessible in a wide variability of materials. A summary of the diverse collections of inorganic fibers is shown in Figure 5.4. Related to common organic fibers, the density of inorganic fibers is high.

Nano wires and fibers signify quasi-one-dimensional (1D) nanostructures that illustrate exceptional electrical, optical, and chemical performances due to dimensional confinement of charge carriers and the existence of surface states. Their performance makes them beneficial as device elements in many applications, such as batteries, piezoelectric energy gleaners, field-effect transistors, sensors, and photovoltaics. Nanofiber networks embody a two-dimensional (2D) conformation fashioned by the entwined measures of a 1D element; this structural design is favorable due to the mechanical constancy and suppleness of the meshes indispensable for device incorporation and the opportunity to activate the functional performance through an external stimulus. Recently, researchers have developed several energy devices using various upcoming materials and inorganic fibers; metal, oxides, and carbon are the prime among them. The application of each fiber has been explained in detail in the coming sections. The electrospinning process, followed by calcination treatment, is the most effective way of manufacturing these fibers.

5.2.2 Metallic Fibers for Energy Devices

Metal fibers exhibit superior physical, electrical, and chemical properties that contribute to the application in energy devices such as capacitors, battery components, fuel cells, and solar cells. Metal nanofibers (NFs) have been occasionally functionalized with numerous constituents, such as polymers, organic functional composite anions, metal nanoparticles (MNPs), carbon nanotubes (CNTs), and other carbon-based materials. Even though the metal oxide-based nanofibers could be manufactured by many new procedures, the spinning process is found to be most apt due to budgetary, modest, and well-used techniques [12]. Various metal fibers, as well as metal oxide fibers that can be exploited for energy devices, have been described in detail.

Mn_2O_3 NF was developed as an electrode with improved constancy and specific capacitance for supercapacitors in six aqueous environments due to their outstanding physical properties, including large explicit surface area and porous structures. The latest research confirms that the amorphous Mn_2O_3 NFs structure has a high specific surface area and excellent capacitive performance, as compared to crystalline Mn_2O_3 [20].

In batteries, metallic nanofibers have been used as both anode and cathode materials. Two primary matters to address are that the fiber electrodes have high conductivity and flexibility. Largely, metal wires with exceptional conductivity are rigid and heavy, which limits their application in this sector. However, using metal nanoparticles to coat flexible textile material is another common practice to overcome this issue. Cobalt nanofiber has been widely used as an anode in Li-ion batteries [21]. The construction of electrospun copper oxide (CuO) nanofibers has been verified as an anode material in lithium-ion batteries [19]. Cavaliere et al. used coaxial electrospinning to make the core-shell $LiCoO_2/MgO$ composite nanofiber. After 40 cycles, $LiCoO_2/MgO$ could reclaim 90% of its initial charge capacity, compared to 52% for the fiber electrode without MgO [22]. Inorganic oxide fibers can be utilized as substrates to fabricate separators for lithium-ion batteries. A new separator has been created by a phase-inversion method by combining inorganic zirconium oxide staple fibers with poly(vinylidenefluoride-co-hexafluoropropylene). The separator features a skinless interface, a highly porous interior structure, and a homogeneous distribution of pore size [23]. The reinforcement from the fiber skeleton can give the separator appropriate porosity, adequate mechanical strength, superior thermal stability, and enhanced electrochemical performance, according to a systematic comparison of a fiber-based separator and a powder-based separator. The notion of integrating inorganic fiber substrates with polymeric media is expected to provide a platform method for fabricating high-safety separators for lithium-ion batteries and other energy storage devices.

5.2.3 CARBON-BASED FIBERS FOR ENERGY DEVICES

Carbon fibers (CF) include fibers with turbostratic carbon layers made up of graphite crystallites or flaking that are oriented all along the fiber axis and contain more than 90% carbon. Due to their unique structure, CFs have outstanding characteristics, such as extraordinary mechanical properties, high temperature stability, low resistance, high electrical conductivity, and low specific density. CNT and graphene fibers have shown significant promise in high-strength fibers, flexible conductors/ electrodes, electrocatalysis, biosensing, and other applications. Carbon-based nanomaterials function as building blocks for flexible energy storage devices due to their unique electrical and mechanical properties. High-performance, scalable, and shape-variable flexible micro-supercapacitors can be prepared in favor of wire-format supercapacitors. Decoration of active carbon nanomaterials, such as commercial ink-carbon, CNTs, and reduced graphenes on pristine CFs, can enhance the electrochemical capacity [24].

Fiber-based supercapacitors using CF/CNTs and CF/MnO_2/CNTs fiber electrodes were developed by dismantling activated CF textiles, which exhibited high areal capacitance of up to 640 mF/cm^2 [25]. In comparison to granular-activated carbons, which are extensively utilized in commercial supercapacitors, CFs and CF fabrics are more promising in the construction of customized supercapacitors because they can have considerably higher flexibility, conductivity, continuity, and strength. They're also helpful in miniature supercapacitors for microcircuits and minuscule devices because of their small diameter and changeable length. Typical configurations of CF-based supercapacitors are illustrated in Figure 5.5.

Carbon fibers are frequently cast off because of their low-slung mass density, robust tensile strength, and large surface area, which improve capacitance and electrical conductivity. Two aligned carbon

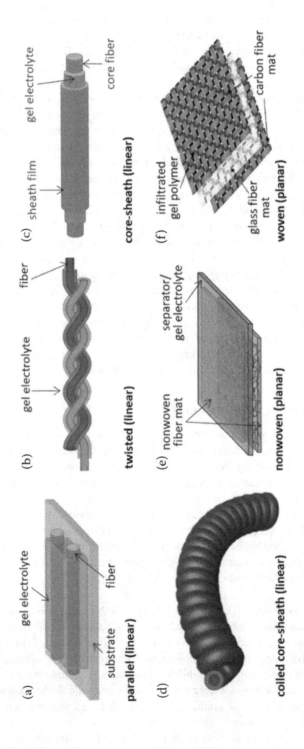

FIGURE 5.5 Archetypal conformations of carbon fiber-based supercapacitors. (A–D) Planar structures (Adapted with permission from [26], Copyright (2013) Wiley and (E) with nonwoven carbon fiber fabric and (F) is with woven carbon fiber fabric. (Adapted with permission from ref. [27], Copyright (2013) Elsevier).

nanotube (CNT) fibers covered with gel electrolyte were twisted to produce a supercapacitor fiber at an early stage; they served as commonly developed candidates. The gadgets that resulted could withstand bending deformations while maintaining a constant specific capacity. In both coaxial and paralleled arrangements, graphene fibers are suitable fiber electrodes for energy storage [28]. CNT/graphene composite fibers were manufactured to couple these properties: the high conductivity of CNT fibers and the high electrocatalytic activity of graphene fibers. To overcome the decreased interfacial area of the above material, hollow-reduced graphene oxide (RGO)/conducting polymer composite fibers with high interfacial areas were developed. As a result, the charge-storage capacity of the supercapacitor is raised to 27.1 μWh/cm^2. Because being stretched is a more common occurrence in most wearable electronics, attaining stretchability becomes the next priority to achieve after fabricating flexible fiber electrodes [29]. Designing inherently stretchy fiber electrodes and electrolytes has recently improved a new approach for stretchable fiber electronics.

Up to this point, the major focus for carbon-based fibers in batteries and energy devices has been on Li batteries. CF-based yarn and fabric electrodes provide many benefits over traditional electrodes and current collectors, including (1) their mesoporous framework allows for good electrolyte infiltration and ion diffusion kinetics; (2) their continuous conductive network allows for quick charge (electrons or holes) transport to active materials and metal ions; and (3) they can be used as freestanding, supple, frivolous, and smooth structural electrodes deprived of the use of a polymer binder or conductive agent. Later, CNT fibers, RGO fibers, and CNT/RGO fibers were used as LIB anodes [30]. Due to exceptional flexibility, high electrical conductivity, tensile strength, and customizable assembly, graphene fibers have shown several recompenses in the fiber-shaped LIB. Furthermore, GFs are made using a wet-spinning technique, allowing for the continuous production of fiber electrodes. As a result, GF-based LIBs have attracted a growing amount of research attention [31]. The low specific capacity, on the other hand, continues to be a key stumbling point. Active functional components like silicon, MoS_2, $Li_4Ti_5O_{12}$, $LiCoO_2$, and titania have been added to improve the electrochemical performance of GFs to overcome this problem. Because of the higher active materials, greater exposed area, and layered structure, the produced GF-based fiber-shape LIB exhibited improved rate capabilities and cycle performance when compared to most fiber electrodes. CNTs have been investigated for use in organic solar cells as a way to decrease unwanted carrier recombination and improve photooxidation resistance. Due to the finding of effective multiple-exciton production at p-n junctions produced inside individual CNTs, CNTs have also been investigated for integration into CNT-Si heterojunctions for possible uses in solar cells. The macroscopic graphene fiber had a high electrical conductivity (127.3 S/cm) and showed energy conversion efficiencies of 3.25% when combined with TiO_2 nanotubes/Ti wire for the dye-sensitized solar cell (DSSC) electrodes [26].

5.2.4 BIOBASED FIBERS FOR ENERGY DEVICES

Biobased fibers can be derived from a variety of sources, such as cellulose, chitin, keratin, biodegradable polymers like poly-(L-lactic acid), poly-(ε-caprolactone), etc. The nature and the physical characteristics of the fibers vary with the source and the processing techniques. Natural fiber or biofiber-based composites have gained much attention these days due to their unique physical properties, cost-effectiveness, renewability, and environmental friendliness, etc. Synthetic polymers are always demanding a substitute because of their polluting nature and toxicity. Biodegradable polymers are an alternative to synthetic ones. They are similar in properties and processing, yet they are decomposable by the microorganism. Among the flexible energy devices, wearables, energy-harvesting materials, and the components in biomedical fields are very demanding. Renewable and waste materials with fibrous morphology can also be cost-effective materials for these applications. Many multicomponent composites are fabricated from natural fibers and have superior electrical and mechanical properties.

5.2.4.1 Cellulose-based fibers

Cellulose is one of the most explored natural fibers for various applications. Cellulose is the most abundant natural polymer. It is chemically composed of glucose monomers, each connected to the other utilizing 1,4-glycosidic linkage. The different forms of materials from cellulose are being employed in composite structures. Nano and microfibers of cellulose always attract interest due to the ease of availability and processing. The nanoscale structure of cellulose offers supple surface behavior, low thermal expansion coefficients, optical transparency, and a high degree of elasticity. Electronic devices such as circuit boards, electrical energy storage devices, batteries, fuel cells, etc., utilize cellulose nanocrystals and nanofibers. When cellulosic materials are incorporated with conductive fillers, such as carbon nanotubes, they excel in their properties. Lokendra Pal and coworkers fabricated a pressure sensor by coating cellulose nanofibers on multiwalled carbon nanotubes, and the material was suggested in potential applications for smart wearables, diagnosis, electronic skins, and biomedical applications [32]. Cellulose in its different forms, especially fibers, has been made applicable in many energy applications. The cellulose fiber present in the paper had a mesoporous structure and was modified by carbon nanotubes to enable them to conduct. These composites were employed in lithium-based battery components [31]. Cellulose could also be used for three-dimensional photoelectrode devices [33]. Compared to the conventional metallic and plastics materials, the cellulose papers were found to be suitable for enhancing the adhesion [34]. Figure 5.6 shows the fabrication steps of a flexible paper-based energy device. The electronic dual-layer frameworks were stripped off and attached mutually at the sides of the cellulose paper, which acted as a mechanical reinforcement as well as a separator membrane. The lamination procedure, as opposed to directly coating the double-layer structure onto the cellulose paper, received great mechanical stability of the electrode materials and avoided leakage via the paper. Output controllable energy devices can be fabricated from three-dimensional cellulosic fiber and gold nanoparticles (interwined on cellulose).

Supercapacitors are an important area in electronics and energy storage. Cellulose-based papers with various other fillers, such as metal oxides and nanotubes, etc., have been employed widely for supercapacitors. A copper oxide-copper-cellulose combination with hierarchical morphology was utilized in the fabrication of supercapacitors. The findings show the promise of hybrid paper as a great enactment electrode medium for lightweight energy storage applications and portable electronics [36]. A schematic representation and the corresponding forest-like morphology of the Cu_2O/Cu based cellulose fiber composites are given in Figure 5.7. The material produced a special type of morphology for each composition. They presented a three-dimensional porous structure to support the forest-like Cu_2O/Cu array design, serve as an electrolyte storage unit to assist ion transport, and serve as an electrical-

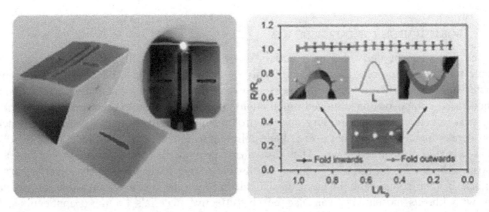

FIGURE 5.6 Flexible paper-based electrode materials (Adapted with permission from ref. [35], Copyright (2018) RSC).

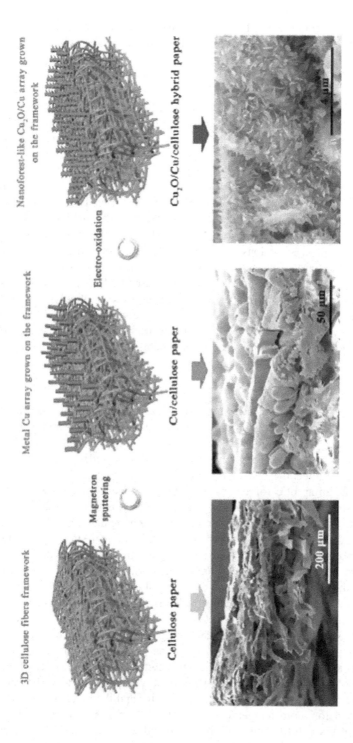

FIGURE 5.7 Forest-like hierarchical morphology for cellulose paper Cu_2O/Cu materials (Reproduced with permission from ref. [37], Copyright (2018) RSC).

insulating separator to build supercapacitor devices. Cu_2O/Cu arrays were grown on the two sides of cellulose paper in this research, for example, to create a symmetric supercapacitor. Positive and negative electrodes were formed by the Cu_2O/Cu combination on both sides.

Pushparaj et al. created cellulose/CNTs composite paper supercapacitors instead of using premade cellulose papers by embedding cellulose solutions into vertically aligned multiwalled carbon nanotube (MWNT) arrays grown on Si substrates. Unaltered plant cellulose was infiltrated into the MWNT with an ionic liquid (1-butyl, 3-methylimidazolium chloride), and the ionic liquid was extracted by solidifying the film on dry ice and indulging it in ethanol for recovery [38]. Cellulose fibers also act as binders in supercapacitors. Yang et al. reported the fabrication of cellulose nanofiber papers, which were flexible and transparent. As the transparent conductive oxide (TCO) layer, they used ALD to develop an aluminum-doped zinc oxide (AZO) thin film on CNF paper. The AZO film had outstanding crystallinity, conformality, and very low surface roughness while being deposited at just 150°C to maintain the structural and chemical integrity of the CNF paper [39].

Nanocellulose can be used as a substrate for supercapacitors. Hu et al. demonstrated the usage of cellulose fibers for the aforementioned purpose. They have also tested various combinations, such as CNTs and MnO_2 along cellulose fibers to find the fittest material for the application [37]. The water-swelling effect of the cellulose fibers can absorb the electrolyte, and the mesoporous internal structure of the fibers can provide pathways for ions to diffuse to the electrochemical energy-storage materials, highlighting the merits of mesoporous cellulose fibers as substrates for supercapacitor electrodes. Wet spinning and in situ polymerization polypyrrole have been used by Nie et al. to create scalable, binder-free, high-performance, fiber-based supercapacitors based on hierarchical polypyrrole-TEMPO-oxidized bacterial cellulose/reduced graphene oxide macrofibers [40]. Similar work explained the fabrication of polypyrrole/cobalt oxyhydroxide/cellulose fiber hybrid electrode in an open device at room temperature using a "liquid step reduction" technique. The electrochemical properties of the composite electrode were fantastic, with a high basic capacitance and capacitance retention [41]. Bacterial cellulose has been exploited by many researchers. The conducting polymers like polypyrrole, PANI, polythiophene, etc., are mainly used as the matrices.

Lithium-ion batteries are the most common secondary cells. Lithium-ion batteries consist of lithium-based electrodes and electrolytes and battery separators. The separators can be fabricated using polymers. Among the biopolymers, cellulose is mostly attracted toward application in lithium-based battery technologies. They have been used as electrodes, electrolytes, separators, and supports. Kuribayashi studied the use of cellulose fibers immersed in a microporous cellulosic matrix saturated inside an aprotic solvent as thin composite cellulosic separators of nearly 39–85 µm thickness for Li-ion batteries [42]. Battery separators for small-scale batteries were developed. This can be done with paper sheets that contain cellulose. Zhang and coworkers designed rice paper separators of about 100 µm having flexibility and electrochemical stability [43,44]. Only limited works have been reported in cellulose-based electrolytes. In one report, cellulose fibers from inexpensive plant resources, such as wheat hays and corn stalks, used the technique of graft-copolymerization with acrylic acid and with the aid of polyvinyl alcohol into thin films. These were swelled by a standard electrolyte [45]. Figure 5.8 given below represents some of the flexible devices fabricated using cellulose fibers.

Leijonmarck suggested a system for fabricating lightweight and solid-state batteries that were assembled into a single flexible paper frame and used nano-fibrillated cellulose (NFC) for both electrode binding and separator fabrication [47]. Carboxymethylcellulose, micro-fibrillated cellulose, and cellulose-dissolved ionic liquid materials were employed in the battery. The coating of a slurry onto a metal foil, which serves as both the current collector and the mechanical substrate in standard electrodes, was needed for room temperature ionic liquid processing of unmodified cellulose and aqueous processing of CMC [46]. Thus, cellulose has indeed been consistently used as a binder in electrodes, a reinforcing ingredient in polymer and gel-polymer electrolytes, and a base material for microporous separators.

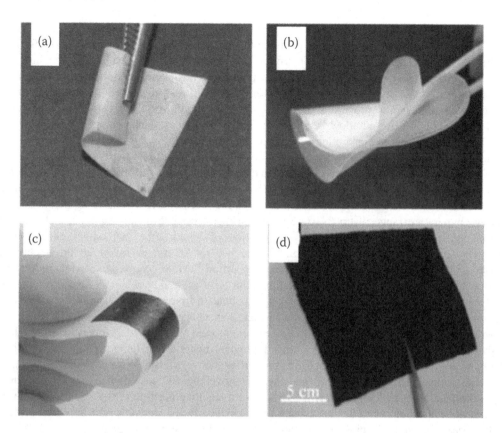

FIGURE 5.8 Different flexible materials fabricated using cellulose. (a) Cellulose reinforced gel-polymer electrolyte membrane (b) Flexible microfibrillar cellulose composite polymer membrane (c) Polymer-coated supercapacitor on paper (d) textile dipped in single-walled carbon nanotube ink (Adapted with permission from ref. [46], Copyright (2003) Elsevier).

5.2.4.2 Keratin and chitin fiber composites

Electromagnetic interference-shielding materials can be fabricated from fiber composites. One such composite was from graphene and lotus leaf fiber, which showed exceptionally high electrical properties and EMI shielding efficiency. They were suggested for energy storage and heat generation [48]. Single material with versatile applications is a widely accepted concept in the present research era. If such materials are fabricated, they can reduce the cost to a great degree. Chicken feather waste-based fibers were incorporated in composite materials, and the resultant materials exhibited piezoelectric behavior. Mukharjee and coworkers fabricated flexible energy devices that could be biocompatible and flexible. The fiber composites were also considered excellent candidates in flexible illuminating LEDs, wearable electronics, and healthcare monitoring devices. The results showed that they were of very low weight (nearly 1 g), provided a piezoelectric output voltage of 10 V, and power density of 6 μW/cm^2 with a current density of 1.8 mA/cm^2. The superior properties were owed to the hydrogen bonding present in the keratin of the chicken feathers. The mechanical strength was attributed to the sulfide linkages [49]. A similar composite was reported, by the inclusion of chicken feather waste, to be used in oriented circuit boards. The inclusion was in the epoxy resin, and the composite materials exhibited all the required properties for a printed circuit board [50]. Piezoelectric nanogenerators can be fabricated even from eggshells, which is surprising. The bio nanogenerators exhibited an output voltage of nearly 27 V, which is fairly high when compared with the above-mentoned report. The efficiency of the membrane was tested with LEDs and circuits; an assembly of five nanogenerators from the eggshell membrane composites could light up to 90 LEDs with an output voltage of 130 V. The membranes were even sensitive toward the minor responses from the human body and serve as a promising material for flexible wearables [51]. Flexible high-strength electrodes and supports can be developed from human hair keratin. Human hair, having high elasticity, along with nickel salt and graphene sheets imparted excellent electrical properties that are suited for portable wearable [52]. Low-k-dielectrics are important materials since they impart dielectric properties. Dielectric insulators with very low k values in the range of 1.7–2.7 can be obtained from hollow keratin fibers and triglycerides, which compete with or overcome the silicon and epoxy-based conventional materials [53].

Wearable electronic components have been successfully manufactured; these include heaters, which are flexible and translucent and have excellent mechanical durability and foldability with minimal improvements in electrical conductivity. The research was done by Son and coworkers, who conclusively proved the success of a flexible heater focused on transparent conductive keratin nanofibers with a broad operating temperature range of up to 65.75°C, a fast recovery time of around 10 s, and low energy consumption. In the work, they used a combination of keratin fiber and silver nanowires [54]. Transparent and flexible photodetectors can be designed on keratin nanofiber textiles incorporating zincoxide-graphene quantum dots [55]. The next-generation flexible energy devices mainly depend on the above-mentioned natural nanofibers and, along with cellulose and keratin, silk fibroin is also used in several studies. Silk fibroin is derived from silk fibers obtained from silkworms and spiders. These fibers are rich in glycine. Silk fibroin is made up of α-helices, β-sheet crystals, and irregular coils, which are put together using hydrogen bonding, hydrophobic interactions, and Van der Waals forces by repeated amino acid sequences. Silk fibroin outperforms other biological materials in terms of biocompatibility and biodegradability, as well as a host of appealing properties such as adjustable water solubility, exceptional optical transmittance, high mechanical robustness, lightweight, and ease of handling, which are only partly present, in other biological materials. Silk fibroin has long been used as a key component in the development of biocompatible flexible electronics, specifically for wearable and implantable devices. They have been known to develop sensors (temperature and pressure), nanogenerators, etc. Chitin is the second-most abundant biopolymer and is utilized for composite materials. Chitin and chitosan-based components are incorporated in electronic devices. Chitin fibers can be used in battery separators and other energy devices [56]. Chitin nanofiber papers are employed in the flexible organic light-emitting diodes (OLEDs) [57].

5.3 ELECTROSPUN FIBERS FOR FLEXIBLE ENERGY DEVICES

Electrospinning is an advanced technique to produce fibrous structures, especially in the nanometer dimensions. In this technique, a high voltage is applied to polymeric solutions of particular viscosity to draw threads. These are collected on a foil and then allowed to condense. The resultant fibers will be of several nanometers. The fibers obtained from this technique have a high surface area, high aspect ratio, flexible surface functionalities, and enhanced mechanical strength. In the direct electrospinning process, the fibers of conductive polymers, such as PEDOT-PSS, P3HT, PVDf-Tr FE, etc., are obtained from solutions of the polymers. The spun fibers are employed in batteries, sensors, textiles, hydrogen storage, supercapacitors, etc. The different fields of application of electrospun fibers are given in Figure 5.9.

Coaxial spinning can be effectively utilized in the production of fiber-based supercapacitors with a thickness of fewer than 100 μm. For the production, an inner poly(3,4-ethylenedioxythiophene) polystyrene sulfonate (PEDOT:PSS) core was coated with an ionically conducting chitosan sheath before being wrapped with a carbon nanotube fiber nanostructure. The superconductors were developed to meet all the requirements of supercapacitors, having flexibility and being lightweight [59]. Nitrogen-enriched carbon nanofiber aerogels with an ultrafine nanofiber network structure were developed using chitin nanofiber aerogels as the precursor. Because of the homogeneous nanofibrous shape and nitrogen-rich structure of chitin nanofiber aerogels, the NCNAs have a large specific surface area. Marine chitin-derived aerogels have a lot of possibilities for energy storage and ecological remediation [60]. Many

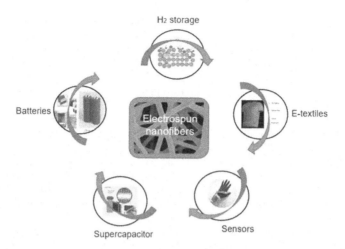

FIGURE 5.9 Applications of electrospun fibers in various energy fields (Adapted with permission from ref. [58], Copyright (A. Mirabedini, Z. Lu, S. Mostafavian, J. Foroughi), some rights reserved; exclusive licensee [MDPI]. Distributed under a Creative Commons Attribution License 4.0 (CC BY)).

polymers are converted to fibers via the electrospinning process to use in wearables. Poly(vinylidene fluoride) (PVDF) fibers were prepared by Huang and coworkers, and the fiber mat was modified by incorporating barium titanate, which is well-known for its piezoelectric properties and graphene sheet. The resultant materials could light up 15 LEDs with a peak voltage of 112 V. This material is an example of the synergistic mechanism of materials with different properties, and these were proposed for energy harvesting devices [58]. PVDF nanofibers membranes with a sandwiched structure exhibited interesting morphology and could light up LEDs [61].

5.4 CONCLUSIONS

Both synthetic and natural fibers are used for a variety of applications. The chapter discusses different types of fibers used for the flexible energy devices. The electronic industry and smart wearables are attracting attention these days. Novel materials and applications are always demanding. Carbon-based fibers are the most commonly employed material for sensors, batteries, LEDs, etc. Cellulose, keratin, and electrospun fibers are used in sensing devices, wearables, and supercapacitors. The fiber composites of different conducting polymers with virgin or surface-modified fibers act as good candidates in flexible devices and other advanced applications.

REFERENCES

1. P. Hassanzadeh, M. Kharaziha, M. Nikkhah, S.R. Shin, J. Jin, S. He, W. Sun, C. Zhong, M.R. Dokmeci, A. Khademhosseini, M. Rolandi, Chitin nanofiber micropatterned flexible substrates for tissue engineering, *J. Mater. Chem. B.* 1 (2013) 4217–4224. 10.1039/c3tb20782j.
2. Kenry.C.T., Lim, Nanofiber technology: Current status and emerging developments, *Prog. Polym. Sci.* 70 (2017) 1–17. 10.1016/j.progpolymsci.2017.03.002.
3. M. Endo, Y.A. Kim, T. Hayashi, K. Nishimura, T. Matusita, K. Miyashita, M.S. Dresselhaus, Vapor-grown carbon fibers (VGCFs): Basic properties and their battery applications, *Carbon N. Y.* 39 (2001) 1287–1297. 10.1016/S0008-6223(00)00295-5.
4. S. Agarwal, A. Greimer, J.H. Wendorff, Electrospinning of manmade and biopolymer nanofibers - Progress in techniques, materials, and applications, *Adv. Funct. Mater.* 19 (2009) 2863–2879. 10.1002/adfm.200900591.

5. L. Feng, N. Xie, J. Zhong, Carbon nanofibers and their composites: A review of synthesizing, properties and applications, *Materials (Basel).* 7 (2014) 3919–3945. 10.3390/ma7053919.

6. T. Iwasaki, Y. Makino, M. Fukukawa, H. Nakamura, S. Watano, Low-temperature growth of nitrogen-doped carbon nanofibers by acetonitrile catalytic CVD using Ni-based catalysts, *Appl. Nanosci.* 6 (2016) 1211–1218. 10.1007/s13204-016-0535-x.

7. J.D. Schiffman, C.L. Schauer, A review: Electrospinning of biopolymer nanofibers and their applications, *Polym. Rev.* 48 (2008) 317–352. 10.1080/15583720802022182.

8. J. Moohan, S.A. Stewart, E. Espinosa, A. Rosal, A. Rodríguez, E. Larrañeta, R.F. Donnelly, J. Domínguez-Robles, Cellulose nanofibers and other biopolymers for biomedical applications. A review, *Appl. Sci.* 10 (2020). 10.3390/app10010065.

9. B. Zhang, W. Wang, M. Tian, N. Ning, L. Zhang, Preparation of aramid nanofiber and its application in polymer reinforcement: A review, *Eur. Polym. J.* 139 (2020). 10.1016/j.eurpolymj.2020.109996.

10. Y. Dai, W. Liu, E. Formo, Y. Sun, Y. Xia, Ceramic nanofibers fabricated by electrospinning and their applications in catalysis, environmental science, and energy technology, *Polym. Adv. Technol.* 22 (2011) 326–338. 10.1002/pat.1839.

11. L. Persano, A. Camposeo, D. Pisignano, Active polymer nanofibers for photonics, electronics, energy generation and micromechanics, *Prog. Polym. Sci.* 43 (2015) 48–95. 10.1016/j.progpolymsci.2014.10.001.

12. X. Shi, W. Zhou, D. Ma, Q. Ma, D. Bridges, Y. Ma, A. Hu, *Electrospinning of nanofibers and their applications for energy devices*, 2015 (2015).

13. T. Huang, C. Wang, H. Yu, H. Wang, Q. Zhang, M. Zhu, Human walking-driven wearable all-fiber triboelectric nanogenerator containing electrospun polyvinylidene fluoride piezoelectric nanofibers, *Nano Energy.* 14 (2014) 226–235. 10.1016/j.nanoen.2015.01.038.

14. J. Chang, M. Dommer, C. Chang, L. Lin, Piezoelectric nanofibers for energy scavenging applications, *Nano Energy.* 1 (2012) 356–371. 10.1016/j.nanoen.2012.02.003.

15. H.B. Kang, J. Chang, K. Koh, L. Lin, Y.S. Cho, High quality Mn-doped (Na,K)NbO3 nanofibers for flexible piezoelectric nanogenerators, *ACS Appl. Mater. Interfaces.* 6 (2014) 10576–10582. 10.1021/am502234q.

16. X. Chen, S. Xu, N. Yao, Y. Shi, 1.6 v nanogenerator for mechanical energy harvesting using PZT nanofibers, *Nano Lett.* 10 (2010) 2133–2137. 10.1021/nl100812k.

17. B. Mahltig, *Introduction to inorganic fibers*, Elsevier Ltd., (2018). 10.1016/B978-0-08-102228-3.00001-3.

18. T. Ishikawa, Advances in inorganic fibers (2005) 109–110. 10.1007/b104208.

19. H. Wu, L. Hu, M.W. Rowell, D. Kong, J.J. Cha, J.R. McDonough, J. Zhu, Y. Yang, M.D. McGehee, Y. Cui, Electrospun metal nanofiber webs as high-performance transparent electrode, *Nano Lett.* 10 (2010) 4242–4248. 10.1021/nl102725k.

20. A.M. Teli, S.A. Beknalkar, D.S. Patil, S.A. Pawar, D.P. Dubal, V.Y. Burute, T.D. Dongale, J. Cheol, P.S. Patil, Applied Surface Science E ff ect of annealing temperature on charge storage kinetics of an electrospun deposited manganese oxide supercapacitor, *Appl. Surf. Sci.* 511 (2020) 145466. 10.1016/j.apsusc.202 0.145466.

21. J.S. Lee, M.S. Jo, R. Saroha, D.S. Jung, Y.H. Seon, J.S. Lee, Y.C. Kang, D. Kang, J.S. Cho, *Hierarchically well-developed porous graphene nanofibers comprising N-doped graphitic C-coated cobalt oxide hollow nanospheres as anodes for high-rate Li-Ion batteries*, 2002213 (2020) 1–14. 10.1002/smll.202002213.

22. E. Environ, S. Cavaliere, S. Subianto, I. Savych, D.J. Jones, J. Rozi, *Environmental science electrospinning: Designed architectures for energy conversion and storage devices* (2011) 4761–4785. 10.1039/c1ee02201f.

23. M. Wang, X. Chen, H. Wang, H. Wu, C. Huang, Improved performances of lithium-ion batteries with a separator based on inorganic fi bers †, *J. Mater. Chem. A Mater. Energy Sustain.* 00 (2016) 1–8. 10.1039/C6TA08404D.

24. Z.D. Caia Xin, Zhanga Chaoqun, Zhanga Shengsen, Fanga Yueping, Materials chemistry A, *J. Mater. Chem. A Mater. Energy Sustain.* (2016). 10.1039/C6TA07868K.

25. L. Dong, C. Xu, Y. Li, C. Wu, B. Jiang, Q. Yang, *Simultaneous production of high-performance flexible textile electrodes and fiber electrodes for wearable energy storage*, (2016) 1675–1681. 10.1002/adma.2015 04747.

26. D.D. Tune, B.S. Flavel, R. Krupke, J.G. Shapter, Carbon nanotube-silicon solar cells, *Adv. Energy Mater.* 2 (2012) 1043–1055. 10.1002/aenm.201200249.

27. N. Shirshova, H. Qian, M.S.P. Shaffer, J.H.G. Steinke, E.S. Greenhalgh, P.T. Curtis, A. Kucernak, A. Bismarck, Structural composite supercapacitors, *Compos. Part A Appl. Sci. Manuf.* 46 (2013) 96–107. 10.1 016/j.compositesa.2012.10.007.

28. S. Chen, L. Qiu, H. Cheng, Carbon-based fibers for advanced electrochemical energy storage devices (2020). 10.1021/acs.chemrev.9b00466.

29. S. Pan, H. Lin, J. Deng, P. Chen, X. Chen, Z. Yang, *Novel wearable energy devices based on aligned carbon nanotube fiber textiles*, (2014) 1–7. 10.1002/aenm.201401438.

30. Q. Xia, H. Yang, M. Wang, M. Yang, Q. Guo, L. Wan, H. Xia, *High energy and high power lithium-ion capacitors based on boron and nitrogen dual-doped 3D carbon nanofibers as both cathode and anode*, 1701336 (2017) 1–9. 10.1002/aenm.201701336.

31. X. Chen, H. Zhu, C. Liu, Y.C. Chen, N. Weadock, G. Rubloff, L. Hu, Role of mesoporosity in cellulose fibers for paper-based fast electrochemical energy storage, *J. Mater. Chem. A*. 1 (2013) 8201–8208. 10.1039/c3ta10972k.

32. H. Zhang, X. Sun, M. Hubbe, L. Pal, Flexible and pressure-responsive sensors from cellulose fibers coated with multiwalled carbon nanotubes, *ACS Appl. Electron. Mater.* 1 (2019) 1179–1188. 10.1021/acsaelm.9b00182.

33. Z. Li, C. Yao, Y. Yu, Z. Cai, X. Wang, Photoelectrodes: Highly efficient capillary photoelectrochemical water splitting using cellulose nanofiber-templated TiO_2 photoanodes, *Adv. Mater.* 26 (2014) 2110. 10.1002/adma.201470089.

34. L. Hu, J.W. Choi, Y. Yang, S. Jeong, F. La Mantia, L.F. Cui, Y. Cui, Highly conductive paper for energy-storage devices, *Proc. Natl. Acad. Sci. U. S. A.* 106 (2009) 21490–21494. 10.1073/pnas.0908858106.

35. Y. Zhang, H. Yang, K. Cui, L. Zhang, J. Xu, H. Liu, J. Yu, Highly conductive and bendable gold networks attached on intertwined cellulose fibers for output controllable power paper, *J. Mater. Chem. A*. 6 (2018) 19611–19620. 10.1039/c8ta08293f.

36. C. Wan, Y. Jiao, J. Li, A cellulose fibers-supported hierarchical forest-like cuprous oxide/copper array architecture as a flexible and free-standing electrode for symmetric supercapacitors, *J. Mater. Chem. A*. 5 (2017) 17267–17278. 10.1039/c7ta04994c.

37. Z. Gui, H. Zhu, E. Gillette, X. Han, G.W. Rubloff, L. Hu, S.B. Lee, G.U.I.E.T. Al, <Nn401818T.Pdf>, (2013) 6037–6046.

38. V.L. Pushparaj, M.M. Shaijumon, A. Kumar, S. Murugesan, L. Ci, R. Vajtai, R.J. Linhardt, O. Nalamasu, P.M. Ajayan, Flexible energy storage devices based on nanocomposite paper, *Proc. Natl. Acad. Sci. U. S. A.* 104 (2007) 13574–13577. 10.1073/pnas.0706508104.

39. X. Wang, C. Yao, F. Wang, Z. Li, Cellulose-based nanomaterials for energy applications, *Small*. 13 (2017). 10.1002/smll.201702240.

40. N. Sheng, S. Chen, J. Yao, F. Guan, M. Zhang, B. Wang, Z. Wu, P. Ji, H. Wang, Polypyrrole@TEMPO-oxidized bacterial cellulose/reduced graphene oxide macrofibers for flexible all-solid-state supercapacitors, *Chem. Eng. J.* 368 (2019) 1022–1032. 10.1016/j.cej.2019.02.173.

41. S. Yang, L. Sun, X. An, X. Qian, Construction of flexible electrodes based on ternary polypyrrole@cobalt oxyhydroxide/cellulose fiber composite for supercapacitor, *Carbohydr. Polym.* 229 (2020). 10.1016/j.carbpol.2019.115455.

42. I. Kuribayashi, Characterization of composite cellulosic separators for rechargeable lithium-ion batteries, *J. Power Sources* 63 (1996) 87–91. 10.1016/S0378-7753(96)02450-0.

43. L.C. Zhang, X. Sun, Z. Hu, C.C. Yuan, C.H. Chen, Rice paper as a separator membrane in lithium-ion batteries, *J. Power Sources* 204 (2012) 149–154. 10.1016/j.jpowsour.2011.12.028.

44. L.C. Zhang, Z. Hu, L. Wang, F. Teng, Y. Yu, C.H. Chen, Rice paper-derived 3D-porous carbon films for lithium-ion batteries, *Electrochim. Acta*. 89 (2013) 310–316. 10.1016/j.electacta.2012.11.042.

45. Z. Yue, I.J. McEwen, J.M.G. Cowie, Novel gel polymer electrolytes based on a cellulose ester with PEO side chains, *Solid State Ionics* 156 (2003) 155–162. 10.1016/S0167-2738(02)00595-7.

46. S.S. Jeong, N. Böckenfeld, A. Balducci, M. Winter, S. Passerini, Natural cellulose as binder for lithium battery electrodes, *J. Power Sources*. 199 (2012) 331–335. 10.1016/j.jpowsour.2011.09.102.

47. S. Leijonmarck, A. Cornell, G. Lindbergh, L. Wågberg, Single-paper flexible Li-ion battery cells through a paper-making process based on nano-fibrillated cellulose, *J. Mater. Chem. A*. 1 (2013) 4671–4677. 10.1039/c3ta01532g.

48. C. Cheng, R. Guo, L. Tan, J. Lan, S. Jiang, Z. Du, L. Zhao, A bio-based multi-functional composite film based on graphene and lotus fiber, *Cellulose* 26 (2019) 1811–1823. 10.1007/s10570-018-2160-1.

49. E. Kar, M. Barman, S. Das, A. Das, P. Datta, S. Mukherjee, M. Tavakoli, N. Mukherjee, N. Bose, Chicken feather fiber-based bio-piezoelectric energy harvester: An efficient green energy source for flexible electronics, *Sustain. Energy Fuels* 5 (2021) 1857–1866. 10.1039/d0se01433h.

50. M. Zhan, R.P. Wool, Design and evaluation of bio-based composites for printed circuit board application, *Compos. Part A Appl. Sci. Manuf.* 47 (2013) 22–30. 10.1016/j.compositesa.2012.11.014.

51. S.K. Karan, S. Maiti, S. Paria, A. Maitra, S.K. Si, J.K. Kim, B.B. Khatua, A new insight towards eggshell membrane as high energy conversion efficient bio-piezoelectric energy harvester, Mater. *Today Energy* 9 (2018) 114–125. 10.1016/j.mtener.2018.05.006.

52. J. Zhao, J. Gong, C. Zhou, C. Miao, R. Hu, K. Zhu, K. Cheng, K. Ye, J. Yan, D. Cao, X. Zhang, G. Wang, Utilizing human hair for solid-state flexible fiber-based asymmetric supercapacitors, *Appl. Surf. Sci.* 508 (2020). 10.1016/j.apsusc.2020.145260.

53. C.K. Hong, R.P. Wool, Low dielectric constant material from hollow fibers and plant oil, *J. Nat. Fibers.* 1 (2004) 83–92. 10.1300/J395v01n02_06.

54. C.H. Lee, Y.J. Yun, H. Cho, K.S. Lee, M. Park, H.Y. Kim, D.I. Son, Environment-friendly, durable, electro-conductive, and highly transparent heaters based on silver nanowire functionalized keratin nanofiber textiles, *J. Mater. Chem. C.* 6 (2018) 7847–7854. 10.1039/c8tc02412j.

55. K.S. Lee, Y.J. Park, J. Shim, H.S. Chung, S.Y. Yim, J.Y. Hwang, H. Cho, B. Lim, D.I. Son, ZnO@graphene QDs with tuned surface functionalities formed on eco-friendly keratin nanofiber textile for transparent and flexible ultraviolet photodetectors, *Org. Electron.* 77 (2020). 10.1016/j.orgel.2019.105489.

56. J.K. Kim, D.H. Kim, S.H. Joo, B. Choi, A. Cha, K.M. Kim, T.H. Kwon, S.K. Kwak, S.J. Kang, J. Jin, Hierarchical chitin fibers with aligned nanofibrillar architectures: A nonwoven-mat separator for lithium metal batteries, *ACS Nano.* 11 (2017) 6114–6121. 10.1021/acsnano.7b02085.

57. J. Jin, D. Lee, H.G. Im, Y.C. Han, E.G. Jeong, M. Rolandi, K.C. Choi, B.S. Bae, Chitin Nanofiber transparent paper for flexible green electronics, *Adv. Mater.* 28 (2016) 5169–5175. 10.1002/adma.201600336.

58. K. Shi, B. Sun, X. Huang, P. Jiang, Synergistic effect of graphene nanosheet and BaTiO3 nanoparticles on performance enhancement of electrospun PVDF nanofiber mat for flexible piezoelectric nanogenerators, *Nano Energy.* 52 (2018) 153–162. 10.1016/j.nanoen.2018.07.053.

59. A. Mirabedini, Z. Lu, S. Mostafavian, J. Foroughi, Triaxial carbon nanotube/conducting polymer wet-spun fibers supercapacitors for wearable electronics, *Nanomaterials* 11 (2021) 1–16. 10.3390/nano11010003.

60. B. Ding, S. Huang, K. Pang, Y. Duan, J. Zhang, Nitrogen-enriched carbon nanofiber aerogels derived from marine chitin for energy storage and environmental remediation, *ACS Sustain. Chem. Eng.* 6 (2018) 177–185. 10.1021/acssuschemeng.7b02164.

61. B. Li, F. Zhang, S. Guan, J. Zheng, C. Xu, Wearable piezoelectric device assembled by one-step continuous electrospinning, *J. Mater. Chem. C.* 4 (2016) 6988–6995. 10.1039/c6tc01696k.

6 3D Printed Flexible Energy Devices

Dongxu He

College of Materials and Chemistry & Chemical Engineering,
Chengdu University of Technology, Chengdu, People's Republic of China

Zheling Li

National Graphene Institute/Department of Materials,
University of Manchester, Manchester, United Kingdom

CONTENTS

6.1 3D printing technologies 99
    6.1.1 Direct ink writing 100
    6.1.2 Fuse deposition modelling 100
    6.1.3 Material jetting 100
    6.1.4 Binder jetting 100
    6.1.5 Directed energy deposition (DED) 102
6.2 Configuration of flexible energy device 102
    6.2.1 Active materials 102
    6.2.2 EES electrodes 102
    6.2.3 Electrolyte and the solid-state devices 104
    6.2.4 Configuration of EES devices 104
6.3 3D printed EES devices 105
    6.3.1 3D printed electrodes 105
        6.3.1.1 Carbon-based electrodes 106
        6.3.1.2 Polymer-based electrodes 108
        6.3.1.3 Others 109
    6.3.2 3D printed electrolytes 111
    6.3.3 3D printed device 112
6.4 Summary and outlook 113
    6.4.1 Precision and resolution of 3D printing 113
    6.4.2 New materials 113
    6.4.3 Integration with multi-materials printing technology and the interface 113
    6.4.4 4D printing 113
References 114

6.1 3D PRINTING TECHNOLOGIES

3D-printing technology generally refers to the construction of 3D objects through a programmed process. It is also named as additive manufacturing (AM), as opposed to the conventional manufacturing process where excess materials are removed through milling, machining, etc. The rapid prototyping technology, which builds 3D models by adding layer upon layer and dates back to the 1980s, is one of the earliest 3D-printing technologies. Almost all the 3D-printing technologies are carried out through a layer-based approach, but they are different regarding how the layers are created and bonded and what materials are used. According to the American Society of Testing and Materials (ASTM) [1], 3D-printing technologies include binder jetting, material jetting (inkjet printing), material extrusion, powder bed fusion, vat photopolymerization (e.g., stereolithography), energy deposition, etc., according to the different physical states of raw materials or the methods used [2]. Each category has several representative technologies. For example, the powder bed fusion includes electron beam melting, selective laser sintering, and selective laser melting [3]. Some of them have already been employed for the fabrication of energy devices, which are briefly introduced below.

6.1.1 DIRECT INK WRITING

Direct ink writing (DIW) and fused deposition modelling (FDM) are two typical techniques belonging to material extrusion [2,4]. The major difference is that in FDM polymer filaments are melted during extrusion, while DIW utilizes inks stored in a syringe. These inks are mixtures of nano- or micro-sized materials and solvents with controlled rheology. Because the DIW enables the creation of arbitrary 3D shapes using a wide range of materials, it has been one of the most versatile AM techniques.

6.1.2 FUSE DEPOSITION MODELLING

As shown in Figure 6.1a, in the FDM process, the melt filaments are squeezed from a micro nozzle and then deposited layer by layer, with the z-axis resolution being lower than other AM techniques (0.25 mm). According to Stratasys Inc. USA white paper [5], more than 10 types of thermoplastics are commercially available to satisfy FDM-supported applications, such as polycarbonate (PC), polyphenylsulfone (PPSF), acrylonitrile butadiene styrene (ABS), and PC-ABS blends. Research is still undertaken for the development of new FDM material having better processability through the construction and higher physical or chemical performance in applications.

6.1.3 MATERIAL JETTING

Material jetting or inkjet printing is an established AM technique that deposits droplets of waxy polymers or liquid photopolymers [6]. It has a high spatial resolution with the drop-on-demand jetting, and then a reaction or evaporation is triggered by an ultraviolet or heating source for solidification (Figure 6.1b). The candidate binder materials for material jetting are typically viscous and capable of forming drops, such as photopolymers or wax-like materials. In addition, in recent years, some kinds of ceramics and metals have also been verified as accessible to the material jetting process [7–10]. In the jetting process, materials are heated to transfer from the solid state to the liquid state, and the reduced viscosity allows the outflow of the jet [6]. Investigations are currently being undertaken to study the effect of parameters, such as print head movement speed, dropping frequency, and droplet velocity. This technology enables the selective deposition of multiple parts [11,12].

6.1.4 BINDER JETTING

In contrast to the material jetting processes, binder jetting processes (Figure 6.1c) only print binder into the powder bed rather than the target material. Typically, binder droplets not only glue the powder particles but also facilitate the bonding among the printed layers. The powder bed will move downward after a new layer finishes printing and then spread a new powder layer [7]. Unbound powder also supports the unconnected parts to allow the formation of internal hollow structures [13]. Comparing with material jetting and some other 3D-printing technologies, binder jetting has a lower cost due to the saving of the heat source. In principle, it can work with any powdered materials as long as they are dimensionally small and can be layered as the only binder is printed.

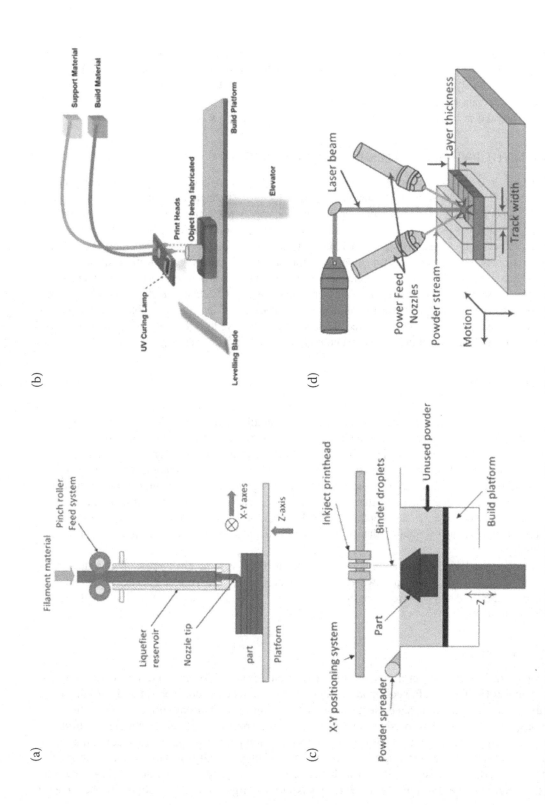

FIGURE 6.1 (a) Material extrusion (DIW and FDM). Reproduced from reference [7]. Copyright (2021) Springer Nature. (b) Material jetting process; Reproduced with permission from ref. [15] – Published by The Royal Society of Chemistry. (c) Binder jetting process. Reproduced from reference [7]. Copyright (2021) Springer Nature. (d) A typical directed energy deposition process. Reproduced from reference [7]. Copyright (2021) Springer Nature.

6.1.5 DIRECTED ENERGY DEPOSITION (DED)

In DED, materials are melted and deposited layer by layer simultaneously. As shown in Figure 6.1d, materials are loaded, possibly as filaments or powders, and the energy source is applied to melt the printed materials. It is predominantly used for metal powders in both research and commercial applications, though it can work for ceramics and polymers as well. First, the feedstock material is melted by a laser or electron beam. Each pass of the filament forms the solidified track that makes up the layer. It is particularly favored to be used for the repairing of an existing part [7,14,15].

6.2 CONFIGURATION OF FLEXIBLE ENERGY DEVICE

Energy devices refer to both energy storage devices and conversion devices, such as supercapacitors, batteries, fuel cells, piezo/triboelectric generators, solar cells, etc. Although some of them, such as piezo/triboelectric generators and solar cells, aim to convert natural energy to electric energy, the low conversation efficiency is a major barrier. More accessible and effective energy storage devices have surged in the past decades in both scientific research and commercial applications, in the fields of wearable devices, vehicles, portable electronics, etc. The unprecedented demand imposed by the emerging applications of flexible and wearable electronics, and the higher requirements for power and energy density, encourage researchers to optimize the 3D structure of electrodes and maximize the capacity of active materials. 3D-printing technology could be a promising solution for this constraint because of its programmable and controllable manufacturing process. Also, it will greatly reduce the cost and improve the efficiency of production. A brief introduction of the fundamental principles and configurations of EES devices is given here.

6.2.1 ACTIVE MATERIALS

In recent decades, many materials have been identified as active materials for energy storage, like carbon materials (graphite, graphene, carbon nanotube, carbon nanofibers, porous carbons, etc.), metal oxide, metal nitrides, metal phosphides (M_xS_y), metals (germanium, tin, phosphorus, indium) and their alloys, polymers (PANi, PPy, PEDOT, etc.) [16–20]. They are different in terms of theoretical capacity, energy storage mechanism, and intrinsic properties, which are based on the intercalation/deintercalation or absorption/desorption of ions in active materials. The shuttling ions with a size comparable to the lattice will cause the large volume change of the nanostructure, especially for the intercalation/deintercalation process, which has deeper penetrated depth. The cracking and pulverization led to the disconnection with a conductive network, and eventually, the isolated parts become dead zones with decreased capacity and cycle stability.

Several strategies have been frequently applied to control the volume changes, such as reducing the particle size and the use of composites, where the host matrix buffers the volume change. When the grain size decreases, the yield and fracture strength of particles increases dramatically because of the decreased dislocations and defects.

6.2.2 EES ELECTRODES

As explained in Figure 6.2a, in a conventional two-dimensional (2D) electrode fabrication method, polymer binders (PVDF, PTFE, sodium alginate, etc) and conductive additives are added to the active materials to form an interconnected electric conduction path to the current collector. Both the short-range transport between active particles and long-range transport throughout the whole electrode are important to their conductivity [23]. However, the tortuous ion pathways (pore structure) in the tightly packed electrodes is another limiting factor to consider. Mathematically, sustaining the same specific capacity (at a certain current density) in a thicker electrode requires proportionally larger currents across the longer charge transport distance. It means more charges should be delivered. The capacity

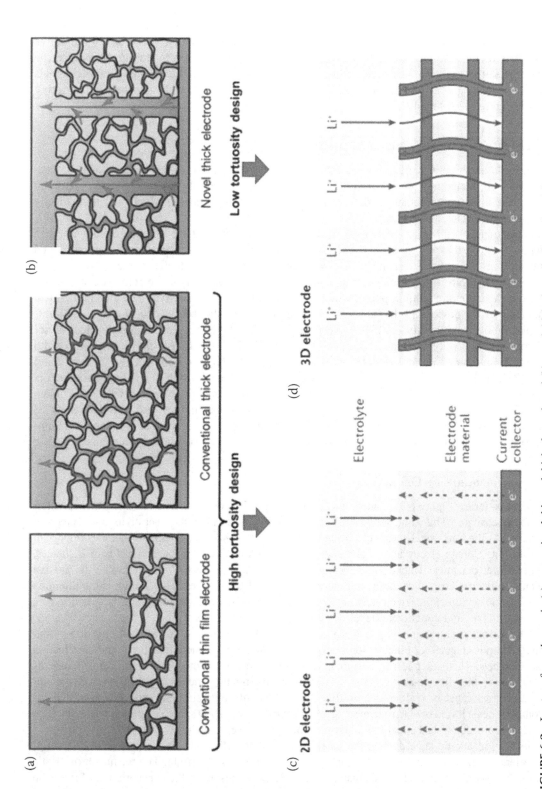

FIGURE 6.2 Ion transfer pathways in (a) conventional thin and thick electrode and (b) novel thick electrode. Reproduced from ref. [21]. Copyright (2019) John Wiley and Sons. Comparison of the charge transfer pathway in a (c) conventional thick electrode and (d) hierarchical 3D electrode. Reprinted by permission from ref. [22]. Copyright (2019), Springer Nature.

degradation becomes more remarkable with the increase of current density and electrode thickness. It's desirable to optimize the electrons' transfer distance from active sites to the current collector and shorten the ions' diffusion pathway from the electrolyte to the active sites to make use of the active materials. The hierarchical structure of the electrode can play a key role in this process since the hierarchical skeleton structure forms the electric conductive network and the coupled porous channels offer the ion diffusion pathway. This outcome can be achieved by scalable and controllable 3D printing techniques (Figure 6.2d) [11–13].

6.2.3 Electrolyte and the Solid-State Devices

Because flexible and wearable EES devices are in proximity to humans, safety issues should be cautiously considered, including flammable and highly volatile liquid electrolytes. It's especially important to consider the long-term mechanical deformation in practical applications, which will dramatically increase the risk of crack and leakage.

An all (or quasi)-solid-state electrolyte enables all the components to be integrated without a membrane separator, which improves both the energy density and the life span. So far, polymer and inorganic options have been developed as two major groups of solid electrolytes. The organic polymer electrolytes are light, flexible, and scalable in nature. Gel polymer electrolytes are solid-state, but they also contain a liquid phase, which enhances the ionic conductivity and electrochemical performance of EES devices [24]. They offer a combination of flexibility, good mechanics, light-weight, high ionic conductivity, and adhesive properties. The latter type of solid-state electrolytes usually consists of a polymer matrix, such as poly (vinyl alcohol) (PVA), poly (ethylene oxide) (PEO), poly (vinylidene fluoride) (PVDF), polyacrylonitrile (PAN), and poly (methyl methacrylate) (PMMA) [24,25]. The solid inorganic electrolytes usually are intrinsic single-ion cationic conductors with higher charge-transport efficiency and lower transference numbers compared with liquid electrolytes. The solid inorganic ceramic electrolytes commonly use oxides and sulfides, such as $Li_{6.5}La_3Zr_{1.5}Ta_{0.5}O_{12}$ [26]. They are compatible with electrodes and the state-of-the-art manufacturing process, and are considered to be excellent routes toward practical requirements [27].

6.2.4 Configuration of EES Devices

Conventional batteries (Figure 6.3a) and supercapacitors typically consist of two electrodes, a separator, and liquid electrolyte. The solid-state electrolyte can be used as both electrolyte and separator (Figure 6.3b). As the ions travel across electrodes and electrolyte, the layer thickness becomes important for ion diffusion. Thinner electrode or (solid-state) electrolyte films have shorter ion diffusion distances, thus higher power densities. Thicker ones increase the mass ratio of the active materials to the passive components (collector, etc.), thus the overall energy density of the devices. The conflict is the dilemma of traditional 2D architectural electrodes, and it partially explains the gap between the performance claimed for lab-level and commercial-level devices [29].

The 3D electrode architecture (Figure 6.3c) contains a 3D scaffold as a current collector and a concomitant 3D porous channel for efficient ion transport. It facilitates the charge transfer by the introduction of porous channels. This structure can potentially overcome the conflict of the conventional structure, as mentioned above; therefore, many groups are currently investigating this structure. However, the improvement of areal capacity is often achieved at the expense of volumetric capacity. Theoretically, tri-continuous nanoscale layers of electrode/electrolyte/electrode folded in 3D space can provide benefits, such as short charge transfer distance and low mass ratio of the passive components (Figure 6.3d) [28].

Additionally, in a solid-state device, the outer surface of the electrode-electrolyte-electrode configuration suffers tensile strain and the inner surface endures compressive strain. The mismatch of elastic modulus and Poisson's ratio of electrode materials and interlayer results in the concentration of stress on the thick electrode or the interface between layers. The exceeded stress over the tolerance limitation will

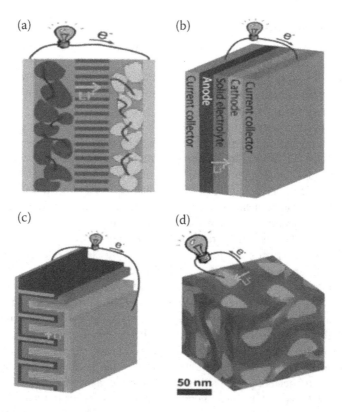

FIGURE 6.3 Illustration of different battery architectures: (a) commonly employed composite, (b) thin film, (c) interdigitated thin film, and (d) 3-D gyroidal designs. Reproduced with permission from ref. [28] – Published by The Royal Society of Chemistry.

destroy structures. The hierarchical 3D structures with ductile porous structures can effectively tackle this issue [22,30].

Despite the proof-of-concept studies, such 3D tri-continuous configurations are still far away from practical applications [31]. Fabricating strictly separated anode and cathode networks into a 3-D continuous architecture is challenging, considering the potential short circuits.

6.3   3D PRINTED EES DEVICES

To solve the problems mentioned above, 3D-printing technologies can be applied. In addition, 3D printing can potentially widen the scope of materials selection with different attributes; for instance, binding, activation, and supporting, etc., can be printed, in principle, in a variety of structures, such as with sandwich-type, in-plane, and 3D tri-continuous configurations. Fillers or additives can be added for optimization. Recent progress in the 3D printing of flexible energy devices and their components is briefly summarised, together with the established performance metrics, such as the evaluation of viscosity, ionic/electric conductivity, cyclability, mechanical properties, flexibility, and operating temperature.

6.3.1   3D PRINTED ELECTRODES

Since the above-mentioned 3D structure is mainly created in the electrode, it is the major part that has been 3D printed. There are some typical categories of materials and the commonly used corresponding structures.

6.3.1.1 Carbon-based electrodes

Due to the chemical processability and tunable viscoelastic properties, carbon materials have been intensively developed with advanced functional components and used to fabricate next-generation 3D-printed structures and devices. The inks are mainly prepared in volatile solutions with active carbon materials as fillers, such as CNTs [32], GO, and carbon blacks. To adjust the viscosity for better printability, the hydrophobic CNTs, carbon black, or graphene generally work with the assistance of polymers.

Graphene is one promising material in the 21st century with superior elasticity, chemical stability, high specific surface area, and excellent physicochemical properties. It can serve as active materials, binders, scaffolds, conductive additives, or forming composites with other active materials for applications in flexible energy storage devices. To further optimize the performance of graphene and graphene composites, an essential premise is the large-scale assembly of 2D building blocks and the transfer of their inherent properties into 3D structures. It can be achieved through chemical reduction, hydrothermal reduction, or cross-linking of the GO sheets. Importantly, the tunable rheological properties of GO dispersion are desirable for the extrusion-based process. GO ink can be reformed and directly used as printable inks to form electrodes.

In 2013, Maher et al. [33] reported a direct laser-writing strategy to fabricate graphene micro-supercapacitors on a large scale. The single-step fabrication technique is similar to selective the laser-sintering technology. On a GO-coated disc, the integrated rGO finger electrodes were reduced and shaped by the DVD burner. After that, Sun et al. [34] fabricated the graphene-based planar micro-supercapacitors via layer-by-layer printing of GO followed by chemical reduction. The thickness can be scaled up and is approximately linearly proportional to the printing cycles, with a decay in performance when the thickness increases.

In 2015, Marcus et al. [35] have demonstrated a DIW strategy for the fabrication of 3D graphene aerogels with designed architectures. With silica filler being the viscosifier, the DIW has a shear thinning behavior to ensure printability, but also with good mechanical properties to maintain the printed structure. The 3D structure was created with the precise deposition of GO inks on a predefined path (Figure 6.4a). This system has been further developed by the authors in 2016 [38]. Besides, Gao's group demonstrated a facile gelation method to fabricate graphene aerogel micro-lattices. The added Ca^{2+} ($CaCl_2$), as cross-linkers, made GO dispersion to hydrogel to maintain the 3D structure. The resistance is orders of magnitude lower than those using only GO as inks since the thickness of the electrodes increases. The excellent ion diffusion enables the interconnected graphene layers and fast transport of ions. The high concentration of GO sheets endows a 3D network. Decreasing the repulsion force or increasing the bonding force between GO sheets in solution will induce GO gelation or participation and reinforce the 3D GO network. Adding cross-linkers was identified as one effective strategy that can realize these purposes (Figure 6.4b). Hydrogen bonds, coordination, and hydrophobic interactions are considered to be the possible driving forces in gelation processes [39]. Additionally, ice-templating and freeze-drying methods were frequently applied to preserve the 3D architecture and restrain shrinkage. For example, Yao et al. [37] fabricated the 3D-printed graphene aerogels as a scaffold (Figure 6.4c) to support the pseudocapacitive MnO_2 (Figure 6.4d). By DIW of GO/hydroxypropyl methylcellulose suspension, the freeze-dried 3D structures were then subjected to a heat treatment process to form a graphene scaffold, with the MnO_2 nanosheets electrodeposited on the scaffold. The high areal capacitance of 44.13 F/cm^2 has been achieved with a 4-mm-thick graphene scaffold with ~180 mg/cm^2 MnO_2 loading.

In addition, electrochemical active materials with different dimensions, such as 0D quantum dots, 1D carbon nanotubes, 2D boron nitride, carbon nitride, or MXenes nanosheets, can be integrated with graphene and act as the inks for 3D printing [40]. In addition to the macrostructures, the microstructure and nanostructure of the electrodes should also be considered for the charge delivery. Studies have demonstrated that the hierarchical porosity is closely linked to their specific capacitance and rate performance [41]. However, restricted by the printing precision and microscaled feedstocks, most current 3D-printing technologies fail to directly shape the nanostructure or microstructure of the electrodes. Therefore, strategies that facilitate this process should be considered, such as the pretreatment of active

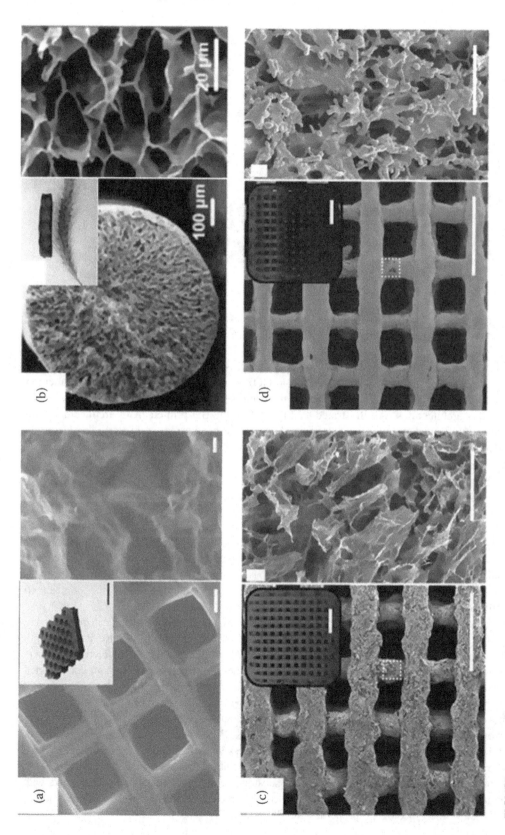

FIGURE 6.4 (a) Structure of the graphene aerogels with 4 wt% R–F after etching, the inset is the 3D-printed honeycomb. scale bar: Left: 200 μm, Right: 100 nm; inset:1 cm. Reproduced from ref. [35]. Copyright (2015) Springer Nature. (b) Morphology and structure of one single fiber of graphene aerogels, scale bar: Left: 100 μm, Right: 20 μm. Adapted from ref. [36] with permission from The Royal Society of Chemistry. The inset is a cubic structure standing on bristlegrass. (c) SEM image of a graphene scaffold by 3D printing. Scale bar: Left: 1 mm; Right 40 μm. The inset is the top-view digital images of 3D-printed graphene aerogel. (d) SEM image of a 3D-printed graphene scaffold after the electrodeposition of MnO_2. Scale bar: Left: 1 mm; Right 40 μm. Reprinted from Yao et al. [37].

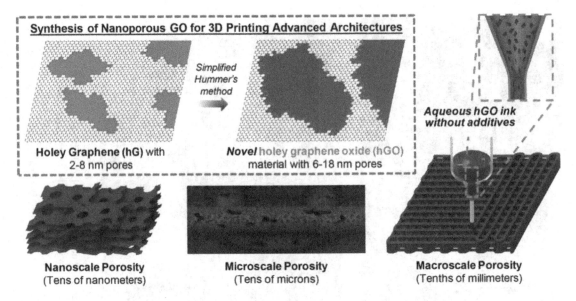

FIGURE 6.5 The process to fabricate 3D architectures with hierarchical porosity (both macroscale and nanoscale). Reproduced from ref. [42]. Copyright (2018) John Wiley and Sons.

materials [41], template-assisted fabrication [38], or post-process of the printed structure [37]. For example, Lacey et al. [42] synthesized the holey GO (hGO) using a controlled air oxidation method followed by liquid-phase oxidation. The hGO was then applied for extrusion-based 3D printing of hierarchical porous architectures. As illustrated in Figure 6.5, the 3D-printed hGO architecture with hierarchical porosity formed many micrometer-sized pores. After thermal reduction, the electrode showed excellent areal capacitance of 13.3 mAh/cm^2 at 0.1 mA/cm^2.

6.3.1.2 Polymer-based electrodes

Polymer-based materials can be 3D printed by different kinds of technologies, such as vat photo-polymerization, powder bed fusion, and selective laser sintering. Since many polymers are electrically insulating or electrochemically inactive, they are only used as templates. Acrylonitrile-butadiene-styrene (ABS) and polylactic acid (PLA) filaments have been widely used in FDM. However, quite a few other combinations have also been demonstrated to be feasible; they include ABS/graphene [43], ABS/carbon black [44], and PLA/graphene [45] [46]. PLA and ABS-based composites are probably not ideal candidates considering their electrochemical inactive properties. Their capacitances were mainly contributed by other electrochemical active components.

Instead, polymers like PANi, PEDOT, or PPy, with their intrinsic electrical conductivity, can be promising for energy storage, especially with the assist of GO additives [47–49]. Still urgently needed is an in-depth understanding of the polymer ink (or raw materials) designs to allow 3D printing of conductive polymers with greater precision, local composition specificity, and electrochemical performance. Among the reported works, Kim et al. [50] obtained a stretched unit by pulling a micropipette filled with a pyrrole solution, followed by oxidative polymerization in air. The wire size can be precisely limited to nanometers by controlling the movement speed. In addition, dense arrays of wires, branches, and bridges have been fabricated. Yuk and co-workers [51] (Figure 6.6) prepared the PEDOT:PPS/DMSO/water mixture inks with a mortar grinder. The relationship between storage modulus and shear stress was investigated and optimized through modulate the ratio of PEDOT:PSS concentration. The inks with low concentrations (1–4 wt%) of PEDOT:PSS conducting polymer spread laterally on the substrate, while blocking occurs with the inks with high concentrations (above 8 wt%) because of the agglomeration. In contrast, the inks with a concentration of around 5–7 wt% show good rheological properties and 3D printability.

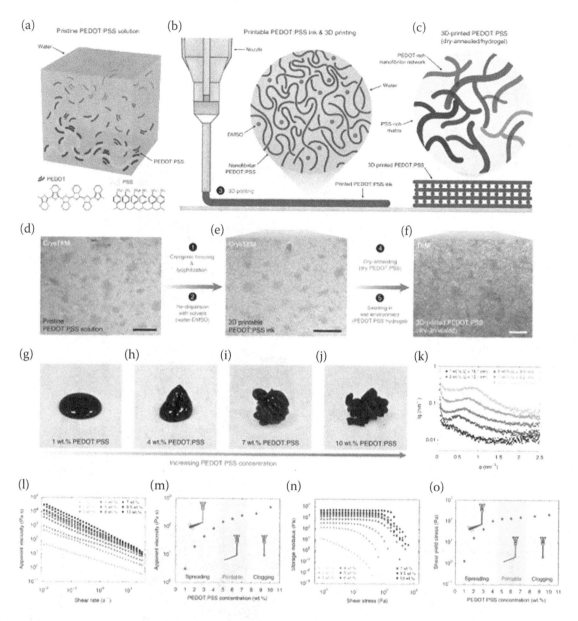

FIGURE 6.6 PEDOT:PSS ink for 3D printability. (d–f) TEM images of the structures, where the scale bars are 100 nm. Reproduced from ref. [51]. Copyright (2020) Springer Nature.

6.3.1.3 Others

Active materials such as LFP, LMFP [52], LTO, V_2O_5 [53], Si [54], SnO_2 [55], etc., are the most commonly used and extensively studied active materials for ion batteries in the past decades. Due to their similarities from conventional coating slurry to printable inks, extensive studies have focused on DIW or inkjet printing. Here, we select the commonly used LFP and LTO as examples, to introduce the reported ink preparation methods, printing technologies, and their performance.

Figure 6.7a shows the relationship between the storage modulus and shear stress of the designed LFP and LTO viscoelastic inks reported by Lewis's group [56]. The addition of hydroxypropyl

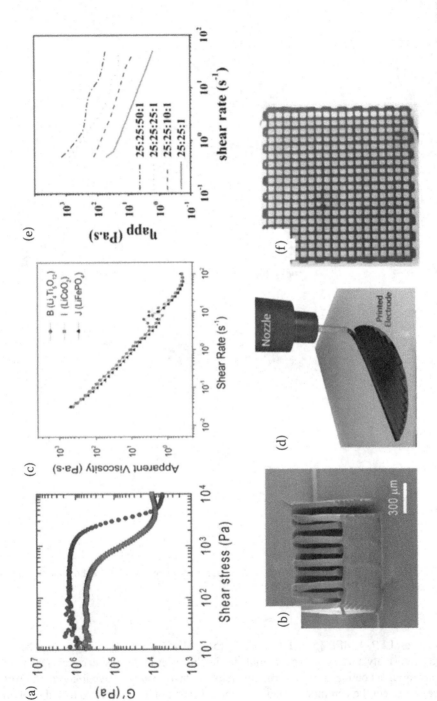

FIGURE 6.7 (a) Storage modulus of the LTO and LFP inks as a function of shear stress; (b) SEM images of the printed LTO-LFP electrode architectures. Reproduced from ref. [56]. Copyright (2013) John Wiley and Sons. (c) Apparent viscosity as a function of shear rate for the composite inks; (d) Printing of the composite electrodes. Reproduced from ref. [57] with permission from The Royal Society of Chemistry. (e) Rheological properties of LFP-CMC-water-1,4 dioxane ink and (f) the printed electrode. Reproduced from ref. [58]. Copyright (2017) MDPI.

cellulose and hydroxyethyl cellulose enabled the printing of thin-walled electrodes (Figure 6.7b). The assembled 3D interdigitated micro-battery shows a high areal energy density of 9.7 J/cm^2 with a power density of 2.7 mW/cm^2. In 2016, Sun et al. used the carbon nanofibers and PVDF as additives for the preparation of LTO and LFP based inks [57]. They identified that each additive plays a significant role: carbon nanofibers improved the conductivity and the porous microstructure for high rate performance, PVDF governed the mechanical properties, and their synergistically improved the rheological properties. The printed electrodes demonstrated excellent cyclability and rate capability. Later, in 2017, Changyong Liu and coworkers [58] optimized the viscosity of the LFP ink through the application of CMC. Results showed that the viscosity increased dramatically with the CMC content (from 1 wt% to 4 wt%) (Figure 6.7c,d). Besides, the apparent viscosity increased with the ratio of 1,4 dioxane until the proportion exceeded 50%. The content of the LFP ink was optimized, and it was printed by direct writing at low temperature, where enhanced rate performance was shown.

Importantly, high viscosity and shearing-thinning are the necessary properties of inks for the high-quality printing. But it still faces the same challenge with conventional electrodes: most of the additives are inactive with no contribution to their gravimetric capacity. Active materials that suit different 3D-printing techniques need to be developed and optimized. Together with new 3D-printing technologies, it is anticipated that the performance of the electrode could be further improved.

6.3.2 3D PRINTED ELECTROLYTES

As introduced before, inorganic ceramic and organic polymers are two main categories for solid-state electrolytes. Current works tend to combine them to synergistically enhance the conductivity and flexibility. Wei et al. [59] fabricated the yttria-stabilized zirconia (YSZ) electrolyte using the digital light processing stereolithography 3D-printing technique. The optimized mixture suspension containing 30 vol% YSZ and photosensitive resin was used as a precursor. A densified microstructure of the YSZ electrolyte was obtained after debinding and sintering. The assembled solid oxide fuel cell with the 3D YSZ electrolyte achieved a power density of 176 mW/cm^2. Progress was made in 2020 by Pesce et al. [60], who 3D printed the electrolyte-supported solid oxide cells with planar and corrugated architectures. The dense and crack-free YSZ electrolytes reach a high ionic conductivity ~3·10^{-2} S/cm. The corrugated cells showed a maximum power density of 410 mW/cm^2 with an increase proportional to the active area.

Cheng and coworkers [61] have reported the direct 3D printing of PVDF-co-HFP based inks with nanosized TiO$_2$ fillers on electrodes at elevated temperatures. The inks exhibited shear-thinning behavior, suitable wettability, and high apparent viscosities. The fabricated devices showed high charge/discharge capacity and good rate performance, and the average coulombic efficiency of around 98.6% was maintained even after 100 cycles. Furthermore, taking advantage of the programmability of 3D printing, greater improvement can be made. Stefanie Zekoll and coworkers [62] applied 3D printing to produce hybrid electrolytes comprising 3D ordered bi-continuous interlocking channels (Figure 6.8): one filled with a Li$_{1.4}$Al$_{0.4}$Ge$_{1.6}$(PO$_4$)$_3$ ceramic electrolyte with channels across the electrolyte, and one filled with an insulating polymer (e.g., epoxy) to improve the mechanical properties. The best performance was achieved with a total ionic conductivity of 1.6·10^{-4} S/cm. Significant improvement in flexural failure strain (up to 500%) and compressive failure strain (up to 28%) in a Li$_{1.4}$Al$_{0.4}$Ge$_{1.6}$(PO$_4$)$_3$ pellet has been achieved, suggesting an improved toughness over the ceramic pellet. Besides, prepared by SLA 3D printing, a flexible 3D solid polymer electrolyte with an Archimedean spiral structure was fabricated for all-solid-state lithium metal batteries (Figure 6.9) [63]. The designed 3D structure shortens the Li-ion transport pathway and reinforces the interface. Compared with the structure-free SPE, the assembled solid-state Li|LFP cell with a mass loading of 5 mg/cm^2 shows a higher specific capacity of 166 mAh/g and better capacity retention. Despite the rapid progress by far, current research on solid-state electrolyte

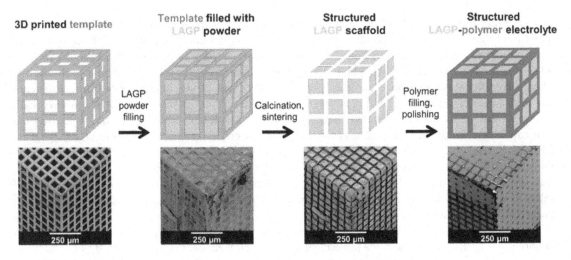

FIGURE 6.8 Schematical and real structure of the 3D printing assisted fabrication of hybrid electrolytes. Reproduced from ref. [62] with permission from The Royal Society of Chemistry.

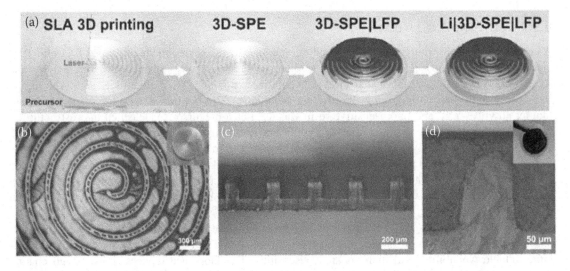

FIGURE 6.9 (a) Fabrication process and (b)(c)(d) morphology of 3D solid polymer electrolyte. Adapted with permission from ref. [63]. Copyright (2020) American Chemical Society.

is still at the early stage. The limited kinds of accessible materials for solid-state electrolytes, unknown ionic transfer mechanisms along with unsatisfied ionic conductivities, are still challenges.

6.3.3 3D PRINTED DEVICE

Although the focus is still on developing new and adaptable materials, for practical applications, it's necessary to fabricate designed electrode and electrolyte structures with better mechanical and electrochemical performance, and direct print fully-packaged devices consisting of integrated functional parts and structural parts . With the rapid development of 3D-printing technologies and the understanding of the energy storage/conversion mechanism, it will become one important direction [2,64–66]. The tri-continuous folded layer structure by 3D printing might be an ideal strategy. It will help retain structural stability, utilize the space, and shorten the ion diffusion distance. Given this potential, integrating the electrode, electrolyte, and current collector into one device remains a challenge. In the hierarchical structure, it is essential that the two opposite electrodes networks are separated in the case of short circuits. The progress in recent years, such as the multimaterial multinozzle 3D printing [67] and multimetal electrohydrodynamic redox 3D printing [68], will probably accelerate the process.

6.4 SUMMARY AND OUTLOOK

AM offers high sustainability and tremendous design flexibility in manufacturing complex 3D structures that will greatly supplement conventional manufacturing technologies in the future. Especially for the flexible energy devices, it could be promising in changing the way future energy devices are fabricated, though many challenges must be tackled.

6.4.1 PRECISION AND RESOLUTION OF 3D PRINTING

Since charge transportation plays a crucial role in EES devices, implementing hierarchical porous structures in electrodes or full devices is vital for achieving high electrochemical performance. But many existing 3D-printing techniques are still not able to construct this with micro- or nano-scale precision. Therefore, the improvement of the fabrication precision is one critical issue. Substantial progress has been made in recent years, such as electrohydrodynamic redox printing (EHD-RP) [68] and aerosol jet technique-based 3D printing [69].

6.4.2 NEW MATERIALS

One of the challenges in 3D printing is the limited material selection, as compared to the variety of materials used in conventional fabrication techniques. The emergence of 3D printing requires researchers to widen the selection of the materials and the ways new materials can be prepared.

6.4.3 INTEGRATION WITH MULTI-MATERIALS PRINTING TECHNOLOGY AND THE INTERFACE

By developing an integrated and hybrid printing chain, raw printing materials can be converted to usable devices. Inkjet printing and DIW could be the candidate choices, but there is still a long way to go before achieving the fully enclosed/packaged device. The application of bifunctional or trifunctional materials to the 3D multi-materials printing for a tri-continuous or multifunctional integrated device with short response time and simplified configuration is an urgent and promising development direction. In the meantime, the interface zone generated between different materials should also be considered: (1) from a mechanical perspective, the mismatch of elastic modulus and Poisson's ratio will result in stress concentration on the interface between layers and the destruction of structures; (2) from electrochemical perspective, the electrode/electrolyte interface will greatly influence the charge transfer.

6.4.4 4D PRINTING

Interestingly, the recent advances in four-dimensional (4D) printing, with time being considered as the fourth dimension, have shown the potential, especially in enabling the structure to be smart and controlled. This goal can be achieved by using smart materials, such as shape memory alloys and polymers [70–72]. There is a reasonable prospect that the development of materials with both electrochemical activity and stimuli-responsive functions, such as shape memory, will help fabricate more smart energy devices. An interdisciplinary investigation is important to overcome those challenges. Due to advantages like sustainability, high speed, low cost, wide material selection, and precise spatial resolution, 3D printing shows the potential in the manufacturing of next-generation EES devices, flexible energy systems, wearable electronics, and even all-in-one smart devices.

REFERENCES

1. Manufacturing Technologies, (2016) Standard Terminology for Additive Manufacturing Technologies, 5–7.
2. Peng Chang, Hui Mei, Shixiang Zhou, Konstantinos G. Dassios, Laifei Cheng, (2019) 3D printed electrochemical energy storage devices, *Journal of Materials Chemistry A*. 7: 4230–4258.
3. Flaviana Calignano, Diego Manfredi, Elisa Paola Ambrosio, Sara Biamino, Mariangela Lombardi, Eleonora Atzeni, (2017) Overview on additive manufacturing technologies, *Proceedings of the IEEE*. 105: 593–612.
4. Victoria G. Rocha, Eduardo Saiz, Iuliia S. Tirichenko, Esther García-Tuñón, (2020) Direct ink writing advances in multi-material structures for a sustainable future, *Journal of Materials Chemistry A*. 8: 15646–15657.
5. Fred Fischer, (2011) White Paper Thermoplastics: The best Choice for 3D printing, *Why ABS is a Good Choice for 3D Printing and When to use another Thermoplastic*, Stratasys, Inc. 1–5.
6. Yee Ling Yap, Chengcheng Wang, Swee Leong Sing, Vishwesh Dikshit, Wai Yee Yeong, Jun Wei, (2017) Material jetting additive manufacturing: An experimental study using designed metrological benchmarks, *Precision Engineering*. 50: 275–285.
7. Ian Gibson, David Rosen, Brent Stucker, Mahyar Khorasani, (2019) Additive Manufacturing Technologies, 3rd Edition.
8. Yinfeng He, Fan Zhang, Ehab Saleh, Jayasheelan Vaithilingam, Nesma Aboulkhair, Belen Begines, (2017) A tripropylene glycol diacrylate-based polymeric support ink for material jetting, *Additive Manufacturing*. 16: 153–161.
9. Yun Lu Tee, Phuong Tran, Martin Leary, Philip Pille, Milan Brandt, (2020) 3D Printing of polymer composites with material jetting: Mechanical and fractographic analysis, *Additive Manufacturing*. 36: 101558.
10. Eduardo Salcedo, Dongcheon Baek, Aaron Berndt, Jong Eun Ryu, (2018) Simulation and validation of three dimension functionally graded materials by material jetting, *Additive Manufacturing*. 22: 351–359.
11. David Bourell, Jean Pierre Kruth, Ming Leu, Gideon Levy, David Rosen, Allison M. Beese, Adam Clare, (2017) Materials for additive manufacturing, *CIRP Annals – Manufacturing Technology*. 66: 659–681.
12. H. Bikas, P. Stavropoulos, G. Chryssolouris, (2016) Additive manufacturing methods and modeling approaches: A critical review, *International Journal of Advanced Manufacturing Technology*. 83: 389–405.
13. Mohsen Ziaee, Nathan B. Crane, (2019) Binder jetting: A review of process, materials, and methods, *Additive Manufacturing*. 28: 781–801.
14. Ian Gibson, David Rosen, Brent Stucker, (2015) Additive manufacturing technologies: 3D printing, rapid prototyping, and direct digital manufacturing, second edition, *Additive Manufacturing Technologies: 3D Printing, Rapid Prototyping, and Direct Digital Manufacturing*, 2nd Edition. 1–498.
15. Merum Sireesha, Jeremy Lee, A. Sandeep Kranthi Kiran, Veluru Jagadeesh Babu, Bernard B.T. Kee, Seeram Ramakrishna, (2018) A review on additive manufacturing and its way into the oil and gas industry, *RSC Advances*. 8: 22460–22468.
16. Ata kamyabi elham kamali heidari, (2018) Electrode materials for lithium ion batteries: A review, *Ultrafine Grainrd and Nanostructured Materials*. 51: 1–12.
17. Zaharaddeen S. Iro, C. Subramani, S.S. Dash, (2016) A brief review on electrode materials for supercapacitor, *International Journal of Electrochemical Science*. 11: 10628–10643.
18. Yuqi Jiang, Jinping Liu, (2019) Definitions of pseudocapacitive materials: A brief review, *Energy & Environmental Materials*. 2: 30–37.
19. Li Zhang, X.S. Zhao, (2009) Carbon-based materials as supercapacitor electrodes, *Chemical Society Reviews*. 38: 2520–2531.
20. Jeffrey W. Fergus, (2010) Recent developments in cathode materials for lithium ion batteries, *Journal of Power Sources*. 195: 939–954.
21. Yudi Kuang, Chaoji Chen, Dylan Kirsch, Liangbing Hu, (2019) Thick electrode batteries: Principles, opportunities, and challenges, *Advanced Energy Materials*. 9: 1–19.
22. Hongtao Sun, Jian Zhu, Daniel Baumann, Lele Peng, Yuxi Xu, Imran Shakir, (2019) Hierarchical 3D electrodes for electrochemical energy storage, *Nature Reviews Materials*. 4: 45–60.
23. Samantha L. Morelly, Nicolas J. Alvarez, Maureen H. Tang, (2018) Short-range contacts govern the performance of industry-relevant battery cathodes, *Journal of Power Sources*. 387: 49–56.

24. Saeideh Alipoori, Saeedeh Mazinani, Seyed Hamed Aboutalebi, Farhad Sharif, (2020) Review of PVA-based gel polymer electrolytes in flexible solid-state supercapacitors: Opportunities and challenges, *Journal of Energy Storage.* 27: 101072.

25. Yun Zheng, Yuze Yao, Jiahua Ou, Matthew Li, Dan Luo, Haozhen Dou, (2020) A review of composite solid-state electrolytes for lithium batteries: Fundamentals, key materials and advanced structures, *Chemical Society Reviews.* 49: 8790–8839.

26. Weiwei Ping, Chengwei Wang, Ruiliu Wang, Qi Dong, Zhiwei Lin, Alexandra H. Brozena, (2020) Printable, high-performance solid-state electrolyte films, 1–10.

27. Qing Zhao, Sanjuna Stalin, Chen Zi Zhao, Lynden A. Archer, (2020) Designing solid-state electrolytes for safe, energy-dense batteries, *Nature Reviews Materials.* 5: 229–252.

28. J.G. Werner, G.G. Rodríguez-Calero, H.D. Abruña, U. Wiesner, (2018) Block copolymer derived 3-D interpenetrating multifunctional gyroidal nanohybrids for electrical energy storage, *Energy and Environmental Science.* 11: 1261–1270.

29. Y. Gogotsi, P. Simon, (2011) True performance metrics in electrochemical energy storage, *Science.* 334: 917–918.

30. Lijuan Mao, Qinghai Meng, Aziz Ahmad, Zhixiang Wei, (2017) Mechanical analyses and structural design requirements for flexible energy storage devices, *Advanced Energy Materials.* 7: 1–19.

31. Christopher P. Rhodes, Jeffrey W. Long, Katherine A. Pettigrew, Rhonda M. Stroud, Debra R. Rolison, (2011) Architectural integration of the components necessary for electrical energy storage on the nanoscale and in 3D, *Nanoscale.* 3: 1731–1740.

32. Wei Yu, Han Zhou, Ben Q. Li, Shujiang Ding, (2017) 3D printing of carbon nanotubes-based micro-supercapacitors, *ACS Applied Materials and Interfaces.* 9: 4597–4604.

33. Maher F. El-Kady, Richard B. Kaner, (2013) Scalable fabrication of high-power graphene micro-supercapacitors for flexible and on-chip energy storage, *Nature Communications.* 4: 1475–1479.

34. Gengzhi Sun, Jia An, Chee Kai Chua, Hongchang Pang, Jie Zhang, Peng Chen, (2015) Layer-by-layer printing of laminated graphene-based interdigitated microelectrodes for flexible planar micro-supercapacitors, *Electrochemistry Communications.* 51: 33–36.

35. Cheng Zhu, T. Yong, Jin Han, Eric B. Duoss, Alexandra M. Golobic, Joshua D. Kuntz, (2015) Highly compressible 3D periodic graphene aerogel microlattices, *Nature Communications.* 6: 1–8.

36. Shijia Yuan, Wei Fan, Dong Wang, Longsheng Zhang, Yue E. Miao, Feili Lai, (2021) 3D printed carbon aerogel microlattices for customizable supercapacitors with high areal capacitance, *Journal of Materials Chemistry A.* 9: 423–432.

37. Bin Yao, Swetha Chandrasekaran, Jing Zhang, Wang Xiao, Fang Qian, Cheng Zhu, (2019) Efficient 3D printed pseudocapacitive electrodes with ultrahigh MnO_2 loading, *Joule.* 3: 459–470.

38. Cheng Zhu, Tianyu Liu, Fang Qian, T. Yong, Jin Han, Eric B. Duoss, (2016) Supercapacitors based on three-dimensional hierarchical graphene aerogels with periodic macropores, *Nano Letters.* 16: 3448–3456.

39. Hua Bai, Chun Li, Xiaolin Wang, Gaoquan Shi, (2011) On the gelation of graphene oxide, *Journal of Physical Chemistry C.* 115: 5545–5551.

40. Xingwei Tang, Han Zhou, Zuocheng Cai, Dongdong Cheng, Peisheng He, Peiwen Xie, (2018) Generalized 3D printing of graphene-based mixed-dimensional hybrid aerogels, *ACS Nano.* 12: 3502–3511.

41. Yuxi Xu, Zhaoyang Lin, Xing Zhong, Xiaoqing Huang, Nathan O. Weiss, Yu Huang, (2014) Holey graphene frameworks for highly efficient capacitive energy storage, *Nature Communications.* 5:4554.

42. Steven D. Lacey, Dylan J. Kirsch, Yiju Li, Joseph T. Morgenstern, Brady C. Zarket, Yonggang Yao, (2018) Extrusion-based 3D printing of hierarchically porous advanced battery electrodes, *Advanced Materials.* 30: 1–9.

43. Xiaojun Wei, Dong Li, Wei Jiang, Zheming Gu, Xiaojuan Wang, Zengxing Zhang, (2015) 3D printable graphene composite, *Scientific Reports.* 5: 1–7.

44. Hairul Hisham Bin Hamzah, Oliver Keattch, Derek Covill, Bhavik Anil Patel, (2018) The effects of printing orientation on the electrochemical behaviour of 3D printed acrylonitrile butadiene styrene (ABS)/carbon black electrodes, *Scientific Reports.* 8: 1–8.

45. D. Vernardou, K.C. Vasilopoulos, G. Kenanakis, (2017) 3D printed graphene-based electrodes with high electrochemical performance, *Applied Physics A: Materials Science and Processing.* 123: 1–7.

46. Di Zhang, Baihong Chi, Bowen Li, Zewen Gao, Yao Du, Jinbao Guo, (2016) Fabrication of highly conductive graphene flexible circuits by 3D printing, *Synthetic Metals.* 217: 79–86.

47. Fuguo Zhou, Sancan Han, Qingren Qian, Yufang Zhu, (2019) 3D printing of free-standing and flexible

nitrogen doped graphene/polyaniline electrode for electrochemical energy storage, *Chemical Physics Letters.* 728: 6–13.

48. Kai Chi, Zheye Zhang, Jiangbo Xi, Yongan Huang, Fei Xiao, Shuai Wang, (2014) Freestanding graphene paper supported three-dimensional porous graphene-polyaniline nanocomposite synthesized by inkjet printing and in flexible all-solid-state supercapacitor, *ACS Applied Materials and Interfaces.* 6: 16312–16319.

49. Muhammad Wajahat, Jung Hyun Kim, Jinhyuck Ahn, Sanghyeon Lee, Jongcheon Bae, Jaeyeon Pyo, (2020) 3D printing of Fe_3O_4 functionalized graphene-polymer (FGP) composite microarchitectures, *Carbon,* 167:278–284.

50. Ji Tae Kim, Seung Kwon Seol, Jaeyeon Pyo, Ji San Lee, Jung Ho Je G. Margaritondo, (2011) Three-dimensional writing of conducting polymer nanowire arrays by meniscus-guided polymerization, *Advanced Materials.* 23: 1968–1970.

51. Hyunwoo Yuk, Baoyang Lu, Shen Lin, Kai Qu, Jingkun Xu, Jianhong Luo, (2020) 3D printing of conducting polymers, Nature *Communications. 11: 4–11.*

52. Jiangtao Hu, Yi Jiang, Suihan Cui, Yandong Duan, Tongchao Liu, Hua Guo, (2016) 3D-printed cathodes of $LiMn_{1-x}Fe_xPO4$ nanocrystals achieve both ultrahigh rate and high capacity for advanced lithium-ion battery, *Advanced Energy Materials.* 6: 1–8.

53. Kai Shen, Junwei Ding, Shubin Yang, (2018) 3D printing quasi-solid-state asymmetric micro-supercapacitors with ultrahigh areal energy density, *Advanced Energy Materials.* 8: 1–7.

54. Stephen Lawes, Qian Sun, Andrew Lushington, Biwei Xiao, Yulong Liu, Xueliang Sun, (2017) Inkjet-printed silicon as high performance anodes for Li-ion batteries, *Nano Energy.* 36: 313–321.

55. Yaomin Zhao, Qin Zhou, Ling Liu, Juan Xu, Manming Yan, Zhiyu Jiang, (2006) A novel and facile route of ink-jet printing to thin film SnO_2 anode for rechargeable lithium ion batteries, *Electrochimica Acta.* 51: 2639–2645.

56. Ke Sun, Teng Sing Wei, Bok Yeop Ahn, Jung Yoon Seo, Shen J. Dillon, Jennifer A. Lewis, (2013) 3D printing of interdigitated Li-ion microbattery architectures, *Advanced Materials.* 25: 4539–4543.

57. Ryan R. Kohlmeyer, Aaron J. Blake, James O. Hardin, Eric A. Carmona, Jennifer Carpena-Núñez, Benji Maruyama, (2016) Composite batteries: A simple yet universal approach to 3D printable lithium-ion battery electrodes, *Journal of Materials Chemistry A.* 4: 16856–16864.

58. Changyong Liu, Xingxing Cheng, Bohan Li, Zhangwei Chen, Shengli Mi, Changshi Lao, (2017) Fabrication and characterization of 3D-printed highly-porous 3D $LiFePO_4$ electrodes by low temperature direct writing process, *Materials.* 10:934.

59. Luyang Wei, Jinjin Zhang, Fangyong Yu, Weimin Zhang, Xiuxia Meng, Naitao Yang, (2019) A novel fabrication of yttria-stabilized-zirconia dense electrolyte for solid oxide fuel cells by 3D printing technique, *International Journal of Hydrogen Energy.* 44: 6182–6191.

60. Arianna Pesce, Aitor Hornés, Marc Núñez, Alex Morata, Marc Torrell, Albert Tarancón, (2020) 3D printing the next generation of enhanced solid oxide fuel and electrolysis cells, *Journal of Materials Chemistry A.* 8: 16926–16932.

61. Meng Cheng, Yizhou Jiang, Wentao Yao, Yifei Yuan, Ramasubramonian Deivanayagam, Tara Foroozan, (2018) Elevated-temperature 3D printing of hybrid solid-state electrolyte for li-ion batteries, *Advanced Materials.* 30: 1–10.

62. Stefanie Zekoll, Cassian Marriner-Edwards, A.K. Ola Hekselman, Jitti Kasemchainan, Christian Kuss, David E.J. Armstrong, (2018) Hybrid electrolytes with 3D bicontinuous ordered ceramic and polymer microchannels for all-solid-state batteries, *Energy and Environmental Science.* 11: 185–201.

63. Yingjie He, Shaojie Chen, Lu Nie, Zhetao Sun, Xinsheng Wu, Wei Liu, (2020) Stereolithography three-dimensional printing solid polymer electrolytes for all-solid-state lithium metal batteries, *Nano Letters.* 20: 7136–7143.

64. Vladimir Egorov, Umair Gulzar, Yan Zhang, Siobhán Breen, Colm O'Dwyer, (2020) Evolution of 3D printing methods and materials for electrochemical energy storage, *Advanced Materials.* 32: 1–27.

65. Feng Zhang, Min Wei, Vilayanur V. Viswanathan, Benjamin Swart, Yuyan Shao, Gang Wu, Chi Zhou, (2017) 3D printing technologies for electrochemical energy storage, *Nano Energy.* 40: 418–431.

66. Bolin Chen, Yizhou Jiang, Xiaohui Tang, Yayue Pan, Shan Hu, (2017) Fully packaged carbon nanotube supercapacitors by direct ink writing on flexible substrates, *ACS Applied Materials and Interfaces.* 9: 28433–28440.

67. Mark A. Skylar-Scott, Jochen Mueller, Claas W. Visser, Jennifer A. Lewis, (2019) Voxelated soft matter via multimaterial multinozzle 3D printing, *Nature.* 575: 330–335.

68. Alain Reiser, Marcus Lindén, Patrik Rohner, Adrien Marchand, Henning Galinski, Alla S. Sologubenko,

(2019) Multi-metal electrohydrodynamic redox 3D printing at the submicron scale, *Nature Communications.* 10: 1–8.

69. Wooik Jung, Yoon ho Jung, Peter V. Pikhitsa, Jooyeon Shin, Kijoon Bang, Jicheng Feng, (2018) Three-dimensional nanoprinting via charged aerosol focusing, *Nature.* 592:54–60.

70. Xiao Kuang, Devin J. Roach, Jiangtao Wu, Craig M. Hamel, Zhen Ding, Tiejun Wang, (2019) Advances in 4D printing: Materials and applications, *Advanced Functional Materials.* 29: 1–23.

71. Ye Zhou, Wei Min Huang, Shu Feng Kang, Xue Lian Wu, Hai Bao Lu, Jun Fu, (2015) From 3D to 4D printing: Approaches and typical applications, *Journal of Mechanical Science and Technology.* 29: 4281–4288.

72. Zhen Ding, Chao Yuan, Xirui Peng, Tiejun Wang, H. Jerry Qi, Martin L. Dunn, (2017) Direct 4D printing via active composite materials, *Science Advances.* 3: e1602890.

7 Environmental Impact of Flexible Energy Devices

H. Bencherif and A. Meddour
University Mostefa Benboulaid Batna, Batna, Algeria

T. Bendib and M. A. Abdi
HNS-RE2SD, Higher National School of Renewable Energy,
Environment & Sustainable Development, Batna, Algeria

LEPCM laboratory, Departement of physics,
University of Batna 1, Batna, Algeria

CONTENTS

7.1 Introduction... 119
7.2 Technical description of flexible devices 121
    7.2.1 Energy conversion devices.. 121
        7.2.1.1 Flexible solar cells... 121
    7.2.2 Energy storage devices.. 124
        7.2.2.1 Flexible supercapacitors 124
        7.2.2.2 Modern designs of lithium-ion batteries... 125
7.3 Flexible materials environmental effects... 126
    7.3.1 Cadmium.. 126
    7.3.2 Amorphous silicon (a-Si)... 126
    7.3.3 Copper indium gallium diselenide (CIGS).. 126
    7.3.4 Lead halide 127
    7.3.5 Carbon-based nanomaterials 127
    7.3.6 Tellurium and indium.. 127
    7.3.7 Toxic flexible substrate 128
7.4 Processing routes and design strategies for safe and sustainable manufacturing........ 128
    7.4.1 Toxic materials replacement 128
        7.4.1.1 Lead-free perovskite 128
        7.4.1.2 Indium 128
    7.4.2 Improving processing routes 129
    7.4.3 Recycling 129
    7.4.4 Encapsulation... 130
7.5 Conclusion 130
References... 130

7.1 INTRODUCTION

The introduction of flexible electronics is seen as a watershed moment that has sparked widespread interest. Flexible electronics are more convenient, light-weight, malleable, and potentially implantable or wearable than traditional electronics [1]. The growth of electronic devices with various

DOI: 10.1201/9781003186755-7

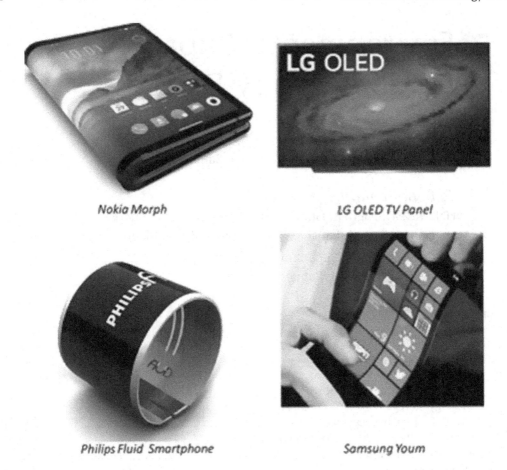

FIGURE 7.1 Nokia Morph Concept, LG OLED TV panel, Philips Fluid flexible smartphone, and Samsung Youm flexible display.

features, counting smart mobile phone devices, roll-up screens, implantable biosensors, and so on, is strongly influenced by these enhanced and improved qualities. LG TV screen OLED (organic light-emitting diode), Nokia Morph model, flexible smartphone from Philips, and flexible Samsung Youm are only a few examples of new ideas and concepts of flexible electronics (Figure 7.1) [1].

Different from the curved-screen LG OLED TV, Apple's new patent application suggests that a wrap-around smartphone display might be on the way [2]. We expect that a significant increase in the use of flexible electronics will occur in the near future. As a result, flexible energy storage and conversion devices as a novel power source category seem to have interesting applications. In recent years, considerable advancement has been realized in the development of high-performance flexible energy conversion and storage technologies, such as solar cells, supercapacitors (SCs), and lithium-ion batteries (LIBs) [3,4].

Even though flexible technology has been around for a while, only recently has it been used to manufacture electronic devices. Many studies are necessary until mass manufacturing can start, but there is also the possibility to include environmental factors into the technological development process from the starting phase. Over 80% of a product's environmental impact is believed to be influenced by decisions taken during the initial stage of design [5]. It is reasonable to believe that prior evaluation of environmental problems has a similar impact on designing new technologies. In today's technological industry, environmental challenges entail more than simply applying the constitution. Manufacturing items' environmental performance may be used in marketing and to improve the company's reputation. Environmental data must be reported transparently to persuade decision-makers that modern technology will not have unintended consequences on the environment. Environmental evaluation of new technology at a broad level is feasible, but it is also difficult because many of the requirements are yet unknown.

This chapter endeavors to discuss the environmental effect of new flexible devices technology. It investigates the impact of different materials and processing techniques on human beings and the environment. Besides, alternative designs strategies and processing routes that pave the way for safe and sustainable materials and manufacturing processes are presented. The presented chapter is arranged as follows. Section 2 is devoted to the technical description of the investigated flexible devices. In Section 3, we highlight the main environmental effects of materials and processing techniques. The new design strategies and manufacturing routes for safe and sustainable flexible devices are showcased in Section 4. In Section 5, we conclude with some remarks and future research directions.

7.2   TECHNICAL DESCRIPTION OF FLEXIBLE DEVICES

In what follows, we introduce briefly the principle devices widely encountered in the flexible electronics field.

7.2.1   Energy conversion devices

7.2.1.1   Flexible solar cells

Figure 7.2 shows the different materials utilized to manufacture a flexible thin-film solar cell which comprises a transparent substrate (right) and an opaque substrate (left).

The optical specification for both electrodes and substrate is determined by the light direction.

The fabrication of thin-film solar cells is made via stacking on a flexible substrate using multiple operational layers and processes, including printing, solution-phase spin-coating, and vacuum-phase deposition. The entire cell is supported mechanically and protected from the environment by a flexible substrate. Figure 7.3 shows some of the most typical flexible substrates for solar cell manufacturing that the literature mentions [6,7]. They can be classified based on the material they are made of, such as metal, ceramic, or plastic substrates.

Two electrodes also gather photoelectric charge carriers and offer electrical connections to an external load. As a top electrode, optical transparency is required to enable sunlight to pass through and permits the absorption by an active semiconductor layer, which assures the conversion of light to photoelectric charge carriers. Figure 7.4 depicts each type of electrode material often employed in solar cells and implemented in a flexible shape, as well as a comparison of their properties.

The active materials, as important elements of flexible solar cells, have a significant influence on power-conversion efficiency. Inorganic, organic, and inorganic-organic hybrid semiconductors are the three main types of active materials. Copper indium gallium diselenide, amorphous silicon, and cadmium telluride are the most prevalent inorganic semiconductors. Organic semiconductors are made up of donor and acceptor organic components, which are further divided into tiny organic molecules and polymers based on their molecular sizes. Metal halide perovskite is the main attractive inorganic-organic semiconductor for flexible solar cell manufacturing. Figure 7.5 shows various types of active materials presently employed in solar cells and possibly applicable in a flexible shape, as well as the matching record certified efficiency.

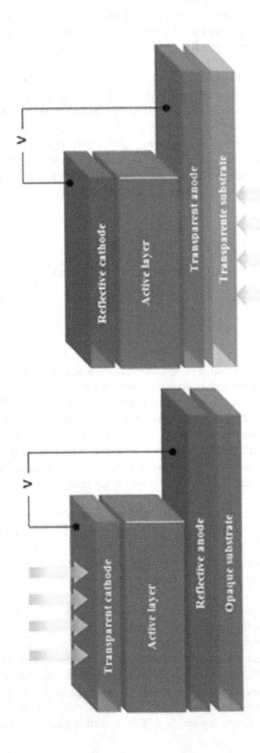

FIGURE 7.2   Flexible solar cell structure including different functional materials.

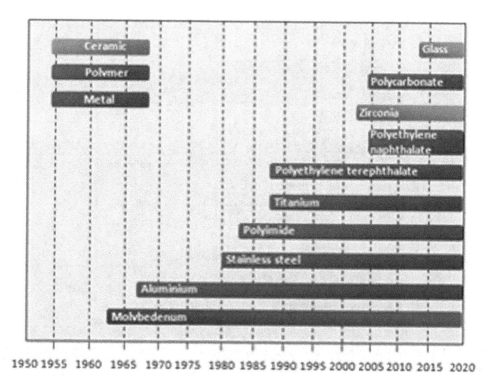

FIGURE 7.3 A timeline of the most widely employed substrates for flexible solar cells, as discussed in the literature. Adapted with permission [6]. Copyright (2021) Materials Reports: Energy.

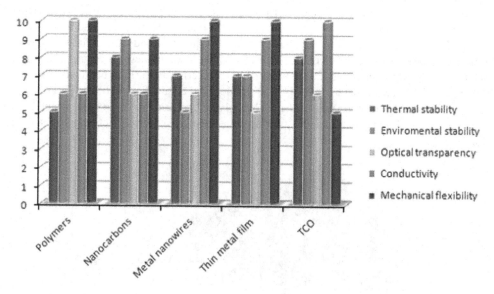

FIGURE 7.4 Physical characteristics comparison of several electrode materials. Adapted with permission [6]. Copyright (2021) Materials Reports: Energy.

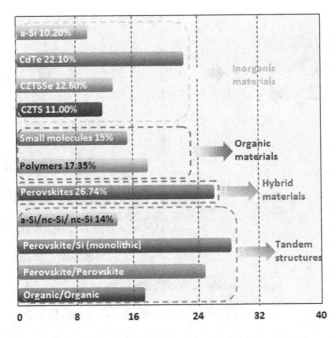

FIGURE 7.5 Highest recorded efficiencies of solar cells produced with different active materials (built on flexible substrates). Adapted with permission [6]. Copyright (2021) Materials Reports: Energy.

7.2.2 ENERGY STORAGE DEVICES

7.2.2.1 Flexible supercapacitors

Flexible solid-state supercapacitors devices commonly include flexible electrodes, a solid-state electrolyte, and a flexible packaging material split by a separator, which is analogous to that of conventional SCs (Figure 7.6). The flexible solid-state SCs differ from ordinary SCs in two aspects; namely, the electrolyte and electrode must have some flexibility, and the electrolyte must be of solid-state nature [8].

Preparing flexible electrodes with superior mechanical qualities and outstanding electrochemical effectiveness is indeed a big issue. Ni foil [9], Ti foil [10], and stainless steel grids [11] have been employed to build flexible electrodes. The aforementioned materials perform as conductive substrates for the combination of capacitive materials deposition, binder, and conducting additives. However, besides the reduced flexibility, the aqueous nature of the electrolytes facilitates the corrosion, which hinders the devices' lifetime. In addition, using metallic existing collectors and other conductive additional materials

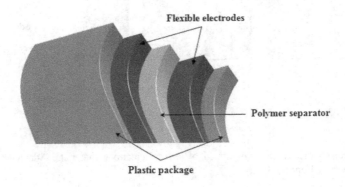

FIGURE 7.6 Schematic views of flexible solid-state supercapacitors.

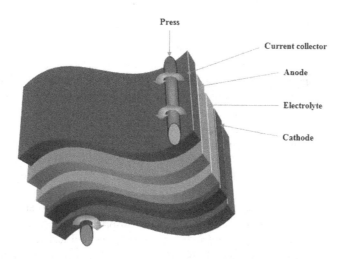

FIGURE 7.7 A schematic view of a flexible LIB.

will significantly raise the weight of SC devices, lowering their energy and specific power densities. As a result, recent studies have concentrated on the conception and production of flexible SC electrodes made of nonmetal materials. Carbon nanomaterials have recently advanced at a breakneck pace, opening up new possibilities toward the manufacturing of high-performance flexible electrodes [12].

Another essential element of flexible SCs is the electrolyte. Solid-state electrolytes are fairly easy to manage compared with liquid electrolytes, which have higher stability and a broader operating temperature range. Likewise, employing a solid-state electrolyte eliminates the risk of leakage, lowering the price of device packaging. Gel polymers are the most often employed solid-state electrolytes in SCs.

7.2.2.2 Modern designs of lithium-ion batteries

The fast growth of flexible electronics has prompted substantial research into how flexible energy and storage devices might be used as power supplies. Flexible lithium-ion batteries (LIBs) recently have shown to be an intriguing and adaptable energy storage technology. Recently, a variety of designs and structures for flexible LIBs have been examined. Remarkable advancement has been made in the development of high-performance electrode and electrolyte materials, as well as innovative structural designs and production processes for flexible LIBs (Figure 7.7).

Flexible electrode concept and manufacturing are both critical for enhancing the performance of flexible LIBs [13]. The useful implementation of flexible LIBs can be aided by a thorough insight into the benefits, disadvantages, and fabrication efficiency of flexible electrodes. A wide range of electrodes has been investigated to design flexible LIBs. These electrodes, which are classed as anodes and cathodes, are commonly made out of soft organic or inorganic materials implanted in a flexible substrate, or their composites.

A high-ionic conductivity liquid electrolyte is a crucial element in the construction of a commercialized Li-ion battery. Nevertheless, liquid electrolyte's drawbacks, such as fluidity, toxicity, flammability, volatility, and the risk of leakage, severely restrict its use in flexible batteries. As a result, greater attention has been dedicated to the development of electrolytes polymer to substitute electrolytes liquid [14]. For the construction of flexible LIBs, materials with a wide electrochemical window, intrinsic flexibility, high safety, and decreased toxicity, such as polymer gel electrolytes and solid-state polymer electrodes, have been used.

Ongoing attempts should be made to fabricate fully flexible batteries to enhance the practical use of flexible LIBs. The flexible LIBs' distinguishing qualities in terms of stability, flexibility, compatibility, simple implantation, and cost are all determined by effective building methodologies.

7.3   FLEXIBLE MATERIALS ENVIRONMENTAL EFFECTS

It is commonly recognized that, depending on the kind of flexible device being constructed, it is necessary to employ further, or a lesser amount of, substantial volumes of acids and solvents for synthesis and cleaning, as well as gases for depositing ultra-thin films of metals and materials. The majority of these compounds are extremely dangerous to the environment and human health. Some of the health and environmental difficulties created by chemical hazards relating to the toxicity, flammability, and carcinogenic characteristics of materials are explored here. An overview of significant environmental and health hazards linked with the manufacturing and employment of various materials is provided below.

7.3.1   CADMIUM

Cadmium, for example, is not biodegradable and hence has significant environmental durability. Cadmium is poisonous to many living creatures. Plants absorb it more readily than other elements do. Cadmium levels in plants might be comparable to those in the soil. Cadmium can also be absorbed from the air. Reduced production, decreased photosynthesis and transpiration, and altered enzyme activity are some of the consequences in plants [15]. Cadmium, being a harmful metal, has ecological impacts in aquatic ecosystems at a certain concentration over the background level.

In terrestrial animals, doses that cause acute effects are infrequent. However, accidental exposure, such as in the workplace, can induce acute consequences in people. Either inhalation or ingestion causes side effects that might range from pulmonary or gastrointestinal inflammation to death, based on the amount consumed. The major worry with cadmium is chronic consequences, rather than acute poisoning, caused by the element's ability to collect in the body. Chronic cadmium poisoning has been reported in Japan, where it has resulted in renal failure and bone structural abnormalities. Individuals who have been exposed to cadmium dust and fumes for a long time have developed a variety of lung ailments, including lung cancer.

7.3.2   AMORPHOUS SILICON (A-SI)

The plasma-enhanced chemical vapor deposition (PECVD) method is commonly employed to produce amorphous silicon solar panels. The major safety danger of this technique is silane gas (SiH_4), which is usually employed as a precursor and is particularly pyrophoric. Based on the carrier gas, it can autonomously flame at smaller doses of 2–3%. Pyrophoric footprints can be detected if mixes are not full, despite the strong pyrophoric tendency of silane, even at doses less than 2% in the carrier gas. For silane concentrations of more than 4.5%, mixtures might be metastable and ignited after a time [16]. Because hydrogen atoms are essential for the material's electrical characteristics, amorphous silicon solar cells have a high concentration of them, roughly 10%. However, since "unhydrogenated amorphous silicon" is of no use in electronic applications, the technique normally refers to "amorphous silicon" rather than "hydrogenated amorphous silicon (a-Si:H)." Because the hydrogen employed in amorphous silicon fabrication is flammable and explosive [16], PV makers must utilize complex gas management systems to reduce, if not eliminate, the danger of fire and explosion. To prevent replacing gas cylinders, one effective technique to prevent these risks is to keep gases like hydrogen and silane in bulk away from tube trailers. Certain toxic gases utilized for doping in a-Si:H production, including phosphine (PH_3), arsine (AsH_3) and germane (GeH_4), may not cause any severe health or environmental risks when employed in modest volumes. These gases' leakage, on the other hand, should be prevented since they pose considerable occupational dangers.

7.3.3   COPPER INDIUM GALLIUM DISELENIDE (CIGS)

Layers deposition can be achieved either by thermally co-evaporation to create CIGS thin films or by rapid deposition and subsequent reaction in a future processing step to generate the final compound [17].

The deposition of a thin cadmium sulfide (CdS) coating as a buffer layer in CIGS TFSCs, using the chemical bath process, is performed. However, cadmium-free CIGS solar cells have previously been efficiently manufactured. Copper, indium, gallium, and selenium are all regarded as nontoxic. Furthermore, basic selenium is essential in human feeding; the absorption of selenium (500–860 g) per day is tolerated for lengthy periods [17]. Despite the fact that pure selenium has low toxicity, hydrogen selenide (H_2Se), which is utilized in the production of CIGS TFSCs, is very poisonous and harmful to health and life. Although its vapor pressure is inferior when compared to arsine, hydrogen selenide has the same impact on the body. To avoid dangers from extremely poisonous H_2Se gas, the production system should be performed under negative pressure and drained via an important control scrubber. Hazardous substances can be reduced by utilizing safer substitutes, such as restriction valves flow and additional safety measures, as Fthenakis [18] explains in detail. CIS and CGS have low systemic toxicity, with no impact on reproduction, liver, ovulation, or kidney, according to certain studies. CIS, on the other hand, has been proven to be less hazardous than CGS and CdTe.

7.3.4 Lead halide

Among the difficult obstacles that limit the solar cell's commercialization, lead toxicity has been identified as the severest, along with cost-effective manufacturing methods and stability issues [19]. The lead (and tin) poisoning dynamics and their impacts on the human health and environment have been extensively analyzed, and they constitute significant dangers that must be addressed, including probable contamination pathways that have been investigated, with a particular emphasis on the development of rain on PSC modules [20]. Contamination from lead is still a concern and a possible roadblock to large industrial PSC adoption, which contributes to the present problem of disposing of electronic equipment containing lead and similar hazardous compounds [19]. Possibilities of lead contamination throughout the manufacturing process by laboratories, researchers, or personnel in future fabrication facilities ought to be considered. According to many studies, the average result is that lead perovskite solar cells do not provide additional difficulties that hinder mass production and implementation, especially in comparison to other photovoltaic technologies in the cradle-to-gate situation. In addition, as will be depicted in the section on recycling difficulties, significant work has been done to create effective recycling methods for lead utilized in perovskite solar cells.

7.3.5 Carbon-based nanomaterials

Various studies on the environmental and health consequences of nanomaterials have been conducted as a result of the introduction and rapid growth of nanotechnology [21]. CNMs are useful in pollution management because they can absorb heavy metals, antibiotics, and toxic chemicals. CNMs, on the other hand, will unavoidably penetrate the environment as they evolve. They are toxic to living creatures in the environment and take a long time to decompose in natural environments. Electrochemical sensors, drug delivery, imaging systems, cancer discovery and therapy, and regenerative medicine are just a few of the medical applications where CNMs have been studied [22,23]. Recent works show the aggregation of nanoparticles in various organs, along with the kidneys, lung, brain, heart, liver, and spleen; this aggregation is caused via ingestion, intravenous injection, and inhalation [24].

7.3.6 Tellurium and indium

Tellurium and indium are uncommon elements with no recognized biological role, and as a result, even at extremely low concentrations, they can be harmful. Regarding their eco-toxicological qualities, very little information exists. Indium can be absorbed through the lungs, causing harm to the lungs, kidneys, and liver [25]. Indium is not extremely poisonous, while tellurium has a hazardous impact when consumed. It is metabolized to dimethyl telluride in the body, giving breath and perspiration a garlic-like odor. Effects range from headaches, metallic taste, nausea, and respiratory arrest to death, depending on the dosage level [26].

7.3.7 TOXIC FLEXIBLE SUBSTRATE

A further essential aspect of flexible solar cells is the substrate material's environmental stability. Throughout external solar cell operation, the substrate should be immune to chemical reactions from the environment. The environmental stability of metal-based flexible substrates is typically good. For instance, stainless steel with a significant chromium concentration (>10 wt%) may withstand acid and base corrosion quite well [27]. Titanium is a reactive metal that, when subjected to air, rapidly forms a thick oxide coating. The created passive oxide protects the titanium foil against further oxidation and makes it immune to chemicals like acid. The environmental stability of ceramic materials is also acceptable. Glass is chemically inert, meaning it can withstand reactions from salt solutions, water, organic compounds, and acids. Nevertheless, it is vulnerable to base and hydrofluoric acid reactions [28]. Solar cells manufactured on ceramic substrates or metal foils can be safe and eco-friendly for routine outdoor use. Plastic substrates, on the other hand, are less environmentally stable when compared to ceramic or metallic substrates. The quick polymer degradation in the face of ambient oxygen is the principal cause of this poor environmental stability. UV irradiation may also induce highly brittle polymers. As a result, UV filters are frequently required to protect plastic substrates.

7.4 PROCESSING ROUTES AND DESIGN STRATEGIES FOR SAFE AND SUSTAINABLE MANUFACTURING

This part focuses on the investigation of processing routes and design techniques, as well as their potential for improvement, with an emphasis on lowering environmental and economic costs. This section provides some guidelines to remedy issues mentioned previously.

7.4.1 TOXIC MATERIALS REPLACEMENT

7.4.1.1 Lead-free perovskite

Even while the outcomes with lead are encouraging, major health and hazardous environmental issues still exist [29]. Though the risk of lead poisoning from perovskite solar cells is less than previously thought, it is worthwhile to investigate lead-free alternatives. Lead-free PSCs had a 6% efficiency at a period when lead-based PSCs had a 17% efficiency; additionally, Sn-based devices are less stable in comparison to Pb counterparts [30]. The replacement of lead by tin is the typical method, although elements have also been used to make halide double perovskites (elpasolites). A plethora of studies, such as life cycle approach (LCA) [31] and further extensive studies comparing lead and tin, were used to assess the environmental advantages of lead-free (or lead-reduced) devices; notably, the substitution of lead did not reduce the environmental consequences in either case. The reason is that the degradation of stability and PCE causes an environmental penalty that outweighs the benefits of lead-free devices. As a result, extensive work is required to resolve such problems.

7.4.1.2 Indium

Indium has been recommended as the most unsustainable metal for the large implementation of organic devices utilizing transparent conducting oxides. It is an abundant and also very costly metal. Nonetheless, indium-tin-oxide (ITO) substrates have been superseded by fluor-doped tin oxide (FTO), which is currently employed as a nearly ubiquitous substrate with an easy manufacturing process and outstanding performance [32]. When compared to ITO-based similar devices, most devices nowadays employ FTO substrates without any noticeable PCE or stability degradation, and the trend demonstrates that almost all contemporary devices tend to employ FTO rather than ITO. Furthermore, due to FTO's superior temperature stability and surface roughness compared to that of ITO, FTO is now diminishing the benefits that ITO had for solar cells in its early progress stages. Particularly in comparison to devices that use pristine FTO, ITO recovery via reuse and recycling has generated cells with just a 0.85% decrease in efficiency.

7.4.2 IMPROVING PROCESSING ROUTES

Despite the lower usage of power, solution-processable ways have a lesser environmental effect than evaporation-based ways. Solvent processing has efficiently replaced the evaporation of metals and oxides in several cases. For example, ZnO ETL has inhibited in perovskite inverted cells with no important degradation of PCE, resulting in a considerable improvement in stability, with a rise of 40%. Processing methods for perovskite active layers are considerably easier; however, for spiro-OMeTAD, processing is quite difficult, including toxic solvents. Therefore, it's necessary to develope replacements for spiroOMeTAD since HTL is a reasonable proposal for a basic processing method that may be accredited to industrial levels.

Low-temperature (70 °C), sequential slot-die coating scalable techniques with outstanding efficiency (10%) have been employed for all layers excluding the electrode in large-scale commercial manufacturing [33]. Besides, the ink-jet printing will help conserve material throughout processing, as evidenced by efficiencies >12% [33]. This approach, with its significant decrease in material consumption (and waste), paves the way to a significant decrease in environmental consequences; an added benefit was the relatively low temperature (50 °C) of the optimum cells. These investigations are particularly noteworthy since they highlight further toxicity issues emerging from the employment of solvents throughout charge transport layer processing, which had previously gone unnoticed. Furthermore, throughout experimental work, a specific worry for toxicity must be addressed because laboratory dangers are mostly caused by the high absorption of toxic lead when handled in solution, notably via the cutaneous and respiratory pathways [34].

7.4.3 RECYCLING

If perovskite solar modules are commercialized and large-scale perovskite photovoltaic systems are installed, significant numbers of encapsulated or bared panels will be processed and stored in a short period, contrasted to certain PV technologies now on the market. This viewpoint necessitates the development of efficient recycling pathways, which facilitate industrial recycling; a particular focal point relates to lead recuperation since lead from the methylammonium halide active layer poses a danger of inadvertent release into the environment [35].

Numerous ways to recycle and reuse materials from PSCs have been presented, with some demonstrating recycling paths that permit for the recovery of all key elements, the preservation of raw materials, and the fabrication of second generation cells without considerable PCE loss, as compared with the original device [36]. Moreover, researchers reported the effective manufacture of perovskite solar cells employing lead from recycled batteries, achieving 9.37% as conversion efficiency; this result is nearly identical to PCE produced using a commercial PbI_2 source, utilizing lead recovered from a single lead-acid vehicle battery [37]. The issue on the prospect of Pb recycling and its applicability to PSC manufacture merits a more complete LCA in the future, based on present and prospective experimental findings.

Three primary material categories have been studied for electrode usage thus far, depending on their charge storage mechanism. Metal oxides, carbonaceous materials, and polymers or polymer composites are among them [38]. Using the double layer adsorption method, carbonaceous materials can store charges. These materials are especially appealing since metal oxide-based supercapacitors suffer from low stability. In addition, these materials are also chemically inert, nontoxic, and stable at high temperatures, readily available, and frequently sustainable [38]. Various sorts of carbon, such as graphene, carbon nanotubes, activated carbons, graphite, carbon aerogel, and carbon nanofibers, have gained a lot of interest owing to their large surface area and porosity [39]. The manufacture of activated carbon from organic matter (biomass) accompanied by chemical or physical activation to be used as supercapacitor electrodes is particularly attractive as it overcomes waste management challenges, offers a less expensive recycling technique, and decreases landfill [40].

7.4.4 ENCAPSULATION

The encapsulation of devices has been shown to improve stability in several applications. The lifetime of standard MAPI/Spiro-OMeTAD cells was increased up to 500 hours. Encapsulation, as a result, is a suitable technique for providing effective immunity against oxygen and moisture deterioration, and it is already consistent with inexpensive and commercial substitutes. Encapsulation techniques should also be developed to prevent the release of harmful compounds into the environment, particularly lead, throughout the device's lifetime or at its terminal phase (recycling or landfilling). Lead is among the highest recycled elements, according to the International Lead Association (ILA), with an 80% recovery rate. This advice is based on a paucity of papers examining encapsulation methods' capacity to capture degradation subproducts toward the end of device life. Supercapacitor liquid electrolytes, on the other hand, need adequate encapsulation to avoid contamination and a separator to evade internal short-circuiting. Solid-state electrolytes that also function as separators have been claimed to be nonflammable, ecologically benign, and shape conformable, but their conductivity and mechanical qualities are inadequate. A suitable electrolyte for flexible power sources should be extremely conductive and flexible, as well as nontoxic, and have appropriate electrode contact. An additional obstacle to surmount is the difficulty in locating suitable electrolytes.

7.5 CONCLUSION

The increased demand for flexible electronics presents both benefits and challenges. The primary objective is to provide extremely flexible energy sources with high power and energy densities, outstanding rate capacity and cycle stability, safe operation, lightweight, reduced cost, and scalable manufacturing. The most difficult aspect of the examination of electrode, electrolyte, active materials, and packaging materials concerns the material choice and fabrication according to the environmental impacts. This chapter studied the environmental effect of storage and conversion flexible devices. Different materials and processing techniques for flexible devices toxicity have been presented. Besides, new strategies and manufacturing routes have been highlighted to make the manufacturing of flexible devices more safe and sustainable. Although recent advances in flexible electronics look very promising, there is still scope for further enhancements, for instance, conducting life-cycle analysis dedicated to perovskite solar cells.

REFERENCES

1. Li L., Wu Z., Yuan S., Zhang X. B., (2014) Advances and challenges for flexible energy storage and conversion devices and systems. *Energy & Environmental Science*, 7(7), 2101–2122.
2. http://www.techienews.co.uk/974063/apples-latest-patent-filing-hits-iphone-6-wrap-around-display/
3. Sharma K., Arora A., Tripathi S. K., (2019) Review of supercapacitors: Materials and devices. *Journal of Energy Storage*, 21, 801–825.
4. Tillman A.-M. (1995) Environmental assessment of photovoltaic technologies. *Energy & Environment*, 6(1), 43–61.
5. Díaz-Ramírez M. C., Ferreira V. J., García-Armingol T., López-Sabirón A. M., Ferreira G. (2020) Environmental assessment of electrochemical energy storage device manufacturing to identify drivers for attaining goals of sustainable materials 4.0. *Sustainability*, 12(1), 342.

6. Li X., Li P., Wu Z., Luo D., Yu H. Y., Lu Z. H. (2020) Review and perspective of materials for flexible solar cells. *Materials Reports: Energy*, 1, 100001.

7. Green M. A., Dunlop E. D., Hohl-Ebinger J., Yoshita M., Kopidakis N., Hao X. (2020) Solar cell efficiency tables (version 56). *Progress in Photovoltaics: Research and Applications*, 28(NREL/JA-5900–77544).

8. Lu X., Yu M., Wang G., Tong Y., Li Y. (2014) Flexible solid-state supercapacitors: Design, fabrication and applications. *Energy & Environmental Science*, 7(7), 2160–2181.

9. Duffy N. W., Baldsing W., Pandolfo A. G. (2008) The nickel–carbon asymmetric supercapacitor-Performance, energy density and electrode mass ratios. *Electrochimica Acta*, 54(2), 535–539.

10. Sivakkumar S. R., Kim W. J., Choi J. A., MacFarlane D. R., Forsyth M., Kim D. W. (2007) Electrochemical performance of polyaniline nanofibres and polyaniline/multi-walled carbon nanotube composite as an electrode material for aqueous redox supercapacitors. *Journal of Power Sources*, 171(2), 1062–1068.

11. Prasad K. R., Miura N. (2004) Electrochemically synthesized MnO_2-based mixed oxides for high performance redox supercapacitors. *Electrochemistry Communications*, 6(10), 1004–1008.

12. Jariwala D., Sangwan V. K., Lauhon L. J., Marks T. J., Hersam M. C. (2013) Carbon nanomaterials for electronics, optoelectronics, photovoltaics, and sensing. *Chemical Society Reviews*, 42(7), 2824–2860.

13. Noerochim L., Wang J. Z., Wexler D., Rahman M. M., Chen J., Liu H. K. (2012) Impact of mechanical bending on the electrochemical performance of bendable lithium batteries with paper-like free-standing V 2 O 5–polypyrrole cathodes. *Journal of Materials Chemistry*, 22(22), 11159–11165.

14. Zhou G., Li F., Cheng H. M. (2014) Progress in flexible lithium batteries and future prospects. *Energy & Environmental Science*, 7(4), 1307–1338.

15. Moulis J. M., Thévenod F. (2010) New perspectives in cadmium toxicity: An introduction. *BioMetals*, 23, 763–768.

16. Britton L. G. (1990) Combustion hazards of silane and its chlorides. *Plant/Operations Progress*, 9(1), 16–38.

17. Zweibel K., Mitchell R. (1990) CuInSe 2 and CdTe scale-up for manufacturing. In *Advances in solar energy* (pp. 485–579). Springer, Boston, MA.

18. Moskowitz P. D., Fthenakis V. M., Zweibel K. (1990) *Health, safety and environmental issues relating to cadmium usage in photovoltaic energy systems (No. SERI/TR-211-3621)*. Solar Energy Research Inst., Golden, CO (USA).

19. Urbina A. (2020) The balance between efficiency, stability and environmental impacts in perovskite solar cells: A review. *Journal of Physics: Energy*, 2(2), 022001.

20. Hailegnaw B., Kirmayer S., Edri E., Hodes G., Cahen D. (2015) Rain on methylammonium lead iodide based perovskites: Possible environmental effects of perovskite solar cells. *The Journal of Physical Chemistry Letters*, 6(9), 1543–1547.

21. Smith S. C., Rodrigues D. F. (2015) Carbon-based nanomaterials for removal of chemical and biological contaminants from water: A review of mechanisms and applications. *Carbon*, 91, 122–143.

22. Biju V. (2014) Chemical modifications and bioconjugate reactions of nanomaterials for sensing, imaging, drug delivery and therapy. *Chemical Society Reviews*, 43(3), 744–764.

23. Panwar N., Soehartono A. M., Chan K. K., Zeng S., Xu G., Qu J., Chen X. (2019) Nanocarbons for biology and medicine: Sensing, imaging, and drug delivery. *Chemical Reviews*, 119(16), 9559–9656.

24. Schrand A. M., Rahman M. F., Hussain S. M., Schlager, J. J., Smith D. A., Syed A. F. (2010) Metal-based nanoparticles and their toxicity assessment. *Wiley Interdisciplinary Reviews: Nanomedicine and Nanobiotechnology*, 2(5), 544–568.

25. Li H., Chen Z., Li J., Liu R., Zhao F., Liu R. (2020) Indium oxide nanoparticles induce lung intercellular toxicity between bronchial epithelial cells and macrophages. *Journal of Applied Toxicology*, 40(12), 1636–1646.

26. Malczewska-Toth B. (2001) Phosphorus, selenium, tellurium, and sulfur. *Patty's Toxicology*, 841–884. doi: 1 0.1002/0471435139.tox044.pub2

27. Ravindranath K., Malhotra S. N. (1995) The influence of aging on the intergranular corrosion of 22 chromium-5 nickel duplex stainless steel. *Corrosion Science*, 37(1), 121–132.

28. Kolli M., Hamidouche M., Bouaouadja N., & Fantozzi G. (2009) HF etching effect on sandblasted soda-lime glass properties. *Journal of the European Ceramic Society*, 29(13), 2697–2704.

29. Green M. A., Ho-Baillie A., Snaith H. J. (2014) The emergence of perovskite solar cells. *Nature Photonics*, 8(7), 506–514.

30. Noel N. K., Stranks S. D., Abate A., Wehrenfennig C., Guarnera S., Haghighirad A. A., Snaith H. J. (2014) Lead-free organic–inorganic tin halide perovskites for photovoltaic applications. *Energy & Environmental Science*, 7(9), 3061–3068.

31. Zhang J., Gao X., Deng Y., Li B., Yuan C. (2015) Life cycle assessment of titania perovskite solar cell technology for sustainable design and manufacturing. *ChemSusChem*, 8(22), 3882–3891.

32. Hu Z., Zhang J., Hao Z., Hao Q., Geng X., Zhao Y. (2011) Highly efficient organic photovoltaic devices using F-doped SnO_2 anodes. *Applied Physics Letters*, 98(12), 66.

33. Hwang K., Jung Y. S., Heo, Y. J., Scholes F. H., Watkins S. E., Subbiah J., Vak D. (2015) Toward large scale roll-to-roll production of fully printed perovskite solar cells. *Advanced Materials*, 27(7), 1241–1247.

34. Sikkema J., de Bont J. A., Poolman B. (1995) Mechanisms of membrane toxicity of hydrocarbons. *Microbiological Reviews*, 59(2), 201–222.

35. Kadro J. M., Hagfeldt A. (2017) The end-of-life of perovskite PV. *Joule*, 1(1), 29–46.

36. Kadro J. M., Pellet N., Giordano F., Ulianov A., Müntener O., Maier J., Hagfeldt A. (2016) Proof-of-concept for facile perovskite solar cell recycling. *Energy & Environmental Science*, 9(10), 3172–3179.

37. Chen P. Y., Qi J., Klug M. T., Dang X., Hammond P. T., Belcher A. M. (2014) Environmentally responsible fabrication of efficient perovskite solar cells from recycled car batteries. *Energy & Environmental Science*, 7(11), 3659–3665.

38. Zequine C., Ranaweera C. K., Wang Z., Dvornic P. R., Kahol P. K., Singh S., Gupta R. K. (2017) High-performance flexible supercapacitors obtained via recycled jute: bio-waste to energy storage approach. *Scientific Reports*, 7(1), 1–12.

39. Jariwala D., Sangwan V. K., Lauhon L. J., Marks T. J., Hersam M. C. (2013). Carbon nanomaterials for electronics, optoelectronics, photovoltaics, and sensing. *Chemical Society Reviews*, 42(7), 2824–2860.

40. Mousavi S. M., Hashemi S. A., Yousefi K., Gholami A., Bahrani S., Chiang W. H. (2017) Historical background and present status of the capacitors and supercapacitor for high bioenergy storage applications. *Carbon*, 175(430). doi:10.1016/B978-0-12-819723-3.00041-X

8 Metal Oxide-Based Materials for Flexible and Portable Fuel Cells: Current Status and Future Prospects

Bincy George Abraham, S. Vinod Selvaganesh, and Raghuram Chetty
Department of Chemical Engineering, Indian Institute of Technology Madras, Chennai, Tamil Nadu, India

P. Dhanasekaran
CSIR-Central Electrochemical Research Institute-Madras Unit, CSIR Madras Complex, Taramani, Chennai, Tamil Nadu, India

CONTENTS

8.1 Introduction 134
8.2 Current architecture and materials for flexible and portable fuel cells 134
8.3 Material challenges for flexible and portable fuel cells 137
8.4 Strategies to tailor metal oxides for fuel cells 138
    8.4.1 Morphological control 138
    8.4.2 Phase structure engineering 141
    8.4.3 Oxygen-vacancy control 141
    8.4.4 Doping 142
    8.4.5 Compositing with carbon/metal-based materials 142
8.5 Current status of metal oxide-based materials in flexible and portable fuel cells 143
    8.5.1 Metal oxide-based catalysts 143
        8.5.1.1 Simple metal oxides as catalysts 143
        8.5.1.2 Perovskites as catalysts 145
        8.5.1.3 Spinel oxides as catalysts 146
    8.5.2 Metal oxide-based co-catalysts 146
    8.5.3 Metal oxide-based catalyst supports 146
    8.5.4 Metal oxide-based electrolytes/membranes 147
    8.5.5 Metal oxide-based bipolar plates and substrates 149
    8.5.6 Metal oxide-based current-collector 149
    8.5.7 Metal oxide-based electrodes 150
8.6 Future avenues for metal oxide systems in empowering flexible and portable fuel cells 150
8.7 Acknowledgments 152
References 152

DOI: 10.1201/9781003186755-8

8.1 INTRODUCTION

Present-day portable and flexible electronic devices have attracted significant attention and market demand because they are integral in making human life convenient, secure, and hassle free. However, with increased built-in functionalities and improved sophistication, these portable devices have become power hungry. Consequently, demand exists for power sources that are power-dense, reliable, and affordable. Though the significant share of energy is still from oil and gas, considerable developments have taken place recently to minimize the dependence on these fossil-based energy sources and thrust toward renewable and clean fuels.

Several electrochemical energy storage and harvesting systems, such as supercapacitors, solar cells, batteries, etc., and their combination have been extensively investigated for application in portable and flexible devices, with batteries receiving the most attention. Fuel cell technology is also quite an attractive and viable source of clean energy and has received broad interest for its enhanced operational flexibility. Like batteries, fuel cells have no moving parts, which allow silent, vibration-free operations, with minimal maintenance requirements. However, batteries store their reactants within the cell, while fuel cells operate with the reactants fed continuously from external sources. Hence, fuel cells provide nearly instantaneous refueling capability and thereby offer prolonged and uninterrupted operation. A variety of fuels, including hydrogen, methanol, bio-mass derived materials, etc., can be employed for fuel cells to generate power. Fuel cells operate on a wide range of temperatures and have fewer environmental issues associated with disposal. Consequently, fuel cell technology can be recognized to be more long-lasting and flexible than batteries in terms of architecture and operation for various applications.

Solid oxide fuel cells (SOFC), polymer electrolyte membrane fuel cells (PEMFC), direct liquid fuel cells (DLFC), and microbial fuel cells (MFC) are the most explored type of fuel cells for flexible and portable applications. Modifications are needed on conventionally rigid fuel cell components, especially bipolar plates, catalyst layer, gas diffusion layer, membranes, etc., to make the fuel cells more favorable for portable and flexible devices. By altering the components with suitable materials, a simpler, light-weight, flexible, compact, and inexpensive fuel cell catering to the requirements of portable applications can be developed.

This chapter briefly describes the architecture and materials employed in several reported flexible and portable fuel cells. The need for metal oxide-based materials in fuel cell components and various methods to beneficially modify metal oxides is also discussed. This information can help design novel fuel cell stack architecture and new/modified materials for cell components that can help reduce stack volume and enhance power output for portable and flexible operations.

8.2 CURRENT ARCHITECTURE AND MATERIALS FOR FLEXIBLE AND PORTABLE FUEL CELLS

Several materials for catalysts and supports have been extensively studied to minimize costs, improve catalyst utilization, and, ultimately, improve the cell performance. The catalyst-layer fabrication techniques on flexible electrodes are also quite challenging. Electrodeposition is an attractive technique to deposit catalysts, but it is suitable only for conductive electrodes. Decal transfer is another well-established technique where inks are deposited onto decal-transfer carrier films, which are then bound to the membrane by heat and pressure. Screen printing is also promising, wherein the ink is forced through a fine fabric screen and flows through the open meshes based on requisite patterns. The need to achieve higher flexibility and quality in membrane electrode assembly (MEA) production has resulted in exploring techniques such as additive-layer manufacturing and inkjet printing. The appropriate method for catalyst-layer fabrication depends on the type of catalyst, type of flexible substrate, as well as formulation, and rheological properties of suitable ink for the corresponding technique.

Significant progress has also been made with the fuel cell architecture with respect to fuel delivery mechanisms, flow field patterns, MEA design, cell configuration, and stacking, depending on the application.

Passive delivery of reactants can help to simplify the fuel cell components. Passive systems only use natural capillary forces, diffusion, or evaporation to feed the fuel without additional power consumption. In air-breathing fuel cells, the natural diffusion of ambient air provides the required oxidant for the cathode reaction. Hence, passive systems are small, less complex, and have the potential for higher fuel utilization and lower cost; these qualities make them favorable for portable and flexible devices.

Planar and tubular architecture are the most explored configurations for portable fuel cells. The tubular configuration offers much higher active area to volume ratios compared to the planar design. Additionally, the tubular configuration is beneficial, especially in air-breathing fuel cells, since a higher cathode surface is available for breathing air. Apart from these, wearable, as well as shape-conforming configurations, are also being explored. Images of various reported portable and flexible fuel cell working models are shown in Figure 8.1. Ning et al. fabricated a flexible PEMFC stack by preparing flexible composite electrodes comprising carbon paper and carbon nanotube (CNT) membrane, which offered a high degree of flexibility without compromising conductivity and permeability (Figure 8.1a). Interestingly, the flexible stack retained nearly 90% of initial performance, even after 600 bendings [1]. Abraham and Chetty designed and fabricated a portable, cylindrical air-breathing direct methanol fuel cell (DMFC) (Figure 8.1b); its unique quick-fit architecture allowed it to easily replace MEAs while holding cell components together firmly, avoiding fuel leakage, and limiting cell resistance [2]. Hsu et al. demonstrated a portable PEMFC having a flexible electrode consisting of metal wires embedded in carbon fiber bunches (Figure 8.1c). The carbon fiber current collector suitably changes the contact characteristics between itself and the electrode [3].

Hsieh and Huang fabricated a planar array module of PEMFC with a unique pin electrode design operating in a series on a standard supporting plate. Copper slims (250 μm) were used as substrates in a microfabrication process employing deep ultraviolet (UV) lithography to obtain Cu flow field plates with serpentine flow channels. For demonstration, a portable electric fan was operated using an eight-cell stack (Figure 8.1d) at a power of nearly 160 mW cm^{-2} for 1 h [4]. Kim et al. fabricated a thin, flexible printed circuit board (PCB) as a current collector for air-breathing PEMFC. The flexible PCB consisted of a single layer of Au coated on a nonconductive polyimide film [5]. Ning and coworkers reported in their work that the operational capability of flexible air-breathing PEMFC could be enhanced up to eightfold by simply optimizing the directions of current collectors. The different performances of various current collection types were attributed to the diverse lengths of the electron transfer pathways [6].

Several flexible MFCs have already been reported. Taghavi and coworkers prepared a tubular flexible MFC in which ultra-flexible carbon sleeves were used as electrodes [10]. Bandodkar et al. designed and fabricated a skin-conformal MFC with active bio-components arranged in a stretchable island-bridge configuration and connected through a lithographically patterned electronic framework (Figure 8.1e). The screen-printed CNT-based bioelectrodes in the MFC offered a high power density of 1.2 mW cm^{-2} [7]. Zhang et al. prepared a flexible and stretchable 3D nanocomposite membrane based on PVAc-g-PVDF coating on surface-modified cotton fabric, which offered good conductivity and low glucose permeability for use in MFC [11]. Wang et al. developed a flexible composite electrode comprising of 3D-reduced graphene oxide (rGO) deposited on nickel foam as an anode for MFC. The composite electrode offered a uniform macro-porous scaffold, providing a large surface area for microbial colonization, as well as effective mass diffusion of the culture medium [12].

Chu et al. explored the synthesis of a flexible electrode comprising PEDOT modified graphene paper on a current collector, which was further decorated with Pt nanoparticles for use in PEMFC [8]. Maiyalagan and coworkers prepared a 3D interconnected graphene monolith as a freestanding electrode support for Pt catalyst, which offered enhanced catalytic activity for methanol oxidation reaction (MOR) due to its 3D interconnected seamless porous structure, high surface area, and high conductivity [9].

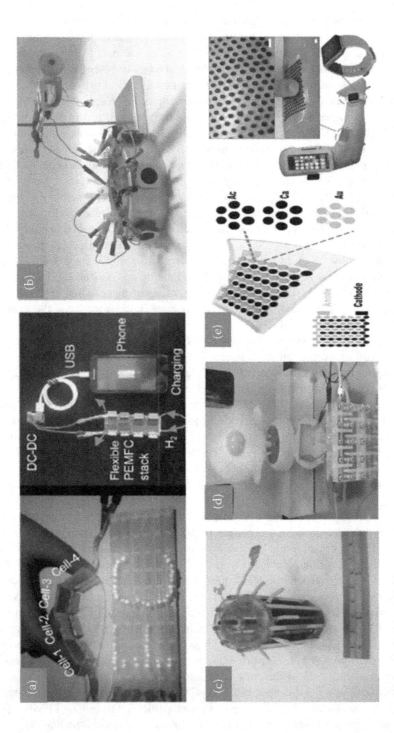

FIGURE 8.1 (a) Photographs of a series of LED lights powered using a bent four-cell PEMFC stack and a mobile phone being charged using a flexible eight-cell PEMFC stack (Reproduced with permission (through Copyright Clearance Central) from ref. [1] Copyright © 2017, American Chemical Society); (b) Photograph of six-cell air-breathing DMFC powering a small fan motor (Reproduced with permission (through Copyright Clearance Central) from ref. [2] Copyright © 2020, Elsevier); (c) Photograph of non-flexible cylindrical portable PEMFC prepared with banded MEA and carbon fiber current collector (Reproduced with permission (through Copyright Clearance Central) from ref. [3] Copyright © 2012, Elsevier); (d) Photograph demonstrating working of 8-cell planar array module of PEMFC (Reproduced with permission (through Copyright Clearance Central) from ref. [4] Copyright © 2013, Elsevier); (e) Schematic representation of various components in a stretchable electronic-skin-based MFC (Ac – Active biocomponents, Au – Gold and Ca – Carbon) along with image of the MFC adorned by a human subject for wearable applications (Reproduced with permission (through Copyright Clearance Central) from ref. [7] Copyright © 2017, The Royal Society of Chemistry).

In an effort toward miniaturization, paper-based fuel cells and on-chip fuel cells are gaining interest. The capillary action inherent in the paper alleviates the need for auxiliary systems to supply reactants. Wang et al. developed a paper-based fuel cell (PBFC) with an innovative structure to store hydrogen. During operation of the PBFC, alkaline electrolyte solution, absorbed into the paper substrate through capillary action, reacts with Al foil embedded inside the paper substrate to generate hydrogen. The remaining components of PBFC, such as electrodes and current collectors are directly prepared on the paper surface, thus forming a simple and flexible fuel cell [13]. Tomanika et al. developed on-chip fuel cells with an air-breathing, membrane-less, and monolithic design, where all the components necessary for power generation were integrated on a single substrate made of flexible cycloolefin polymer films [14].

Developing better multifunctional materials for fuel cell components in the flexible and portable fuel cell is an ongoing process. Recent breakthroughs in chemistry and materials science have allowed the synthesis and exploration of alternative nanomaterials for these functional components. These advances in nanostructuring and material science can help take full advantage of the material's properties and easily prepare composites with other materials that enhance the overall properties of the parent material.

8.3 MATERIAL CHALLENGES FOR FLEXIBLE AND PORTABLE FUEL CELLS

Flexible power sources are expected to offer safe operation due to their high energy, power densities, and lightweightness. The selection of materials for the components is quite challenging because the functional materials need to have properties such as good ionic and electronic conductivity, high electrochemical performance, efficiency, and being mechanically stable and durable to suit the requirements for portable and flexible applications. They should be of low cost and appropriate for scalable production. To achieve this need, the material selection and fabrication/synthesis technique employed are of paramount importance. With the increasing importance of flexible power sources, the various challenges involved must be extensively explored.

- Electrolytes should possess a suitable balance between mechanical stability and ionic transport. Composites are often employed in solid electrolytes to operate in conditions close to ambient conditions. Moreover, fillers are often incorporated in solid electrolytes to improve flexibility and control permeability and the crossover of reactants. However, the processing of these electrolytes can become more complex. Liquid electrolytes require a dimensionally stable matrix whose porosity and pore size are tuned for ideal contact with the electrode material and gas. The electrolytes should have good integrity and compatibility to serve safely, irrespective of the power source configuration.
- Electrodes must enable fast electrochemical reactions along with providing uniform and facile transport of reactants and products. The catalyst layer in electrodes should offer fast reaction kinetics, resistance to catalytic poisoning from reaction intermediates, thermal stability, and corrosion resistance. To overcome these issues, several strategies, such as alloying of noble metals with cheaper metals/metal-based composites, etc., are explored to minimize the cost without compromising the electrocatalysts' electrochemical performance and durability. It is also critical to adopt large-scale fabrication techniques, such as inkjet and screen printing, etc., to prepare cost-effective electrodes and catalysts.
- High surface-area support materials are essential to anchor the supported metal catalyst and enable the catalysts to withstand the fuel cell operating conditions, which include high potentials, continuous voltage transients, etc. Conventionally employed carbon supports have excellent conductivity and offer high surface area but are susceptible to carbon corrosion, thereby limiting the durability of the catalysts. Alternative materials with conductivity and surface area comparable to carbon supports continue to be an obstacle.

- Owing to the nature of flexible electronics, the high density of conventional rigid metal current collectors and the tendency for fatigue failure under constant bending conditions make them less suitable for application in flexible power sources. Several nanostructured materials, such as metal oxide composites and conducting polymers, are being explored as flexible and lightweight current collectors. It is crucial to optimize the materials for stability in the operating conditions of the fuel cell without compromising on conductivity and flexibility.
- Cell structural design and assembly is also integral component that will dictate the performance of the fuel cell. Additionally, electrodes should be modified and cell structure should be optimized to minimize reliance on auxiliary components for water and heat management during operation. Materials offering some "feel good" features, such as stretchability and optical transparency, can help to make the devices attractive in the market.

Strategic use of material science and engineering can help to overcome the above-mentioned challenges. Exploiting suitable synthesis techniques for size, morphology, and composition control can help to prepare tunable materials with beneficial structural and electrochemical properties, thereby opening new possibilities to design performance materials.

8.4 STRATEGIES TO TAILOR METAL OXIDES FOR FUEL CELLS

Among the various materials explored, metals and metal oxides are extensively employed as fuel cell components. Metal oxides offer a clear superiority over other metal compounds due to their structural tunability, consequently offering desirable electronic and chemical properties. Pure metal oxides have received interest primarily because their surfaces can carry more hydroxyl groups, which are helpful for electrochemical systems. The thermodynamic and electrochemical stability of metal oxides in acidic conditions and highly oxidative conditions make them particularly attractive for fuel cell applications. However, their poor electrical properties have limited their application, especially as electrocatalysts.

Advances in synthesis techniques have allowed for the preparation of high-quality materials with desirable features, such as size and morphology control, high surface-to-volume ratio, and improved conductivity. These enhanced physical and electrical properties help nanostructured metal oxides exhibit acceptable performance while reducing the requirement for expensive high-performance materials. Some of the strategies explored to effectively exploit the various functionalities of metal oxides are discussed briefly. These strategies can help to optimize the electronic structure, thereby enhancing the conductivity and increasing the catalytically active sites.

8.4.1 MORPHOLOGICAL CONTROL

The size, shape, and morphology of metal oxides have a significant impact on their electrochemical properties. Morphology-controlled synthesis of metal oxides is often used to prepare 1D, 2D, and 3D structures that offer a larger active area, electron mobility, and tailor catalytic activity compared to irregular-shaped metal oxide structures. The benefits of 1D metal oxide nanostructures, such as nanorods, nanowires, nanotubes, etc., are extensively explored. Selvaganesh et al. prepared TiO_2 nanowires (Figure 8.2a) to improve the performance and durability of Pt catalysts for PEMFC [15]. Dhanasekaran et al. explored pristine TiO_2 nanorods (Figure 8.2b) and Cu, N dual-doped TiO_2 nanorods as effective and durable support for Pt toward oxygen reduction reaction (ORR) [16]. Abraham et al. reported that Pd supported on titania nanotube (TNT) (Figure 8.2c), prepared by anodization of titanium substrate, offered a tenfold increase in peak current for formic acid oxidation compared to Pd on bare titanium [17].

Few reports have also explored 2D nanosheets for electrochemical systems. Wu et al. synthesized flower-like ultrathin Co_3O_4 nanosheets (Figure 8.2d), which exhibited better stability and higher ORR

FIGURE 8.2 Morphological control.FE-SEM images of (a) titania nanowires (Reproduced with permission from ref. [15] Copyright © 2017, Walter de Gruyter GmbH. The article was printed under a CC-BY license), (b) titania nanorods (Reproduced with permission (through Copyright Clearance Central) from ref. [16] Copyright © 2017, The Royal Society of Chemistry), (c) titania nanotubes (Reproduced with permission (through Copyright Clearance Central) from ref. [17] Copyright © 2016, Wiley-VCH Verlag GmbH & Co. KGaA, Weinheim), (d) flower-like Co_3O_4 nanosheets (Reproduced with permission (through Copyright Clearance Central) from ref. [18] Copyright © 2017, The Royal Society of Chemistry), (e) nanocube Cu_2O and (f) nanoporous Cu_2O with higher magnification image as inset (Reproduced with permission (through Copyright Clearance Central) from ref. [19] Copyright © 2013, American Chemical Society); SEM images of 3-D TiO_2 nanostructures viz. (g) cross-linked nanorods, (h) nanowire microspheres, and (i) mesoporous nanospindles (Reproduced with permission (through Copyright Clearance Central) from ref. [20] Copyright © 2016, Wiley-VCH Verlag GmbH & Co. KGaA, Weinheim).

activity in alkaline solution than Co_3O_4 nanoparticles [18]. The 3D nanomaterials formed by spatial organizations of a group of 0D, 1D, 2D, or 3D nanostructures are also quite beneficial. Li et al. investigated the relation of morphology to the ORR activity of Cu_2O nanocrystals by preparing a series of Cu_2O nanoparticles compared with nanocubes (Figure 8.2e), nanoporous morphology (Figure 8.2f), etc., via a solution chemistry route. Cu_2O nanocubes, which consisted predominantly of {100} crystal planes, exhibited the highest activity and four-electron selectivity for ORR, compared to others [19]. Sun et al. studied controlled solution growth of TiO_2 nanostructures, such as cross-linked nanorods (Figure 8.2g), 3D dendritic nanowire spheres (Figure 8.2h), mesoporous nanospindles, etc., (Figure 8.2i) by rational design of the solutions and synthesis parameters [20].

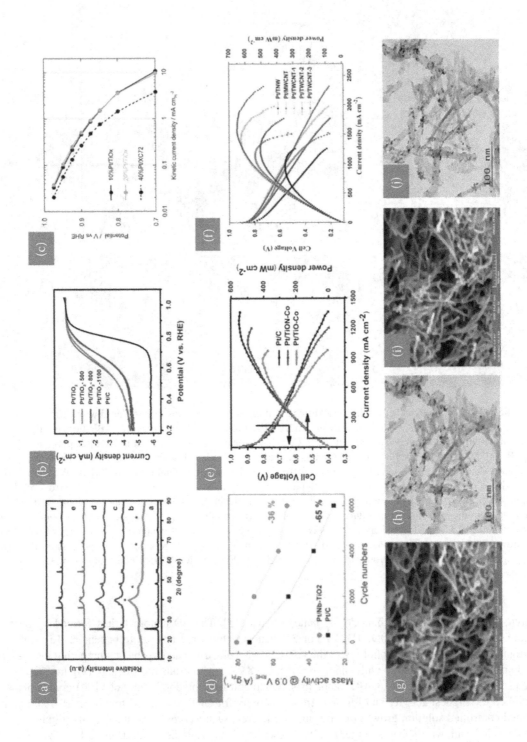

8.4.2 PHASE STRUCTURE ENGINEERING

The crystalline structure of catalysts significantly dictates the properties offered by metal oxides. Dhanasekaran et al. explored the impact of phase purity and the crystalline structure of TiO_2 prepared using the annealing method towards ORR and full cell performance of PEMFC. X-ray diffraction (XRD) studies (Figure 8.3a) clearly showed the various crystalline phases formed during the annealing of TiO_2 at various temperatures. Electrochemical studies toward ORR indicated that, although ORR activity of the prepared Pt/TiO_2 catalysts was lower compared to Pt/C, Pt deposited on rutile TiO_2 formed by heat-treating at 800 and 1100 °C showed better ORR activity (Figure 8.3b) compared to anatase TiO_2 (P-25) supported Pt catalyst [21]. According to Cao et al., the ORR activity was strongly dictated by the crystalline structure of manganese oxides, with δ-MnO_2 significantly outperforming λ-MnO_2 [22]. Dhanasekaran et al. carried out stability tests for rutile TiO_2 supported Pt catalyst, which retained higher electrochemical surface area than anatase TiO_2 supported Pt catalyst and Pt/C after 20,000 potential cycles (from 0.6 to 1.2 V (vs. RHE)). The authors observed lower performance due to larger particle size and reduced reactive sites in the TiO_x surface formed subsequest to the high-temperature (1100 °C) treatment [21]. This result clearly shows that temperatures used to prepare crystalline phases could also impact the performance of the catalysts.

8.4.3 OXYGEN-VACANCY CONTROL

Controlling oxygen vacancies can help design functional properties of oxides by inducing and compensating for electronic charges through ionic defects. Disordered structure forms in the metal oxide lattice when oxygen atoms are removed from metal oxides. The two electrons formerly bound to the oxygen ion remain in the oxide and populate defect states, thereby manipulating their donor densities, ultimately influencing metal oxides' electronic and chemical properties. Controlled chemical reduction (using $NaBH_4$, hydrazine, etc.), thermal treatment in reducing atmospheres, electrochemical reduction, irradiation techniques, etc., are various methods explored to prepare oxygen-deficient metal oxides. Oxygen vacancies can influence the electronic properties at the active site. They also help in forming OH_{ads} species and incorporating strong metal-support interactions that manipulate the bonding interactions. Oxygen vacancies can also create lots of defects on the surface, producing highly active metal ion redox pairs and consequently increasing the conductivity several manifolds.

FIGURE 8.3 **Phase structure engineering**: (a) XRD pattern and (b) LSV measurements for Pt/TiO_2 and Pt/C electrocatalysts as a function of TiO_2 support calcination temperature: a. TiO_2 (P25), b. Pt/C, c. Pt/TiO_2, d. Pt/TiO_2–500, e. Pt/TiO_2–800, f. Pt/TiO_2–1100 (Reproduced with permission (through Copyright Clearance Central) from ref. [21] Copyright © 2016, Springer Nature); **Oxygen vacancy control**: (c) Tafel plots for ORR on 10 wt % Pt/TiO_x, 20 wt % Pt/TiO_x and 40 wt % Pt/XC-72 in O_2-saturated 0.1 M $HClO_4$ at 25 °C (Reproduced with permission (through Copyright Clearance Central) from ref. [24] Copyright © 2011, IOP Publishing); **Doping**: (d) Retention rate of mass activity of Pt nanoparticles @ 0.9 V (vs. RHE) during stability test of 6000 cycles for Pt/Nb-TiO_2 nanofiber catalyst and Pt/C catalyst (Reproduced with permission (through Copyright Clearance Central) from ref. [26] Copyright © 2017, The Author(s). The article was printed under a CC-BY license); (e) Steady-state polarization curves comparing Pt/C catalyst, Pt on Co-doped TiO_2 and Pt on Co, N dual doped-TiO_2 (Reproduced with permission (through Copyright Clearance Central) from ref. [25] Copyright © 2016, The Royal Society of Chemistry); **Compositing with carbon/metal-based materials**: (f) Steady-state polarization curves comparing Pt on various supports viz. MWCNT, titania nanowires (TNW) and composites TWCNT-1, TWCNT-2 and TWCNT-3 prepared from MWCNT and TNW of 1:1, 2:1 and 3:1 weight ratios of MWCNT to TiO_2, respectively. (g) SEM image of TWCNT-2 and (h) TEM image of Pt/TWCNT-2 (Reproduced with permission (through Copyright Clearance Central) from ref. [15] Copyright © 2017, Walter de Gruyter GmbH. The article was printed under a CC-BY license); (i) SEM image of spherical TiN nanoparticles incorporated in TiO_2 nanowires (TiO_2:TiON composite nanowires) and (j) TEM image of Pt/TiO_2:TiON composite nanowires (Reproduced with permission (through Copyright Clearance Central) from ref. [27] Copyright © 2017, The Royal Society of Chemistry).

The presence of oxygen vacancies facilitates oxygen ion diffusion, which is beneficial in SOFCs. Increased oxygen vacancies in SOFC cathodes can allow molecular oxygen reduction to occur, not just in the vicinity of the electrode-electrolyte interface but also throughout the electrode. This eventually increases the effective active sites and delocalizes the triple-phase boundary (TPB) for SOFC. Oxygen vacancy in metal oxides is also explored for synthesizing ORR electrocatalysts. Li et al. prepared rutile-type β-MnO_2 with oxygen vacancies introduced through heat treatment. MnO_2 with oxygen vacancies exhibited enhanced ORR performance compared to its pristine counterpart [23]. Ioroi et al. explored the formation of the magneli phase of TiO_x via the UV pulse laser method. The specific ORR activities (Figure 8.3c) were reported to be more than twofold higher for TiO_x supported Pt matrix compared to Pt/C [24].

8.4.4 DOPING

Suitable metals or heteroatoms can be effectively used to dope metal oxides and improve their conductivity. Doping also allows support to donate more electron density to the catalyst and enhance catalyst durability and activity toward electrochemical reactions. During the doping process, charge compensation occurs in anion or cation sites, resulting in increased electrical conductivity. Besides, defect states created via doping also help in the adsorption and desorption of reactant species on catalyst surfaces [25].

Kim et al. prepared Nb-TiO_2 nanofibers using the electrospinning technique and investigated the effect of calcination temperature and Nb doping level to the conductivity of TiO_2 nanofibres, which are utilized as support for Pt catalysts toward ORR. Doping TiO_2 with Nb up to 25% was shown to improve the electronic conductivity significantly. However, further doping of Nb diminished the conductivity due to the significant formation of Nb oxides. The authors reported that the prepared Pt on Nb-TiO_2 nanofiber catalyst exhibited higher durability than the Pt/C catalyst since the former lost only 36% of initial ORR activity compared to 65% in the case of the latter during the stability test of 6000 cycles (Figure 8.3d) [26].

Dhanasekaran et al. prepared Pt on cobalt and nitrogen dual-doped TiO_2 and compared the catalytic performance with Pt on Co-doped TiO_2 in PEMFC. The Co, N-doped TiO_2 exhibited a maximum power density of nearly 500 mW cm^{-2}, nearly 25% higher than that exhibited by Co-doped TiO_2 supported Pt cathode electrocatalyst (Figure 8.3e) [25]. Hsieh et al. prepared Pt on dual-doped TiO_2 with nitrogen as the anion and tungsten or niobium as the cation. Dual doping of Nb or W, along with N on TiO_2 support, offered excellent electronic conductivity, which was several orders of magnitude higher than single-doped (Nb/W/N) TiO_2. Additionally, Pt on dual-doped TiO_2 support showed excellent ORR activity and higher durability compared to Pt/C and Pt on single-doped TiO_2. The authors attributed the enhanced performance to the synergistic effect of dual doping and SMSI between Pt and support [28].

8.4.5 COMPOSITING WITH CARBON/METAL-BASED MATERIALS

Hybrid nanocomposites consisting of metal oxides with various carbon-based materials, such as CNT, graphene, etc., and metal-based materials, such as metal nitrides, carbides, etc., are extremely useful as cocatalysts and supports. Properties of metal oxides, such as increased acid stability, improved resistance to corrosion, and enhanced catalytic activity, operate in synergy with carbon/metal-based materials, which offer better conductivity and a large surface area. Xia et al. designed composite support composed of TiO_2 nanosheets grafted on CNT backbone for Pt electrocatalyst. The as-prepared Pt/CNT@TiO_2 electrocatalysts exhibited higher electrocatalytic activity with greatly improved stability as compared with conventional Pt/C or Pt/CNT electrocatalysts. The enhancement is attributed to the synergetic effect of TiO_2 and CNT [29].

Selvaganesh et al. investigated the optimum composition of TiO_2 nanowires (TNW) and graphitic MWCNTs as hybrid support (TWCNT) for Pt as cathode catalyst. The authors compared the PEMFC performance of Pt/MWCNT, Pt/TNW, and Pt/TWCNT of various MWCNT to TiO_2 ratios

(Figure 8.3f). The results showed that TWCNT-2, which has MWCNT to TiO_2 ratio of 2 (Figure 8.3g), effectively supported Pt catalysts. The Pt/TWCNT-2 (Figure 8.3h) showed the highest performance of 810 mA cm^{-2} at 0.6 V (vs. DHE) compared to Pt/MWCNT (600 mA cm^{-2}) and Pt/TNW (350 mA cm^{-2}) [15]. Metal nitrides and carbides are often explored in combination with metal oxides as electrocatalyst support. The difference in valence electrons for nitrogen and carbon induces different beneficial geometric and electronic properties for nitrides and carbides. To ameliorate the electrical and physical properties of TiO_2, Dhanasekaran et al. prepared a combination of TiO_2 and TiN to form highly stable support of TiO_2:TiN composite nanowires (Figure 8.3i). Pt/TiO_2:TiN catalysts (Figure 8.3j) showed improved ORR activity compared to Pt/TiO_2 nanowires and Pt/TiN catalysts, as well as better stability compared to Pt/C [27].

8.5 CURRENT STATUS OF METAL OXIDE-BASED MATERIALS IN FLEXIBLE AND PORTABLE FUEL CELLS

Metal oxides are employed either as the main constituent or as additives in fuel cell components, such as catalyst, co-catalyst, catalyst support, bipolar plate, current collector plates, etc. Thus, metal oxide systems have diverse applications in portable and flexible fuel cell systems.

8.5.1 METAL OXIDE-BASED CATALYSTS

There is a lot of scope for improvement in the activity and durability of catalysts, especially PEMFC and DLFC. The research focus for catalysts has shifted from expensive Pt and its alloys toward transition metal compounds, such as metal oxides, chalcogenides, carbides, phosphates, and nitrides. Among the above-mentioned compounds, metal oxides are particularly attractive for acidic and alkaline electrolytes owing to their abundance and tunability of electronic and morphological characteristics toward selectivity. With most of the electrochemical reactions being catalyst structure-sensitive, surface engineering of metal oxide could tune the facets exposed and thereby enhance fuel cell performance. Metal oxide systems play a vital role in the oxidation of adsorbed reaction intermediates, especially CO_{ads}, which act as a catalytic poison of noble metal catalysts. Metal oxides can interact with noble metal catalysts through hypo-hyper-d-electronic interaction and alter the interfacial charge-transfer properties, leading to improved reaction kinetics. Metal oxides generally contain complex compositions and various surface element valances. It is, therefore, crucial to find the structure-activity relationship to direct their further design. The various forms in which metal oxides are employed as catalysts are briefly described.

8.5.1.1 Simple metal oxides as catalysts

Metal oxide-based electrocatalysts are being extensively explored as cost-effective alternatives to costly noble metal-based materials. Iron oxides are quite beneficial as ORR catalysts. They are often supported on conductive supports to restrict uncontrollable growth and agglomeration of iron oxides that often occur during the synthesis of iron oxides. Zhao et al. prepared nano-Fe_3O_4/graphene (Figure 8.4a) and FeO(OH) nanoflake/graphene composites. They reported that the nano-Fe_3O_4/graphene composite catalyst displayed more efficient catalytic activity for ORR than FeO(OH) nanoflake/graphene composite catalyst. The enhancement was attributed to the synergism between Fe_3O_4 and graphene [30]. Similarly, copper oxides supported on conductive supports are extensively investigated as catalyst systems. Zhou et al. synthesized CuO/N-rGO nanocomposite, which exhibited high current density and more positive onset potential with transfer of four electrons during the ORR reaction. The authors attributed the enhancement to the synergic effect in the Cu–N interaction [31]. Li et al. reported that Cu_2O nanocrystals exhibited enhanced ORR durability and better methanol tolerance than the commercial Pt/C materials [19].

Molybdenum oxides and manganese oxides are attractive as catalysts due to their versatile crystallographic structures, short diffusion path lengths, and multiple valence states of Mo and Mn.

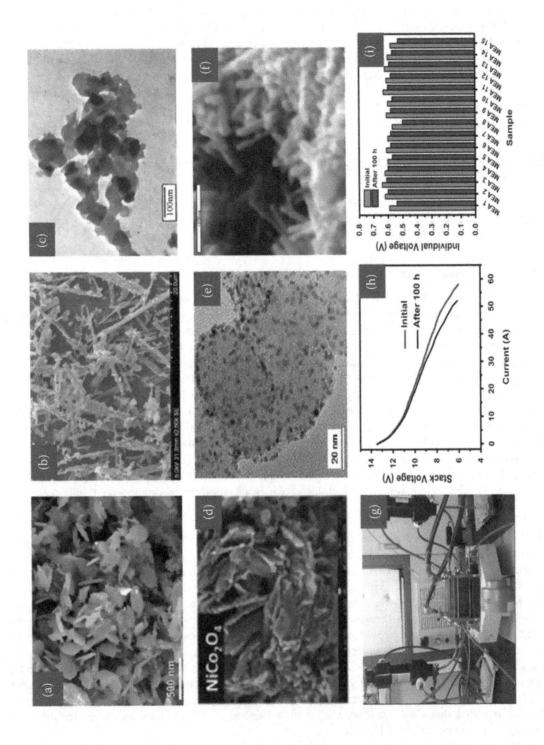

MoO_2 composited with polypyrrole (Ppy) was explored by Karthick et al. as a cathode catalyst in MFCs. The MoO_2-polypyrrole composite (Figure 8.4b) coated carbon cloth exhibited comparable performance to

Pt catalyst coated carbon cloth [32]. Dessie et al. prepared an amorphous Mn_xO_y-NiO_x hybrid, which was reported to exhibit superior activity toward ORR due to the synergistic effect of manganese and nickel oxides [33].

8.5.1.2 Perovskites as catalysts

Perovskites are oxide-based materials that have an ABO_3 structure, in which the A-site is usually made up of rare earth metals and the B-site with transition metals. Commonly, B, the transition metal cation, has six-fold coordination with oxygen anions to give an octahedral complex, while the A cation has twelve-fold coordination with respect to the oxygen anions that are located in between octahedral complexes of the structure. Perovskite oxides have attracted considerable attention in various areas due to their fascinating chemical, physical, and catalytic properties, as well as their low cost and environmental friendliness [34].

Perovskite materials such as strontium-doped lanthanum manganite oxide $(LaSr)(Mn)O_3$ or LSM, etc., are commonly used as SOFC cathode. The Sr^{2+} and La^{3+} on the A-site of the perovskite structure allow manganese to adopt both Mn^{3+} and Mn^{4+} states within the perovskite structure. The presence of multivalent manganese in the perovskite helps to promote electron transfer through the material via a hopping mechanism, ultimately resulting in high electronic conductivity. However, below 800 °C, the ionic and electronic conductivity of LSM is low. Hence, alternative perovskites, such as $(La,Sr)(Fe,Co)O_3$ and $(Ba,Sr)(Fe,Co)O_3$, are extensively explored for improved conductivity at lower temperatures. Similarly, $(La,Sr)(Ga,Mn)O_3$, $(La,Sr)(Cr,Mn)O_3$, etc., have also shown promise as potential anode catalysts in SOFCs due to their high resistance to coking and sulfur tolerance, as well as high ionic conductivity. Increased ionic conductivity of these materials can increase the TPB, where the oxide ions, the fuel, and the active catalytic sites come into contact.

Suntivich et al. extensively explored several transition metal oxide perovskites with various A-site and B-site substitutions for ORR activity [34]. ORR activity for perovskite oxides could be further improved by creating an A-site cation deficiency, considering that the additional oxygen vacancies produced from the A-site cation deficiency can facilitate the transport of oxygen ions and thus improve the ORR activity [35]. Similarly, perovskites have been explored as superior oxidation catalysts, especially for MOR. The oxygen vacancies in perovskites facilitate oxygen ion transport, which is quite efficient in removing oxidation intermediates. Yu et al. synthesized Sr substituted lanthanum cobalt/copper oxide (Figure 8.4c) and reported increased MOR activity, as evidenced by the onset values for

FIGURE 8.4 **Simple metal oxide catalysts**: FE-SEM image of (a) Fe_3O_4/graphene nanocomposite (Reproduced with permission (through Copyright Clearance Central) from ref. [30] Copyright © 2015, American Chemical Society) and (b) MoO_2-polypyrrole composite (Reproduced with permission (through Copyright Clearance Central) from ref. [32] Copyright © 2020, Elsevier); **Perovskite catalysts**: (c) TEM image of Sr substituted lanthanum copper oxide perovskite nanoparticles calcined at 600 °C in pure O_2 atmosphere (Reproduced with permission (through Copyright Clearance Central) from ref. [36] Copyright © 2004, Elsevier); **Spinel oxide catalysts**: (d) FE-SEM image of $NiCo_2O_4$ spinel oxides (Reproduced with permission (through Copyright Clearance Central) from ref. [38] Copyright © 2015, Wiley-VCH Verlag GmbH & Co. KGaA, Weinheim); **Metal oxide supports**: (e) TEM image of Pt supported on MoO_3/C (Reproduced with permission (through Copyright Clearance Central) from ref. [39] Copyright © 2011, Elsevier); (f) FE-SEM image of WO_3 nanorods (Reproduced with permission (through Copyright Clearance Central) from ref. [40] Copyright © 2007, Elsevier); (g) Photograph of 15 cell stack assembly employing Pt catalyst supported on carbon semi-coated on titania nanorods, (h) stack polarization before and after hold at 35 A for 100 h and (i) comparison of voltage from different individual cells of the 15-cell stack before and after hold at 35 A for 100 h (Reproduced with permission (through Copyright Clearance Central) from ref. [41] Copyright © 2018, Elsevier).

oxidation [36]. $SrRuO_3$-Pt and $LaRuO_3$-Pt hybrids, fabricated by directly incorporating Pt during the solution combustion synthesis process, were found to exhibit high MOR activity [37].

8.5.1.3   Spinel oxides as catalysts

Spinel oxides are another interesting class of metal oxide structure being exploited for ORR and MOR in fuel cell applications. Spinel oxides have a general formula of AB_2O_4 and usually possess a cubic close-packed structure with eight tetrahedral and four octahedral sites. Spinel structures such as Fe_3O_4, Co_3O_4, as well as Zn, Cu, and Ni-ferrites ($ZnFe_2O_4$, $CuFe_2O_4$, and $NiFe_2O_4$) are being extensively explored as catalysts for ORR [38].

Li et al. employed solution-based synthesis to structurally control the formation of a wide compositional range of $Co_{3-x}Mn_xO_4$ spinel oxides. The systematic structure–performance study of the spinel catalysts showed that a cubic phase and a high Mn concentration in CoMnO spinels favored intrinsic ORR activity [42]. Toh et al. investigated various nanocrystal spinel oxides, namely $ZnCo_2O_4$, $NiCo_2O_4$, $NiMn_2O_4$, and $ZnMn_2O_4$ as ORR catalysts. Among the spinel oxides studied, $NiCo_2O_4$ (Figure 8.4d) exhibited excellent electrocatalytic activity toward the ORR [38]. In addition, spinel oxides were also explored as support for Pt electrocatalysts. Mohanraju et al. synthesized Mn_3O_4 spinels through the co-precipitation method and employed them as composite support along with carbon for Pt electrocatalyst. The prepared catalysts exhibited enhanced ORR mass and specific activity compared to Pt/C [43].

8.5.2   METAL OXIDE-BASED CO-CATALYSTS

Catalysts are prone to dissolution in the fuel cell operating conditions. The addition of stable metal oxides with Pt has also been investigated for enhancing the stability of catalysts. Metal oxides, such as TiO_x, WO_x, MoO_x, etc., are very stable in acidic media, while MnO_x and perovskite oxides are stable in alkaline media [44]. TiO_2 is extensively explored as a cocatalyst because it can anchor metal nanoparticles, disperse catalysts, and alter the electronic properties of metal nanocatalysts [45]. TiO_2 nanorods, nanofibers, nanosheets, and aligned TNT have been explored to enhance activity and stability. TNT is quite attractive as alcohol oxidation catalysts because they provide several catalyst-promoting effects, such as the bifunctional mechanism, electronic effects, etc., while offering linear pathways for facile electron transfer compared to TiO_2 nanoparticles [45].

Ruthenium oxides, having various oxidation states and hydrous states, promote electrochemical activity through the bifunctional mechanism. RuO_2 is chemically more stable under corrosive or oxidizing conditions than its pristine metal counterpart. The negatively charged RuO_2 acts like a solid acid, attracting dissolved cationic Pt-ion, thereby performing as a protective layer limiting the diffusion of Pt-ion into the electrolyte. Inhibition of carbon corrosion could also be due to the strong metal-support-interaction (SMSI). Accordingly, RuO_2 can help to improve the durability of Pt in Pt-RuO_2/C under stressful potential transients [46]. Tin oxides are also explored as cocatalysts since they offer high tolerance to CO poisoning based on bifunctional or electronic mechanisms. Pt/SnO_2 has been found to exhibit better selectivity toward ORR, MOR, ethanol oxidation reaction (EOR), etc., because of its high electronic conductivity and stability in low-temperature fuel cell working conditions [47].

8.5.3   METAL OXIDE-BASED CATALYST SUPPORTS

Various types of metal oxide systems, such as AO_2, ABO_3, etc., are well investigated as alternative stable supports for the corrosion-prone carbon-based support because they bring about a stabilizing effect through a stronger interaction with the supported catalyst, referred to as strong metal-support interaction (SMSI), along with improved corrosion resistance. Justin et al. investigated MoO_3 incorporated Vulcan carbon as support for Pt catalysts toward MOR. Enhanced activity and CO stability observed on Pt-

MoO$_3$/C (Figure 8.4e) were attributed to the formation of hydrogen molybdenum bronze (H$_x$MoO$_3$), which results from hydrogen intercalation/de-intercalation within MoO$_3$. H$_x$MoO$_3$ provides a proton spillover effect on Pt sites, resulting in efficient desorption of the adsorbed CO from Pt surface [39]. Tungsten oxide supports are quite effective for alcohol oxidation since the oxophilic tungsten oxides can absorb water to form adsorbed OH species, enhancing the oxidation of adsorbed intermediates at lower potentials. Rajeswari et al. reported that WO$_3$ in nanorod morphology (Figure 8.4f) offered better stability in sulphuric acid media. Pt-loaded WO$_3$ nanorods showed enhanced performance toward MOR when compared to Pt deposited on bulk WO$_3$ [40].

Huang et al. prepared Pt deposited on porous TiO$_2$ and observed that TiO$_2$-supported Pt could retain its particle size during corrosion testing by holding the electrodes at 1.2 V (vs. NHE). In contrast, corrosion of carbon support occurred in Pt/C during the stability test, resulting in the migration of Pt particles and a significant increase in particle size. This indicated that TiO$_2$ is resistant to corrosion and can strongly anchor the catalyst particles by interacting with Pt, consequently restricting its migration and agglomeration [48]. Dubau et al. compared stability studies for niobium-doped tin dioxide (NTO) and antimony-doped tin dioxide (ATO) supports for Pt cathode catalyst in air-breathing PEMFC systems. Though Pt/ATO cathodes showed better cell performance than Pt/NTO due to a lower electronic conductivity of the NTO support, niobium was found to be more stable to dissolution than antimony in the harsh environment of PEMFC [49]. Dhanasekaran et al. prepared carbon semi-coated on titania nanorods (CCT) as support for Pt electrocatalyst and investigated its durability in a 15-cell PEMFC stack assembly (Figure 8.4g). Pt/CCT catalyst retained 85% of fuel cell performance, even after 100 h (Figure 8.4h,i), compared to 74% for Pt/C. The minimal loss in performance of Pt/CCT after 100 h of durability test indicated improved corrosion resistance of the catalyst, which could be attributed to the synergetic interaction of CCT with Pt [41].

8.5.4 METAL OXIDE-BASED ELECTROLYTES/MEMBRANES

Solid electrolytes used in SOFCs consist of oxygen ion-conducting ceramics such as yttria-stabilized zirconia (YSZ), etc., while PEMFCs, DLFCs, etc., employ proton conducting membranes. Flexible ceramic components are being explored to provide SOFCs with thermomechanical shock tolerance and relieve any stress during stacking of SOFCs, thereby achieving long-lasting operation [50]. The proton conducting membranes, however, are inherently flexible. Among the proton conducting materials explored to date, the most popular is the perfluorosulfonic acid (PFSA), commercially known as Nafion ionomer. Though it is widely used due to its thermal and mechanical stability, unique micro-morphological structure, and ionic conductivity, the Nafion membrane has limitations, such as membrane synthesis difficulties, the permeability of reactant fuels, and high cost. Several alternate membranes, such as polyether sulphones, polybenzimidazole, sulphonated polyimide, etc., have been studied as an alternative to PFSA based membranes. Recently, Yoon and coworkers prepared polyphenylenes/fluorosulfonyl imide-based blend polymer electrolytes, which were thermally and chemically stable with conductivity close to that of Nafion [51].

Various strategies are being explored to improve durability, proton conductivity, flexibility, and lower reactant permeability of membranes. Adjemian and co-workers studied various metal oxides viz. SiO$_2$, TiO$_2$, Al$_2$O$_3$, and ZrO$_2$ to prepare metal oxide-recast Nafion composite membranes for PEMFC. The authors concluded that composite membranes that incorporated TiO$_2$ or SiO$_2$ offered lower ionic resistivities at low RH operation and high operational temperature than a simple Nafion-based fuel cell [52]. Similarly, Ketpang et al. synthesized mesoporous TiO$_2$, CeO$_2$, and ZrO$_{1.95}$ nanotubes as fillers to prepare Nafion composite membranes for operating PEMFCs under both dry and fully humid conditions. Among the composite membranes explored, the Nafion-TNT composite membrane (Figure 8.5a) demonstrated enhanced performance compared to commercial Nafion 212 when operated at 80 °C under 100% RH, as well as 18% RH. The enhancement was ascribed to the fact that the mesoporous TNT helped in retaining water and lowering the ohmic resistance [53].

Metal oxide-based fillers in polymeric membranes were found to be attractive in DMFC systems since they can reduce the methanol crossing over the membrane under osmotic drag, thereby improving the

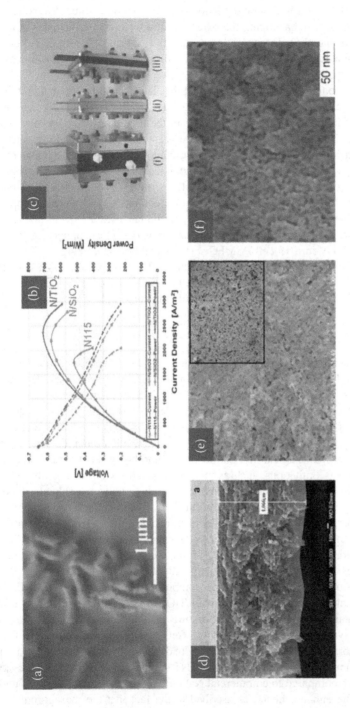

FIGURE 8.5 **Metal oxide-based membranes:** (a) Cross-sectional FE-SEM image of Nafion–TNT composite membrane (Reproduced with permission (through Copyright Clearance Central) from ref. [53] Copyright © 2014, American Chemical Society); (b) Polarization curves comparing Nafion 115-based MEA to Nafion/TiO$_2$ as well as Nafion/SiO$_2$ composite membrane-based MEA employed in a DMFC at 95 °C fed with 1 M methanol (Reproduced with permission (through Copyright Clearance Central) from ref. [54] Copyright © 2016, Elsevier); **Metal oxide-based bipolar plates:** (c) Photograph showing comparison of single-cell PEMFC with bipolar plates made of (i) graphite, (ii) pure titanium, and (iii) titanium sintered with IrO$_2$ (Reproduced with permission (through Copyright Clearance Central) from ref. [56] Copyright © 2013, The American Society of Mechanical Engineers); (d) FE-SEM image of V$_2$O$_3$ films grown on SS bipolar plates (Reproduced with permission (through Copyright Clearance Central) from ref. [57] Copyright © 2011, Elsevier); **Metal oxide-based GDL:** (e) HR-SEM image of TiO$_2$ coated GDL with inset showing pristine GDL (Reproduced with permission (through Copyright Clearance Central) from ref. [62] Copyright © 2009, Elsevier); (f) SEM image of highly porous IrO$_2$ coated on macro-porous Ti sheet (Reproduced with permission (through Copyright Clearance Central) from ref. [63] Copyright © 2013, Elsevier).

cell performance. Ercelik et al. investigated SiO_2 and TiO_2 incorporated polymeric membranes for DMFC systems. The authors reported that the composite membranes performed better than their pristine counterparts, especially under high-temperature operation. At 95 °C, Nafion/TiO_2 composite membrane-based MEA exhibited higher power density (710 W/m^2) in comparison to Nafion/SiO_2 (660 W/m^2) and Nafion 115 membranes, (520 W/m^2) as seen in Figure 8.5b [54].

8.5.5 METAL OXIDE-BASED BIPOLAR PLATES AND SUBSTRATES

Bipolar plate stability and thickness play a vital role in the size and architecture of the PEMFC stack. Therefore, significant importance is directed toward research on materials, designs, and fabrication techniques for bipolar plates. Metal-based systems, such as metal nanostructures, metal mesh, metal foam, etc., and other materials, such as conductive polymers, graphene sheets, etc., are explored as substrates for flexible applications. Though metal-based plates offer better mechanical strength, satisfactory electrical conductivity, and cost-effective production, they are plagued with concerns such as oxide formation and ion dissolution due to an aggressive fuel cell environment and stack operating conditions. One strategy is to incorporate additional materials in the metal to resist corrosion [55]. Ti is often alloyed with Nb and Ta to prepare viable bipolar plate materials because the resistivity of their surface oxides is lower than that of pure Ti. Chromium is often added to stainless steel (SS) bipolar plates because Cr can form a thin layer of Cr_2O_3, which prevents any further corrosion.

Preventive metal oxide coatings over the bipolar plates are considered another major strategy. Metal oxide coatings impart thermal and electrochemical stability to the bipolar plates, thereby improving their life. Temperature differentials that bipolar plates may experience should be factored while choosing to coat; the coating layer may not expand or contract at the same rate as the metallic base plate. Thermal expansion differences and maintenance of the coating devoid of microcracks or micropores can be controlled by introducing intermediate coating between layers. Indium tin oxide and lead oxide are often employed as coatings on SS bipolar plates through electron beam evaporation, vapor deposition, or sputtering techniques, etc. Wang et al. sintered IrO_2 on Ti substrates (Figure 8.5c) to improve corrosion resistance; these were explored as bipolar plates in portable applications due to their lightweight and low sintering cost [56]. Vanadium oxide (V_2O_3) thin films consisting of large vermicular grains of V_2O_3 (Figure 8.5d) were grown on 316L SS bipolar plates to exploit the negative temperature coefficient offered by V_2O_3, which can help minimize parasitic power loss and eliminate any necessity for external equipment to supply heat in freezing conditions for the cold start of fuel cell vehicles [57]. Numerical models and experimental studies were carried out by Peng et al. to explore the feasibility of fabricating thin metallic sheet bipolar plates, which can help to suitably reduce stack volume in portable fuel cell systems [58].

Additionally, extensive research is also being focused on developing mechanically resilient and shape-conformable electroactive materials. Ultrathin plastic substrates, such as polyimide, polyethylene terephthalate, polyurethane, polydimethylsiloxane (PDMS), etc., are being explored as electrodes for stretchable electronic skin applications due to their ink compatibility and bendability, as well as their temperature stability [59].

8.5.6 METAL OXIDE-BASED CURRENT-COLLECTOR

Current collectors are also exposed to the thermal and electrochemically corrosive environment of fuel cells. Usually, copper plates coated with a thin layer of gold are employed as current collector plates owing to the electrochemical stability and higher conductivity of gold. However, during continuous assembly and disassembly, the coating can be deformed/peeled off, leading to electrochemical corrosion of exposed copper plates and ultimately compromising the stability and lifetime. Given this result, metal oxide coatings are being explored as an alternative and efficient strategy to stabilize the existing current collector plates and improve their electrochemical and mechanical stability.

Thin and flexible printed circuit boards made of fiberglass and coated with Cu, Au, or Ag are used as a current collector and bipolar plates. Itagaki et al. used electrophoretic deposition (EPD) to coat metal wire meshes with the electrode materials, such as LSM and NiO/YSZ, for use as SOFC current collectors. The EPD coating of the metal mesh increased the effective contact area between the wire mesh and electrode surface [60]. Sarikaya et al. reported that incorporation of LSM oxide particles to the Ag matrix led to the formation of porous composite Ag–LSM current collector, which offered stable electrical properties in SOFC for over 5000 hours at 800 °C in the air [61]. Abraham and Chetty employed Ti mesh as a current collector, which also served as catalyst support and reactant distributor to the catalyst layer, thereby effectively decreasing the overall thickness of the MEA [2].

8.5.7 METAL OXIDE-BASED ELECTRODES

The gas diffusion layer (GDL) is another integral fuel cell component, especially for PEMFC and DLFCs. The conventionally used carbon-based materials, such as graphitic nanofibers and amorphous carbon, offer high conductivity, high surface area, and high porosity for reactant transport, while preventing water accumulation. A microporous layer (MPL) is often coated over the macroporous GDLs, which decreases the interfacial contact resistance, protects the membrane from being punctured by GDL fibers, and promotes water/gas transport. The rugged nature of the MPL and GDL helps to retain enough water to keep membranes hydrated. However, MPLs and GDLs are susceptible to corrosion under fuel cell operating conditions. Hence, it is mandatory to stabilize the MPL and GDL to significantly improve the performance, durability, and stability of the fuel cell. Moreover, they are brittle and inflexible for flexible applications.

Metal oxides are considered an alternative or additional material in MPL and GDL to counter the risk of oxidation or carbon corrosion; metal oxides are electrochemically and thermodynamically more stable than amorphous carbon materials. Composite GDLs are being explored via the wet-laying process by utilizing various alternative materials, for instance, using natural cellulose as the binder to increase the flexibility without compromising porosity and torturous structure. Hydrophilic materials are being explored as an additional coating onto traditionally hydrophobic GDL, which could help to effectively remove condensed water from reaction sites toward the gas flow channels. Additionally, an ad-layer of thin metal oxide coating between the MPL and catalyst is often employed to help the catalyst layer, and consequently, the membrane, to retain water in its layer, which is necessary when the stack is operated under low-RH conditions.

Cindrella et al. investigated the preparation of self-humidifying MEA by coating a hydrophilic layer with TiO_2, SiO_2, Al_2O_3, etc., on GDL and explored the cell performance at various humidity conditions. Results indicated that the GDL with TiO_2 coating (Figure 8.5e) offered superior performance at RH conditions of 80–100%, while at low RH condition (50–70%), the GDL containing Al_2O_3 or SiO_2 layer was found to be most suitable compared to its pristine counterpart consisting of carbon nanofibers and porous carbon black with porous structure (Figure 8.5e**inset**) [62]. Wang et al. prepared a SiO_2-PDMS composite MPL layer that had a super-hydrophobic surface with hydrophilic pores inside. The hydroxyl groups on the SiO_2 surface offered hydrophilic pores to maintain uniform water distribution in the GDL, while the super-hydrophobic surface quickly removed water from the catalyst layer-MPL interface [64]. IrO_2 strongly coated on the inner and outer surfaces of a macro-porous titanium sheet substrate was investigated as GDL for PEMFC by Takasu et al. The highly porous IrO_2 in the IrO_2-coated macroporous Ti sheet (Figure 8.5f) functioned effectively as the catalyst for both oxidation and reduction reactions without the need for Pt catalyst or carbon support [63].

8.6 FUTURE AVENUES FOR METAL OXIDE SYSTEMS IN EMPOWERING FLEXIBLE AND PORTABLE FUEL CELLS

Future efforts can be focused on the following aspects to expedite the development of flexible and portable fuel cells for commercial applications.

a. **Tailoring the cost, activity, and durability of the electrode assembly**

Metal oxides are cost-effective and relatively easy to synthesize. Hence, they could simultaneously address performance, cost, and reliability, which are the deciding factors of fuel cell stacks for commercial applications. Advanced characterization techniques can help to understand the influence of the metal oxide systems on corresponding electrocatalytic mechanisms. An overall approach in catalyst engineering employing morphological, compositional, and structural control of the electrocatalysts, along with suitable modification with metal oxide systems based on the underlying reaction mechanism, is required to improve catalyst utilization and enhance the intrinsic activity for practical applications. For example, employing a 1D structured metal oxide supports/ co-catalyst such as nanorods, nanowires, and nanoribbons, etc., could significantly enhance the transport properties and improve catalyst utilization in the catalyst layer while mitigating parasitic reactions. Such 1D metal oxide structures should be further extensively explored to effectively bring down the loading of precious metal catalysts, as well as the overall cost of the fuel cell stack.

b. **Engineering the cell components for flexible cell architecture**

Bendable substrates are integral components for flexible cell architecture. In this regard, metal oxides are particularly intriguing, considering the various forms metal oxide systems can be easily engineered into while retaining the corrosion-resistant behavior as well as electrochemical and thermal stability under fuel cell operating conditions. Lightweight and flexible metal oxide mesh and foam can be further explored as substrates to increase the stack power to volume ratio significantly.

Metal oxides can be integral components of membranes and bipolar plates. More efforts are necessary on metal oxide-based systems to evaluate the architectures, compatibility with composite materials, and their relation to fuel cell performance. Furthermore, the conductivity and permeability of metal oxide-based membranes and the durability of metal oxide-modified bipolar plates should be evaluated with advanced testing protocols to determine suitability in flexible and portable applications.

c. **Realization of a self-sustaining reliable fuel cell stack**

For low-temperature fuel cells, several research groups have reported that metal oxides, especially TiO_2, can retain water given its spillover effect, which could make the catalyst layer, or composite membrane, function even under dry gas feed conditions, making the fuel cell stack self-sustaining. Metal oxide-based catalyst layers, including metal oxide-based catalyst, co-catalyst, or gas-diffusion layer, could be further explored to allow fuel cell operation in ambient conditions, thereby helping to reduce dependency on auxiliary systems and making them more suitable for portable and flexible applications.

d. **Miniaturization and optimization of fuel cell technologies for wearable and shape conforming applications**

Application-oriented fuel cell architecture is a challenge, since traditionally, most fuel cell technologies use rigid components for various applications. However, portable and flexible applications necessitate the need for nonrigid architecture. Most wearable and shape-conforming applications are on the human body. They use paper-based fuel cells and MFCs that employ microorganisms, enzymes, and precious metals as catalysts and make it possible to harvest energy from human bodily fluids, such as saliva, urine, sweat, blood, etc., through electrochemical mechanisms. Novel and modified fuel cell technologies and architectures should be explored to meet application-side requirements, such as miniaturization, compatibility, simplicity, and flexibility, while running on cost-effective metal oxide-based systems and simple, readily available fuels.

To summarize, metal oxide-based materials are revealed to be suitable materials to empower flexible and portable fuel cell applications. Low-cost, highly efficient, and robust metal oxide-based fuel cell components are developed, which, when combined with suitable architecture, advanced materials, and superior manufacturing technologies, can enable the further application of fuel cells as sustainable power sources in portable and flexible applications.

8.7   ACKNOWLEDGMENTS

Dr. P. Dhanasekaran (Scientist's Pool Scheme-9123-A) and Dr. S. Vinod Selvaganesh (Scientist's Pool Scheme-9178-A) would like to thank CSIR for the Senior Research Associateship funding under Scientist Pool Scheme. Prof. Raghuram Chetty and Dr. Bincy George Abraham would like to acknowledge the Ministry of New and Renewable Energy (MNRE Grant No: 102/61/2009-NT), Government of India.

ABBREVIATIONS

ATO	Antimony-doped tin dioxide
CCT	Carbon semi-coated on titania nanorods
CNT	Carbon nanotube
DHE	Dynamic Hydrogen Electrode
DLFC	Direct liquid fuel cell
DMFC	Direct methanol fuel cell
EOR	Ethanol oxidation reaction
EPD	Electrophoretic deposition
GDL	Gas diffusion layer
LSM	Strontium-doped lanthanum manganite oxide
MEA	Membrane electrode assembly
MFC	Microbial fuel cell
MOR	Methanol oxidation reaction
MPL	Microporous layer
NTO	Niobium-doped tin dioxide
ORR	Oxygen reduction reaction
PBFC	Paper-based fuel cell
PCB	Printed circuit board
PDMS	Polydimethylsiloxane
PEDOT	Poly(3,4-ethylenedioxythiophene)
PEMFC	Polymer electrolyte membrane fuel cell
PFSA	Perfluorosulfonic acid
Ppy	Polypyrrole
PVAc-g-PVDF	Polyvinyl acetate-graft-polyvinylidene fluoride
rGO	Reduced graphene oxide
RH	Relative humidity
RHE	Reverible Hydrogen Electrode
SMSI	Strong metal support interaction
SOFC	Solid oxide fuel cell
SS	Stainless steel
TNT	Titania nanotube
TNW	TiO_2 nanowires
TPB	Triple-phase boundary
UV	Ultraviolet
YSZ	Yttria-stabilized zirconia

REFERENCES

1. F. Ning *et al.*, "Flexible and lightweight fuel cell with high specific power density," *ACS Nano*, vol. 11, no. 6, pp. 5982–5991, 2017, doi: 10.1021/acsnano.7b01880.

2. B. G. Abraham and R. Chetty, "Design and fabrication of a quick-fit architecture air breathing direct methanol fuel cell," *Int. J. Hydrogen Energy*, vol. 46, no. 9, pp. 6845–6856, Feb. 2021, doi: 10.1016/j.ijhydene.2020.11.184.

3. F. K. Hsu, M. S. Lee, C. C. Lin, Y. K. Lin, and W. T. Hsu, "A flexible portable proton exchange membrane fuel cell," *J. Power Sources*, vol. 219, pp. 180–187, 2012, doi: 10.1016/j.jpowsour.2012.07.054.

4. S. S. Hsieh and C. F. Huang, "Design, fabrication and performance test of a planar array module-type micro fuel cell stack," *Energy Convers. Manag.*, vol. 76, pp. 971–979, 2013, doi: 10.1016/j.enconman.2013.08.062.

5. S. H. Kim, H. Y. Cha, C. M. Miesse, J. H. Jang, Y. S. Oh, and S. W. Cha, "Air-breathing miniature planar stack using the flexible printed circuit board as a current collector," *Int. J. Hydrogen Energy*, vol. 34, no. 1, pp. 459–466, 2009, doi: 10.1016/j.ijhydene.2008.09.088.

6. F. Ning *et al.*, "Critical importance of current collector property to the performance of flexible electrochemical power sources," *Chinese Chem. Lett.*, vol. 30, no. 6, pp. 1282–1288, 2019, doi: 10.1016/j.cclet.2019.02.032.

7. A. J. Bandodkar *et al.*, "Soft, stretchable, high power density electronic skin-based biofuel cells for scavenging energy from human sweat," *Energy Environ. Sci.*, vol. 10, no. 7, pp. 1581–1589, 2017, doi: 10.1039/c7ee00865a.

8. C. Y. Chu, J. T. Tsai, and C. L. Sun, "Synthesis of PEDOT-modified graphene composite materials as flexible electrodes for energy storage and conversion applications," *Int. J. Hydrogen Energy*, vol. 37, no. 18, pp. 13880–13886, 2012, doi: 10.1016/j.ijhydene.2012.05.017.

9. T. Maiyalagan, X. Dong, P. Chen, and X. Wang, "Electrodeposited Pt on three-dimensional interconnected graphene as a free-standing electrode for fuel cell application," *J. Mater. Chem.*, vol. 22, no. 12, pp. 5286–5290, 2012, doi: 10.1039/c2jm16541d.

10. M. Taghavi *et al.*, "High-Performance, Totally Flexible, Tubular Microbial Fuel Cell," *ChemElectroChem*, vol. 1, no. 11, pp. 1994–1999, 2014, doi: 10.1002/celc.201402131.

11. B. Zhang, Y. Jiang, and J. Han, "A flexible nanocomposite membrane based on traditional cotton fabric to enhance performance of microbial fuel cell," *Fibers Polym.*, vol. 18, no. 7, pp. 1296–1303, 2017, doi: 10.1007/s12221-017-7191-y.

12. H. Wang *et al.*, "High power density microbial fuel cell with flexible 3D graphene-nickel foam as anode," *Nanoscale*, vol. 5, no. 21, pp. 10283–10290, 2013, doi: 10.1039/c3nr03487a.

13. Y. Wang *et al.*, "A flexible paper-based hydrogen fuel cell for small power applications," *Int. J. Hydrogen Energy*, vol. 44, no. 56, pp. 29680–29691, 2019, doi: 10.1016/j.ijhydene.2019.04.066.

14. S. Tominaka, H. Nishizeko, J. Mizuno, and T. Osaka, "Bendable fuel cells: On-chip fuel cell on a flexible polymer substrate," *Energy Environ. Sci.*, vol. 2, no. 10, pp. 1074–1077, 2009, doi: 10.1039/b915389f.

15. S. V. Selvaganesh, P. Dhanasekaran, and S. D. Bhat, "TiO2-nanowire/MWCNT composite with enhanced performance and durability for polymer electrolyte fuel cells," *Electrochem. Energy Technol.*, vol. 3, no. 1, pp. 9–26, 2018, doi: 10.1515/eetech-2017-0002.

16. P. Dhanasekaran, S. V. Selvaganesh, and S. D. Bhat, "Enhanced catalytic activity and stability of copper and nitrogen doped titania nanorod supported Pt electrocatalyst for oxygen reduction reaction in polymer electrolyte fuel cells," *New J. Chem.*, vol. 41, no. 21, pp. 13012–13026, 2017, doi: 10.1039/C7NJ03463F.

17. B. G. Abraham, K. K. Maniam, A. Kuniyil, and R. Chetty, "Electrocatalytic performance of palladium dendrites deposited on titania nanotubes for formic acid oxidation," *Fuel Cells*, vol. 16, no. 5, pp. 656–661, Oct. 2016, doi: 10.1002/fuce.201600023.

18. H. Wu, W. Sun, J. Shen, D. W. Rooney, Z. Wang, and K. Sun, "Role of flower-like ultrathin Co3O4 nanosheets in water splitting and non-aqueous Li-O2 batteries," *Nanoscale*, vol. 10, no. 21, pp. 10221–10231, 2018, doi: 10.1039/c8nr02376j.

19. Q. Li *et al.*, "Structure-dependent electrocatalytic properties of Cu_2O nanocrystals for oxygen reduction reaction," *J. Phys. Chem. C*, vol. 117, no. 27, pp. 13872–13878, 2013, doi: 10.1021/jp403655y.

20. Z. Sun, T. Liao, L. Sheng, L. Kou, J. H. Kim, and S. X. Dou, "Deliberate design of TiO_2 nanostructures towards superior photovoltaic cells," *Chem. – A Eur. J.*, vol. 22, no. 32, pp. 11357–11364, 2016, doi: 10.1002/chem.201601546.

21. P. Dhanasekaran, S. V. Selvaganesh, L. Sarathi, and S. D. Bhat, "Rutile TiO_2 supported Pt as stable electrocatalyst for improved oxygen reduction reaction and durability in polymer electrolyte fuel cells," *Electrocatalysis*, vol. 7, no. 6, pp. 495–506, 2016, doi: 10.1007/s12678-016-0329-7.

22. Y. L. Cao, H. X. Yang, X. P. Ai, and L. F. Xiao, "The mechanism of oxygen reduction on MnO2-catalyzed air cathode in alkaline solution," *J. Electroanal. Chem.*, vol. 557, pp. 127–134, 2003, doi: 10.1016/S0022-0728(03)00355-3.

23. L. Li *et al.*, "Insight into the effect of oxygen vacancy concentration on the catalytic performance of MnO_2," *ACS Catal.*, vol. 5, no. 8, pp. 4825–4832, 2015, doi: 10.1021/acscatal.5b00320.

24. T. Ioroi, T. Akita, S. Yamazaki, Z. Siroma, N. Fujiwara, and K. Yasuda, "Corrosion-resistant PEMFC cathode catalysts based on a magnéli-phase titanium oxide support synthesized by pulsed UV laser irradiation," *J. Electrochem. Soc.*, vol. 158, no. 10, p. C329, 2011, doi: 10.1149/1.3622297.

25. P. Dhanasekaran, S. V. Selvaganesh, and S. D. Bhat, "A nitrogen and cobalt co-doped titanium dioxide framework as a stable catalyst support for polymer electrolyte fuel cells," *RSC Adv.*, vol. 6, no. 91, pp. 88736–88750, 2016, doi: 10.1039/c6ra18083c.

26. M. Kim, C. Kwon, K. Eom, J. Kim, and E. Cho, "Electrospun Nb-doped TiO_2 nanofiber support for Pt nanoparticles with high electrocatalytic activity and durability," *Sci. Rep.*, vol. 7, no. February, pp. 1–8, 2017, doi: 10.1038/srep44411.

27. P. Dhanasekaran, S. V. Selvaganesh, and S. D. Bhat, "Preparation of TiO_2:TiN composite nanowires as a support with improved long-term durability in acidic medium for polymer electrolyte fuel cells," *New J. Chem.*, vol. 41, no. 8, pp. 2987–2996, 2017, doi: 10.1039/c7nj00374a.

28. B. J. Hsieh *et al.*, "Platinum loaded on dual-doped TiO2 as an active and durable oxygen reduction reaction catalyst," *NPG Asia Mater.*, vol. 9, no. 7, 2017, doi: 10.1038/am.2017.78.

29. B. Y. Xia, S. Ding, H. Bin Wu, X. Wang, and X. Wen, "Hierarchically structured Pt/CNT@TiO_2 nanocatalysts with ultrahigh stability for low-temperature fuel cells," *RSC Adv.*, vol. 2, no. 3, pp. 792–796, 2012, doi: 10.1039/c1ra00587a.

30. B. Zhao *et al.*, "Multifunctional iron oxide nanoflake/graphene composites derived from mechanochemical synthesis for enhanced lithium storage and electrocatalysis," *ACS Appl. Mater. Interfaces*, vol. 7, no. 26, pp. 14446–14455, 2015, doi: 10.1021/acsami.5b03477.

31. R. Zhou, Y. Zheng, D. Hulicova-Jurcakova, and S. Z. Qiao, "Enhanced electrochemical catalytic activity by copper oxide grown on nitrogen-doped reduced graphene oxide," *J. Mater. Chem. A*, vol. 1, no. 42, p. 13179, 2013, doi: 10.1039/c3ta13299d.

32. S. Karthick and K. Haribabu, "Bioelectricity generation in a microbial fuel cell using polypyrrole-molybdenum oxide composite as an effective cathode catalyst," *Fuel*, vol. 275, no. April, p 117994, Sep. 2020, doi: 10.1016/j.fuel.2020.117994.

33. Y. Dessie, S. Tadesse, R. Eswaramoorthy, and B. Abebe, "Recent developments in manganese oxide based nanomaterials with oxygen reduction reaction functionalities for energy conversion and storage applications: A review," *J. Sci. Adv. Mater. Devices*, vol. 4, no. 3, pp. 353–369, 2019, doi: 10.1016/j.jsamd.2019.07.001.

34. J. Suntivich, H. A. Gasteiger, N. Yabuuchi, H. Nakanishi, J. B. Goodenough, and Y. Shao-Horn, "Design principles for oxygen-reduction activity on perovskite oxide catalysts for fuel cells and metal-air batteries," *Nat. Chem.*, vol. 3, no. 7, pp. 546–550, 2011, doi: 10.1038/nchem.1069.

35. H. Wang, X. Chen, D. Huang, M. Zhou, D. Ding, and H. Luo, "Cation deficiency tuning of lacoo 3 perovskite as bifunctional oxygen electrocatalyst," *ChemCatChem*, vol. 12, no. 10, pp. 2768–2775, May 2020, doi: 10.1002/cctc.201902392.

36. H. C. Yu, K. Z. Fung, T. C. Guo, and W. L. Chang, "Syntheses of perovskite oxides nanoparticles La 1-xSr xMO 3-δ (M = Co and Cu) as anode electrocatalyst for direct methanol fuel cell," *Electrochim. Acta*, vol. 50, no. 2-3 SPEC. ISS., pp. 811–816, 2004, doi: 10.1016/j.electacta.2004.01.121.

37. A. Lan and A. S. Mukasyan, "Complex SrRuO3-Pt and LaRuO3-Pt catalysts for direct alcohol fuel cells," *Ind. Eng. Chem. Res.*, vol. 47, no. 23, pp. 8989–8994, 2008, doi: 10.1021/ie8000698.

38. R. J. Toh, A. Y. S. Eng, Z. Sofer, D. Sedmidubsky, and M. Pumera, "Ternary transition metal oxide nanoparticles with spinel structure for the oxygen reduction reaction," *ChemElectroChem*, vol. 2, no. 7, pp. 982–987, 2015, doi: 10.1002/celc.201500070.

39. P. Justin and G. Ranga Rao, "Methanol oxidation on MoO_3 promoted Pt/C electrocatalyst," *Int. J. Hydrogen Energy*, vol. 36, no. 10, pp. 5875–5884, 2011, doi: 10.1016/j.ijhydene.2011.01.122.

40. J. Rajeswari, B. Viswanathan, and T. K. Varadarajan, "Tungsten trioxide nanorods as supports for platinum in methanol oxidation," *Mater. Chem. Phys.*, vol. 106, no. 2–3, pp. 168–174, 2007, doi: 10.1016/j.matchemphys.2007.05.032.

41. P. Dhanasekaran, B. Saravanan, S. V. Selvaganesh, and S. D. Bhat, "Addressing LT-PEFC 15 cell stack durability using carbon semi-coated titania nanorods-Pt electrocatalyst," *Int. J. Hydrogen Energy*, vol. 44, no. 3, pp. 1940–1952, 2019, doi: 10.1016/j.ijhydene.2018.11.097.

42. C. Li, X. Han, F. Cheng, Y. Hu, C. Chen, and J. Chen, "Phase and composition controllable synthesis of cobalt manganese spinel nanoparticles towards efficient oxygen electrocatalysis," *Nat. Commun.*, vol. 6, pp. 1–8, 2015, doi: 10.1038/ncomms8345.

43. K. Mohanraju, P. S. Kirankumar, L. Cindrella, and O. J. Kwon, "Enhanced electrocatalytic activity of Pt decorated spinals (M3O4, M = Mn, Fe, Co)/C for oxygen reduction reaction in PEM fuel cell and their evaluation by hydrodynamic techniques," *J. Electroanal. Chem.*, vol. 794, pp. 164–174, 2017, doi: 10.1016/j.jelechem.2017.04.011.

44. Z. Zhang, J. Liu, J. Gu, L. Su, and L. Cheng, "An overview of metal oxide materials as electrocatalysts and supports for polymer electrolyte fuel cells," *Energy Environ. Sci.*, vol. 7, no. 8, pp. 2535–2558, 2014, doi: 10.1039/C3EE43886D.

45. B. G. Abraham, K. K. Maniam, A. Kuniyil, and R. Chetty, "Electrocatalytic performance of palladium dendrites deposited on titania nanotubes for formic acid oxidation," *Fuel Cells*, vol. 16, no. 5, 2016, doi: 10.1002/fuce.201600023.

46. S. V. Selvaganesh, G. Selvarani, P. Sridhar, S. Pitchumani, and A. K. Shukla, " A durable RuO_2-carbon-supported Pt catalyst for PEFCs: A cause and effect study," *J. Electrochem. Soc.*, vol. 159, no. 5, pp. B463–B470, 2012, doi: 10.1149/2.jes113440.

47. H. Zhang, C. Hu, X. He, L. Hong, G. Du, and Y. Zhang, "Pt support of multidimensional active sites and radial channels formed by SnO2 flower-like crystals for methanol and ethanol oxidation," *J. Power Sources*, vol. 196, no. 10, pp. 4499–4505, 2011, doi: 10.1016/j.jpowsour.2011.01.030.

48. S. Y. Huang, P. Ganesan, S. Park, and B. N. Popov, "Development of a titanium dioxide-supported platinum catalyst with ultrahigh stability for polymer electrolyte membrane fuel cell applications," *J. Am. Chem. Soc.*, vol. 131, no. 39, pp. 13898–13899, 2009, doi: 10.1021/ja904810h.

49. L. Dubau *et al.*, "Durability of alternative metal oxide supports for application at a proton-exchange membrane fuel cell cathode—comparison of antimony- And niobium-doped tin oxide," *Energies*, vol. 13, no. 2, pp. 1–14, 2020, doi: 10.3390/en13020403.

50. O. S. Jeon, H. J. Hwang, O. Chan Kwon, J. G. Lee, and Y. G. Shul, "Next-generation flexible solid oxide fuel cells with high thermomechanical stability," *J. Mater. Chem. A*, vol. 6, no. 37, pp. 18018–18024, 2018, doi: 10.1039/c8ta03573c.

51. S. Yoon *et al.*, "Flexible blend polymer electrolyte membranes with excellent conductivity for fuel cells," *Int. J. Hydrogen Energy*, vol. 45, no. 51, pp. 27611–27621, 2020, doi: 10.1016/j.ijhydene.2020.07.076.

52. K. T. Adjemian *et al.*, "Function and characterization of metal oxide-nafion composite membranes for elevated-temperature H 2/O 2 PEM fuel cells," *Chem. Mater.*, vol. 18, no. 9, pp. 2238–2248, 2006, doi: 10.1021/cm051781b.

53. K. Ketpang, K. Lee, and S. Shanmugam, "Facile synthesis of porous metal oxide nanotubes and modified nafion composite membranes for polymer electrolyte fuel cells operated under low relative humidity," *ACS Appl. Mater. Interfaces*, vol. 6, no. 19, pp. 16734–16744, 2014, doi: 10.1021/am503789d.

54. M. Ercelik, A. Ozden, Y. Devrim, and C. O. Colpan, "Investigation of Naf ion based composite membranes on the performance of DMFCs," *Int. J. Hydrogen Energy*, vol. 42, no. 4, pp. 2658–2668, 2017, doi: 10.1016/j.ijhydene.2016.06.215.

55. N. Kumar, G. P. Shaik, S. Pandurangan, B. Khalkho, L. Neelakantan, and R. Chetty, "Corrosion characteristics and fuel cell performance of a cost-effective high Mn–Low Ni austenitic stainless steel as an alternative to SS 316L bipolar plate," *Int. J. Energy Res.*, vol. 44, no. 8, pp. 6804–6818, 2020, doi: 10.1002/er.5422.

56. S. H. Wang, W. B. Lui, J. Peng, and J. S. Zhang, "Performance of the iridium oxide (IrO2)-modified titanium bipolar plates for the light weight proton exchange membrane fuel cells," *J. Fuel Cell Sci. Technol.*, vol. 10, no. 4, pp. 1–6, 2013, doi: 10.1115/1.4024565.

57. H. M. Jung and S. Um, "An experimental feasibility study of vanadium oxide films on metallic bipolar plates for the cold start enhancement of fuel cell vehicles," *Int. J. Hydrogen Energy*, vol. 36, no. 24, pp. 15826–15837, 2011, doi: 10.1016/j.ijhydene.2011.09.008.

58. L. Peng, D. Liu, P. Hu, X. Lai, and J. Ni, "Fabrication of metallic bipolar plates for proton exchange membrane fuel cell by flexible forming process-numerical simulations and experiments," *J. Fuel Cell Sci. Technol.*, vol. 7, no. 3, pp. 0310091–0310099, 2010, doi: 10.1115/1.3207870.

59. J. Chen *et al.*, "Polydimethylsiloxane (PDMS)-based flexible resistive strain sensors for wearable applications," *Appl. Sci.*, vol. 8, no. 3, 2018, doi: 10.3390/app8030345.

60. Y. Itagaki, F. Matsubara, M. Asamoto, H. Yamaura, H. Yahiro, and Y. Sadaoka, "Electrophoretically coated wire meshes as current collectors for solid oxide fuel cell," *ECS Trans.*, vol. 7, no. 1, pp. 1319–1325, Dec. 2007, doi: 10.1149/1.2729235.

61. A. Sarikaya, V. Petrovsky, and F. Dogan, "Silver composites as highly stable cathode current collectors for solid oxide fuel cells," *J. Mater. Res.*, vol. 27, no. 15, pp. 2024–2029 2012, doi: 10.1557/jmr.2012.175.

62. L. Cindrella, A. M. Kannan, R. Ahmad, and M. Thommes, "Surface modification of gas diffusion layers by inorganic nanomaterials for performance enhancement of proton exchange membrane fuel cells at low RH conditions," *Int. J. Hydrogen Energy*, vol. 34, no. 15, pp. 6377–6383, 2009, doi: 10.1016/j.ijhydene.2009.05.086.

63. Y. Takasu, H. Fukunaga, H. S. Yang, T. Ohashi, M. Suzuki, and W. Sugimoto, "A gas-diffusion cathode coated with oxide-catalyst for polymer electrolyte fuel cells using neither platinum catalyst nor carbon catalyst-support," *Electrochim. Acta*, vol. 105, pp. 224–229, 2013, doi: 10.1016/j.electacta.2013.04.133.

64. Y. Wang, S. Al Shakhshir, and X. Li, "Development and impact of sandwich wettability structure for gas distribution media on PEM fuel cell performance," *Appl. Energy*, vol. 88, no. 6, pp. 2168–2175, 2011, doi: 10.1016/j.apenergy.2010.12.054.

9 Flexible Fuel Cells Based on Microbes

Hamide Ehtesabi
Faculty of Life Sciences and Biotechnology,
Shahid Beheshti University, Tehran, Iran

CONTENTS

9.1 Introduction 157
9.2 Basics of MFCs 158
    9.2.1 Instrumental bases 159
    9.2.2 Two-compartment MFCs 159
    9.2.3 Single-compartment MFCs 159
9.3 Flexible MFCs 162
    9.3.1 Electrodes 162
        9.3.1.1 Carbonaceous material 162
        9.3.1.2 Bacterial cellulose 163
        9.3.1.3 Graphene sheet 163
        9.3.1.4 Polypyrrole (PPy) 164
    9.3.2 Membrane 164
    9.3.3 Microorganism 165
    9.3.4 Fabrication 166
    9.3.5 Applications 166
        9.3.5.1 Energy harvesting 166
        9.3.5.2 Treatment of wastewater 167
        9.3.5.3 Sensors and portable power machines 167
9.4 Future aspect 167
    9.4.1 Large-scale uses 167
    9.4.2 Anode manipulation 167
    9.4.3 Membrane-free MFC 168
9.5 Conclusion 168
References 168

9.1 INTRODUCTION

Natural gas, petroleum, and coal are conventional fossil fuels that were generated from buried plants and animals compressed for millions of years through upper sediments. Such fossil fuels can generate high power to support vehicles' engines, electronic appliances, and daily life. Conversely, they are considered unsustainable sources of energy because they are finite. Therefore, a requirement exists for natural and enduring fuels to replace fossil fuels and improve up-to-date industrial civilization [1, 2].

MFCs is considered as a biological-electrochemical instrument that transfers chemical energy into electrical energy through the activities of the microbes. Using organic materials like wastewater within MFC makes it an eco-friendly instrument that provides mutual advantages of waste management and bioelectricity production. The microbes (bacteria) work as biocatalysts that activate the organic

DOI: 10.1201/9781003186755-9

materials' degradation to generate electrons; these electrons move to the cathode side by the electric circuit. Such bacteria are named "exoelectrogens" (exo- stands for exocellular and "electrogens," on the basis of the capability to move electrons directly to a chemical or material that is not considered an immediate electron acceptor) [3]. A representative MFC is composed of two anode and cathode chambers. To simplify the flow of electrons from the anode chamber toward the cathode one, as well as an electrolytic medium between both chambers to permit positive ions to distribute toward the cathode, the external resistance is implemented between both electrodes. The MFC anode is exhibited to electrogenic bacteria, and oxidation takes place at cathode reduction.

Bacteria and enzymatic protein as a microorganism and biomacromolecules, respectively, are considered to be the major electrochemical reactors used in anode chamber [4]. Several kinds of organic waste were implemented as fuel for microbial anodes; however, electroactive bacteria kinetics continue to be poor. It is still unclear why electrodes and bacteria interact with each other. In addition, no description relates to interacting and/or coexisting the electron transfer mechanisms between solid electrodes and bacteria, particularly in complicated areas where a wide range of microbial types (electroactive or not) were found on the electrodes [2]. Eventually, because of the difficulty of coupling the complex procedures of microbial electrochemistry and the available imaging technology, it is unclear why the microbial cells are attracted toward the electrodes, why the anode surface and biofilm forms, why inter-species cooperation and interaction occurs, as well as why environmental parameters affect microbial colonization [5]. One of the perfect substrates for bioelectricity generation is organic waste, which is composed of low-strength wastewaters and lignocellulose biomass. An ample amount of low-strength wastewater and many kinds of organic compositions are available in municipal sewage, which is comparable to food compounds, such as carbohydrates (25–50%), protein (40–60%), and fats (10%).

New health-monitoring technologies have a strong dependency on textile-based wearable electronics. The permanent and efficient connection of the nonuniform and highly mobile human skin with monitoring devices depends on the flexibility and stretchable nature of electronic parts. For this reason, wearable and flexible electronic devices are widely used in real-time monitoring of biological parameters in the human body. However, achieving these permanent and stand-alone flexible devices, which generate their electrical energy, is in progress [6]. In addition to rigidity and high weight, common battery-based systems do not ensure long-term devices' advanced functionality because these batteries need to be changed continuously [7]. Emerging energy-harvesting systems raise hopes to obtain flexible and wearable devices that have low power consumption. This technology uses sun, body motion, natural human heat, and other costless energy sources to produce electrical power. But such energy sources are also finite and not permanent. For example, both sun energy and body actions stop at night. A recent candidate to overcome these challenges is biological fuel cells (BFCs). BFCs utilize enzymes and microorganisms to produce biochemical energy; they usually use sweat, saliva, blood, or other organic sources. In this bio-system, energy producers are living organisms that present a self-maintaining and self-repairing system in combination with nonorganic components of a sensing device. Biodegradable components of BFCs make them more environmental friendly as devices, permitting self-assembly, self-repair, and self-maintenance properties. The present study aimed to introduce instrumental of MFCs and electrochemical foundation. After that, membrane, electrodes, and fabrication of flexible MFCs are introduced. At the end, the usages and future dimensions of such MFCs are explained.

9.2 BASICS OF MFCS

Transferring electrons from cell to circuit results in exploring MFC by the metabolic potential of microbes for converting the organic substrate into electricity. In an anodic chamber, the substrate oxidation (oxidation of acetate substrate to the carbon dioxide) when oxygen is not present by respiratory bacteria generates protons and electrons that are transferred onto terminal e^- acceptor [O_2, nitrate, or Fe (III)] by an electron transport chain (ETC). But, if e^- acceptor in an MFC is absent, some electrons are passed onto anode by microorganisms. An effective electron shuttle to anode may be obtained either by spontaneous (direct) or electron-shuttling mediators. Electrons can be directly transferred to the anode through bacteria, which need some physical contact with the electrode for the current production. A plunge lineup between anode surface and bacteria consists of putative conductive pili or outer membrane-bound cytochromes called nanowires. In these mediator-less MFCs, a biofilm is created on the anode surface as the final electron acceptor [8]. Regarding mediated electron transfer machinery, microbes generate their own natural-synthetic exogenous mediators (dye or metallorganic) or soluble redox compounds (quinones and flavin) to alternate electrons between anode surface and terminal respiratory enzyme. These mediators interact with the outer cell membrane, divert electrons from the respiratory chain, and move toward shuttle electrons to the electrode. The substrate consumed by microbes has a key number of produced electrons. MFCs that have less mediator and more commercial potential as mediators are costly and are sometimes poisonous to microorganisms. The electrode is reacted in an MFC compartment as follows [9]. Because of its abundance and high reduction potential at the cathode chamber, oxygen has mainly been applied as the oxidant. Moreover, it was found that the probability of using metallic oxidants can be decreased to a less toxic oxidation state. One of the main bottlenecks of this technology is the oxygen reduction reaction (ORR), which is encountered because of the high overpotentials and low kinetics (Figure 9.1).

9.2.1 Instrumental bases

A typical MFC device comprises two anode and cathode chambers produced by the glass, polycarbonate, or plexiglass with electrodes made of graphite felt, carbon paper, graphite, carbon-cloth, Pt black, or Pt. These two chambers are separated by nafion or ultrex. Organic substrates sill the anodic chamber, which is metabolized by microbes for energy production and growth while producing protons and electrons. For completing the circuit, a cathode is filled with a maximum potential electron acceptor. A typical electron acceptor should not interfere with microbes and should be a sustainable combination having no toxic impact. Oxygen is used as an ideal electron acceptor because of its nontoxic impact and is preferred as an oxidizing reagent because it simplifies the use of an MFC, or else, standard media with an appropriate electron acceptor like ferricyanide can also be implemented for increasing power density. Based on the assembly of cathode and anode chambers, a simple prototype of MFC can be chambered either doubly or singularly. In addition to these two conventional designs, the prototype of MFC design and structure has been adapted [10].

9.2.2 Two-compartment MFCs

In two-compartment MFCs, a membrane completely separates the anode and cathode chambers (Figure 9.2). Both anode and cathode must be electron conductive; additionally, anode needs to have biocompatible properties to prove efficient attachment of microorganisms on the anode surface and allow them a permanent growth on its surface. Microorganisms attach to the anode surface and create a biofilm. Then, they oxidize their substrate, followed by the production of electrons and protons. Electrons pass an external circuit toward the cathode chamber to reduce oxygen (ORR) there. On the other side, protons migrate across the membrane and react with oxygen to produce water. Such systems are typically used for treating waste with simultaneous power generation. Intensifying two-chamber MFCs to an industrial size is tough. In addition, periodic inflation of cathodic compartments restricts the usage spectrum of double compartment MFCs.

9.2.3 Single-compartment MFCs

In the single-compartment MFCs, anodic chamber and porous cathode are integrated with a gas diffusion layer (Figure 9.3). To complete the circuit, the electrons are passed to a porous cathode. A restricted need for periodic recharging with oxidative media and aeration results in the single chamber microbial fuel cell

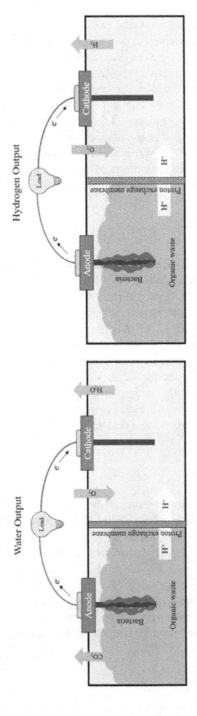

FIGURE 9.1 The basics of a microbial fuel cell.

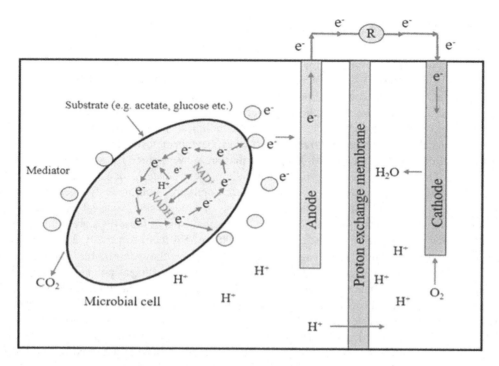

FIGURE 9.2 Two-compartment MFCs.

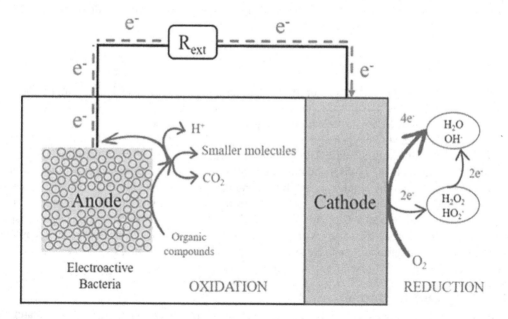

FIGURE 9.3 Single-compartment MFCs.

system becoming more versatile. Among various benefits, single-chamber MFC consists of decreased setup expenses (because of not presenting cathodic chambers and expensive membranes), which allow flexible uses in power generation and wastewater treatment. This flexibility eliminates ion-exchange membranes in the single-chamber MFCs, decreased internal system resistance and undesirable pH gradient.

9.3 FLEXIBLE MFCS

Practical construction of flexible MFCs first happened using enzymatic BFCs (EFCs); researchers have established enzymes on the flexible and stretchable substrates to harvest energy from human sweat or tears [11]. In the EFCs, oxidizing of blood and tear metabolites is catalyzed by anode chamber enzymes. However, enzymes' performance as an energy producer presents a challenge as they are unstable, and their activity reduces power and electron-transferring potency to the anode fluctuate under condition changes [12]. Protective techniques, such as immobilization, applying mediators, or providing direct electron transfer pathways, have been tested to improve durability and practical feasibility of EFCs, but achieving a self-sustaining power source for wearable electronics is a long way off. On the other side, microbial fuel cells present concerns due to their microbial cytotoxicity. However, if we consider 3.8×10^{13} bacterial cells that continuously live with a human body compared to 3.0×10^{13} of our cells [13], the use of noninvasive bacterial cells in wearable electronics as a power generator is not a concern. Microorganisms in the MFCs are complete living cells with various complicated enzymatic pathways, which help them to rapidly adapt to environmental shocks and react against harsh conditions [14]. Therefore, they are reliable, permanent, self-assembling, self-repairing, and self-maintaining agents that have sufficient potency to continuously generate green energy for flexible fuel cells [15]. In addition to what was mentioned about MFCs as a power generator, they have a low output power density for several reasons, including low efficiency of extracellular electron transfer (EET) to the electrode (redox-active outer membrane proteins, electron shuttles, and conductive bacterial pili) and low microbial-loading capacity on the anode surface [16], which restricted their practical competence in wearable and flexible MFCs [17]. Progresses in flexible MFCs can create a great revolution in novel technologies, such as in human-machine interface [18], geometrically compatible military, environmental sensing [19], and wearable biomedical care [20]. Figure 9.4 shows common materials used in flexible MFCs. Also, Table 9.1 listed the main properties and applications of flexible MFCs.

9.3.1 ELECTRODES

Microbes metabolize organic molecules in the MFCs and create an electro-active biofilm on the anode surface. This biofilm releases oxidized molecules in form of protons and electrons via a redox reaction. The solution and external circuit pass electrons and protons to the cathode, respectively, to participate in a reduction reaction related to oxygen at the cathode surface. Electrodes play a critical role in this cycle. The cathode will significantly increase MFC efficiency and also lower costs. In addition, the essential elements are the active anodic biofilm and anode electrode. The type of microbes, substrate, and anode material are key factors in the efficiency, durability, and activity of anodic biofilms. Broad surface area, chemical stability, good biocompatibility, and high electrical conductivity are all requirements for the electrode material used in MFC. Clogging (particularly with graphite fiber brushes), large resistance, high prices, and fragility are all issues with electrode materials. For better electrode efficiency, some scholars use heat treatment and catalysts to alter electrode materials [26]. Conversely, these changes increase the cost of the MFC reactor's design in ref. [27].

9.3.1.1 Carbonaceous material

Activated carbon is considered as a carbonaceous substance with a wide internal surface area and well-developed porous structure that results from the high-temperature processing of raw materials. Activated carbon contains 87–97% carbon, but it can also contain other elements that depend on the manufacturing method and raw materials used. The chemical activation process is used to prepare activated carbon. In this approach, reactants and activators undergo thermal decomposition; therefore, carbonization and activation reactions occur simultaneously [21]. The activated carbon can be coated on the surface using the doctor blade technique to prepare electrodes for MFCs.

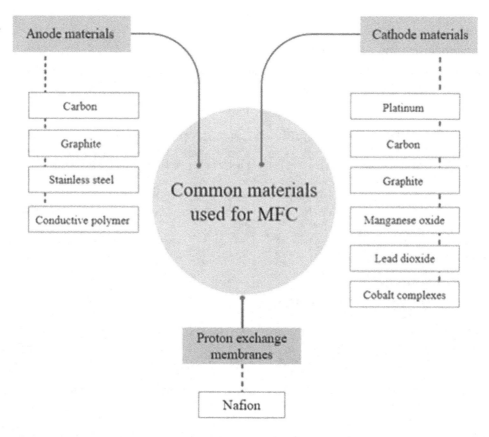

FIGURE 9.4 Common materials used in flexible MFCs.

9.3.1.2 Bacterial cellulose

Bacterial cellulose (BC) is utilized in flexible MFCs electrodes not only due to its stretchable and flexible properties, but also because of its low-cost production, hydrophilicity, biocompatibility, eco-friendly, and porous structure. In addition to conductive polymers, such as polypyrrole and polyaniline (PANI), carbon-based nanomaterial like CNTs and graphene are used to coat BC to modulate its conductivity. This binder-less coating method is easy and low cost in comparison with common bindery coating by polytetrafluoroethylene (PTFE) or nafion. To synthesize CNT-coated-polyaniline-bacterial cellulose (BC-CNT-PANI) bioanode, Mashkour et al. placed a sheet of wet BC on the glass filter of an Erlen Bochner flask. A vacuum-filtering approach was selected to coat the BC anode with CNT and incubated at room temperature for the drying process [23].

9.3.1.3 Graphene sheet

Mink JE et al. introduced a thin, flexible anode for the construction of microsized microbial fuel cells. For this process, the chemical vapor deposition technique (CVD) was applied to deposit nickel metal on the silicon oxide surface, and then, a thin layer of graphene sheets was grown on nickel. The final flexible anode was obtained via etching the silicon. The obtained flexible MFC consist of a rubber anode chamber with a remarkable available area for bacterial growth (25 mm^2 and 1 mm thickness), while the total volume of the device was 25 ml. This microsized flexible MFC utilized a thin film of graphene sheets on the nickel to achieve anodes without concern of rupturing and polymer-based contamination [24].

TABLE 9.1

The main properties and application of flexible MFCs

Electrodes		Membranes		Microorganism	Application	Ref.
Precursor	Final structure	Precursor	Final structure			
Date pits	Carbon Composite (activated charcoal) coated on the cotton	Salt and agar solution	PVC pipe containing salt and agar	Microbes present in activated sludge	Wastewater treatment and bioelectricity generation	[21]
—	Stainless steel electrodes	Polyvinylidene fluoride, vinyl acetate	PVAc-PVDF coated Cotton Fabric	Palm Oil sludge	Wastewater treatment and bioelectricity generation	[22]
Bacterial cellulose, Aniline	Polyaniline modified bacterial cellulose	Free membrane		Anaerobic sludge of biogas plant	Supercapacitive MFC	[23]
Graphene, Nickle	Thin film graphene on nickel anode	Free membrane		Domestic waste water	Low-cost power source for lab-on-chip applications and water purification	[24]
Fabric, Silver oxide, Ethylene glycol, Polystyrene sulfonate	Anodic and cathodic chambers on the single-layer textile	Free membrane		*Pseudomonas aeruginosa* (PAO1)	Textile-based biobatteries, wearable electronics	[25]

9.3.1.4 Polypyrrole (PPy)

Various approaches were examined to improve the practical performance of flexible MFCs in recent years [28]. One of the most capable techniques is the modification of electrodes using polymers and nanomaterials, such as graphene and carbon nanotubes (CNTs). As a high conductive polymer, polypyrrole is utilized in anode electrode structures. Other remarkable properties that make polypyrrole suitable as a key material in flexible MFCs are its paper-like structure, chemical stability, biocompatibility, flexibility, and production low-cost. Zhao and his coworkers embedded PPy-nanotube into the anode material to promote practical power generation in flexible MFCs. In comparison with typical carbon paper anodes, the paper-like PPy coated anode has given the flexible MFC about six times higher power density [16].

9.3.2 MEMBRANE

The main parameters in determining the flexible efficiency of MFCs are the types of utilized microorganisms, membrane, and electrodes. In addition, the electrical resistance of the whole system, available space of each chamber, anode-to-cathode distance, and ionic strength of the solution have their influence on the MFC performance. The ion-exchange membrane is responsible for a remarkable part of the internal resistance in MFCs that waste a significant part of the system's power generation [29]. Up to now, a wide variety of materials are explored to enhance ion-exchange membrane performance, for example, bipolar membranes, polystyrene, nafion, divinylbenzene, ultrex, dialyzed membrane, glass wool, micro and nano filters [30]. A typical ion-exchange membrane is nafion, which has its deficiencies, including the high cost of synthesis, leakage of ions during transferring, cation transport, and biofouling.

In recent years, flexible substrates and fabric-based materials have been introduced as powerful tools to increase the practical potential of ion-exchange membranes. These materials are soft, porous, and have a three-dimensional structure. Furthermore, their synthesis techniques are easy and simple, for example, pressing or weaving of natural or synthetic fibers. Among various fabric materials, cotton is ubiquitous, low-price, and eco-friendly. Cotton has complicated structures, including crystalline, amorphous, microfibrillar, cavities, and the pores. Zhang et al. prepare an ion-exchange membrane based on traditional cotton to improve the performance of flexible MFCs. In this report, they first cleaned a piece of cotton fabric with acetone and deionized water to remove impurities. They then immersed cotton into the solution of polyvinyl acetate-polyvinylidene fluoride (PVAc-PVDF) and incubated it for 3 minutes. In the next part, the PVAc-PVDF-coated cotton fabric was thoroughly rinsed in 70% ethanol solution for 24 h and washed with deionized water. Finally, the product cotton was rinsed in 1 mol/l HCl solution for 8 hours. The introduced cotton membrane shows low glucose permeability. In addition, prepared flexible MFC based on this cotton-fabric membrane shows good electrical power and current generation potential (400 ± 10 mW/m^2 and 92 mA/m^2, respectively) [22].

9.3.3 MICROORGANISM

Fortunately, a wide range of microorganisms, from all three domains of life, have the potential to generate electrical current and transfer electrons to the anode electrode of MFCs. In other words, in desirable conditions, lots of microbes from yeasts to archaea can generate high densities of electrical power. Among these exoelectrogen microorganisms, bacteria with iron-reducing properties, such as *geobacter sulfurreducens,* have remarkable competence for electrical energy generation. Electrotrophic microorganisms can play a role as electron acceptors in cathodes. The incredible variety of electroactive microorganisms has provided the construction possibility of efficient flexible MFC devices [31]. Figure 9.5 shows typical microorganism applied in MFCs.

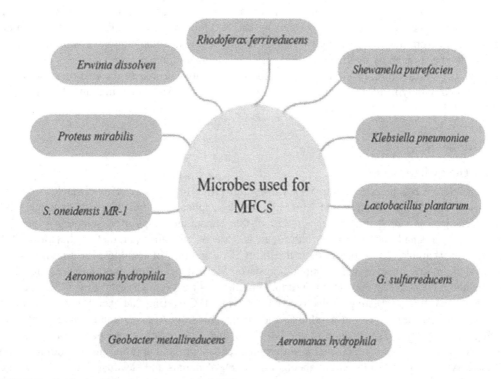

FIGURE 9.5 Typical microorganisms used in MFCs.

9.3.4 FABRICATION

Both top-down and bottom-up methods are utilized to obtain flexible and wearable MFCs. In these approaches, all parts of the system are precisely placed on the fiber or yarn to achieve the desired flexibility in the MFCs structure. In the top-down method, yarn fibers, which are usually physically and chemically functionalized, are weaved into the electronic part of a bulk textile. Notably, batteries and fuel cells usually contain liquid electrolytes, which add more challenges to fabricate flexible devices. The bottom-up approaches are in progress and need more attention to achieve their desired level. Flexible MFCs collect electrons from microorganisms; so, in addition to flexibility, the selected yarn must have a porous and hydrophilic structure to adsorb the cells [32]. Double chamber H-type MFC was created using two polyethylene terephthalate plastic bottles as chambers. A union joint was used to connect the two chambers, with salt and agar solution stored in PVC pipe mounted as a proton exchange membrane in the center of the reactor. The anode chamber also had an inlet and outlet valve. Before starting the reactor, the anode chamber was properly maintained as anaerobic by removing all leaks and purging nitrogen. Using an Arduino board microcontroller and temperature, pH, and voltage sensors, real-time data was saved to a memory card. Installing an aquarium air pump to provide oxygen/air kept the cathode chamber aerobic. Greywater was treated in the anode chamber, while DI water was used in the cathode chamber [29]. Mashour et al. prepared supercapacitive flexible MFC based on electro-polymerized polyaniline modified as conductive bacterial cellulose anode. Fabrication of this flexible MFC is followed by incubation of plexiglass in a sludge of biogas plant phosphate buffer solution and potassium acetate. The whole system consists of anodes and membrane-less air-breathing cathodes, as well as a layer of activated carbon coated on the stainless steel mesh to form a cathode electrode. To prevent leakage, the obtained air cathode was sandwiched between two silicone frames. BC sheets were placed on the stainless steel mesh to prepare an anode electrode and then pressed into the plastic meshes. The as-prepared MFCs incubated in open-circuit voltage (OCV) overnight. In this study, power density of flexible MFCs improved by increasing the active anodic surface [23]. In another example, Justine et al. constructed a flexible micrometer-size MFC for sustainable power generation and water treatment applications. The air cathode was prepared by using carbon cloth and a thin layer of platinum to utilize ambient oxygen from the air. This system has no membrane, and cathode/anode electrodes are on the sides of nafion because applying the air cathode removes the membrane part and provides the possibility of one-chamber construction. In addition, this technique obviously decreases the cost of the total system when compared with typical MFCs and generates a high power density by eliminating the ion-exchange membrane [24].

9.3.5 APPLICATIONS

9.3.5.1 Energy harvesting

Significant internal resistance prevents MFCs from generating high power. This obstacle is one of the main reasons for the low exposure of MFCs in various applications, but they can generate this low electrical energy for a long time (over 9 months); that encourages hope of developing a long-lasting power supply for small wearable electronic devices. One of the other parameters that prove MFCs' application in wearable electronics is their flexibility. Construction of flexible MFCs is discussed and demonstrated in this chapter; achieving a long-term MFC depends on energy-harvesting abilities. Adekunle and coworkers have explored optimum conditions of energy harvesting by MFCs. According to the results, real-time harvesting of electrical energy with MFC, storing that energy in a capacitor in the optimal charge/discharge cycle, is a reliable and sustainable way to accumulate energy without jeopardizing the long-term performance of the MFC. In that report, Sphagnum peat, composted cattle, humus, and maple sawdust were elected as an organic recourse to find appropriate metabolite for power generation. According to results, simultaneous utilization of humus and sawdust shows the best power generation [33].

9.3.5.2 Treatment of wastewater

To progress along the technological readiness level (TRL) scale, MFCs have earned less funding. However, many studies of MFC implementation in a variety of applications demonstrate the technology's practical value. Several groups have investigated MFCs as wastewater treatment reactors, ranging in size from lab-scale (less than 1 L) to thousands of liters. Rather than energy recovery, the main target in such instances was the identification of compounds or the degradation of organics. Zheng and Zhen discuss the practical aspects of MFC water treatment. One of the few long-term field trials of large-scale MFC systems treating real wastewater has been demonstrated. Though its efficiency was not as good as that of laboratory MFC systems, it did show that it could effectively treat the primary effluent by combining bioelectrochemical and aerobic treatment. For a modularized framework, optimizing the communication between MFC modules for optimal output will be a major challenge. The difference between theoretical energy recovery and that of the real world has been discovered by an examination of the onsite application of the emitted energy. The MFC system may have a capital cost compared to small-scale wastewater treatment facilities, with the potential for more cost reductions thanks to the production of low-cost membrane materials. The findings of this study are expected to pique the interest of peer researchers to conduct similar or larger-scale studies of MFC systems to advance MFC technology for long-term wastewater treatment [38].

9.3.5.3 Sensors and portable power machines

Lab-on-a-chip sensors and micro-meter devices have attracted recent attention due to their small size, the smaller sample required, fast analysis capability, and ability to perform the analytical test in a spatial condition that is more comparable with biochemical reaction in real life. Another reason micro-size devices are portable is because of their small size and low weight, but it is possible if they have a micro-size energy source. Therefore, flexible and wearable MFC technology plays a key role in the fabrication of practical portable devices and sensors.

9.4 FUTURE ASPECT

9.4.1 LARGE-SCALE USES

Although lots of experimental researches confirm the remarkable potential of MFCs in various applications, such as power generation and water treatment, few reports are available to assess their large-scale application; scale-up of MFCs has its challenges [34], including low performance, high cost, energy-intensive operation, and scaling bottlenecks. Up to now, a proven area for utilizing MFCs is small-scale batteries technology. These MFC-based batteries possess superior properties, such as being self-standing, having low cost, disposability, simplicity, and long-term power generation ability; these qualities make them suitable for flexible smart textile applications [35], for instance, biofuels [36], water desalination [37], bioremediation [38], and toxicity detection [39]. Although the large-scale capabilities of MFCs are ambiguous, this area of technology is still attractive and may be the only alternative for electricity generation in harsh and shortage situations because, in MFCs, electricity is obtained from approximately limitless sources.

9.4.2 ANODE MANIPULATION

Several organic compounds from various urban and industrial wastewater forms have been successfully investigated. These investigations show that BES can generate power while simultaneously degrading contaminants, making it a viable alternative technology for water cleaning with nil or a positive energy budget. Anode benefits from hydrophilicity, excellent conductivity, durability, low cost, large surface area, biocompatibility, and high capacitance are the benefits. Despite the wide variety of materials tested for the synthesis of anodes having the mentioned properties, this scope of MFCs technology requires more study.

9.4.3 MEMBRANE-FREE MFC

A very important step toward the practical utility of flexible MFCs in medical, wearable, and portable electronics is the resolution of the low power density of these devices; this progress is achievable by eliminating the ion-exchange membrane. In this way, the internal resistance of the system decreases, and the amount of electrical power generation increases. Bilayer MFC has an intricate design, difficult assembly, and expensive construction, which restrict their entrance to the novel technologies. In the membrane-based MFC, the additional distance between the anode and cathode chambers causes higher resistance versus proton traveling and finally decreases the MFC performance. Besides, the wax-based membranes may retract biological reactions because poor proton transfer produces higher proton concentration at the anodes and interrupts the pH of the anodic chamber. In addition, the mechanical deformation of the previous textile MFC decreases device performance due to the misalignment and vertical discontinuity of layers. Furthermore, increasing the stretching cycles causes mechanical resistance to degradation [25].

9.5   CONCLUSION

We summarized the progress made toward creating energy-producing devices based on the MFC. While they have shown early promise, electroactive microbes have yet to deliver on their promise of revolutionizing renewable energy generation. Low power density, nonlinear increases of performance with the system scaling up, and high cost of fabrication and maintenance in large scale hinder MFC progress in practical applications. Small-scale, lightweight, autonomous, and disposable flexible MFCs that consume low power offer an innovative, but reasonable solution. Carbon-based materials with high specific surface area, conductivity, chemical stability, and biocompatibility have been commonly used as flexible anodes. Small system footprints, high versatility, and impressive power output were presented in novel cable-type or fiber-type designs. Despite this, their high fabrication costs continue to present a challenge. Textile-based MFCs have the added benefits of increased mechanical power, flexibility, and wearability, making them ideal for use in other wearable electronic applications. Yarn-based MFCs are especially promising because they can be easily woven or knitted into smart fabric and textiles. Finally, future advancement in this area will be built on the foundations of the aforementioned works, but as our knowledge grows, new focal points are being discovered. Fossil resources will be exhausted in the immediate future. Microbial fuel cells (MFC) are bioelectrochemical devices that use microorganisms as a biocatalyst to turn chemical energy into electrical energy and are one of the green alternatives.

REFERENCES

1. Santoro C, Arbizzani C, Erable B, Ieropoulos I Microbial fuel cells: From fundamentals to applications. A review. *J Power Sources*. 2017;356:225–244.
2. Pandey P, Shinde VN, Deopurkar RL, Kale SP, Patil SA, Pant D Recent advances in the use of different substrates in microbial fuel cells toward wastewater treatment and simultaneous energy recovery. *Appl Energy*. 2016;168:706–723.
3. Obileke K, Onyeaka H, Meyer EL, Nwokolo N Microbial fuel cells, a renewable energy technology for bio-electricity generation: A mini-review. *Electrochem Commun*. 2021;107003.
4. Tee P-F, Abdullah MO, Tan IAW, Amin MAM, Nolasco-Hipolito C, Bujang K Effects of temperature on wastewater treatment in an affordable microbial fuel cell-adsorption hybrid system. *J Environ Chem Eng*. 2017;5(1):178–188.
5. Blanchet E, Erable B, De Solan M-L, Bergel A Two-dimensional carbon cloth and three-dimensional carbon felt perform similarly to form bioanode fed with food waste. *Electrochem commun*. 2016;66: 38–41.
6. Yetisen AK, Qu H, Manbachi A, Butt H, Dokmeci MR, Hinestroza JP, et al. Nanotechnology in textiles. *ACS Nano*. 2016;10(3):3042–3068.

7. Bandodkar AJ, You J-M, Kim N-H, Gu Y, Kumar R, Mohan AMV, et al. Soft, stretchable, high power density electronic skin-based biofuel cells for scavenging energy from human sweat. *Energy Environ Sci.* 2017;10(7):1581–1589.

8. Wrighton KC, Coates JD Microbial fuel cells: Plug-in and Power-on Microbiology-These devices already prove valuable for characterizing physiology, modeling electron flow, and framing and testing hypotheses. *Microbe.* 2009;4(6):281.

9. Bennetto HP, Stirling JL, Tanaka K, Vega CA Anodic reactions in microbial fuel cells. *Biotechnol Bioeng.* 1983;25(2):559–568.

10. Pant D, Van Bogaert G, Diels L, Vanbroekhoven K A review of the substrates used in microbial fuel cells (MFCs) for sustainable energy production. *Bioresour Technol.* 2010;101(6):1533–1543.

11. Bandodkar AJ, Wang J Wearable biofuel cells: A review. *Electroanalysis.* 2016;28(6):1188–1200.

12. Rasmussen M, Abdellaoui S, Minteer SD Enzymatic biofuel cells: 30 years of critical advancements. *Biosens Bioelectron.* 2016;76:91–102.

13. Sender R, Fuchs S, Milo R Revised estimates for the number of human and bacteria cells in the body. *PLoS Biol.* 2016;14(8):e1002533.

14. Osman MH, Shah AA, Walsh FC Recent progress and continuing challenges in bio-fuel cells. Part II: Microbial. *Biosens Bioelectron.* 2010;26(3):953–963.

15. Pang S, Gao Y, Choi S Flexible and stretchable microbial fuel cells with modified conductive and hydrophilic textile. *Biosens Bioelectron.* 2018;100:504–511.

16. Zhao C, Wu J, Kjelleberg S, Loo JSC, Zhang Q Employing a flexible and low-cost polypyrrole nanotube membrane as an anode to enhance current generation in microbial fuel cells. *Small.* 2015;11(28):3440–3443.

17. Yang W, Li J, Lan L, Li Z, Wei W, Lu JE, et al. Facile synthesis of Fe/N/S-doped carbon tubes as high-performance cathode and anode for microbial fuel cells. *ChemCatChem.* 2019;11(24):6070–6077.

18. Cao R, Pu X, Du X, Yang W, Wang J, Guo H, et al. Screen-printed washable electronic textiles as self-powered touch/gesture tribo-sensors for intelligent human–machine interaction. *ACS Nano.* 2018;12(6):5190–5196.

19. Whitesides GM Soft robotics. *Angew Chemie Int Ed.* 2018;57(16):4258–4273.

20. Herbert R, Kim J-H, Kim YS, Lee HM, Yeo W-H Soft material-enabled, flexible hybrid electronics for medicine, healthcare, and human-machine interfaces. *Materials (Basel).* 2018;11(2):187.

21. Mubeen A, Arain M, Memon AA, Ahmed D, Tehseen M Microbial fuel cell with carbon composite coated flexible electrode. Retrieved from http://icsdc.muet.edu.pk/wp-content/uploads/2020/06/240.pdf

22. Zhang B, Jiang Y, Han J A flexible nanocomposite membrane based on traditional cotton fabric to enhance performance of microbial fuel cell. *Fibers Polym.* 2017;18(7):1296–1303.

23. Mashkour M, Rahimnejad M, Mashkour M, Soavi F Electro-polymerized polyaniline modified conductive bacterial cellulose anode for supercapacitive microbial fuel cells and studying the role of anodic biofilm in the capacitive behavior. *J Power Sources.* 2020;478:228822.

24. Mink JE, Qaisi RM, Hussain MM Graphene-based flexible micrometer-sized microbial fuel cell. *Energy Technol.* 2013;1(11):648–652.

25. Pang S, Gao Y, Choi S Flexible and stretchable biobatteries: Monolithic integration of membrane-free microbial fuel cells in a single textile layer. *Adv Energy Mater.* 2018;8(7):1702261.

26. Choudhury P, Prasad Uday US, Bandyopadhyay TK, Ray RN, Bhunia B Performance improvement of microbial fuel cell (MFC) using suitable electrode and bioengineered organisms: A review. *Bioengineered.* 2017;8(5):471–487.

27. Choudhury P, Uday USP, Mahata N, Tiwari ON, Ray RN, Bandyopadhyay TK, et al. Performance improvement of microbial fuel cells for waste water treatment along with value addition: A review on past achievements and recent perspectives. *Renew Sustain Energy Rev.* 2017;79:372–389.

28. Kirchhofer ND, Chen X, Marsili E, Sumner JJ, Dahlquist FW, Bazan GC The conjugated oligoelectrolyte DSSN+ enables exceptional coulombic efficiency via direct electron transfer for anode-respiring Shewanella oneidensis MR-1—a mechanistic study. *Phys Chem Chem Phys.* 2014;16(38):20436–20443.

29. Gude VG Wastewater treatment in microbial fuel cells: An overview. *J Clean Prod.* 2016;122:287–307.

30. Angioni S, Millia L, Bruni G, Tealdi C, Mustarelli P, Quartarone E Improving the performances of Nafion™-based membranes for microbial fuel cells with silica-based, organically-functionalized mesostructured fillers. *J Power Sources.* 2016;334:120–127.

31. Logan BE, Rossi R, Saikaly PE Electroactive microorganisms in bioelectrochemical systems. *Nat Rev Microbiol.* 2019;17(5):307–319.

32. Shin S, Kumar R, Roh JW, Ko D-S, Kim H-S, Kim SIl, et al. High-performance screen-printed thermo-electric films on fabrics. *Sci Rep.* 2017;7(1):1–9.

33. Adekunle A, Raghavan V, Tartakovsky B Carbon source and energy harvesting optimization in solid anolyte microbial fuel cells. *J Power Sources.* 2017;356:324–330.

34. Zebda A, Alcaraz J-P, Vadgama P, Shleev S, Minteer SD, Boucher F, et al. Challenges for successful implantation of biofuel cells. *Bioelectrochemistry.* 2018;124:57–72.

35. Gao Y *Flexible and Stretchable Bacteria-Powered Biobatteries.* State University of New York at Binghamton; 2020.

36. Choi O, Sang B-I Extracellular electron transfer from cathode to microbes: Application for biofuel pro-duction. *Biotechnol Biofuels.* 2016;9(1):1–14.

37. Chen S, Liu G, Zhang R, Qin B, Luo Y Development of the microbial electrolysis desalination and chemical-production cell for desalination as well as acid and alkali productions. *Environ Sci Technol.* 2012;46(4): 2467–2472.

38. He Y-R, Xiao X, Li W-W, Cai P-J, Yuan S-J, Yan F-F, et al. Electricity generation from dissolved organic matter in polluted lake water using a microbial fuel cell (MFC). *Biochem Eng J.* 2013;71:57–61.

39. Yang W, Wei X, Fraiwan A, Coogan CG, Lee H, Choi S Fast and sensitive water quality assessment: A µL-scale microbial fuel cell-based biosensor integrated with an air-bubble trap and electrochemical sensing functionality. *Sensors Actuators B Chem.* 2016;226:191–195.

10 Flexible Silicon Photovoltaic Solar Cells

Pratik Deorao Shende, Krishna Nama Manjunatha, Iulia Salarou, and Shashi Paul
Emerging Technologies Research Center, De Montfort University,
The Gateway, Leicester, United Kingdom

CONTENTS

10.1 Introduction..171
10.2 Classification of flexible photovoltaic solar cells 172
    10.2.1 Inorganic flexible photovoltaic solar cells..174
    10.2.2 Organic flexible photovoltaic solar cells 174
    10.2.3 Hybrid flexible photovoltaic solar cells...175
10.3 Flexible silicon (Si) photovoltaic solar cells...175
    10.3.1 Flexible crystalline silicon solar cells 176
        10.3.1.1 Recent progress in flexible crystalline silicon solar cells.........176
    10.3.2 Flexible thin-film amorphous silicon solar cells.....................................176
        10.3.2.1 Recent progress in flexible amorphous silicon solar cells........178
    10.3.3 Silicon nanostructures for flexible solar cells 179
        10.3.3.1 Silicon nanowire flexible solar cells...180
        10.3.3.2 Silicon nanopyramid solar cells..183
        10.3.3.3 Silicon nanoparticles for solar cells..185
        10.3.3.4 Silicon ink-based solar cells..187
10.4 Outlook and conclusions..189
Acknowledgement 191
References...191

10.1 INTRODUCTION

Today, approximately 38% of global electricity generation still depends on nonrenewable energy sources, such as conventional fossil fuels; these dominate renewable energy sources, such as wind energy, solar energy, tidal energy, hydro energy, and geothermal energy, etc. [1]. However, renewable energy sources have an indisputable potential to replace conventional fossil fuels currently in use, and they have added advantages, such as a low carbon footprint, which produces a positive impact on fauna and flora, as well as the reduction, or complete elimination, of greenhouse gases (GHGs) and reassurance & sustainability of a healthier environment [2]. Moreover, an International Energy Agency (IEA) anticipates that within a few years, renewable energy sources will surpass coal capacity and solar energy, becoming the most prominent source of global electricity power supplier among all other renewable energy sources by 2030 [3].

The sun's light is an excellent alternative source since it is the most abundant and clean source of renewable energy [5]. The earth receives about 1.8×10^{11} MW from the sun every day, which is sufficiently higher than the present global energy needs [6,7]. Hence, the sun has a key advantage over the conventional power generators; photovoltaic (PV) solar systems use solar energy to produce electricity directly from the sunlight [8]. Interestingly, the overall global PV solar system capacity has increased from 107 GW in 2020 to 117 GW in 2021; see Figure 10.1 [4]. The reduced prices and versatility of solar photovoltaic systems make it one of the key players in the renewable energy market. The use of the photovoltaic effect for the conversion of photons to electricity is the most direct use of solar energy. Furthermore, many early experiments have documented the photovoltaic effect as a phenomenon, with the experiment of Alexandre-Edmond Becquerel in 1839 being the best known [9].

Silicon (Si) is the second most abundant element in the earth's crust, comprising 25% percent of the crust's mass [10]. Considering its adoption in a photovoltaic module, silicon is a nontoxic, biocompatible, and effectively stable material compared to other semiconductor materials [11]. Owing to these significant properties, versatility, and improved industrial and technological development, Si has been identified as a nucleus of the advanced semiconductor industry [12]. As a result, the conventional wafer-based single or mono-crystalline and multi or poly-crystalline silicon photovoltaic solar cells currently constitute nearly 92% of the photovoltaic solar cells sector, followed by a thin film based amorphous silicon (a-Si) photovoltaic solar cells ~1% of the PV sector [13]. Silicon nanostructure-based photovoltaic solar cells are presently at the exploratory stage, which could be commercialized in the near future. Other photovoltaic technologies, such as cadmium telluride (CdTe), copper indium gallium selenide (CIGS), dye-sensitized (DSSC), perovskite, and combination of these with silicon (hybrid solar cells), represent the remaining share. According to R. Singh [14] and R. Arora et al. [15], silicon still remains a dominant photovoltaic solar cell material in the 21st century compared to other existing and emerging materials.

However, Goetzberger et al. [16] indicated that the conventional silicon-wafer-based photovoltaic solar cells are generally rigid, heavier, bulkier, and breakable under certain conditions. Moreover, silicon wafer-based solar cells require more material; there is a minimum wafer thickness (500 μm) so that the wafer can be handled by a robotic arm without damaging it. The average penetration depth of photon into silicon depends on the incident wavelength (e.g., the penetration depth is 100 μm for 980 nm wavelength, and it decreases as the wavelength of incident light is decreased), [17] and thus, a remaining material is not active in the generation of carriers.

Additionally, these solar cells demand advanced fabrication facilities to transform them into flexible photovoltaic solar cells; as a result, they are extremely expensive to manufacture [18]. Currently, thin film based a-Si, ultra-thin crystalline silicon (c-Si), and silicon nanostructures based photovoltaic solar cells are fabricated at research labs aiming to achieve flexible, and lightweight, solar cells. Flexible a-Si solar cells are commercially available in the market, and they are suitable for both indoor and outdoor applications [19,20]. Moreover, flexible photovoltaic cells can provide power to flexible and portable electronic appliances, such as flexible biomedical devices and flexible wearable gadgets, and they can be integrated with the curved surfaces of vehicles or buildings [21,22]. Consequently, Si-based thin-film a-Si, ultrathin c-Si, and nanostructured silicon-based flexible photovoltaic solar cells have been further investigated by many reearchers, and they can be integrated at a lower cost compared to conventional wafer-based silicon photovoltaic solar cells, with some compromise in the efficiency.

10.2 CLASSIFICATION OF FLEXIBLE PHOTOVOLTAIC SOLAR CELLS

The classification of flexible photovoltaic solar cells is generally based on the semiconductor materials (absorber) used in the solar cell. These materials include inorganic materials, such as silicon, gallium arsenide (GaAs), cadmium telluride (CdTe), copper indium gallium selenide (CIGS), [23] and organic materials, such as polymers, small molecules [24], and a combination of both organic and inorganic materials, which have been referred as hybrid materials [25]. Furthermore, the flexible photovoltaic solar cells in this review article are classified into three categories: inorganic flexible photovoltaic solar cells, organic flexible photovoltaic solar cells, and organic-inorganic (hybrid) flexible photovoltaic solar cells; see Figure 10.2.

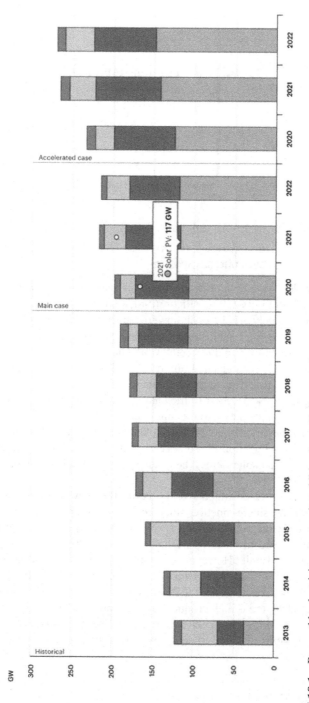

FIGURE 10.1 Renewable electricity net capacity addition by technology, main, and accelerated cases 2013 to 2022 [4]. "IEA, Renewable electricity net capacity additions by technology, main and accelerated cases" IEA, Paris, 2013–2022. [Online]. Available: https://www.iea.org/data-and-statistics/charts/renewable-electricity-net-capacity-additions-by-technology-main-and-accelerated-cases-2013–2022.

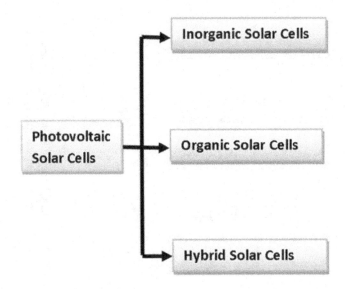

FIGURE 10.2   Classification of photovoltaic solar cells based on absorber materials.

10.2.1   INORGANIC FLEXIBLE PHOTOVOLTAIC SOLAR CELLS

Silicon constitutes the major element of silicon-wafer-based inorganic flexible photovoltaic solar cells, which include thin-film and ultrathin crystalline silicon (c-Si) photovoltaic solar cells. Flexible silicon solar cells have shown approximately ~1–19% of power conversion efficiency (PCE) [26], with an improved transformation technique and nanostructure geometry compared to conventional mono-crystalline and poly-crystalline silicon-based solar cells, which has demonstrated dominant efficiency of ~18–27% [27,28]. Furthermore, the hydrogenated amorphous silicon (a-Si:H), copper indium gallium selenide (CIGS), and cadmium telluride (CdTe) are the main materials for inorganic flexible photovoltaic solar cells. By demonstrating nearly ~10–23% of efficiency [23,29], such types of flexible photovoltaic solar cells reduce production cost since they avoid using thicker silicon wafers and use less material as compared to the conventional wafer-based photovoltaic solar cells. Inorganic flexible photovoltaic solar cells consist of new emerging single-junction PV solar cell technologies, such as quantum dot (QD) solar cells and silicon nanostructure-based solar cells. These single-junction solar cells have demonstrated a high efficiency of ~10–18% [30,31].

Multi-junction solar cells have demonstrated world-record efficiency for those produced in the research labs as compared with the single-junction solar cell. This efficiency is attributed to a higher absorption coefficient in the wide range of the solar spectrum. As defined by W. Shockley et al. [32], single-junction solar cells are limited to maximum efficiency of 33% by radiative recombination, also known as the Shockley-Queisser Limit. However, by using an innovative multi-junction solar cell technology (using a combination of different semiconductors), above 38% of efficiency has been achieved [33]. Multi-junction solar cells, based on III–V semiconductors, have fewer commercial applications due to the increased material and production cost compared to the other thin-film based photovoltaic solar cells. This technology has been referred to as the most advanced photovoltaic solar cells technology to date and is generally used in advanced applications, such as space exploration (satellites) and the military [34].

10.2.2   ORGANIC FLEXIBLE PHOTOVOLTAIC SOLAR CELLS

Certain organic polymers or molecules show semiconductor behavior with the excitation of carriers (electrons) when exposed to light; these organic materials are used as absorber layers in the flexible organic solar cells (OSCs). Such types of OSCs are inexpensive [24]. OSCs are unique and have numerous potential benefits, including simple processability, flexibility, low material cost, and low fabrication cost for large-scale production [35]. Regardless of their benefits, they also have several limitations. One such limitation is that while commercial inorganic wafer-based Si photovoltaic solar cells may last for approximately 25 years, organic-based solar cells barely last a handful of years [36]. This limitation is attributed to significant degradation when exposed to the ambient atmosphere (moisture and air) with sunlight, which reduces efficiency over time [36,37].

These OSCs typically include a bulk heterojunction comprising a solution mix of an electron acceptor, an electron donor, an n-conjugated polymer semiconductor including poly 3-hexylthiophene (P3HT), and a fullerene such as [6,6]-phenyl C61-butyric acid methylester (PCBM) in a photoactive layer [38]. The fabrication process starts with a substrate, typically, a plastic sheet such as polyethylene terephthalate (PET), and a roll-to-roll (R2R) manufacturing technique is used to reduce the cost [39,40]. Such a fabrication process significantly reduces the overall duration involved in the fabrication. It is efficient as far as the manufacturing process is concerned, but the efficiency of the solar cell is low. Notably, a recent review of polymer organic solar cells shows that their efficiency is limited to ~ 10% [41]. This limitation means that their efficiency is lower than that of the flexible silicon-based solar cells. Therefore, the inorganic material-based flexible photovoltaic solar cells are comparatively better when compared with the organic polymer-based flexible photovoltaic solar cells, and the interest in exploring silicon-based flexible solar cell further continues with a good reason.

10.2.3 HYBRID FLEXIBLE PHOTOVOLTAIC SOLAR CELLS

The hybrid photovoltaic solar cells are the combination of both organic and inorganic materials, e.g., an organic polymer layer matrix with a combination of silicon or GaAs or perovskite-based multi-junction solar cells. As a result, they provide the advantages of both organic-inorganic materials, such as efficiency, stability, durability, flexibility, efficient synthesis, and versatile manufacturing [42]. However, due to a combination of several material properties and complex device structures, various additional fabrication processes are incorporated; this results in a longer manufacturing period and higher material and production costs compared to thin-film single-junction solar cells [43]. Consequently, these hybrid solar cells have been explored due to tuneable absorption spectra and enhanced efficiency [44]. Ronald et al. [25] demonstrated hydrogenated amorphous silicon (a-Si:H) and organic-polymer-based hybrid multi-junction flexible solar cell with a power-conversion efficiency of 11.7%. Khang, D. Y. et al. [45] incorporated silicon nanostructures, rather than the standard silicon wafer with organic material (PEDOT:PSS), as an alternative to obtain flexible hybrid solar cells; however, the efficiency of such solar cells is reduced to ~6%. The efficiency of hybrid solar cells is predominately affected by the absorption of moisture during the deposition. He et al. [46] have investigated and resolved moisture-associated problems by fabricating these hybrid solar cells using a glove box in an inert atmosphere (argon gas). Further research efforts are needed to improve the stability and efficiency of organic-inorganic hybrid solar cells.

10.3 FLEXIBLE SILICON (SI) PHOTOVOLTAIC SOLAR CELLS

For all these reasons, silicon photovoltaic solar cells show better stability and durability, have a longer carrier lifetime, and involve a mature, industrially accepted fabrication process. Therefore, Si-based photovoltaic solar cells outnumber other nonsilicon photovoltaic solar cells [47]. Such silicon-based solar cells, where they offer flexible solar cell production possibilities, open new paths and integrate seamlessly into the next generation of flexible electronics, such as IoT (Internet of Things) devices. Additional sections will provide further information on recent developments, manufacturing techniques, and the benefits of flexible silicon solar cells, which are central to this review chapter.

10.3.1 FLEXIBLE CRYSTALLINE SILICON SOLAR CELLS

Monocrystalline and polycrystalline silicon has been used for more than 50 years in commercial solar panels that already show the strength of silicon as a promising semiconductor for photovoltaic applications. Due to the high demand for flexible solar cells in the PV market, because they offer lightweight, good conformal capability and are cost-effective, it is already possible to manufacture semiflexible c-Si solar cells with efficiencies in the range of 14–19% (depending on the crystallinity and purity of silicon). Such semi-flexible SCs are fabricated by reducing the thickness of bulk silicon wafers at the micro-scale to obtain ultrathin wafers during the ingot slicing process [48]. However, such a production process results in a 50% waste of silicon from the ingot during the slicing and polishing process [49]. As a result, about 50% of the original single crystal of silicon is effectively used for the fabrication of solar cells. That challenge indicates, typically, growing the crystalline silicon ingots and then slicing them into wafers, a process that requires a timeframe of about one month, which contributes toward 50% production cost of using a silicon substrate for crystalline solar cells [50,51]. Consequently, due to limited bendability (bending radius less than 10 cm) [52], large wafer thickness, high production cost, and inadequate flexibility, crystalline silicon-based flexible solar cells are still far away from other emerging solar cells technologies for commercialization [21].

10.3.1.1 Recent progress in flexible crystalline silicon solar cells

The development of flexible crystalline solar cells was considerably challenging due to the higher cost and the complexity in the manufacturing process, which uses flexible nonsilicon substrates (glass, plastic, metal sheets) on large-scale production [53]. However, in 2008, a lightweight, bendable, and flexible integrated circuit (ICs) has been successfully demonstrated by using transfer printing technology (TPT) that utilizes monocrystalline silicon with nano-band array connections [54]. Later in the same year, for the first time in the photovoltaic solar cell industry, the monocrystalline flexible solar cell was manufactured by employing conventional methods of photoetching and TPT on a plastic substrate [55]. This method utilizes a reduced amount of silicon as an absorbent material, showing excellent light absorption that translates to the power-conversion efficiency of 4% with an individual microcell [56]. Such a process is promising for manufacturing an inexpensive, industrially compatible manufacturing of monocrystalline silicon-based flexible solar cells, also called "flat flexible solar cells" [56]. Michael R. et al. demonstrated a novel free-standing transfer process to obtain an ultrathin monocrystalline Si solar cell made from 50 μm thick wafer with a PCE of 17%. This free-standing transfer process uses a layer transfer technique (LTP) and TPT by providing a simple process and eliminates the complex process of transferring Si onto a glass substrate with an epoxy resin. The solar cell fabricated by this process not only shows better absorption but also provides higher efficiency [57]. Recently, there is further progress in the transformation of traditional wafer-scale interdigitated back contact (IBC) based monocrystalline silicon cells into flexible solar cells. Such solar cells have achieved a global record of ultra-stretchability of 95% and an efficiency of 19% by employing a novel laser-patterning-based corrugation technique [52]. This method relies on the patterning of the hard mask to obtain the islands of silicon, which can be readily interconnected via the IBC grid, as illustrated in Figure 10.3. Moreover, this technique demonstrates incredibly negligible degradation in electrical characteristics of a solar cell as fabricated; however, cracks were developed in the hard mask, with the smaller bending radius of solar cells leading to damage in the electrical connections. Table 10.1 illustrates the reported efficiency (η), fill factor (FF), open-circuit voltage (V_{OC}), short circuit current density (J_{SC}), thickness, and chosen substrate for the fabrication of flexible crystalline silicon solar cells.

10.3.2 FLEXIBLE THIN-FILM AMORPHOUS SILICON SOLAR CELLS

In the recent decade, there has been significant development and interest in the flexible thin-film amorphous silicon (a-Si) solar cells; as a result, they have increasingly received global attention [59]. Mostly, hydrogenated amorphous silicon (a-Si:H) has been used in commercial devices. a-Si is non-crystalline

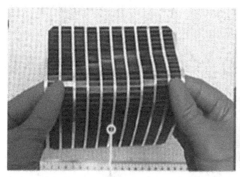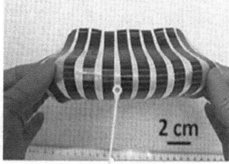

FIGURE 10.3 Image of monocrystalline flexible silicon solar cell. IBC device structure was used for this solar cell. Reproduced with permission from ref. [52] Copyright (2021), John Wiley & Sons, Inc.

TABLE 10.1

Selective summary of flexible crystalline silicon solar cells performance

Type	Substrate	Thickness (μm)	Efficiency (%)	Jsc (mA/cm^2)	FF (%)	Voc (V)	Ref.
Mono/c-Si	Plastic	100	4	–	–	–	[56]
Mono/c-Si	Glass	47	17	33.5	78.7	0.64	[57]
Mono/c-Si	Elastomer	50	19	38.75	75	0.63	[52]
Ultra-thin c-Si	Ultrathin Si wafer	14	14.9	–	–	–	[58]

silicon, and, with the use of hydrogen, defect density, and recombination of carriers are minimized [48]. Thin film of hydrogenated amorphous silicon (a-Si:H) solar cells is possible to be fabricated on flexible substrates and provides opportunities to use such solar cells in flexible electronic devices. Flexible a-Si-H solar cells are fabricated on transparent and nontransparent flexible substrates, such as plastic (polyethylene-naphthalene -PEN, polyethylene terephthalate – PET, other polymers) and metal sheets (aluminium, copper, titanium, stainless steel) [26].

Söderström, T. et al. [59] mentioned that, among various available solar cell technologies, Si-based solar cells have, over the years, dominated the photovoltaic market since the vast majority of commercial solar cell modules used today are based on crystalline Si; a rising interest exists in the thin-film silicon for its use in solar cells, owing to its mature and cheaper fabrication process. Amorphous silicon-based solar cells have dominated the market due to interesting features, including flexibility, lightweight, and reasonable power density with a competitive cost per watt. However, amorphous silicon has an indirect bandgap, higher defect density, and suffers from light-induced degradation [48]. Therefore, the challenge in this technology is to achieve higher power-conversion efficiency.

Hydrogenated amorphous silicon (a-Si:H) solar cells have also shown their uses in the indoors (artificial light) and under poor light conditions by generating better power as compared with the crystalline silicon solar cells. The added advantage of a:Si:H, unlike other thin-film solar cells, is that a range of flexible substrates, with economics in mind, can be used to realize a-Si:H solar cells. These solar cells can be fabricated as a single module with an area of approximately one square foot [60]. Thin-film a-Si:H solar cells continue to improve efficiently; it has been predicted that they would overtake the typical inflexible/rigid photovoltaic technologies that were in use during the mid-20th century and become more effective in generating electricity where other photovoltaic cells cannot be used, for instance, on curved surfaces [61]. This is achieved with the use of tandem device architecture comprised of absorber layers

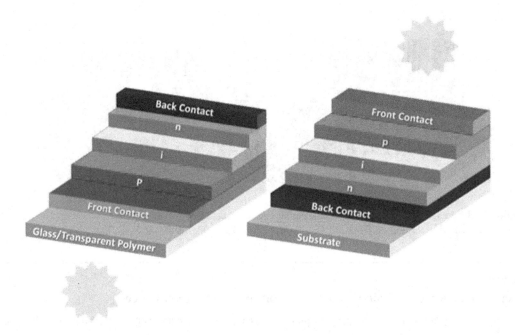

FIGURE 10.4   The standard commonly used device structures of a-Si:H solar cell. This image is reused under the Attribution 4.0 International (CC BY 4.0) [26].

such as a-Si:H, a-SiGe:H, and a-Ge:H. As reported by Ramanujam, J. et al. [13], flexible substrates are more advantageous compared to rigid substrates for the fabrication of futuristic electronic devices. a-Si:H further identified some of the advantages of flexible substrates, which include the suitability of solar cells on curved surfaces and building-integrated photovoltaics (BIPV). a-Si:H flexible modules are much cheaper compared to the conventional c-Si solar cell, and 1% of Si (amount of material) is used as compared with the c-Si solar cells. Moreover, metal frames and glass cover used in Si modules are not used in the flexible a-Si:H solar cells; this not only reduces the cost, but also reduces the weight of the solar cells.

10.3.2.1   Recent progress in flexible amorphous silicon solar cells

In 1976, a first thin-film amorphous silicon a-Si solar cell with ~$1\mu m$ thickness was fabricated using hydrogenated amorphous silicon (a-Si:H) by Carlson and Wronski [62] of RCA (Radio Corporation of America) Laboratory in the United States, which was then referred as a new era of thin-film solar cell technologies. Since then, different experiments were conducted to improve the efficiency of a-Si:H solar cells, including fabrication on various flexible substrates. The structure of (a-Si:H) solar cell depends on the substrate used, for metal substrates n-i-p structure is employed and substrate-like flexible glass p-i-n structure is employed, as shown in Figure 10.4.

Seung Y.M. et al. has demonstrated flexible a-Si:H solar cell with 9.3% power conversion efficiency using dual-junction p-i-n type a-Si:H with hydrogenated silicon fabricated on flexible 100um thin glass substrate using low-pressure chemical vapor deposition technique (LPCVD) [63]. It is reported that with the use of textured transparent fluorine-doped tin oxide (FTO) as a front electrode, absorption of light is increased due to the light scattering. Zhang et al. [64] claimed 8.17% efficiency for a solar cell fabricated on polymer substrate using a sol-gel-based nonprinting method with indium tin oxide (ITO) as a top electrode. The fabrication process reported here seems promising and efficient compared with the previous solar cell fabricated on glass demonstrating similar comparable efficiency. Additionally, nanoholes and nanocones films were added to the device, which served as an

TABLE 10.2

Selective summary of flexible amorphous silicon solar cells performance fabricated on different flexible substrates

Substrate	Efficiency (%)	Jsc (mA/cm^2)	FF (%)	Voc (V)	Year	Ref.
Cellulose	5.5	11.5	47.2	0.86	2017	[66]
Glass	9.3	10.2	73.7	0.82	2014	[63]
Polymer	8.17	15.57	0.64	0.834	2017	[64]
Metal foil	8.05	14.6	63.9	0.86	2016	[65]

antireflection layer and resulted in the increased efficiency due to reduced reflection of light. Although results were promising for the pristine device, ~17% of degradation in the efficiency was observed when solar cells were subjected to 100,000 cycles of 180° bending angle. Lin, Yinyu et al. [65] demonstrated, dual-layer pattern nanostructured-based thin-film a-Si solar cell on flexible titanium foil, which has shown the efficiency of 8.05% under AM 1.5 irradiation. The aforementioned devices have shown the successful fabrication of solar cells on flexible substrates. However, the researchers need to know that the solar cells fabricated on flexible glass can provide additional features, like high-temperature resistance and increased transparency for the solar cells. Similarly, solar cells fabricated on flexible plastic substrates provide the possibility of fabricating lightweight solar cells as compared with the solar cells fabricated on flexible glass and metal sheets. It is also equally important to mention that ultra-thin glass substrate has limited flexibility, which makes it unrollable and requires careful handling to avoid cracks in the solar cells. Table 10.2 illustrates the reported efficiency (η), fill factor (FF), open circuit voltage (V_{OC}), short circuit density (J_{SC}), thickness and chosen substrate for the fabrication of flexible amorphous silicon solar cells.

10.3.3 SILICON NANOSTRUCTURES FOR FLEXIBLE SOLAR CELLS

The most conventional monocrystalline and polycrystalline silicon photovoltaic solar cells are breakable, bulkier, require an additional antireflection coating to absorb more light, and are in-flexible and thicker (~500μm) to achieve a stable and efficient solar cell that promotes ease in the fabrication process [67]. Conventional bulk crystalline silicon has a partial degree of freedom in the material design due to an indirect bandgap that cannot be altered. However, energy bandgap engineering, by reducing the size of materials into the nanoscale, provides opportunities to alter the properties of bulk silicon, e.g., direct bandgap semiconductors can be obtained from the silicon nanostructures [68]. Therefore, for the flexible application of photovoltaic solar cells, nanostructured silicon has opened a new era of solar energy-conversion devices by demonstrating flexibility and superior light-trapping capabilities [69,70]. The nanostructured silicon provides supra-indirect-gap absorption [71] along with excellent optical (reduced reflection/increased absorption) and electrical properties [72]. Light incident on a nanostructured silicon surface (nanowires, nanopyramid, nanopillars, nanovoids, nanodents, nanowells, and nanoholes) shows a significant increase in the light absorption owing to surface plasmon resonance, gradual change in the effective refractive index, and light-trapping capabilities. Moreover, nanostructured silicon shows significant light-trapping capability without the need for an additional anti-reflection layer [12]. Several nanostructured silicon, such as silicon nanowire [73], silicon nanopyramids [74], silicon nanohole [75], and silicon nanoparticles [76], are discussed in the later sections, including their possible integration in flexible photovoltaic devices.

10.3.3.1 Silicon nanowire flexible solar cells

According to Tsakalakos et al. [77] silicon nanowire-based solar cells are a promising class of nanostructured photovoltaic devices for immediate commercialization. These solar cells determine an immediate route for increasing the efficiency due to the geometry of the nanostructures that show better absorption of light compared to other nanostructures [78,79]. Over the past decade, many studies have been conducted extensively on silicon nanowire (SiNW) solar cells and their potential use in the next generation photovoltaic technologies, with an aim of high-power conversion efficiency and reduced overall cost [80,81]. As described by Li G. et al. [82], one-dimensional silicon nanowires show promising applications in gas sensors, drug delivery, solar cells, lithium-ion batteries, medical diagnostics, and photocatalysis, among others with a diameter in the range of 1–50 nm. These nanowires enhance the efficiency of solar cells as a result of the better electrical and optical properties, and such required properties can be tailored by tuning their physical dimensions. Such nanostructures create unique opportunities for improved performance in other devices, such as sensors and optoelectronic devices (including photovoltaic devices) [83]. Garnett, Erik et al. [71] stated that for a solar cell that may be regarded as an optoelectronic device, the nanowire geometry has many advantages: strong light absorption, better efficiency in the carrier collection and separation owing to the short direct path for the charge transportation. Manjunatha et al. [84] have recently shown the elimination of toxic gases that are used as dopants for the fabrication of SiNWs solar cells; thus, the fabrication process is sustainable and minimizes carbon footprints in the environment. Other advantages of the nanowire solar cells include lattice strain relaxation, which enables a greater degree of freedom to design efficient solar cells, and control over material properties, such as interface gradient and dopant composition. The study of solar cell properties in the nanoscale geometrics serves as an effective platform for the advancement in the physical limits posed by the standard bulk material-based solar cells with the adoption of new device concepts [80].

Accordingly, the device structure of a silicon nanowire solar cell relates to its physical and geometrical aspects; this can be further classified into two different device architectures of silicon nanowire solar cells; these include radial and axial junction silicon nanowire solar cells. Such solar cell structures can be further explored with heterojunction and homojunction solar cells, with a combination of different materials (e.g., Si and Ge) and/or change in the properties of similar material depending upon their crystallinity and size (e.g., c-Si and a-Si) [80]. SiNWs solar cells have shown a rapid collection of photogenerated charge carriers before recombination; charge carriers travel radially between the core of the silicon nanowires in the radial junction solar cell, and the charge carriers travel longitudinally along the silicon nanowires in the axial junction solar cell.

As demonstrated by Tsakalakos et al. [77], fabrication of silicon nanowire-based flexible solar cells has been carried out by growing p-type silicon nanowires with a diameter 109 nm on low cost, flexible stainless steel (SS) metal sheet using a novel chemical vapor deposition (CVD) technique at 650° C. Solar cells fabricated in the aforementioned report use the p-n junction device structure by depositing a thin layer of hydrogenated amorphous silicon (a-Si:H) over the silicon nanowire array using PECVD. The device structure of silicon nanowire-based solar cells is shown in Figure 10.5. The electrical properties of such devices were reported: power-conversion efficiency of ~0.1% with FF = 28% and Voc = 0.13V. Furthermore, Xiaobing Xie et al. claimed 3.56% of efficiency with radial n-i-p SiNWs based flexible solar cell fabricated on stainless steel substrate using a PECVD technique [85]. With the use of n-i-p as compared with the aforementioned n-p device structure, solar cell electrical parameters have significantly improved with Voc = 0.62V and Jsc = 13.36 mA/cm^2. Increased power-conversion efficiency is additionally attributed to low shunt resistance and high series resistance in the solar cell. However, more work needs to be undertaken to alter the desirable contact resistance, shunt resistance by optimizing nanowire geometry, and reduce the recombination of charge carriers, thereby increasing the power-conversion efficiency (PCE) of silicon nanowire-based flexible solar cells.

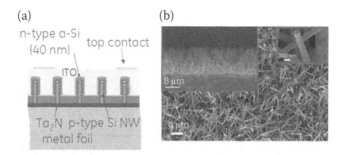

FIGURE 10.5 The device structure of silicon nanowire-based solar cell and SEM image of silicon nanowires grown on metal sheets. Reprinted from ref. [77], Tsakalakos, L., J. Balch, J. Fronheiser, B. A. Korevaar, O. Sulima, and J. Rand. "Silicon nanowire solar cells." Applied physics letters 91, no. 23 (2007): 233117, with the permission of AIP Publishing.

According to Xiaolin Sun et al. [86], the high power-to-weight ratio (PTWR) is an important factor to be considered for research focused on flexible solar cells. Owing to PTWR, flexible SiNWs based solar cell is fabricated using a three-dimensional radial p-i-n junction solar cell on a 15 μm aluminum sheet with a negligible compromise on the electrical properties of the solar cell tested at a bending radius of 5 mm. This solar cell demonstrated an efficiency of 5.6%. The fabrication process and device structure are shown in Figure 10.6.

Fabrication of SiNWs based radial junction flexible solar cells on the aluminum foil eliminates the use of an extra surface-texturing step on the TCO layer that is commonly used in conventional a-Si:H solar cells [64]. Every standard thin-film-based solar cell uses TCO as the bottom electrode, and Ag is deposited over the TCO to minimize the resistive losses. Not only the thin-film devices, but also wafer-based electronic devices use Ag to minimize the resistive losses. Morphology of TCO in such solar cells that are used as a bottom electrode is altered to obtain more back scattering of light into the device to absorb more light. However, it is possible to eliminate the use of TCO and Ag in SiNWs solar cells fabricated on unpolished Al foil; as a result, ~46% of the cost is minimized that accounts for the cost of Ag and TCO electrodes as shown in Figure 10.7. Textured TCO is replaced with the unpolished Al foil that provides a similar back scattering of light into the device to absorb more light. Moreover, light-harvesting capabilities have been achieved with the use of SiNWs and unpolished Al foil that enables the elimination of additional antireflection layers; this is evidenced by observing the improvement in J_{SC} of 14.2 mA/cm^2 as compared with the planar device. SiNWs based solar cell discussed here has shown outstanding PTWR of >1300 W/kg, which signifies the advantage of employing SiNWs flexible solar cells for commercialization as compared with the thin-film a-Si:H solar cells.

Further work by Zhang et al. [87] demonstrates the increased efficiency of flexible radial tandem junction (RTJ) SiNWs solar cell from 5.6% to 6.6% with the use of additional hydrogenated amorphous silicon germanium (a-SiGe:H) layer that has narrow bandgap and promoting broader external quantum efficiency (EQE) response. Other absorber layers, electrodes, and the substrate in this device remained the same, including the device architecture Figure 10.8. Along with the increase in efficiency, the performance of solar cells negligently changed up to a bending radius of 10 mm as compared with their previous device that was tested up to a bending radius of 5 mm.

Incredibly, this RTJ-3D structured SiNWs based flexible solar cell demonstrated PTWR of ~1628 W/kg, which is the highest compared to other reported work and tabulated in Table 10.3. In the future, low-temperature fabrication techniques can be implemented on less expensive polymer or cardboard substrates to achieve flexible thin-film solar cells at an extremely low cost without any compromise in performance. Employing wider bandgap material such as uc-SiOx:H, which has higher transparency and less absorption loss compared to a-Si:H [88], or employing wide-bandgap

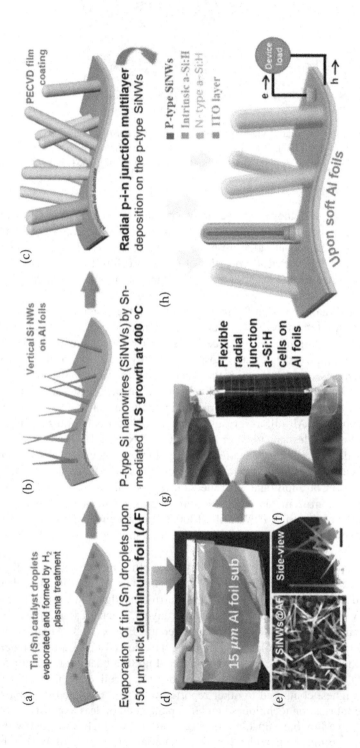

FIGURE 10.6 (a–c) The fabrication process of SiNW based a-Si:H radial junction solar cell on aluminum foil, (d) aluminum foil with thickness 15 μm, (e–f) top and side-view of the SiNWs, (g) solar cells deposited on EVA polymer, (h) p-i-n multilayer device structure of the radial junction solar cell. Reprinted from ref. [86]. Firmly standing three-dimensional radial junctions on soft aluminium foils enable extremely low-cost, flexible thin-film solar cells with very high power-to-weight performance, Nano Energy 53 (2018): 83–90, Sun, Xiaolin, Ting Zhang, Junzhuan Wang, Fan Yang, Ling Xu, Jun Xu, Yi Shi, Kunji Chen, Pere Roca i Cabarrocas, and Linwei Yu, Copyright (2021), with permission from Elsevier.

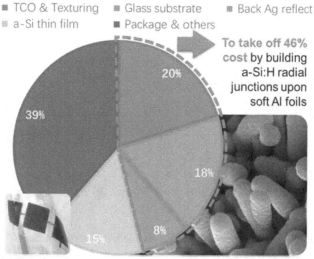

COST BREAK-DOWN IN A-SI:H SOLAR CELLS

■ TCO & Texturing   ■ Glass substrate   ■ Back Ag reflect
■ a-Si thin film   ■ Package & others

To take off 46% cost by building a-Si:H radial junctions upon soft Al foils

20%

39%

18%

8%

15%

FIGURE 10.7 The fabrication cost reduced by ~46% by employing Al foil as a substrate compared to other flexible substrate-based SiNWs and a-Si:H solar cells. Reprinted from ref. [86]. Firmly standing three-dimensional radial junctions on soft aluminium foils enable extremely low-cost, flexible thin-film solar cells with very high power-to-weight performance, Nano Energy 53 (2018): 83–90, Sun, Xiaolin, Ting Zhang, Junzhuan Wang, Fan Yang, Ling Xu, Jun Xu, Yi Shi, Kunji Chen, Pere Roca i Cabarrocas, and Linwei Yu, Copyright (2021), with permission from Elsevier.

perovskite thin film with a-SiGe:H hybrid solar cell structure [89], there is a possibility to achieve a superior efficiency with this RTJ-3D structured using SiNWs. Table 10.3 illustrates the reported efficiency (η), fill factor (FF), open circuit voltage (V_{OC}), short circuit density (J_{SC}), bending radius and chosen substrate for the fabrication of flexible silicon nanowire solar cells.

10.3.3.2   Silicon nanopyramid solar cells

A randomly textured surface over c-Si and polycrystalline silicon wafers with morphology similar to micro-size pyramids (typically, 3–10 μm) have been used for reducing the reflections from the solar cells. The reduction in light reflection (or in other words – trapping of light) is due to the strong light-trapping phenomena exhibited by the Si wafers containing nano- and micro-sized pyramid structures at the surface of silicon wafers [91,92]. Some bulk crystalline and polycrystalline silicon-based photovoltaic solar cells that are extremely thin (<50 μm) do not have pyramid structures at the surface owing to the complexity of the etching process while handling the wafers during the fabrication process. As a result, various light-trapping silicon nanostructures have been explored, such as silicon nanowire [93,94]. Furthermore, due to the large surface area of the solar cells obtained by texturing pyramidal structures on the surface of the silicon substrate, surface recombination of charge carriers is increased [95], which adversely reduces the efficiency of solar cells. Approximately 2% reduction in the standard silicon wafer-based solar cells is observed with the texturing (Figure 10.9) of silicon wafers without surface passivation layers [93]. This is also the limiting factor in the solar cells fabricated by the use of silicon nanowires and silicon nanoholes, and hence, these solar cells have not surpassed efficiency beyond 10% [75,96]. However, by establishing an inverted nanopyramid structure, the surface recombination losses can be minimized, which precedes an enhancement of light absorption capabilities, and such technique is claimed to be cost-effective [97]. Several advanced techniques, such as interference lithography (IL) [98], wet Si etching [99], ultraviolet (UV), and nanoimprint lithography (UV-NIL) [100,101] are used to obtain nanopyramid structure at the surface of silicon wafers.

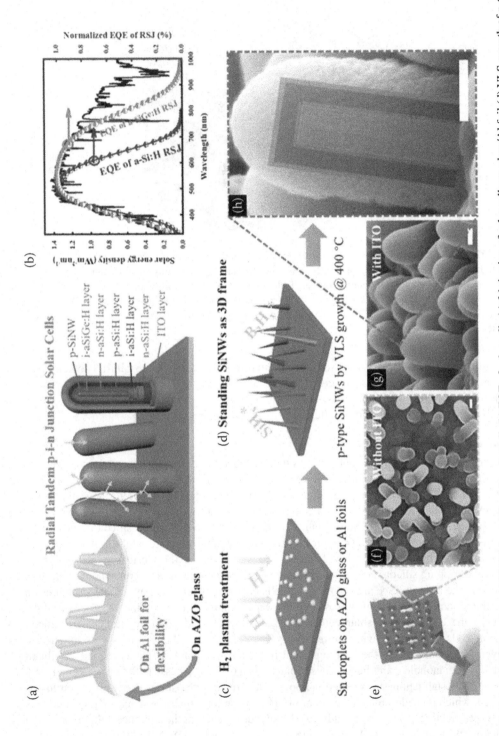

FIGURE 10.8 (a) RTJ structure of SiNWs based flexible solar cell. (b) EQE of solar cells (c) fabrication of solar cells on glass/Al foil (d) VLS growth of p-type c-SiNWs (e) photograph of actual RTJ solar cell, (f) The SEM images of SiNWs (h) side-view of SiNWs with ITO top electrod Reprinted from ref. [87]. Highly flexible radial tandem junction thin film solar cells with excellent power-to-weight ratio, Nano Energy 86 (2021): 106121, Zhang, Shaobo, Ting Zhang, Zongguang Liu, Junzhuan Wang, Linwei Yu, Jun Xu, Kunji Chen, and Pere Roca i Cabarrocas, Copyright (2021), with permission from Elsevier.

TABLE 10.3

Selective summary of flexible silicon nanowire solar cells performance fabricated on different flexible substrates

Substrate	Efficiency (%)	Bending radius	Jsc (mA/cm2)	P-W ratio (W/Kg)	FF (%)	Voc (V)	Year	Ref.
SS	~0.1	–	–	–	0.28	~0.13		[77]
SS	3.56	–	13.36	–	0.43	0.62		[85]
PEN	4	–	14.2	~80	–	0.60		[90]
Al foil	5.6	5 mm	14.2	1382	56	0.71		[86]
Al foil	6.6	10 mm	11.1	1628	65	1.1		[87]

In 2014, Li, Guijun, et al. [74] proposed ultrathin amorphous silicon (a-Si) and ultrathin crystalline silicon (c-Si) based flexible tandem photovoltaic solar cell with nanopyramid structure (Figure 10.9) and obtained 13.3% efficiency with a significant improvement in light-trapping capabilities up to 48% as compared with the planar surface. Recent studies demonstrated that the power conversion efficiency up to 17% [102] (Table 10.4) is achieved with a cost-effective Ag assisted chemical-etching process followed by a post nanostructure rebuilding (NSR) method to manufacture an ultrathin c-Si-based flexible photovoltaic solar cell. The fabrication process of nanostructured ultrathin c-Si flexible solar cell is shown in Figure 10.10. This method includes a high-temperature phosphorus diffusion followed by antireflection surface passivation (SiNx) layer deposited by the PECVD technique to minimize the surface recombination of charge carriers [102,103]. Solar cells fabricated in this study have shown reduced reflection (<12%) in the wavelength region of 300–1100 nm due to efficient light absorption shown by surface nano-texturing of the wafer coupled with the antireflection layer. However, the theoretical simulation analysis has predicted efficiency up to 26% of inverted nanopyramids surface texturing coupled with surface passivation. Table 10.4 illustrates the reported efficiency (η), fill factor (FF), open circuit voltage (V_{OC}), short circuit density (J_{SC}), and chosen substrate for the fabrication of flexible thin and ultrathin crystalline silicon solar cells.

10.3.3.3 Silicon nanoparticles for solar cells

Dielectric nanoparticles can trap/scatter the light of different wavelengths close to the device's active layer. Dielectric nanoparticles are preferred over metal nanoparticles because they are inefficient in absorbing light and chemically inert, and so they do not interact with other underlying layers of the solar cell during the fabrication, ensuring overall stability [104]. Another drawback of metal nanoparticle is that quenching of the electron-hole pair is commonly observed when these NPs touch the surface of absorber layers, leading to reduced I_{SC} and increased recombination of charge carriers at the interface. Silicon nanoparticles are dielectric, and hence, aforementioned properties can be exploited in the solar cells with the use of silicon nanoparticles. Mostly, silicon nanoparticles are used in the device where the absorber layer is extremely thin and thickness cannot be increased. As the thickness cannot be increased, absorption of light will be compromised [104]; hence, dielectric nanoparticles (Si) are used to increase the absorption where the thickness of the absorber layer is extremely thin (<500 nm). With the use of an ultrathin layer of silicon nanoparticles (1 nm blue luminescent and 2.85 nm red luminescent silicon nanoparticles) directly deposited on the polycrystalline silicon wafer, the absorption of light is improved in the polysilicon solar cell [105]. According to Di Vece et al. [106], implementing the bottom-top approach to deposit silicon nanoparticles results in an enhancement in the power-conversion efficiency of a solar cell fabricated over the flexible substrate. Crystalline silicon solar cells have additionally used zinc oxide (ZnO) heterostructure to prevent carrier recombination in the solar cell containing silica nanoparticles as plasmonic nanoparticles that were used to obtain far-field light scattering [107,108]. Gribov et al. [76] has indicated that solar cells containing silicon nanoparticles may offer increased

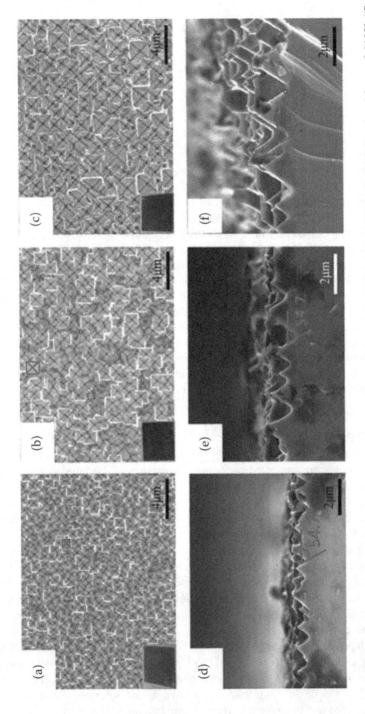

FIGURE 10.9 SEM image of nanopyramid structure treated with different temperature (a)(d) 25°C (b)(e) 40°C (c)(f) 60°C. Reprinted from ref. [102]. "Superiority of random inverted nanopyramid as efficient light trapping structure in ultrathin flexible c-Si solar cell." Renewable Energy 133 (2019): 883–892, Tang, Quntao, Honglie Shen, Hanyu Yao, Kai Gao, Ye Jiang, Yufang Li, Youwen Liu, Lei Zhang, Zhichun Ni, and Qingzhu Wei, Copyright (2021), with permission from Elsevier.

TABLE 10.4

Selective summary of flexible thin and ultrathin crystalline silicon solar cells with nanopyramid structure performance

Substrate	Efficiency (%)	Jsc (mA/cm2)	FF (%)	Voc (V)	Ref.
Thin c-Si	>1	37.5	N/A	N/A	[97]
Thin c-Si	2.73	16.92	46	0.54	[26]
Thin c-Si	13.3	13.4	58.7	0.89	[74]
Ultrathin c-Si	17	36.6	70.78	0.65	[102]
Ultrathin c-Si	17.44	34.8	76.4	0.65	[103]

absorption, but such solar cell will require an additional passivation layer to improve power-conversion efficiencies of solar cells.

10.3.3.4 Silicon ink-based solar cells

Currently, silicon-based solar cell technology relies on vapor-phase deposition methods for commercial applications. Not only that these methods are expensive, but they also create a large carbon footprint, including a large volume of wasted raw materials. However, this technology approach does not fit with the current trend to manufacture flexible, bendable electronic devices. The solution processing of silicon thin films has the potential to dramatically change the current phase and to open a new avenue of flexible and large-area deposition of silicon thin films, particularly with the realization of flexible silicon solar cells. In this section, few solar cells are discussed that are not flexible, but there is a potential to fabricate flexible solar cells due to the compatibility of the fabrication process on flexible substrates without altering the device structure.

So far, only a few articles are published on silicon solar cells that utilize liquid silicon as a source/precursor for the deposition of silicon absorber layer dedicated toward photovoltaic applications. Masuda et al. [109] have demonstrated the first proof of concept solution-processed p-i-n silicon solar cell, where the spin coating/cast deposition technique was used to obtain solution-processed silicon thin films. Cyclopentasiloxane (CPS: Si_5H_{10}) was used as a silicon precursor and amorphous hydrogenated silicon thin film (a-Si:H) was obtained with the use of polydihydrosilane synthesized by photo-induced polymerization. Solution-processed solar cells fabricated with the use of the above techniques and materials have shown PCE of 0.31–0.51% under AM-1.5G illumination. Follow-up work by Masuda et al. [110] used the conventional method, i.e., thermal chemical vapor deposition for making the hydrogenated amorphous silicon (a-Si:H) film; but, in this work, the silicon reactant gas was generated from liquid cyclopentasilane (CPS) instead of conventional silane-based gas precursor. Additionally, the authors reported a method of making p-type doping using decaborane and n-type doping by white phosphorus, with both chemicals dissolved in liquid CPS. This approach was mainly considered to eliminate vacuum-based deposition methods and to eliminate the use of conventional toxic gas, such as diborane and phosphine. In this work, two different device structures are considered; Glass/AZO/n-i-p/ITO and Glass/AZO/p-i-n/ITO solar cells were fabricated and investigated [110]. Using the same solution-based precursors with an alternative deposition technique, the PCE of solar cells was increased from 0.5% to 2.6%. Bronger et al. used neopentasilane (NPS) as an alternative silicon precursor for the fabrication of solution-based thin-film silicon solar cells. It is reported that use of NPS compared with the CPS will provide better processing efficiency, including improved material quality. Bronger et al. reported, for the first time, the use of NPS as an alternative silicon precursor for the fabrication of solar cells [111]. The silicon-based inks for three functional layers, i.e., n-layer, intrinsic absorber, and p-layer, were formulated and investigated in this work. The n-i-p structure solar cells were fabricated with the use of NPS

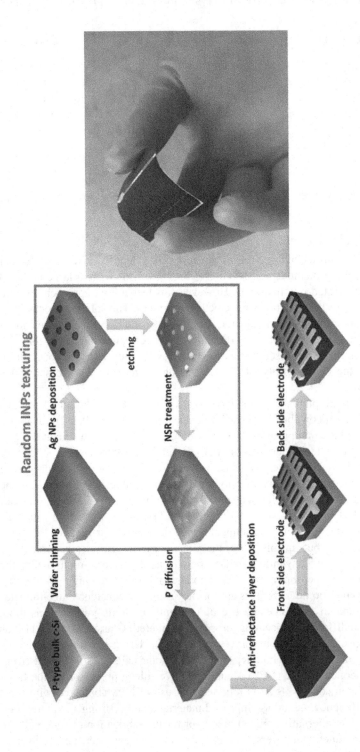

FIGURE 10.10 Process schematics of the random inverted nanopyramids textured flexible ultrathin c-Si solar cell fabrication. Reprinted from ref. [102], "Superiority of random inverted nanopyramid as efficient light trapping structure in ultrathin flexible c-Si solar cell." Renewable Energy 133 (2019): 883–892, Tang, Quntao, Honglie Shen, Hanyu Yao, Kai Gao, Ye Jiang, Yufang Li, Youwen Liu, Lei Zhang, Zhichun Ni, and Qingzhu Wei, Copyright (2021), with permission from Elsevier.

modified silicon inks deposited by spin/slot die coating to further increase the efficiency to 3.5% as compared with the CPS silicon precursor based solar cells. After successful demonstration of solution-processed amorphous silicon for application in photovoltaic devices, as discussed above, Sontheimer and co-workers reported the fabrication of crystalline silicon solar cells from a precursor solution containing silicon with the use of slot die- and/or spin-coating process [112]. The liquid silicon solution was formulated using a branched molecule of NPS with all processes, starting from the NPS precursor to the final step to obtain crystalline silicon film is illustrated in Figure 10.11. The liquid amorphous Si film is converted into hydrogenated amorphous silicon (a-Si:H) at elevated temperatures between 350–550°C in a hydrogen atmosphere. A crystallization process involved annealing a-Si:H layers in a nitrogen atmosphere at 600°C to obtain liquid-to-solid phase transition. With the use of all these additional techniques, n-i-p silicon solar cell was fabricated with an efficiency of 0.7%. This attempt seems to be unsuccessful, with increased cost requiring additional thermal treatments that increase the carbon footprint even more. Importantly, solar cell efficiency was significantly lower, as compared with the solution-processed amorphous silicon discussed above (Table 10.5).

10.4   OUTLOOK AND CONCLUSIONS

Silicon-based photovoltaic solar cells have been studied for more than 50 years. This comprehensive review provides a detailed investigation of emerging silicon-based flexible photovoltaic solar cells. In this review, advancements and progress in the areas of silicon-based flexible and stretchable solar cells are discussed and highlighted, with a focus on various fabrication techniques, suitability of fabrication on flexible substrates, use of nanomaterials to overcome the limitations posed by bulk materials, and practical aspects, including challenges. Flexible solar cells in this review article have mainly focused on low-dimensional and bulk silicon, including thin films for their use in photovoltaic applications. Presently, most of the PV market is acquired by silicon-based solar cells. Among various types of silicon solar cells, only a-Si: H-based solar cells are commercially available and fabricated on flexible substrates. However, lab-scale research has shown that flexible silicon solar cells are achievable by employing emerging fabrication techniques, novel device structures, novel nanomaterials, and flexible substrates to fabricate solar cells from crystalline, polycrystalline, and amorphous silicon. c-Si-based flexible solar cells are manufactured successfully using novel laser-patterning based corrugation technique with ultra-stretchability of 95%, and power-conversion efficiency of 19% is achieved. However, although these flexible solar cells promise higher efficiencies, the fabrication techniques involved are either complex or expensive. This complexity and cost further inhibit commercialization, so these cells may be used only in niche applications or remain as a lab research device, with the limited purpose of contributing to the knowledge. Therefore, it is essential to develop new cost-effective techniques that will allow commercialization.

Comparatively, thin-film-based a-Si:H flexible solar cells offer more flexibility, and these devices are cost-effective. The highest efficiency achieved so far from flexible a-Si:H solar cell is 9.3%, when fabricated on a flexible glass substrate. The power-conversion efficiency of these solar cells is still far behind as compared with the rigid commercial crystalline silicon solar cells. Nevertheless, the cost of production of flexible thin-film silicon solar cells can be reduced when fabricated on substrates; these are easily integrated, on windows for instance. In the future, the market demand for flexible a-Si-based solar cells will depend mainly on the material cost and stability under different harsh environmental conditions.

The future perspectives are to obtain high-performance flexible solar silicon solar cells, which could demonstrate the power-conversion efficiencies up to 40% surpassing the Shockley–Queisser limit. This is achievable using nanomaterials of silicon that exhibit superior photo-response and increased absorption of the light, including controllable bandgap. That said, these nanostructured silicon-based photovoltaic solar cells show better light absorption capabilities in single-junction solar cells; due to increased recombination of charge carriers at the surface (presence of surface dangling bonds), the power

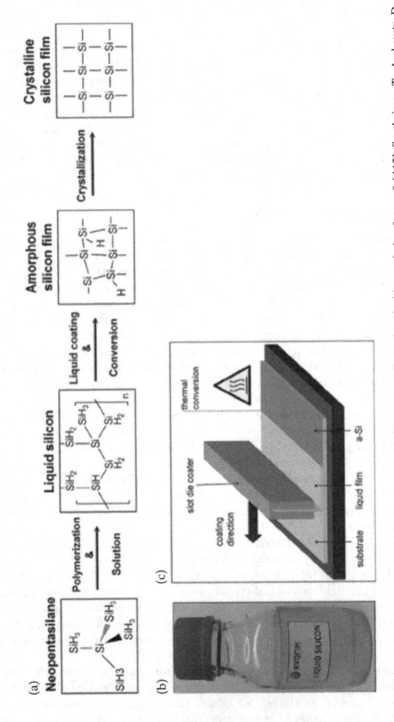

FIGURE 10.11 Synthesis of crystalline silicon film from the NPS precursor. Reproduced with permission from ref. [112]. Sontheimer, T., Amkreutz, D., Schulz, K., Wöbkenberg, P. H., Guenther, C., Bakumov, V...... & Rech, B. (2014). Solution-Processed Crystalline Silicon Thin-Film Solar Cells. Advanced Materials Interfaces, 1(3), 1300046, Copyright 2021, John Wiley & Sons, Inc.

TABLE 10.5

Selective summary of silicon solar cells fabricated with the use of silicon containing inks

Device structure	Liquid silicon pre-cursor ink	Efficiency (%)	J_{sc} (mA/cm^2)	FF (%)	V_{oc} (V)	Fabrication techniques	Ref.
P-I-N	cyclopentasilane (CPS)	0.5	2.46	35	0.59	Spin Coating	[109]
P-I-N	cyclopentasilane (CPS)	2.36	4.97	58	0.82	Non-vacuum CVD	[110]
N-I-P	cyclopentasilane (CPS)	2.34	5.29	56	0.79	Non-vacuum CVD	[110]
N-I-P	neopentasilane (NPS)	3.5	9.0	53.8	0.73	Spin coating/ slot die	[111]
N$^+$-I-P$^+$	neopentasilane (NPS)	0.7	3.7	55	0.34	Spin coating/ slot die	[112]

conversion efficiency is not directly translated from the increased absorption of light. In the future, for the production of commercial silicon-based flexible solar cells, a metal substrate such as aluminium foil will be commonly used due to its cost, excellent flexibility, and thermal stability. Furthermore, it is necessary to develop roll-to-roll manufacturing technologies that are commonly used in DSSCs for the bulk production of flexible silicon solar cells. This is possible to be achieved with the use of silicon inks in the fabrication of solar cells.

In short, future advances hold promise for flexible silicon-based solar cells. Now researchers and industries need to narrow and/or bridge the gap between the efficiency of rigid solar cells and flexible photovoltaic devices by using novel cheap fabrication techniques and utilizing the strengths of nanomaterials/nanostructures of silicon.

ACKNOWLEDGEMENT

The author (Shashi Paul) would like to thank the EPSRC (#EP/P020518/1) for their financial support.

REFERENCES

1. Li, Xiaoyue, Peicheng Li, Zhongbin Wu, Deying Luo, Hong-Yu Yu, and Zheng-Hong Lu, "Review and perspective of materials for flexible solar cells," *Materials Reports: Energy 1*, pp. (2021) 100001.
2. Panwar, N. L., S. C. Kaushik and Surendra Kothari , "Role of renewable energy sources in environmental protection: A review," *Renewable and Sustainable Energy Reviews 15, no. 3*, pp. (2011) 1513–1524.
3. Malinowski, Mariusz, Jose I. Leon, and Haitham Abu-Rub, "Solar photovoltaic and thermal energy systems: Current technology and future trends," *Proceedings of the IEEE 105, no. 11*, pp. (2017) 2132–2146.
4. "IEA, Renewable electricity net capacity additions by technology, main and accelerated cases," IEA, Paris, 2013–2022. [Online]. Available: https://www.iea.org/data-and-statistics/charts/renewable-electricity-net-capacity-additions-by-technology-main-and-accelerated-cases-2013–2022.
5. Guangul, Fiseha Mekonnen, and Girma T. Chala, "Solar energy as renewable energy source: SWOT analysis," In *2019 4th MEC International Conference on Big Data and Smart City (ICBDSC)*, pp. 1–5. IEEE, (2019).
6. Parida, Bhubaneswari, Selvarasan Iniyan, and Ranko Goic, "A review of solar photovoltaic technologies," *Renewable and Sustainable Energy Reviews 15.3*, pp. (2011) 1625–1636.
7. Kim, Sangmo, Hoang Van Quy, and Chung Wung Bark, "Photovoltaic technologies for flexible solar cells: Beyond silicon," *Materials Today Energy 19*, p. (2020) 100583.
8. V. Sivaram, *Taming the sun: Innovations to harness solar energy and power the planet*, MIT Press, (2018).
9. Becquerel, M. E., "Memoir on the electrical effects produced under the influence of solar rays," *Weekly Reports of the Sessions of the Academy of Sciences 9*, pp. (1839) 561–567.
10. Snyder, George H., Vladimir V. Matichenkov, and Lawrence E. Datnoff., "Silicon," *In Handbook of plant nutrition*, CRC Press, pp. 567–584 (2016).

11. Luceño-Sánchez, José Antonio, Ana María Díez-Pascual, and Rafael Peña Capilla, "Materials for photovoltaics: State of art and recent developments," *International journal of molecular sciences 20, no. 4,* p. (2019) 976.

12. Sun, Baoquan, Mingwang Shao, and Shuitong Lee, "Nanostructured silicon used for flexible and mobile electricity generation," *Advanced Materials 28, no. 47,* pp. (2016) 10539–10547.

13. Ramanujam, J., Bishop, D. M., Todorov, T. K., Gunawan, O., Rath, J., Nekovei, R., Artegiani, E., and Romeo, A., "Flexible CIGS, CdTe and a-Si:H based thin film solar cells: A review," *Progress in Materials Science 110,* p. (2019) 100619.

14. R. Singh, "Why silicon is and will remain the dominant photovoltaic material," *Journal of Nanophotonics 3, no. 1,* p. (2009) 032503.

15. R. Arora and D. Agrawal, "Will silicon remain the dominant electronics material for the entire 21st century? [The way I see it]," *IEEE Potentials, 30, no. 6,* pp. (2011) 7–29.

16. Goetzberger, Adolf, Joachim Knobloch, and Bernhard Voss, *Crystalline silicon solar cells. Vol. 1.* Chichester: Wiley (1998).

17. Breitenstein, O., Frühauf, F., & Bauer, J., "Advanced local characterization of silicon solar cells," *Physica status solidi (a), 214, no. 12,* p. (2017) 1700611.

18. Tao, Meng, "Inorganic photovoltaic solar cells: silicon and beyond," *The Electrochemical Society Interface 17, no. 4,* p. (2008) 30.

19. Cao, Shuangying, Dongliang Yu, Yinyue Lin, Chi Zhang, Linfeng Lu, Min Yin, Xufei Zhu, Xiaoyuan Chen, and Dongdong Li, "Light propagation in flexible thin-film amorphous silicon solar cells with nanotextured metal back reflectors," *ACS Applied Materials & Interfaces 12, no. 23,* pp. (2020), 26184–26192.

20. Lee, Youngseok, Yeeun Woo, Doh-Kwon Lee, and Inho Kim, "Fabrication of quasi-hexagonal Si nanostructures and its application for flexible crystalline ultrathin Si solar cells," *Solar Energy 208,* pp. (2020): 957–965.

21. Blakers, A. W., and T. Armour, "Flexible silicon solar cells," *Solar Energy Materials and Solar Cells 93, no. 8,* pp. (2009), 1440–1443.

22. Águas, Hugo, Tiago Mateus, António Vicente, Diana Gaspar, Manuel J. Mendes, Wolfgang A. Schmidt, Luís Pereira, Elvira Fortunato, and Rodrigo Martins, "Thin film silicon photovoltaic cells on paper for flexible indoor applications," *Advanced Functional Materials 25, no. 23,* pp. (2015), 3592–3598.

23. Wu, Jyh-Lih, Yoshiaki Hirai, Takuya Kato, Hiroki Sugimoto, and Veronica Bermudez, "New world record efficiency up to 22.9% for Cu (In, Ga)(Se, S) 2 thin-film solar cells," In *7th World Conference on Photovoltaic Energy Conversion (WCPEC-7), Waikoloa, Hawaii, USA* (2018).

24. Wöhrle, Dieter, and Dieter Meissner, "Organic solar cells," *Advanced Materials 3, no. 3,* pp. (1991), 129–138.

25. Roland, Steffen, Sebastian Neubert, Steve Albrecht, Bernd Stannowski, Mark Seger, Antonio Facchetti, Rutger Schlatmann, Bernd Rech, and Dieter Neher, "Hybrid organic/inorganic thin-film multijunction solar cells exceeding 11% power conversion efficiency," *Advanced Materials 27, no. 7,* pp. (2015), 1262–1267.

26. El-Atab, Nazek, and Muhammad M. Hussain, "Flexible and stretchable inorganic solar cells: Progress, challenges, and opportunities," *MRS Energy & Sustainability 7,* (2020), 19.

27. Verlinden P. J., et al., "Will we have >22% efficient multi-crystalline Silicon solar Cell?," *PVSEC 26, Singapore,* 24–28 October (2016).

28. Haase, Felix, Christina Hollemann, Sören Schäfer, Agnes Merkle, Michael Rienäcker, Jan Krügener, Rolf Brendel, and Robby Peibst, "Laser contact openings for local poly-Si-metal contacts enabling 26.1%-efficient POLO-IBC solar cells," *Solar Energy Material 186,* (2018): 184–193.

29. Mahabaduge, H. P., W. L. Rance, J. M. Burst, M. O. Reese, D. M. Meysing, C. A. Wolden, J. Li et al., "High-efficiency, flexible CdTe solar cells on ultra-thin glass substrates," *Applied Physics Letters 106, no. 13,* p. (2015), 133501.

30. Zhang, Xiaoliang, Viktor A. Öberg, Juan Du, Jianhua Liu, and Erik MJ Johansson, "Extremely lightweight and ultra-flexible infrared light-converting quantum dot solar cells with high power-per-weight output using a solution-processed bending durable silver," *Energy & Environmental Science 11, no. 2,* pp. (2018), 354–364.

31. Q. Tang et al, "Investigation of optical and mechanical performance of inverted pyramid based ultrathin flexible c-Si solar cell for potential application on curved surface," *Applied Surface Science 504,* p. (2020), 144588.

32. W. Shockley and H. J. Queisser, "Detailed balance limit of efficiency of p-n junction solar cells," *Journal of Applied Physics, 32, no. 3,* pp. (1961), 510–519.

33. M. Green et al., "Solar cell efficiency tables (version 57)," *Progress in Photovoltaics: Research and Applications, 29, no. 1*, pp. (2021), 3–15.

34. Bett, Andreas W., Frank Dimroth, Wolfgang Guter, Raymond Hoheisel, Eduard Oliva, Simon P. Philipps, Jan Schöne et al., "Highest efficiency multi-junction solar cell for terrestrial and space applications," *Space 25, no. 25.8*, pp. (2009), 30–36.

35. Bagher, Askari Mohammad, "Introduction to organic solar cells," *Sustainable Energy 2, no. 3*, pp. (2014), 85–90.

36. Krebs, F. C. and Norrman, K., "Failure mechanism for a stable organic photovoltaic during 10 000 h of testing," *Analysis of the Progress in Photovoltaics: Research and Applications 15, no. 8*, pp. (2007), 697–712.

37. Gevorgyan, S. A., Madsen, M. V., Dam, H. F., Jørgensen, M., Fell, C. J., Anderson, K. F., Duck, B. C., Mescheloff, A., Katz, E. A., Elschner, A. and Roesch, R., "Interlaboratory outdoor stability studies of flexible roll-to-roll coated organic photovoltaic," *Modules: Stability over 10,000 h. Solar Energy Materials and Solar Cells, 116*, pp. (2013), 187–196.

38. Fung, Dixon DS, and Wallace CH Choy, "Introduction to organic solar cells," *Organic Solar Cells.* Springer, London, pp. 1–16, (2013).

39. Petsch, Tino, Jens Haenel, Maurice Clair, Bernd Keiper, and Christian Scholz, "Laser processing of organic photovoltaic cells with a roll-to-roll manufacturing process," *In Laser-based Micro-and Nanopackaging and Assembly V*, vol. 7921, p. 79210U, International Society for Optics and Photonics (2011).

40. Krebs, Frederik C., "Fabrication and processing of polymer solar cells: A review of printing and coating techniques," *Solar Energy Materials and Solar Cells 93, no. 4*, pp. (2009), 394–412.

41. Mori, Shigehiko, Haruhi Oh-Oka, Hideyuki Nakao, Takeshi Gotanda, Yoshihiko Nakano, Hyangmi Jung, Atsuko Iida et al., "Organic photovoltaic module development with inverted device structure," *MRS Online Proceedings Library (OPL) 1737*, pp. (2015), 26–31.

42. Günes, Serap, and Niyazi Serdar Sariciftci, "Hybrid solar cells," *Inorganica Chimica Acta 361 no. 3*, pp. (2008), 581–588.

43. Michaels, Hannes, Iacopo Benesperi, and Marina Freitag, "Challenges and prospects of ambient hybrid solar cell applications," *Chemical Science 12, no. 14*, pp. (2021), 5002–5015.

44. Wright, Matthew, and Ashraf Uddin, "Organic—inorganic hybrid solar cells: A comparative review," *Solar Energy Materials and Solar Cells 107*, pp. (2012), 87–111.

45. Khang, Dahl-Young, "Recent progress in Si-PEDOT: PSS inorganic–organic hybrid solar cells," *Journal of Physics D: Applied Physics 52, no. 50*, p. (2019), 503002.

46. He, W. W., K. J. Wu, K. Wang, T. F. Shi, L. Wu, S. X. Li, D. Y. Teng, and C. H. Ye, "Towards stable silicon nanoarray hybrid solar cells," *Scientific Reports 4, no. 1*, pp. (2014), 1–7.

47. Orgil, Khulan, "Comparison of organic and inorganic solar photovoltaic systems," (2018). https://digitalcommons.calpoly.edu/cgi/viewcontent.cgi?article=1467&context=eesp.

48. Chen, Ying, Ye Jiang, Yin Huang, and Xue Feng, "IV group materials-based solar cells and their Flexible Photovoltaic Technologies," *Inorganic Flexible Optoelectronics: Materials and Applications 6*, p. (2019), 8.

49. Moeller, Hans Joachim, "Wafering of silicon," *In Semiconductors and Semimetals*, vol. 92, pp. 63–109, Elsevier, (2015).

50. Hecini, M., Drouiche, N. and Bouchelaghem, O., "Recovery of cutting fluids used in polycrystalline silicon ingot slicing," *Journal of Crystal Growth 453*, pp. (2016), 143–150.

51. Green, Martin A., "The future of crystalline silicon solar cells," *Progress in Photovoltaics: Research and Applications 8, no. 1*, (2000), 127–139.

52. El-Atab, Nazek, Nadeem Qaiser, Rabab Bahabry, and Muhammad Mustafa Hussain, "Corrugation enabled asymmetrically ultrastretchable (95%) monocrystalline silicon solar cells with high efficiency (19%)," *Advanced Energy Materials 9, no. 45*, p. (2019), 1902883.

53. Polman, Albert, Mark Knight, Erik C. Garnett, Bruno Ehrler, and Wim C. Sinke, "Photovoltaic materials: Present efficiencies and future challenges," *Science 352, no. 6283* (2016). doi:10.1126/science.aad4424

54. Kim, Dae-Hyeong, Jong-Hyun Ahn, Won Mook Choi, Hoon-Sik Kim, Tae-Ho Kim, Jizhou Song, Yonggang Y. Huang, Zhuangjian Liu, Chun Lu, and John A. Rogers, "Stretchable and foldable silicon integrated circuits," *Science 320, no. 5875*, (2008), pp. 507–511.

55. Fan, Zhiyong, and Ali Javey, "Solar cells on curtains," *Nature Materials 7, no. 11*, pp. (2008), 835–836.

56. Zou, Dechun, Dan Wang, Zengze Chu, Zhibin Lv, and Xing Fan, "Fiber-shaped flexible solar cells," *Coordination Chemistry Reviews 254, no. 9–10*, pp. (2010), 1169–1178.

57. Reuter, Michael, Willi Brendle, Osama Tobail, and Jürgen H. Werner, "50 µm thin solar cells with 17.0% efficiency," *Solar Energy Materials and Solar Cells 93, no. 6–7*, pp. (2009), 704–706.

58. Cruz-Campa, Jose L., Gregory N. Nielson, Paul J. Resnick, Carlos A. Sanchez, Peggy J. Clews, Murat Okandan, Tom Friedmann, and Vipin P. Gupta, "Ultrathin flexible crystalline silicon: Microsystems-enabled photovoltaics," *IEEE Journal of Photovoltaics 1*. doi:10.1109/PVSC.2011. 6186687

59. Söderström, T., Haug, F.-J., Terrazzoni-Daudrix, V., and Ballif, C., "Optimization of amorphous silicon thin film solar cells for flexible photovoltaics," *Journal of Applied Physics 103, no. 11*, p. (2008), 114509.

60. Tao, Ke, Jin Wang, Hongkun Cai, Dexian Zhang, Yanpin Sui, Yi Zhang, and Yun Sun, "Low-temperature preparation of flexible a-Si: H solar cells with hydrogenated nanocrystalline silicon p layer," *Vacuum 86, no. 10*, pp. (2012), 1477–1481.

61. Burgess, D., "Thin-film solar cell" (2020).

62. Carlson, David E., and Cristopher R. Wronski, "Amorphous silicon solar cell," *Applied Physics Letters 28, no. 11*, pp. (1976), 671–673.

63. Myong, Seung Yeop, and Seong Won Kwon, "Superstrate type flexible thin-film Si solar cells using flexible glass substrates," *Thin Solid Films 550*, pp. (2014), 705–709.

64. Zhang, Chi, Ye Song, Min Wang, Min Yin, Xufei Zhu, Li Tian, Hui Wang et al., "Efficient and flexible thin film amorphous silicon solar cells on nanotextured polymer substrate using sol–gel based nanoimprinting method," *Advanced Functional Materials 27, no. 13*, p. (2017), 1604720.

65. Lin, Yinyue, Zhen Xu, Dongliang Yu, Linfeng Lu, Min Yin, Mohammad Mahdi Tavakoli, Xiaoyuan Chen et al., "Dual-layer nanostructured flexible thin-film amorphous silicon solar cells with enhanced light harvesting and photoelectric conversion efficiency," *ACS Applied Materials & Interfaces 8, no. 17*, pp. (2016), 10929–10936.

66. Smeets, M., K. Wilken, K. Bittkau, H. Aguas, L. Pereira, E. Fortunato, R. Martins, and V. Smirnov., "Flexible thin film solar cells on cellulose substrates with improved light management," *Physica status solidi (a) 214, no. 8*, p. (2017), 1700070.

67. Saga, Tatsuo, "Advances in crystalline silicon solar cell technology for industrial mass production," *Npg asia materials 2, no. 3*, pp. (2010), 96–102.

68. Priolo, Francesco, Tom Gregorkiewicz, Matteo Galli, and Thomas F. Krauss, "Silicon nanostructures for photonics and photovoltaics," *Nature Nanotechnology 9, no. 1*, pp. (2014), 19–32.

69. Han, Sang Eon, and Gang Chen, "Toward the Lambertian limit of light trapping in thin nanostructured silicon solar cells," *Nano Letters 10, no. 11*, pp. (2010), 4692–4696.

70. Haug, F.-J., and C. Ballif, "Light management in thin film silicon solar cells," *Energy & Environmental Science 8, no. 3*, pp. (2015), 824–837.

71. Garnett, Erik, and Peidong Yang, "Light trapping in silicon nanowire solar cells," *Nano Letters 10, no. 3*, pp. (2010), 1082–1087.

72. Guo, Chun-Sheng, Lin-Bao Luo, Guo-Dong Yuan, Xiao-Bao Yang, Rui-Qin Zhang, Wen-Jun Zhang, and Shuit-Tong Lee. "Surface passivation and transfer doping of silicon nanowires," *Angewandte Chemie 121, no. 52*, pp. (2009), 10080–10084.

73. Pang, Chunlei, Hao Cui, Guowei Yang, and Chengxin Wang, "Flexible transparent and free-standing silicon nanowires paper," *Nano Letters 13, no. 10*, pp. (2013), 4708–4714.

74. Li, Guijun, He Li, Jacob YL Ho, Man Wong, and Hoi Sing Kwok, "Nanopyramid structure for ultrathin c-Si tandem solar cells," *Nano Letters 14, no. 5*, pp. (2014), 2563–2568.

75. Peng, Kui-Qing, Xin Wang, Li Li, Xiao-Ling Wu, and Shuit-Tong Lee, "High-performance silicon nano-hole solar cells," *Journal of the American Chemical Society 132, no. 20*, pp. (2010), 6872–6873.

76. Gribov, B. G., K. V. Zinov'ev, O. N. Kalashnik, N. N. Gerasimenko, D. I. Smirnov, V. N. Sukhanov, N. N. Kononov, and S. G. Dorofeev, "Production of silicon nanoparticles for use in solar cells," *Semiconductors 51, no. 13*, pp. (2017), 1675–1680.

77. Tsakalakos, L., Balch, J., Fronheiser, J., Korevaar, B. A., Sulima, O. and Rand, J., "Silicon nanowire solar cells.," *Applied Physics Letters, 91, no. 23*, p. (2007), 233117.

78. Sivakov, V., G. Andrä, A. Gawlik, A. Berger, J. Plentz, F. Falk, and S. H. Christiansen, "Silicon nanowire-based solar cells on glass: Synthesis, optical properties, and cell parameters," *Nano Letters 9, no. 4*, pp. (2009), 1549–1554.

79. Gabrielyan, Nare, Konstantina Saranti, Krishna Nama Manjunatha, and Shashi Paul, "Growth of low temperature silicon nano-structures for electronic and electrical energy generation applications," *Nanoscale Research Letters 8, no. 1*, pp. (2013), 1–8.

80. Sahoo, M. K. and Kale, P., "Integration of silicon nanowires in solar cell structure for efficiency enhancement: A review," *Journal of Materiomics.* (2018). doi:10.1016/j.jmat.2018.11.007

81. Manjunatha, Krishna Nama, Iulia Salaoru, William I. Milne, and Shashi Paul, "Comparative study of silicon nanowires grown from Ga, In, Sn, and Bi for energy harvesting," *IEEE Journal of Photovoltaics 10, no. 6*, pp. (2020), 1667–1674.

82. Li G., Kwok H. S., "Silicon nanowire solar cells," In: Ikhmayies S. (ed) *Advances in Silicon Solar Cells.* Springer, Cham. 10.1007/978-3-319-69703-1_10, (2018).

83. Song, T., Shuit-Tong L., and Baoquan S., "Silicon nanowires for photovoltaic applications: The progress and challenge," *Nano Energy 1, no. 5*, pp. (2012), 654–673.

84. Manjunatha, Krishna Nama, and Shashi Paul, "In-situ catalyst mediated growth and self-doped silicon nanowires for use in nanowire solar cells," *Vacuum 139*, pp. (2017), 178–184.

85. Xie, Xiaobing, Xiangbo Zeng, Ping Yang, Hao Li, Jingyan Li, Xiaodong Zhang, and Qiming Wang, "Radial n–i–p structure silicon nanowire-based solar cells on flexible stainless steel substrates," *Physica status solidi (a) 210, no. 2*, pp. (2013), 341–344.

86. Sun, Xiaolin, Ting Zhang, Junzhuan Wang, Fan Yang, Ling Xu, Jun Xu, Yi Shi, Kunji Chen, Pere Roca i Cabarrocas, and Linwei Yu, "Firmly standing three-dimensional radial junctions on soft aluminum foils enable extremely low cost flexible thin film solar cells with very high power-to-weight performance," *Nano Energy 53*, pp. (2018), 83–90.

87. Zhang, Shaobo, Ting Zhang, Zongguang Liu, Junzhuan Wang, Linwei Yu, Jun Xu, Kunji Chen, and Pere Roca i Cabarrocas, "Highly flexible radial tandem junction thin film solar cells with excellent power-to-weight ratio," *Nano Energy 86*, p. (2021), 106121.

88. Qian, Shengyi, Soumyadeep Misra, Jiawen Lu, Zhongwei Yu, Linwei Yu, Jun Xu, Junzhuan Wang et al., "Full potential of radial junction Si thin film solar cells with advanced junction materials and design," *Applied Physics Letters 107, no. 4*, p. (2015), 043902.

89. Chapa, Manuel, Miguel F. Alexandre, Manuel J. Mendes, Hugo Águas, Elvira Fortunato, and Rodrigo Martins, "All-thin-film perovskite/C–Si four-terminal tandems: interlayer and intermediate contacts optimization," *ACS Applied Energy Materials 2, no. 6* (2019).

90. Jung, Kwang Hoon, Sun Jin Yun, Seong Hyun Lee, Yoo Jeong Lee, Kyu-Sung Lee, Jung Wook Lim, Kyoung-Bo Kim, Moojin Kim, and R. E. I. Schropp, "Double-layered Ag–Al back reflector on stainless steel substrate for a-Si: H thin film solar cells," *Solar Energy Materials and Solar Cells 145*, pp. (2016), 368–374.

91. Campbell, Patrick, and Martin A. Green, "Light trapping properties of pyramidally textured surfaces," *Journal of Applied Physics 62, no. 1*, pp. (1987), 243–249.

92. Campbell, Patrick, and Martin A. Green, "High performance light trapping textures for monocrystalline silicon solar cells," *Solar Energy Materials and Solar Cells 65, no. 1–4*, pp. (2001), 369–375.

93. Peng, Kuiqing, Ying Xu, Yin Wu, Yunjie Yan, Shuit-Tong Lee, and Jing Zhu, "Aligned single-crystalline Si nanowire arrays for photovoltaic applications," *Small 1, no. 11*, pp. (2005), 1062–1067.

94. Subramani, Thiyagu, Chen-Chih Hsueh, Hong-Jhang Syu, Chien-Ting Liu, Song-Ting Yang, and Ching-Fuh Lin, "Interface modification for efficiency enhancement in silicon nanohole hybrid solar cells," *RSC Advances 6, no. 15*, pp. (2016), 12374–12381.

95. Kelzenberg, M. D. et al., "High-performance Si microwire photovoltaics," *Energy & Environmental Science 4*, pp. (2011), 866–871.

96. Muskens, Otto L., Jaime Gómez Rivas, Rienk E. Algra, Erik PAM Bakkers, and Ad Lagendijk, "Design of light scattering in nanowire materials for photovoltaic applications," *Nano Letters 8, no. 9*, pp. (2008), 2638–2642.

97. Mavrokefalos, Anastassios, Sang Eon Han, Selcuk Yerci, Matthew S. Branham, and Gang Chen, "Efficient light trapping in inverted nanopyramid thin crystalline silicon membranes for solar cell applications," *Nano Letters 12, no. 6*, pp. (2012), 2792–2796.

98. Savas, T. A., M. L. Schattenburg, J. M. Carter, and Henry I. Smith, "Large-area achromatic interferometric lithography for 100 nm period gratings and grids," *Journal of Vacuum Science & Technology B: Microelectronics and Nanometer Structures Processing, Measurement, and Phenomena 14, no. 6*, pp. (1996), 4167–4170.

99. Kovacs, Gregory TA, Nadim I. Maluf, and Kurt E. Petersen, "Bulk micromachining of silicon," *Proceedings of the IEEE 86, no. 8*, pp. (1998), 1536–1551.

100. Battaglia, Corsin, Jordi Escarré, Karin Söderström, Lukas Erni, Laura Ding, Grégory Bugnon, Adrian Billet et al., "Nanoimprint lithography for high-efficiency thin-film silicon solar cells," *Nano Letters 11, no. 2*, pp. (2011), 661–665.

101. Amalathas, Amalraj Peter, and Maan M. Alkaisi, "Efficient light trapping nanopyramid structures for solar cells patterned using UV nanoimprint lithography," *Materials Science in Semiconductor Processing 57*, pp. (2017), 54–58.

102. Tang, Quntao, Honglie Shen, Hanyu Yao, Kai Gao, Ye Jiang, Yufang Li, Youwen Liu, Lei Zhang, Zhichun Ni, and Qingzhu Wei. "Superiority of random inverted nanopyramid as efficient light trapping structure in ultrathin flexible c-Si solar cell," *Renewable Energy 133*, pp. (2019), 883–892.

103. Q. Tang et al., "Investigation of optical and mechanical performance of inverted pyramid based ultrathin flexible c-Si solar cell for potential application on curved surface," *Applied Surface Science, vol. 504*, p. (2020), 144588.

104. Furasova, Aleksandra, Emanuele Calabró, Enrico Lamanna, Ekaterina Tiguntseva, Elena Ushakova, Eugene Ubyivovk, Vladimir Mikhailovskii, Anvar Zakhidov, Sergey Makarov, and Aldo Di Carlo, "Resonant silicon nanoparticles for enhanced light harvesting in halide perovskite solar cells," *Advanced Optical Materials 6, no. 21*, p. (2018), 1800576.

105. Stupca, M., M. Alsalhi, T. Al Saud, A. Almuhanna, and M. H. Nayfeh, "Enhancement of polycrystalline silicon solar cells using ultrathin films of silicon nanoparticle," *Applied Physics Letters 91, no. 6*, p. (2007), 063107.

106. Di Vece, Marcel, "Using nanoparticles as a bottom-up approach to increase solar cell efficiency," *KONA Powder and Particle Journal, 36*, pp. (2019), 72–87.

107. Hemaprabha, Elangovan, Upendra K. Pandey, Kamanio Chattopadhyay, and Praveen C. Ramamurthy, "Doped silicon nanoparticles for enhanced charge transportation in organic-inorganic hybrid solar cells," *Solar Energy 173*, pp. (2018), 744–751.

108. Roy, Arijit Bardhan, Sonali Das, Avra Kundu, Chandan Banerjee, and Nillohit Mukherjee, "c-Si/n-ZnO-based flexible solar cells with silica nanoparticles as a light trapping metamaterial," *Physical Chemistry Chemical Physics 19, no. 20*, pp. (2017), 12838–12844.

109. Masuda, Takashi, Naoya Sotani, Hiroki Hamada, Yasuo Matsuki, and Tatsuya Shimoda, "Fabrication of solution-processed hydrogenated amorphous silicon single-junction solar cells," *Applied Physics Letters 100, no. 25*, p. (2012), 253908.

110. Masuda et al., "Phosphorus- and boron-doped hydrogenated amorphous silicon films prepared using vaporized liquid cyclopentasilane," *Thin Solid Films, 589*, p. (2015), 221, 10.1016/J.TSF.2015.05.040.

111. T. Bronger, P. H. Wöbkenberg, J. Wördenweber, S. Muthmann, U. W. Paetzold, V. Smirnov, S. Traut, Ü. Dagkaldiran, S. Wieber, M. Cölle, A. Prodi-Schwab,O. Wunnicke, M. Patz, M. Trocha, U. Rau, R. Carius, "Solution-based silicon in thin film solar cells," *Advanced Energy Materials 4*, p. (2014), 1301871.

112. T. Sontheimer, D. Amkreutz, K. Schulz, P. H. Wöbkenberg, C. Guenther, V. Bakumov, J. Erz, C. Mader, S. Traut, F. Ruske, M. Weizman, A. Schnegg, M. Patz, M. Trocha, O. Wunnicke, B. Rech, "Solution-processed crystalline silicon thin-film solar cells," *Advanced Materials Interfaces 1*, p. (2014), 1300046, 10.1002/admi.201300046.

11 Flexible Solar Cells Based on Metal Oxides

Soner Çakar

Zonguldak Bulent Ecevit University, Department of Chemistry, Zonguldak, Turkey

Sakarya University, Biomaterials, Energy, Photocatalysis, Enzyme Technology, Nano & Advanced Materials, Additive Manufacturing, Environmental Applications and Sustainability Research & Development Group (BIOENAMS R & D Group), Sakarya, Turkey

Mahmut Özacar

Sakarya University, Faculty of Science & Arts, Department of Chemistry, Sakarya, Turkey

Sakarya University, Biomaterials, Energy, Photocatalysis, Enzyme Technology, Nano & Advanced Materials, Additive Manufacturing, Environmental Applications and Sustainability Research & Development Group (BIOENAMS R & D Group), Sakarya, Turkey

Fehim Findik

Sakarya University, Biomaterials, Energy, Photocatalysis, Enzyme Technology, Nano & Advanced Materials, Additive Manufacturing, Environmental Applications and Sustainability Research & Development Group (BIOENAMS R & D Group), Sakarya, Turkey

Sakarya University of Applied Science, Faculty of Technology, Department of Metallurgical and Materials Engineerings, Sakarya, Turkey

CONTENTS

11.1 Introduction.. 198
11.2 Substrate materials in flexible solar cells 198
11.3 Flexible dye-sensitized solar cells based on metal oxides........................... 198
11.4 Flexible organic solar cells based on metal oxides 201
11.5 Flexible perovskite solar cells based on metal oxides................................. 203
11.6 Other flexible solar cells based on metal oxides 205
11.7 Conclusion 207
References... 208

DOI: 10.1201/9781003186755-11

11.1 INTRODUCTION

With the rapid increase in the world's population, the need for energy is constantly increasing. On the other hand, because fossil fuels will be depleted in the next 50 years, interest in renewable energy sources is enhancing to meet the increasing need [1]. Amid the renewable energy sources, solar energy is the most remarkable because it is an endless energy source. Solar cells, as also known photovoltaics, are apparatus that alter sunlight into electrical energy. Solar cells are divided into three generations; first-generation solar cells – silicon solar cells, second-generation solar cells – thin-film solar cells, and third-generation solar cells – emerging technologies [2]. The early studies on the first-generation solar cell technology were carried out by Bell Laboratories in 1940; the crystalline silicon technology is still widely used commercially today. Second-generation solar cells (thin-film solar cells) that can be an alternative to the first-generation solar cells were developed in the middle of the 20th century, and these are based on gallium arsenide (GaAs), cadmium telluride (CdTe), copper indium gallium selenide (CuInGaSe), and amorphous silicon (a-Si:H). Third-generation solar cell technology has developed at the end of the 20th century, and these solar cells are classified into dye-sensitized solar cells (DSSCs), organic/polymer solar cells (OSCs), perovskite solar cells (PSCs), and quantum dots solar cells (QDSCs) [3]. They have been studied extensively owing to their low cost and environmentally friendly nature. With the development of third-generation solar cell technology, flexible solar cells are now being investigated. The biggest challenge in the commercialization of third-generation solar cells is that they have lower efficiencies than first- and second-generation solar cells. However, their commercialization can be actualized by using third-generation solar cells, especially as flexible solar cells in applications where small power is needed [4]. In recent years, flexible solar cells have attracted great interest due to their flexibility, ease of fabrication, cost-effectiveness, etc. This chapter presents a comprehensive review of metal oxides used in FSCs.

11.2 SUBSTRATE MATERIALS IN FLEXIBLE SOLAR CELLS

The structure of flexible solar cells is divided into two main groups: i) polymer substrates and ii) metallic substrates. Polyethylene naphthalate (PEN) and polyethylene terephthalate (PET) polymers are the most commonly used substrates in flexible solar cells [1]. These polymeric substrates are generally coated with $In:SnO_2$ (ITO) or $F:SnO_2$ (FTO) to gain conductivity. ITO or FTO coated polymer substrates have about 10–15 ohm/cm^2 sheet resistance. Additionally, these ITO or FTO coated polymer substrates have high flexibility and transparency in VIS-NIR region. However, the properties of ITO or FTO, such as compressive stress, brittle structure, and tensile sensitivity, may reduce the conductivity of flexible substrates. Therefore, different materials like conductive polymers, graphene, and carbon nanotubes have been studied to displace ITO or FTO. Moreover, plastic substrates are less resistant to oxygen and water than glass substrates [5]. Thin metal foils can be alternatives to polymer substrate since they provide excellent conductivity compared to polymers and ITO or FTO with great flexibility. Various metals like copper, aluminum, titanium, stainless steel, zinc, and some alloys have been examined for such applications [4]. Additionally, metal mesh and metal wires are also used in flexible solar cells. However, these metal-based substrates have lower corrosion resistance and stability than polymeric substrates.

11.3 FLEXIBLE DYE-SENSITIZED SOLAR CELLS BASED ON METAL OXIDES

Metal oxides are extensively utilized as photoanodes in DSSCs. Semiconductors, such as TiO_2, SiO_x, ZnO, Zn_2SnO_4 [6,7], etc., are used on flexible substrates. In addition, the use of Co_3O_4 as a counter electrode also stands out recently for flexible DSSCs. The power-conversion efficiency of flexible stainless steel DSSC, where $Co_3O_4@Co_3S_4$ was used as the counter electrode, was found to be 5.30%.

The conversion-efficiency value obtained in the DSSC using Pt counter electrode under similar conditions was 5.23%. Also, it is observed that this flexible solar cell can maintain 93% of cell efficiency after a 15-day aging test and different blending angles [8]. The conversion efficiencies of DSSCs formed by coating 10 μm thick TiO_2 films on different flexible metallic substrates (W, Zn, Ti, stainless steel, Co, Pt, Ni, Al) of ~100 μm thickness were investigated. While conversion-efficiency values of 3.60, 3.32, 2.79, 2.20% were obtained for Ti, W, stainless steel, Zn metallic substrates, respectively, very low cell efficiencies were found for Pt, Co, Al, Ni metallic substrates. It has also been reported that the SiO_x layer coated as an insulator in the $Al/ITO/SiO_x/TiO_2$ electrode prepared on the Al flexible substrate prevents corrosion caused by iodine [9]. In another study where the effect of the SiO_x insulator layer on flexible substrates was examined, different device configurations, such as bare, ITO/metal, and ITO/SiO_x/metal prepared using Ti, W, stainless steel metallic substrates, were examined by electrochemical impedance spectroscopy, and the efficiencies of these cells were determined. While the SiO_x insulator layer on the stainless steel substrate increases the solar cell efficiency, it has been seen that this layer decreases the conversion efficiency in Ti and W substrates. The reason for this decrease is that although SiO_x insulator layer increases the J_{SC} values, the conversion efficiencies decrease as it decreases the FF values [10].

Titanium oxide nanotube (TNT) was formed by anodization process on flexible metallic Ti foil substrate, and TiO_2 nanoparticles were coated on it. The TNT served as electron transfer paths and as a photoanode. The conversion efficiency of the DSSC where TNT was used as the electron transfer layer was 6.68%, which was 5.55% higher than that achieved for the DSSC with only TiO_2 nanoparticles [11]. In studies performed on DSSCs with TiO_2 photoanodes on flexible stainless steel mesh electrodes, it was reported that the stainless steel mesh electrodes can be more effective and efficient in large area models [12]. The influence of $ZnSn_2O_4$ semiconductor on DSSCs with flexible stainless steel mesh electrodes was examined, and conversion efficiency of 1.15% was obtained in 10 cm × 5 cm large area flexible DSSCs [13]. The effects of monolithic back contact DSSCs (mBC-DSSCs) that can create an alternative instead of conventional DSSCs on Ti foil metallic substrates were investigated. In mBC-DSSCs structures, two co-planar charge collection electrodes were discreted with a ZrO_2 porous spacer. These type mBC-DSSCs can be applied by roll-to-roll printing techniques on flexible metal substrates of cost-effective and transparent conductive DSSCs [14]. In the study performed by coating ZnO, another metal oxide widely used in DSSCs, on stainless steel mesh electrodes, a conversion efficiency of 2.71% was obtained [15]. Generally, ZnO provides a lower conversion efficiency than TiO_2 based solar cells due to its solubility and aggregation problems. When the efficacy of TiO_2 photoanotes on DSSCs with stainless steel metallic foils was examined, efficiency of 5.51% was obtained. This type of stainless steel-essenced solar cells can form potential alternatives that can preserve their efficiency properties under bending in rigid or plastic substrates [16]. Metal oxides can also be used in DSSCs by coating on metal mesh electrodes with the electrodeposition method. In the study performed by a coating of ZnO nanorods on stainless steel mesh electrode with electrodeposition, more homogeneous and smooth coatings of ZnO were obtained and a conversion efficiency value of 1.81% was achieved [7].

Flexible woven metal wires are utilized as flexible substrates in DSSCs. An efficiency value of 2.302% was obtained with $ZnSn_2O_4$ nanostructures coated on these metal wires. In particular, it is reported that in adding N to $ZnSn_2O_4$ nanostructures, the conversion efficiency is further increased [6]. Although wire-shaped DSSCs show low efficiencies, they are the most groundbreaking studies on flexible DSSCs. Wire-based DSSCs are also prepared by twisted the working and counter electrodes. Optical and SEM images of wire-based DSSCs can be seen in Figure 11.1 [17]. When two fiber-like electrodes are twisted together, the entire cell is still a flexible wire. In wire-based DSSCs, J_{SC} values are especially in direct proportion to the length. In metal wire-based solar cells, in the coating of ZnO and TiO_2 semiconductors, techniques such as thermal amplification, atomic layer deposition, etc., by adding seeds coating, are generally used [18]. In cylindrically shaped Pt coated Ti rod-based DSSCs, solar cells prepared with TiO_2 semiconductors have lower conversion efficiencies than traditional DSSCs. The reason for this is due to optical losses like transmittance and

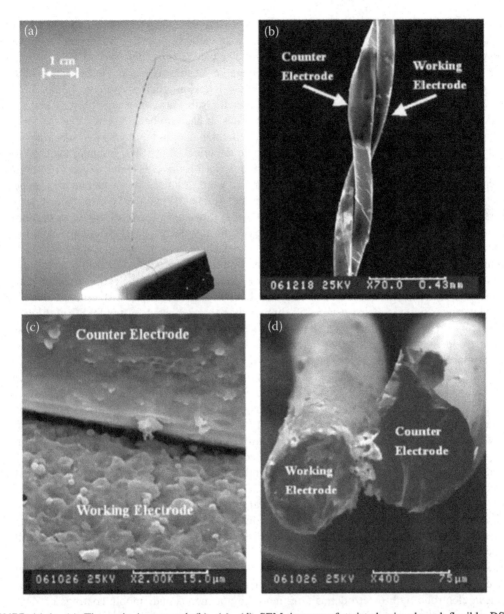

FIGURE 11.1 (a) The optic image and (b), (c), (d) SEM images of twisted wire-shaped flexible DSSCs. Reproduction from ref. [19].

reflection [19]. In a study conducted with TiO_2 blocking layer coated on flexible substrate using anodic aluminum oxide template in flexible DSSCs, a relatively high efficiency of 5.19% was observed for flexible solar cells [20]. In another study, TiO_2 semiconductors are coated with laser-assisted nanoparticles deposition systems in flexible DSSCs. DSSCs prepared with this special coating system were integrated into the power manager unit device and used in sensor applications. With this power manager unit developed, a new application area is brought to DSSCs thanks to the use of DSSCs as small and flexible solar cells in integrated circuits [21]. Some studies on flexible DSSCs are given in Table 11.1.

TABLE 11.1

Some DSSCs Based on Metal Oxide

Substrate	Oxide	PA or CE[1]	Jsc (mA/cm^2)	Voc (V)	FF (%)	η (%)	Ref.
Ti	TiO$_2$	PA	8.14	0.62	71	3.60	[9]
Ti	TiO$_2$ SiO$_x$	PA	8.46	0.62	68	3.54	[10]
SS wire	TiO$_2$	PA	0.06	0.61	38	0.014	[17]
SS mesh	TiO$_2$	PA	5.32	0.57	55	1.68	[12]
SS mesh	Zn$_2$SnO$_4$	PA	6.5	0.5	35.9	1.15	[13]
Metal wire	ZnO	PA	7.89	0.53	64.3	2.70	[18]
Ti metal	ZrO$_2$, TiO$_2$	PA	8.2	0.79	64	4.2	[14]
Ti foil	TiO$_2$	PA	10.71	0.68	71	5.19	[20]
SS mesh-Ti sputtered	TiO$_2$ rod	PA	9.62	0.73	54	3.80	[19]
PET	TiO$_2$ nanolaser	PA	3.82	0.73	68.58	1.92	[21]
SS mesh	ZnO	PA	9.60	0.50	56	2.71	[15]
SS	TiO$_2$	PA	15.82	0.67	52	5.51	[16]
SS metal wires	Zn$_2$SnO$_4$	PA	8.734	0.51	51.7	2.302	[6]
SS mesh	ZnO	PA	6.67	0.497	55	1.81	[7]

SS: stainless steel PA: photoanode CE: counter electrode, PET: polyethylene terephthalate.

11.4 FLEXIBLE ORGANIC SOLAR CELLS BASED ON METAL OXIDES

Metal oxides are mostly utilized as electron transporter layers in OSCs, while metal oxides such as ZnO, MoO$_3$, WO$_3$ [22,23] are also often utilized as hole transporter layers. ZnO has lofty electron mobility (200–100 cm^2 V^{-1} s^{-1}), great excitation binding energy (60 mV), and a broad bandgap (3.37 eV) [24]. The efficiency values of 1.65–2.15% were obtained in OSCs formed by ZnO coating with the help of acetylacetone and monoethanolamine organic precursors onto PET-ITO flexible substrate. It has been reported that the ZnO electron transporter layer prepared with the aid of acetylacetone increases the efficiency by forming an electron collection layer [22]. When the n-type ZnO layer on the PEN-ITO flexible substrate was prepared by the photoactivated method, an efficiency value of 2.71% was obtained. When similar solar cells were prepared on a glass substrate, an efficiency value of 3.60% was found [25]. ZnO/N-doped Cu/ZnO electrode with oxide/metal/oxide (OMO) configuration has been developed to increase the electron transport effect (Figure 11.2). A stable ZnO/N-doped Cu/ZnO type OSCs that can be prepared at low cost have shown good transmittance, high conductivity, long-term stability against corrosion, and high-efficiency features. The conversion efficiency value of 7.1% was obtained with bendable OSC, which has higher efficiency than conventional ITO-based OSC. Furthermore, it is also reported that this 2D Cu (N) ultrathin film can be successfully applied to different polymer substrates with industrial-scale roll coaters techniques [26]. Ag/Al:ZnO is utilized as a nano-thickness transparent electron transporter layer on PET electrodes. Metals like Al and Ag added to ZnO to upsurge the effect and efficiency of the ZnO electron transporter layer to provide high electrical conductivity, mechanical ductility, transparency, and antireflection properties. The efficiency value of 2.6% was observed in flexible OSC prepared using Ag/Al: ZnO electrode which is very close to the efficiency value of ITO-based OSC (2.9%) [27]. ZnO/AgOx/ZnO electrode system was used to demonstrate higher efficiency values of ZnO electron transporter layers. Here, oxygen-doped silver electrode was coated as ultrathin between the electron transporter layers, and its application in flexible OSCs was examined. This developed ZnO/AgOx/ZnO electron transporter layer was coated on PET substrate, and thanks to its lofty transmittance (95%) and low

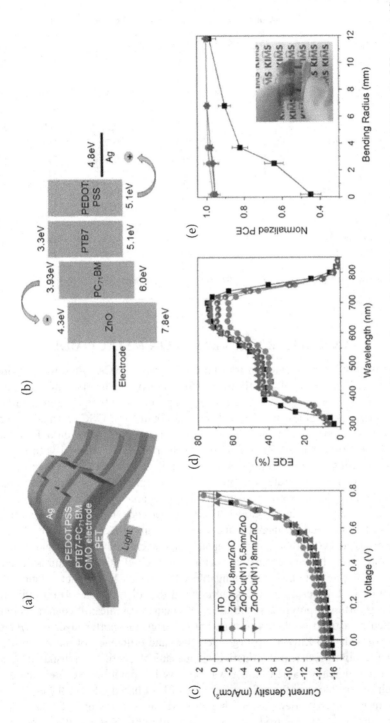

FIGURE 11.2 (A) The schematic diagram of flexible OSCs, (b) energy diagram of example OSCs, the characterizations of OSCs (c) J-V curves, (d) IPCE measurements, (e) Normalized PCE curves. Reproduction from ref. [26].

TABLE 11.2

Some Organic Solar Cells Based on Metal Oxide

Substrate	Oxide	ET or HT[1]	J_{SC} (mA/cm^2)	V_{OC} (V)	FF (%)	η (%)	Ref.
PET	ZnO	ET	7.33	0.55	53	2.15	[22]
PET	Ag/Al:ZnO	ET	6.6	0.76	53	2.6	[27]
PEN	TeO$_2$/Ag	ET	10.39	0.86	65	5.78	[32]
PET	Ta$_2$O$_5$/Ag/WO$_{3-x}$	ET	7.4	0.63	61	3.0	[23]
PET	ITO/AgO$_x$/ITO	ET	14.62	0.67	60.33	5.88	[33]
PET	ZnO/AgO$_x$/ZnO	ET	16.39	0.68	57.93	6.34	[28]
PET	NiO/Ag/NiO	ET	12.91	0.71	61	5.55	[34]
PEI	VO$_x$	HT	10.16	0.86	64	5.58	[35]
PET	ZnO/Cu/ZnO	ET	15.40	0.74	62.15	7.10	[26]
PET	Al-doped ZnO	ET	25.05	0.832	72.97	15.21	[29]

SS foil: stainless steel foil, PET: polyethylene terephthalate, PEN: polyethylene naphthalate, PEI: polyethylenimine, ET: electron transport layer, HT: hole transport layer.

resistance (20 Ω/sq) properties, high conversion efficiency (6.34%) was obtained [28]. It has been stated that Ag nanowires/Al-doped ZnO based electrode (Em-Ag/AgNWS:Al:ZnO:SG) designed in the concept of "welding" flexible transparent electrode in flexible OSCs, show the highest efficiency. This also provided outstanding optoelectronic features, smooth surface, flexibility, and enhanced mechanical properties and conversion efficiency of 15.21% [29].

Although MoO$_3$ is especially used as hole transport layers in OSCs, it can be also used as an electron transport material. For this purpose, the electron transfer efficiency was increased by using Ag nanofilm and MoO$_3$ together. The use of MoO$_3$/Ag/MoO$_3$ multilayer electrodes as electron transporter layers in OSCs has been investigated, and it is reported that these electrodes show high adherence and flexibility on PET surfaces [30]. In the study performed with MoO$_3$/LiF/MoO$_3$/Ag/MoO$_3$ transparent electrode where MoO$_3$ was used as multilayer, an efficiency of 2.77% was obtained due to the increase in transmittance. This alternative transparent electrode showed higher efficiency than ITO-based OSCs due to its low reflective index and electrode/organic interface interaction [31]. It was aimed to provide zero reflection conditions in a broad wavelength range with the formed multilayer electrode by adding Ta$_2$O$_5$ in MoO$_3$/Ag. In the multilayer electrode formed for this purpose, Ta$_2$O$_5$ has outer dielectric and zero reflection characteristics, and a conversion efficiency value of 2.9% was obtained in OSCs [23]. Apart from these, some metal oxides such as TeO$_x$, AgO$_x$, NiO, VO$_x$, and V$_2$O$_5$ [32–35] are also utilized as electron transporter layers in OSCs. Some studies on flexible OSCs that used various metal oxides are given in Table 11.2.

11.5 FLEXIBLE PEROVSKITE SOLAR CELLS BASED ON METAL OXIDES

Perovskite solar cells were first developed in 2009 by Miyasaka et al. [36]. The PSCs, known for their structure similar to DSSCs, have been observed to have 3.8% efficiency values in this first study with mesoporous TiO$_2$ electron collector and thin perovskite layer (CH$_3$NH$_3$PbBr$_3$ and CH$_3$NH$_3$PbI$_3$). Shortly after this first study, PSCs became most attractive to researchers because they provided a conversion efficiency of 20% [37]. Very recently, 25% conversion efficiency values have been achieved in PSCs [38]. Although PSCs have reached high efficiency, major challenges in their future commercialization are their low environmental stability (compared to crystalline solar cells) and use of toxic materials. Some studies on flexible PSCs are given in Table 11.3. PSCs are made up of transparent

TABLE 11.3

Some PSCs Based on Metal Oxides

Substrate	Oxide	ET or HT[1]	Jsc (mA/cm^2)	Voc (V)	FF (%)	η (%)	Ref.
PET	Au or Ag-In$_2$O$_3$ ZnO	TCO	12	0.96	66	7.8	[39]
PET	Ag NW/ZnO:F	TCO	12.2	0.685	39.5	3.29	[40]
PEN/ITO	Zn$_2$SnO$_4$	ETL	21.6	1.05	67	15.3	[44]
ITO/PEN	Li/SnO$_2$	ETL	23.27	1.106	70.71	14.78	[43]
PET	Cu$_2$CrO$_2$	HTL	20.48	1.07	70	15.30	[48]
Ti plate	TiO$_2$	ETL	17.9	1.09	74	14.7	[42]
PEN/ITO	ZnO/NiO$_x$	ETL/ HTL	20.6	1.021	73	15.3	[45]
ITO/PEN	a-Al:ZnO/AgNW/ Al:ZnO/PES	TCO	17.4	1.05	74.4	13.56	[41]
PET/ITO	NiO	HTL	23.5	0.985	77.6	18.0	[46]
PEN	NiO	HTL	20.5	1.05	65.6	13.8	[47]
ITO/SiO$_2$/invar	SiO$_2$	TCO	18.12	1.11	47	9.70	[5]

Notes

[1] ET: electron transporter layer, HT: hole transporter layer, TCO: transparent conductive oxide, MCL: metal contact layer.

conductive oxide substrate, electron transporter layer, perovskite sensitizers, hole transporter layer, and metal contact components. In PSCs prepared onto flexible substrates, various metal oxides were used in all components except sensitizers. Especially, metal oxide essenced electron transporter layers are widely utilized in flexible PSCs. Substrates utilized in conventional solar cells are generally ITO and FTO layers. In flexible solar cells, studies have been carried out for the use of layers that can create an alternative to SiO$_2$ layer. For this purpose, Au or Ag/In$_2$O$_3$, AgNWS/ZnO:F, a-Al:ZnO/AgNW/Al:ZnO and SiO$_2$ were utilized as transparent conductive oxides in flexible solar cells. The conversion efficiency of PET/M-In$_2$O$_3$/ZnO/CH$_3$NH$_3$PbI$_3$/Spiro-OMeTAD/Ag type PSCs in flexible solar cells was reported as 7.8% [39]. Ag NWs/ZnO:F composite developed by coating on flexible PET substrate for PSCs exhibited 90.4% transmittance at 550 nm wavelength, and 23 Ω/sq sheet resistance [40]. In the study conducting by the coating of a-Al:ZnO layers on Ag NWs based transparent electrodes as protective layers especially to increase the chemical stability in flexible solar cells, the 11.23% cell efficiency was obtained from PSC with Al:ZnO/AgNW/Al:ZnO/ZnO/CH$_3$NH$_3$PbI$_3$/spiro-OMeTAD/Au cell configuration. It is especially reported that this protective layer (a-Al:ZnO/AgNW/Al:ZnO) has maintained 94% of its initial solar cell efficiency in 12.5 mm blending angle and 400 blending cycles, whereas the ITO electrode is only safeguarded at 42% under similar conditions [41]. It has been reported that when SiO$_2$ insulator layers are used between ITO films coated with sputtering on flexible invar metal foils, they show high reflectance properties. The conversion efficiency value of 9.70% was obtained in the ITO layer coated with a 150 nm thick invar substrate/SiO$_2$/ITO transparent conductive oxide electrode developed for this purpose. In this study, it was stated that invar flexible substrates could be used as an alternative instead of flexible polymer substrates [5].

Metal oxides such as TiO$_2$ and ZnO, which have a wide bandgap, are widely utilized as the electron transporter layer in PSCs. Extensive studies have also been performed on the usage of metal oxides as electron transporter layers in flexible PSCs, and these studies have focused on the development of alternative metal oxide layers in recent years. The conversion efficiency of PSC, used TiO$_2$ electron transport electrode prepared by thermal oxidation on a flexible Ti metal plate, was found to be 14.9%. In this study, the TiO$_2$ layer's oxygen vacancy amount was adjusted in a controlled manner. The agglomeration that occurs in these oxygen vacancies causes low resistance in the transport of the electron and thus decreases shunt-resistance values of PSC. The decreasing shunt resistance causes the alternating current path to arise for the current generated by the transformation of light and power losses occurring in PSCs. Although the Ti metal plates/TiO_2 highly bendable flexible PSCs are applied 1000 times the blending test, they do not show any decrease in solar cell efficiencies [42]. In the study on the use of SnO_2 electron transporter layers in flexible PSCs, the changes in the PSCs efficiencies depending on the SnO_2 sintering temperatures were investigated. The fact that the electron transporter layers used in PSCs do not contain pinholes has caused the conversion efficiency to be high. For this reason, intensive studies are carried out to optimize the sintering temperatures and times of the electron transporter layers. Various solar cell efficiencies such as 16.50%, 18.1%, 16.20% were obtained with nanocrystalline SnO_2 electron transporter layers on PET surface. It has been stated that the conversion efficiency of flexible PSCs fabricated by the SnO_2 electron transporter layer decreases almost very little at 1000 blending cycles. Studies were also performed on the development of metal-doped SnO_2 electron transporter layers used in flexible PSCs. Flexible and wearable PSCs with Li doped SnO_2 electron transporter layer have been developed, and the potential availability as both a flexible and wearable energy source is reported with an efficiency of 14.78% [43]. A binary metal oxide electrode, such as Zn_2SnO_4, was utilized as an electron transporter layer on PEN/ITO, and the photoconversion efficiency value was determined as 14.85%. It is stated that Zn_2SnO_4 is a suitable structure for flexible substrates with its n-type semiconductor, wide bandgap (3.8eV), low reflective index feature [44].

Nickel oxide (NiO_x) is utilized as hole transporter material in flexible PSCs. With the NiO_x hole transporter materials, photoconversion efficiency values of 16.6%, 16,7%, and 15.9% were obtained in PSCs. Polymeric and organic hole transporter materials like Spiro-MeOTAD, P3HT, PTAA, PPy, PEDOT:PSS, CuSCN, etc., are also widely used in PSCs. In PSCs, good electrical properties and more stable devices are considered to be developed when hole transporter materials are compatible with back contact. In flexible solar cells, under thermal stress (when temperature ranges from 45 to 100 °C), especially photoconversion efficiency of NiO_x keeps > 80% of the initial performance for > 1000 hours [45]. It was also reported that cell stability is increased when NiO_x hole transport materials are used together with PEDOT: PSS [46]. NiO_x hole transporter layer coating onto plastic substrate ensured the attachment of more NiOOH groups on the surface and thus improved the perovskite layer quality with the increase of hole collection. Cell efficiency of 15.9% was obtained when this metal oxide was used as hole transporter material in flexible PSCs [47]. In an important study where binary copper-chromium metal oxide ($Cu_yCr_zO_2$) was used as a hole transporter layer in flexible PSCs, efficiency values of 12–15% were observed by changing the composition ratios of copper or cobalt. The highest efficiency was obtained by using the $Cu_yCr_zO_2$ (2:1) hole transporter layer with a 15.30% efficiency. This new $Cu_yCr_zO_2$ hole transporter material, with properties such as forming a uniform perovskite film, being compatible and wettable, has shown that it can be an alternative to mono-metal oxide hole transporter layers [48]. The various flexible perovskite solar cells are shown in Figure 11.3.

11.6 OTHER FLEXIBLE SOLAR CELLS BASED ON METAL OXIDES

Apart from the extensive studies on the use of various metal oxide structures in OSCs, DSSCs and PSCs, a limited number of studies have been conducted on their use in second- and third-generation solar cells. One of these studied the effect of Al:ZnO thin films on flexible amorphous silicon (a-Si:H) thin-film solar cells. Al:ZnO layers of different thicknesses in a-Si:H thin-film solar cell were coated with the radio frequency magnetron sputtering method, and the highest light absorption data and cell efficiency (6.36%) were obtained with a 100 nm spacer layer. The efficiencies of solar cells, depending on the blending angle, have been specially investigated [50]. In a-Si:H thin-film solar cells, high-row V-shaped anodic aluminum oxide molds as anti-reflective layers were coated on polyimide foils by sol-gel method, and it was stated that the efficiency increased from 5.50% to 8.17%. The variations in efficiency depending on the bending angle were specifically investigated, and cells showed excellent efficiency performance in 100,000 blending cycles [51].

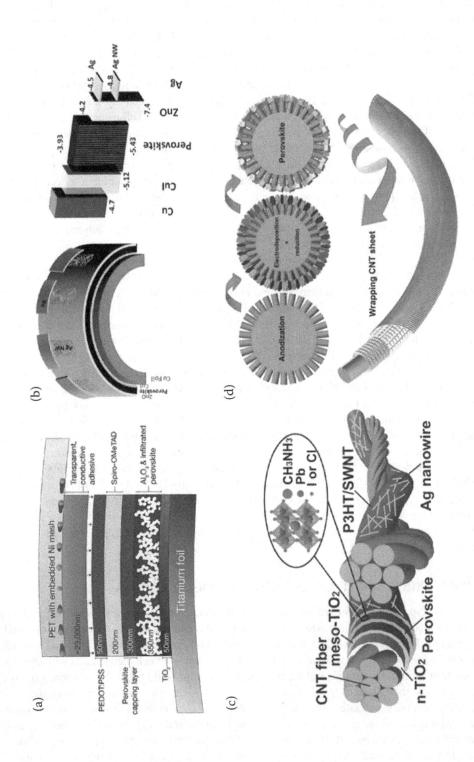

FIGURE 11.3 The various shaped PSCs based on TiO$_2$ or ZnO electron transporter layers. (a) planar, 9b) highly bendable, (c) wire shaped, and (d) fiber shaped of PSCs. Reproduction from ref. [49].

In gallium arsenide (GaAs) type solar cells, another thin-film solar cell, CeO_x structures, were coated on a flexible substrate with epitaxial growth. In this study, MgO homoepitaxial layer was also used as hole transporter material. High purity single crystal (001) GaAs layer was grown epitaxially on a CeO_2 flexible film, and this cell showed shunted diode feature. It has been reported that the CeO_2 layer reduces the cost of GaAs-type solar cells and increases the surface roughness with simple manufacturing processes [52]. Studies on the effects of Al_2O_3 barrier layers in flexible Cu(In, Ga)Se$_2$ (CIGSe) thin-film solar cells are also performed. While 7.9% efficiency was obtained with CIGSe thin-film solar cells without barrier layer prepared on stainless steel surfaces, 14.4% efficiency was obtained when Al_2O_3 barrier layers were used [53]. For a good conversion efficiency, iron diffusion from steel to CIGSe material should be prevented. In a similar study, it has been shown that the Al_2O_3 barrier layer reduces the diffusion of iron into the chalcopyrite absorbent layer. In addition, the interaction and role of sodium used in the preparation of chalcopyrite solar cells with the Al_2O_3 barrier layer were also debated [54].

The properties of the sandwich structure like roughness, density, layer resistance, and transparency were investigated for Al:ZnO/Ag NWs/Al:ZnO sandwich conductors in CuIGSe solar cells. It was observed that Ag nanowires showed efficient charge carrier extraction and collection features under strenuous mechanical bending in these flexible solar cells and maintained 95% of power efficiency at 1000 bending cycles with the hybrid electrode used [55]. Another important study performed on flexible solar cells examined the effect of plasmonic Au-SiO_2 core-shell nanoparticles on CuIGSe thin-film solar cells. It showed that with the plasmonic Au-SiO_2 core-shell nanoparticles contribution, the conversion efficiency increased from 9.28% to 10.88% in flexible solar cells. It was reported that the Au-SiO_2 core-shell nanoparticles increased scattering and efficiency with plasmonic effect [56]. The effect of MgZnO type composite nanocrystals on third-generation quantum dot solar cells has been investigated. MgZnO nanocrystals were used as electron transporter layer on PET/ITO substrate and 9.4% conversion efficiency was obtained. It has been reported that the MgZnO electron transporter layer, synthesized at low temperatures, shows efficient, light, stable properties and will be a good interconnector layer in tandem solar cells [57].

11.7 CONCLUSION

Many studies in the literature use various metal oxides for several purposes in different types of flexible solar cells, and those with important results were concisely summarized in this chapter. Generally, the metal oxide's nature is the main factor to impacts the efficiency of flexible solar cells. In many studies, metal oxides, such as TiO_2, ZnO in DSSCs, MoO_3, ZnO in OSCs, and SnO_2 in PSCs, were generally used as an electron transporter layer and partially hole transporter material. When fabricating flexible solar cells, PET, PEN, stainless steel, and various metallic substrates are widely used as flexible substrates, along with metal oxides. Different techniques have been used in both preparation and coating the metal oxides onto flexible substrates. In OSCs where flexible metal oxides are used, solar cell efficiency values usually range from 1% to 4%, but rarely, 7% efficiency has been observed in OSCs. The major challenge of using metal oxide layers in flexible OSCs, is their low efficiency values and the problems that occur after bending. In DSSCs with flexible metal oxide, the conversion efficiency values have varied the range of 2–7%. One of the major drawbacks of these solar cells is ensuring the compatibility between flexible materials and metal oxides. Conversion efficiencies in the range of 3–18% have been found in PSCs using various metal oxides. High-efficiency values have been achieved in PSCs, but difficulties are encountered in their commercialization because they do not have long-term stability compared to crystalline solar cells, and environmentally friendly materials are not used in their production. Although many studies have been done on flexible solar cells and significant results have been achieved, they can still be developed to increase their efficiency and service life.

REFERENCES

1. Abujarad, S. Y., Mustafa, M. W., Jamian, J. J. Recent Approaches of Unit Commitment in the Presence of Intermittent Renewable Energy Resources: A Review. *Renew. Sustain. Energy Rev.*, 2017, *70* (October 2015), 215–223. 10.1016/j.rser.2016.11.246.

2. Sugathan, V., John, E., Sudhakar, K. Recent Improvements in Dye Sensitized Solar Cells: A Review. *Renew. Sustain. Energy Rev.*, 2015, *52*, 54–64. 10.1016/j.rser.2015.07.076.

3. Raj, C. C., Prasanth, R. A Critical Review of Recent Developments in Nanomaterials for Photoelectrodes in Dye Sensitized Solar Cells. *J. Power Sources*, 2016, *317*, 120–132. 10.1016/j.jpowsour.2016.03.016.

4. Balasingam, S. K., Kang, M. G., Yongseok Jun. Metal Substrate Based Electrodes for Flexible Dye-Sensitized Solar Cells: Fabrication Methods, Progress and Challenges. *Chem. Commun.*, 2013, *49*, 11457–11475. 10.1039/C3CC46224B.

5. Seok, H. J., Kim, H. K. Study of Sputtered ITO Films on Flexible Invar Metal Foils for Curved Perovskite Solar Cells. *Metals (Basel).*, 2019, *9* (2). 10.3390/met9020120.

6. Li, Z., Ma, Q., Li, Y., Liu, R., Yang, H. Flexible Woven Metal Wires Supported Nanosheets and Nanoparticles Double-Layered Nitrogen-Doped Zinc Stannate toward Enhanced Solar Energy Utilization. *Ceram. Int.*, 2018, *44* (1), 905–914. 10.1016/j.ceramint.2017.10.021.

7. Yang, S., Sha, S., Lu, H., Wu, J., Ma, J., Wang, D., Sheng, Z. Electrodeposition of Hierarchical Zinc Oxide Nanostructures on Metal Meshes as Photoanodes for Flexible Dye-Sensitized Solar Cells. *Colloids Surfaces A Physicochem. Eng. Asp.*, 2020, *594* (February), 124665. 10.1016/j.colsurfa.2020.124665.

8. Li, Z., Chen, L., Yang, Q., Yang, H., Zhou, Y. Compacted Stainless Steel Mesh-Supported Co3O4 Porous Nanobelts for HCHO Catalytic Oxidation and Co3O4@Co3S4 via In Situ Sulfurization as Platinum-Free Counter Electrode for Flexible Dye-Sensitized Solar Cells. *Appl. Surf. Sci.*, 2021, *536* (February 2020), 147815. 10.1016/j.apsusc.2020.147815.

9. Man, G. K., Park, N. G., Kwang, S. R., Soon, H. C., Kim, K. J. Flexible Metallic Substrates for TiO2 Film of Dye-Sensitized Solar Cells. *Chem. Lett.*, 2005, *34* (6), 804–805. 10.1246/cl.2005.804.

10. Jun, Y., Kang, M. G. The Characterization of Nanocrystalline Dye-Sensitized Solar Cells with Flexible Metal Substrates by Electrochemical Impedance Spectroscopy. *J. Electrochem. Soc.*, 2007, *154* (1), B68. 10.1149/1.2374943.

11. Lin, L. Y., Yeh, M. H., Lee, C. P., Chen, Y. H., Vittal, R., Ho, K. C. Metal-Based Flexible TiO2 Photoanode with Titanium Oxide Nanotubes as the Underlayer for Enhancement of Performance of a Dye-Sensitized Solar Cell. *Electrochim. Acta*, 2011, *57* (1), 270–276. 10.1016/j.electacta.2011.03.065.

12. Vijayakumar, V., Du Pasquier, A., Birnie, D. P. Electrical and Optical Studies of Flexible Stainless Steel Mesh Electrodes for Dye Sensitized Solar Cells. *Sol. Energy Mater. Sol. Cells*, 2011, *95* (8), 2120–2125. 10.1016/j.solmat.2011.03.010.

13. Li, Z., Zhou, Y., Bao, C., Xue, G., Zhang, J., Liu, J., Yu, T., Zou, Z. Vertically Building Zn 2SnO 4 Nanowire Arrays on Stainless Steel Mesh toward Fabrication of Large-Area, Flexible Dye-Sensitized Solar Cells. *Nanoscale*, 2012, *4* (11), 3490–3494. 10.1039/c2nr30279a.

14. Fu, D., Lay, P., Bach, U. TCO-Free Flexible Monolithic Back-Contact Dye-Sensitized Solar Cells. *Energy Environ. Sci.*, 2013, *6* (3), 824–829. 10.1039/c3ee24338a.

15. Li, Z., Liu, G., Zhang, Y., Zhou, Y., Yang, Y. Porous Nanosheet-Based Hierarchical Zinc Oxide Aggregations Grown on Compacted Stainless Steel Meshes: Enhanced Flexible Dye-Sensitized Solar Cells and Photocatalytic Activity. *Mater. Res. Bull.*, 2016, *80*, 191–199. 10.1016/j.materresbull.2016.04.005.

16. Salehi Taleghani, S., Zamani Meymian, M. R., Ameri, M. Interfacial Modification to Optimize Stainless Steel Photoanode Design for Flexible Dye Sensitized Solar Cells: An Experimental and Numerical Modeling Approach. *J. Phys. D. Appl. Phys.*, 2016, *49* (40), 405601. 10.1088/0022-3727/49/40/405601.

17. Fan, X., Chu, Z., Wang, F., Zhang, C., Chen, L., Tang, Y., Zou, D. Wire-Shaped Flexible Dye-Sensitized Solar Cells. *Adv. Mater.*, 2008, *20* (3), 592–595. 10.1002/adma.200701249.

18. Dai, H., Zhou, Y., Chen, L., Guo, B., Li, A., Liu, J., Yu, T., Zou, Z. Porous ZnO Nanosheet Arrays Constructed on Weaved Metal Wire for Flexible Dye-Sensitized Solar Cells. *Nanoscale*, 2013, *5* (11), 5102–5108. 10.1039/c3nr34265d.

19. Usagawa, J., Pandey, S. S., Ogomi, Y., Noguchi, S., Yamaguchi, Y., Hayase, S. Transparent Conductive Oxide-less Three-dimensional Cylindrical Dye-sensitized Solar Cell Fabricated with Flexible Metal Mesh Electrode. *Prog. PHOTOVOLTAICS Res. Appl.*, 2011, *21* (4), 517–524. 10.1002/pip.1223.

20. Kim, K. P., Lee, S. J., Kim, D. H., Hwang, D. K. Effect of Anodic Aluminum Oxide Template Imprinting on TiO$_2$ Blocking Layer of Flexible Dye-Sensitized Solar Cell. *J. Nanosci. Nanotechnol.*, 2013, *13* (3), 1888–1890. 10.1166/jnn.2013.6983.

21. Lee, J. W., Choi, J. O., Jeong, J. E., Yang, S., Ahn, S. H., Kwon, K. W., Lee, C. S. Energy Harvesting of Flexible and Translucent Dye-Sensitized Solar Cell Fabricated by Laser Assisted Nano Particle Deposition System. *Electrochim. Acta*, 2013, *103*, 252–258. 10.1016/j.electacta.2013.04.050.

22. Kuwabara, T., Nakashima, T., Yamaguchi, T., Takahashi, K. Flexible Inverted Polymer Solar Cells on Polyethylene Terephthalate Substrate Containing Zinc Oxide Electron-Collection-Layer Prepared by Novel Sol-Gel Method and Low-Temperature Treatments. *Org. Electron.*, 2012, *13* (7), 1136–1140. 10.1016/j.orgel.2012.03.015.

23. Ham, J., Kim, S., Jung, G. H., Dong, W. J., Lee, J. L. Design of Broadband Transparent Electrodes for Flexible Organic Solar Cells. *J. Mater. Chem. A*, 2013, *1* (9), 3076–3082. 10.1039/c2ta00946c.

24. Çakar, S., Güy, N., Özacar, M., Fehim, F. Investigation of Vegetable Tannins and Their Iron Complex Dyes for Dye Sensitized Solar Cell Applications. *Electrochim. Acta*, 2016, *209*, 407–422. 10.1016/j.electacta.2016.05.024.

25. Lee, J. W., Lee, S. H., Kim, Y. H., Park, S. K. Low-Temperature Photo-Activated Inorganic Electron Transport Layers for Flexible Inverted Polymer Solar Cells. *Appl. Phys. A Mater. Sci. Process.*, 2014, *116* (4), 2087–2093. 10.1007/s00339-014-8407-2.

26. Zhao, G., Kim, S. M., Lee, S. G., Bae, T. S., Mun, C. W., Lee, S., Yu, H., Lee, G. H., Lee, H. S., Song, M., et al. Bendable Solar Cells from Stable, Flexible, and Transparent Conducting Electrodes Fabricated Using a Nitrogen-Doped Ultrathin Copper Film. *Adv. Funct. Mater.*, 2016, *26* (23), 4180–4191. 10.1002/adfm.201600392.

27. Ghosh, D. S., Chen, T. L., Formica, N., Hwang, J., Bruder, I., Pruneri, V. High Figure-of-Merit Ag/Al:ZnO Nano-Thick Transparent Electrodes for Indium-Free Flexible Photovoltaics. *Sol. Energy Mater. Sol. Cells*, 2012, *107*, 338–343. 10.1016/j.solmat.2012.07.009.

28. Wang, W., Song, M., Bae, T. S., Park, Y. H., Kang, Y. C., Lee, S. G., Kim, S. Y., Kim, D. H., Lee, S., Min, G., et al. Transparent Ultrathin Oxygen-Doped Silver Electrodes for Flexible Organic Solar Cells. *Adv. Funct. Mater.*, 2014, *24* (11), 1551–1561. 10.1002/adfm.201301359.

29. Chen, X., Xu, G., Zeng, G., Gu, H., Chen, H., Xu, H., Yao, H., Li, Y., Hou, J., Li, Y. Realizing Ultrahigh Mechanical Flexibility and >15% Efficiency of Flexible Organic Solar Cells via a "Welding" Flexible Transparent Electrode. *Adv. Mater.*, 2020, *32* (14), 1–10. 10.1002/adma.201908478.

30. Abachi, T., Cattin, L., Louarn, G., Lare, Y., Bou, A., Makha, M., Torchio, P., Fleury, M., Morsli, M., Addou, M., et al. Highly Flexible, Conductive and Transparent MoO₃/Ag/MoO₃ Multilayer Electrode for Organic Photovoltaic Cells. *Thin Solid Films*, 2013, *545*, 438–444. 10.1016/j.tsf.2013.07.048.

31. Chen, S., Dai, Y., Zhao, D., Zhang, H. ITO-Free Flexible Organic Photovoltaics with Multilayer MoO₃/LiF/MoO₃/Ag/MoO₃ as the Transparent Electrode. *Semicond. Sci. Technol.*, 2016, *31* (5), 55013. 10.1088/0268-1242/31/5/055013.

32. Salinas, J. F., Yip, H. L., Chueh, C. C., Li, C. Z., Maldonado, J. L., Jen, A. K. Y. Optical Design of Transparent Thin Metal Electrodes to Enhance In-Coupling and Trapping of Light in Flexible Polymer Solar Cells. *Adv. Mater.*, 2012, *24* (47), 6362–6367. 10.1002/adma.201203099.

33. Yun, J., Wang, W., Bae, T. S., Park, Y. H., Kang, Y. C., Kim, D. H., Lee, S., Lee, G. H., Song, M., Kang, J. W. Preparation of Flexible Organic Solar Cells with Highly Conductive and Transparent Metal-Oxide Multilayer Electrodes Based on Silver Oxide. *ACS Appl. Mater. Interfaces*, 2013, *5* (20), 9933–9941. 10.1021/am401845n.

34. Xue, Z., Liu, X., Zhang, N., Chen, H., Zheng, X., Wang, H., Guo, X. High-Performance NiO/Ag/NiO Transparent Electrodes for Flexible Organic Photovoltaic Cells. *ACS Appl. Mater. Interfaces*, 2014, *6* (18), 16403–16408. 10.1021/am504806k.

35. Back, H., Kong, J., Kang, H., Kim, J., Kim, J. R., Lee, K. Flexible Polymer Solar Cell Modules with Patterned Vanadium Suboxide Layers Deposited by an Electro-Spray Printing Method. *Sol. Energy Mater. Sol. Cells*, 2014, *130*, 555–560. 10.1016/j.solmat.2014.07.053.

36. Kojima, A., Teshima, K., Shirai, Y., Miyasaka, T. Organometal Halide Perovskites as Visible-Light Sensitizers for Photovoltaic Cells. *J. Am. Chem. Soc.*, 2009, *131* (17), 6050–6051.

37. Zhou, H., Chen, Q., Li, G., Luo, S., Song, T., Duan, H.-S., Hong, Z., You, J., Liu, Y., Yang, Y. Interface Engineering of Highly Efficient Perovskite Solar Cells. *Science (80-.).*, 2014, *345* (6196), 542 LP–546. 10.1126/science.1254050.

38. Jeong, M., Choi, I. W., Go, E. M., Cho, Y., Kim, M., Lee, B., Jeong, S., Jo, Y., Choi, H. W., Lee, J., et al. Stable Perovskite Solar Cells with Efficiency Exceeding 24.8% and 0.3-V Voltage Loss. *Science (80-.).*, 2020, *369* (6511), 1615 LP–1620. 10.1126/science.abb7167.

39. Poorkazem, K., Liu, D., Kelly, T. L. Fatigue Resistance of a Flexible, Efficient, and Metal Oxide-Free Perovskite Solar Cell. *J. Mater. Chem. A*, 2015, *3* (17), 9241–9248. 10.1039/c5ta00084j.

40. Han, J., Yuan, S., Liu, L., Qiu, X., Gong, H., Yang, X., Li, C., Hao, Y., Cao, B. Fully Indium-Free Flexible Ag Nanowires/ZnO:F Composite Transparent Conductive Electrodes with High Haze. *J. Mater. Chem. A*, 2015, *3* (10), 5375–5384. 10.1039/c4ta05728g.

41. Lee, E., Ahn, J., Kwon, H. C., Ma, S., Kim, K., Yun, S., Moon, J. All-Solution-Processed Silver Nanowire Window Electrode-Based Flexible Perovskite Solar Cells Enabled with Amorphous Metal Oxide Protection. *Adv. Energy Mater.*, 2018, *8* (9), 1–11. 10.1002/aenm.201702182.

42. Han, G. S., Lee, S., Duff, M. L., Qin, F., Lee, J. K. Highly Bendable Flexible Perovskite Solar Cells on a Nanoscale Surface Oxide Layer of Titanium Metal Plates. *ACS Appl. Mater. Interfaces*, 2018, *10* (5), 4697–4704. 10.1021/acsami.7b16499.

43. Park, M., Kim, J. Y., Son, H. J., Lee, C. H., Jang, S. S., Ko, M. J. Low-Temperature Solution-Processed Li-Doped SnO_2 as an Effective Electron Transporting Layer for High-Performance Flexible and Wearable Perovskite Solar Cells. *Nano Energy*, 2016, *26*, 208–215. 10.1016/j.nanoen.2016.04.060.

44. Shin, S. S., Yang, W. S., Noh, J. H., Suk, J. H., Jeon, N. J., Park, J. H., Kim, J. S., Seong, W. M.; Seok, S. Il. High-Performance Flexible Perovskite Solar Cells Exploiting Zn_2SnO_4 Prepared in Solution below 100 °C. *Nat. Commun.*, 2015, *6* (May), 1–8. 10.1038/ncomms8410.

45. Najafi, M., Di Giacomo, F., Zhang, D., Shanmugam, S., Senes, A., Verhees, W., Hadipour, A., Galagan, Y., Aernouts, T., Veenstra, S., et al. Highly Efficient and Stable Flexible Perovskite Solar Cells with Metal Oxides Nanoparticle Charge Extraction Layers. *Small*, 2018, *14* (12), 1–10. 10.1002/smll.201702775.

46. Hou, L., Wang, Y., Liu, X., Wang, J., Wang, L., Li, X., Fu, G., Yang, S. 18.0% Efficiency Flexible Perovskite Solar Cells Based on Double Hole Transport Layers and $CH_3NH_3PbI_3$-XCl_x with Dual Additives. *J. Mater. Chem. C*, 2018, *6* (32), 8770–8777. 10.1039/c8tc02906g.

47. Hou, C. H., Shyue, J. J., Su, W. F., Tsai, F. Y. Catalytic Metal-Induced Crystallization of Sol-Gel Metal Oxides for High-Efficiency Flexible Perovskite Solar Cells. *J. Mater. Chem. A*, 2018, *6* (34), 16450–16457. 10.1039/c8ta05973j.

48. Qin, P. L., He, Q., Chen, C., Zheng, X. L., Yang, G., Tao, H., Xiong, L. Bin; Xiong, L., Li, G., Fang, G. J. High-Performance Rigid and Flexible Perovskite Solar Cells with Low-Temperature Solution-Processable Binary Metal Oxide Hole-Transporting Materials. *Sol. RRL*, 2017, *1* (8), 1–10. 10.1002/solr.201700058.

49. Li, L., Zhang, S., Yang, Z., Berthold, E. E. S., Chen, W. Recent Advances of Flexible Perovskite Solar Cells. *J. Energy Chem.*, 2018, *27* (3), 673–689. 10.1016/j.jechem.2018.01.003.

50. Xiao, H., Wang, J., Huang, H., Lu, L., Lin, Q., Fan, Z., Chen, X., Jeong, C., Zhu, X., Li, D. Performance Optimization of Flexible A-Si: H Solar Cells with Nanotextured Plasmonic Substrate by Tuning the Thickness of Oxide Spacer Layer. *Nano Energy*, 2015, *11*, 78–87. 10.1016/j.nanoen.2014.10.006.

51. Zhang, C., Song, Y., Wang, M., Yin, M., Zhu, X., Tian, L., Wang, H., Chen, X., Fan, Z., Lu, L., et al. Efficient and Flexible Thin Film Amorphous Silicon Solar Cells on Nanotextured Polymer Substrate Using Sol–Gel Based Nanoimprinting Method. *Adv. Funct. Mater.*, 2017, *27* (13). 10.1002/adfm.201604720.

52. Mehrotra, A., Freundlich, A., Majkic, G., Wang, R., Sambandam, S. Epitaxial Growth of (100) GaAs on CeOx Coated Flexible Metal Substrates. *2012 38th IEEE Photovolt. Spec. Conf.*, 2012, 002571–002574.

53. Gledhill, S., Zykov, A., Allsop, N., Rissom, T., Schniebs, J., Kaufmann, C. A., Lux-Steiner, M., Fischer, C. H. Spray Pyrolysis of Barrier Layers for Flexible Thin Film Solar Cells on Steel. *Sol. Energy Mater. Sol. Cells*, 2011, *95* (2), 504–509. 10.1016/j.solmat.2010.09.010.

54. Gledhill, S. E., Zykov, A., Rissom, T., Caballero, R., Kaufmann, C. A., Fischer, C. H., Lux-Steiner, M., Efimova, V., Hoffmann, V., Oswald, S. The Role of the Spray Pyrolysed Al_2O_3 Barrier Layer in Achieving High Efficiency Solar Cells on Flexible Steel Substrates. *Appl. Phys. A Mater. Sci. Process.*, 2011, *104* (1), 407–413. 10.1007/s00339-010-6176-0.

55. Tsai, W. C., Thomas, S. R., Hsu, C. H., Huang, Y. C., Tseng, J. Y., Wu, T. T., Chang, C. H., Wang, Z. M., Shieh, J. M., Shen, C. H., et al. Flexible High Performance Hybrid AZO/Ag-Nanowire/AZO Sandwich Structured Transparent Conductors for Flexible $Cu(In,Ga)Se_2$ Solar Cell Applications. *J. Mater. Chem. A*, 2016, *4* (18), 6980–6988. 10.1039/c5ta09000h.

56. Chen, C. W., Chen, Y. J., Thomas, S. R., Yen, Y. T., Cheng, L. T., Wang, Y. C., Su, T. Y., Lin, H., Hsu, C. H., Ho, J. C., et al. Enhanced Power Conversion Efficiency in Solution-Processed Rigid CuIn(S,Se)2 and Flexible Cu(In,Ga)Se2 Solar Cells Utilizing Plasmonic Au-SiO2 Core-Shell Nanoparticles. *Sol. RRL*, 2019, *3* (5), 1–8. 10.1002/solr.201800343.

57. Zhang, X., Santra, P. K., Tian, L., Johansson, M. B., Rensmo, H., Johansson, E. M. J. Highly Efficient Flexible Quantum Dot Solar Cells with Improved Electron Extraction Using MgZnO Nanocrystals. *ACS Nano*, 2017, *11* (8), 8478–8487. 10.1021/acsnano.7b04332.

12 Inorganic Materials for Flexible Solar Cells

Mozhgan Hosseinnezhad
Department of Organic Chemistry, Institute for Color Science
and Technology, Tehran, Iran

Zahra Ranjbar
Department of Surface Coatings and Novel Technologies-Institute
for Color Science and Technology, Tehran, Iran

Center of Excellence for Color Science and Technology, Tehran, Iran

CONTENTS

12.1   Introduction.. 211
12.2   Inorganic photoactive devices... 211
12.3   Cu(In,Ga)Se$_2$ (CIGS) solar cells.. 212
12.4   Cu$_2$ZnSn(S,Se)$_4$ solar cells.. 213
12.5   CdTe solar cells.. 216
12.6   Sb$_2$Se$_3$ solar cells 218
12.7   CsPb(I$_{1-x}$Br$_x$)$_3$ solar cells 219
12.8   Environmental and economic concerns... 223
12.9   Conclusion 223
References.. 224

12.1   INTRODUCTION

Electrical energy consumption is increasing worldwide, and people have different electricity consumption based on their needs. The International Energy Agency monitors the annual energy consumption plans of different countries. Today, around the world, energy is supplied through fossil fuels. The use of fossil and coal resources has created complex environmental problems. The important thing is that the use of renewable energy protects the environment. The use of energy products by the sun creates a clean and accessible source of energy in the world. This clean energy produced can be used in a wide range of applications, such as photovoltaic devices and thermo-electronic devices, photoelectron-chemical tools, and photo-catalytic tools. The research shows that photovoltaic devices are an important smart technology due to direct production of electrical energy from sunlight [1]. Today, the energy consumption of new technologies, such as mobile phones, has increased and the availability of recharging technology is very important. Among today's sources of electricity generation, flexible solar cells (FSCs) are promising for a variety of reasons. It is very important to attend to the performance power efficiency of these devices, and the use of durable materials reinforces this feature. One of these materials is inorganic compounds that have shown excellent performance in these devices from the very beginning [2,3]. The purpose of this chapter is to provide an overview of inorganic-based flexible solar devices.

12.2   INORGANIC PHOTOACTIVE DEVICES

Maximum energy production and long life in environmental conditions are very important for customers of solar cell devices (PV technologies). Today, the performance of solar cell panels in standard laboratory conditions is evaluated and presented. These standard laboratory conditions include direct irradiation (1000 Wm^{-2}) and temperature of 25°C, which may not be met under real conditions [4]. Various components, such as organic, organic-inorganic, inorganic, and hybrid compounds, are used in these optical devices and can serve as a basis for their classification. Thus, inorganic photovoltaic components have received a lot of attention due to their excellent optical performance, low cost, and good validity [5].

Reported studies have introduced new inorganic components with high adsorption coefficient and usability in solar cells. These components can layer on flexible substrates, so they can be used in flexible solar devices. Therefore, due to the high efficiency of these components in solar devices, they can replace silicone devices. Among inorganic components for PV technologies, Cu(In, Ga)Se$_2$ (CIGSe) and CdTe show excellent optical efficiency and have expanded quickly. The CIGSe and CdTe (as inorganic components) optical devices have obtained good power-conversion efficiency of about 22%, and they may be a good candidate for commercialization [6].

12.3   Cu(In,Ga)Se$_2$ (CIGS) SOLAR CELLS

For the first time in 1970, basic research and characterization of CuInSe$_2$-based solar cells were reported. All research was conducted at Bell Labs. These devices are suitable for replacing silicon solar devices due to flexibility, tunable bandgap, and a good absorption coefficient. From this type of structure, two types of solar cells, Cu(In$_x$,Ga$_{1-x}$)Se$_2$ (CIGS) and Cu(In$_x$,Ga$_{1-x}$)(S$_y$,Se$_{1-y}$) (CIGSSe), were introduced, which showed the best performance. The use of flexible substrates is more important than hard substrates due to the variety of applications. In the following, CIGS will be introduced as an introduction to environmentally friendly energy production devices. There are requirements for providing flexible devices and employing flexible substrates, including chemical inertness, thermal stability, and surface smoothness. Therefore, before producing flexible solar cells and introducing new inorganic materials, these points should be considered [7].

Different types of CIGSe-based solar device substrates are PI, ZrO$_2$, PTFE, paper, PET, MS, SS, Ti, Al, Mo, Kovar, and Cu. The performance of each device and the year of its development are given in Figure 12.1. The use of polyimide, polyethylene terephaltalte, or ceramic foils as substructures are suitable for the development of flexible solar cell devices; metal substructures cannot be used for this

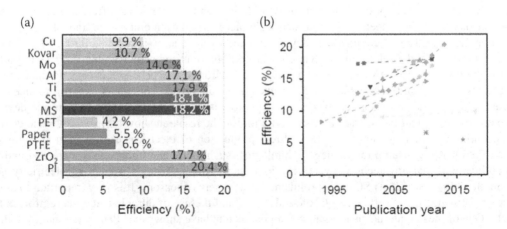

FIGURE 12.1   (a) Yields and (b) efficiency progress of flexible solar cells-CIGS on different substrates [8].

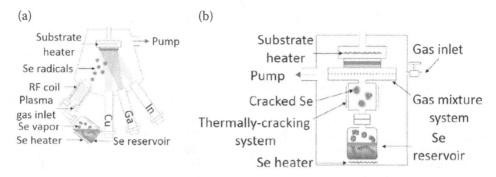

FIGURE 12.2   The scheme of CIGS MBE with (a) Se beam and (b) thermally cracked Se selenization system [8].

purpose. But polyimide has been used in most research as a flexible substrate because of its prominent attributes like the independence from impurity diffusion and lightweight. Of course, polyimide film also has a limitation property that should not be used at temperatures above 450 °C. However, the power-conversion efficiency of flexible and rigid solar cells CIGSe vary, and only rarely are the same or close values obtained. The use of glass and SS substrates has been used to prepare commercial CIGSe-based solar cell devices. But commercial models with flexible substrates have not yet been introduced. The largest reported power-conversion efficiency from the commercial sample of these devices is about 17%.

The development of solar cell (CIGSe kind) construction and production depends on several factors, including preparation of single-crystalline $CuInSe_2$, etc. These properties prevent the production of films with a thickness of less than 1 µm, which is necessary for producing solar cells. The first $CuInSe_2$ solar cell device with a glass substrate was introduced in the early 1980s and had an efficiency of 9.4%, attracting a lot of attention. With the development of research and the use of new technologies, the efficiency of this device has increased by about 22%. The Se beam from CIGS MBE technique (Figure 12.2a) was employed for the preparation of Se radicals to enhance the reactivity and controllability of the Se source. Se radicals increased grain sizes, but they did not significantly change the solar cell efficiency. Then, the RF-cracked Se technology as a thermal-cracking system (Figure 12.2b) was utilized to improve the results. The uniformity of the solar cell-CIGS prepared by the high cracked-selenium (HC-Se) method is greater than the low cracked-selenium (LC-Se) method [8,9].

CIGSe solar cells are a promising option for a wide range of commercial applications due to their excellent photovoltaic properties, large tolerance of compositional deviation, and benign grain boundaries. However, the cost of making these devices is usually very high, which is the main limitation to replacing solar devices based on silicon. The flexible solar cells were prepared using a zirconia sheet as substrate with a thickness of 50 µm. The absorber layers were employed using alkali doped of CIGS and CIS. The presence of doped layers increases the efficiency by 20%, enhancing absorption and decreasing nominal bandgap energy [10]. The Na-doped molybdenum bi-layer was prepared and used for flexible solar cells (flexible CIGS). This layer acts as a diffusion barrier and increases the efficiency by up to 38%. It is very important to pay attention to the exact percentage of sodium because its excessive use causes a recombination phenomenon [11]. Routes that can be considered for the development and advancement of this technology are: employing of Cd-free buffer layer, tandem configuration, and study of bandgap CIGS devices [8,12].

12.4 $Cu_2ZnSn(S,Se)_4$ SOLAR CELLS

Katagiri et al. were prepared CZTSSe solar and the efficiency and photovoltage (V_{OC}) of the prepared device were 0.66% and 400 mV, respectively. The CZTSSe film was prepared via sequential high vacuum evaporation technique. In another study using the deposition method, the power-conversion efficiency was increased up to 2%. It is important to note that the natural resources of In and Ga are limited and predicted to affect the development of this generation of cells (CIGSe solar cells). Due to the requests for low-cost, green components and ternary configuration to more complex quaternary I-II-IV-VI configuration, employing cross substitution cation excitation prepares many elections. The study of CIGSe led to the introduction of a new generation of flexible solar cells, namely CZTSSe. In this generation, indium or galium metals were replaced by zinc and tin. A common structure of this solar cell is: flexible layer/Mo/CZTSSe/CdS/i-zinc oxid/AZO(ITO). Route of CIGSe solar cells development can inspire CZTSSe device progress. Thermal dynamically stable kieserite Cu_2ZnSnS_4 and related $Cu_2ZnSn(S,Se)_4$ solid media were identified as the best options due to suitable bandgap (1.0–1.5 eV) and high absorption coefficient ($>10^4$ cm^{-1}). However, intrinsic defects and the controlling phase are complicated due to the complex quaternary CZTSSe system [13]. The phase-stable areas of kesterite CZTSSe were illustrated by the black region. During the synthesis of CZTSSe, the secondary phases were created by annealing to 500°C. The amount of tin at a temperature above 300°C decreases due to Sn-S(Se) phase volatility, and this percentage reduction occurs faster in a vacuum. Prepared species often have extensive network defects. This phenomenon has a direct impact on the lifetime of the flexible solar cell. However, the development of these compounds is slow due to significant network defects and difficult control.

Although no efficiency of more than 12.6% has been achieved for these devices so far, there is hope to solve the problems by recognizing the properties of the materials and the complete preparation process. With careful study and preparation of numerous solar cells, no higher efficiency has been achieved, with most reports publishing efficiency of around 11%. The maximum performance of pure sulfur CZTS devices prepared by co-sputtering and a post-sulfurization method is reported to be about 11%. By changing the manufacturing technique and using the sputtering and post-selenization method, the efficiency increases by about 12.5%. The most important disadvantage of CZTSSe devices is the lack of high photovoltage. Resolving this weakness is essential to achieving a high return of about 15% [14].

The parasite of CdS buffer reduces photocurrent value and thus reduces power-conversion efficiency. In CZTS devices with pure sulfide, the amount of photovoltage can be brought to an acceptable level using other buffer components with appropriate conduction band edge. Figure 12.3a illustrated the optimum "spike"-like CBO with (Zn,Cd)S/CZTS heterojunction containing the highest photovoltage about 100 mV. However, researchers have explored the idea of a Cd-free buffer layer to reduce toxicity. For this purpose, the atomic layer deposition (ALD) method has been used for (Zn,Sn)O buffer layers production and application in flexible photovoltaic devices, and 10% efficiency has been obtained. Passivation of the surface of the CZTSSe absorber is an appropriate process for enhancing the junction interface. To improve efficiency, the surface-etching process with KCN or $(NH_4)_2S$ was proposed. Combining the two methods, the etching process and ammonia soaking during CBD treatment have the best results in this case and for preparation of the secondary phase of detrimental $Cu_{2-x}(S,Se)$ on the surface. Other groups researched the Al_2O_3 layers for surface passivation. As shown in Figure 12.3b, the photovoltage increased due to fewer local potential fluctuations, which is due to the introduction of Al_2O_3, the charged point defects of CZTS. The redispensation of sodium and elements inter-diffusion passivated the nonradiative recombination point and created a steady conduction band transition at the heterojunction interface (Figure 12.3c), creating the 11% power conversion efficiency CZTS devices based on pure sulfide. But using the toxic metal Cd has prevented much $Cu_2(Zn,Cd)Sn(S,Se)_4$ development. The band adjustment between other kesterite quaternary components for gradient bandgap designing is illustrated in Figure 12.3d. Ag-based kesterite components illustrate lower valence and conduction band, which is appropriate for the front bandgap gradient. But gallium-based components behave in the opposite way to Ag [12].

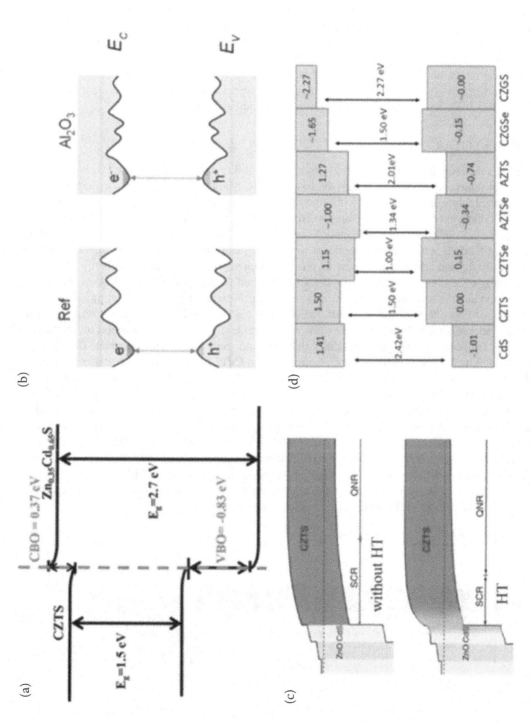

FIGURE 12.3 Schematic of the band alignment (a) at CZTS/$Zn_{0.35}Cd_{0.65}S$ interface, (b) CZTS with and without the HT process and (d) Calculated band alignment between various CZTSSe [12].

The next generation of flexible inorganic solar cells, the Kesterite CZTSSe based semiconductors, will be due to low-cost assembling, environmental-friendly constituents, and considerable potential for excellent performance. Sun et al. investigated the effect of using Cd-doped layer in flexible CZTSSe. This layer was prepared using the evaporating technique and applied on a flexible solar device. The maximum photovoltage of the device was 342.52 mV, and its efficiency was improved about 50% [15]. Yu et al. studied the effect of using In-doped layer in flexible CZTSSe. The efficiency of the flexible device was estimated about 7.19%, and the photovoltage value was improved about 62% [16]. This knowledge of formation and controlling of the defects and passivation procedure is necessary to provide an efficient, flexible device with maximum power-conversion efficiency [10,12,17].

12.5 CDTE SOLAR CELLS

CdTe has demonstrated the highest scalability and reproducibility among thin films so far, as reported by the successful industrial production within the last 10 years. The high absorption coefficient allows the production of highly efficient devices with ultra-thin absorbers, and values exceeding 10% have been obtained with thickness below 1 μm and around 10% for only 0.5 μm [18]. The CdTe thickness reduction could be a crucial point for increasing the efficiency and stability of flexible devices. Both rigid and flexible CdTe flexible-optical devices perform well when made in superstrate configuration, and this happens mainly for two reasons: (i) the special process required with the so-called "activation treatment" where typically a Cd chloride ($CdCl_2$) thin layer is deposited on top of the CdTe and then the stack is annealed in the air; the activation treatment acts as a step for increasing the electrical properties of the absorber and for improving the junction between CdS and CdTe. So if it is made after CdTe deposition on CdS, the performance is higher. (ii) the back contact generally requires the addition of Cu for high performance. In substrate configuration, Cu is deposited on the back of the solar cell. During CdTe deposition (with high substrate temperature), Cu reacts with CdTe film, which causes degradation in the film. The $CdCl_2$ treatment promotes recrystallization of the CdTe and enhances grains growth; it also improves the CdTe/CdS junction properties by enhancing inter-diffusion of sulfur into the CdTe and reducing the lattice mismatch between the two layers. So the treatment has a double purpose, which makes it necessary to be used after the CdTe deposition. When CdTe is deposited in substrate configuration, $CdCl_2$ must be applied two times, a first time before and a second time after CdS deposition. CdTe solar cells in superstrate configuration have reached the superior efficiency of 22.1%, reported by First Solar Inc. So superstrate configuration is still preferable also for flexible devices. On the other hand, this requires particular substrates that have to fulfill transparency and flexibility requirements simultaneously. Configurations of CdTe solar cells are shown in Figure 12.4.

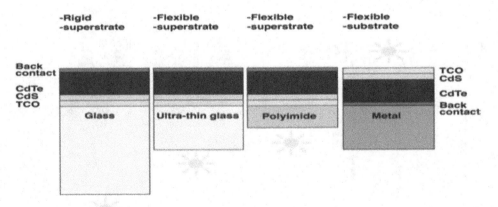

FIGURE 12.4   Schematic of superstrate and substrate CdTe devices [19].

TABLE 12.1

Best Efficiency of Flexible CdTe Solar Cells [19]

Structure	Substrate	Substrate (T °C)	η (%)	V_{OC} (mV)	J_{SC} (mAcm^{-2})
Superstrate	Flex glass	> 550	16.4	831	25.5
Superstrate	PI film	< 450	13.6	846	22.3
Substrate	Mo film	< 450	11.5	821	21.8

The paths for fabricating efficient CdTe solar cells have been divided into two main cases: (i) to optimize a fabrication process for substrate configuration CdTe solar cells by using a nontransparent substrate such as metal-based or polymer-based layers, and (ii) to fabricate and optimize flexible CdTe cells on transparent substrate capable of withstanding high process temperatures [20].

The devices were made by close space sublimation at temperatures above 600°C. After CdCl$_2$ vapor treatment and deposition of ZnTe:Cu back contact, a record efficiency of 14.05% was obtained. This value has been overcome by researchers from the same group, by the application of a sputtered CdS:O instead of the prior chemical bath deposited CdS, gaining in transparency, which resulted in a higher response in the blue light region with higher efficiency of 16.4%. A summary of the best efficiency for CdTe flexible solar cells is presented in Table 12.1.

CdTe solar cells, whether built-in superstrate or substrate configuration, need long-term stability. Generally, CdTe is a very robust and stable material, and no particular problems have been addressed for stability when suitable encapsulation against moisture is applied. In earlier times, an age-old issue of performance stability was due to copper diffusion. However, this has been practically solved by strongly reducing the copper amount and adding a copper barrier at the back contact, such as ZnTe that acts as a barrier to Cu, as previously mentioned. Other solutions to this problem also have been introduced, such as a barrier or As$_2$Te$_3$ or the incorporation of copper in the form of a chlorine salt, which allows introducing an extremely low amount of copper by combining it with the CdTe back surface. However, now CdTe-based solar cells on glass have demonstrated optimal stability for long-time performance, as also demonstrated by the large module production.

For flexible solar cells, new issues must be considered in terms of stability. The bending and stretching of the module should not affect the device's performance. Only a few reports are available for the stability of flexible devices. Rance et al. studied the effects of tensile and compressive stresses for flexible cells on ultrathin glass. The devices were fastened with PVC tubes so it was possible to bend them. Different curvatures were experimented and tested with, and the resulting efficiencies were measured. Analogous data for cells on polymers were not reported. However, the change in efficiencies (in some cases, efficiency increases when cells are bent) is not attributed to the type of substrate but the stress in the CdTe itself; similar results would be expected for the cells on polymer substrates [21].

Contrarily to other thin-film materials, CdTe solar cells have come to large mass production, for the moment, concentrated on one company. The simple stoichiometry and high reproducibility reduce problems with inhomogeneity, which are more frequent for other high-efficiency thin-film flexible devices. Thus, the large mass production of Si wafers has put production costs of thin-film modules at the same level as Si modules. Actually, thin films have a higher potential for cost reduction if we can take advantage of their ability to adapt to different shapes. This adaptability will allow integrating and substituting conventional roofs with energy producing roofs. Application of CdTe to new substrates must be studied; CdTe is a straightforward material to be grown, it is not affected by the different substrates, it has reduced issues in inhomogeneity for large scale, and it has a significant potential of mass production in a flexible configuration. Superstrate devices are the easiest to fabricate. They would be optimal for windows and flexible glass, while substrate devices would be preferred on tiles and ceramic substrates [19,22]. Salavei et al. investigated the effect of different substrates on the performance of flexible CdTe devices. The deposition was prepared on low-temperature procedure. The results show that the polyimide with low transmittance after layering presents the best performance [23]. Teoleken et al. produced flexible CdTe solar cells using flexible ultrathin glass with an efficiency of 17.7%. The results illustrated that this substrate is a new approach in building flexible optical devices with optimal performance [24].

12.6 Sb_2Se_3 SOLAR CELLS

In 1950, an isostructure component was developed from Sb_2S_3, Sb_2Se_3. Due to the long history of research on Sb_2S_3, it became much easier to understand the physical properties of Sb_2Se_3 [25]. In the 1960s, many of the physical properties of Sb_2Se_3 composition were studied and identified. The study of the optical properties of Sb_2Se_3 showed that the crystalline structure or noncrystalline (amorphous) directly affects the physical properties. From the 1970s to the 1990s, research on Sb_2Se_3 and its physical properties and their thin films gradually expanded and reported.

The Sb_2Se_3 has unique and diverse features, such as electrical, electronic, chemical structure, magnetic, and photonic properties. Sb_2Se_3 was first used in solar cell devices as a supporting element or passive component. Although Sb_2Se_3 was first made in the 1950s, the first solar cell using the compound was introduced in 1982 with an efficiency of 1%. However, active studies on Sb_2Se_3 solar cells began in 2014, maybe after published papers on TiO_2-sensitized inorganic-organic solar cells ($\eta = 3.2\%$), TiO_2/Sb_2Se_3 and CdS/Sb_2Se_3 thin-film devices ($\eta = 2.26\%$). Figure 12.5a illustrated the power-conversion efficiency of Sb_2Se_3 and other Sb-containing thin-film devices. But today, studies on the use of this compound are growing significantly (Figure 12.5b) [26].

Excellent thin layers are prepared from the Sb_2Se_3 using the low-temperature vacuum-based deposition technique for CdTe solar cell devices. This achievement is due to the characteristics of the low melting point of Sb_2Se_3 and its high vapor pressure of Sb_2Se_3. Experimental studies illustrated that polycrystalline Sb_2Se_3 with stoichiometric ratio has a bandgap (E_g) of about 1.03 eV indirect and 1.17 eV direct, which is appropriate for using Sb_2Se_3 as a photovoltaic absorber layer. Of course, achieving a good power-conversion efficiency is also influenced by the high charge-carrier transport properties. Various researchers have extensively studied the electronic structure of Sb_2Se_3. The results show that the Sb_2Se_3 is an indirect-bandgap semiconductor. The properties of crystal structure and chemical composition have a direct effect on the optical properties of Sb-Se components. In most of the

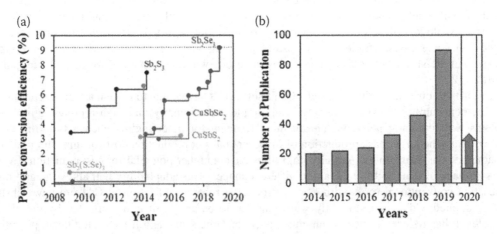

FIGURE 12.5   (a) The record efficiency for Sb-related devices, (b) annually published papers about Sb_2Se_3-based devices [26].

reports, polycrystalline Sb_2Se_3 films with stoichiometric composition illustrated E_g within the range of 1.0–1.3 eV. The method of preparing the thin film has changed the amount of E_g variation in poly-crystalline [27].

The superstrate system is often used to supply flexible Sb_2Se_3-devices. In these systems, CdS layers transmit transports electrons to the Sb_2Se_3 layer, which is capable of producing electron-hole pairs. The energy of the photons is very important to enter this process and should be less than 2.4 eV. Doing this process accurately can cause p-n heterojunction to form between the Sb_2Se_3/CdS interface and all-electron transfer into the TE component. However, holes are possessed by the p-type hole-transport layer and assembled by the back contact. It is important to note that all solar cell layers are critical to high power-conversion efficiency. The Sb_2Se_3–based solar cell device is illustrated in Figure 12.6. It also illustrates the technical subjects that affect the efficiency of the device. Therefore, some issues such as sputtering failure, the action of the crystal orientation (in Sb_2Se_3), and rollover effects (at the interface and surface of Sb_2Se_3) need to be carefully studied [26].

The characteristics of low price and low toxicity of Sb_2Se_3, compared to CdTe, have caused more attention. Two other unique properties of Sb_2Se_3 that have a great impact on the construction of the flexible solar cell are: high absorption coefficient of about 10^5 cm^{-1} and optimal bandgap of 1.1 eV. As mentioned earlier, Sb_2Se_3 has been around for over 50 years, but its use in solar cells is very short-lived. Sb_2Se_3 band-gap value is lower than CdTe, but its photocurrent amount is higher due to ex-panding absorption spectra [28]. Of course, in addition to high power-conversion efficiency, features such as low toxicity, low cost, and high stability are very important for the preparation and com-mercialization of solar cells. The high stability of solar cell devices under light, heat, and ultraviolet radiation is essential because the devices are installed outdoors. Several Sb_2Se_3 papers reported these subjects from this point of view. The highest stability of about 100 days was achieved on the glass/FTO/CdS/Sb_2Se_3/Au structure. The advantages of the 1D crystal-structural Sb_2Se_3 structure are su-perior flexibility and high bandgap, but its disadvantages include complex defect chemistry and highly anisotropic carrier transport. To achieve high power-conversion efficiency, these problems must be solved. Li et al. investigated the optical properties of ultra-flexible Sb_2Se_3 devices. The prepared flexible solar cell illustrated high efficiency of 6.13% and kept 96% efficiency after 1000-time use. This research presents a new concept for efficient flexible Sb_2Se_3 devices [29]. Wen et al. utilized Sb_2Se_3 as a natural abundant component for the preparation of flexible solar cells. The prepared flexible device illustrated high efficiency of 7.15% and kept 90% responsibility after 1000-time using [30].

12.7 $CsPb(I_{1-x}Br_x)_3$ SOLAR CELLS

New photovoltaic components with ABX_3 chemical formula as lead-halide perovskite were prepared where A is an inorganic or organic cation, B is a metal ion usually lead, or tin and X is a halide-like I, Br or Cl, or a mixture of these. These components were introduced in 2009 for the first time and used for the preparation of flexible photovoltaic devices [31]. The advantages of this new technology are low manufacturing cost and solution processability. These properties predict a promising future for this material and flexible solar cells. For this purpose, their most important defect, namely, the considerably lower thermal decomposition temperature, must be eliminated. To this end, a solution is a displacement of the organic ion in the A motion by an inorganic cation like the cesium (Cs^+) organization perovskite like the $CsPb(I_{1-x}Br_x)_3$ (X = 0–1). Newly prepared compounds have high thermal stability. One of the important features of this compound is its dependence on its crystal structure on temperature. Figure 12.7 illustrates four crystalline structures of $CsPbI_3$ as (i) cubic (α-phase), (ii) tetragonal (β-phase), (iii) or-thorhombic (γ-phase), and hexagonal (δ-phase). All structures, namely "black phase," have photovoltaic properties. Among these configurations, the highest temperature stability is related to the α-phase, and other phases occur as the temperature decreases [32].

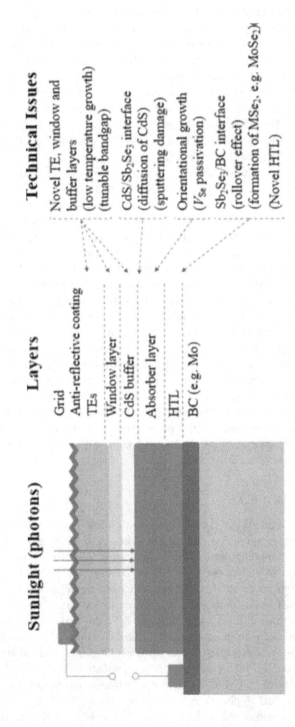

FIGURE 12.6   Structure of flexible devices and each layer and/or interface [26].

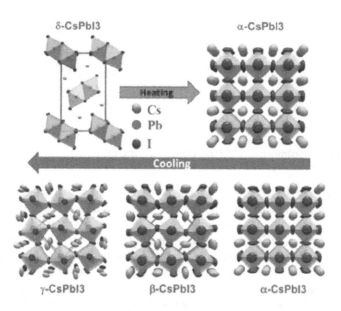

FIGURE 12.7 Scheme of $CsPb(I_{1-x}Br_x)_3$ temperature dependence [12].

The band gap of $CsPb(I_{1-x}Br_x)_3$ is adjustable by changing the I and Br ratios, e.g., $CsPbI_3$ and $CsPbBr_3$ show a bandgap of 1.73 and 2.36 eV, respectively. So the amount of absorption grows due to an increase in Br content [33]. Of course, the amount of bromine must be controlled since an increasing percentage of bromine can limit the use of the final compound due to the excessive width of the absorption spectrum. The best results are observed by using both halides $(CsPb(I_{1-x}Br_x)_3)$. The first report on inorganic cesium lead halides was published in 1893. However, their use in solar cell technology took years and was first applied in 2015 in solar cells as $CsPbBr_3$ solar cell devices. The results show that the $CsPbBr_3$ is a competitive analog for the hybrid perovskite delivering a power-conversion efficiency of 5.9%. In the same year, another group used $CsPbI_3$ in the solar cell structure and achieved an efficiency of 2.9%. The rate of cesium led halides PSCs development in the solar cell structure then increased due to narrower bandgap. The thermal co-evaporation techniques of CsI and PbI_2 were employed in the preparation of highly efficient flexible $CsPbI_3$ devices, with an efficiency of about 10%. In 2019, a b-$CsPbI_3$ PSC device was assembled with power-conversion efficiency exceeding. Now, using di-methylammonium iodine (DMAI) into $CsPbI_3$ causes the enhancing the efficiency about 19.0%, which is the highest power-conversion efficiency published so far for this kind of inorganic device [34]. The important point is that after the formation of the perovskite system, no organic additive is observed in it, but this material plays an essential role in film preparation. These important achievements and high efficiency predict a promising concept for this technology (flexible $CsPbI_3$ devices). However, high power-conversion efficiency $CsPbI_2Br$ device with an efficiency of 16.79% was obtained.

When flexible $CsPbBr_3$ devices were first introduced, they were not expected to make much progress due to wide bandgap. But today, this technology has provided flexible devices with 10% efficiency and very high stability. The two determinants of the highest $CsPb(I_{1-x}Br_x)_3$ PSCs efficiency are smooth morphology and uniformity of thin films. Studies show that dopants, processing conditions, and deposition methods have the greatest impact on the performance of $CsPb(I_{1-x}Br_x)_3$ devices. The main reasons for the development of this technology are its ease of preparation using the spin-coating deposition method and low cost.

Two methods, such as one-step and two-step processes, can be used for the application of the spin-coating method and preparation of thin layers. Using the one-step technique, a solution of lead halide and

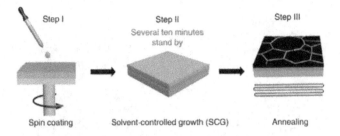

FIGURE 12.8 Schematic illustration of $CsPb(I_{1-x}Br_x)_3$ preparation [12].

cesium halide is prepared in the DMF or DMSO (or a mixture of them) and used for the coating layer for preparation of $CsPb(I_{1-x}Br_x)_3$ perovskite. To produce a uniform film that has the best performance, the solvent evaporation rate must be carefully monitored and the concentration kept carefully (Figure 12.8). Using the two-step technique, lead halide and cesium halide are layered separately and respectively for preparation of $CsPb(I_{1-x}Br_x)_3$ perovskite. The second technique (two-step) effectively overcame the solubility restriction of $PbBr_2$ and was used in many studies; it developed more rapidly, such as to employ the spray technique for the assemby of the $CsPbIBr_2$ device [35].

The phase stability of $CsPb(I_{1-x}Br_x)_3$ material is crucial for solar device development. Reducing the crystal size increases the surface energy, which leads to increased black-phase stability at ambient conditions. The best result of this method is obtained for perovskite nanocrystals. Two important factors in the production and transfer of charge are HTL and/or electrons transport layer (ETL) and interfaces between inorganic perovskite absorbers. These two factors have a direct impact on flexible device performance [36]. As a result, it's important to thoroughly study the two factors of passivation process defects and lower the energy barrier. Introducing a buffer layer between inorganic components and charge transport layers (CTLs) is the main challenge. In some studies, C60 and graphene have been employed to increase electron transfer from $CsPb(I_{1-x}Br_x)_3$ perovskite to the ETL of semiconductor components like TiO_2 or SnO_2. Effective technology uses a layer of ZnO between SnO_2 ETL and $CsPbI_2Br$ perovskite. This layer facilitates charge flow and effective level alignment (Figure 12.9a). There are several benefits to using SnO_2/ZnO bi-layered ETL, including decreasing the charge re-combination, enhancing the charge preparation, obtaining high energy, increasing the photovoltage value (Figure 12.9b). In other words, the SnO_2/ZnO bi-layered ETL plays the role of accelerating hole transport. Fadla et al. introduced the inorganic halide phase in the flexible solar cell as $CsPbI_{3-x}Br_x$. The

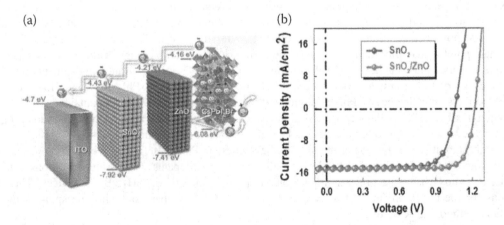

FIGURE 12.9 (a) Energy diagrams of $CsPbI_2Br$ PSC and (b) J-V characteristics of $CsPbI_2Br$ PSC with SnO_2/ZnO bilayered ETL and SnO_2 ETL [12].

optical and electronic properties of this device were investigated. Used layered were prepared in 320°C and assembled devices illustrated high and perfect performance [36].

Two essential properties of $CsPb(I_{1-x}Br_x)_3$ solar cells, namely, thermal stability and excellent photovoltaic performance, have made these devices an essential option for commercialization. $CsPb(I_{1-x}Br_x)_3$ devices are suitable for many applications due to the bandgap tunable feature. But their most important defect, the phase stability issue, must be fixed. Fixing this limitation can lead to the development of this technology and significantly increase efficiency [32,35].

12.8 ENVIRONMENTAL AND ECONOMIC CONCERNS

Most of the new technology uses toxic (Cadmium) and less toxic (indium) metals. Indium and tin are used to prepare indium tin oxide (ITO), which is widely used in solar cell technology for the preparation of conductive substrates due to its excellent performance. Indium resources are expected to run out in 12–13 years, according to global estimates. Its global consumption (about 6000 metric tons) is currently very high and is still increasing. Therefore, recycling indium will be the only option to access indium. Another critical point is the cost of indium, which is rising rapidly. The reason for the high slope of cost is the imbalance of production and supply with market needs. In 2002, market demand increased by 40% compared to the previous year, but production increased by only 27%. As a result, the market for innovative technologies such as CIS and CIGS with heavy metals may be limited [36]. Rare-earth metals used in solar cell technology increase the price of solar cell production. This limitation prevents many users from developing and using this technology.

Much research has focused on improving photovoltaic properties to achieve maximum efficiency of solar devices. Many aspects of getting an optimal photoanode still need to be reconsidered and studied. With more restrictions removed and new advances made, it can be hoped that the large-scale commercialization of solar cells will occur at high speed and volume. Some points to consider are: (1) silicon solar cells are more efficient than second- and third-generation technologies; if charge recombination is reduced, it is possible to produce higher efficiency and better performance; (2) to achieve higher efficiency, the photo-anode component of this technology needs to be further developed and perform better; (3) sensitizers have a great impact on the performance of the solar cell, and the choice of dyes with a wide absorption spectrum or co-sensitization method can increase the efficiency; (4) it is necessary to conduct research on the knowledge of how the dye relates to the structure of the semiconductor and the electrolyte and the rotation of the electron produced to be accurate and to identify the factors affecting the facilitation of the process; (5) the sensitizer absorption spectrum and its expansion and the amount of semiconductor bandgap should be analyzed and optimized; (6) liquid electrolytes are a major constraint in the development of this technology for various reasons, such as evaporation, the possibility of linkage, concentration change, and it is necessary to introduce efficient solid and semi-solid electrolytes; (7) the application of new compounds in various components, such as hybrid perovskite and new target molecule for the electrolyte, as well as the production of flexible devices, can be the main topic of future research; and (8) another important factor in this technology is the effective lifetime and response time. Flexible devices are expected to have a longer lifetime [1,37,38]. Therefore, to achieve effective photovoltaic devices based on inorganic materials especially flexible ones that are still commercially available, a huge amount of researches and studies are still needed. Examples for an objective comparison of performance, effective factors, and differences between the different types of inorganic-based solar cells are presented in Table 12.2.

12.9 CONCLUSION

Solar cells and clean energy production are a definite approach for the future because energy resources are running out and energy demand is increasing. To commercialize this technology, efficiency and longevity must be increased and costs reduced. One of the important things that needs to be corrected

TABLE 12.2

Comparison of Various Properties of Different Types of Inorganic-Based Solar Cells [12]

PV material	Crystal structure	Band gap (ev)	Absorption coefficient	Best PCE
$Cu(In,Ga)Se_2$	Tetragonal	1.02–1.67	$> 10^5$ cm^{-1}	23.35 %
$Cu_2ZnSn(S,Se)_4$	Tetragonal	1.0–1.5	$> 10^4$ cm^{-1}	12.6 %
CdTe	Cubic	1.45	$> 10^5$ cm^{-1}	22.1 %
Sb_2Se_3	Orthorhombic	1.1	$> 10^5$ cm^{-1}	9.2 %
$CsPb(I_{1-x}Br_x)_3$	Cubic	1.73–2.36	$> 10^4$ cm^{-1}	19.03 %

is the substrate; the initial options to consider are ceramics and plastics. To reduce the cost and increase the application time, the use of durable substrates such as flexible substrates is inevitable. The use of these substrates in solar cell technology requires careful and comprehensive study. In reported studies, new inorganic components with high adsorption coefficients and usability in solar cells have been introduced. These components can layer on flexible substrates, so they can be used in flexible solar devices. Among inorganic components for PV technologies, Cu(In, Ga)Se$_2$ (CIGSe) and CdTe with outstanding photoelectric performance have practiced rapid expansion. CIGSe solar cells are a promising option for a wide range of commercial applications due to their excellent photovoltaic properties, large tolerance of compositional deviation, and benign grain boundaries. CdTe has demonstrated the highest scalability and reproducibility among thin films so far, as reported by the successful industrial production within the last 10 years. The advantages of the 1D crystal-structural Sb$_2$Se$_3$ structure are superior flexibility and high bandgap, but its disadvantages include complex defect chemistry and highly anisotropic carrier transport.

REFERENCES

1. N. Tomar et al., Ruthenium complexes based dye sensitized solar cells: Fundamentals and research trends, *Solar Energy*, 207, 2020, 59–76.
2. G. Li et al., Engineering flexible dye-sensitized solar cells for portable electronics, *Solar Energy*, 177, 2019, 80–98.
3. M. Hosseinnezhad et al., Acid azo dyes for efficient molecular photovoltaic: study of dye-sensitized solar cells performance, *Prog. Color Colorants Coat*, 9, 2016, 61–70.
4. Mario Pagliaro et al., *Flexible Solar Cells*, Wiley-VCH Pub., 2008, Germany, p. 55–69.
5. M.A. Green et al., The emergence of perovskite solar cells, *Nat. Photon.*, 8, 2014, 506–517.
6. Y. Wang et al., The Role of Dimethylammonium Iodide in CsPbI3 Perovskite Fabrication: Additive or Dopant? *Angew. Chem. Int. Ed.*, 58, 2019, 16691.
7. W. Xiang, A. Hagfeldt, Phase stabilization of all-inorganic perovskite materials for photovoltaics, *Curr. Opin. Electrochem.*, 11, 2018, 141.
8. Y. Wang et al., A critical review on flexible Cu(In, Ga)Se2 (CIGS) solar cells, *Mater. Chem. Phys.*, 234, 2019, 329–344.
9. P. Jackson et al., Effects of heavy alkali elements in Cu(In,Ga)Se2 solar cells with efficiencies up to 22.6%, *Phys. Status Solidi RRL*, 10, 2016, 583.
10. S. Ishizuka et al., Development of high-efficiency flexible Cu(In,Ga)Se2 solar cells: A study of alkali doping effects on CIS, CIGS, and CGS using alkali-silicate glass thin layers, *Current Appl. Phys.*, 10, 2010, S154.
11. S.T. Kim et al., Effect of Na-doped Mo layer as a controllable sodium reservoir and diffusion barrier for flexible Cu(In,Ga)Se2 solar cells, *Energy Report*, 7, 2021, 2255.
12. F. Liu et al., Emerging inorganic compound thin film photovoltaic materials: Progress, challenges and strategies, *Mater. Todays*, 41, 2020, 120–142.
13. D.H. Kim et al., Bimolecular additives improve wide-band-gap perovskites for efficient tandem solar cells with CIGS, *Joule*, 3, 2019, 1734.

14. H.X. Deng et al., Origin of the distinct diffusion behaviors of Cu and Ag in covalent and ionic semiconductors, *Phys. Rev. Lett.*, 117, 2016, 165901.

15. L. Sun et al., Effect of evaporated CdS layer on formation and performance enhancement of flexible Cu2ZnSn(S,Se)4 solar cells, *Vacuum*, 187, 2021, 110098.

16. X. Yu et al., Efficient flexible Mo foil-based Cu2ZnSn(S, Se)4 solar cells from In-doping technique, *Sol. Energy Mater. Sol. Cells*, 209, 2020, 110434.

17. J. Li et al., Cation substitution in earth-abundant kesterite photovoltaic materials, *Adv. Sci.*, 5, 2018, 1700744.

18. S.H. Demtsu, J.R. Sites, Effect of back-contact barrier on thin-film CdTe solar cells, *Thin Solid Films*, 510, 2006, 320.

19. J. Ramanujam et al., Flexible CIGS, CdTe and a-Si:H based thin film solar cells: A review, *Prog. Mater. Sci.*, 110, 2020, 100619.

20. J.M. Burst et al., CdTe solar cells with open-circuit voltage breaking the 1 V barrie, *Nat. Energy*, 1, 2016, 16015.

21. S.G. Kumar, K.S.R.K. Rao, Physics and chemistry of CdTe/CdS thin film heterojunction photovoltaic devices: fundamental and critical aspects, *Energy Environ. Sci.*, 7, 2014, 45.

22. B. K. Ghosh et al., mcSi and CdTe solar photovoltaic challenges: Pathways to progress, *Optik*, 206, 2020, 164278.

23. A. Salavei et al., Comparison of high efficiency flexible CdTe solar cells on different substrates at low temperature deposition, *Sol. Energy*, 139, 2016, 13–18.

24. A.C. Teloeken et al., Effect of bending test on the performance of CdTe solar cells on flexible ultra-thin glass produced by MOCVD,*Sol. Energy Mater. Sol Cells*, 211, 2020, 110552.

25. Y. Zhou et al., Solution-processed antimony selenide heterojunction solar cells, *Adv. Energy Mater.*, 4, 2014, 1301846.

26. A. Mavlonov et al., A review of Sb2Se3 photovoltaic absorber materials and thin-film solar cells, *Sol. Energy*, 201, 2020, 227–246.

27. C. Chen et al., 6.5% certified efficiency Sb2Se3 solar cells using PbS colloidal quantum dot film as hole-transporting layer, *ACS Energy Lett.*, 2, 2017, 2125.

28. Y. Zhouet al., Thin-film Sb2Se3 photovoltaics with oriented one-dimensional ribbons and benign grain boundaries, *Nat. Photonics*, 9, 2015, 409.

29. K. Li et al., One-dimensional Sb2Se3 enabling ultra-flexible solar cells and mini-modules for IoT applications, *Nano Energy*, 86, 2021, 106101.

30. X. Wen et al., Efficient and stable flexible Sb2Se3 thin film solar cells enabled by an epitaxial CdS buffer layer, *Nano Energy*, 85, 2021, 106019.

31. G.E. Eperon, et al., Inorganic caesium lead iodide perovskite solar cells, *J. Mater. Chem. A*, 3, 2015, 19688.

32. A. Marronnier, et al., Anharmonicity and disorder in the black phases of cesium lead iodide used for stable inorganic perovskite solar cells, *ACS Nano*, 12, 2018, 3477.

33. R.E. Beal, et al., Cesium lead halide perovskites with improved stability for tandem solar cells, *J. Phys. Chem. Lett.*, 7, 2016, 746.

34. L.A. Frolova et al., Highly efficient all-inorganic planar heterojunction perovskite solar cells produced by thermal coevaporation of CsI and PbI2, *J. Phys. Chem. Lett.*, 8, 2017, 67.

35. P. Wang et al., Solvent-controlled growth of inorganic perovskite films in dry environment for efficient and stable solar cells, *Nat. Commun.*, 9, 2018, 2225.

36. M.A. Fadla et al., Insights on the opto-electronic structure of the inorganic mixed halide perovskites γ-CsPb (I1-xBrx)3 with low symmetry black phase, *J. Alloy Compound*, 832, 2020, 154847.

37. H. Yuan et al., All-inorganic CsPbBr3 perovskite solar cell with 10.26% efficiency by spectra engineering, *J. Mater. Chem. A*, 6, 2018, 24324.

38. C. Gao et al., Review on transition metal compounds based counter electrode for dye-sensitized solar cells, *J. Energy Chem.*, 27, 2018, 703–712.

13 Efficient Metal Oxide-Based Flexible Perovskite Solar Cells

Subhash Chander and Surya Kant Tripathi
Centre for Advanced Study in Physics, Department of Physics, Panjab University Chandigarh, Chandigarh, India

CONTENTS

13.1 Introduction.. 227
13.2 Requirement for alternate energy resources.. 228
13.3 Metal oxide nanostructures 228
13.4 Metal oxide based flexible solar cells 231
13.5 Metal oxides based flexible perovskite solar cells.. 235
13.6 Recent advancement in metal oxide-based flexible perovskite solar cells 235
13.7 Summary and outlook 237
Acknowledgments.. 238
References... 238

13.1 INTRODUCTION

Ever-rising energy demand has augmented the risk to the existence of humanity. On the one side, increased energy demands directly impace pollution and global warming, while on the other side, it circuitously stimulates political and strategic conflict over the stock of fossil fuels. Although the renewable energy sector developments significantly counteract this scenario, many challenges must be addressed. Renewable energies contribute to less than 20% of the global energy demand [1]. Many factors determine the growth of the renewable energy sector, including investments, government policies, industrial growth and demand, and public awareness, etc. Many countries, developed and developing ones, are trying to boost renewable energy production, but the appetite for renewable energy is lower because of factors including cost, portability, and utility. Human resources and money are poured into renewable energy research and development to square the challenges. Amid the renewable energies, photovoltaic energy is one of the main concentrations; because surplus resources are available, energy can be directly converted into electricity without any intermediate state, this energy is stable, as is the life cycle of the solar cells. In connection with the rapid development of electric vehicles by leading manufacturers like Tesla, Inc., and Mahindra, hydrocarbon fuels will be overrun by the electricity-based systems in a decade or two. The production of electricity for automobiles through fossil fuels will be meaningless. Only photovoltaic electricity production will solve the problem. The amorphous Si thin-film-based solar cell was first used in the late 1970s, while the lightweight and flexible thin-film-solar-cells are equated with traditional crystalline Si-based solar cells. The mechanically hard and thus far flexible modules based on thin films offer tremendously striking energy source solutions [2]. With swift advancements in recent years in new material schemes, such as organic semiconductors and metal halide perovskites, flexible solar cell panels are expected to be commercialized in many more future marketable products.

DOI: 10.1201/9781003186755-13

13.2 REQUIREMENT FOR ALTERNATE ENERGY RESOURCES

The energy demand is rising at an extraordinary rate to compete with the stride of technical advancements and the upsurging world population. At present, the three prime fossil fuels (oil, natural gas, and coal) fulfill more than 80% of world energy demands [3,4]. At present, yearly world energy consumption is 15 TW, which is anticipated to upsurge to 30 TW by 2050; of this 30 TW, 75–80% of the demand will be fulfilled by these fossil fuels, with the remainder fulfilled by nuclear power, hydro, and renewable energy sources. This augmented energy demand is one side of the story, while the other side encompasses the exhaustion of natural resources, augmented production cost, and deep ecological apprehensions, such as global warming owing to unwarranted usage of fossil fuels. We know that fossil fuels yield greenhouse gases upon ignition and are accountable for global climate change due to global warming, thus putting life nourishment at an augmented risk on the earth. These conventional energy resources are inadequate, and by the present rate of consumption, the oil, natural gas, and coal resources are anticipated to be exhausted in the coming 50, 60, and 120–150 years [1,5]. While it may appear that fossil fuels will continue to control the energy economy, one also must consider the high risks that fossil fuels levy against the planet's atmosphere, such as a 35% escalation in the CO_2 levels, resulting in an increase in global temperature by 0.8°C, which will cause the sea level to rise [6]. Current consciousness of fossil fuels' environmental and health risks has led to an urgency to explore alternate clean and renewable energy resources. This understanding a few decades back inspired sincere efforts to develop such renewable energy resources. The present commercial solar cell modules based on crystalline and amorphous Si demonstrated an efficiency of 25.1%, but these are too costly. The high cost and difficulty of fabrication are the chief problems to the commercial production of solar cells. Consequently, thin-film-based cadmium-telluride (CdTe) and copper-indium-gallium-selenide/sulfide (CIGS/Se) solar cells are broadly studied and commercially developed with proven photovoltaic cell efficiencies and cheaper and less material consumption [7,8]. Table 13.1 summarizes the efficiencies of different single-junction solar cells.

The highest attainable efficiencies of single-junction Si and thin-film solar cells lie between 20% and 25%, which is considerably lower than the supreme hypothetical efficiency edge of 33%, as projected by Shockley and Queisser. Therefore, superficial and affordable methods to develop solar cells to achieve significant efficiencies currently are an urgent need. Recent years have seen a rapid advancement in new material systems (namely, organic semiconductors and metal halide perovskites), and flexible solar cell panels are expected to be available on the market in coming years.

13.3 METAL OXIDE NANOSTRUCTURES

Metal oxides have persisted as an attractive material owing to the wide diversity of physicochemical and optoelectronic properties. Metal oxides are beneficial as equated to orthodox semiconducting materials because oxides have additional degrees of freedom; consequently, the optoelectronic and other properties can be altered by manipulating these degrees of freedom. For illustration, changing the oxygen-octahedral inclines in perovskite oxides can dramatically affect the properties of the oxides. Among the different classes of oxides, transition metal oxides (TMOs) have attracted widespread consideration because of their great applications in sensors, solar cells, fuel cells, etc. [9]. TMOs are compounds made of transition metals and oxygen atoms, and these can occur in the form of monoxide, dioxide, trioxide, etc., owing to the multivalent nature of transition metals. Figure 13.1 demonstrates 5d-orbitals, which play an essential part in several applications of TMOs.

Metal-oxide nanostructures (MONs) for several specialized applications are one of the blistering areas in the field of oxide electronics. Researchers have developed MONs with an extensive range of morphologies through different chemical or physical evolution methods in the past three decades. In photovoltaics, MONs can be used as a scaffold layer to dye-sensitized solar cells (DSSCs) and organic-inorganic hybrid perovskite solar cells (OIHPSCs). These MONs can also be used as electron and hole transport layers in DSSCs and organic solar cells (OSCs). The function of a scaffold is to enable charge separation and charge transport, while the purpose of transport layers is to transform one kind of charge-carrier block to the other kind. Consequently, modifying their properties is inevitable to develop high-performing and flexible photovoltaic devices using these layers [11]. This work emphasizes the different metal oxides, such as TiO_2, ZnO, WO_3, CuO, Cu_2O, etc., for their use in flexible solar cell devices. These materials are the most common and predictable because they are profitable, stable, effective, and eco-friendly. This work also highpoints the photophysical and physiochemical properties of such metal oxides, recent progress, trials, and amendments made on these metal oxides to overcome their limits and exploit their performance in photovoltaic applications. The applications of some of the representative metal oxide-based materials prepared in different nanostructures are tabulated in Table 13.2.

MONs have been reported, ranging from nanoparticles, nanowires, nanotubes, nanosheets, nano-whiskers, nanospheres, among others. These materials can be employed in several applications, such as catalysis, microelectronic circuits, transistors, energy storage and conversion, biomedicines, and sensors [9]. Pimentel and coworkers [13] reported the synthesis of TiO_2-based nanostructured arrangements on the flexible substrates using different solvents (Figure 13.2). SEM microstructure revealed that the different solvents mostly affected the density of the nanostructured films, along with their shape and size (Figure 13.3). They used the optimized results to fabricate a flexible solar cell device, but the efficiency was not great.

TABLE 13.1

The Power Conversion Efficiencies of Different Single-Junction Photovoltaic Technologies at 25°C with Global AM1.5 Spectrum (1000 Wm^{-2}). Adapted and Modified with Permission from Ref. [7], Copyright (2021), Wiley

Classification	V_{OC} (V)	J_{SC} (mA/cm^2)	Fill factor (%)	Efficiency (%)
Crystalline cell	0.738	42.65	84.9	26.7
Large-crystalline cell	0.694	41.58	83.3	24.0
GaAs (thin film cell)	1.127	29.78	86.7	29.1
GaAs (multi-crystalline)	0.994	23.2	79.78	18.4
InP (crystalline cell)	0.939	31.15	2.6	24.2
GaInP	1.493	16.31	87.7	21.4
Thin Film Solar Cells				
CIGS	0.992	35.70	77.6	22.9
CIGSS (Cd-free)	0.685	39.91	76.4	20.9
CdTe	0.887	31.69	78.5	22.1
CZTSS	0.513	35.21	69.8	12.6
CZTS	0.730	21.74	65.1	11.1
Amorphous/microcrystalline				
Amorphous silicon cell	0.896	16.36	69.8	10.2
Microcrystalline silicon cell	0.550	29.72	75.0	11.9
Other PV technology				
Dye-sensitized cell	0.744	22.47	71.2	11.9
Organic cell	0.896	25.72	78.9	18.2
Perovskite cell	1.189	25.74	83.2	25.5

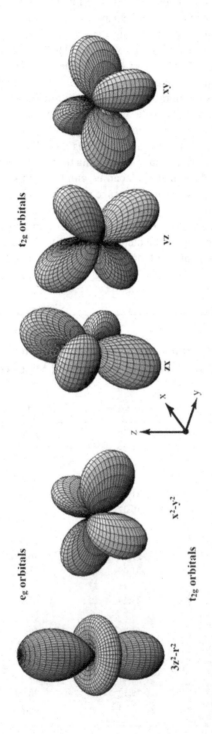

FIGURE 13.1 The five 'd' orbitals. In the cubic crystal field, this fivefold degeneracy is lifted to two e_g and three t_{2g} orbitals. Adapted with permission from ref. [10]. Copyright (2000), The American Association for the Advancement of Science (AAAS).

TABLE 13.2

Applications of Some Representative Metal Oxide-Based Materials with Various Structures Adapted with permission from ref. [12], Copyright (2020), MDPI. Copyright The Authors, some rights reserved; exclusive licensee [MDPI]. Distributed under a Creative Commons Attribution License 3.0 (CC BY)

Metal-oxides based nanostructures	Structural design	Synthesis method	Application(s)
ZnO-nanosheets	3D hierarchical flower-like architectures	Solvothermal	Adsorption of triphenylmethane dyes
Fe_3O_4UiO-66 composite	Cubical NPs arranged	Sonication	Adsorption
Fe_3O_4MIL-100(Fe) Core-Shell Bionanocomposites	Core-shell structure with Fe_3O_4 core, immobilized on P. Putida	Sonication followed by attaching the NPs on bacteria	Adsorption
ZnO-TiO_2/clay	TiO_2 and ZnO NPs mounted on the clay surface	Sol-gel method	Degradation of MG
Cu/ZnO/Al_2O_3	Cu and ZnO impregnated γ-Al_2O_3	Impregnation method	CO removal from reformed fuel
Co^{2+}, Ni^{2+} doped Fe_3O_4 NPs	Cubic lattice	Co-precipitation method	Photodegradation of Carbol Fuchsi
Ce/Fe bimetallic oxides (CFBO)	Flower-like 3D hierarchical architecture	No-template hydrothermal method	As(V) and Cr(VI) Remediation
Perovskite titanates ($ATiO_3$, A = Sr, Ca and Pb)	Leaf-architectured 3D Hierarchical structure	Combination of biosynthesis from Cherry Blossom, heating, grinding, and photo-deposition	Artificial Photosynthetic System for photoreduction of CO_2
TiO_2 polypyrrole	Core-shell nanowires (NWs)	Seed-assisted hydrothermal method	Flexible supercapacitors on carbon cloth
Fe_3O_4/WO_3	Hierarchical Core-shell Structure	Solvothermal growth + oxidation route	Photodegradation of organic-dye materials

Park and coworkers [14] reported the hydrothermal synthesis of rutile TiO_2 nanorod films to be integrated on flexible perovskite solar cells. The efficiency was found to be reliant on the nanorod's length and current density, and the open-circuited voltage decreased with increasing nanorod's length.

13.4 METAL OXIDE BASED FLEXIBLE SOLAR CELLS

The thin-film solar cells (TFSCs), quantum dot solar cells (QDSCs), DSSCs, OSCs, and PSCs are some of the most gifted renewable energy resources for sustainable progress of the future. These solar cells have received widespread consideration because they are easy to fabricate, low-priced, and have tunable optical properties [8,15]. The flexible perovskite solar cells (PSCs) have shown remarkable efficiency improvement, which is considerably increased from 3.8% to more than 25% in the past few years; this is comparable with the Si-based solar cells. The high power-conversion efficiency is achieved using a metal-oxide-based electron transport layer, such as TiO_2, ZnO, SnO_2, NiO, Al_2O_3, V_2O_5, ZrO_2, CeO_2, Nb_2O_5, etc. These devices have accredited all-embracing consideration owing to low cost, high efficiency, and ease of fabrication procedure [16]. The metal-oxide heterojunction devices have also attracted researchers owing to their stability and low cost.

The metal oxides are intrinsic semiconductors, and they can either be n-type (viz. TiO_2, Fe_2O_3, ZnO) or p-type (viz. CuO, Cu_2O, Co_3O_4) [17,18]. Metal-oxide semiconductors having wide bandgaps are cast off as the window layer, while metal oxides having small bandgaps can be used as an absorber layer. The metal-oxides-based transport layer in PSCs can act as a compact layer to avoid any short circuits [19]. TiO_2 and ZnO are excellent choices for transparent window layers because they have a broad bandgap greater than 3 eV and are transparent to visible light. ZnO is a direct bandgap material used widely in metal-oxides-based flexible solar cells [20]. TiO_2 primarily exists as indirect bandgap semiconductors and is extensively used in dye-sensitized solar cells [21]. Among various p-type metal oxides, copper oxides are the most popular materials for solar cells. Stable oxides of copper, Cu_2O, and CuO are p-type semiconductors having a direct bandgap of ~2.1 eV and 1.2 eV. Both semiconductors have high absorption in the visible region and good minority charge-carrier diffusion-length, making them ideal candidates as an absorber layer [22]. Another promising p-type semiconductor with good absorption in the visible region is cobalt oxide. Co_3O_4 is a compound with mixed-valence states and has a spinel structure. CoO is a more stable state of cobalt oxide with a very high bandgap (~5 eV); consequently, Co_3O_4 is the most preferred phase for photovoltaic applications. Numerous metal-oxides-based flexible solar cells fabricated using various techniques in combinations of these metal-oxides are reported in the literature, among which $ZnO-Cu_2O$ based heterojunction solar cells are the most occurring combination; however, it shows very low efficiencies, though it has the theoretical efficiency of 18% [23]. Regardless of the strategies researchers adopt, the improvement in efficiencies is still very low. Among several such reported works, a few remarkable efforts are scrutinized herein.

Wei and coworkers [23] have fabricated three types of $ZnO-Cu_2O$ flexible solar cells by coating ZnO as thin films, nanowires, and nanotubes with a thin film of copper oxides. Using ZnO as thin films yielded very low efficiencies of 0.02%, while ZnO wires and tubes showed slightly higher efficiencies of around 0.12% when compared to the films. In nanotubes and nanowires solar cells, the improved efficiency can be attributed to the increase in the heterojunction area between the Cu_2O and ZnO. Similar solar cells were reported where $ZnO-Cu_2O$ heterojunction solar cells were fabricated using nanorods and nanotubes of ZnO with Cu_2O films [24]. The ZnO nanorods showed a photovoltaic efficiency of 0.4%, whereas the ZnO nanotubes showed 0.8% efficiency. The efficiency was doubled in nanotubes compared to nanorods because of the higher surface area offered by the ZnO nanotubes. In

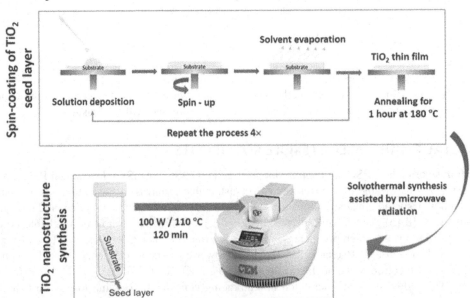

FIGURE 13.2 TiO_2 seed layer production and nanostructured synthesis on flexible substrates. Adapted with permission from ref. [12], Copyright (2016), IntechOpen Ltd. Adapted with permission from ref [12] Copyright The Authors, some rights reserved; exclusive licensee [IntechOpen Ltd]. Distributed under a Creative Commons Attribution License 3.0 (CC BY).

FIGURE 13.3 SEM images and their cross-section of TiO$_2$ nanostructures films synthesized using different solvents. Adapted with permission from ref. [13], Copyright (2016), IntechOpen Ltd. Adapted with permission from ref. [13] Copyright The Authors, some rights reserved; exclusive licensee [IntechOpen Ltd]. Distributed under a Creative Commons Attribution License 3.0 (CC BY).

this route, Zhang's group [25] fabricated flexible solar cells using solution-processed 3D square-patterned nanorod arrays of n-type ZnO and Cu$_2$O thin film. These cells demonstrated the efficiency of 1.52% with a current density of 9.89 mA/cm^2, which revealed a significant improvement compared to that of cells without patterns. Few other researchers tried to improve the efficiency of the Cu$_2$O heterojunction solar cells by varying the deposition methods. Wee and coworkers [26] have deposited Cu$_2$O thin films on the textured metallic substrate employing a pulsed-layer deposition. As shown in the schematic in Figure 13.4a, they fabricated a flexible solar cell structure with an epitaxial Cu$_2$O absorber layer by introducing a thick epitaxial SrRuO$_3$ conductive oxide layer (~100 nm) between Cu$_2$O and SrTiO$_3$ layers as well as n-type transparent conductive oxides ZnO and Al-doped ZnO layers on top of Cu$_2$O layer. This pure epitaxial phase of Cu$_2$O had produced an efficiency of 1.65% (Figure 13.4b). Pławecki et al. [27] invented flexible ZnO/Cu$_2$O solar cells exploiting the electro-deposition method, and they reported the effect of Cu$_2$O thickness on the photovoltaic efficiency of the solar cell layers and achieved efficiencies as high as 2.7%.

Chatterjee et al. [28] reported another technique, namely successive ionic-layer adsorption and reaction (SILAR), where they coated an additional NiO layer over the Cu2O as hole transport and an electron-block layer, and SnO$_2$ on the ZnO side (Figure 13.5).

The prepared NiO/Cu$_2$O/ZnO/SnO$_2$ heterojunction solar cell devices having staircase-like energy levels facilitated the efficient flow of electrons and holes, yielding a photon-conversion efficiency of 1.12%. High efficiency of 3.83% is reported by Minami et al. [29] for high-purity Cu$_2$O layers prepared by thermal oxidation. The high-purity copper sheets were oxidized to Cu$_2$O by heating at 1010°C under a very controlled atmosphere, while ZnO layers were doped with aluminum.

From these explanations, it is understood that the purity of the copper-oxide plays an important role in improving the efficiency of the solar cells, but the preparation of such high-purity Cu$_2$O makes the solar cell fabrication process complex and costly. Contrasting Cu$_2$O, there are only a few reports on photophysical properties of CuO, and the achieved efficiency was lower than 2% owing to minor carrier concentration, but it has an excellent bandgap to construct the flexible solar cell. CuO-based solar cells fabricated with n-type silicon showed 0.36% efficiency [30], whereas TiO$_2$-CuO based solar cells manufactured using electroplating exhibited efficiencies of around 1.6% [31].

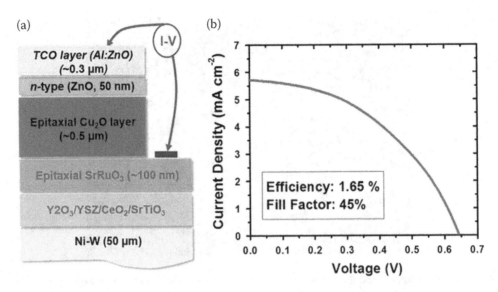

FIGURE 13.4 (a) Schematic design and (b) Current-voltage characteristics of a flexible solar cell device based on epitaxial Cu_2O layer on the textured metallic substrate. Adapted with permission from ref. [26], Copyright (2015), Nature Research. Adapted with permission from ref. [26] Copyright The Authors, some rights reserved; exclusive licensee [Nature Research]. Distributed under a Creative Commons Attribution License 3.0 (CC BY).

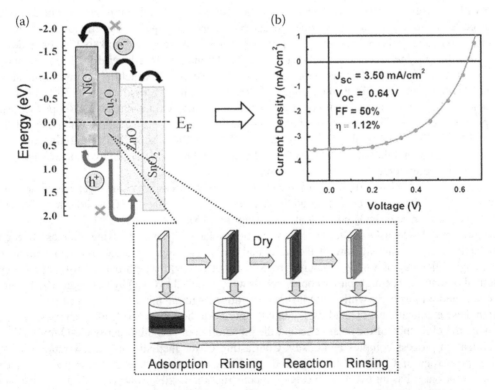

FIGURE 13.5 Staircase-like energy levels and current-voltage characteristics of $NiO/Cu_2O/ZnO/SnO_2$ heterojunction. The dashed-line denotes the Fermi-energy after contact. Adapted with permission from ref. [28], Copyright (2016), Elsevier.

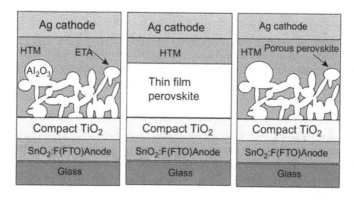

FIGURE 13.6 Typical configuration structures of a perovskite solar cell. Adapted with permission from ref. [32], Copyright (2018), Elsevier.

13.5 METAL OXIDES BASED FLEXIBLE PEROVSKITE SOLAR CELLS

MONs are considered to be a fascinating functional material, and hence, they are an active research topic nowadays. The metal oxides nanostructured in PSCs can act as a compact layer to avoid any short circuit conditions, while the porous-scaffold layer underneath the perovskite active layer may help to crystallize the technology or increase the surface-active-area [19]. Typical PSCs usually have a sandwiched architecture having two electrodes and a photoactive layer (Figure 13.6).

The three conventional configurations of PSCs are (i) semiconductor meso-porous-structured PSCs, where any solution-processed semiconducting material as a porous-scaffold-layer to deliver a meso-super-structured solar cell, (ii) thin-film based PSCs, where no porosity is needed, and perovskite active layer is sandwiched between p- and n-type charge-extracting-contacts, and (iii) p-n heterojunction solar cells, perovskite active layer is porous [32]. The first semiconductor properties of perovskites with the first crystallographic studies of cesium lead halides ($CsPbX_3$, where X = Cl/Br/I) had been introduced in 1958 by Danish scientist Christian Møller [33]. After two decades, Dieter Weber [34–36] invented new three-dimensional (3D) organic-inorganic hybrid perovskites in 1978, where he replaced cesium with methylammonium cations. On the other hand, Zhai and colleagues were the first to report the organometallic halide perovskites having good performances [37]. In the organic-inorganic hybrid PSCs, the morphologies are strongminded by the different crystalline phases, such as tetragonal, orthorhombic, and cubic, which are dependent on the temperature and processing solvent, while the electronic properties can be altered by the change in the morphology [38].

Acting as a photoactive layer in flexible solar cells, perovskite demonstrates numerous extraordinary aids since its tunable bandgap can be altered within the range of 1.24–3.55 eV by exploiting mixed ions for single-junction flexible solar cell devices [2]. The first attempt to use perovskite as an active layer in flexible solar cells was made in 2013, where preliminary efficiency was around 2.6% [39]. A sharp upsurge in the efficiency for flexible PSCs is attained in the last few years, which is discussed in the next section.

13.6 RECENT ADVANCEMENT IN METAL OXIDE-BASED FLEXIBLE PEROVSKITE SOLAR CELLS

Perovskite solar cells are meticulous; they are the top contender for next-generation high-efficient solar cell technologies, being well-matched with low-priced, flexible substrates and large-area construction progressions. High-efficient PSCs are made of perovskites having a ABX_3 three-dimensional structure and are generally made of an organic or inorganic mono-valent cation, A = ($CH_3NH_3^+$, $CH(NH_2)_2^+$, Cs^+, Rb^+), a divalent metal cation, B = (Pb^{2+}; Sn^{2+}), and halide anion motif X_3 = (I^-; Br^-,Cl^-). Perovskite attracted scientists' interest owing to their easy fabrication methods based on solution precipitation or evaporation, such as one-step deposition, two-step sequential deposition, and vapor-assisted solution processing. The process of deposition determines the quality of perovskite films. Methylammonium-lead-iodide (MAPbI$_3$) perovskite has been intensively cast off as light-harvesting material, while organic-inorganic hybrid perovskites (OIHPs) have been developed significantly during the past few years as they demonstrated the photo-conversion efficiencies (PCEs) comparable to the Si, CdTe, and CIGS solar cells [7,38]. In a systematic construction of a PSC, the most extensively used ETL is TiO$_2$, which is usually attained by sintering a precursor solution at 450°C or 500°C. This practice not only upsurges engineering costs, but it also limits the applications of flexible plastic substrates. Consequently, other substitutes as inorganic ETLs have been established, for example, ZnO, SnO$_2$, Fe$_2$O$_3$, In$_2$S$_3$, WO$_x$, etc. Very early on, a few ternary metal-oxides such as BaSnO$_3$, SrTiO$_3$, Zn$_2$SnO$_4$, Nb:SnO$_2$, W(Nb)O$_x$, etc., were also investigated as ETLs in PSCs [40].

In 2009, the organic-inorganic perovskite was first used as a sensitizer in liquid-state DSSCs by Miyasaka's group [41], where they developed the new solar cells on a DSSCs configuration with a thin perovskite layer such as CH$_3$NH$_3$PbI$_3$ (MAPbI$_3$) and CH$_3$NH$_3$PbBr$_3$ (MAPbBr$_3$) on the mesoporous TiO$_2$ and delivered PCEs of 3.81% and 3.1%, respectively. However, the stability of these solar cells was inferior due to the corrosive properties of the liquid electrolyte. In 2011, NG Park's group [42] developed a solar cell with the same perovskite-sensitized configuration, where they used perovskite quantum dots as sensitizers and achieved an improved PCE of 6.5%. Later, in 2012, his group invented solid-state mesoscopic PSCs using solid-state spiro-MeOTAD as hole-transport material (HTM) after investigating the corrosion phenomena associated with liquid electrolytes to enhance the cell stability [43]. They used one-step fabrication of PSC with a solid-state electrolyte by a solution-based technique and achieved a PCE of 9.7%, having reasonable durability because this HTM layer can dissolve in organic solvents (chlorobenzene and toluene) in which hybrid perovskite do not dissolve and are easily deposited on the top of hybrid perovskite.

Snaith's group [44] fabricated meso-superstructure solar cells and achieved PCE of 10.9% for the first time by replacing TiO$_2$ conducting layer with the mesoporous alumina scaffold layer (Al$_2$O$_3$). Because of the large bandgap of Al$_2$O$_3$, it cannot assist the electron extraction, which suggests that perovskites possess effective electrical transport properties without the use of a nanoporous layer (the whole purpose of the nanoporous layer is to collect electrons that cannot be transported to the thin film by creating more interfaces of metal oxide and perovskite) [45]. The PSCs fabricated employing sequential deposition methods demonstrated a PCE of 15% with high reproducibility [46]. Later, Snaith's group used the vapor-deposition technique to fabricate planar heterojunction PSC for further development in PCE up to 15.4% [47]. Further, Seok's group [48] cast-off combined planar and scaffold layer of mixed halide CH$_3$NH$_3$Pb(I$_{1-x}$Br$_x$)$_3$, which improved the PCE to 17.9%. Here, they also varied the thickness of layers as well as the chemical compositions.

The first attempt to use perovskite as an active layer in flexible solar cells was made by the Snaith group [39], where they reported preliminary efficiency of 2.6% [39]. The flexible PSC with a triple-cation composition demonstrated an efficiency of 18.6%, while it was extremely stabilized at 17.7%. Over ligand-free and high-crystalline oxide-transport-layer, the flexible PSC attained a certified efficiency of 17.3%, while flexible all-perovskite tandem solar cells demonstrated the efficiency of 21.3% [49,50].

In 2021, with several developments in this field, such as altering the characteristics of perovskite through interface engineering, chemical composition, developing effective growth methods, and hole-transport materials optimization, the PCE of PSCs has increased significantly to 25.5% within a short period, as demonstrated in Figure 13.7. The high efficiency basically benefits from the good-looking optoelectrical properties of the perovskites; for instance, the high absorption-coefficient, higher electron/hole mobility, excellent lifetime and long diffusion-length of the charge carriers, etc. Thus, the rapid enhancement can be very close to the theoretical efficiency level by combining the design and development of the perovskite materials.

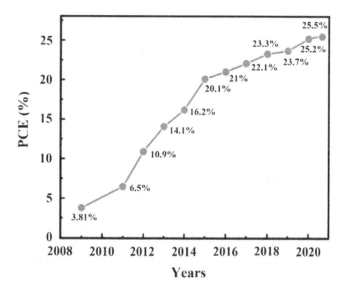

13.FIGURE 13.7 The power-conversion efficiency (PCE) evolution of metal-oxide-based perovskite solar cells.

The sturdiness of PSCs can be administrated by various issues like the instability of halide-perovskites, interfaces, and also charge-transport-layers (ETLs and HTLs). Since the stability of PSCs is always verified by the white-light-emitting-diode or UV light-filtered solar simulators. The fabrication of double-layer or amalgamated ETL is established to be a beneficial method to improve performance. Hence, PSCs are fascinating enormous attention as next-generation thin-film-based solar cells with the expectation to develop economically practicable power supplies to advance our eco-friendly sustainability in the coming years. Nevertheless, numerous problems that presently hamper the commercialization of PSCs require painstaking efforts, mainly their hysteresis, reproducibility, durability, and stability at ambient conditions.

13.7 SUMMARY AND OUTLOOK

The role of metal-oxides-based nanostructures (MONs) on diverse energy-related applications has been excellently reviewed, especially flexible solar cells. Global resource consumption is a genuine concern to humanity and is depleting rapidly. Furthermore, the use of these resources has a significant influence on global warming and climate change. Exploiting MONs attracted the attention of numerous scientists owing to their low-priced, ecofriendly, and real-world features. We have observed exponential growth in research activities on different metal oxides (TiO_2, ZnO, WO_3, SnO_2, NiO, Al_2O_3, V_2O_5, ZrO_2, CuO, Cu_2O, CeO_2, Nb_2O_5) based nanostructures, targeting its exploitation for flexible PSC applications. These MONs may arise in different shapes and sizes, which will impact the performance of the concerned devices. The use of MONs has presented a superior performance in devices due to their high surface area to volume ratio, optical effects, advantageous electron transport properties, and confinement effects, properties that are strongly impacted by their microstructural formations. The development of new synthesis strategies is of great importance to boosting, even more, the properties of MONs.

The extraordinary performance of perovskites as light-harvester for solar cell applications are ascribed to outstanding material properties such as direct bandgap, high absorption-coefficient, abrupt optical band-edge, large charge-carrier diffusion-length, and low-exciton binding energy. The metal-oxides based flexible PSCs show impressive competitiveness with other existing photovoltaic systems owing to their exclusive advantages, such as (a) low-priced, earth-abundant, and relaxed fabrication, (b) near-perfect crystallinity at low-temperature, (c) large charge-carrier diffusion-length, and (d) lower value of loss-in-potential in a solar cell. The metal-oxides-based PSCs are attracting huge interest as next-generation thin-film photovoltaics with the hope of developing economically feasible power supplies to improve our environmental sustainability soon. Nevertheless, numerous problems that presently hamper the commercialization of PSCs require painstaking responses, mainly their hysteresis, reproducibility, durability, and stability at ambient conditions. But, perovskites also have few inexorable drawbacks because these materials are enormously vulnerable to oxygen and water vapor, which causes the crystal structure to break down and dissolve the salt-like perovskite. Consequently, the long-term stability of the PSCs is hampered. So preparation of perovskite thin films should be performed in an inert atmosphere. Another issue is related to the environment since the lead is most castoff in PSCs, which is perilous and could leak out of the solar panel onto rooftops or underneath soil. The most promising approaches to improving degradation and phase/polymorphism stability in PSCs are introducing the mixing of A (binary, tertiary, or even quaternary mixed position) site-cations and X site halide-anions. The chemical-composition engineering or alloying approach has been barely exploited for lead-free perovskite systems, and additional investigation for a magical composition is predictable to bring subsequent years of vital results for the photovoltaic community stimulating further interests of both academia and industry. Because of the merits and demerits mentioned above, some novel approaches can be innovative to progress the performance of PSCs further: (a) the tunable bandgap of perovskites, consent absorption toward 940 nm, which would significantly augment the efficiency, (b) using high mobility and low-cost hole-transport materials would advance the fill factor and reduction in the cost, (c) refining the excellence of the contact-layers by cutting-edge growth procedures among the photoanode or perovskite or hole-transporters may also lead to improved fill factor, and (d) considering that the PSC matches well with Si and other thin-film technologies (viz. CdTe, CIGS, & CZTS/Se), a tandem solar cell merging the PSC with a crystalline Si/CIGS/CZTSSe would additionally advance the efficiency till 29.6%. The perovskite-based technologies would permit the mass-production of high-efficient solar cells and moderately low-slung temperatures, which would explanation for a substantial cost reduction. In the time being, this technology may lead to high-throughput device construction owing to the modest deposition method required. It is believed that these new-generation PSCs will find extensive applications and will ultimately lead to devices that competing standard Si-based solar cells.

ACKNOWLEDGMENTS

SC would like to acknowledge the University Grants Commission (UGC), New Delhi, to provide financial assistance through Dr. D.S. Kothari Post-Doctoral Fellowship (DSKPDF (PH/19–20/0089)).

REFERENCES

1. Annual Energy Outlook (2020) *US Energy Information Administration (EIA)*. Washington, DC, United States.
2. Li X, Li P, Wu Z, Luo D, Yu H-Y, Lu Z-H (2021) Review and perspective of materials for flexible solar cells. *Mater. Reports: Energy* 1:100001.
3. International. Energy Agency (2011) *Report on 'Key World Energy Statistics'*, Paris France.
4. Energy Information Administrator (2006) *Report on 'World Consumption of Primary Energy by Energy Type and Selected Country'*. Washington, DC, United States.
5. Luque A, Hegedus S (2003) *Handbook of Photovoltaic Science and Engineering*, John Wiley & Sons Ltd, Chichester, West Sussex, United Kingdom.
6. Alexander LV, Zhang X, Peterson TC, et al. (2006) Global observed changes in daily climate extremes of temperature and precipitation. *J. Geophys. Res. Atmos.* 111:D05109.
7. Green MA, Dunlop ED, Hohl-Ebinger J, Yoshita M, Kopidakis N, Hao X (2021) Solar cell efficiency tables (Version 57). *Prog. Photovolt: Res. Appl.* 29:3–15.
8. Chander S (2020) Advancement in $CdIn_2Se_4$/CdTe based photoelectrochemical solar cells, In: *Advances in Energy Materials. Advances in Material Research and Technology* (Ikhmayies S (eds)). Springer Nature, Switzerland.

9. Ogale SB, Venkatesan TV, Blamire M (2013) *Functional Metal Oxides: New Science and Novel Applications*, Wiley-VCH, Singapore.

10. Tokura Y, Nagaosa N (2000) Orbital physics in transition-metal oxides. *Science* 288:462–468.

11. Elumalai NK, Vijila C, Jose R, Uddin A, Ramakrishna S (2015) Metal oxide semiconducting interfacial layers for photovoltaic and photocatalytic applications. *Mater Renew Sustain Energy* 4:11.

12. Danish MSS, Bhattacharya A, Stepanova D, Mikhaylov A, Grilli ML, Khosravy M, Senjyu T (2020) A systematic review of metal oxide applications for energy and environmental sustainability. *Metals* 10:1604.

13. Pimentel A, Nunes D, Pereira S, Martins R, Fortunato E (2016) Photocatalytic activity of TiO$_2$ nanostructured arrays prepared by microwave-assisted solvothermal method. In: *Semiconductor Photocatalysis – Materials, Mechanisms and Applications* (Cao W (eds)). IntechOpen Ltd, London, UK.

14. Kim HS, Lee JW, Yantara N, Boix PP, Kulkarni SA, Mhaisalkar S, Grätzel M, Park NG (2013) High efficiency solid-state sensitized solar cell-based on submicrometer rutile TiO$_2$ nanorod and CH$_3$NH$_3$PbI$_3$ perovskite sensitizer. *Nano. Lett.* 13:2412–2417.

15. Chander S, Dhaka MS (2019) Exploration of CdMnTe thin film solar cells. *Sol. Energy* 183:544–550.

16. Yu Z, Hagfeldt A, Sun L (2020), The application of transition metal complexes in hole-transporting layers for perovskite solar cells: Recent progress and future perspectives. *Coord. Chem. Rev.* 406:213143.

17. Meyer BK, Polity A, Reppin D, et al. (2012) Binary copper oxide semiconductors: From materials towards devices. *Phys. Status Solidi B* 249:1487–1509.

18. Danish MSS, Estrella LL, Alemaida IMA, Lisin A, Moiseev N, Ahmadi M, Nazari M, Wali M, Zaheb H, Senjyu T (2021) Photocatalytic applications of metal oxides for sustainable environmental remediation. *Metals* 11:80.

19. Mahmood K, Sarwar S, Mehran MT (2017) Current status of electron transport layers in perovskite solar cells: Materials and properties. *RSC Adv.* 7:17044–17062.

20. Huang J, Yin Z, Zheng, Q (2011) Applications of ZnO in organic and hybrid solar cells. *Energy Environ. Sci.* 4:3861–3877.

21. Ahmed MS, Pandey AK, Rahim, NA (2017) Advancements in the development of TiO$_2$ photoanodes and its fabrication methods for dye sensitized solar cell (DSSC) applications: A review. *Renew. Sustain. Energy Rev.* 77:89–108.

22. Wong T, Zhuk S, Masudy-Panah S, Dalapati G (2016) Current status and future prospects of copper oxide heterojunction solar cells. *Mater.* 9:271-.

23. Wei H, Gong H, Wang Y, Hu X, Chen L, Xu H, Liu P, Cao B (2011) Three kinds of Cu$_2$O/ZnO heterostructure solar cells fabricated with electrochemical deposition and their structure-related photovoltaic properties. *CrystEngComm.* 13:6065–6070.

24. Abd-Ellah M, Thomas JP, Zhang L, Leung KT (2016) Enhancement of solar cell performance of p-Cu$_2$O/n-ZnO-nanotube and nanorod heterojunction devices. *Sol. Energy Mater. Sol. Cells* 152:87–93.

25. Chen X, Lin P, Yan X, Bai Z, Yuan H, Shen Y, Liu Y, Zhang G, Zhang Z, Zhang Y (2015) Three-dimensional ordered ZnO/Cu$_2$O nano-heterojunctions for efficient metal-oxide solar cells. *ACS Appl. Mater. Interfaces* 7:3216–3223.

26. Wee SH, Huang PS, Lee JK, Goyal, A (2015) Heteroepitaxial Cu$_2$O thin film solar cell on metallic substrates. *Sci. Rep.* 5:16272.

27. Pławecki M, Rówiński E, Mieszczak Ł (2016) Zinc oxide/cuprous(I) oxide-based solar cells prepared by electrodeposition. *Acta Phys. Pol. A* 130:1144–1146.

28. Chatterjee S, Saha, SK, Pal, AJ (2016) Formation of all-oxide solar cells in atmospheric condition based on Cu$_2$O thin-films grown through SILAR technique. *Sol. Energy Mater. Sol. Cells* 147:17–26.

29. Minami T, Nishi Y, Miyata T, Nomoto, JI (2011) High-efficiency oxide solar cells with ZnO/Cu$_2$O heterojunction fabricated on thermally oxidized Cu$_2$O sheets. *Appl. Phys. Exp.* 4:062301.

30. Masudy-Panah S, Dalapati GK, Radhakrishnan K, Kumar A, Tan HR, Elumalai NK, Vijila C, Tan CC, Chi, D (2015) p -CuO/n-Si heterojunction solar cells with high open circuit voltage and photocurrent through interfacial engineering. *Prog. Photovolt.: Res. App.* 23:637–645.

31. Rokhmat M, Wibowo E, Sutisna, Khairurrijal, Abdullah M (2017) Performance improvement of TiO$_2$/CuO solar cell by growing copper particle using fix current electroplating method. *Procedia Eng.* 170:72–77.

32. Guo W, Xu Z, Li T (2018) Metal-based semiconductor nanomaterials for thin-film solar cells. In: *Multifunctional Photocatalytic Materials for Energy* (Lin Z, Ye M, Wang M (eds.)), Woodhead Publishing, Cambridge, UK.

33. Møller CK (1958) Crystal structure and photoconductivity of cesium plumbohalides. *Nat.* 182:1436-1436.

34. Weber D (1978) $CH_3NH_3PbX_3$, ein Pb(II)-System mit kubischer Perowskitstruktur/$CH_3NH_3PbX_3$, a Pb(II)-system with cubic perovskite structure. *Z. Naturforsch. B* 33:1443–1445.

35. Chen J, Zhou S, Jin S, Li H, Zhai T (2016) Crystal organometal halide perovskites with promising optoelectronic applications. *J. Mater. Chem. C* 4:11–27.

36. Munir R, Sheikh AD, Abdelsamie, M, Hu H, Yu L, Zhao K, Kim T, Tall OE, Li R, Smilgies DM, Amassian A (2017) Hybrid perovskite thin film photovoltaics: In-situ diagnostics and importance of the precursor solvate phases. *Adv. Mater.* 29:1604113.

37. Ono LK, Juarez-Perez EJ, Qi Y (2017) Progress on perovskite materials and solar cells with mixed cations and halide anions. *ACS Appl. Mater. Interfaces* 9:3019730246.

38. Wang W, Shao Z (2020) Perovskite materials in photovoltaics. In: *Revolution of Perovskite* (Arul NS, Nithya VD (eds.)), Springer Nature, Singapore.

39. Docampo P, Ball JM, Darwich M, Eperon GE, Snaith HJ (2013) Efficient organometal trihalide perovskite planar-heterojunction solar cells on flexible polymer substrates. *Nat Commun.* 4:2761.

40. Dong J, Wu J, Jia J, Fan L, Lin Y, Lin J, Huang M (2017) Efficient perovskite solar cells employing a simply-processed CdS electron transport layer. *J. Mater. Chem. C* 5:10023.

41. Kojima A, Teshima K, Shirai Y, Miyasaka T (2009) Organometal halide perovskites as visible-light sensitizers for photovoltaic cells. *J. Am. Chem. Soc.* 131:6050–6051.

42. Im JH, Lee CR, Lee JW, Park SW, Park NG (2011) 6.5% efficient perovskite quantum-dot-sensitized solar cell. *Nanoscale* 3:4088–4093.

43. Kim HS, Lee CR, Im JH, Lee KB, Moehl T, Marchioro A, Moon SJ, Humphy-Baker R, Yum JH, Moser JE, Grätzel M, Park NG (2012) Lead iodide perovskite sensitized all-solid-state submicron thin film mesoscopic solar cell with efficiency exceeding 9%. *Sci. Rep.* 2:591.

44. Lee MM, Teuscher J, Miyasaka T, Murakami TN, Snaith HJ (2012) Efficient hybrid solar cells based on meso-superstructured organometal halide perovskites. *Sci.* 338:643–647.

45. Yin WJ, Yang JH, Kang J, Yan Y, Wei SH (2015) Halide perovskite materials for solar cells: a theoretical review. *J. Mater. Chem. A* 3:8926–8942.

46. Burschka J, Pellet N, Moon SJ, Humphry-Baker R, Gao P, Nazeeruddin MK, Grätzel M (2013) Sequential deposition as a route to high-performance perovskite-sensitized solar cells. *Nature* 499:316–319.

47. Liu M, Johnston MB, Snaith HJ (2013) Efficient planar heterojunction perovskite solar cells by vapour deposition. *Nature* 501:395–398.

48. Jeon NJ, Noh JH, Kim YC, Yang WS, Ryu S, Seok SI (2014) Solvent engineering for high-performance inorganic-organic hybrid perovskite solar cells. *Nat Mater* 13:897–903.

49. Liu C, Zhang L, Zhou X, et al. (2019) Hydrothermally treated SnO_2 as the electron transport layer in high-efficiency flexible perovskite solar cells with a certified efficiency of 17.3%. *Adv. Funct. Mater.* 29:1807604.

50. Palmstrom AF, Eperon GE, Leijtens T, et al. (2019) Enabling flexible all-perovskite tandem solar cells. *Joule* 3:2193–2204.

14 Flexible Solar Cells Based on Chalcogenides

Kulwinder Kaur, Nisika, and Mukesh Kumar
Functional and Renewable Energy Materials Laboratory Indian
Institute of Technology Ropar, Punjab, India

CONTENTS

14.1   Introduction.. 242
14.2   Merits of flexible solar cells 243
14.3   Progress and development on different substrates 244
    14.3.1   CIGS 244
        14.3.1.1   Polyimide 244
        14.3.1.2   Metal foils 245
        14.3.1.3   Ceramic and other materials 248
    14.3.2   CdTe 249
        14.3.2.1   Metal Foils 249
        14.3.2.2   Polymer 249
        14.3.2.3   Ceramics 250
    14.3.3   CZTS/CZTS(Se) 250
        14.3.3.1   Metal foils 250
        14.3.3.2   UTG 251
        14.3.3.3   Polymer and other materials 252
    14.3.4 Sb_2Se_3 252
14.4   Fabrication issues and challenges with flexible solar cells 252
    14.4.1   Crack initiation 253
    14.4.2   Performance degradation under bending 254
    14.4.3   Substrate choice 255
    14.4.4   Electrodes issues 255
    14.4.5   Stability and scalability issues 256
14.5   Future prospects and strategies for further advancements 256
    14.5.1   Absorber optimization 256
    14.5.2   New chalcogenide materials 256
    14.5.3   Optimizing every layer of solar module 257
    14.5.4   Material database and machine learning algorithms 257
    14.5.5   Rigorous testing.. 257
    14.5.6   Development of transparent/semitransparent solar cells 257
    14.5.7   Integration with existing technologies 258
14.6   Conclusion 258
References... 258

DOI: 10.1201/9781003186755-14

14.1 INTRODUCTION

Renewable energy resources are the only reliable sources of energy for continuously inflating energy demands in the near future. It is frightening that 38% of the world's total electricity production through 2019 still depends on the burning of coal [1]. Solar energy has long been touted as the most reliable and attractive option for the cost-effective origin of clean energy; thus, solar cells have been explored and developed over time. Si-based photovoltaic (PV) technology has been the main spotlight since the 1950s; this technology currently possesses 80–85% PV market share, globally. In the late 1970s, amorphous thin-film silicon solar cells (TFSCs) were used as a power source for a handheld calculators for the first time [2].

Interest in alternative materials has been accelerated over the last two decades for device processing due to Si's rigidity, weight, low absorption coefficient, and expensive wafer, and the processing limitations of Si-based PV technology [3]. This further directed extensive research and development of thin-film PV to overcome these limitations. Chalcogenide compounds with high absorption coefficients, such as $Cu(In,Ga)Se_2$ (CIGS), CdTe, CZTS(Se), and Sb_2Se_3, can be utilized as very thin films (2–3 μm) to absorb nearly all incidental solar light. Apart from a decrease in required optically active material per solar cell, TFSCs also cater to low-cost deposition on a large and flexible substrate. The most inviting characteristic of flexible TFSCs is the high potential to reduce the production cost; roll-to-roll manufacturing on flexible substrates permits the use of compact-size deposition equipment, high throughput, and low thermal expense [4]. New opportunities have evolved with the rapid progress of solar PV on flexible substrates. For example, flexible lightweight modules promote easy-to-install features for portable consumer devices, automobile industry, roofs of buildings, wearable electronics, and integrated photovoltaics with lower transport and installation costs. Moreover, power supply equipment to be installed in the Internet of Things (IoT), for instance, demands that solar energy be efficiently harnessed from the environment sustainably, which is possible with flexible TFSCs [5].

Chalcogenide-based flexible solar cells have emerged as promising energy resources over the past decade. CIGS, the most researched chalcogenide compound for PV technology, have already achieved 20.8% efficiency on a flexible substrate, which is close to its record efficiency of 23.4% on a rigid substrate [6]. CdTe PV offers minimum energy payback time among all PV technologies [1]. Likewise, CZTS(Se) and the newest Sb_2Se_3 chalcogenide have already shown their potential on flexible substrates and achieved efficiencies as high as on rigid substrates [5,7]. However, they still are being explored for further advancements in efficiency to compete with their counterparts, CIGS and CdTe. Chalcogenide-based flexible solar cells have been demonstrated on a variety of substrates featuring distinct advantages. To find a suitable flexible substrate, certain criteria need to be fulfilled before addressing the technical aspects of the flexible solar cell fabrication process. Prerequisites involving different physical and chemical properties, such as chemical inertness, vacuum, humidity barrier function, surface smoothness, thermal stability, and suitable coefficient of thermal expansion (CTE) must be satisfied [3,8]. Other than that, stability and compatibility of the substrate material bearing the whole fabrication steps and prolonged operational hours of PV modules are mandatory. The comparison of some of the properties of metal, ceramic, and plastic substrates used for chalcogenide flexible solar cells is represented in Figure 14.1. Despite having many advantages, every substrate has distinct disadvantages as well. Metal substrates hold good thermal and environmental stability, while they lack stability related to portability and adequate optical properties required for flexible solar cells. Likewise, ceramic substrates, having good thermal stability, environmental stability, and optical properties, however, have portability and flexibility issues. On the other hand, plastic substrates offer good flexibility, portability, and optical properties, while they mainly suffer from thermal and environmental stability issues. Therefore, a suitable substrate may be selected depending on the application requirement. The progress and development of chalcogenide-based flexible solar cells over the years is represented with the help of Figure 14.2. Notably, CIGS, being the oldest among all, exist in the topmost range of efficiencies, followed by CdTe, CZTS(Se), and Sb_2Se_3.

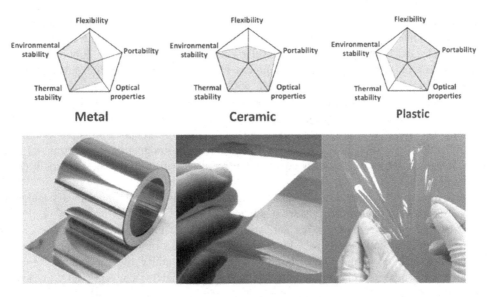

FIGURE 14.1 Representing comparison of different properties of metal, ceramic, and plastic substrate. Below is the picture of individual substrates, i.e., stainless steel (SS), thin corning glass, polyimide sheet (from left to right). Reproduced with permission from ref. [2]. Copyright (2020) Elsevier.

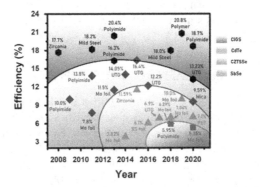

FIGURE 14.2 Progress and development of different chalcogenide flexible solar cells on different substrates over the years. *UTG (ultra-thin glass).*

However, the efficiencies are still limited by certain factors and challenges, which need to be addressed for future advancements of these solar cells. This chapter reviews the development of chalcogenide-based flexible solar cells on different substrates and challenges imposed during fabrication processes due to substrate limitations. In addition, it discusses in detail different viable strategies with their future prospects to push the efficiency bar further.

14.2 MERITS OF FLEXIBLE SOLAR CELLS

Thin-film chalcogenide-based flexible solar cells have grabbed enormous attention due to their wide variety of applications compared with solar cells fabricated on rigid substrates. Some of the merits of flexible solar cells are listed as follows, which make them more advantageous than rigid substrate (SLG) solar cells.

- Flexible light-weight modules offer easy installation features on portable devices, automobiles, roofs, and integrated photovoltaics with a small budget for transportation and installation [8].

- They are perfect for rooftop applications where glass-covered typical Si modules fail in roof tests due to additional weight and technical and structural issues [4].
- The flexible and lightweight of flexible solar cells enable their easy integration with wearable electronics, making way for the development of self-powered electronic devices. Also, these flexible solar cells can be easily integrated with daily use items, such as umbrellas, cars, bags, clothes, etc., which in turn can act as an additional sustainable and renewable power source for charging personalized devices [9].
- A major advantage is the reduced production cost. A roll-to-roll deposition is possible for flexible modules; this deposition is revolutionary in managing the production cost compared to that of rigid substrates. Moreover, a glass cover is not required, which saves the additional cost to the module [3].
- Flexible modules do not require any hard labor due to less installation time and easy integration; no glass cover, racking assembly, etc. are needed [3].
- Glass-based rigid solar modules are delicate and fragile, while flexible modules are robust and unbreakable; they can be easily rolled up, offer ease of transportation, and handle comfortably [3].
- Cell-to-cell interconnections in monolithic integration simplify the production steps to form modules with no soldering required or the lay-up essential for conventional Si-based module assembly [8].
- Chalcogenide-based active materials, i.e., CIGS, CZTS(Se), CdTe, and Sb_2S_3 for flexible modules are much economical than bulk Si, glass cover frames used in conventional Si modules [7,8,10,10,11].
- CIGS absorber material is known to be stable under harsh environments/different irradiations; therefore, CIGS flexible modules are very attractive for space applications, unlike rigid substrate solar cells [3].

14.3   PROGRESS AND DEVELOPMENT ON DIFFERENT SUBSTRATES

Substrate plays a pivotal role in defining the device properties; however, it requires meeting certain qualifications before addressing the technological characteristics of a flexible solar cell fabrication. Stability and compatibility of the flexible substrate with the whole manufacturing and fabrication process flow and extended operational lifetime are the foremost demands [8]. Numerous types of metal, thin glass, and polymer substrates have been employed for flexible PV technology. This section will focus on the evolution of flexible chalcogenide based solar cells using different substrates.

14.3.1   CIGS

$CuInGaSe_2$ (CIGS), being the oldest of other chalcogenide materials for solar cells, is still an interesting and attractive candidate for flexible solar cells. Reduced absorber thickness, along with deposition on flexible substrate, is a technically feasible strategy to produce cost-effective CIGS solar modules. CIGS solar cells have been testified to many flexible substrates, including polymers, metal foils, ceramics, etc., for a wide variety of applications.

14.3.1.1   Polyimide

Polyimide (PI) substrates have inviting features, such as high flexibility and being lightweight. Polymeric substrates are advantageous due to their electrically insulating characteristic, which permits the direct monolithic integration of solar modules through sequential patterning of sheets during the different processing steps [8,12]. However, it suffers primarily from poor stability at a high temperature, which restricts the temperature to around 450 °C [12]. Methodologies manifesting low-temperature fabrication schemes have struggled for many years to reach efficiency as high as 15%, until 2000 [3]. After a decade of fact-finding in detail, the efficiency had remained below 16%, since 2010 [3]. Recently, 20.8%

efficiency, a record high on a flexible (PI) substrate, was reported by Romain et al. [6]. But analysis of this drastic advancement in CIGS flexible photovoltaic technology is required.

Back in 1996, Basol et al. fabricated flexible $CuInSe_2$ (CIS) solar cells on a lightweight PI substrate for the first time [13]. A decent efficiency of 9.3% is achieved, which has attracted interest toward the flexible thin-film PV technology [13]. Moreover, the high stability standard of CIS solar cells on PI formed the base for the development of CIGS for terrestrial applications. Soon after, the research on flexible and lightweight CIGS PV technology was on the front foot. Various approaches have been adopted for the efficiency enhancement of flexible CIGS solar cells. In 2005, Rudmann and his coworkers recorded 14.1% efficiency for flexible CIGS solar cells on PI, which represented the highest efficiency reported for any solar cell grown on polymer films until that time [3]. The highest efficiencies are reported on a conventional soda-lime glass substrate (SLG), which is generally ascribed to the beneficial traits of Na diffusion in CIGS from the substrate. To achieve the same effect in PI, Na is incorporated by predeposition of NaF using evaporation on Mo-coated PI film; 12.2% efficiency was achieved [3]. However, this efficiency was observed to be less compared to the efficiency obtained by NaF postdeposition treatment by Rudmann et al. Na content was varied to establish optimum growth conditions at low temperature (400 °C), which is refrained due to the use of PI. This provides an impulsion toward the investigation of the alkali treatment effect on CIGS solar cells. PI-coated SLG substrates have been widely used to incorporate Na from the substrate, as well as by post-deposition treatment (PDT) with NaF and KF. Reinhard et al. fabricated the first CIGS solar cells using both NaF and KF PDT on flexible PI film that led to efficiencies reaching 20% [8,14]. The flexible device on PI can also be fabricated by using a lifted-off process from SLG substrate, which enables the fabrication of flexible CIGS solar cells without altering their properties, as shown in Figure 14.3a. Figure 14.3b represents the J-V curve of the device before and after the lift-off process, where efficiency as high as 11.8% is achieved [3,8]. This efficiency was further improved to 15% and 18% after NaF and KF PDT, respectively, by the same group [15]. The effects of NaF and KF-PDT on the morphology (SEM images in Figure 14.3c–k) indicate that the improved V_{OC} for KF-PDT devices is related to the presence of small holes on the CIGS surface. It is actually ascribed to the formation of K-In-Se based passivation layer on the CIGS surface. Similarly, with advanced alkali treatment, efficiencies have reached as high as 20.8% for CIGS on flexible PI, reported by Romain et al. in 2019 [6]. Apart from the baseline process of NaF PDT, three technological improvements have successively improved efficiencies via RbF PDT treatment, co-evaporation of NaF during the third stage, and RbF co-evaporation during cap deposition of In and Se over CIGS.

Advancements are done in terms of altering the deposition of Cu, In, Ga, and Se during different deposition stages with different substrate temperatures, followed by NaF PDT then followed by RbF PDT, represented as steps A and B, respectively (Figure 14.4a). Step C involves the evaporation of NaF during In-Ga-Se deposition starting from the stoichiometry point. Finally, step D demonstrates the RbF co-evaporation during the capping layer deposition of In-Ga-Se before PDT. Figure 14.4b represents the cross-section SEM image of the final solar cell structure of flexible CIGS. Figure 14.4c reports the efficiency output for the four technological steps. The efficiency represented for every sample is calculated from the average of 18 individual cells fabricated without antireflection coating, where CGI depicts the [Cu/(In+Ga)] ratio [6]. Figure 14.4d represents the plot of voltage loss vs. CGI. RbF PDT noticed to yield a V_{OC} gain of 15–20 meV for a given CGI, the same as observed for KF. Na addition at the third stage improves V_{OC} by 15 meV for comparable CGI. Conversely, RbF treatment during capping marginally affects V_{OC}. The improvement in efficiency is independent of CGI ratio rather than primarily dependent on alkali treatment.

14.3.1.2   Metal foils

For the development of flexible solar cells on metal foils, International Solar Electric Technology (ISET Chatsworth, CA) had tested Mo, Al and Ti foils for the first time in 1992 [3]. The best

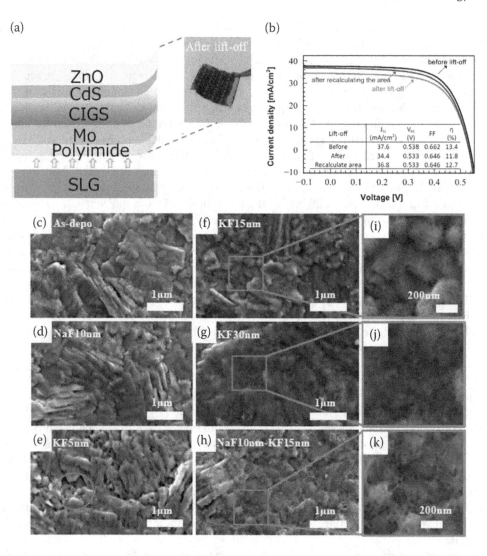

FIGURE 14.3 (a) Structure of flexible CIGS solar cell fabricated on PI supported by SLG, (inset) solar cell after lift-off process, (b) efficiencies compared before and after the lift-off process. Reproduced with permission from ref. [16]. Copyright (2015) Institute of Physics. The SEM image of (c) as deposited CIGS, (d) with 10 nm NaF PDT, (e) with 5 nm KF PDT, (f) with 15 nm KF PDT, (g) with 10 nm NaF PDT, (h) with 10 nm NaF-15 nm KF PDT, enlarged image of red marked region for (f), (g) and (h) is represented by (i), (j) and (k), respectively. Reproduced with permission from ref. [15]. Copyright (2017) Institute of Physics.

efficiency on Mo foil was reported to be 8.3% on 0.09 cm^2 cell, using e-beam evaporation for CIGS deposition by Kapur et al. [3,8]. The same group reported an efficiency of 11.7% on Mo foil by adopting an ink-based deposition process in 2003 [8]. The efficiency was reported to be further improved to 14.6% on 0.481 cm^2 cell area by deploying a three-stage co-evaporation process [8,17]. Thereafter, Mo foil did not receive much attention due to its mismatched coefficient of thermal expansion (CTE) that directly affects the CIGS growth. On the other side, Ti had received great attention owing to its lightweight and well-matched CTE (8.4×10^{-6} K^{-1}). Developments on Ti foil were also progressing when ZSW, Baden-Wurttemberg, Germany had reported 12% efficiency on Ti foil in 2009 [8]. SiOx: Na-based barrier layers were used to induce Na doping, which led to improved

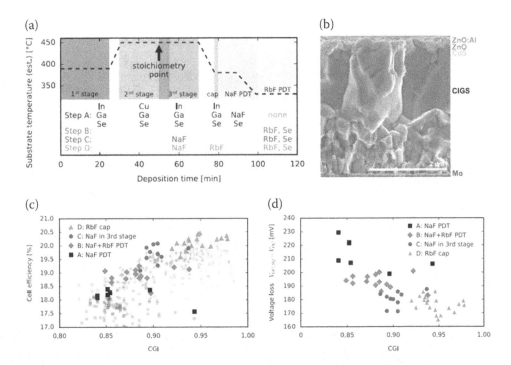

FIGURE 14.4 (a) Schematic representation of CIGS deposition processes step A (black), technological steps B (red), C (green), and D (cyan), (b) cross-section SEM of a final solar cell structure of CIGS on PI, (c) represents the efficiency vs. CGI plot and (d) voltage loss plot as a function of CGI, for all the four technological steps. Reproduced with permission from ref. [6]. Copyright (2019) John Wiley and Sons.

efficiency of 13.1% on Ti foil [8]. The performance was substantially improved by Aoyama Gakuin University (AGU, Tokyo, Japan) to 17.9% through a three-stage co-evaporation process and removing the conventional CdS buffer layer [3,8].

Nevertheless, the cost of Ti remains higher than other flexible substrate choices, such as stainless steel (SS). SS of grade 400 series has better matched CTE (e.g., 430 type 11×10^{-4} K^{-1}), therefore being the constantly studied flexible substrate material. A 17.4% [8] efficiency was reported on SS by the three-stage co-evaporation synthesis method, which a year later was improved to 17.5% in 2000 [8]. A 11.65% efficiency demonstrated for roll-to-roll co-evaporation process, including Na diffusion externally, in 2002 [3,8]. Diffusion barrier layers were also introduced to avoid impurities during the synthesis process. Herz et al. achieved an efficiency of 10.9% by using an Al_2O_3 diffusion barrier layer on SS foil without Na doping [18]. In 2012, 17.7% efficiency was demonstrated on SS/Ti/Mo/Mo substrates without barrier layer at a low substrate temperature of 475°C and NaF-PDT [8]. Apart from that, more economical mild steel (MS) with Cr and Ni-based corrosion protective layers were also used as one of the flexible substrates. Efficiencies as high as 17.6% and 18.2% were demonstrated by using enamel films as barrier layers on MS substrate, in 2012 and 2013, respectively [3]. This high performance was ascribed to the alkali sources, such as Na_2O and K_2O from the enamel layer. Other low-cost metal foils such as Cu and Al have also been investigated as flexible substrates. Recently, efficiency reaching 10% is demonstrated on Cu foil using graphene as a hole-transport layer (HTL) [19], as shown in Figure 14.5a. Figure 14.5b illustrates the band diagram of CIGS/graphene solar cells on Cu foil at zero bias, showing the generation and collection of charge carriers. Figure 14.5c compares the J–V characteristics of the CIGS solar cell fabricated on graphene/Cu foil to that on Mo/STS, while inset image shows the actual flexible solar cell.

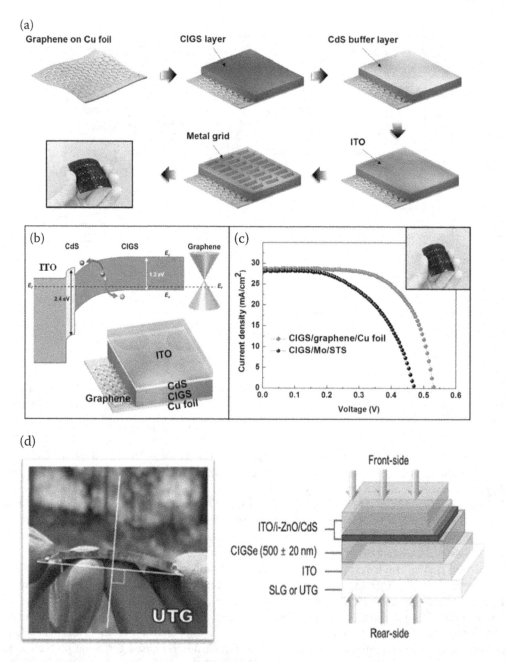

FIGURE 14.5 (a) Schematic of step-by-step fabrication of CIGS solar cell on Cu foil using Graphene HTL, (b) Band diagram of the flexible CIGS solar cell, (c) comparison of J-V plots of the device on Cu foil and SS. Reproduced with permission from ref. [19]. Copyright (2018) Elsevier. (d) Photograph of CIGSe solar cell fabricated on UTG, in bending position, (e) Illustration of flexible UTG CIGSe bifacial solar cell structure. Reproduced with permission from ref. [20]. Copyright (2020) John Wiley and Sons.

14.3.1.3 Ceramic and other materials

Other heat-resistant ceramics (e.g., ultrathin glass (UTG), ZrO_2) have also been explored for the development of flexible CIGS solar cells. ZrO_2 has been investigated to achieve an efficiency of 17.7% using MBE at 550°C [8]. Arnaud et al. reported 11.2% efficiency on UTG and successfully examined it for different bending tests [3]. The benefit of using UTG as a flexible substrate is the Na diffusion from the substrate itself. The decreased performance of the UTG-based cells under bending is ascribed to the deterioration of the top ZnO:Al (AZO) contact under tensile stress. Recently, flexible UTG is reported to exhibit an efficiency of 13.23%, paving the way toward bifacial photovoltaic solar cells, by Kim et al. [20]. Figure 14.5d represents the photograph of a bent CIGS solar cell. The utilization of thin flexible glass allows illumination from both the front and rear sides, as illustrated in Figure 14.5e. UTG has proven to be advantageous over SLG, not just because of flexibility and high transmittance, but also because of its indemnity to produce larger grain size, better crystallinity, precise control over the Na diffusion, and, most importantly, reduced thickness of the interfacial GaO_x layer, which elevates the device performance and is irreplaceable. Other substrates, such as paper, polyethylene terephthalate (PET), and Kovar (Fe-Ni alloy) have also been investigated arbitrarily. However, the efficiencies reported are less for these substrates, due to the inefficient fabrication procedures, such as crack formation during the mechanical lift-off poses technical issues.

14.3.2 CdTe

14.3.2.1 Metal Foils

CdTe solar cells fabricated over metal foils offer various advantages owing to their low-cost, lightweight, wide range of temperature operations and ability to be integrated with oriented or curved surfaces for various space and terrestrial applications. Mo is most interesting due to its matched CTE (5.0×10^{-6} K^{-1}) to that of CdTe (4.8×10^{-6} K^{-1}), out of all metal foils that can be used [10]. Thus, CdTe remains stable even at high temperatures of 550 °C without forming any blisters or bubbles on Mo substrate. Great progress has been made in developing CdTe flexible solar cells on metallic foils, following the initial efforts by Matulionis et al. who have achieved 7.8% efficiency on Mo foil, in 2001 [21]. Various strategies have been employed to improve the performance of CdTe solar cells based on metallic foils, which primarily involves the investigations of Cu doping. In 2013, the effect of Cu doping in CdTe has been successfully demonstrated to enhance hole density and carrier lifetime to attain 11.5% efficiency, Kranz et al. [1] (maximum to date on Mo foil). Additionally, integration of the ZnTe layer between CdTe and back contact effectively reduces the ohmic resistivity at CdTe/back contact interface [10]. The high-temperature processibility and robustness of these substrates enable roll-to-roll manufacturing to industrialize solar cells fabrication based on flexible metallic foils. However, solar cells fabrication in superstrate configuration (which offers better solar harvesting efficiencies by trapping light) is not possible using these metallic foils, due to their opaque nature.

14.3.2.2 Polymer

Polymeric substrates are not only much more flexible, cheap, and easily processable, but they are also mostly transparent to visible and NIR irradiation, making them viable for fabricating CdTe flexible solar cells in superstrate configuration. Until 2006, solar harvesting efficiency was found to increase substantially when fabricated in superstrate over substrate configuration, with efficiency boost to 11% from 7.3% with superstrate configuration on PI and substrate configuration on metal sheet, respectively (Figure 14.6) [22].

To further improve device performance, various approaches, such as $CdCl_2$ annealing treatment for junction activation and recrystallization, using different back-contact layers, modifying n-type layer have been employed to boost device efficiency. The integration of anti-reflective coating on top of cells also effectively increases the efficiency of the solar cell from 12.4% to 12.7% [23], which was later enhanced to 13.8% by the same group in 2012. However, despite various advantages, PI substrates suffer from the problem of material degradation at high-temperature processing, hindering their commercialization.

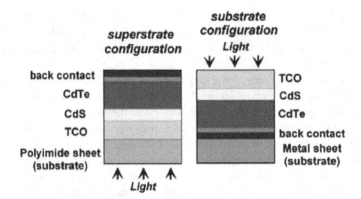

FIGURE 14.6 (a) Actual CdTe flexible solar cell deployed using metal foil substrate, (d) superstrate and substrate configurations followed by Romeo et.al for their investigations. Reproduced with permission from ref. [22]. Copyright (2006) Elsevier.

14.3.2.3 Ceramics

Recently, CdTe flexible solar cell using mica has also been fabricated, with solar harvesting efficiency reaching 9.6% [24]. UTG has also been investigated as one of the flexible substrates in this category. Comparison of CdTe solar cells fabricated on various flexible substrates reveals less stress generation for devices fabricated on UTG as compared to PI substrates [25]. Moreover, transparency of PI substrates was found to decline after $CdCl_2$ treatment (for improving crystal structure), which was not the case using UTG substrates. Taking notice of various advantages of UTG over various other flexible substrates, considerable research efforts have been devoted to increasing the efficiency of solar cells using UTG substrates. In 2014, CdTe solar cells based on UTG with efficiency surpassing 14% have been developed with relatively small efficiency drop undergoing different static-bending testing [26]. Later, the efficiency was boosted to 16.4% efficiency by the same group using a modified synthesis procedure in 2015 [27].

14.3.3 CZTS/CZTS(Se)

14.3.3.1 Metal foils

The quaternary kesterite semiconductors: Cu_2ZnSnS_4 (CZTS), $Cu_2ZnSn(SSe)_4$ (CZTSSe) have attracted great attention among the research community since the late 2000s, owing to their earth-abundant constituents, optimum bandgap ($E_g = 1.0$–1.5 eV) and high absorption coefficient ($>10^4$ cm^{-1}) [28–30]. Strategies employed the process of modifying back contact and different alkali metal doping in parallel to significantly improve CZTSe device (synthesized over ferritic steel substrates) PCE from 2.2% to 6.1% [31]. Champion solar cell with PCE value of 7.04% was fabricated over flexible Mo foils by properly optimizing Na doping and stacking order of CZTS precursors, in 2018 [32]. In 2019, careful investigations on the role of preparative parameters by Jo et al. were successful in demonstrating a record efficiency of ~8% [33], surpassing the previous record of 7.04% for Na incorporated CZTSSe sputter-based TFSCs. Moreover, the application of thin and multilayered precursor structure not only improves device performance but also enhances performance homogeneity among the same sample, promoting commercialization of flexible CZTSSe solar modules. This fabrication strategy effectively controls the Sn-related donor defects leading to mitigation of local V_{OC} losses, to reach PCE value of over 10% for CZTSSe layers fabricated over Mo foils [34]. Recently, efficiency as high as 10.30% is reported on flexible SS foil using ZnO and SiO_2 diffusion barrier layers, as shown in Figure 14.7a–b [35]. Depth profile of Fe and Cr impurities in SS obtained from time-of-flight secondary ion mass spectrometry (TOF-SIMS)

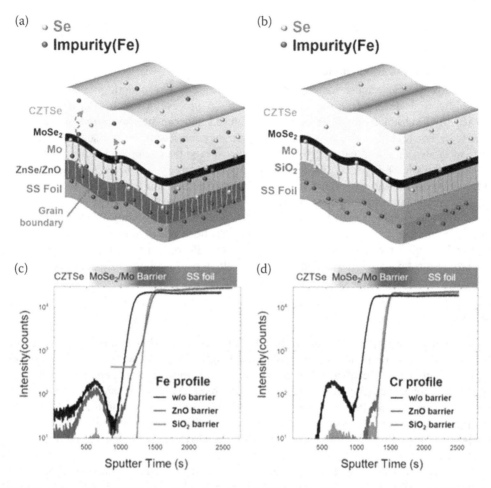

FIGURE 14.7 Schematic of CZTSe flexible solar cell on SS foil using diffusion barrier layer of (a) ZnO, (b) SiO$_2$, (c) and (d) respectively, shows the Fe and Cr impurity depth profiles obtained from TOF-SIMS. Reproduced with permission from ref. [35]. Copyright (2019) Royal Society of Chemistry.

indicate that SiO$_2$ acts as an efficient barrier hindering the impurity diffusion (Figure 14.7c–d). Most of the efficiencies for flexible CZTS(Se) solar cells on metal foil substrates have been reported after 2015, which indicates that this particular area is still an open quest.

14.3.3.2 UTG

In addition to metallic substrates, CZTSSe solar cells have also been developed UTG offering distinct advantages and disadvantages. For instance, UTG can sustain high temperatures essential for crystallization of CZTSSe films. In 2014, Peng et al. demonstrated that the etching of sputtered CZTS films with HCl and NaCN solutions can effectively increase device PCE value surpassing 3% [36]. Kesterite solar cells fabricated on SLG substrates offers higher solar conversion efficiencies because it is alkali (Na) doped, which gets diffused to the absorber layer, increasing device efficiency. However, the millimeter thick SLG substrate not only makes the solar cell rigid but also makes the solar module very heavy to be installed in remote locations. Na doping in CZTS(Se) through optimized NaF deposition over UTG has demonstrated an efficiency of 6.9% [37].

Nevertheless, CZTS(Se) based flexible solar cells are yet to be explored in detail for further development.

14.3.3.3   Polymer and other materials

To overcome low melting temperature issues of polymeric substrates hindering processing of Kesterite solar cells, Min et al. developed a unique material transfer process in which solar cells are deposited over normal SLG substrates, after which they are etched and can be transferred to various flexible substrates [38]. To check the efficacy of this approach, they fabricated flexible solar cells over various polymeric substrates with outstanding efficiency. They achieved the champion efficiency of 7.1%, 6.7%, and 5.9% over flexible PET, paper, and cloth substrates, respectively. However, synthesizing solar cells on a different substrate and then its transfer can be very cumbersome and consume tremendous resources. To overcome this shortocming, researchers have developed kesterite solar cells over ceramic substrates, which are highly flexible, light, and can sustain high processing temperature. In fact, the maximum PCE value of 11.5% for CZTSSe solar cells has been achieved over zirconia substrates [39]. Also, the high melting temperature of ceramic substrates allows a rapid rampup of temperatures, to decrease processing time. Despite having the highest efficiency reported, not much research has been invested in developing kesterite solar cells over ceramic substrates.

14.3.4 Sb_2Se_3

Antimony selenide (Sb_2Se_3) is the newest of all chalcogenides; it has received wide attention as a potential solar cell absorber because it has demonstrated high PCE (>9%) within a small span of research efforts [40]. Additionally, electrically benign boundaries are formed upon bending, which do not introduce any additional recombination centers, and they require low processing temperatures, making Sb_2Se_3 an ideal candidate for flexible solar cell modules. The first report demonstrating the fabrication of flexible Sb_2Se_3 based solar cells achieved PCE value of 5.95% over an ultrathin PI substrate, by Li et al. in 2018 [41]. Additionally, the devices demonstrated excellent stability undergoing various bending and flexibility tests.

Using two-step treatment, first etching by HCl and then annealing, they enhanced the performance of Sb_2Se_3 solar cells on flexible Mo foils from 1.6% to 5.35% [11]. Interfacial defects can greatly reduce device performance because they serve as centers for recombination. To overcome this barrier, Wen et al. demonstrated the use of an epitaxial CdS buffer layer to effectively reduce the number of interfacial defects. This in turn enhances the PCE value of Sb_2Se_3 solar cell fabricated on mica substrate to 7.15% (which is the highest reported value to the date) [42]. Recently, in 2021, Sb_2Se_3 flexible solar cells were tested for 6.14% efficiency bearing a high power-per-weight of 2.04 $W.g^{-1}$ with mini-modules (area 25 cm^2) applied for the power supply of IoT sensors [5]. Figure 14.8a–c represents the schematic diagram, cross-sectional SEM image indicating all the interfacial layers and dense morphology with large grain sizes, respectively. J–V curves (Figure 14.8d) indicate better performance on a flexible substrate (PI) than glass. EQE of both PI and glass devices (Figure 14.8e) almost overlap from 550 nm to 1050 nm. Therefore, major energy loss below 550 nm is ascribed to intense parasitic absorption of PI substrate (Figure 14.8f). The flexible Sb_2Se_3 mini-module is represented in Figure 14.8g; while connected to the flower monitor, environment information can be extracted from the smartphone via Bluetooth in real-time (Figure 14.8h). As Sb_2Se_3 is relatively a new photovoltaic semiconductor, much effort is needed to optimize the composition of Sb_2Se_3 and various interfacial layers to enhance the performance of flexible solar cells based on Sb_2Se_3.

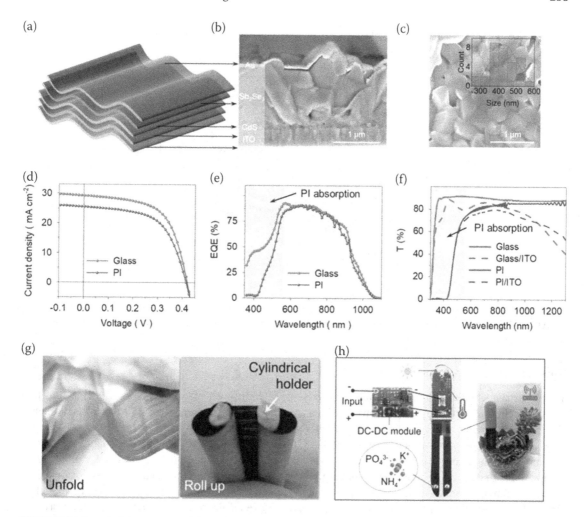

FIGURE 14.8 (a) Schematic structure, (b) SEM cross-section and (c) surface morphology from SEM of flexible Sb_2Se_3 solar cell, (d) J-V plots and (e) EQE on PI and glass substrates, (f) transmission of glass, glass/ITO, PI and PI/ITO, (g) photographs of actual flexible Sb_2Se_3 mini-module during unfolding and roll-up position and (h) illustration of Sb_2Se_3 mini-module and flower monitor. Reproduced with permission from ref. [5]. Copyright (2021) Elsevier.

14.4 FABRICATION ISSUES AND CHALLENGES WITH FLEXIBLE SOLAR CELLS

Over the past decade, great success has been made in increasing the efficiency of flexible solar cells based on chalcogenide materials, while minimizing manufacturing costs, environment deterioration, and scaling manufacturing to meet industrial standards. But as these chalcogenide materials are composed of various elements, many problems associated with material composition, interfacial defects, and elemental inter-diffusion hinder the potential of solar modules based on these materials. In addition to substrate choice, the lattice mismatch between different layers of a solar cell, material degradation under bending/folding tests, etc., as shown in Figure 14.9, are other limiting factors of flexible solar cells based on chalcogenide materials. All these factors have been detailed in the section below.

FIGURE 14.9 Representing various existing challenges hindering the advancements in chalcogenide-based flexible solar cells.

14.4.1 CRACK INITIATION

Flexible chalcogenides solar cell modules comprising different layers (absorber, buffer, charge transport layers, electrodes, substrate, window layer), which are made of different materials. For better performance, the lattice parameter of all the layers should be very close to ensure proper adhesion between different layers. However, not just the lattice parameter, but CTE is another important feature that should have parity among different layers. Therefore, when these flexible solar modules are subjected to various loads (thermal, pressure, etc.), different layers undergo deformation to a different extent. When the load exceeds the yield strength, the material undergoes plastic deformation, which is not recoverable, and material fails if further loading is applied [43]. When the load is removed, even if the rest of the layers retain their original shape, the one layer that sustains a permanent deformation will act as a nucleation site for crack generation, which will ultimately lead to a failure of material [44]. Even the substrate itself can generate a crack undergoing bending/twisting, which is expected to hold solar cells for its operation life period (20–25 years) [45]. Ceramic substrates are usually brittle, so they can directly fail without warning when the applied load exceeds the yield strength. Also, chalcogenides materials are inorganic, meaning they are not highly flexible. Moreover, solar cell modules have to sustain wide temperature fluctuations during their operation period [46]. These temperature fluctuations can cause solar cell material to expand or contract accordingly. As the solar cell module is composed of different materials with different expansion coefficients, each layer deforms to a different degree for each thermal cycle. Also, the substrate used for flexible solar (plastics, polymers) cells has a comparatively higher thermal expansion coefficient than rigid glass substrates [47]. This in turn can lead to generation of stress among adjoining layers. Therefore, care must be taken while fabricating flexible solar cell modules so that the thermal expansion coefficient of different layers is comparable to each other to avoid material failure due to thermal creep.

14.4.2 PERFORMANCE DEGRADATION UNDER BENDING

The electronic bandgap of material greatly influences the solar harvesting capability of the material, which can be changed by the application of various external factors [48]. Application of strain has been proven very useful to improve carrier dynamics for improved solar-harvesting applications [49].

However, it may also serve as a bottleneck for the commercialization of flexible solar cell modules. Solar cell materials are selected very carefully to ensure proper band alignment between different layers to enable efficient separation and transportation of electrons and holes to respective ends of a solar cell. However, when this system is subjected to bending/twisting, this induces strain in the system, which in turn influences the band structure of individual materials. This, in turn, can alter the bandgap alignment between the individual layers, which is destructive for the performance of the solar cell. Not much work has been done on tracking how the performance of chalcogenide-based flexible solar cells degrade under bending or twisting. But researchers have shown that PCE value can reduce by as much as 20% going under bending to 50 mm radius [36]. In real-life conditions, bending to a greater degree (lower bent radius) is required with a much lower decrease in efficiency to induce monetary benefits of chalcogenides-based flexible solar cells. New Sb_2Se_3 based flexible solar cells show great promise since they retain 97% of the original PCE value even going after 1000 cycles of bending (2.5 mm radius) [11]. However, much work is needed to lift the efficiency of these solar cells to achieve a decent market share.

14.4.3 SUBSTRATE CHOICE

Flexible substrates can be divided primarily into three categories: metal, ceramic, and polymer substrates. Various factors, such as cost, flexibility, optical properties, roughness, environmental stability, etc., need to be considered for choosing the best substrate for the required application. For instance, if these substrates are compared based on their flexibility, polymeric substrates have the upper edge. Young's modulus is the measure of the stiffness of the material, and elastic elongation determines the maximum strain the material can endure in an elastic deformation zone. These two properties greatly dictate the flexibility of the material. Young's modulus of PET is 3 GPa and elastic elongation is $\sim 10^{-2}$, while soda-lime glass has Young's modulus of 70 GPa and elastic elongation of $\sim 10^{-3}$ [50,51]. This in turn makes glass stiffer and less deformable in the elastic deformation range. Metallic and ceramic substrates have even higher Young's modulus and lower elastic elongation [52]. The cost of the substrate is another crucial factor. Ceramic and metallic substrates are usually expensive, while polymeric substrates are cheaply available due to their easy processibility. As chalcogenide materials usually require high processing temperatures, the substrate should be able to handle extreme temperatures. Metallic substrate like stainless steel sustains up to 900°C and ceramic substrates like glass can sustain 650°C, which makes them adequate for high temperature processing [43], while polymeric substrates usually start to fail around 200°C [53]. Therefore, there is a trade-off between the substrates due to distinct properties such as flexibility and thermal stability. Additionally, metallic and glass substrates are, however, very reluctant to permeate water and moisture; thus, they can operate smoothly for years without damage. Also, depending on solar cell architecture: substrate or superstrate, the substrate must be chosen with high transmittance or reflective properties. For substrate configuration, the substrate should be highly reflective, where metal substrates can offer better application [54], while for superstrate configuration, glass and polymeric substrates are a better choice [55]. Due to the distinct merits and demerits of an individual substrate, it is difficult to standardize the substrate choice for upcoming fabricating processes.

14.4.4 ELECTRODES ISSUES

Each solar cell comprises two electrode layers, whether it is fabricated in substrate or superstrate configuration. One of these electrode layers is usually transparent, while the other is reflective, depending on solar cell configuration. Both these electrode layers should have optimum bandgap to allow efficient extraction of charge carriers (electrons-holes). But in the case of flexible solar cells, an additional requirement is there: electrodes should be highly flexible. Transparent conducting oxides (TCOs) are widely used for solar cell applications. The most commonly used material is ITO, which has high electrical conductivity and is highly transparent (>90%) to solar irradiation. However, due to its brittle ceramic nature, cracks start to form when ITO is subjected to even small strain rates [56]. It is usually observed that crack initiation takes place when bending curvature is lowered by 10 mm radius [57]. Some techniques, like fabricating heterogeneous electrodes of ITO/Ag NWs [58] or ITO/PEDOT:PSS [49], were fabricated to increase electrical conductivity and bending ability of the electrodes, which in turn complicate the fabrication process and the cost of the overall solar module. Notably, indium's limited supply is a concern for industrialists; they have to compete against TV, smartphone, display makers to obtain indium. A metallic film with thickness >100 nm is opaque. They can be made transparent if their thickness is reduced below 20 nm, to be served as front electrodes for solar cell application. But because chalcogenide material requires high processing temperatures, these metallic electrodes can diffuse to an adjoining absorber or buffer layer, serving as recombination centers, thus decreasing the overall performance of solar cell [60]. Carbon nanomaterials, such as carbon nanotubes (CNTs) and graphene show great promise to serve as a viable replacement for flexible electrodes [61], but much work is needed in the direction before concrete assertions can be made.

14.4.5 STABILITY AND SCALABILITY ISSUES

Lab-manufactured flexible solar cells have very small active surface areas. The efficiency drops significantly when they are scaled up to large areas. For instance, the PCE value of flexible CZTSSe solar cells decrease by 2% (from 10% to 8%) as the active area is increased from 0.5 cm^2 to 2 cm^2 [34]. This in turn poses a huge challenge to commercialize flexible chalcogenides solar cell technology. Moreover, these solar cells are tested under lab conditions, rather than harsh real-life conditions. Therefore, efficiency may further fall, when they would be set to operate in real-life conditions. The other issue hindering technological advancements is associated with the long-term stability of these devices. This provides an impetus to boost the efficiency of these flexible devices for proper commercialization.

14.5 FUTURE PROSPECTS AND STRATEGIES FOR FURTHER ADVANCEMENTS

Chalcogenide-based flexible solar cells show great potential for high efficiency, low cost, and high throughput to be used as mobile power and product integration. Great progress has been achieved in optimizing the composition and efficient synthesis of chalcogenide-based flexible solar modules to increase the solar harvesting capability. Material transfer processes have also been developed to enable the proper transfer of absorber material over flexible substrates without requiring processing at high temperatures. Still, there are several concerns like performance degradation under bending, insufficient energy conversion values, and scalability issues, which need to be properly addressed to ensure the legitimate establishment of chalcogenide-based flexible solar modules. In this direction, future prospects and certain strategies are proposed for the augmentation of chalcogenide flexible solar cells.

14.5.1 ABSORBER OPTIMIZATION

Chalcogenide materials are associated with various compositional and interfacial defects owing to their complex crystal structure comprising of various elements. For instance, CZTS solar cells have theoretical attainable SQL efficiency of 33.7%; however, PCE value of just 12.6% has been achieved to date [62]. This huge deficit between experimental and theoretical values is generally associated with a nonradiative recombination, which limits device performance. To mitigate this energy loss, a deeper understanding of carrier dynamics inside chalcogenide absorbers at the nanoscale is required. Real-time or in situ material characterizations would be critical for the evolution of material properties, which can provide useful insights to modify synthesis routes to maximize solar harvesting capacity. Moreover, the integration of ultrafast spectroscopic techniques with conventional material characterization tools may provide a better comprehension of the influence of crystal lattice on carrier dynamics.

14.5.2 New chalcogenide materials

Commercialized chalcogenide material technologies (CIGS, CdTe solar cells) use rare-earth or environment hazardous materials. Environment-friendly kesterite solar cells have attained limited solar-conversion efficiencies, after investing years of research. This stimulated the need to develop new chalcogenide materials that are both environmentally friendly and can attain higher solar-conversion efficiencies. Sb_2Se_3 is a relatively new chalcogenide material that has achieved a PCE value of over 9% in a short time [40]. Just a few reports are demonstrating the efficacy of Sb_2Se_3 based flexible solar cells. Thus, more efforts should be addressed in bringing up the efficiency of this new chalcogenide material. Moreover, various new chalcogenide materials have been proposed computationally for various photovoltaic and photocatalytic applications [63]. Also, modifying crystal lattice with other materials, such as chalcogenide perovskite materials, also shows high promise for environment-friendly photovoltaics [64]. Accordingly, all these new chalcogenide materials deserve appropriate attention, which may play a pivotal role in the further development of chalcogenide flexible solar cells.

14.5.3 Optimizing every layer of solar module

Even if one layer is mismatched in terms of the thermal expansion coefficient, linear expansion coefficient, yield strength, or band alignment, the performance of the solar cell device can reduce drastically. Therefore, each component of flexible solar cell, i.e., electrodes, absorber and buffer layer, window layer, and the substrate, should be properly optimized to match the lattice coefficient of adjacent layers to a great extent to ensure proper heterojunction formation. Moreover, more attention should be ascribed to selecting material for proper band alignment at interfaces, to mitigate energy loss subjected to bending or twisting.

14.5.4 Material database and machine learning algorithms

Many combinations of substrate, absorber, buffer layer, and electrodes are being investigated to maximize the performance of chalcogenide-based flexible solar cells. Various material processing and treatment strategies are being developed; therefore, it is resilient to keep track of different methods and materials to enhance device performance. So, there is an immediate need to develop a robust material database that can be updated regularly to keep a record of various methods. In addition, machine-learning algorithms can easily quantify the combinations of various absorber/buffer layers, electrodes, and substrates (given appropriate dataset) to maximize the solar-conversion capability of flexible solar cell devices. These programs can perform experiments virtually to mitigate the number of undesirable combinations, which in turn saves a huge amount of time and resources. Along with that, useful insights can be drawn from various trends of different combinations, allowing researchers to try new materials and sequences to fill up the gaps to maximize device efficiency.

14.5.5 Rigorous testing

CIGS-based flexible solar cells have achieved a maximum efficiency of 20.8% on polymer substrate to date [6]. It is noteworthy that this efficiency is achieved under lab conditions and over a small (~0.5 cm^2) surface area. These numbers will reduce drastically when subjected to real-time environmental conditions and scaling to meet industrial standards. Thus, flexible solar modules need to be tested under real-life conditions, such as rain, prolonged light exposure, extreme temperature fluctuations, pressure, varying relative degrees of humidity, etc., to check the actual efficacy of these solar modules. The performance may also reduce drastically when subjected to continuous bending. Therefore, testing must be performed robustly under varying degrees of deformation and environmental factors altogether, specifically for the application of wearable electronics. It urges to build simulators mimicking environmental conditions while undergoing rigorous deformation.

14.5.6 DEVELOPMENT OF TRANSPARENT/SEMITRANSPARENT SOLAR CELLS

Flexible solar cells provide an edge over conventional rigid solar modules because they can be easily integrated with existing infrastructure, without any requirement of providing extra supports. This, in turn, mitigates one of the biggest challenges associated with solar energy production, i.e., huge landmasses, without destroying or disturbing natural habitats. This progress can be accelerated even further, with the development of transparent or semitransparent flexible solar modules. These solar panels can be easily integrated with the various buildings to their glassy exterior without hindering the normal working lifestyle of people. As these solar modules may have less PCE, the cost of these solar modules has to be reduced drastically for justification of their integration with existing infrastructure.

14.5.7 INTEGRATION WITH EXISTING TECHNOLOGIES

Flexible solar cells are very lightweight, making them a perfect candidate to be integrated with various wearable electronics. However, this progress has been very slow as researchers have been struggling to push the efficiency of chalcogenide-based flexible solar cells over the years. Proper steps must be initiated to increase the efficiency of flexible solar cells while demonstrating their efficacy when integrated with several daily-use items. Wearable electronics would consume a huge amount of power, which is going to increase exponentially with changing lifestyles. Thus, wearable electronics powered by these flexible solar cells can make a substantial impact in meeting the world's soaring energy demands.

14.6 CONCLUSION

Chalcogenide-based flexible solar cells offer a viable solution to meet the world's ever-increasing energy demands, both sustainably and practically. However, more efforts are needed to boost the efficiency and efficacy of the solar modules for their economical mass production, outreach to common people, and effective market capture. The above-mentioned steps can provide great assistance in achieving aspired performances; however, unanimous endeavors of physicists, chemists, materials scientists, industrialists, and policymakers would be required for the promotion of chalcogenide-based flexible solar cells and modules.

REFERENCES

1. Kranz, L., et al., Doping of polycrystalline CdTe for high-efficiency solar cells on flexible metal foil. *Nature Communications*, 2013. 4(1): p. 1–7.
2. Li, X., et al., Review and perspective of materials for flexible solar cells. *Materials Reports: Energy*, 2020. 1(1): p. 100001.
3. Reinhard, P., et al. Review of progress toward 20% efficiency flexible CIGS solar cells and manufacturing issues of solar modules. in *2012 IEEE 38th Photovoltaic Specialists Conference (PVSC) PART 2*. 2012. IEEE.
4. Ramanujam, J., et al., Flexible CIGS, CdTe and a-Si:H based thin film solar cells: A review. *Progress in Materials Science*, 2020. 110: p. 100619.
5. Li, K., et al., One-dimensional Sb2Se3 enabling ultra-flexible solar cells and mini-modules for IoT applications. *Nano Energy*, 2021. 86: p. 106101.
6. Carron, R., et al., *Advanced alkali treatments for high-efficiency Cu(In,Ga)Se2 solar cells on flexible substrates. Advanced Energy Materials*, 2019. 9(24): p. 1900408.
7. Yang, K.-J., et al., Flexible Cu 2 ZnSn (S, Se) 4 solar cells with over 10% efficiency and methods of enlarging the cell area. *Nature Communications*, 2019. 10(1): p. 1–10.
8. Wang, Y.-C., T.-T. Wu, and Y.-L. Chueh, A critical review on flexible Cu (In, Ga) Se2 (CIGS) solar cells. *Materials Chemistry and Physics*, 2019. 234: p. 329–344.
9. Hashemi, S. A., S. Ramakrishna, and A. G. Aberle, Recent progress in flexible–wearable solar cells for self-powered electronic devices. *Energy & Environmental Science*, 2020. 13(3): p. 685–743.
10. Aliyu, M., et al., Recent developments of flexible CdTe solar cells on metallic substrates: Issues and prospects. *International Journal of Photoenergy*, 2012. 2012.

11. Wang, C., et al., Efficiency improvement of flexible Sb2Se3 solar cells with non-toxic buffer layer via interface engineering. *Nano Energy*, 2020. 71: p. 104577.

12. Reinhard, P., et al., Flexible Cu (In, Ga) Se2 solar cells with reduced absorber thickness. *Progress in Photovoltaics: Research and Applications*, 2015. 23(3): p. 281–289.

13. Başol, B.M., et al., Flexible and light weight copper indium diselenide solar cells on polyimide substrates. *Solar Energy Materials and Solar Cells*, 1996. 43(1): p. 93–98.

14. Reinhard, P., et al., Features of KF and NaF postdeposition treatments of Cu (In, Ga) Se2 absorbers for high efficiency thin film solar cells. *Chemistry of Materials*, 2015. 27(16): p. 5755–5764.

15. Sadono, A., et al., Efficiency enhancement of flexible Cu (In, Ga) Se2 solar cells deposited on polyimide-coated soda lime glass substrate by alkali treatment. *Japanese Journal of Applied Physics*, 2017. 56(8S2): p. 08MC15.

16. Sadono, A., et al., Flexible Cu (In, Ga) Se2 solar cells fabricated using a polyimide-coated soda-lime glass substrate. *Japanese Journal of Applied Physics*, 2015. 54(8S1): p. 08KC16.

17. Niki, S., et al., CIGS absorbers and processes. *Progress in Photovoltaics: Research and Applications*, 2010. 18(6): p. 453–466.

18. Herz, K., et al., Diffusion barriers for CIGS solar cells on metallic substrates. *Thin Solid Films*, 2003. 431: p. 392–397.

19. Sim, J.-K., et al., Implementation of graphene as hole transport electrode in flexible CIGS solar cells fabricated on Cu foil. *Solar Energy*, 2018. 162: p. 357–363.

20. Kim, D., et al., Flexible and semi-transparent ultra-thin CIGSe solar cells prepared on ultra-thin glass substrate: A key to flexible bifacial photovoltaic applications. *Advanced Functional Materials*, 2020. 30(36): p. 2001775.

21. Matulionis, I., et al., Cadmium telluride solar cells on molybdenum substrates. *MRS Online Proceedings Library (OPL)*, 2001. 668.

22. Romeo, A., et al., High-efficiency flexible CdTe solar cells on polymer substrates. *Solar Energy Materials and Solar Cells*, 2006. 90(18): p. 3407–3415.

23. Perrenoud, J., et al., Fabrication of flexible CdTe solar modules with monolithic cell interconnection. *Solar Energy Materials and Solar Cells*, 2011. 95: p. S8–S12.

24. Wen, X., et al., Epitaxial CdTe thin films on mica by vapor transport deposition for flexible solar cells. *ACS Applied Energy Materials*, 2020. 3(5): p. 4589–4599.

25. Salavei, A., et al., Comparison of high efficiency flexible CdTe solar cells on different substrates at low temperature deposition. *Solar Energy*, 2016. 139: p. 13–18.

26. Rance, W.L., et al., 14%-efficient flexible CdTe solar cells on ultra-thin glass substrates. *Applied Physics Letters*, 2014. 104(14): p. 143903.

27. Mahabaduge, H. P., et al., High-efficiency, flexible CdTe solar cells on ultra-thin glass substrates. *Applied Physics Letters*, 2015. 106(13): p. 133501.

28. Kumar, M., et al., Strategic review of secondary phases, defects and defect-complexes in kesterite CZTS–Se solar cells. *Energy & Environmental Science*, 2015. 8(11): p. 3134–3159.

29. Kaur, K., N. Kumar, and M. Kumar, Strategic review of interface carrier recombination in earth abundant Cu–Zn–Sn–S–Se solar cells: Current challenges and future prospects. *Journal of Materials Chemistry A*, 2017. 5(7): p. 3069–3090.

30. Nisika, K. Kaur, and M. Kumar, Progress and prospects of CZTSSe/CdS interface engineering to combat high open-circuit voltage deficit of kesterite photovoltaics: a critical review. *Journal of Materials Chemistry A*, 2020. 8(41): p. 21547–21584.

31. López-Marino, S., et al., Alkali doping strategies for flexible and light-weight Cu2ZnSnSe4 solar cells. *Journal of Materials Chemistry A*, 2016. 4(5): p. 1895–1907.

32. Yang, K.-J., et al., The alterations of carrier separation in kesterite solar cells. *Nano Energy*, 2018. 52: p. 38–53.

33. Jo, E., et al., 8% efficiency Cu2ZnSn (S, Se) 4 (CZTSSe) thin film solar cells on flexible and lightweight molybdenum foil substrates. *ACS Applied Materials & Interfaces*, 2019. 11(26): p. 23118–23124.

34. Yang, K.-J., et al., Flexible Cu2ZnSn(S,Se)4 solar cells with over 10% efficiency and methods of enlarging the cell area. *Nature Communications*, 2019. 10(1): p. 2959.

35. Ahn, K., et al., Flexible high-efficiency CZTSSe solar cells on stainless steel substrates. *Journal of Materials Chemistry A*, 2019. 7(43): p. 24891–24899.

36. Peng, C.-Y., et al., Fabrication of Cu2ZnSnS4 solar cell on a flexible glass substrate. *Thin Solid Films*, 2014. 562: p. 574–577.

37. Brew, K.W., et al., Improving efficiencies of Cu2ZnSnS4 nanoparticle based solar cells on flexible glass substrates. *Thin Solid Films*, 2017. 642: p. 110–116.

38. Min, J.-H., et al., Flexible high-efficiency CZTSSe solar cells on diverse flexible substrates via an adhesive-bonding transfer method. *ACS Applied Materials & Interfaces*, 2020. 12(7): p. 8189–8197.

39. Todorov, T., et al. Flexible kesterite solar cells on ceramic substrates for advanced thermal processing. in *2015 IEEE 42nd Photovoltaic Specialist Conference (PVSC)*. 2015.

40. Li, Z., et al., 9.2%-efficient core-shell structured antimony selenide nanorod array solar cells. *Nature Communications*, 2019. 10(1): p. 125.

41. Li, K. and J. Tang. Super-flexible Sb2Se3 solar cell. in *The International Photonics and Optoelectronics Meeting (POEM)*. 2018. Wuhan: Optical Society of America.

42. Wen, X., et al., Efficient and stable flexible Sb2Se3 thin film solar cells enabled by an epitaxial CdS buffer layer. *Nano Energy*, 2021. 85: p. 106019.

43. Wong, W.S. and A. Salleo, Flexible electronics: Materials and applications. Vol. 11. 2009: Springer Science & Business Media.

44. Méndez-Hernández, J. M., et al., Effects of mechanical deformations on P3HT:PCBM layers for flexible solar cells. *Mechanics of Materials*, 2021. 154: p. 103708.

45. Cabral, J.T., et al., Bending fatigue study of sputtered ITO on flexible substrate. *Journal of Display Technology*, 2011. 7(11): p. 593–600.

46. Pont, S., J. R. Durrant, and J. T. Cabral, Dynamic PCBM:Dimer population in solar cells under light and temperature fluctuations. *Advanced Energy Materials*, 2019. 9(19): p. 1803948.

47. MacDonald, W. A., et al., Latest advances in substrates for flexible electronics. *Journal of the Society for Information Display*, 2007. 15(12): p. 1075–1083.

48. Chaves, A., et al., Bandgap engineering of two-dimensional semiconductor materials. *NPJ 2D Materials and Applications*, 2020. 4(1): p. 29.

49. Zhu, C., et al., Strain engineering in perovskite solar cells and its impacts on carrier dynamics. *Nature Communications*, 2019. 10(1): p. 815.

50. Amborski, L. E. and D. W. Flierl, Physical properties of polyethylene terephthalate films. *Industrial & Engineering Chemistry*, 1953. 45(10): p. 2290–2295.

51. Kese, K. O., Z. C. Li, and B. Bergman, Influence of residual stress on elastic modulus and hardness of soda-lime glass measured by nanoindentation. *Journal of Materials Research*, 2004. 19(10): p. 3109–3119.

52. Rösler, J., H. Harders, and M. Bäker, Mechanical behaviour of engineering materials: Metals, ceramics, polymers, and composites. 2007: Springer Science & Business Media.

53. Huang, J. M., P. P. Chu, and F. C. Chang, Conformational changes and molecular motion of poly(ethylene terephthalate) annealed above glass transition temperature. *Polymer*, 2000. 41(5): p. 1741–1748.

54. Karlsson, T. and A. Roos, Optical properties and spectral selectivity of copper oxide on stainless steel. *Solar Energy Materials*, 1984. 10(1): p. 105–119.

55. Ni, H.-j., et al., A review on colorless and optically transparent polyimide films: Chemistry, process and engineering applications. *Journal of Industrial and Engineering Chemistry*, 2015. 28: p. 16–27.

56. Valla, A., et al., Understanding the role of mobility of ITO films for silicon heterojunction solar cell applications. *Solar Energy Materials and Solar Cells*, 2016. 157: p. 874–880.

57. Lewis, J., Material challenge for flexible organic devices. *Materials Today*, 2006. 9(4): p. 38–45.

58. Wang, C., et al., Protective integrated transparent conductive film with high mechanical stability and uniform electric-field distribution. *Nanotechnology*, 2019. 30(18): p. 185303.

59. Po, R., et al., The role of buffer layers in polymer solar cells. *Energy & Environmental Science*, 2011. 4(2): p. 285–310.

60. Li, W., et al., Effect of a TiN alkali diffusion barrier layer on the physical properties of Mo back electrodes for CIGS solar cell applications. *Current Applied Physics*, 2017. 17(12): p. 1747–1753.

61. Zhang, Z., et al., All-carbon electrodes for flexible solar cells. *Applied Sciences*, 2018. 8(2): p. 152.

62. Wang, W., et al., Device characteristics of CZTSSe thin-film solar cells with 12.6% efficiency. *Advanced Energy Materials*, 2014. 4(7): p. 1301465.

63. Woods-Robinson, R., et al., Wide band gap chalcogenide semiconductors. *Chemical Reviews*, 2020. 120(9): p. 4007–4055.

64. Swarnkar, A., et al., Are chalcogenide perovskites an emerging class of semiconductors for optoelectronic properties and solar cell? *Chemistry of Materials*, 2019. 31(3): p. 565–575.

15 Perovskite-Based Flexible Solar Cells

Rushi Jani

Department of Electrical and Computer Science Engineering, Institute of Infrastructure, Technology, Research and Management, Ahmedabad, Gujarat, India

Kshitij Bhargava

Department of Electrical and Computer Science Engineering, Institute of Infrastructure, Technology, Research and Management, Ahmedabad, Gujarat, India

CONTENTS

15.1   Introduction.. 262
15.2   Device structure and development of FPSCs... 263
    15.2.1   Device structure of FPSC 263
    15.2.2   Development of FPSCs... 264
15.3   FPSC fabrication methods 265
    15.3.1   Laboratory scale fabrication methods... 265
        15.3.1.1   Spin coating.. 265
        15.3.1.2   Thermal evaporation.. 268
    15.3.2   Large scale fabrication methods 269
        15.3.2.1   Inkjet printing 269
        15.3.2.2   Blade coating.. 269
        15.3.2.3   Spray coating.. 270
        15.3.2.4   Slot-die coating.. 270
15.4   Materials for FPSCs.. 271
    15.4.1   Perovskite absorber layer.. 271
    15.4.2   Charge transport layers 272
        15.4.2.1   Electron transport layer 272
        15.4.2.2   Hole transport layer... 273
    15.4.3   Flexible substrates 273
        15.4.3.1   Polymer (or plastic) substrates... 273
        15.4.3.2   Metal substrates 274
        15.4.3.3   Fiber shaped PSCs.. 274
        15.4.3.4   Other flexible substrates 274
    15.4.4   Transparent conducting layer.. 275
    15.4.5   Encapsulation.. 275
15.5   Recycling of FPSCs 275
15.6   Challenges and future perspectives 278
    15.6.1   Environmental stability 278
    15.6.2   Mechanical stability 280

DOI: 10.1201/9781003186755-15

        15.6.3   High manufacturing cost...280
        15.6.4   Large-area fabrication 280
        15.6.5   Toxicity...281
15.7   Applications of FPSCs...282
15.8   Conclusion...282
References...282

15.1   INTRODUCTION

In the present era, energy is a crucial commodity and is very much intertwined with human development. The growing population has led to an increase in per capita energy consumption. Currently, the primary source of energy is fossil fuels, i.e., oil, coal, and natural gas. These conventional sources of energy have finite capacity and environmental ramifications, including climate change, increase in carbon footprint, etc. There is a dire need for renewable energy sources, such as solar energy, wind energy, hydro energy, etc. Moreover, these renewable energy resources are sustainable, have a low carbon footprint, and help confront the urgent issue of global warming. Among the various renewable energy sources, solar energy had shown colossal potential owing to the reduced levelized cost of electricity (LCOE), which is comparable to that produced through fossil fuels. Although the photovoltaic (PV) market is burgeoning, a huge scale-up in installation is necessary to compete with fossil fuels and cater to the global energy demand. In the near future, independent and distributed energy sources for electric vehicles, mobiles, sensors, IoT devices, electronic textiles, wearable devices, etc., will be required. In such a scenario, flexible solar cells prove to be a promising solution; that is, they can be integrated and implemented efficiently in extraterrestrial applications, vehicle-integrated photovoltaics, building-integrated photovoltaics (BIPV), wearable electronics, and other versatile applications. Therefore, in the past few years, flexible solar cells are being extensively investigated. The flexible solar cells can be manufactured via a roll-to-roll (R2R) process at low temperatures, which expedites mass production and also decreases production costs. Also, flexible solar cells are lightweight, thin, easily transportable, bendable, and offer economic installation.

The PV technology can be categorized into three generations: The first generation includes silicon-based solar cells, such as polycrystalline and monocrystalline solar cells. The second generation includes thin-film solar cells, such as cadmium telluride (CdTe), copper indium gallium selenide (CIGS), amorphous silicon (a-Si), and gallium arsenide (GaAs). The third generation involves emerging technologies, such as perovskite solar cells (PSC), quantum dot solar cells, organic solar cells, and dye-sensitized solar cells (DSSC). Table 15.1 compares the performance of various flexible PV technologies. Among the mentioned flexible PV technologies, perovskite solar cells have been extensively explored in the last decade. The PSC technology incorporates perovskite material as a photoactive absorber layer. The first-ever PSC was reported in 2009; it demonstrated efficiency of 3.8%. Furthermore, within a decade, the efficiency has ameliorated from 3.8% to 25.2% and has emerged as a potential replacement of silicon-based PVs. The promising performance of PSCs is attributed to high charge carrier mobility, lower exciton binding energy, higher absorption edge, higher carrier diffusion length, economic fabrication, and easily available precursor materials making it suitable for mass fabrication through solution route [8–10]. The flexible perovskite solar cells (FPSCs) are fabricated via low-temperature solution-processing techniques on flexible substrates. The first-ever FPSC was reported by Kumar et al. in 2013; it demonstrated power-conversion efficiency (PCE) of 2.62% [11]. In subsequent years, the PCE of FPSC has improved to 19.9%. With an aim to ramp up the performance of FPSCs, extensive research has been done on the fabrication of superior quality perovskite thin films, development of novel transparent electrodes, and low-temperature deposition methods of electron and hole transport layer. The aforementioned studies have enabled roll-to-roll processed large-area fabrication at a reduced cost. FPSCs lack pragmatic applications owing to some vital factors, namely, environmental stability, manufacturing cost, large-area fabrication, toxicity, and mechanical stability.

TABLE 15.1

Comparison of Various Flexible PV Technologies

Flexible PV technology	Certified efficiency (%)	Merits	Demerits
Silicon solar cells	19 [1]	• Exhibits air stability • Better efficiency • Lesser efficiency degradation while upscaling	• More thick • Development of cracks even at smaller bending radius
Cadmium Telluride solar cells	13.6 [2]	• Compatible to several fabrication techniques. • High efficiency	• Expensive raw materials • Manufacturing is expensive • Contains toxic elements
Copper indium gallium selenide solar cells	20.8 [2]	• Longer lifetime • Exhibits air stability • High efficiency	• Manufacturing process is cost-intensive • Contains toxic elements
Gallium arsenide solar cells	29.10 [3]	• Able to function in case of weak-light • Exhibits thermal stability • Very high efficiency	• Lifetime reduces rapidly • Encapsulation method is very intricate • Its assembly is cost-intensive
Perovskite solar cells	19.9 [4]	• High power to weight ratio • High efficiency • Involves low cost	• Poor mechanical robustness • Performance is prone to environmental factors • Efficiency degrades while up-scaling
Quantum dot solar cells	12.3 [5]	• Exhibits better flexibility • High power to weight ratio • Its manufacturing is cost intensive	• Performance severely degrades upon exposure to UV and under humid environment • Contains toxic elements • Difficult to control particle size
Dye sensitized solar cells	7.09 [6]	• Able to function in case of weak light • Involves low cost • Higher efficiency is achieved at elevated temperature.	• Meagre efficiency • Under low temperature or at elevated temperature liquid electrolyte might freeze or evaporate
Organic solar cells	15.21 [7]	• Involves low cost • Exhibits better flexibility • Lightweight	• Inferior lifespan • Lower efficiency

15.2 DEVICE STRUCTURE AND DEVELOPMENT OF FPSCS

15.2.1 DEVICE STRUCTURE OF FPSC

A typical FPSC device consists of the following layers: flexible substrate, transparent conducting layer (or front contact), hole-transport layer (HTL), perovskite absorber layer, electron transport layer (ETL), and back contact. Generally, the FPSC can be categorized into two structures: normal structure (n-i-p) and inverted structure (p-i-n), where p: p-type HTL, i: intrinsic perovskite absorber layer, n: n-type ETL. Typically, FPSC is based on a normal structure (n-i-p) in which HTL is placed between the metal back contact and the perovskite absorber layer, and ETL is placed between the perovskite absorber layer and transparent conducting layer, as shown in Figure 15.1a. In the case of inverted p-i-n structure, the position of HTL and ETL is swapped, as shown in Figure 15.1b.

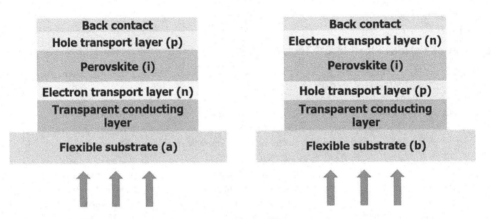

FIGURE 15.1   Types of FPSCs architecture: normal (n-i-p) (a) and inverted (p-i-n) (b).

The absorption of incident light in the perovskite absorber layer leads to the formation of excitons, which are bounded electron-hole pairs with exciton binding energy of nearly 26 meV, which is similar to room-temperature thermal energy. If the thermal energy is greater than exciton binding energy, free electrons and holes will be generated, and if thermal energy is less than exciton binding energy, it will require some external energy to form free electrons and holes. The photogenerated electrons and holes will travel toward their respective electrodes via the charge transport layer, i.e., electrons will travel through ETL and holes will travel through HTL. Hence, the incident photons are converted into useful electrical power. The quality of interfaces, namely perovskite/ETL, HTL/perovskite, and other interfaces between different materials, severely affects the performance of FPSCs; this is because recombination of charge carriers generally occurs at these interfaces, which subsequently reduces the value of voltage and current, thus curtailing the PCE.

In FPSCs, the commonly used flexible substrates are polymer, paper, metal foils, and fibers. From the aforementioned materials, polymer-based flexible substrates, such as polyethylene 2,6-naphthalate (PEN), polyethylene terephthalate (PET), and polyimide (PI), have been extensively explored. To be precise, PET is extensively used as a flexible substrate for n-i-p FPSCs. Conventionally, the transparent conducting layer is made up of indium-doped tin oxide (ITO) and fluorine-doped tin oxide (FTO). The combination of PET and ITO is widely used in normal and inverted FPSCs. Simultaneously, n-type organic materials, such as C_{60}, PCBM, BCP, etc., and inorganic materials, such as Zn_2SnO_4, SnO_2, ZnO, TiO_2, etc., are used as ETL. For HTL, p-type organic materials are used, such as PEDOT:PSS, spiro-OMeTAD, PTAA, etc., and inorganic materials, such as Cu_2O, NiO_x, etc. Generally, in normal structure, (2,2′,7,7′-tetrakis(N,N-di-p-methoxyphenylamine)-9,9′-spirobifluorene) (Spiro-OMeTAD) and inorganic materials, such as SnO_2, TiO_2, ZnO, are widely used as HTL and ETL, respectively. For inverted structure, PEDOT:PSS and NiO_x are commonly used as HTL owing to their higher transparency, while PCBM and its derivatives are used as ETL owing to their higher opacity [12]. The back contact is generally made of metals, including aluminum, gold, silver, and copper.

15.2.2   Development of FPSCs

The prior research and advancement of PSCs was based on rigid glass substrates, which are heat re-silient, highly transparent, and adhere well with the transparent conductive layer. Howbeit, flexible substrates have several benefits as compared to rigid glass substrates; for instance, they are stretchable, lightweight, foldable, and bendable to a certain extent. Furthermore, flexible films are compatible with the roll-to-roll (R2R) fabrication technique, which enables a higher production rate. Also, as compared to rigid PSC, FPSC fabrication involves low-temperature energy-efficient processes that further reduce the fabrication cost. The PCE and stability of FPSCs are lesser as compared to rigid PSCs, which is mainly attributed to the material properties of the substrate. For instance, the fabrication of FPSC on polymer substrates must be done at a temperature less than 150 °C as the polymer substrates cannot sustain high temperatures. The different layers involved in FPSC must undergo high-temperature processing to attain superior grade semiconductor thin film with good electromechanical properties. Furthermore, another reason for the low efficiency of FPSC is that the optical transmittance of flexible substrates (metal foil and plastic substrates) is lesser as compared to rigid substrates. Also, the deposition of indium tin oxide (ITO) layer over the flexible substrate, which is used as a transparent conducting layer, involves low-temperature processing as compared to a rigid glass substrate. This low-temperature processed ITO deposited over flexible substrates has lower crystallinity and hence, lesser conductivity. Ideally, the FPSC substrate must possess a lower water vapor transmission rate (WVTR), low coefficient of thermal expansion, high flexibility, low cost, lightweight, high transmittance and offer less sheet resistance.

The mediocre power-conversion efficiency of FPSCs has been augmented from 2.62% to 19.9% within a decade. Throughout the history of FPSCs, the normal structure devices have attained high performance. Tables 15.2 and 15.3 show the noteworthy development of normal and inverted structures FPSC, respectively.

15.3 FPSC FABRICATION METHODS

The FPSC fabrication methods are divided into two categories, namely, laboratory-scale methods and large-scale methods, which are described as follows.

15.3.1 LABORATORY SCALE FABRICATION METHODS

15.3.1.1 Spin coating

Spin coating is a straightforward method and is extensively used for the deposition of a pin-hole-free and uniform perovskite thin film. The spin coating process involves the following three steps. In the first step, a definite amount of solution precursor is dispensed onto the center of the substrate. In the second step, the substrate is spun at a high angular speed to spread the dispensed solution precursor by centrifugal force. In the third step, the volatile precursor solvent starts evaporating during the rotation. As a result, a dried film with definite thickness is achieved. The thickness and quality of the resultant film are dependent on the following parameters: duration of rotation, solution precursor concentration, the centrifugal force experienced, which is proportional to angular speed, the viscosity of solution precursor, and the dispensed volume of solution [23].

Generally, the spin-coating process is followed by heat treatment (annealing) to confirm the formation of perovskite crystals, and depending on the perovskite composition, the heat treatment (annealing) temperature is decided. In this context, perovskite, which involves heat treatment (annealing) temperature greater than 150 °C, is not compatible with plastic substrates. There are two popular methods for deposition of perovskite thin-film via spin coating: the one-step method and the two-step method. The one-step spin-coating method comprises three stages: preparation of precursor, spin coating, and heat treatment (annealing). In the first stage, the precursor is prepared by mixing metal halides, such as lead chloride ($PbCl_2$)/lead iodide (PbI_2), and methyl ammonium (MA)/formamidinium (FA) halides, such as MAX or FAX (X = Br, I, Cl) in 1:1 stoichiometry or 1:3 in organic solvents like dimethyl sulfoxide (DMSO), gummabutyrolactone (GBL), N,N-dimethylformamide (DMF) or their mixture. In the second stage, the prepared precursor is spin-coated on the substrate. In the final stage, the resultant spin-coated film undergoes heat treatment (annealing) to remove residual solvent, improve perovskite film coverage, and confirm the formation of perovskite crystals. Furthermore, to produce highly uniform perovskite thin films, an anti-solvent is usually introduced during the spin-coating process of the one-step method. The incorporation of anti-solvent expedites the formation of uniform perovskite crystals and facilitates the segregation of perovskite crystals from precursor solvent. For

TABLE 15.2

Development of Normal (n-i-p) Type FPSCs

Normal structure	V_{oc}	J_{sc}	FF	PCE	Key remark	Bendability
Au / Spiro-OMeTAD / Perovskite / C-ZnO/ZnO / PET-ITO	0.8	7.52	43.14	2.62 [11]	Deposition of zinc oxide (ZnO) via electrodeposition	–
Ag / Spiro-OMeTAD / Perovskite / ZnO / PET-ITO	1.03	13.4	73.9	10.2 [13]	Deposition of ZnO nanoparticles via spin coating	Retained 90% of its initial efficiency at bending radius $(B_r) = 1.60$ mm
Ag / Spiro-OMeTAD / Perovskite / TiO$_x$ / PEN-ITO	0.95	21.4	60	12.2 [14]	Deposition of TiO$_x$ via atomic layer deposition	Retained 95% of its initial efficiency after 1000 bending cycles at $B_r = 10$ mm
Au / PTAA / Perovskite / Zn$_2$SnO$_4$ / PEN-ITO	1.05	21.6	67	15.3 [15]	Deposition of Zn$_2$SnO$_4$ nanoparticles via spin coating	Retained 95% of its initial efficiency after 300 bending.
Au / Spiro-OMeTAD / Perovskite / SS-IL / PET-ITO	1.07	22.72	66.2	16.09 [16]	Incorporating solid state ionic liquid as ETL	Retained 91% of its initial efficiency after 300 bending cycles at $B_r = 5$ mm
Au / Spiro-OMeTAD / Perovskite / Nb$_2$O$_5$ / PET-ITO	1.103	22.48	74.2	18.4 [17]	Preparation of perovskite thin film using dimethyl sulphide additive	Retained 83% of its initial efficiency after 5000 bending cycles at $B_r = 4$ mm
Au / Spiro-OMeTAD / Perovskite / Zn$_2$SnO$_4$/SnO$_2$ / PEN-ITO	1.19	21.9	76.3	19.9 [4]	Porous planar ETL	Retains 15% of its initial efficiency after 1000 bending cycles at $B_r = 10$ mm

TABLE 15.3

Development of Inverted (p-i-n) Type FPSCs

Inverted structure	V_{oc}	J_{sc}	FF	PCE	Key remark	Bendability
Ag / PCBM/Bis-C$_{60}$ / Perovskite / PEDOT:PSS / PET-ITO	0.87	13.91	59	7.14 [18]	Fabrication of FPSC via blade coating instead of screen printing	Retained 80% of its initial efficiency after 100 bending cycles at $B_r = 10$ mm
Ga-In alloy / PCBM / Perovskite / PEDOT:PSS / NOA 63	0.92	16.58	70.5	10.75 [19]	Utilizing shape recoverable polymer (NOA 63) as flexible substrate	Retained 90% of its initial efficiency after 1000 bending cycles at $B_r = 1$ mm
Ag / C$_{60}$/BCP / Perovskite / PEDOT:PSS / PET-ITO	0.97	21.76	80	16.13 [20]	Optimization of mixed cation perovskite, anti-solvents and additive	Retained >90% of its initial efficiency after 1000 bending cycles at $B_r = 6$ mm
Au / PCBM/BCP / Triple cation perovskite / NiO$_x$ / Glass-ITO	1.02	23.6	71.5	17.2 [21]	Utilizing all inkjet printed layers	–
Ag / PCBM/BCP / Perovskite / PEDOT:PSS/PTAA / PEN-ITO	1.09	21.98	81	19.41 [22]	Morphology control and interface design	Retained ≈100% of its initial efficiency after 1000 bending cycles at $B_r = 3$ mm

instance, chlorobenzene, diethyl ether, and toluene are extensively used anti-solvents. In particular, chlorobenzene is highly preferred to fabricate flexible and rigid perovskite solar cells. Nevertheless, the addition of an anti-solvent requires very accurate control, and hence, it is difficult to produce large-area FPSC. Cao et al. fabricated FPSC in which quintuple cation perovskite thin film was deposited via one-step spin coating and yielded a PCE of 19.11% [24].

The two-step method does not involve anti-solvents and comprises three stages, namely deposition of metal halide, spin coating, and heat treatment (annealing). The first stage involves the deposition of a metal halide layer (PbI$_2$) on the substrate. The second stage involves spin coating of a methylammonium (MA)/formamidinium (FA) layer over a metal halide layer; later, these layers fuse into each other. The final stage involves heat treatment (annealing) of the resultant film, which aids the crystallization process. The two-step method proves to be more effective than the one-step method owing to superior interface control and morphology. Despite the aforementioned benefits, it suffers from partial conversion of PbI$_2$ into perovskite film and hence lacks large-scale manufacturing productivity of FPSCs with higher reproducibility. Nejand et al. fabricated FPSC in which MAPbI$_3$ perovskite thin film was deposited via two-step spin coating and achieved a PCE of 12.8% [25]. Both the methods lead to high waste of materials and hence are limited to laboratory scale [9,12,23].

Advantages

i. Simple operation and efficient method
ii. Cost-effective
iii. Faster drying
iv. Very fine control over thin film thickness over a small area and hence possible to achieve a wide range of thin-film thickness

Disadvantages

i. Waste of precursor material, hence limited to laboratory scale
ii. Upon increasing fabrication area, it is prone to pinhole, uncovered area, and surface defects, which hamper the efficiency of FPSC
iii. Unable to produce a thin film with uniform thickness over a large area
iv. Not suitable for batch production

15.3.1.2 Thermal evaporation

It is a popular laboratory scale method to fabricate uniform and smooth textured perovskite thin films. This method is performed inside a vacuum chamber and involves the following two steps: (1) The perovskite precursor material is evaporated by heating at high temperature. (2) The vapor particles produced reach the substrate where it solidifies, and hence, a perovskite thin film gets deposited. This method is suitable for the fabrication of FPSC because it does not involve high-temperature processes during deposition. However, this method is complex and the setup cost is high due to the vacuum process involved.

This method can be categorized into two types, namely the dual-source and single-source thermal evaporation method. The dual-source thermal evaporation method involves simultaneous evaporation of organic and inorganic materials. Liu et al. reported the first-ever perovskite thin-film deposition using the dual-source thermal evaporation method [26]. In M. Xu et al. fabricated FPSC (Glass/SU-8/MoO$_3$/Au-thin film/PEDOT:PSS/perovskite/Ca/Ag), the perovskite thin film was deposited using the dual-source thermal evaporation method, which yielded a power-conversion efficiency of 9.05% [27]. The single-source thermal evaporation method involves evaporation of a single source. Li et al. fabricated FPSC (PEN/ITO/C$_{60}$/perovskite/spiro-OMeTAD/Au) in which perovskite thin film was deposited using a single-source thermal evaporation method that yielded a power-conversion efficiency of 13.9% [28]. The single-source thermal evaporation method is energy efficient and has a simple procedure, unlike the dual-source evaporation method.

Advantages

i. Lesser damage to the underlying layer and hence suitable for fabricating perovskite-based tandem solar cells
ii. The resultant thin film is pin-hole free
iii. Accurate control of thin-film thickness
iv. Higher thin-film deposition rate

Disadvantages

i. The substrate temperature is difficult to control
ii. Complex procedure
iii. Difficult to improve shadowing (or step coverage)

15.3.2 LARGE SCALE FABRICATION METHODS

Roll-to-Roll (R2R) processing is a proven technique for manufacturing solar cells. Generally, R-2-R processing comprises various steps, such as drying, cutting, printing/coating, lamination, UV-curing, etc. This section focuses on printing/coating methods, including inkjet printing, blade coating, spray coating, and slot-die coating which are compatible with R2R processing.

15.3.2.1 Inkjet printing

This method primarily involves the ejection of a definite quantity of ink droplets on a substrate through a nozzle. It is a promising method for fabricating FPSC, although it was originally utilized for fabricating dye-sensitized solar cells and organic solar cells. This method is not capable of using anti-solvents during the process; therefore, to produce thin films with homogenous morphology, it requires ancillary steps (such as vacuum annealing), precise optimization of the ink composition, and substrate temperature. The inkjet printing technology can be categorized into three distinct categories: field or flow-induced tip streaming, drop on demand, and continuous inkjet. It is also capable of fabricating three-dimensional (3D) electronic parts, optical devices, electronics components (capacitor, resistor, transistor), and polymer solar cells [29]. Recently, Schackmar et al. demonstrated all-inkjet-printed PSC (glass/ITO/NiO$_x$/Triple-cation perovskite/PCBM/BCP/Au) and achieved a PCE >17% [21].

Advantages

    i. Lesser production of waste
    ii. Higher resolution and hassle-free computerized fabrication
    iii. Cost-effective
    iv. Non-contact and maskless printing

Disadvantages

    i. Requires precise optimization of the ink composition
    ii. Nozzle clogging is a serious issue
    iii. Difficult to fabricate thin films with homogenous morphology
    iv. Thin-film formation is a complex process

15.3.2.2 Blade coating

Initially, blade coating was utilized for fabricating dye-sensitized solar cells. This method involves suitably placing a sharp blade (0.01–0.5 μm) above the substrate and spreading the coating solution in front of the blade. The blade is then moved at a uniform speed (generally several cm/s) over the substrate, depositing a film of definite thickness [23]. Note that the operating temperature is relatively low and hence suitable for fabricatating FPSC. This method is nearly identical to the one-step spin-coating method, although the blade-coating method consumes less material. For instance, to fully deposit a thin film over an area of 225 mm^2, the blade-coating method consumes 0.01–0.02 mL of precursor solution, whereas the one-step spin-coating method consumes 0.05–0.1 mL of the precursor solution.

In 2015, Yang et al. demonstrated the first-ever blade-coated FPSC with a configuration of (Glass/ITO/PEDOT:PSS/Perovskite/Bis-C$_{60}$/Ag), all layers other than silver electrode were fabricated via blade coating. The fabricated FPSC showcased a PCE of 7.14 and retained 80% of its initial PCE after 100 bending cycles at a bending radius (B$_r$) of 10 mm [18]. Later on, Wang et al. fabricated FPSC with a configuration of (PEN/ITO/PEDOT:PSS/PTAA/MAPbI$_3$/PCBM/BCP/Ag), which illustrated a PCE of 19.41%. Moreover, the fabricated FPSC almost retained its initial PCE after 1000 bending cycles at a B$_r$ of 3 mm [22].

Advantages

  i. Lesser wastage of precursor solution
  ii. Compatible to fabricate FPSC and rigid PSC
  iii. High throughput and highly efficient
  iv. Fabrication area has lesser impact on thin-film uniformity

Disadvantages

  i. Lower reproducibility
  ii. Process is more prone to contamination
  iii. Not suitable for fabricating ultra-thin films

15.3.2.3 Spray coating

Spray coating is a highly suitable and contact-free method to fabricate FPSC owing to its low cost, ability to produce large-area FPSC, high throughput, and superior robustness. This method involves the conversion of perovskite solution into ultrafine droplets, ejection of ultrafine droplets from the air-driven spray nozzle, and deposit on a substrate. This method is not suitable for fabrication over polymer substrates because evaporation of perovskite solution into gaseous phase involves high temperature. Das et al. reported a high throughput ultrasonic spray coating process to fabricate FPSC with configuration (PET/ITO/TiO_2/Perovskite/Spiro-OMeTAD/Ag) and yielded a maximum PCE of 8.1%. The fabricated FPSC retained nearly 80% of maximum efficiency after 1000 bending cycles at a bending radius of 7 mm [30].

Advantages

  i. Faster fabrication
  ii. Lesser production of waste
  iii. Lesser damage to the underlying layer
  iv. Patterning is possible with this technique

Disadvantages

  i. Difficult to achieve ultra-smooth surface
  ii. Incongruous for thin-film deposition over a flexible polymer substrate
  iii. Process parameters are difficult to control

15.3.2.4 Slot-die coating

This contact-free method comprises a coating head, a moving substrate, and a pump that feeds ink to the coating head. The ink flows through the coating head and is directly deposited onto the moving substrate, which is placed underneath the coating head. Initially, the slot-die coating method and spin-coating method were jointly used to manufacture FPSC. This method is cost-effective and has higher reproducibility owing to negligible evaporation of ink. Schmidt et al. fabricated FPSC based on glass/ITO substrate via one-step slot-die roll coating and two-step slot dies roll coating process, which yielded PCE of 4.9% for a one-step slot-die roll-coating process and 2.6% for a two-step slot-die roll coating process [31]. Hwang et al. demonstrated first-ever fully printed PSC via slot-die coating method and large area R2R processing under ambient conditions, which yielded PCE of 11.96% [32]. Remeika et al. fabricated flexible slot-die coated PSC under an ambient condition with configuration (PET/ITO/ZnO-np/Perovskite/P3HT/Au), which yielded PCE of 3.6% with negligible hysteresis [33].

Advantages

  i. Noncontact method which avoids scratches on the substrate
  ii. High coating speed

iii. Negligible evaporation loss of perovskite precursor/solution

iv. Process is completely sealed therefore no leakage of noxious perovskite precursor

Disadvantages

i. A complex process due to the requirement of optimization of multiple parameters

ii. Involves very high initial setup cost

iii. Involves large supporting infrastructure to house the equipment

iv. Difficult to analyze the origins of defects owing to the complexity of the process

15.4   MATERIALS FOR FPSCS

In this section, we will discuss the various materials involved in the manufacturing of FPSCs. An appropriate selection of materials plays a pivotal role in the performance of FPSCs. While choosing the materials for FPSCs, the following criteria are considered: durability, flexibility, optoelectronic properties, mechanical properties, and performance metrics. Generally, the following materials are utilized during the preparation of FPSCs.

i. Perovskite absorber layer

ii. Charge transport layers

iii. Flexible substrate

iv. Transparent conducting layer

v. Encapsulation

15.4.1   Perovskite absorber layer

The quality of perovskite thin film is crucial to realize high-efficiency FPSCs. The perovskite thin films are required to possess properties such as homogeneity, compactness, mechanical durability (resilience toward the formation of cracks occurring due to bending cycles, which hampers the performance), and defect-free. Methylammonium lead tri-iodide ($MAPbI_3$) perovskite absorber layer has been extensively investigated since it was first reported in 2009. Recently, Wang et al. fabricated $MAPbI_3$ based FPSC with a configuration of (PEN /ITO/PEDOT:PSS/PTAA/$MAPbI_3$/ PCBM/BCP/Ag), which illustrated a PCE of 19.41%. Despite demonstrating huge prospects of PV application, the perovskite layers suffer from pitfalls like lower performance, thermal stability, low mobility lifetime product, and lower absorption edge. Therefore, formamidinium lead iodide ($FAPbI_3$) was introduced owing to benefits such as narrow bandgap, superior mobility lifetime product, and lesser thermodynamic degradation. Recently, Wu et al. fabricated $FAPbI_3$ based FPSC with configuration (PEN/ITO/SnO_2/$FAPbI_3$/Spiro-OMeTAD/Au), which demonstrated PCE of 19.38% [34]. Despite the aforementioned benefits, $FAPbI_3$ based FPSCs lack realization due to low resilience toward humidity, the phase transition from photoactive black-colored phase to photo-inactive yellow-colored phase at room temperature and post-heating temperature is 150 °C while 100 °C for $MAPbI_3$ due to the presence of photoinactive yellow phase at annealing temperature <150 °C; hence, it is difficult to fabricate over plastic substrates. To counter these issues and improve their performance, several methods were proposed, including addition of methylammonium (MA^+) and formamidinium (FA^+) cations, i.e., mixed cation perovskite, addition of bromide (Br^-) anions, i.e., mixed anion perovskite. Pandey et al. fabricated mixed cation perovskite-based FPSC with configuration (PET/ITO/PEDOT:PSS/$FA_{0.2}MA_{0.8}PbI_3$/C_{60}/BCP/Ag), which achieved a maximum PCE of 16.13% [20]. Cao et al. studied mixed cation and mixed anion system, i.e., fabricated quintuple cation and mixed bromide anion perovskite-based FPSC with configuration (PEN/ITO/ $Rb_{5-x}K_xCs_{0.05}FA_{0.83}MA_{0.17}PbI_xBr_{3-x}$/Spiro-OMeTAD/Au), which achieved a PCE of 19.11% [24].

Nowadays, researchers are exploring lead-free (tin or bismuth) perovskite-based FPSCs due to the toxicity of lead-based FPSC. Rao et al. fabricated formamidinium tin iodide (FASnI$_3$) based flexible and wearable perovskite solar cell with configuration (PDMS/hc-PEDOT:PSS/PEDOT:PSS/FASnI$_3$/C$_{60}$/BCP/Ag), which yielded PCE of 8.56%. Furthermore, they applied graphite phase-C$_3$N$_4$ as a crystalline template into FASnI$_3$ perovskite [35].

15.4.2 CHARGE TRANSPORT LAYERS

Free charge carriers (electrons and holes) are generated inside the perovskite absorber layer upon absorption of incident photons. Later, these photogenerated free charge carriers will diffuse toward their respective electrodes, i.e., holes will diffuse toward the cathode and electrons will diffuse toward the anode. Charge transport layers are of two types: an electron transport layer (ETL), which aids diffusion of electrons, and a hole-transport layer (HTL), which aids diffusion of holes. Usually, certain aspects are considered while selecting charge transport layers; these include the value of electron and hole mobility of ETL and HTL, respectively, sufficient mechanical durability, the ability of ETL and HTL to block holes and electrons respectively, electrical properties, optoelectronic properties, processing temperature, and environmental stability, i.e., in case of flexible plastics, the WVTR is high as compared to rigid substrates; therefore, the charge transport layers should be resilient toward corrosion.

15.4.2.1 Electron transport layer

The conduction band maximum (CBM) of ETL should match well with the CBM of perovskite absorber and anode, which facilitates the diffusion of electrons and thereby ameliorates the performance. Generally, utilized inorganic ETL materials are titanium dioxide (TiO$_2$), zinc oxide (ZnO), tin oxide (SnO$_2$), niobium pentoxide (Nb$_2$O$_5$), zinc stannate (Zn$_2$SnO$_4$), etc. TiO$_2$ and ZnO are ubiquitous ETL materials used in preparing FPSC.

TiO$_2$ is widely used as ETL in FPSC owing to relatively high stability, suitable alignment with CBM of the perovskite layer, and excellent mechanical properties. In 2013, Docampo et al. synthesized TiO$_x$ at 130 °C via solution process and for the first time, they successfully fabricated FPSC with a bilayer ETL, i.e., PCBM/TiO$_x$, which achieved a PCE of ~6% [36]. Recently, Wang et al. fabricated FPSC with room-temperature processed fullerene/TiO$_2$ nanocomposite as ETL, which achieved an efficiency of 18.06% [37]. Despite being a fundamental ETL material, TiO$_2$ requires high-temperature annealing (~500 °C) and hence is incompatible on flexible plastic substrates. Thus, efficient alternatives such as ZnO, Zn$_2$SnO$_4$, SnO$_2$ have been introduced. ZnO is a promising replacement for TiO$_2$ due to low-temperature processing, long diffusion coefficient, and superior electron mobility. In 2013, Kumar et al. reported the first-ever FPSC with ZnO nanorods grown by chemical bath deposition and ZnO compact layer deposited via electrodeposition, which yielded PCE of 2.62% [11]. Heo et al. fabricated FPSC with ZnO as ETL, which yielded a PCE of 15.5% [38]. Furthermore, other novel inorganic ETL materials, such as solid-state ionic liquid, tungsten oxide, transition metal dichalcogenide materials (MoS$_2$, SnS$_2$), etc., are also being investigated.

In addition, organic materials are also utilized as ETL due to low-temperature processing. Currently, fullerene and its derivatives, i.e., C$_{60}$, PCBM, PCBC6, etc., are being extensively explored as ETL. PCBM, a derivative of fullerene is being extensively used due to suitable energy-level alignment with perovskite, better electron extraction, and solution processability. Recently, Meng et al. fabricated FPSC with PCBM, which yielded a PCE of 19.87% [39]. Also, as compared to PCBM, C$_{60}$ demonstrates superior conductivity and electron mobility. In a recent study, Tavakoli et al. fabricated FPSC with C$_{60}$ as ETL, which yielded a PCE of 18.68%. Furthermore, other novel organic ETL materials, such as C$_{60}$ pyrrolidine tris acid (CPTA), a thin layer of polyethylenimine-ethoxylated (PEIE) over a transparent conducting layer for better electron extraction, perylene diimide, etc., are being studied.

15.4.2.2 Hole transport layer

It is desirable that the valence band maximum (VBM) of HTL should match well with the VBM of perovskite absorber and cathode, which facilitates diffusion of holes and thereby ameliorates the performance. Generally, utilized inorganic HTL materials are nickel oxide (NiO_x), copper iodide (CuI), copper oxide (CuO), copper phthalocyanine (CuPc), etc. Also, inorganic HTL materials are low cost and more environmentally stable as compared to their organic counterpart. Among the aforementioned inorganic HTL, NiO_x is being extensively employed owing to better alignment with VBM of the perovskite absorber layer, good optical transparency, better hole extraction, and efficient electron blocking. Hou et al. fabricated efficient FPSC by employing a double HTL design of PEDOT:PSS/NiO. The fabricated FPSC yielded a PCE of 18%, which is the highest PCE of NiO_x HTL based FPSC. Howbeit, deposition of NiO_x requires high-temperature processing and hence is less suitable for flexible plastic substrates. Furthermore, other novel inorganic HTL materials such as molybdenum oxide (MoO_x), copper chromium binary metal oxide ($CuCrO_2$), cobalt nitride (CoN) nano-films, etc. are being investigated.

Furthermore, organic HTL materials are often employed owing to their low-temperature solution method and well-aligned VBM with perovskite absorber. Currently, organic HTL materials such as Spiro-OMeTAD, PEDOT:PSS, PTAA, and P3HT are often used. Among the aforementioned organic HTLs, Spiro-OMeTAD is most commonly used. Recently, Chung et al. fabricated FPSC with Spiro-OMeTAD and achieved a certified efficiency of 19.9%, which is the highest PCE reported to date [4]. However, Spiro-OMeTAD suffers from poor crystallinity, is prone to degradation, and is expensive. Therefore, PEDOT:PSS is proposed as a potential substitute to Spiro-OMeTAD owing to its excellent hole mobility, low cost, hassle-free fabrication, better mechanical flexibility, and low-temperature processing. Recently, Ma et al. reported efficient FPSC by incorporating PEDOT:PSS/N,N′-Bis-(1-naphthalenyl)-N,N′-bis-phenyl-(1,1′-biphenyl)-4,4′-diamine (NPB) bi-layer as HTL and realized a highest PCE of 14.4% [40]. However, the acidic and hygroscopic nature of PEDOT:PSS limits its application. High acidity leads to corrosion of ITO electrode, and along with its hygroscopic nature, it hampers device stability. Therefore, other novel organic materials, including polyTPD, PhNa-1T, etc., are being explored.

15.4.3 FLEXIBLE SUBSTRATES

The flexibility and performance of FPSC essentially rely on the substrate material utilized. A good substrate should possess high flexibility (larger bending radii), low sheet resistance, high transparency, resilience toward corrosion, and thermal stability. The substrates can be categorized into two categories: polymer (or plastic) and metal substrates. Also, other substrates such as willow glass, paper, mica, etc. are being investigated. Furthermore, novel heat resistant and weavable substrates such as fiber-shaped PSCs are being explored [12].

15.4.3.1 Polymer (or plastic) substrates

Polymer materials, namely polyethylene 2,6-naphthalate (PEN), polyethylene terephthalate (PET), and polyimide (PI) are ubiquitous substrate materials owing to their low cost, lightweight, ease of availability, mechanical robustness, high optical transparency, high flexibility, and hassle-free transportation. Feng et al. utilized PET as a flexible substrate and realized a PCE of 18.4%. Furthermore, the FPSC retained 83% of its initial efficiency after 5000 bending cycles at B_r = 4mm [17]. To date, the highest certified PCE of 19.9% is realized on PEN substrate. Moreover, the FPSC retained 15% of its initial efficiency after 1000 bending cycles at B_r = 10 mm [4]. Howbeit, the above-mentioned PET and PEN substrates have low heat tolerance, i.e., PET and PEN have low glass transition temperature (T_G), $T_G \approx 105\,°C$ (for PET), and $T_G \approx 125\,°C$ (for PEN) [10]. Therefore, it is cumbersome to deposit metal oxide-based charge transport layers because it usually involves high-temperature processing ($\geq 400\,°C$), which would further damage the substrate.

Recently, polyimide (PI) has proved to be a favorable substitute to PET/PEN, owing to its higher heat tolerance ($T_G > 200\ °C$). However, PI has an inherently light brown color with lower optical transparency [10]. Park et al. utilized colorless polyimide (C-PI) as a flexible substrate and realized a PCE of 15.5%. Furthermore, as compared to PET substrate, it offered low sheet resistance (57.8 Ω sq^{-1}), high optical transmittance (83.6%), and superior mechanical flexibility [41]. Other novel polymer substrates were also investigated. For instance, Park et al. utilized a shape recoverable polymer, noland optical adhesive 63 (NOA 63) as a flexible substrate. The fabricated FPSC yielded a PCE over 10% and retained 90% of its initial efficiency after 1000 bending cycles at $B_r = 1$ mm [19].

15.4.3.2   Metal substrates

As opposed to the polymer substrates, metal substrates have higher heat tolerance, i.e., they can withstand high-temperature processing. Furthermore, metal substrates have higher conductivity, moderate flexibility, and lower sheet resistance. Titanium (Ti) foils are being widely used due to their high strength-to-weight ratio and outstanding resilience toward corrosion. Also, Ti foils are helpful to develop TiO$_2$ film as TiO$_2$ being the fundamental ETL. In metal substrate-based FPSC, the counter electrode (or top electrode) should be transparent since metal substrates are usually opaque. Han et al. employed Ti foils as flexible substrate with an ultrathin gold (Au)/copper (Cu) metal film as a transparent counter electrode (or transparent conducting layer). The configuration of fabricated FPSC was Au/Cu/Spiro-OMeTAD/MAPbI$_3$/TiO$_2$/Ti, which yielded a PCE of 14.9%. Furthermore, it retained PCE even after 1000 bending cycles at $B_r = 4$ mm [42].

In addition, Cu foils are also employed as flexible substrates in FPSC. Nejand et al. employed Cu foil as a flexible substrate and silver (Ag) nanowires as a transparent conducting layer. The configuration of fabricated FPSC was (Cu foil/CuI/MAPbI$_3$/ZnO/Ag-nanowires), which yielded a PCE of 12.8% [25].

15.4.3.3   Fiber shaped PSCs

As compared to conventional planar structures, fiber-shaped PSCs are one-dimensional linear structures in the form of coaxial fiber. The novel fiber-shaped PSCs can be employed in wearable electronics and electronics textiles owing to their weavability, high flexibility, and lightweight nature. In fiber-shaped PSCs, all the layers are deposited coaxially on metal or conductive carbon fiber. Furthermore, its distinguishing one-dimensional structures enable uninterrupted power output, which is independent of a solar incidence angle. Qiu et al. reported the first-ever fiber-shaped perovskite solar cell by depositing various layers coaxially over stainless-steel fiber (anode) via the dip-coating method. The layers were deposited in the following sequence, from the inner core to outer shell, compact TiO$_2$ (c-TiO$_2$) layer, mesoporous TiO$_2$ layer, MAPbI$_3$ absorber layer, Spiro-OMeTAD (HTL), and transparent carbon nanotube (CNT) as a cathode. The fabricated FPSC yielded a PCE of 3.3% and retained 95% of its initial PCE after 50 bending cycles [43].

15.4.3.4   Other flexible substrates

Apart from conventional polymer and metal substrates, other novel flexible substrates are succinctly described here. The willow glass or ultrathin flexible glass is a promising flexible substrate due to its compatibility with R2R processing, high optical transparency, malleability, mechanical robustness, hermeticity, and low-thermal expansion coefficient. Tavakoli et al. utilized willow glass to fabricate FPSC with configuration (willow glass/ITO/ZnO/MAPbI$_3$/Spiro-OMeTAD/Au). The fabricated FPSC realized a PCE of 12.06% and retained 96% of its initial PCE after 200 bending cycles with $B_r = 40$ mm [44]. Earth-abundant and eco-friendly mica is also a promising alternative to polymer and metal substrates due to its high heat tolerance, resilience toward moisture, and low thermal expansion coefficient. Additionally, paper is also a favorable substitute to nonbiodegradable polymer substrates owing to its cost-effectiveness, superior flexibility, biodegradability, and nontoxicity. However, the PCE of paper-based FPSC is still below 14%.

15.4.4 TRANSPARENT CONDUCTING LAYER

The role of the transparent conducting layer (TCL) is to collect electrons/holes from charge transport layers. Ideally, a TCL should possess the following crucial properties: high optical transparency, superior corrosion resistance, high conductivity, and mechanical robustness. ITO demonstrates its prevalence in PSCs and FPSCs due to its wide bandgap, high transmittance, and superior conductivity. The PET/ITO and PEN/ITO are widely implemented combinations for FPSCs. Recently, the highest certified PCE of 19.9% was realized by utilizing PEN/ITO flexible substrate. Despite the aforementioned advantages, ITO suffers from the following pitfalls: it is brittle, which leads to mechanical instability, and it comprises indium, which is costly and scanty. Therefore, other TCLs like PEDOT:PSS, Ag nanowires, graphene, etc. are being explored. PEDOT:PSS is widely used as TCL due to tunable conductivity, excellent transmittance, and good mechanical stability. Rao et al. fabricated $FASnI_3$ based flexible and wearable perovskite solar cells by utilizing highly conductive PEDOT:PSS as TCL and achieved a PCE of 8.56% [35]. Furthermore, metal nanostructures are also promising TCL for FPSC. For example, Ag nanowires (NW) demonstrates high optical transmittance, better electrical conductivity, and good mechanical flexibility. Nejand et al. fabricated $MAPbI_3$ based FPSC by using Ag-NW as TCL and achieved a PCE of 12.8% [25].

15.4.5 ENCAPSULATION

Encapsulation is necessary to improve the stability and lifetime of FPSCs because perovskite materials are prone to oxygen and humidity. The encapsulation techniques employed in rigid PSCs cannot be directly implemented in FPSCs. Therefore, research is being done to develop novel encapsulation materials and methods. An efficient encapsulant should possess the following qualities: high optical transmittance, low WVTR, superior strength and flexibility, adhesiveness, high chemical inertness, low oxygen transmission rate (OTR), and resistance to ultra-violet [9].

For encapsulation of FPSCs, several techniques are employed, including atomic-layer deposition (ALD), physical-vapor deposition (PVD), plasma-enhanced chemical vapor deposition (PECVD), etc. The encapsulation of FPSCs comprises two main methods, i.e., single-layer and multiple-layer encapsulation. The first method uses a single layer of organic/inorganic material for encapsulation of FPSCs. However, the encapsulant surface formed by this method is more prone to defects and pinholes, which expedites the penetration of moisture into the device. As a result, the performance and stability of FPSCs worsen. The second method utilizes two layers, one layer comprising of organic material and the other layer comprising of inorganic material. The resultant encapsulant surface is observed to be less prone to defects and hence better device stability. Some of the popular polymer encapsulant materials are ethylene vinyl acetate (EVA), polyvinyl butyral (PVB), polyisobutylene (PIB), UV-cured epoxy, thermoplastic polyurethane (TPU), etc. Table 15.4 summarizes the materials commonly used in the manufacturing of FPSCs.

15.5 RECYCLING OF FPSCS

Although the performance of FPSCs has been improving expeditiously, it still poses serious concern due to the presence of lead (Pb), which is highly toxic and soluble in water. The mainstream FPSCs contain toxicants like lead and other environmentally hazardous components, which lead to environmental pollution and also have deleterious effects on living bodies, which retain these noxious elements long term. Furthermore, the costly TCLs (such as ITO, FTO, and Au) account for 15–20% of the total fabrication cost of FPSCs. It leads to significant consumption of scarce materials, such as indium and gold. The onetime usage of the aforementioned materials leads to a substantial increase in the fabrication cost of FPSCs, which limits its pragmatic applications. Figure 15.2 illustrates the recycling process for FPSCs. Consequently, it is essential to recycle critical components of FPSCs to minimize consumption

TABLE 15.4

Summary of Common Materials Generally Used for Manufacturing FPSCs

Layers	Material	Features
Flexible Substrates	PET	• Widely used substrate • Low $T_G \approx 105°C$ and hence unsuitable to high temperature processing • Resilient toward penetration of water and gases
	PEN	• Properties identical to PET • Higher $T_G \approx 125°C$ and dimensional stability as compared to PET
	PI	• $T_G > 200°C$ and hence higher heat tolerance than PET and PEN • It is inherently light brown in color and hence lower optical transmission
	Metal foils (Cu,Ti)	• Excellent mechanical stability • Higher heat tolerance, i.e., it can withstand high temperature processing • As it is opaque, the counter electrode or back contact should be transparent
	Willow glass or ultra-thin flexible glass	• Promising substitute to polymer substrates • Compatible with R2R processing • High transparency and malleable • Low thermal expansion coefficient
	Mica	• Low thermal expansion coefficient • Earth abundant and eco-friendly • Resilient toward moisture • High heat tolerance
	Paper	• Cost-effective • Superior flexibility • Nontoxic and biodegradable
	Fibre shaped	• High flexibility and lightweight • Weavable and hence able to be utilized in wearable electronics and electronic textiles
Transparent conducting layer	Transparent conducting oxides (FTO, ITO, AZO etc.)	• High optical transparency and good conductivity • Brittle in nature, i.e., after certain bending cycles, cracks are developed)
	Carbonaceous materials (graphene and CNT)	• Low fabrication cost • High optical transparency, flexibility and conductivity • Chemically stable
	Ag NW	• It is not prone to cracks • High optical transparency and conductivity • A barrier layer is required to prevent its reaction with iodide present in perovskite
ETL	TiO_2	• It is being extensively used due to its good energy band alignment with perovskite, high bandgap, and high optical transparency. • The main issues are low-bulk electron mobility, requires a high processing temperature, and UV instability.
	ZnO	• High electron mobility (≈ 200 cm^2/Vs) • Suitable ETL since it does not require high temperature processing

TABLE 15.4 (Continued)

Summary of Common Materials Generally Used for Manufacturing FPSCs

Layers	Material	Features
	SnO_2	• The main issues are: hydroxyl residues and hygroscopic nature, which degrades perovskite • It is a promising replacement of TiO_2, owing to its superior electron mobility (≈ 250 cm^2/Vs) and wide bandgap (3.8eV) • Better UV stability • Less hygroscopic in nature
	Zn_2SnO_4	• Decent electron mobility • Suitable energy-level matching • Good stability • However, FPSCs based on Zn_2SnO_4 demonstrates poor performance
	PCBM	• Better electron extraction • It does not require high temperature processing • Suitable energy-level alignment
	C_{60}	• Superior conductivity and electron mobility • Does not require high temperature processing • The main issue is parasitic absorption due to its narrow bandgap (≈ 1.7 eV). • It is utilized as bilayer with other metal oxide ETLs, owing to its defect passivation characteristics.
HTL	Spiro-OMeTAD	• It is a fundamental HTL material owing to its excellent hole extraction, compatible with solution processing and suitable energy-level matching. • However, it is costly and lacks long-term stability.
	PEDOT:PSS	• It is a promising replacement of Spiro-OMeTAD due to its low-temperature processing, low cost, excellent hole mobility and provides better mechanical stability • However, it is acidic and hygroscopic.
	NiO_x	• Better hole extraction and efficient electron blocking ability • Better energy band alignment with perovskite • Cost-effective
Perovskite absorber layer	$MAPbI_3$	• It is a fundamental perovskite absorber layer owing to its low defect density and high mobility. • However, it lacks long-term stability and lower absorption edge due to wide bandgap (≈ 1.55eV).
	$FAPbI_3$	• It is a promising substitute for $MAPbI_3$ due to better thermal stability, high absorption edge due to narrow bandgap (≈ 1.45eV), and lower defect density.
	Lead-free perovskite ($MASnI_3$, $FASnI_3$, etc.)	• Perovskites based on nontoxic elements, such as Sn, Ge and Bi, are being explored. However, they lack efficiency and long-term stability.
	Mixed cation and/or mixed halide perovskites	• To achieve desired bandgap and optoelectronic properties, A-site cations such as methylammonium (MA^+), formamidinium (FA^+), Caesium (Cs) are mixed with halides (Cl,Br,I) in a definite proportion. • It shows better stability.
Back contact	Metals (Ag,Au)	• High conductivity

(Continued)

TABLE 15.4 (Continued)
Summary of Common Materials Generally Used for Manufacturing FPSCs

Layers	Material	Features
		• However, the main issues are that they are expensive and require a barrier layer to prevent reaction between iodide present in perovskite and metal (Au), which leads to corrosion of back contact.
	Carbonaceous materials (graphene and its derivatives, CNTs)	• High conductivity and thermal stability • Low-cost and excellent mechanical stability • However, carbon-based FPSCs lacks efficiency
Encapsulant	EVA	• Good optical transmittance (91%), adhesiveness and flexibility • WVTR ≈ 15–30 g m^{-2} day^{-1} • Low cost • However, the main issues are less UV resistance and high curing temperature
	PVB	• High transparency • Decent heat tolerance • WVTR ≈ 30 g m^{-2} day^{-1}
	PIB	• Very low WVTR (< 0.001 g m^{-2} day^{-1}) • Lower cost than EVA • Low chemical reactivity with perovskite • It requires relatively low temperature processing and does not require UV curing.

of scarce materials, fabrication cost to enable large-scale applications, and environmental pollution. Yang and coworkers have demonstrated the recycling of TCL, toxic lead components, ETL, HTL, and metal electrodes [45].

15.6 CHALLENGES AND FUTURE PERSPECTIVES

Despite the remarkable advancements in the area of FPSCs, it still lacks pragmatic applications owing to below mentioned major drawbacks [9,10,12].

  i. Environmental stability
 ii. Mechanical stability
iii. High manufacturing cost
 iv. Large-area fabrication
  v. Toxicity

15.6.1 ENVIRONMENTAL STABILITY

Long-term environmental stability is a serious challenge that needs to be addressed for pragmatic applications. It refers to the degradation of different layers of FPSCs, attributed to the presence of humidity, air, UV radiation, and variation in the surrounding temperature. The widely used polymer substrates have very high WTVR and OTR as compared to glass substrates and hence much sensitivity toward moisture

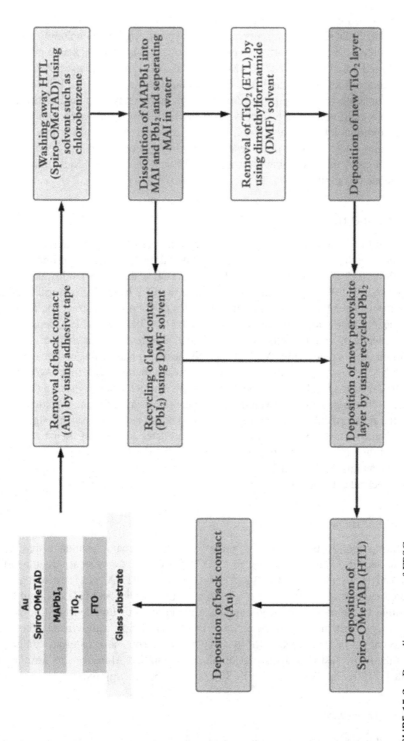

FIGURE 15.2 Recycling process of FPSCs.

and oxygen. This is a major cause for the lack of environmental stability of FPSCs. So, novel flexible substrates having low WVTR and OTR are being developed. Apart from polymer substrate issue, other various factors such as perovskite composition, selection of ETLs, HTLs and electrodes, additive engineering, encapsulation techniques, and interfacial engineering need to be addressed for improving the environmental stability of FPSCs. Therefore, several strategies were employed, including utilizing mixed cation and/or mixed halide perovskite absorber layer; using novel environmental stable materials such as naphthalene diimide based polymer in lieu of PCBM and metal oxide-based nanoparticles as ETLs; NiO_x nanoparticles, and graphene oxide (GO) as HTLs and carbon-based materials as TCL and metal back contact due to its hydrophobic nature; utilizing additives such as dimethyl sulfide and 1,8-diiodooctane (DIO) and H_2O (double additives), which improves crystallinity and forms larger grains and thereby reduces moisture penetration into the device and further develops an effective, low-cost and low-temperature encapsulation technique. Also, research is being conducted to develop novel environmentally stable encapsulants possessing ultra-low WVTR and OTR, high optical transparency, high compactness, and chemical stability.

15.6.2 MECHANICAL STABILITY

It refers to the ability of FPSCs to retain their performance under various physical deformations (such as twisting, bending, and stretching). Currently, the TCL and metal back contact are prime reasons for the low mechanical stability of FPSCs. The ITO is a widely used TCL with polymer substrates. However, in the case of ITO/PET and ITO/PEN, cracks are developed in the ITO layer after several bending cycles owing to the difference between the value of Young's modulus of polymer substrate and ITO. Therefore, to overcome this issue, the following strategies are employed: (1) utilizing suitable encapsulant materials having an optimal thickness, and (2) utilizing non-ITO electrodes such as indium oxide (In_2O_3), PEDOT:PSS, aluminum-doped zinc oxide (AZO), Cu or Ni mesh, Ag nanowires, carbonaceous materials (such as graphene, CNT), etc. Although, these materials showcase better mechanical stability than ITO, those non-ITO electrodes based FPSCs show poor PCE. In this respect, research is being done to elevate PCE of non-ITO electrodes based FPSCs, which further enables large-scale manufacturing.

Now, regarding the replacement of conventional metal back contact (such as Cu, Ag, Au, Al, etc.), graphene and its derivatives have emerged as promising candidates because of their better mechanical and environmental stability and suitable work function (~5.1eV). However, its conductivity is observed to be less than that of metals. Hence, an investigation is being conducted to overcome this issue and fabricate all-carbon electrodes based FPSCs.

15.6.3 HIGH MANUFACTURING COST

The total manufacturing cost, to be precise, the material cost, is a major challenge that needs to be addressed for the successful commercialization of FPSCs. Currently, for normal structure FPSCs, expensive electrodes and HTLs (such as Au, Ag, Spiro-OMeTAD, PTAA, etc.) are being extensively utilized, which increases the manufacturing cost. So, low-cost electrodes and HTLs should be explored as a possible substitute to the aforementioned expensive materials. For inverted FPSCs, expensive organic ETLs, such as C_{60}, derivatives of fullerene, and PCBM are being widely used. Therefore, other low-cost novel ETLs should be investigated; for instance, carbonaceous nanomaterials can be promising replacement due to their low cost, abundantly available, and stability. Moreover, to further reduce the manufacturing cost, HTL-free and/or ETL-free FPSCs are also being explored.

15.6.4 LARGE-AREA FABRICATION

A high PCE is realized under a small fabrication area (~0.1 cm^2). With the increasing fabrication area, the PCE drops significantly, mainly for the following reasons: (1) increase in sheet resistance (R_{sh})

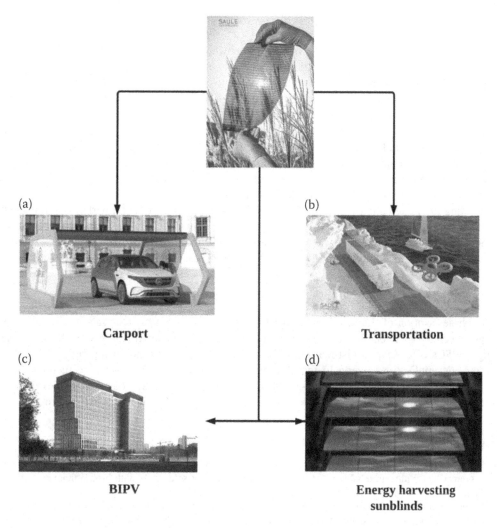

FIGURE 15.3 Applications of FPSCs (Adapted with permission from ref. [46], Saule Technnologies).

offered by TCL (also referred to as ohmic loss), (2) increase of pinholes/defects in the perovskite absorber layer, (3) the uniformity and compactness of the ETL, HTL, and perovskite absorber layer are compromised as its fabrication processes are solution-based. Therefore, to counter the aforementioned issues, several strategies have been proposed, including to serially connect individual FPSC, develop novel fabrication processes, such as vacuum flashing, gas-assisted nucleation, and solvent engineering, which enables large scale production of FPSCs.

15.6.5 TOXICITY

Even though the performance of FPSCs has been improving rapidly, they still pose a serious threat to the environment due to the presence of lead (Pb), which is highly toxic and soluble in water. The high PCE FPSCs are generally lead-based, which is not eco-friendly. Therefore, lead-free FPSCs such as germanium (Ge), tin (Sn), and bismuth (Bi) based are being investigated. However, these lead-free FPSCs have low stability and efficiency. So, the perovskite layer comprising a combination of various nontoxic elements is being developed.

15.7   APPLICATIONS OF FPSCS

FPSCs are mechanically robust, ultrathin, have a low power-to-weight ratio, and are easily transportable and lightweight. With the increasing per capita energy consumption, these features enable pragmatic applications of FPSCs in the modern world. FPSCs have a plethora of applications owing to their distinctive feature of flexibility, which cannot be realized with rigid PSC. Some of the noteworthy applications are shown in Figure 15.3 [46].

- Carport: It is a photovoltaic integrated self-sufficient charging station that does not require any external power supply.
- Transportation: Solar cells are integrated into the vehicle's body to reduce carbon emission, for instance, in solar-powered yachts, solar-powered trucks, etc.
- Building-integrated photovoltaics (BIPV): FPSC modules are an integral part of building elements, such as facade and windows.
- Energy-harvesting sunblinds: The energy harvesting sunblinds are equipped with FPSC modules. It can be operated manually or automatically to control room temperature and optimize energy gain.

15.8   CONCLUSION

A thorough description of the development of FPSCs in the context of materials used, manufacturing techniques, technological challenges, future prospects, and real-time applications of FPSCs is presented. Although FPSCs offer several distinct applications that cannot be realized using rigid PSCs, they still lack pragmatic applications owing to the aforementioned challenges. Moreover, to increase their versatility, semi-transparent FPSCs are being developed. Considering the burgeoning development of perovskite PV, it is believed that FPSCs will demonstrate versatile applications in the future PV market.

REFERENCES

1. El-Atab, Nazek, Nadeem Qaiser, Rabab Bahabry, and Muhammad M. Hussain. 2019. "Corrugation enabled asymmetrically ultrastretchable (95%) monocrystalline silicon solar cells with high efficiency (19%)." *Adv. Energy Mater.* 9 (45): 1902883. doi:10.1002/aenm.201902883.
2. Ramanujam, Jeyakumar, Douglas M. Bishop, Teodor K. Todorov, Oki Gunawan, Jatin Rath, Reza Nekovei, Elisa Artegiani, and Alessandro Romeo. 2020. "Flexible CIGS, CdTe and a-Si:H based thin film solar cells: A review." *Progress Mater. Sci.* 110: 100619. doi:10.1016/j.pmatsci.2019.100619.
3. Green, Martin A., Keith Emery, Yoshihiro Hishikawa, Wilhelm Warta, and Ewan D. Dunlop. 2015. "Solar cell efficiency tables (version 46)." *Prog. Photovolt: Res. Appl.* 23 (7): 805–812. doi:10.1002/pip.2637.
4. Chung, Jaehoon, Seong S. Shin, Kyeongil Hwang, Geunjin Kim, Ki W. Kim, Seul Da Lee, Wansun Kim et al. 2020. "Record-efficiency flexible perovskite solar cell and module enabled by a porous-planar structure as an electron transport layer." *Energy Environ. Sci.* 13 (12): 4854–4861. doi:10.1039/D0EE02164D.
5. Hu, Long, Qian Zhao, Shujuan Huang, Jianghui Zheng, Xinwei Guan, Robert Patterson, Jiyun Kim et al. 2021. "Flexible and efficient perovskite quantum dot solar cells via hybrid interfacial architecture." *Nature Commun.* 12 (1): 466. doi:10.1038/s41467-020-20749-1.
6. Prasad, Saradh, D. Devaraj, Rajender Boddula, Sunitha Salla, and Mohamad S. AlSalhi. 2019. "Fabrication, device performance, and MPPT for flexible dye-sensitized solar panel based on gel-polymer phthaloylchitosan based electrolyte and nanocluster CoS_2 counter electrode." *Mater Sci Energy Technol* 2 (2): 319–328. doi:10.1016/j.mset.2019.02.004.
7. Chen, Xiaobin, Guiying Xu, Guang Zeng, Hongwei Gu, Haiyang Chen, Haitao Xu, Huifeng Yao, Yaowen Li, Jianhui Hou, and Yongfang Li. 2020. "Realizing ultrahigh mechanical flexibility and 15% efficiency of flexible organic solar cells via a 'welding' flexible transparent electrode." *Advanced Materials* 32 (14): 1908478. doi:10.1002/adma.201908478.
8. Mujahid, Muhammad, Chen Chen, Wei Hu, Zhao-Kui Wang, and Yu Duan. 2020. "Progress of high-throughput and low-cost flexible perovskite solar cells." *Sol. RRL* 4 (8): 1900556. doi:10.1002/solr.201900556.

9. Zhang, Jing, Wei Zhang, Hui-Ming Cheng, and S. R. P. Silva. 2020. "Critical review of recent progress of flexible perovskite solar cells." *Materials Today* 39: 66–88. doi:10.1016/j.mattod.2020.05.002.

10. Mishra, Snehangshu, Subrata Ghosh, and Trilok Singh. 2021. "Progress in materials development for flexible perovskite solar cells and future prospects." *ChemSusChem* 14 (2): 512–538. doi:10.1002/cssc.202002095.

11. Kumar, Mulmudi H., Natalia Yantara, Sabba Dharani, Michael Graetzel, Subodh Mhaisalkar, Pablo P. Boix, and Nripan Mathews. 2013. "Flexible, low-temperature, solution processed zno-based perovskite solid state solar cells." *Chem Commun (Cambridge, England)* 49 (94): 11089–11091. doi:10.1039/c3cc46534a.

12. Zeng, Peng, Wenbin Deng, and Mingzhen Liu. 2020. "Recent advances of device components toward efficient flexible perovskite solar cells." *Sol. RRL* 4 (3): 1900485. doi:10.1002/solr.201900485.

13. Liu, Dianyi, and Timothy L. Kelly. 2014. "Perovskite solar cells with a planar heterojunction structure prepared using room-temperature solution processing techniques." *Nature Photon* 8 (2): 133–138. doi:10.1038/nphoton.2013.342.

14. Kim, Byeong J., Dong H. Kim, Yoo-Yong Lee, Hee-Won Shin, Gill S. Han, Jung S. Hong, Khalid Mahmood et al. 2015. "Highly efficient and bending durable perovskite solar cells: toward a wearable power source." *Energy Environ. Sci.* 8 (3): 916–921. doi:10.1039/c4ee02441a.

15. Shin, Seong S., Woon S. Yang, Jun H. Noh, Jae H. Suk, Nam J. Jeon, Jong H. Park, Ju S. Kim, Won M. Seong, and Sang I. Seok. 2015. "High-performance flexible perovskite solar cells exploiting Zn_2SnO_4 prepared in solution below 100 °C." *Nature Commun.* 6: 7410. doi:10.1038/ncomms8410.

16. Yang, Dong, Ruixia Yang, Xiaodong Ren, Xuejie Zhu, Zhou Yang, Can Li, and Shengzhong F. Liu. 2016. "Hysteresis-suppressed high-efficiency flexible perovskite solar cells using solid-state ionic-liquids for effective electron transport." *Advanced Mater.* 28 (26): 5206–5213. doi:10.1002/adma.201600446.

17. Feng, Jiangshan, Xuejie Zhu, Zhou Yang, Xiaorong Zhang, Jinzhi Niu, Ziyu Wang, Shengnan Zuo, Shashank Priya, Shengzhong F. Liu, and Dong Yang. 2018. "Record efficiency stable flexible perovskite solar cell using effective additive assistant strategy." *Advanced materials* 30 (35): e1801418. doi:10.1002/adma.201801418.

18. Yang, Zhibin, Chu-Chen Chueh, Fan Zuo, Jong H. Kim, Po-Wei Liang, and Alex K.-Y. Jen. 2015. "High-performance fully printable perovskite solar cells via blade-coating technique under the ambient condition." *Adv. Energy Mater.* 5 (13): 1500328. doi:10.1002/aenm.201500328.

19. Park, Minwoo, Hae J. Kim, Inyoung Jeong, Jinwoo Lee, Hyungsuk Lee, Hae J. Son, Dae-Eun Kim, and Min J. Ko. 2015. "Mechanically recoverable and highly efficient perovskite solar cells: Investigation of intrinsic flexibility of organic-inorganic perovskite." *Adv. Energy Mater.* 5 (22): 1501406. doi:10.1002/aenm.201501406.

20. Pandey, Manish, Gaurav Kapil, Kazuhiko Sakamoto, Daisuke Hirotani, Muhammad A. Kamrudin, Zhen Wang, Kengo Hamada et al. 2019. "Efficient, hysteresis free, inverted planar flexible perovskite solar cells via perovskite engineering and stability in cylindrical encapsulation." *Sustain. Energy Fuels* 3 (7): 1739–1748. doi:10.1039/C9SE00153K.

21. Schackmar, Fabian, Helge Eggers, Markus Frericks, Bryce S. Richards, Uli Lemmer, Gerardo Hernandez-Sosa, and Ulrich W. Paetzold. 2021. "Perovskite solar cells with all-inkjet-printed absorber and charge transport layers." *Adv. Mater. Technol.* 6 (2): 2000271. doi:10.1002/admt.202000271.

22. Wang, Zhen, Linxiang Zeng, Cuiling Zhang, Yuanlin Lu, Shudi Qiu, Chuan Wang, Chong Liu et al. 2020. "Rational interface design and morphology control for blade-coating efficient flexible perovskite solar cells with a record fill factor of 81%." *Adv. Funct. Mater.* 30 (32): 2001240. doi:10.1002/adfm.202001240.

23. Yang, Zhichun, Shasha Zhang, Lingbo Li, and Wei Chen. 2017. "Research progress on large-area perovskite thin films and solar modules." *J. Materiom.* 3 (4): 231–244. doi:10.1016/j.jmat.2017.09.002.

24. Cao, Bingbing, Longkai Yang, Shusen Jiang, Hong Lin, Ning Wang, and Xin Li. 2019. "Flexible quintuple cation perovskite solar cells with high efficiency." *J. Mater. Chem. A* 7 (9): 4960–4970. doi:10.1039/C8TA11945G.

25. Abdollahi Nejand, B., P. Nazari, S. Gharibzadeh, V. Ahmadi, and A. Moshaii. 2017. "All-inorganic large-area low-cost and durable flexible perovskite solar cells using copper foil as a substrate." *Chem. Commun.* 53 (4): 747–750. doi:10.1039/C6CC07573H.

26. Liu, Mingzhen, Michael B. Johnston, and Henry J. Snaith. 2013. "Efficient planar heterojunction perovskite solar cells by vapour deposition." *Nature* 501 (7467): 395–398. doi:10.1038/nature12509.

27. Xu, Ming, Jing Feng, Zhao-Jun Fan, Xia-Li Ou, Zhen-Yu Zhang, Hai-Yu Wang, and Hong-Bo Sun. 2017. "Flexible perovskite solar cells with ultrathin Au anode and vapour-deposited perovskite film." *Solar Energy Mater. Solar Cells* 169: 8–12. doi:10.1016/j.solmat.2017.04.039.

28. Li, Kunpeng, Junyan Xiao, Xinxin Yu, Tianhui Li, Da Xiao, Jiang He, Peng Zhou et al. 2018. "An efficient, flexible perovskite solar module exceeding 8% prepared with an ultrafast PbI2 deposition rate." *Scien. Rep.* 8 (1): 442. doi:10.1038/s41598-017-18970-y.

29. Basaran, Osman A., Haijing Gao, and Pradeep P. Bhat. 2013. "Nonstandard inkjets." *Annu. Rev. Fluid Mech.* 45 (1): 85–113. doi:10.1146/annurev-fluid-120710-101148.

30. Das, Sanjib, Bin Yang, Gong Gu, Pooran C. Joshi, Ilia N. Ivanov, Christopher M. Rouleau, Tolga Aytug, David B. Geohegan, and Kai Xiao. 2015. "High-performance flexible perovskite solar cells by using a combination of ultrasonic spray-coating and low thermal budget photonic curing." *ACS Photonics* 2 (6): 680–686. doi:10.1021/acsphotonics.5b00119.

31. Schmidt, Thomas M., Thue T. Larsen-Olsen, Jon E. Carlé, Dechan Angmo, and Frederik C. Krebs. 2015. "Upscaling of perovskite solar cells: fully ambient roll processing of flexible perovskite solar cells with printed back electrodes." *Adv. Energy Mater.* 5 (15): 1500569. doi:10.1002/aenm.201500569.

32. Hwang, Kyeongil, Yen-Sook Jung, Youn-Jung Heo, Fiona H. Scholes, Scott E. Watkins, Jegadesan Subbiah, David J. Jones, Dong-Yu Kim, and Doojin Vak. 2015. "Toward large scale roll-to-roll production of fully printed perovskite solar cells." *Advanced Mater.* 27 (7): 1241–1247. doi:10.1002/adma.201404598.

33. Remeika, Mikas, Luis K. Ono, Maki Maeda, Zhanhao Hu, and Yabing Qi. 2018. "High-throughput surface preparation for flexible slot die coated perovskite solar cells." *Organic Electronics* 54: 72–79. doi:10.1016/j.orgel.2017.12.027.

34. Wu, Cuncun, Duo Wang, Yuqing Zhang, Feidan Gu, Ganghong Liu, Ning Zhu, Wei Luo et al. 2019. "FAPbI$_3$ flexible solar cells with a record efficiency of 19.38% fabricated in air via ligand and additive synergetic process." *Adv. Funct. Mater.* 29 (34): 1902974. doi:10.1002/adfm.201902974.

35. Rao, Li, Xiangchuan Meng, Shuqin Xiao, Zhi Xing, Qingxia Fu, Hongyu Wang, Chenxiang Gong et al. 2021. "Wearable tin-based perovskite solar cells achieved by a crystallographic size-effect." *Angewandte Chemie (International ed. in English)*. doi:10.1002/anie.202104201.

36. Docampo, Pablo, James M. Ball, Mariam Darwich, Giles E. Eperon, and Henry J. Snaith. 2013. "Efficient organometal trihalide perovskite planar-heterojunction solar cells on flexible polymer substrates." *Nature Commun.* 4: 2761. doi:10.1038/ncomms3761.

37. Wang, Ping-Cheng, Venkatesan Govindan, Chien-Hung Chiang, and Chun-Guey Wu. 2020. "Room-temperature -processed fullerene/TiO$_2$ nanocomposite electron transporting layer for high-efficiency rigid and flexible planar perovskite solar cells." *Sol. RRL* 4 (10): 2000247. doi:10.1002/solr.202000247.

38. Heo, Jin H., Min H. Lee, Hye J. Han, Basavaraj R. Patil, Jae S. Yu, and Sang H. Im. 2016. "Highly efficient low temperature solution processable planar type CH 3 NH 3 PbI 3 perovskite flexible solar cells." *J. Mater. Chem. A* 4 (5): 1572–1578. doi:10.1039/C5TA09520D.

39. Meng, Xiangchuan, Zheren Cai, Yanyan Zhang, Xiaotian Hu, Zhi Xing, Zengqi Huang, Zhandong Huang et al. 2020. "Bio-inspired vertebral design for scalable and flexible perovskite solar cells." *Nature Commun.* 11 (1): 3016. doi:10.1038/s41467-020-16831-3.

40. Ma, Shuang, Xuepeng Liu, Yunzhao Wu, Ye Tao, Yong Ding, Molang Cai, Songyuan Dai, Xiaoyan Liu, Ahmed Alsaedi, and Tasawar Hayat. 2020. "Efficient and flexible solar cells with improved stability through incorporation of a multifunctional small molecule at PEDOT:PSS/perovskite interface." *Solar Energy Mater. Solar Cells* 208: 110379. doi:10.1016/j.solmat.2019.110379.

41. Park, Jeong-Il, Jin H. Heo, Sung-Hyun Park, Ki I. Hong, Hak G. Jeong, Sang H. Im, and Han-Ki Kim. 2017. "Highly flexible InSnO electrodes on thin colourless polyimide substrate for high-performance flexible CH3NH3PbI3 perovskite solar cells." *J. Power Sources* 341: 340–347. doi:10.1016/j.jpowsour.2016.12.026.

42. Han, Gill S., Seongha Lee, Matthew L. Duff, Fen Qin, and Jung-Kun Lee. 2018. "Highly bendable flexible perovskite solar cells on a nanoscale surface oxide layer of titanium metal plates." *ACS App. Mater. Interfaces* 10 (5): 4697–4704. doi:10.1021/acsami.7b16499.

43. Qiu, Longbin, Jue Deng, Xin Lu, Zhibin Yang, and Huisheng Peng. 2014. "Integrating perovskite solar cells into a flexible fiber." *Angew. Chem.* 126 (39): 10593–10596. doi:10.1002/ange.201404973.

44. Tavakoli, Mohammad M., Kwong-Hoi Tsui, Qianpeng Zhang, Jin He, Yan Yao, Dongdong Li, and Zhiyong Fan. 2015. "Highly efficient flexible perovskite solar cells with antireflection and self-cleaning nanostructures." *ACS Nano.* 9 (10): 10287–10295. doi:10.1021/acsnano.5b04284.

45. Yang, Fengjiu, Shenghao Wang, Pengfei Dai, Luyang Chen, Atushi Wakamiya, and Kazunari Matsuda. 2021. "Progress in recycling organic–inorganic perovskite solar cells for eco-friendly fabrication." *J. Mater. Chem. A* 9 (5): 2612–2627. doi:10.1039/D0TA07495K.

46. https://sauletech.com/.

16 Quantum Dots Based Flexible Solar Cells

Sandeep Kumar and Prashant Kumar
Department of Energy Engineering, Central University of Jharkhand,
Brambe, Ranchi, Jharkhand, India

Arup Mahapatra and Basudev Pradhan
Department of Energy Engineering, Central University of Jharkhand,
Brambe, Ranchi, Jharkhand, India
Centre of Excellence (CoE) in Green and Efficient Energy Technology (GEET),
Central University of Jharkhand, Brambe, Ranchi, Jharkhand, India

CONTENTS

16.1   Introduction... 285
16.2   Theoretical background of QDs... 288
    16.2.1   Quantum size effect 288
    16.2.2   Multiple exciton generation 290
    16.2.3   Ultrafast charge transfer... 291
16.3   Synthesis and characterization of QDs... 292
    16.3.1   Colloidal synthesis 292
    16.3.2   Surface engineering... 295
16.4   QDs based flexible heterojunction solar cell.. 295
16.5   QD based flexible sensitized solar cells.. 298
16.6   QDs based flexible perovskite solar cells 299
16.7   Flexible QD-silicon hybrid solar cells.. 300
16.7   Conclusion 302
References... 303

16.1   INTRODUCTION

Flexible quantum dots (QDs) solar cells have attracted pronounced research attention due to their competitive pricing, solution processability, and flexible substrate compatibility. Even so, the photovoltaic market is brimming with solar cells that are dominantly silicon wafer-based; besides, they are also bulky and rigid. Moreover, the requirement of the energy-intensive manufacturing process involved in wafer-based technology, such as Czochralski and float-zone methods starting from sand to solar-grade silicon, increases the overall cost of silicon-based solar modules and makes it less attractive on the marketing front. With the intention of lowering the tariff, as well as adding flexibility, amorphous silicon-based cells were introduced. The lowering of cost and the addition of flexibility both were achieved to some extent, but the price reduction was not enough to improve its market acceptability. On the other hand, the flexibility had a brittle nature, which prevents its application for next-generation devices. This rationale drove the leading researchers in the field of solar photovoltaics to work toward the third-generation solar cells established on thin-film technology.

DOI: 10.1201/9781003186755-16

The properties of flexibility and lightweight solar photovoltaic devices have always been intriguing and desirable. The colloidal QDs solar cells have a distinct advantage in this direction owing to their unique nanocrystal character and due to quantum confinement. This enables QDs solar cells to convert energy, even in the situation of extreme deformation. The QDs also have an edge over other techniques due to the dependence of optoelectronic properties on the size of QDs, solution processability, and broad light absorption covering the whole solar spectral range [1,2]. Furthermore, some other exciting phenomena, such as possibilities of multiple exciton generation (MEG), might further push the efficacies of solar cell devices beyond the Shockley–Quessier efficiency, which is limited to 32.9% for single absorber solar cells [3]. These fascinating properties have led to various developments entailing stretchable and foldable optoelectronic devices, roll-to-roll (R2R) printing processes, and self-healing material strategies. Over the few last decades, QDs solar cells have received huge attention in the third-generation photovoltaic research domain. Low-temperature synthesis and fabrication involved in device fabrication enable it to be deposited over flexible polyethylene terephthalate (PET) substrates or other similar flexible transparent conducting surfaces. The QDs fabrication also incorporates capping agents, which act as a protective layer strengthening its air stability. Innovative and simple routes of fabrication, such as solvent engineering, have helped achieve highly efficient QDs thin films. Moreover, the utilization of solution-processing technology in manufacturing makes it one of the finest candidates for use in flexible electronic devices since these can be printed over conducting flexible substrates of large sheets using continuous R2R manufacturing approaches with spraying or printing techniques. These distinguishing qualities convey higher prospects of very high through-put solar cell production and reduced expenditure in solar cell fabrication cost. The QDs film manifests high mechanical stability, as exhibited by the ability to bend or stretch multiple times; this is one of the key features for flexible optoelectronics devices, by the virtue of the small size of the QDs and tunable inter QDs distance without distorting QDs crystal structure. The improvised performance in QDs solar cells was enhanced mainly by the advancement of QDs surface chemistry and energy-band engineering, and accompanied by optimization of the device structure over the last few years. In a solar cell, several semiconducting QDs have been probed, including cadmium sulfide (CdS), cadmium telluride (CdTe), indium phosphide (InP), cadmium selenide (CdSe), zinc selenide (ZnSe), zinc sulfide (ZnS), lead(II) sulfide (PbS), lead selenide(PbSe). These include core-shell types, such as CdSe/ZnS, ZnSe/ZnS, CdSe/CdS/ZnS core-shell–shell QDs [4], and most recently, metal halide perovskites, such as $CsPbX_3$QDs, have joined the game. In a decade of extensive research, the power-conversion efficiency(PCE) of PbS based QDs solar cells plunged to 14% and 16.6% for perovskite QDs solar cells in traditional device structures on a glass substrate, as reported by Kim et al. [5] and Hao et al. [6], respectively. On account of these advancements, the researchers commenced making QDs based solar cells on a flexible substrate. The first flexible QDs based solar cell was demonstrated by Kramer et al.; it had a PCE of 7.2% by spray coating PbS QDs on PET substrate [7]. Thereafter, it attracted the scientific community all over the world and vast research was carried out. Table 16.1 shows the summary of the development of flexible QDs based solar cell and their performance to date. Zang et al. has reported the highest PCE of 9.9% using PbS QDs on silver nanowire-coated flexible substrate [8]. On the other hand, very recently, Hu et al. achieved the PCE of 12.3% using cesium lead halide perovskite ($CsPbI_3$) QDs on flexible substrate showed quite impressive open-circuit voltage (V_{OC}) of 1.24 V [9].

The requirement of high open-circuit voltage is vital to power the Internet of hings (IoT) based low-powered wearable devices. For many wearable electronic applications, flexibility and durability, together with the power output per weight (specific power of unit watts/gram), are more crucial compared with the energy conversion per unit area. The specific power output of various lightweight solar cells has been summarized in Table 16.2. Traditional ultrathin silicon or copper indium gallium selenide (CIGS) based solar cells have a power-per-weight output of <3.5 Wg^{-1}. Considering the fabrication of lightweight devices, it is always advisable to use a thin layer of absorber material having a strong absorption

TABLE 16.1

Summary of Photovoltaic Performance of Flexible Quantum Dot Solar Cells

Device architecture	V_{OC} (V)	J_{SC} (mA/cm^2)	FF	PCE (%)	Ref.
PET/ITO/ZnO/PbS–MPA/P3HT/Au	0.56	12.1	0.50	3.4	[10]
PET/TiO$_2$/CdS/CdSe sensitizer with Cu$_2$S CE	0.58	9.28	0.65	3.49	[11]
PEN/ITO/TiO$_2$/CdSe sensitizer with Cu$_2$S/Ni CE	0.581	9.95	0.61	3.55	[12]
Ti/ZnO/ZnSe/CdSe sensitizer with carbon/Ti CE	0.638	15.32	0.52	5.08	[13]
PET/SWCNTs/PDOT:PSS/CQDs/PCBM/Ag	0.55	18.7	0.55	5.6	[14]
PET/MoO$_3$/Au/MoO$_3$/PbS – MPA/ZnO/Al	0.59	19.5	0.59	6.8	[15]
PET/ITO/PbS–MPA/MoO$_3$/Au	0.59	22.9	0.54	7.2	[7]
PET/ITO/MZO/PbS–TBAI/PbS–EDT/Au	0.62	24.5	0.62	9.4	[16]
PEN/AgNWs/AZO/PbS-PbX$_2$/PbS–EDT/Au	0.64	24.6	0.63	9.9	[8]
PET/ITO/SnO$_2$/ CsPbI$_3$QD:PCBM/QDs/ PTB7/MoO$_3$/Ag	1.23	13.6	0.74	12.3	[9]
Ultrathin Silicon/Cd$_{0.5}$Zn$_{0.5}$S/ZnS core-shell QDs	0.495	33.05	0.76	12.37	[17]

TABLE 16.2

Photovoltaic Parameters and Specific Power (Watt/gram) of Different Lightweight Solar Cells

Solar cells	V_{OC} (V)	J_{SC} (mA/cm^2)	FF	PCE (%)	Specific power (Wg^{-1})	Ref.
α–Si/ nc Si	1.26	10.04	0.64	8.12	0.13	[18]
CdTe	0.765	20.9	0.71	11.4	2	[19]
CIGS	0.712	34.8	0.76	18.7	3.04–3.12	[20]
Mono c–Si	0.696	42.0	0.84	24.4	0.39	[21]
Poly c–Si	0.654	38.1	0.79	19.8	0.32	[22]
InP	0.62	29.6	0.55	10.2	2	[23]
Organic	0.58	11.9	0.61	4	10	[24]
Perovskite	0.926	17.6	0.73	11.9	22.88	[25]
Quantum Dot	0.64	24.6	0.63	9.9	15.2	[8]

coefficient in the broadband spectrum. In general, QDs based solar cells are very lightweight because they require less material to achieve similar kinds of efficiency due to the high absorption coefficient of QDs materials. For instance, in PbS QDs, which depict strong absorption needs about 350–450 nm thick layer of solid QDs thin film for lightweight solar cell fabrication, which shows a specific power output of 15.2 Wg^{-1} Another important issue is the weight of different contact electrodes used in the solar cell fabrication process; they need to be accounted for in weight reduction. Zang et al. has used highly conducting and transparent Ag NW network electrodes in comparison with the ultrathin metal mesh film and carbon electrode to reduce the weight [8]. Similar performance of the QDs solar cell fabricated on the different substrates, including glass and flexible polyethylene naphthalate (PEN) with the Ag NWs as a front electrode, also offers a great possibility for better device performance. This opens the door to the selection of substrate, according to our choice of application without compromising the high power output of the devices. In this chapter, the detailed review of the development of high-efficiency flexible QDs solar cell has thoroughly been discussed, which has a significant possibility for future commercial production in near future.

16.2 THEORETICAL BACKGROUND OF QDS

16.2.1 QUANTUM SIZE EFFECT

Quantum dots are tiny forms of nanomaterials having one dimension in the nanometer range of the order 1–100 nm. The QDs often exhibit a very unique optoelectronic property, which is intermediate between their bulk counterparts, and discrete atoms or molecules, which attract a great deal of research interest in nanoelectronic devices. The optoelectronic properties of these QDs change as a function of both size and shape. In general, based on the degree of freedom of excitons (loosely bound electron-hole pair), materials exhibit different electronic properties. In the case of bulk semiconductors, electrons and holes (excitons) are free to move in all three directions, giving very close energy levels in a way that energy bands are formed. The highest occupied energy band maximum is called the valence band, whereas the lowest unoccupied band minima is the conduction band. The energy gap between the valence band maxima and the conduction band minima is called the bandgap of the material E_g(eV). An exciton is formed when photon energy is absorbed, giving a hole in the valence band and an electron in the conduction band. The separated electron and hole system, due to photon absorption, forms a hydrogen-like structure. The separation between electron and hole is called the Bohr's radius. In bulk semiconductor materials, the density of states is proportional to $(E - E_{c/v})^{1/2}$ (Figure 16.1a). When the charge carriers (i.e., electrons and holes) are only free to move in two dimensions restricting the movement in third dimension (two-dimensional system), this gives a thin-film form of the semiconductor material having a step-like density of states (Figure 16.1b). Nanowires are nanomaterials in which electrons and holes have one-dimensional confinement and have a density of states as a function of $(E - E_{c/v})^{-1/2}$ as shown in Figure 16.1c. Quantum dot is a zero-dimensional structure since charge carriers (electron and hole) are confined in all three dimensions, hence having an atomic-like density of states. The density of states for QDs is mathematically represented by delta function $\delta (E - E_{c/v})$, as illustrated in Figure 16.1d.

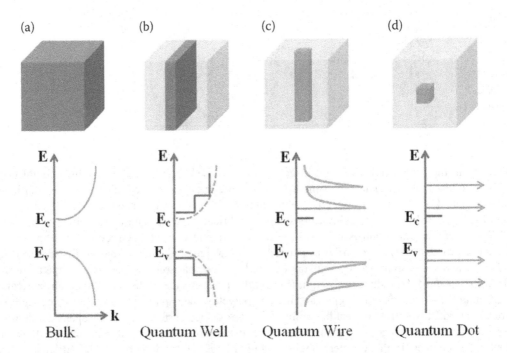

FIGURE 16.1 Schematic illustration of the density of states (DOS) in quantum-confined systems: (a) bulk, (b) quantum well, (c) quantum wire, and (d) quantum dot.

If any dimension of a nanomaterial is close to, or smaller than, the Bohr radius of the exciton, which is the natural length scale of an exciton in a bulk semiconductor, then the electronic levels and optical properties show size dependency. This phenomenon is known as the quantum confinement effect.

$$a_B = \varepsilon_r \left(\frac{m_e}{\mu^*} \right) a_0 \tag{16.1}$$

Where ε_r is the dielectric constant, m_e is the mass of a free electron, μ^* is the effective reduced mass of an exciton, and $a_0 = 0.053$ nm is the Bohr radius of hydrogen atoms (an exciton is an electron-hole pair bound by Coulomb force, which can be approximated by the *hydrogen atom model*). The quantum confinement effect originates from the size-dependent electronic structure of materials. In bulk semiconductor crystals, the electrons and holes can move in three dimensions and have three degrees of freedom in which the electronic structures can be solved by treating the semiconductor as an infinite crystal with a periodic lattice potential. For a small crystal, the boundary conditions for solving the Schrödinger equations have to be modified since the electron and hole wave-functions are "confined" by the physical size of the crystal, which deviates from the infinite crystal assumption. It can be shown that using the simplified *particle in a box* model, the ground-state energy in a small crystal is higher than that in the bulk. This energy difference is called the confinement energy $E_{confinement}$ and is inversely proportional to the square of the size L.

$$E_{confinement} = \frac{\hbar^2 \pi^2}{2\mu^* L^2} \tag{16.2}$$

In addition, the density of states (DOS) in quantum-confined systems takes different functional forms depending on the number of confined dimensions (Figure 16.1). The DOS in bulk materials is proportional to $E^{1/2}$. In 2-D *quantum well* structures, electrons and holes have two degrees of freedom (1-D confinement) and the DOS is independent of energy. Materials with 2-D confinement are called *quantum wires,* and their DOS is proportional to $E^{1/2}$, whereas materials with 3-D confinement are called *quantum dots* whose DOS are delta functions in energy. In reality, the DOS of quantum-confined materials are further modified by the homogeneity, lattice potential, and the shape of the nanocrystal. Overall, the observed bandgap or the band-edge absorption energy in a quantum-confined system can be described by Eq. (16.1–16.3):

$$E_{band\ edge} \approx E_{g,bulk} + \frac{\hbar^2 \pi^2}{2\mu^* L^2} - E_X \tag{16.3}$$

Where $E_{g,bulk}$ is bandgap energy of the bulk material, and E_x is the exciton binding energy (E_x is also a function of the size, and the size-dependent on dielectric constant). The significant quantum-confinement effect in any material can be observed when the size of the material is comparable to the exciton Bohr radius, which depends on material-specific properties as indicated by Eq (16.1) and is on the order of 10 nm. For example, the exciton Bohr radii are approximately 1 nm for Copper(I) chloride (CuCl), 6 nm for CdSe, 20 nm for PbS, and 54 nm for Indium antimonide (InSb). Within this quantum-confinement range, the bandgap of a semiconductor nanocrystal increases as the size decreases. It should be noted that all nanomaterials are quantum-confined materials. For example, a 10 nm (in diameter) PbS crystal is considered as a quantum dot, while a 10 nm CuCl crystal is not a quantum dot and is termed as a nanocrystal or nanoparticle.

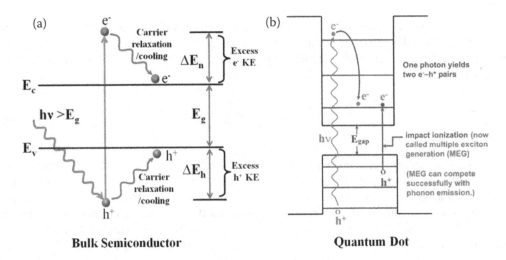

FIGURE 16.2 Thermalization of hot carriers in (a) bulk semiconductor, (b) quantum dots. Adapted with permission from ref. [27], Copyright (2002), Elsevier.

16.2.2 MULTIPLE EXCITON GENERATION

In a conventional solar cell, a single electron-hole pair, or exciton, is created due to the absorption of a single photon by the semiconductor. If the energy of the incident photon has higher energy than the bandgap, the excess energy of that photon is dissipated as heat energy through electron-phonon scattering, as shown in Figure 16.2a. But in the case of quantum dots, theoretically, more than one exciton can be generated due to single-photon absorption; this phenomenon is commonly known as multiple exciton generation (MEG). The MEGs can strengthen the power-conversion efficiency of QDs solar cells beyond the Shockley–Queisser limit in a single junction, which was thoroughly investigated by Nozik and his coworker [26]. In bulk semiconductors such as crystalline silicon, this phenomenon is not possible due to the continuous energy level. But, QDs can generate multiple excitons after collision with one photon of energy exceeding the bandgap, which is demonstrated in Figure 16.2b. The absorption of photons possessing energy more than bandgap in bulk semiconductors makes an electron jump from valance band to a higher level in the conduction band; these electrons holding higher energies are often designated as hot carriers. Since these hot carriers or excited electrons tend to attain lower energy levels, they often undergo several nonradiative relaxations, such as thermalization, multi-phonon emission before attaining the bottom energy levels of the conduction band. Even so, in a quantum dot, the hot carrier experiences an impact ionization process (carrier multiplication). The rate of electron relaxation via electron-phonon interaction in QDs is reduced due to the discrete nature of the e^-–h^+ spectra. At the same time, for a higher photogenerated current, rate of carrier cooling, electron transfer of a hot electron, and Auger recombination needs to be minimized. On the other hand, because of the carrier confinement, the rate of reverse Auger recombination process of exciton multiplication and e^-–h^+ Coulomb interaction is increased. The schematic presentation of MEG in QDs is shown in Figure 16.2b. Consequently, the absorption of a single photon in QDs gives rise to multiple electron-hole pairs. Note that the absorption of UV photons in QDs creates more electron-hole pairs compared to near-infrared photons.

The MEG phenomenon was experimentally reported in different QDs. It was observed that post-photonic absorption, the time taken by hot carriers for impact ionization is 0.1 ps(picosecond), after which it takes nearly 2 ps for the carriers to relax. Following, it holds nearly 20 ps duration for Auger recombination and at last 2 ps for cooling down and becoming available for new excitation. The phenomenon of carrier multiplication within quantum dots in which one photon can generate multiple excitons through impact ionization or reverse of Auger recombination was studied experimentally,

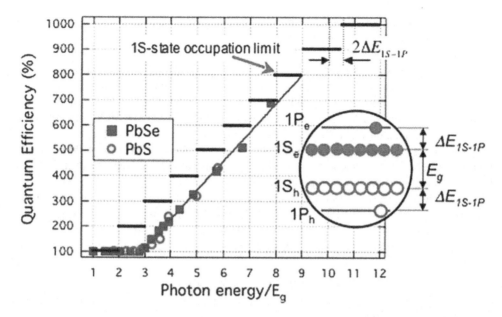

FIGURE 16.3 Experimental confirmations of quantum efficiency (QE) for photon-to-exciton conversion in PbSe and PbS nanocrystals [29]. The black bars represent the "ideal" QEs derived from energy conservation, whereas blue solid squares and red open circles indicate the experimental QEs measured for PbSe and PbS NCs, respectively, as a function of pump-photon energy normalized by an energy gap. Adapted with permission from ref. [29], Copyright (2006), American Chemical Society.

employing transient absorption spectroscopy. In this technique, laser pulses are aimed at the samples separately, but for both excitation and absorption. The measurement of relaxation time helps in stipulating whether radiative recombination in a single exciton is taking place or biexcitons recombination through the Auger recombination process. Studies have revealed that a linear proportionality exists between absorption changes for a number of generated electron-hole pairs, which is less than three. These insights are affirmed by the investigations in PbSe and PbS QDs, in PbSe QDs where 300% photo-generated carrier density(quantum yield) is found using a photon with four times energy to that of the energy spacing between highest occupied molecular orbital (HOMO)-lowest unoccupied molecular orbital (LUMO) (E_g) of the QD itself, as illustrated in the Figure 16.3 [28]. Furthermore, Schaller and his coworkers also reported carrier multiplication efficiencies of 700% of photogenerated carrier density by using photons eight times higher than the E_g of QDs [29]. As part of this study, it was observed that the onset of multiplication starts when photon energy reaches three times the multiple of the bandgap (E_g), as shown in Figure 16.3, which indicates complete filling of the 1S quantized state (eight electron-hole pairs). It is evident from the figure that as photon energy increases, the number of excitons (from one to seven) generation per single-photon increases as the amplitude of Auger-decay component enhancement. The slower QE growth for energy greater than nine times of bandgap is expected by populating 1P states of QDs, which require photon energy of $E_g + 2\Delta E_{1S-1P}$ as represented in the inset, where ΔE_{1S-1P} is the energy separation between the 1S and 1P states.

16.2.3   ULTRAFAST CHARGE TRANSFER

Nanostructured-based devices can harvest and exploit sunlight efficiently, depending upon the profound understanding of the fundamental microscopic principles that control the underlying photophysical conversion processes. This whole operation occurs on an extremely short femto-second time scale. The real-time investigation regarding the dynamics-of-charge transfer due to the application of light is very crucial. The QDs also have the advantage of ultrafast charge transfer due to the spectral tunability of QDs. The fast carrier transport is essential for better performance of solar cells, which arises due to wave function overlap between donor and acceptor materials in the bulk heterojunction device structure. A.J. Goodman et al. [30] has studied the samples having the fastest charge-transfer rates (≤50 fs) in CdSe QDs donor having heterojunction with tungsten sulfide (WS_2) layer, the results manifested astonishingly strong electronic coupling between QDs and transition metal dichalcogenide interface. They also have demonstrated the usefulness of time-resolved second harmonic generation for exploring ultrafast electronic-vibrational dynamics. In a similar study, Zhang et al. have examined charge-transfer dynamics in PbS QDs/ WSe_2 system, which shows tunable hole transfer rate from ~1 ns to <100 fs by a factor of more than four orders of magnitude and also observed transition to strong coupling from weak coupling regime due to quantum confinement [31]. In another study, Proppe et al. have measured experimentally the highest diffusion lengths of ~300 nm in metal halide exchanged PbS QDs with stronger interdot coupling, which shows PCE of 12% [32]. The charge carrier took 8 ps for interdot hopping. PbS QDs with smaller diameter (d = ~3.2 nm) exhibit faster interdot carrier transfer rates up to five times with 10 times lower trap-state density as compared to the QDs having a larger diameter (d = ~5.5 nm).

16.3 SYNTHESIS AND CHARACTERIZATION OF QDS

16.3.1 COLLOIDAL SYNTHESIS

The synthesis of uniform high-quality QDs is vitally important to achieving the best performances in any device application. More often, the QDs are synthesized by wet chemical methods due to their easy processability and convertibility of these high-quality QDs into high-quality dense thin films. The key challenges to realize high-quality QDs based solar cells include QDs packing, surface passivation, and the possibility of different materials that can be employed as a thin layer for effective transport of holes and electrons across the ends of the device. The chemical synthesis of QDs mainly includes two steps: (1) rapid nucleation, (2) slow growth after nucleation [33]. The command over to control the size and shape of QDs can be carried out using the thermal decomposition method. This method is not only considered effective in generating high-quality QDs but also economical [33–35]. In the synthesis of QDs, the hot injection or the heat-up method marks the separation of the growth phase from the nucleation phase. In the hot injection method, the precursor(s) is injected rapidly, usually with the help of a syringe into the hot solution containing a high boiling point surfactant. (Figure 16.4a). Since temperature plays a key role in the synthesis, it becomes crucial to consider the injection temperature because the decomposition of the precursor(s) relies on it. Upon injection, nucleation initiates due to induced supersaturation conditions in the solution. The injection of a room temperature solution will decrease the overall reaction temperature, hence terminating the nucleation stage and commencing the growth stage. In most cases, for nanoparticles stabilization, surplus surfactants are commonly used to start the crystal growth in solution. For QDs synthesis, surfactants like oleic acid (OA), trioctylphosphine oxide (TOPO), 1-hexadecylamine (HAD), tetradecylphosphonic acid (TDPA), etc. are commonly used. The size of the QDs strongly depends on the injection speed of precursor(s). A high degree of supersaturation solution is created due to fast injection within a very short period, resulting in fast nucleation. These nuclei slowly grow into the larger crystal. As nucleation and crystal growth are two separate processes, the QDs synthesized through this hot injection method are always monodispersed and have narrow size distribution. Alternatively, the noninjection heat-up approach involves steady heating of a mixture of precursor(s) and ligand in a single pot (Figure 16.4b). It has been observed that this approach leads to poly-dispersed particle-size distribution; however, the attainment of monodispersed QDs is also possible using precursors with the right level of reactivity. The heat-up method is very useful for large-scale synthesis of QDs, it also provides better control over the reaction parameters accompanied by full automation control. Another advantage associated with this process is that it allows the use of continuous flow synthesis, which includes the separation of nucleation and growth stage steps by the use of a dual

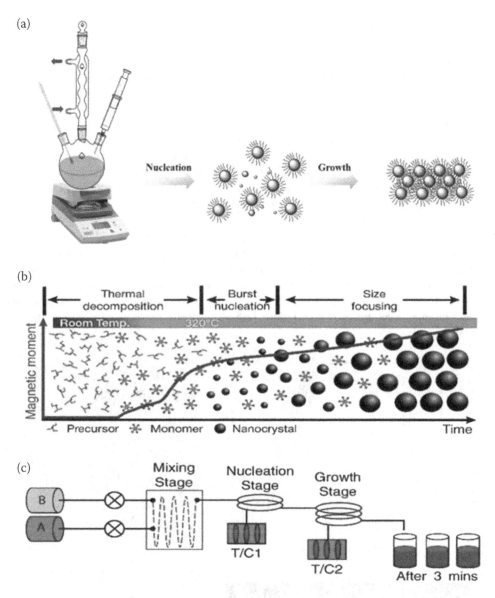

FIGURE 16.4 Schematic diagrams of QDs synthesis techniques. (a) Hot-injection synthesis showing nucleation and crystal growth phase. Adapted with permission from ref. [36], Copyright (2020), Elsevier. (b) Noninjection heat-up method. All precursors are heated to initiate nucleation and growth stages. Adapted with permission from ref. [35], Copyright (2007), American Chemical Society. (c) Flow reactor synthesis in a multistage, low-volume, continual flow synthesis reactor. Adapted with permission from ref. [37], Copyright (2015), American Chemical Society.

temperature stage-flow reactor. In a recent study, it has been observed that this method helps in improving the process control along with a narrower QD size distribution. A schematic representation of the dual-flow reactor synthesis method has been shown in (Figure 16.4c).

To fabricate highly efficient solar cells, the QDs need to undergo the purification stage. The purification of QDs becomes vital to discard the excess surfactants and unreacted precursors, which obstruct the charge carriers' transfer in solar cell devices [38,39]. The surfactant capped QDs are completely dissolved in nonpolar solvent after the chemical reaction. So, usually, sequential precipitation is employed for purification of QDs in which nonsolvent and solvent are used for precipitation and dispersion, respectively. The QDs thus purified can be dispersed into the solvent of our choice in the form of colloidal suspensions, otherwise also known as inks. Investigations in this purification technique depicted reduced photoluminescence quantum yield in QDs (PLQY) [40]. In another study, similar reductions in PLQY have also been observed for the recently used purification method using gel permeation chromatography (GPC) [41]. These reductions in PLQY can be mainly attributed to the loss of ligands, which can also be restored if the removed ligands can be reintroduced. In the quest of improving the PLQY, other methods were investigated, such as successive ionic-layer adsorption and reaction (SILAR). SILAR is an in situ method of synthesizing QDs. The SILAR method has proven to be effective, up to a certain extent, by providing effective control over the stoichiometry and size, thus improving the surface ligand coverage; however, it also suffers from disadvantages of corrosion. The solar cell devices fabricated using this method have been observed to have lower efficiencies compared to ex situ synthesized QD devices.

The photophysical study is very important to characterize the QDs. The photoluminescence (PL) spectroscopy is carried out to assess the size distribution of synthesized QDs. A sharp peak in photoluminescence spectra has small width of the peak at half maxima, which shows a more uniform size distribution of QDs, whereas the broad peak represents nonuniform size distribution. Composition-dependent PL spectra of $CsPb(Cl_{1-x}Br_x)_3$ QDs under 360 nm excitation wavelength with different chlorine and bromine ions are shown in Figure 16.5 [42]. In the reaction process, chlorine ions are

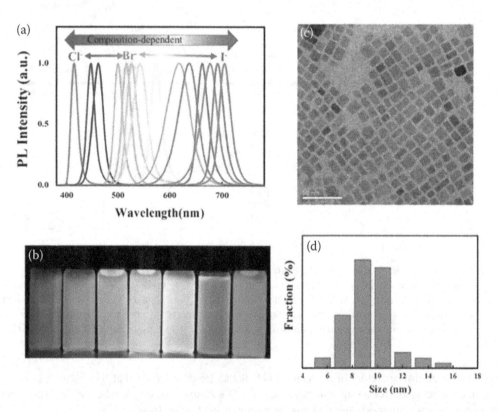

FIGURE 16.5 (a) Represents the various PL-spectra of $CsPbX_3$ quantum dots, depending upon their composition. (b) The figure shows photoluminescence properties of QDs in seven different colors under UV-exposure (violet, blue, cyan, green, yellow, orange, and red from left to right). The TEM images and size distribution of the yellow color light-emitting perovskite QDs are shown in (c) and (d), respectively. Adapted with permission from ref. [42], Copyright (2017), The Optical Society.

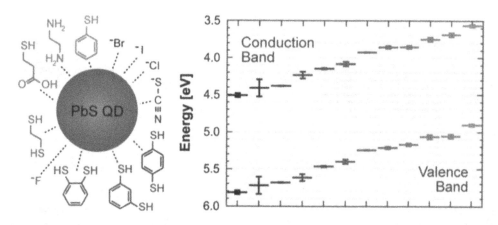

FIGURE 16.6   Various ligands used to exchange with the long-chain oleic acid of PbS QDs and the corresponding energy levels with respect to vacuum (right). Adapted with permission from ref. [43], Copyright (2014), American Chemical Society.

injected into the solution. As the chlorine ion concentration increases, the PL spectra peak shifts to higher energy. Alternatively, the iodine ions can shift the PL peak to the lower energy direction due to the presence of more iodine ions. From Figure 16.5a, it is clear that PL spectra can be tuned within 402 nm to 740 nm by adjusting the chlorine and iodine ions concentration. Figure 16.5b shows seven typical color photographs of samples in solution under 360 nm UV exposures. The microstructural properties of QDs are studied by a high-resolution transmission electron microscope (HRTEM) at the atomic level. The HRTEM image of monodisperse yellow light-emitting perovskite QDs is shown in Figure 16.5c; the size distribution of QDs has an average size of 9.28 nm (Figure 16.5d).

16.3.2   SURFACE ENGINEERING

As synthesized QDs are generally capped with long electronically insulating aliphatic chains, such as oleic acid $(CH_3(CH_2)_7CH=CH(CH_2)_7COOH)$ called as ligand. These long alkyl chains provide colloidal stability of QDs in the solution phase in an organic solvent. But, due to these long chains, inter-dot separation is large, which impedes charge transport between adjacent QDs in a thin film. These ligands also produce a tunneling barrier joining neighboring QDs and in between the QDs layer and the outer electrodes. Therefore, for the fabrication of QDs based solar cells, solid-state ligand exchange is performed to replace these long-chain ligands with short-chain ligands. After the exchange with short ligands, the conduction and valence band energy level are also modified, as shown in Figure 16.6. By performing ultraviolet photon spectroscopy, a shift up to 0.9 eV is achieved by tuning the different chemical ligands in PbS QDs [43]. Hence, surface engineering provides a great impact in a shift of the energy levels, which in turn enhances the electronics properties of colloidal QDs films.

16.4   QDS BASED FLEXIBLE HETEROJUNCTION SOLAR CELL

Most high-performance traditional QDs solar cells are fabricated on a glass substrate in bulk heterojunction architecture. In bulk heterojunction structure, n-type and p-type semiconductor material forms a three-dimensional interpenetrating network where each photogenerated exciton gets a charge-separating interface within exciton diffusion length, which is also quite efficient when compared to other types of solar cells. In the case of QDs based bulk heterojunction solar cells, a

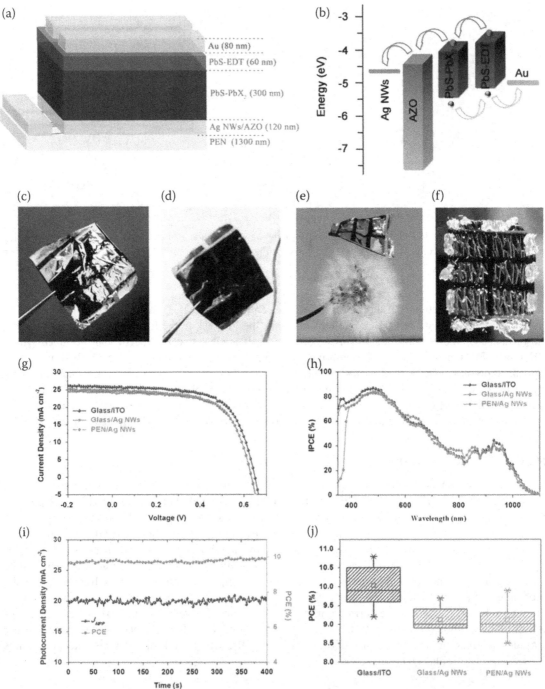

FIGURE 16.7 (a) Device architecture of lightweight and flexible PbS QDs heterojunction solar cell. (b) Schematic energy band diagram of the QDs solar cell with device architecture of PEN/Ag NWs/AZO/PbS-PbX$_2$/PbS-EDT/Au. Photograph shows (c) back side and (d) front side of the flexible solar cell. (e) To demonstrate the lightweight, the flexible solar cell placed on a dandelion without any deformation of the dandelion. (f) The flexible solar cell under the extreme mechanical stress. (g) J–V characteristic and (h) IPCE spectra of the QDs solar cell on glass/ITO, glass/Ag NWs and flexible PEN/Ag NWs substrate, respectively. (i) Steady-state power-conversion efficiency and photocurrent density of the flexible solar cell at the MPP. (j) Device-to-device variation of PCE of different solar cells fabricated on a different substrate. Figure 16.7 adapted with permission from ref. [8]. Copyright The Authors, some rights reserved; exclusive licensee [Royal Society of Chemistry]. Distributed under a Creative Commons Attribution License 3.0 (CC BY).

bilayer of QDs thin film is sandwiched between n-type wide-bandgap metal oxide semiconductors, such as TiO_2 or ZnO, etc and metal contact. The different materials interface enhances the junction deletion region and helps to incorporate more QDs materials, which ultimately increases both absorption and carrier collection simultaneously. However, in QDs based solar cells, transportation of minority carriers is a major issue in the bilayer device structure. This problem is supplemented by the bulk heterojunction concept. In the bulk heterojunction structure, a large number of interfaces is available throughout the active layer for exciton dissociation as compared to bilayer structure, but at the same time, it also introduces a larger number of defect sites, which increases the bimolecular recombination. Practically, it reduces the device's built-in voltage, which is reflected in device open-circuit voltage. To overcome this issue, conduction band offset at QDs and metal oxide interface is required to reduce the back-recombination. Additionally, practical steps must be taken to prevent shunt resistance in highly structured devices. The various QDs, such as CdS, CdSe, PbSe, Ag_2S Ag_2Se, $CuInS_2$ $AgBiS_2$ including PbS, have been studied in QDs based solar cells [44]. Among these QDs, PbS QDs shows the highest power-conversion efficiency (PCE) of 13.3% [45] in traditional bulk heterojunction structure (ITO/ZnO as the electron transport layer/PbS QDs film as the active layer/thin PbS QDs film treated with 1,2-ethanethiol (EDT) as hole-transport layer/Au on glass substrate). PbS QDs possess a wide light-absorption spectrum with a bandgap in the range of 0.8–1.6 eV, which can be achieved by tuning QDs size. Due to significant improvement in device performance of QDs based solar cell, more attention is being given to the development of QDs based solar cell on flexible substrate.

Researchers Zhang et al. from Uppsala University have demonstrated flexible QDs based solar cells with PCE of 9.4% using PbS QDs with stable MgZnO nanocrystal as electron transport layer [16]. Later on, the same research group has reported so far the highest PCE of 9.9% using the same QDs with improved mechanical properties, and it works during large compression-stretching deformation cycles [8]. A highly conducting silver nanowire (Ag NW) network was used as a transparent conducting electrode for this device fabrication on 1.3 μm-thick PEN substrate at a low temperature. The 5 mg/ml of Ag NW ink in isopropanol was spin-coated on the PEN film with various speeds (700–4000 rpm) under ambient conditions followed by annealed at 150°C for 20 min under nitrogen atmosphere to achieve Ag NW network of different transparency and resistance. Figure 16.7a shows the schematic device structure of PEN/Ag NWs/AlZnO (AZO)/PbS-PbX_2 (X = I and Br)/PbS-1,2-ethanedithol (EDT)/Au. With the help of spin coating, the AZO nanoparticle layer was deposited on the Ag NW network film and was subsequently annealed at 140°C for 30 min under ambient conditions. The oleic acid ligand capped PbS QDs was synthesized by the commonly used hot injection method, as mentioned in Section 3. Prior to the QDs, solid film deposition solution-phase ligand exchange was done. The n-type PbS QDs solution was prepared by treating with PbX_2 and p-type PbS QDs by EDT through ligand exchange. PbS-PbX_2 QDs ink was prepared through solution-phase ligand exchange, and PbS-PbX_2 QDs thin film was deposited by a spin-coating method. This was followed by EDT-treated PbS QDs that was deposited on it and acts as the hole-transport layer, and the metal gold (Au) contact was deposited by a thermal evaporation method. The six solar cells are fabricated with a flexible substrate of $\sim2 \times 2$ cm^2. The corresponding energy band diagram of the complete device is shown in Figure 16.7b. The thickness of the complete solar cell is under 2 μm with an extremely low weight of 6.5 g m^{-2}, which is resulting in a high power-per-weight output of ~15 Wg^{-1}.

Both sides of fabricated flexible QDs solar cells are shown in Figure 16.7c and 7d. No deformation in dandelion was observed when kept on it, as shown in Figure 16.7e, which demonstrates that the cell is extremely lighter in weight. Figure 16.7f shows the device's durability under compress state when it adheres to the pre-stretched elastomer. The photocurrent density–voltage (J–V) characteristic of the flexible QDs solar cells fabricated on the different substrates is shown in Figure 16.7f under AM 1.5G 100 mW/cm^2 illumination conditions. The solar cell shows PCE of

10.8% and 9.7%, which was fabricated on glass/ITO and glass/Ag NWs substrate, respectively. The flexible QDs solar cell fabricated on PEN/Ag NWs substrates reflects PCE of 9.9% with open-circuit voltage (V_{oc}) of 0.64 V, a short-circuit current density (J_{sc}) of 24.6 mA cm^{-2}, and a fill factor (*FF*) of 0.63. Figure 16.7h represents the corresponding incident photon-to-current efficiency (IPCE) spectra of these solar cells. The IPCE spectra is low at UV region for the device on PEN substrate, which is due to the UV absorption of PEN substrate itself. The flexible device stability is also tested under continuous AM 1.5G 100 mW cm^{-2} illumination within a short-term illumination of 400s, which indicates stable steady-state photocurrent density (J_{MPP}) and a slight increase in efficiency at the maximum power point (MPP), as presented in Figure 16.7i. The statistical analysis data of different devices fabricated on a different substrate is shown in Figure 16.7j, which shows the high reproducibility of flexible QDs solar cells. This result indicates that the QDs are very promising for future commercialization of efficient, flexible, and extremely lightweight devices in various advanced applications.

16.5 QD BASED FLEXIBLE SENSITIZED SOLAR CELLS

In the recent past, the QDs have also been explored as a photosensitizer alternative to organic dye in dye-sensitized solar cells structure. Yitan Li et al. have demonstrated CdSe/CdS QDs co-sensitized solar cells by growing uniform zinc oxide (ZnO) nanosheet arrays on woven titanium wires. In this study, the low-temperature hydrothermal synthesis method is used to prepare photoanodes by depositing CdS and CdSe quantum dots on the ZnO nanosheet arrays by successive ionic layer adsorption and reaction (SILAR) technique [46]. They achieved an overall PCE of 0.98% for flexible CdS/CdSe co-sensitized solar cells, where conventional Pt foils are used as counter electrodes. To further enhance the performance of the flexible solar cells, they have introduced PbS and Cu_2S counter electrodes. The most common polysulfide electrolyte is used; it is composed of 1 M sulfur, 1 M Na_2S, and 0.1 M NaOH in an aqueous solution. The highest PCE of CdS/CdSe co-sensitized flexible solar cells incorporating Cu_2S and PbS counter electrodes were 3.4% and 2.5%, respectively, which is a noteworthy enhancement in performance from those achieved in the commonly used Pt and Au electrodes. In 2014, Que et al. also demonstrated flexible QDs sensitized solar cells by decorating CdS and CdSe QDs on ZnO nanowire on ITO/PET substrate with cobalt sulfide (CoS) nanorod arrays (NRAs) deposited on graphite papers (GPs) (CoS NRAs/GP) as the counter electrode [47]. An absolute PCE of 2.7% was reported by this flexible device structure. Du et al. also reported flexible CdSe QDs sensitized solar cells with PCE of 3.55% using copper sulfide (Cu_2S)/nickel as counter electrode [12]. Ma et al. recently demonstrated CdS/CdSe sensitized solar cells with PCE of 3.49% on TiO_2 nanoparticles (P25) coated ITO/PET substrate [11]. To increase the TiO_2 slurry adhesion on the flexible substrate surface, tert-butanol was incorporated as an additive. The quantum efficiency retention rate of flexible CdS/CdSe QDs solar cell is 72.7% after 500 times of bending, which proclaims its good bending stability. Du et al. reported the highest PCE reached so far of 5.08% using CdSe QDs on to ZnO/ZnSe nanocrystal and mesoporous carbon supported on titanium (Ti) mesh as counter electrode [13]. The schematic device structure and real photograph of the working device are depicted in Figures 16.8a and 8b, respectively. The photocurrent density-voltage curve of the flexible QDs solar cell with only ZnO/ZnSe nanosheets and ZnO/ZnSe/CdSe nanosheets after the ion exchange process in Cd^{2+} solution for different periods are shown in Figure 16.8c. As a matter of fact, a solar cell fabricated using ZnO/ZnSe nanosheets shows poor PCE of 0.32%, after growing CdSe QDs on to the ZnO/ZnSe nanosheets for four hours by ion exchange with Cd^{2+} ions, J_{SC} improved from initial value of 3.59 mA cm^{-2} to 15.27 mA cm^{-2} and V_{OC} from 0.187 to 0.638V, respectively, resulting in the so far highest PCE of 5.08% to date. The improved performance is due the formation CdSe QDs passivation layer on to ZnSe layer, which prevents direct contact between electron collecting layer and electrolyte. The recombination of charge carrier at the photoanode

(a)   (b)   (c)

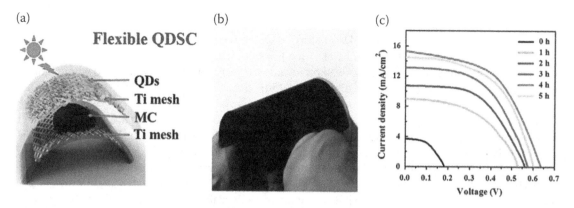

FIGURE 16.8 (a) The schematic device structure of ZnO/ZnSe/CdSe nanosheets on flexible titanium mesh. (b) Real device photograph of the corresponding solar cell. (c) J–V characteristic of the flexible QDs solar cell based on ZnO/ZnSe nanosheets and ZnO/ZnSe/CdSe nanosheets after the cadmium ion exchange for different time periods. Adapted with permission from ref. [13], Copyright (2018), Royal Society of Chemistry.

interface is also reduced, which enhances the open circuit voltage of the solar cell. Various researchers have also explored the use Cu-based QDs sensitizer for solar cell, which is environmentally friendly as compared to the most commonly used toxic Cd- and Pb-based materials. Peng et al. has demonstrated PCE of 3.19% using copper indium sulfide ($CuInS_2$) QDs as sensitizer on ITO/PEN substrate with Cu_2S/brass foil as counter electrode and curving stability of 98% [48]. Besides these, lots of other semiconductor QDs materials are also explored as sensitizers in solar cell application and continuing.

16.6   QDS BASED FLEXIBLE PEROVSKITE SOLAR CELLS

Perovskite solar cells have sprung up as the forefront contender in the realm of photovoltaics. Recent developments of these cells have pushed PCE beyond 25% within just over 11 years of R&D research all over the world. Despite having high power-conversion efficiency, long-term device stability is a growing concern that is impeding the market commercialization of perovskite-based modules. These perovskite materials degrade very fast under ambient environmental stressors, which are due to phase segregation, ion migration, and high-temperature operation; further thorough understanding is required to study the durability of the devices. On the other hand, perovskite quantum dots (PQDs) have emerged as an alternative to their bulk counterpart for photovoltaic applications due to fascinating optoelectronic properties, such as multiple excitons generation, high absorption coefficient, photoluminescence quantum yield, tunable light absorption spectra, high defect tolerance, and high extinction coefficients over a broad absorption range. Broadly, the perovskite materials are used in optoelectronics applications having an ABX_3 structure, A site is large organic or alkali metal cation such as methylammonium ($CH_3NH_3^+$) (MA), formamidinium ($NH_2CHNH_2^+$) (FA), and cesium (Cs^+), B site is a divalent cation (e.g., Pb^{2+}, Sn^{2+}) and X site is halide anion(e.g., Cl^-, Br^-, I^-). These PQDs materials show very good photophysical properties alike other conventional chalcogenide quantum dots with flexible compositional control, and crystalline strain benefits, making them suitable for efficient photovoltaic application. The first PQDs based solar cell with $CsPbI_3$ QD arrays was reported in 2018 with a PCE of 10.77% [49]. So far, the highest reported $Cs_{0.5}FA_{0.5}PbI_3$ based PQDs solar cell achieved Newport Corporation certified PCE of 16.6% in conventional structure [6]. $CsPbI_3$ and $FAPbI_3$ are the most commonly used PQDs in photovoltaics as compared to other perovskites. $FAPbI_3$ also displayed the highest PCE of 13.2%.

With the use of various common nonpolar organic solvents such as hexane, octane, or toluene at room temperature, wet chemical method PQDs are easily synthesized and processed, whereas toxic polar solvents, such as N,N dimethylformamide, are used to synthesize thin-film perovskites.

The first successful methylammonium lead bromide (MAPbBr$_3$) PQDs are synthesized with a long or medium alkyl chain to stabilize the PQDs in 2014 [50]. By and large two popular methods, (1) hot injection, and (2) ligand-assisted reprecipitation (LARP), are used to synthesize the PQDs. in the LARP method, the nucleation and growth of PQDs took place in mixture anti solvents in the presence of oleylamine and oleic acid as the capping ligand at room temperature. Whereas in the case of all inorganic perovskite PQDs, such as CsPbI$_3$, the hot injection method is quite popular, and this ignited major research effort of PQDs solar cell research. These PQDs can be easily deposited on any substrate by a solution process to make a thin film at room temperature, which is essential for flexible devices. These PQDs also show higher mechanical flexibility in contrast to their thin-film counterpart due to nanoscale grain boundaries. Very recently, Hu et al. demonstrated the highest PCE of 12.3% in flexible CsPbI$_3$ PQDs based solar cells with V_{OC} of 1.24 V, J_{SC} of 13.6 mA/cm^2, and FF of 73% [9]. CsPbI$_3$ PQDs based perovskite solar cells show very high V_{OC}, which is distinctly larger than that in CsPbI$_3$ polycrystalline thin-film solar cells. For the improvement of the device performance, a thin hybrid interfacial architecture (HIA) was introduced by incorporating phenyl-C61-butyric acid methyl ester (PCBM) into the CsPbI$_3$ PQDs layer in PET/ITO/SnO$_2$/ CsPbI$_3$/PTB7/MoO$_3$/Ag device configuration. This hybrid interface enhances the charge transfer and also promotes exciton dissociation at both PQDs hetero-interfaces, as well as PQDs/electron transport layer interfaces. Figure 16.9a shows the J–V characteristics with and without HIA-treated depicted as controlled and target device, respectively, with PCE distribution inset figure.

Figure 16.9b shows the normalized PCE change of CsPbI$_3$ PQDs solar cell and CsPbI$_2$Br thin-film solar cell as a function of bending cycles. The real and schematic devices of flexible PQDs solar cells are shown in the inset of Figure 16.9b. The HIA-treated flexible PQDs device retained 90% of the initial value, while only ~75% of initial efficiency value for and CsPbI$_2$Br thin-film solar cell after 100 bending cycles at the radius of 0.75 cm. Figure 16.9c represents the schematic diagram of the bending mechanical testing of PQDs and thin-film on a flexible substrate. A mechanical durability test using SEM was carried out to further investigate the reason for the internal stress build-up after bending of CsPbI$_3$ QDs and CsPbI$_2$Br thin film. The study also included widely studied PbS QDs film and high-efficiency Cs$_{0.05}$MA$_{0.85}$- MA$_{0.1}$PbI$_{2.55}$Br$_{0.45}$ mixed perovskite thin film deposited on sample flexible PET/ITO substrate, which is represented in Figure 16.9d–g. No significant change in morphology was observed after 1000 bending cycles in the case of both CsPbI$_3$ (Figure 16.9d) and PbS (Figure 16.9e) QD films. But noticeable cracks in CsPbI$_2$Br thin film, as well as mixed perovskite thin film, was observed after 100 bending cycle with a gradual increase in the number of bending cycles, as displayed in Figure 16.9f–g. So, it is quite clear from this study that PQDs are better suitable for flexible surfaces in comparison to the extensively studied thin-film materials for solar cell application.

16.7 FLEXIBLE QD-SILICON HYBRID SOLAR CELLS

Researchers from Hanyang University, S. W. Baek et al. [17], tried to implement core-shell QDs as a performance booster for flexible silicon solar cells. Their work established that by the introduction of core-shell QD of Cd$_{0.5}$Zn$_{0.5}$S on the flexible silicon solar cells, the PCE enhanced by 0.7%, thus explaining an energy-down-shift effect, as shown in Figure 16.10. The device structure schematic is depicted in Figure 16.10a. The fabricated solar cell reveals very good flexible and twistable characteristics, which is essential to install solar cells on curved surfaces of a building or many applications (Figure 16.10b, c). These solar cells with an energy down-shifter did not show any light emission when it is exposed under visible light, whereas visible light gets emitted as a result of UV light irradiation on the solar cell surface, as shown in Figure 16.10d. The enhancement of the external quantum efficiency,

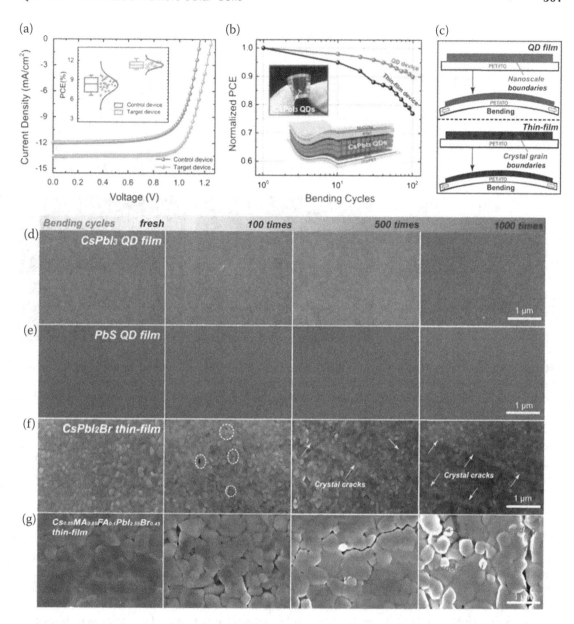

FIGURE 16.9 (a) Current density-voltage (*J–V*) plots of best flexible controlled and target device measured under standard condition (inset: schematic representation of flexible device architecture and PCE distribution of the flexible devices). (b) The variation of PCE of CsPbI$_3$ QDs based solar cell and CsPbI$_2$Br thin-film solar cell as a function of bending cycles. Inset shows the photograph of the flexible QDs based solar cell. (c) Schematic representation of mechanical banding testing of QDs film and thin-film on a flexible substrate. (d)–(g) Top-view SEM images of the CsPbI$_3$ QDs film, PbS QDs film, CsPbI$_2$Br, and mixed perovskite thin-film fabricated on PET/ITO (2.5 × 2.5 cm) having almost similar thickness with mechanical bending (curvature radius of 0.75 cm). Figure 16.9 adapted with permission from ref. [9]. Copyright (2021) the Authors, some rights reserved; exclusive licensee [Springer Nature]. Distributed under a Creative Commons Attribution License 4.0 (CC BY).

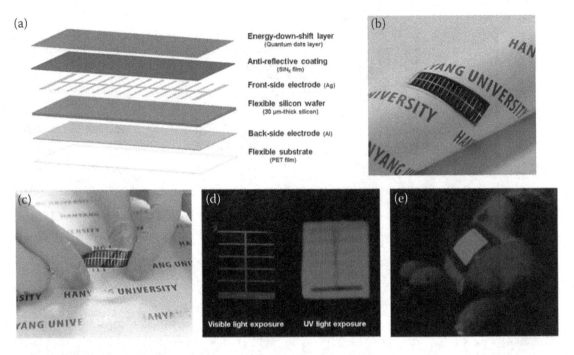

FIGURE 16.10 Effect of energy down-shifter QDs layer in ultrathin silicon solar cells. (a) Schematic device structure. (b) Image demonstrating bendability. (c) twistability of flexible solar cell. (d) Image of flexible solar cell under UV and visible light exposure. (e) Image represents the flexibility of the cells. Adapted with permission from ref. [17], Copyright (2015), Royal Society of Chemistry.

especially in the UV light, can be attributed to the absorbance of UV light by the core/shell QDs and emission of visible light (blue light), undergoing reabsorption into the cell. Also, the cells recorded exceptional bending fatigue performance (Figure 16.10e) as PCE remained at 12.37% under a strain of 5.72% after 5000 bending cycles.

16.7 CONCLUSION

Flexibility and lightweight QDs solar cells are needed in the current advanced photovoltaic market, along with a great deal of utility in powering the IoT-based wearable electronic devices. With continuous progress in PCE of flexible PbS QDs and perovskite QDs based solar cells, achieving efficiency above 15% is only a matter of time. Despite this impressive progress, a further improvement in flexible QDs solar cells is unavoidable, which can be achieved by optimizing device performances and developing solution-processed flexible conducting electrodes. Several nanowires-based thin films, or graphene-based thin films, with comparable conductivity and transparency similar to the conventional ITO-based electrode, which demonstrates superior mechanical properties, can lead to economical and lightweight devices. The long-term device stability of QDs based solar cells is also one of the major growing concerns that demands significant research efforts. The QDs based solar cells need to be highly stable due to their continuous exposure to light under ambient conditions for commercial application. At the same time, flexible QDs solar cells should maintain competitive photovoltaic performance under acute deformation conditions, and they must show robust performance after repetitive compression-stretching cyclic. An extensive theoretical study is also necessary to improve the performance of QDs based solar cells. Overall, the further breakthrough in QDs research will guide lightweight and ultra-flexible QDs based solar cell performance toward commercial realization in the near future.

REFERENCES

1. Carey GH, Abdelhady AL, Ning Z, Thon SM, Bakr OM, Sargent EH (2015) Colloidal quantum dot solar cells. *Chem. Rev.* 115(23):12732–12763.
2. Kirmani AR, Luther JM, Abolhasani M, Amassian A (2020) Colloidal quantum dot photovoltaics: Current progress and path to gigawatt scale enabled by smart manufacturing. *ACS Energy Lett.* 5(9): 3069–3100.
3. Nozik AJ, Beard MC, Luther JM, Law M, Ellingson RJ, Johnson JC (2010) Semiconductor quantum dots and quantum dot arrays and applications of multiple exciton generation to third-generation photovoltaic solar cells. *Chem. Rev.* 110(11):6873–6890.
4. Xuan T-T, Liu J-Q, Li H-L, Sun H-C, Pan L-k, Chen X-H, et al. (2015) Microwave synthesis of high luminescent aqueous CdSe/CdS/ZnS quantum dots for crystalline silicon solar cells with enhanced photovoltaic performance. *RSC Adv.* 5(10):7673–7678.
5. Kim HI, Baek SW, Cheon HJ, Ryu SU, Lee S, Choi MJ, et al. (2020) A tuned alternating D–A copolymer hole-transport layer enables colloidal quantum dot solar cells with superior fill factor and efficiency. *Adv. Mater.* 32(48):2004985.
6. Hao M, Bai Y, Zeiske S, Ren L, Liu J, Yuan Y, et al. (2020) Ligand-assisted cation-exchange engineering for high-efficiency colloidal $Cs_{1-x}FA_xPbI_3$ quantum dot solar cells with reduced phase segregation. *Nat. Energy.* 5(1):79–88.
7. Kramer IJ, Moreno-Bautista G, Minor JC, Kopilovic D, Sargent EH (2014) Colloidal quantum dot solar cells on curved and flexible substrates. *Appl. Phys. Lett.* 105(16):163902.
8. Zhang X, Öberg VA, Du J, Liu J, Johansson EM (2018) Extremely lightweight and ultra-flexible infrared light-converting quantum dot solar cells with high power-per-weight output using a solution-processed bending durable silver nanowire-based electrode. *Energy Environ. Sci.* 11(2):354–364.
9. Hu L, Zhao Q, Huang S, Zheng J, Guan X, Patterson R, et al. (2021) Flexible and efficient perovskite quantum dot solar cells via hybrid interfacial architecture. *Nat. Commun.* 12(1):1–9.
10. Zhang X, Zhang J, Liu J, Johansson EM (2015) Solution processed flexible and bending durable heterojunction colloidal quantum dot solar cell. *Nanoscale* 7(27):11520–11524.
11. Ma F, Deng Y, Ni X, Hou J, Liu G, Peng S (2021) Flexible CdS/CdSe quantum dots sensitized solar cells with high performance and durability. *Nano Select.* 10.1002/nano.202000270.
12. Du Z, Tong J, Guo W, Zhang H, Zhong X (2016) Cuprous sulfide on Ni foam as a counter electrode for flexible quantum dot sensitized solar cells. *J. Mater. Chem. A* 4(30):11754–11761.
13. Du Z, Liu M, Li Y, Chen Y, Zhong X (2017) Titanium mesh based fully flexible highly efficient quantum dots sensitized solar cells. *J. Mater. Chem. A.* 5(11):5577–5584.
14. Zhang X, Aitola K, Hägglund C, Kaskela A, Johansson MB, Sveinbjörnsson K, et al. (2017) Dry-deposited transparent carbon nanotube film as front electrode in colloidal quantum dot solar cells. *ChemSusChem.* 10(2):434–441.
15. Zhang X, Johansson EM (2016) Utilizing light trapping interference effects in microcavity structured colloidal quantum dot solar cells: A combined theoretical and experimental approach. *Nano Energy.* 28: 71–77.
16. Zhang X, Santra PK, Tian L, Johansson MB, Rensmo Hk, Johansson EM (2017) Highly efficient flexible quantum dot solar cells with improved electron extraction using MgZnO nanocrystals. *ACS nano.* 11(8): 8478–8487.
17. Baek S-W, Shim J-H, Ko Y-H, Park J-S, Lee G-S, Jalalah M, et al. (2015) Low-cost and flexible ultra-thin silicon solar cell implemented with energy-down-shift via $Cd_{0.5}Zn_{0.5}S/ZnS$ core/shell quantum dots. *J. Mater. Chem. A.* 3(2):481–487.
18. Rath J, Brinza M, Liu Y, Borreman A, Schropp R (2010) Fabrication of thin film silicon solar cells on plastic substrate by very high frequency PECVD. *Sol. Energy Mater. Sol. Cells.* 94(9):1534–1541.
19. Romeo A, Khrypunov G, Kurdesau F, Arnold M, Bätzner D, Zogg H, et al. (2006) High-efficiency flexible CdTe solar cells on polymer substrates. *Sol. Energy Mater. Sol. Cells.* 90(18-19):3407–3415.
20. Chirilă A, Buecheler S, Pianezzi F, Bloesch P, Gretener C, Uhl AR, et al. (2011) Highly efficient Cu (In, Ga) Se 2 solar cells grown on flexible polymer films. *Nat. Mater.* 10(11):857–861.
21. Zhao J, Wang A, Altermatt P, Green M (1995) Twenty-four percent efficient silicon solar cells with double layer antireflection coatings and reduced resistance loss. *Appl. Phys. Lett.* 66(26):3636–3638.
22. Zhao J, Wang A, Green MA, Ferrazza F (1998) 19.8% efficient "honeycomb" textured multicrystalline and 24.4% monocrystalline silicon solar cells. *Appl. Phys. Lett.* 73(14):1991–1993.

23. Lee K, Shiu K-T, Zimmerman JD, Renshaw CK, Forrest SR (2010) Multiple growths of epitaxial lift-off solar cells from a single InP substrate. *Appl. Phys. Lett.* 97(10):101107.

24. Kaltenbrunner M, White MS, Głowacki ED, Sekitani T, Someya T, Sariciftci NS, et al. (2012) Ultrathin and lightweight organic solar cells with high flexibility. *Nat. Commun.* 3(1):1–7.

25. Shin SS, Yang WS, Noh JH, Suk JH, Jeon NJ, Park JH, et al. (2015) High-performance flexible perovskite solar cells exploiting Zn_2SnO_4 prepared in solution below 100 C. *Nat. Commun.* 6(1):1–8.

26. Nozik AJ (2008) Multiple exciton generation in semiconductor quantum dots. *Chem. Phys. Lett.* 457(1–3): 3–11.

27. Nozik AJ (2002) Quantum dot solar cells. Physica E Low Dimens. *Syst. Nanostruct.* 14(1–2):115–120.

28. Ellingson RJ, Beard MC, Johnson JC, Yu P, Micic OI, Nozik AJ, et al. (2005) Highly efficient multiple exciton generation in colloidal PbSe and PbS quantum dots. *Nano Lett.* 5(5):865–871.

29. Schaller RD, Sykora M, Pietryga JM, Klimov VI (2006) Seven excitons at a cost of one: Redefining the limits for conversion efficiency of photons into charge carriers. *Nano Lett.* 6(3):424–429.

30. Goodman AJ, Dahod NS, Tisdale WA (2018) Ultrafast charge transfer at a quantum dot/2D materials interface probed by second harmonic generation. *J. Phys. Chem. Lett.* 9(15):4227–4232.

31. Zhang C, Lian L, Yang Z, Zhang J, Zhu H (2019) Quantum confinement-tunable ultrafast charge transfer in a PbS quantum dots/WSe$_2$ 0D–2D hybrid structure: transition from the weak to strong coupling regime. *J. Phys. Chem. Lett.* 10(24):7665–7671.

32. Proppe AH, Xu J, Sabatini RP, Fan JZ, Sun B, Hoogland S, et al. (2018) Picosecond charge transfer and long carrier diffusion lengths in colloidal quantum dot solids. *Nano Lett.* 18(11):7052–7059.

33. Thanh NT, Maclean N, Mahiddine S (2014) Mechanisms of nucleation and growth of nanoparticles in solution. *Chem. Rev.* 114(15):7610–7630.

34. Talapin DV, Lee J-S, Kovalenko MV, Shevchenko EV (2010) Prospects of colloidal nanocrystals for electronic and optoelectronic applications. *Chem. Rev.* 110(1):389–458.

35. Kwon SG, Piao Y, Park J, Angappane S, Jo Y, Hwang N-M, et al. (2007) Kinetics of monodisperse iron oxide nanocrystal formation by "heating-up" process. *J. Am. Chem. Soc.* 129(41):12571–12584.

36. Luo Q (2020) Nanoparticles inks: Edited byG. K. Zheng Cui, *Solution Processed Metal Oxide Thin Films for Electronic Applications*, Elsevier.

37. Pan J, El-Ballouli AaO, Rollny L, Voznyy O, Burlakov VM, Goriely A, et al. (2013) Automated synthesis of photovoltaic-quality colloidal quantum dots using separate nucleation and growth stages. *ACS nano.* 7(11):10158–10166.

38. Tagliazucchi M, Tice DB, Sweeney CM, Morris-Cohen AJ, Weiss EA (2011) Ligand-controlled rates of photoinduced electron transfer in hybrid CdSe nanocrystal/poly (viologen) films. *Acs Nano.* 5(12): 9907–9917.

39. King LA, Riley DJ (2012) Importance of QD purification procedure on surface adsorbance of QDs and performance of QD sensitized photoanodes. *J. Phys. Chem. C.*116(5):3349–3355.

40. Kalyuzhny G, Murray RW (2005) Ligand effects on optical properties of CdSe nanocrystals. *J. Phys. Chem. B.* 109(15):7012–7021.

41. Shen Y, Gee MY, Tan R, Pellechia PJ, Greytak AB (2013) Purification of quantum dots by gel permeation chromatography and the effect of excess ligands on shell growth and ligand exchange. *Chem. Mater.* 25(14): 2838–2848.

42. Zhang W, Yang W, Zhong P, Mei S, Zhang G, Chen G, et al. (2017) Spectral optimization of color temperature tunable white LEDs based on perovskite quantum dots for ultrahigh color rendition. *Opt. Mater. Express.* 7(9):3065–3076.

43. Brown PR, Kim D, Lunt RR, Zhao N, Bawendi MG, Grossman JC, et al. (2014) Energy level modification in lead sulfide quantum dot thin films through ligand exchange. *ACS nano.* 8(6):5863–5872.

44. Yuan M, Liu M, Sargent EH (2016) Colloidal quantum dot solids for solution-processed solar cells. *Nat. Energy.* 1(3):1–9.

45. Choi M-J, de Arquer FPG, Proppe AH, Seifitokaldani A, Choi J, Kim J, et al. (2020) Cascade surface modification of colloidal quantum dot inks enables efficient bulk homojunction photovoltaics. *Nat. Commun.* 11(1):1–9.

46. Li Y, Wei L, Wu C, Liu C, Chen Y, Liu H, et al. (2014) Flexible quantum dot-sensitized solar cells with improved efficiencies based on woven titanium wires. *J. Mater. Chem. A.* 2(37):15546–15552.

47. Que M, Guo W, Zhang X, Li X, Hua Q, Dong L, et al. (2014) Flexible quantum dot-sensitized solar cells employing CoS nanorod arrays/graphite paper as effective counter electrodes. *J. Mater. Chem. A.* 2(33): 13661–13666.

48. Peng Z, Liu Z, Liu Y, Chen J, Li C, Li W, et al. (2018) Improving on the interparticle connection for performance enhancement of flexible quantum dot sensitized solar cells. *Mater. Res. Bull.* 105: 91–97.
49. Swarnkar A, Marshall AR, Sanehira EM, Chernomordik BD, Moore DT, Christians JA, et al. (2016) Quantum dot–induced phase stabilization of α-CsPbI$_3$ perovskite for high-efficiency photovoltaics. *Science.* 354(6308):92–95.
50. Schmidt LC, Pertegás A, González-Carrero S, Malinkiewicz O, Agouram S, Minguez Espallargas G, et al. (2014) Nontemplate synthesis of CH$_3$NH$_3$PbBr$_3$ perovskite nanoparticles. *J. Am. Chem. Soc.* 136(3): 850–853.

17 A Method of Strategic Evaluation for Perovskite-Based Flexible Solar Cells

Figen Balo
Department of Industrial Engineering, Firat University, Elazığ, Turkey

Lutfu S. Sua
Department of Management and MKTG, Southern University and A&M College, Louisiana, USA

CONTENTS

17.1 Introduction... 307
17.2 Perovskite-based solar cells' working mechanism.. 313
    17.2.1 The future of perovskite-based solar cells 314
17.3 Methodology.. 315
    17.3.1 AHP analysis 315
17.4 Conclusions... 317
References.. 320

17.1 INTRODUCTION

Power is a key factor for our life quality. The fossil-based energy sources are being depleted quicker than ever before, and yet they still remain as the world's main power resource. Thus, finding alternatives for these fossil-based energy sources has become even more essential, and renewable power proposes a solution. The investment is evaluated as the continuing capital expenditure in power-providing capacity, power effectiveness, and the increased expenditure on more effective goods and equipment. "Fuel providing" contains all investments related to the transformation, provision, and production of liquid, gaseous, and solid energy sources to users; these investments occur primarily in coal, gas, and oil, but they also contain bio-fuels and other less-carbon fuels. "Energy industry" contains the capital expenditure on overall energy production technics, as well as continuing investments in storage and grids. "Power end utilization and effectiveness" contains the investment in effective developments across all end-utilization industries, as well as end-utilization implementations for sustainable heat. The worldwide total power investment has dropped by 20% between 2019 and 2020, while the investment in renewable energy-productivity on a global scale and its share in overall investment has increased [1]. In the latest scenario, the need for innovation and the need for energy of eco-friendly technologies and renewable technologies are of primary significance. Solar power, geothermal, tidal, biomass, and wind energy are arising as alternative resources of power for our power-deprived planet. It is the clean and sustainable form of energy that addresses the growing issue of fossil fuel-sourced greenhouse gases and global warming.

In recent years, the study of sun power has permitted the improvement of photovoltaic cells that can access about 0.20 power transformation performance when sold traditionally and, at minimum, pair while being analyzed in labs. By applying this industry into the commercial production of photovoltaic cells, sun power is a reasonable change for fossil-based sources as the world's main source of power.

DOI: 10.1201/9781003186755-17

The fund investment through chosen and major companies in novel designs outside gas and oil providing has surpassed the 2 billion USD by 2019 [1]. The photovoltaic panels are one of the most attractive alternatives to compensate for the continuous power requisition while meeting the considerable electrical decarbonization. Perovskite-based photovoltaic solar cell performances at current market status are displayed in the Perovskite Handbook [2,3]. Photovoltaic cells are usually grouped into four generations relying on categories and time of materials that are utilized for their production. The most widespread photovoltaic cells existing in the market are the first-generation photovoltaic cells, which contain multi- and single-crystalline silicon. Second-generation photovoltaic cells were presented as feedback to silicon photovoltaic cell cost and high material usage. To decrease the material utilization, the maximal cell thickness for this production was evaluated from tens of micrometers to several nanometers. In the meantime, many investigators have tested light administration concepts utilizing nano-pattering dye-sensitized photovoltaic cells, nano-structuring, perovskite, organic photovoltaic cells, and photochemical cells. Fourth-generation photovoltaic cells fall in the conjectural production group made of composites. The classification of photovoltaic cells based upon the principal energetic material is given in Figure 17.1.

Energy conversion from solar photovoltaics is based on the idea that electricity is produced when a solar film is illumined through sun rays, producing the photovoltaic impact, an event that transforms light into electricity. The amount of electricity produced is determined by a variety of factors, involving film size (the greater the film size, the more individual films turn into either more current or more voltage), film material (such as semiconductor compounds, silicon, and so on), or the light source's quality and strength. For this reason, photovoltaics are typically categorized on the basis of either the energetic materials utilized for the solar films (such as the light-absorber materials) or the overall system structures. Lately, classifications on the basis complication of material have been suggested [4]. Photovoltaics are classified into thin-film and wafer-based technologies from the view point of system structure and design. Whereas wafer-based photovoltaics are developed from semiconducting wafers' ingot-sourced slices [5], thin-film films follow a fundamentally distinct strategy in which insulation substrates, such as elastic plastics or glass, are utilized to deposit semiconducting material layers that shape the structure of the device [4,6].

Generally, photovoltaic industries can be split into two primary groups: first-generation photovoltaics (also known wafer-based photovoltaic) and thin-film photovoltaic solar cells. Conventional crystalline silicon-based solar cells (both multi-crystalline and single-crystalline silicon) and gallium arsenide-based solar films as they pertain to wafer-based photovoltaics, with crystalline silicon cells dominating today's photovoltaic market share (around 90%) and gallium arsenide indicating the maximum performance. The thin-film solar films generally absorb light more than silicon solar films, enabling the eminently thin films' utilization. With more than 17.5% module productivity and 20% film productivity records, the cadmium telluride industry has been meaningfully commercialized, and such films recently hold around 5% of the global market share. The copper indium gallium di-selenide and hydrogenated amorphous silicon solar films are other commercial industries of thin-film, with about 2% market share today. The technology of copper zinc tin sulfide has been subject to extensive research and development for many years and is likely to take some time before actual commercialization. The perovskites are a group of materials that share a common structure, indicating numerous interesting features, such as magnetoresistance, superconductivity, and so on.

The future of solar cells is considered to be these simply modified materials since their characteristic structure makes them ideal for allowing effective and low-expense photovoltaics. They are also expected to play a significant role in lasers, sensors, future-generation electric vehicle batteries, and so on. A developing thin-film photovoltaic group is being studied, also named third-generation photovoltaics. This applies to photovoltaics-utilizing industries that have the ability, or are focused on new materials, to surpass the existing performance and efficiency limits. This third generation of photovoltaics includes perovskite photovoltaic, organic photovoltaics, quantum dot photovoltaic, and dye-sensitized solar cell. A perovskite-based flexible film is a kind of solar cell. As the light-harvesting energetic sheet, it

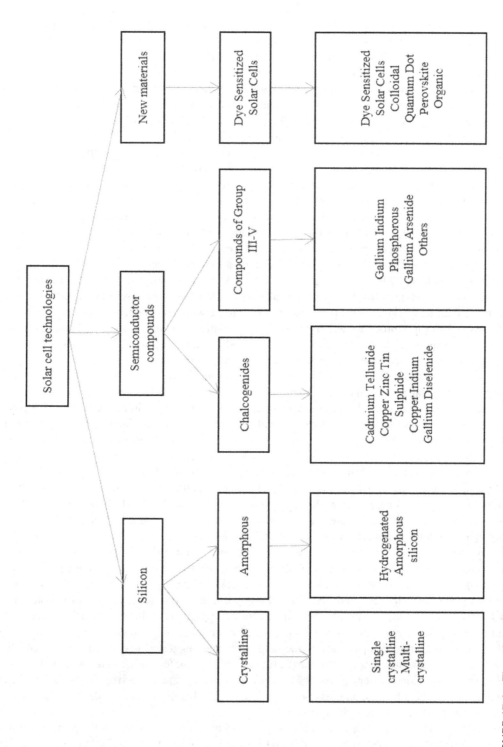

FIGURE 17.1 The solar cell sorting based on the main energetic material.

concludes a perovskite build combination, mostly a hybrid tin halide-sourced material or inorganic-organic lead. Perovskite-based materials are relatively simple to produce and cheap to manufacture, such as methylammonium lead halides. They have intrinsic features that make them attractive materials for solar cells at solid-state, such as fast-charge separation, long carrier separation lifetime, long transport distance of holes and electrons, broad absorption spectrum, and so on. Without a doubt, they are gaining interest in the photovoltaics field. With their capability to adsorb light across almost all noticeable wavelengths, their exceptional energy-transformation performances already passing 20% in the lab, and relative convenience of manufacturing, they are causing excitement within the photovoltaic energy sector. Perovskite-based films still face a few difficulties, but many studies are focusing on them, and some firms are already talking about commercializing them in the next years [7].

To deposit a sheet of perovskite, a few fabrication methods are available [8]. The three prime groups can be: vapor process, solution process, and hybrid (Figure 17.2) [9–10]. Through a tempering operation to type perovskite in steam operation, the pioneers are vaporized subsequently [11–12]. In co-vaporization, the pioneers are vaporized on a substrate together and, in successive operations, the pioneers are vaporized one after another. In resolution operation, the pioneers that are utilized to build up perovskite are melted in an organic solvent, generally in spin covered on the substrate and dimethyl formamide to consist of perovskite. The resolution operation can be of various kinds: a sole resolution operation in which the pioneers are combined in a sole resolution to spin-cover the perovskite sheet [13–14] or a successive operation in which the pioneers are covered one after another, with the perovskite subsequently creating the tempering operation [15]. Liu and his coworkers have also improved a methodology for developing perovskite grain size as a hoover flash-supported resolution operation [16]. Through perovskite tempering, both operations are associated with perovskite tempering. The hybrid operation, which is a vapor process and a mixture of the resolution, can also place perovskite, where the 104 pioneers are vaporized while the other is spin-covered on the substrate [17]. The most prevalent methodology utilized is sole resolution operation with co-evaporation and anti-solvent methodology.

Future-generation wearable and portable electronics are anticipated to be self-employed through compatible power storage tools that can supply power performance whenever required. The arising energy storage and harvesting interconnected system in an elastic assembly, in this regard, has proposed a favorable solution. Nonetheless, formidable difficulties with regard to the inadequate power density restricted all effectiveness, and low-performance voltage of the predominant interconnected energy resources is still present. Because of their flawless opto-electrical features, like strong absorptivity, long diffusion lengths, good carrier mobilities, tunable bandgaps, and small defect densities, perovskites have attracted important attention as favorable option absorber materials for opto-electronic tools specially for utilize in perovskite photovoltaics, where the effectiveness has risen to 22.70% [18] and 18.36% [19] on elastic-polymer and rigid-glass substrates, respectively. This result can be compared to the effectiveness of copper-indium-gallium-selenium and single-crystal silicon films.

Nowadays, numerous studies on these materials have been published. A perovskite-based flexible solar cell's characteristic structure is a clear conductive oxide substrate tracked through the perovskite, a metal electrode, electron move sheet, and a hole-move sheet. The latest rise in effectiveness is attributed to alterations to the tool structure and the material compositions' optimization. Given the potency of perovskite-based flexible solar cells for low-expense power production, the results of researches are significant [20–21]. Hwang and his coworkers compared a sequence of perovskite-based flexible films with differently sized SiO_2 nanoparticles as the help sheet and suggested that when the size of SiO_2 nanoparticles was 50 nm, the power-conversion efficiency was 11.45%, which was moderately higher than the power-conversion efficiency (10.29%) of the devices with the same-sized TiO_2 nanoparticles [22]. Japanese researchers Kojima and his coworkers discovered that dyes were similar to organic metal halide perovskite and were capable of absorbing sunlight.

As sensitizers in photo electrochemical films, Kojima and his coworkers affirmed the primary organic metal halide combinations $CH_3NH_3PbI_3$ and $CH_3NH_3PbBr_3$. They gauged a developed power-conversion performance of 3.13% for the $CH_3NH_3PbI_3$ and 3.81% for the $CH_3NH_3PbBr_3$-sourced tools, respectively

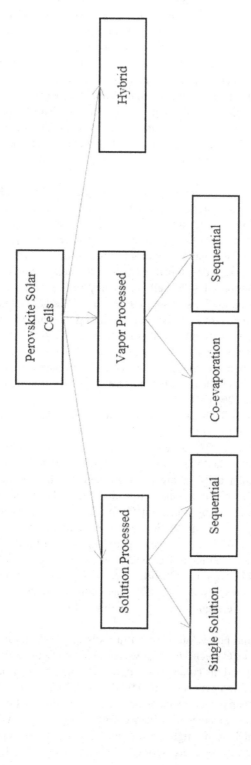

FIGURE 17.2 Diverse production methods for perovskite.

[23]. In 2011, the same team recorded a 6.5% power-conversion efficiency, still utilizing liquid electrolyte contact with $CH_3NH_3PbI_3$-sourced iodide, but with developed preparation terms. In *Nano Letters*, Sang Seok and his colleagues published encouraging results from the utilization of blended halide perovskite-based films by halide optimization in the combination $CH_3NH_3Pb(I1-xBrx)_3$ [24]. This paves the route for perovskite-based film performance's unheard-of growth throughout the years, with 12.3% of energy transformation performance supplied through greatly effective solid status films [25], 15% in 2013, 19.3% in the first half of 2014 [26,27]. Through the Korean Research Institute of Chemical Technology, a perovskite-based solar cell performance of 20.1% was recorded, through Yang et al. utilizing $FAPbI_3$ as an energetic sheet, which was confirmed since 2014 through National Renewable Energy Laboratory [28,29]. The overall solid-status $MAPbI_3$ films based upon the porous nano-TiO_2's mesoscopic structure were recorded through Kim et al. as well as Burschka and his coworkers. The perovskite-based film performances were reported to be 15.0% and 9.7%, and the filling factors were 0.73 and 0.62 [30,31]. The substantial development is primarily because of the high-standard perovskite-based films provided through the process of two-step solution deposition.

Burschka and his colleagues thought that once the two ingredients (PbI_2 and MAI) come into contact, the transformation takes place inside the nano-porous host, allowing much greater control over the morphology of the perovskite [31]. Yang and his coworkers recorded a perovskite-based solar cell efficiency larger than 20% in late 2014 utilizing mesoscopic TiO_2/$FAPbI_3$ perovskite-based fim architectonics [32]. Centered on a 1D TiO_2 nano-arraymesoscopic formation with $MAPbI_2Br$ as the light-adsorbing sheet, Qiu and his colleagues developed overall-solid-status perovskite-based films and found a V_{oc} of 0.82 V and a perovskite-based solar cells efficiency of 4.87% [33]. Qiu and his colleagues displayed that with the 1D TiO_2 nanowire's raising the length of the perovskite-based solar cells, efficiencies primarily raised and then diminished, therefore verifying a balance between perovskite loading and electron transport. The sintering with high-heat is involved in the perovskite-based films depending on mesoscopic TiO_2, and so many researchers have searched the structure of perovskite cells using Al_2O_3 as the mesoscopic material. Through utilizing polypolystyrene sulfonate to replace the traditional dense TiO_2 material, Malinkiewicz and his colleagues made $MAPbI_3$ perovskite-based films and accomplished a power-conversion efficiency of 12% [34].

Through utilizing yttrium-doped TiO_2 as the electron move material, Zhou and his colleagues obtained higher electron mobility and changed the ITO to minimize the working function preferred by injecting electrons from TiO_2 until the ITO electrode. The device's short-circuit current and open-circuit voltage were significantly improved, and the efficiency of power conversion was up by 19.3% [35]. To further develop the perovskite solar cells' stability due to their excellent durability, especially at high temperatures, the integration of overall-inorganic perovskites into solar-sourced tools has been offered. Swarnkar and his colleagues made stable quantum $CsPbI3$ films and dots that showed stability for 60 days when the films were exposed to the ambient atmosphere and displayed a V_{oc} of 1.23 V and an efficiency of power conversion of 10.77% or more in perovskite-based films [36]. $CsPbBr_3$ perovskite-based films with a power-conversion efficiency of 5.95% were reported by Kulbak and his colleagues, and the thermal durability of $MAPbBr_3$ and $CsPbBr_3$ were compared, corroborative that $CsPbBr_3$ can withstand much greater temperatures upwards of 580°C and showed greater stability in two weeks [37]. Though the studies described above mentioned high durability for solar tools, organic HTM was still present in the structures.

The design of overall inorganic perovskite-based film tools in which organic HTM and metal electrodes were entirely removed and a C electrode sheet was covered on the $CsPbBr_3$ sheet was recorded by Jin and his colleagues [38]. Nam and his colleagues introduced potassium cations into the $CsPbI_2Br$ and stated that the perovskite-based films depend on $Cs0.925K0.075PbI_2Br$ and exhibited 9.1% and 10.0% maximum and mean power-conversion efficiencies and showed higher stability than those depend on $CsPbI_2Br$ [39]. Marshall and his colleagues identified $CsSnI_3$ inorganic perovskite-based materials and signified that the addition of $SnCl_2$ to the light-absorbing sheet was most important for durability improvement and could not decrease the energy-conversion efficiency [40]. Liang and his colleagues suggested successive deposition operation within this second stage [41]. As the pioneer solution for spin-covering on the TiO$_2$ substrate, a saturated methanol solution of PbI$_2$ is utilized in this technique.

Roes and colleagues [42]; Espinosa and colleagues [43]; Anctil and colleagues [44]; and all of the above researches used a process approach to LCA, neglecting other upstream operations because of system boundary truncation recognized as a methodological limit for operation-sourced LCA at perovskite-based flexible solar cells. Anctil and coworkers [44], Valverde-García and coworkers [19], Roes and coworkers [42], Parisi and coworkers [45], and Espinosa and coworkers [43] conducted Life Cycle Assessment utilizing an operation strategy, neglecting other upstream events owing to systems boundary truncation, which has been noted as a methodological restriction for operation-based Life Cycle Assessment at perovskite-based flexible films. Perovskite-based solar cells' working mechanism is explained in the following section. Based upon the photovoltaic impact, solar films are among the best approaches to converting sun power into electrical power. The solar cell working mechanism is dependent on the separation, light absorption, and charge career set of the respective electrodes, which determines the potency variation to the other side of the p-n junction. The variation in voltage induced findings in the electrical power generation.

17.2 PEROVSKITE-BASED SOLAR CELLS' WORKING MECHANISM

Solar films are also referred to as first- and second-generation solar films, based upon thin-film and crystalline silicon industries. Restricted supply and the expense of silicon are the demerits. Developing photovoltaics as 3D generation was improved as an alternative. These contain dye-sensitized solar cells, quantum dots, organic photovoltaic, and perovskite-based solar film. Among these, perovskite-based solar films hold the top transformation effectiveness spot and are derived from dye-sensitized solar cells. CH$_3$NH$_3$PbI$_3$ has mostly been used as a sunlight absorber in organolead halide perovskite. To have maximum energy transformation effectiveness, perovskite was utilized to sensitize mesoporous TiO$_2$ solar cells. Ease of manufacturing, synthetic viability, and charge-recombination ratios are perovskite's essential aspects. Solid status sensitized films based upon dye-sensitized solar cells are the fundamental industry for perovskite photovoltaics. The key downside of dye-sensitized solar cells is the liquid electrolyte's utilization that learns from consistent issues, and when solid-state hole-transfer material is utilized, an open-circuit voltage may also be predicted to increase. The perovskite materials serve as electrolytes that excite holes and electrons to absorb sunlight. The electron transport material's electron acts as a semiconductor of the n-type, and the hole transport material's hole in acts as a semiconductor of the p-type. The fluorine-based tin oxide doped is utilized as an anode and as cathode silver or gold. These materials are placed as a solar cell fulfillment of the cell structure on top of the perovskite. Application demonstration and configurations of perovskite-based solar cells and perovskite-based flexible solar film samples are provided in the literature.

Elastic perovskite-based solar cells have attracted important interest because they are proper for mass generation through the roll-to-roll production methods, encouraging big potency for functional implementations in wearable, portable, and light electronic tools. Both the accessibility and low cost of the pioneer materials for the lamellas' simple formation and the perovskites through low-temperature methodologies give essential aid for the perovskites' utilization in elastic films. Unmistakably, the good attribute of the perovskite-based flexible films, with the inclusion of full coverage, uniform morphology, on a large scale focused crystalline design, and appropriate band hole, are preconditions for good effectiveness elastic perovskite-based films. On the basis of the above mentioned, a few diverse types of methodologies have been developed to optimize the perovskite-based elastic solar cells' features, such as through changing the production methods or through tuning the incorporated components' proportion in the perovskite build. Very recently, perovskite-based quantum points have been improved as an efficient absorbent material owing to some particular parameters and a high certain superficies field, which creates perovskite-based quantum points favorable nominees for optoelectronic and photovoltaic implementations [46].

So far, elastic nonvolatile memory-based perovskite-based quantum points [47], elastic light-emissive diodes [36], have been affirmed with encouraging efficiency. Especially, overall inorganic cesium lead halide perovskite-based quantum points have been meaningfully implemented in rigid perovskite-based solar cells, with effectiveness of over 13%. Although there has been no affirming on quantum-point-based elastic perovskite-based photovoltaics, processing with its low temperature creates a feasible option for elastic perovskite-based solar cells. Aside from the materials of the absorber, the elastic interface sheet and the electrodes also play essential roles in developing the devices' performance. Investigators have made progress in endeavoring to improve hole-transporting layers, electron move sheets, and electrode materials, and important accomplishments have been made, with the effectiveness raised to over 18% in a short time. All the same, three topics still challenge the further improvement of elastic perovskite-based films for practical implementation: inconvenient twisting stability, reduced effectiveness, and limited flexibility on a big scale. Traditional elastic perovskite devices are produced on polyethylenen aphthalate or polyethylene terephthalate substrates due to the bonding's lack and high layer resistance between the plastic substrates and electrode sheet. All the same, due to the lack of bonding and high layer resistance between the plastic substrates and the electrode sheet's rigid characteristic activates flaws in the lamella, which is the primary cause for tool malfunction after twisting twice, and most checks for twisting are performed with a twisting radius above 1 mm, or for restricted twisting cycles, none of which is adequate for allegating well twisting tolerance.

The identical troubles are observed in the elastic perovskite solar cells based upon other metal foil substrates or metal oxides. Thus, C materials like carbon nanotubes and graphene, and organic materials like polystyrene sulfonate, were utilized to stay away from these dehiscences troubles. Though the effectiveness is poorer than that of tools used due to the lack of bonding and high layer resistance between the plastic substrates and electrode sheet, they supply an encouraging means to develop twisting consistently, which is an important characteristic for elastic perovskite-based solar cells' practical implementations. Until today, almost all affirmed elastic perovskite-based solar cells have little fields. The photoconversion efficiency is decreased when the field of the tools is raised to big-scale, owing to the homogeneity inevitably lost in the lamellas. In consequence, the deposition procedure for the lamellas at big scale immediately identifies the efficiency of the big-field elastic perovskite-based solar cells. Thus, big-field methods need to be improved for the production of all the sheets in elastic films. The alternative vacuum relic industry should be presented into the roll-to-roll scheme to further decrease the production expense.

17.2.1 THE FUTURE OF PEROVSKITE-BASED SOLAR CELLS

By means of a simple manufacturing process, perovskite-based films can be low-expense materials. The perovskite-based solar cells have made substantial advances in performance, achieving high productivities in laboratory cells, and they will proceed to gain sophistication in maturity as consistency is improved. Generally, perovskite-based solar cells have both a fast payback period and a low cost of materials for energy (<0.22 years). The key difficulty for perovskite-based solar cells at present is to accomplish module-level long-term consistency and to provide sturdy outdoor activity. Within the settlement, it is important for the perovskite photovoltaic group to establish an opinion on the consistent measurement protocols and emit normed assay protocols. To speed up the aging operation under operating situations and accurately estimate the perovskite-based solar cell life, novel protocols should be developed. While adequate encapsulations should inhibit the perovskite absorber's degradation caused through oxygen or humidity in ambient air, the performance deterioration caused through heat and illumination should be addressed through interfaces and building stable materials. Compositional activities in engineering should aim to optimize energy creation and would possibly benefit from the robotic synthesis and evolving algorithms of machine learning. A decrease in performance is observed when the perovskite-based solar cell field raises [48]. It is predicted that the effectiveness aperture between industrial modules and laboratory cells will be reduced and will reach a grade equivalent to that of other photovoltaic industries, with proceeded endeavors by industrial and research societies toward the scaling up of perovskite-based solar cells. Expense per kWh is predominated by solar panel effectiveness and lifespan. The abundance of low-expense feedstocks and the easy processing of perovskite-based solar cells suggest ways of decreasing fabrication expenses below-mentioned traditional photovoltaic industries grades. On the long view, the lower cost of production will provide a more renewable decrease in the overall cost of the panel, rendering solar cells based on perovskite economic competitiveness.

17.3 METHODOLOGY

Many intangibles need to be incorporated into the decisions. To do so, they must be assessed alongside tangibles whose metrics must also be assessed as to how well the decision-goals makers are served. To reproduce prioritize scales, analytic hierarchy process is a theory by pairwise comparisons and depends upon the evaluations of specialists. It is these measures that, in notional terms, calculate abstracts. Comparisons are accomplished utilizing a measure of actual decisions that indicates how much more, with regard to a given quality, one variable beats another. The decisions may be incoherent, and how to calculate incoherence and develop the decisions, when feasible to find better coherence, is an interest of the analytic hierarchy process. The reproduced precedence measures are synthesized by multiplying them by their parent node's precedence and adding them to all of those nodes.

It is important to break down the decision into the below stages to decide on a structured route to produce priorities.

1. It identifies the issue and decides the form of expertise sought.
2. It constructs the judgment hierarchy with the purpose of the determination from the top, and then identifies the goals from a wide framework, by the intermediate norms to the least norm.
3. It constructs group matrices for pairwise comparison. To equate the elements in the norm directly following on this basis, each of the elements in the top norm is utilized.
4. It uses the precedence obtained from the comparisons to automatically weigh the precedence at the stage below. For every part, it does this. Then it applies its weighted values to each of the elements at the level below and obtains its all or global precedence. This adding and weighing process continues till the last preferences of the options at the lowest norm are found [49].

17.3.1 AHP ANALYSIS

For the purpose of evaluating perovskite alternatives, a set of characteristics are determined, as listed below (Table 17.1):

- PCE (photoconversion efficiency) (%)
- E_{loss} (energy loss) (eV)
- FF (fill factor) (%)
- Jsc (short-circuit current density) (mA cm^{-2})
- Vo (open circuit voltage) (V)

These characteristics are not expected to have the same impact on the overall score of alternatives. Thus, the relative priorities of these characteristics need to be determined based on expert evaluations. Experts are asked to rate the relative weights of these characteristics through a pair-wise comparison based on the scale of comparison in Table 17.2.

Based on the pair-wise comparison, the decision matrix presented in Table 17.3 is constructed.

TABLE 17.1

Characteristics [50]

Perovskite	Film	PCE (%)	E_{loss} (eV)	FF (%)	J_{SC} (mA cm^{-2})	V_{oc} (V)	Ref.
$CsPbI_3$	NC film	10.77	0.52	65	13.47	1.23	Swarnkar et al., 2016
$CsPbI_3$	NC film	13.4	0.55	78	14.37	1.20	Sanehira et al., 2017
$CsPbBr_3$	NC film	5.4	0.88	62	5.6	1.5	Akkerman et al., 2016
$CsPbBr_3$	NC film	9.72	0.93	82.1	8.12	1.45	Duan et al., 2018
$CsPbBr_3$	NC film	5.6	0.96	53	7.01	1.42	Hoffman et al., 2017
$CsPbBr_3$	NC film	10.6	0.77	–	–	1.61	Zhao et al., 2019
$CsPbBr_3$	NC film	4.57	0.96	36	9.41	1.34	This work
$CsPbBrI_2$	NC film	12.02	0.5	70	13.13	1.32	Zeng et al., 2018
$CsPbBrI_2$	NC film	12.39	0.73	80	12.93	1.19	Zhang J. et al., 2018

TABLE 17.2

The Comparison's Scale

Intensity	Definition	Explanation
1	Equal significance	Two events take part evenly to the target
2	Slight or weak	
3	Moderate significance	Judgement and experience insignificantly favor one event over another
4	Moderate positive	
5	Strong significance	Judgement and experience firmly favor one event over another
6	Strong positive	
7	Too firm or indicated significance	One event is favored too firmly over another; its superiority indicated practically
8	Very, very strong	
9	Extreme significance	The proof favoring one event over another is of the maximum feasible order of assertion
Reciprocals referred to above	If event i as compared to activity j, has one of the above nonzero numbers assigned to it, then j has the reciprocal value compared to I	An acceptable assertion
1.1–1.9	If the events are too close	

TABLE 17.3

The Decision Matrix

Matrix		PCE (%)	E_loss (eV)	FF (%)	J_sc (mA cm−2)	V_oc (V)
		1	2	3	4	5
PCE (%)	1	1	7	5	2	2
E_loss (eV)	2	1/7	1	1/2	1/5	1/3
FF (%)	3	1/5	2	1	3	3
J_sc (mA cm−2)	4	1/2	5	1/3	1	2
V_oc (V)	5	1/2	3	1/3	1/2	1

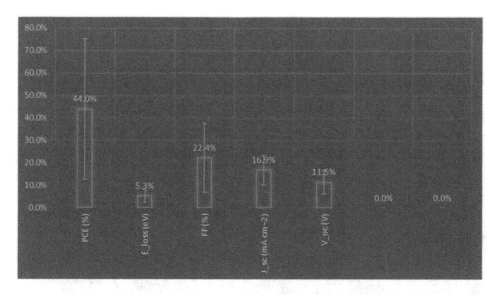

FIGURE 17.3 The criteria's relative priorities.

Figure 17.3 indicates the relative priorities of the characteristics used. Based on the relative priorities shown in Figure 17.1, PCE is the characteristic with the highest impact.

For the purpose of this study, nine alternatives are determined. The scores of these alternatives are provided in Figure 17.4.

Figure 17.5 further presents the overall scores, as indicated in the last row of the table. The results indicate that *CsPbI3* has the highest score (0.1319) among all other alternatives. Figure 17.6 presents the schematic overview of the applied method.

17.4 CONCLUSIONS

One of the most desirable clean-power resources is solar cells. Perovskite-based flexible solar cells have made unprecedented strides during the last decades, to the point that they now demonstrate exceptional maximum efficiency. Thus, perovskite-based flexible solar films are noted to be an outstanding nominee for replacing the common solar cells based on Si that currently predominate the photovoltaic market. Future-generation portable and wearable electronics are predicted to be self-energized through compatible power storage tools that can supply power performance when necessary. In this respect, the emerging interconnected system for energy storage and harvesting in an elastic assembly has provided an encouraging resolution. Nonetheless, daunting difficulties with regard to the inadequate power density restricted all effectiveness and low throughput voltage of the predominant interconnected energy resources.

Due to numerous products with different properties, the perovskite materials' commercial application and proper methodologies have become essential to assess perovskite materials mentioned in all ways. The goal of this search is to shed light on the comparison of the solar films produced in different types according to the main operational characteristics and evaluating the optimum photovoltaic option, among others. Therefore, it is a valuable contribution from three perspectives: assessing a series of criteria needed in the assessment of perovskite-based flexible solar cells, defining their relative precedence based upon expert options, and supplying a simple numerical examination. In further investigations, it is prospected to raise the variable number and change their comparative precedence based upon new appearances.

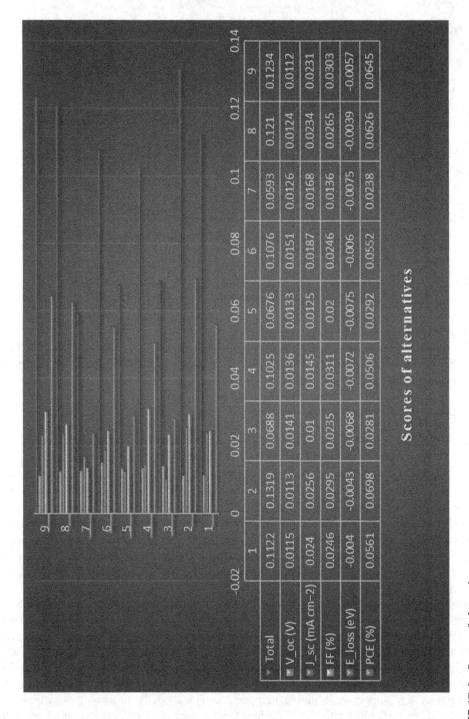

Scores of alternatives

	1	2	3	4	5	6	7	8	9
Total	0.1122	0.1319	0.0688	0.1025	0.0676	0.1076	0.0593	0.121	0.1234
V_oc (V)	0.0115	0.0113	0.0141	0.0136	0.0133	0.0151	0.0126	0.0124	0.0112
J_sc (mA cm−2)	0.024	0.0256	0.01	0.0145	0.0125	0.0187	0.0168	0.0234	0.0231
FF (%)	0.0246	0.0295	0.0235	0.0311	0.02	0.0246	0.0136	0.0265	0.0303
E_loss (eV)	−0.004	−0.0043	−0.0068	−0.0072	−0.0075	−0.006	−0.0075	−0.0039	−0.0057
PCE (%)	0.0561	0.0698	0.0281	0.0506	0.0292	0.0552	0.0238	0.0626	0.0645

FIGURE 17.4 Scores of alternatives.

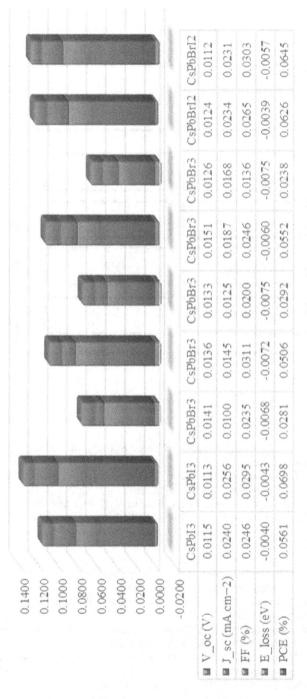

	CsPbI3	CsPbI3	CsPbBr3	CsPbBr3	CsPbBr3	CsPbBr3	CsPbBr3	CsPbBrI2	CsPbBrI2
V_oc (V)	0.0115	0.0113	0.0141	0.0136	0.0133	0.0151	0.0126	0.0124	0.0112
J_sc (mA cm−2)	0.0240	0.0256	0.0100	0.0145	0.0125	0.0187	0.0168	0.0234	0.0231
FF (%)	0.0246	0.0295	0.0235	0.0311	0.0200	0.0246	0.0136	0.0265	0.0303
E_loss (eV)	−0.0040	−0.0043	−0.0068	−0.0072	−0.0075	−0.0060	−0.0075	−0.0039	−0.0057
PCE (%)	0.0561	0.0698	0.0281	0.0506	0.0292	0.0552	0.0238	0.0626	0.0645

■ PCE (%)   ■ E_loss (eV)   ■ FF (%)   ■ J_sc (mA cm−2)   ■ V_oc (V)

FIGURE 17.5 Resulting scores.

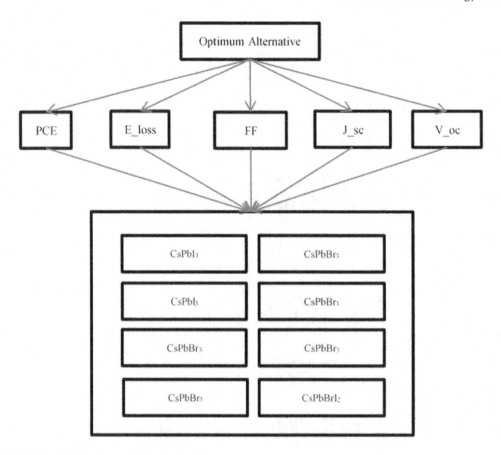

FIGURE 17.6   Schematic overview.

REFERENCES

1. World Energy Investment, 2020, IEA (International Energy Agency), July 2020.
2. The Perovskite Handbook, 2020 edition.
3. T. Ibn-Mohammeda, S. C. L. Kohab, I. M. Reaneyc, A. Acquayed, G. Schileoc, K. B. Mustaphae, and R. Greenoughf, *Renew Sust Energ Rev*, 80, 1321–1344 (2017).
4. Jean J., Brown P. R., Jaffe R. L., Buonassisi T., and Bulović V. *Energy Environ Sci*, 8, 1200–1219 (2015).
5. Zhou Y. *Eco- and renewable energy materials*. Springer; 2015.
6. Kilner J. A., Skinner S., Irvine S., and Edwards P. *Functional materials for sustainable energy applications*. Elsevier; 2012.
7. The Perovskite Handbook, January 2021 edition.
8. S. M. Istiaque Hossain, Performance and stability of perovskite solar cells. *Graduate Theses and Dissertations*, 16379 (2018). https://lib.dr.iastate.edu/etd/16379
9. W. Tress, N. Marinova, T. Moehl, S. M. Zakeeruddin, Mohammad Khaja Nazeeruddin and M. Grätzel, *Energy Environ Sci*, 8, 995–1004 (2015).
10. Hisham A. Abbas, Ranjith Kottokkaran, Balaji Ganapathy, Mehran Samiee, Liang Zhang, Andrew Kitahara, Max Noack and Vikram L. Dalal, *APL Materials*, 3, 016105 (2015).
11. Zhengguo Xiao, Yongbo Yuan, Yuchuan Shao, Qi Wang, Qingfeng Dong, Cheng Bi, Pankaj Sharma, Alexei Gruverman and Jinsong Huang, *Nat Mater*, 14, 193–198 (2015).
12. Yehao Deng, Zhengguo Xiao and Jinsong Huang, *Adv Energy Mater* (2015), doi: 10.1002/aenm.201570107.
13. Wanyi Nie, Hsinhan Tsai, Reza Asadpour, Jean-Christophe Blancon, Amanda J. Neukirch, Gautam Gupta, Jared J. Crochet, Manish Chhowalla, Sergei Tretiak, Muhammad A. Alam, Hsing-Lin Wang and Aditya D. Mohite, *Science*, 347(6221), 522–525 (2015).

14. Nam-Gyu Park, J. *Phy Chem Lett*, 4, 2423–2429 (2013).
15. Nam Joong Jeon, Jun Hong Noh, Woon Seok Yang, Young Chan Kim, Seungchan Ryu, Jangwon Seo and Sang Il Seok, *Nature*, 517, 476 (2015).
16. Dianyi Liu and Timothy L. Kelly, *Nat Photonics*, 8, 133–138 (2014).
17. E. L. Unger, E. T. Hoke, C. D. Bailie, W. H. Nguyen, A. R. Bowring, T. Heumüller, M. G. Christoforo and M. D. McGehee, *Energy Environ Sci*, 7, 3690–3698 (2014).
18. García-Valverde R., Cherni J. A., and Urbina A., *Progress Photovolt: Res Appl*, 18, 535–558 (2010).
19. C. Wang, L. Guan, D. Zhao, Y. Yu, C. R. Grice, Z. Song, R. A. Awni, J. Chen, J. Wang, X. Zhao, and Y. Yan, *ACS Energy Lett.*, 2(9), 2118–2124 (2017).
20. Patwardhan S., Cao D. H., Hatch S., Farha O. K., Hupp J. T., Kanatzidis M. G., et al., *J Phys Chem Lett*, 6, 251–255 (2015).
21. Snaith H. J., *J Phys Chem Lett*, 4, 3623–3630 (2013).
22. H. Zhou, Q. Chen, G. Li, S. Luo, T. B. Song, H. S. Duan, Z. Hong, J. You, Y. Liu, and Y. Yang, *Science*, 345, 542–546 (2014).
23. Im J.-H., Lee C.-R., Lee J.-W., Park S.-W., and Park N.-G., *Nanoscale*, 3(10), 4088–4093 (2011).
24. Noh J. H., Im S. H., Heo J. H., Mandal T. N., and Seok S. I., *Nano Lett*, 13(4), 1764–1769 (2013).
25. Assadi M. K., Bakhoda S., Saidur R., and Hanaei H., *Renew Sustain Energy Rev*, 81, 2812–2822 (2017).
26. Tong X., Lin F., Wu J., and Wang Z. M., *Adv Sci*, 3, 5 (2016).
27. Jung H. S., and Park N. G., *Small*, 11(1), 10–25 (2015).
28. Green M. A., and Ho-Baillie A., *ACS Energy Lett*, 2, 822–830 (2017).
29. Bhatt M. D., and Lee J. S., *New J Chem*, 41(19), 10508–10527 (2017).
30. H.-S. Kim, C.-R. Lee, J.-H. Im et al., *Scientific Reports*, 2, article 591 (2012).
31. J. Burschka, N. Pellet, S. Moon et al., *Nature*, 499(7458), 316–319 (2013).
32. W. S. Yang, J. H. Noh, N. J. Jeon et al., *Science*, 348(6240), 1234–1237 (2015).
33. J. Qiu, Y. Qiu, K. Yan et al., *Nanoscale*, 5(8), 3245–3248 (2013).
34. O..Malinkiewicz, A. Yella, Y. H. Lee et al., *Nature Photonics*, 8(2), 128–132 (2014).
35. H. Zhou, Q. Chen, and G. Li, *Science*, 345(6196), 542–546 (2014).
36. A. Swarnkar, A. R. Marshall, E. M. Sanehira et al., *Science*, 354(6308), 92–95 (2016).
37. M. Kulbak, S. Gupta, N. Kedem et al., *J Phy Chem Lett*, 7(1), 167–172 (2016).
38. J. Liang, C..Wang, Y..Wang et al., *J Am Chem Soc*, 139(7), 2852 (2017).
39. J. K. Nam, S. U. Chai, W. Cha et al., *Nano Lett*, 17(3), 2028–2033 (2017).
40. K. P. Marshall, M. Walker, R. I. Walton, and R. A. Hatton, *Nat Energy*, 1, 16178 (2016).
41. K. Liang, D. B. Mitzi, and M. T. Prikas, *Chem Mater*, 10(1), 403–411 (1998).
42. Roes A., Alsema E., Blok K., and Patel M. K. *Progress Photovolt: Res Appl*, 17, 372–393 (2009).
43. Espinosa N., García-Valverde R., Urbina A., Lenzmann F., Manceau M., Angmo D., et al., *Sol Energy Mater Sol Cells*, 97, 3–13 (2012).
44. Anctil A., Babbitt C., Landi B., and Raffaelle R. P. Life-cycle assessment of organic solar cell technologies. Photovoltaic Specialists Conference (PVSC), 2010 35th IEEE; 2010. pp. 000742–000747.
45. Parisi M. L., Maranghi S., and Basosi R., *Renew Sustain Energy Rev*, 39, 124–138 (2014).
46. K. Yang, F. Li, C. P. Veeramalai, and T. Guo, *Appl. Phys. Lett.*, 110, 083102 (2017).
47. Y. Li, Y. Lv, Z. Guo, L. Dong, J. Zheng, C. Chai, N. Chen, Y. Lu, and C. Chen, *ACS Appl Mater Interfaces*, 10, 15888–15894 (2018).
48. Z. Li et al., *Nat Rev Mater*, 3, 18017 (2018).
49. O. S. Vaidya, and S. Kumar, *Eur J Operat Res*, 169, 1–29 (2006).
50. Lei Zhang, Tianle Hu, Jinglei Li, Lin Zhang, Hongtao Li, Zhilun Lu and Ge Wang, *Front Mater*, 6, 330 (2020).
51. Henry J. Snaith, Antonio Abate, James M. Ball, Giles E. Eperon, Tomas Leijtens, Nakita K. Noel, Samuel D. Stranks, Jacob Tse-Wei Wang, Konrad Wojciechowski, and Wei Zhang, J. *Phy Chem Lett*, 5(9), 1511–1515 (2014).
52. Hui-Seon Kim and Nam-Gyu Park, J., *Phy Chem Lett*, 5(17), 2927–2934 (2014).
53. Mingzhen Liu, Michael B. Johnston and Henry J. Snaith, *Nature*, 501, 395 (2013).
54. Zhao Y., and Zhu K. *J Phys Chem Lett*, 5, 4175–4186 (2014).

18 Flexible Batteries Based on Li-Ion

Solen Kinayyigit
Nanocatalysis and Clean Energy Technologies Laboratory, Institute of
Nanotechnology, Gebze Technical University, Gebze, Kocaeli, Turkey

Emre Bicer
Battery Research Laboratory, Faculty of Engineering and Natural Sciences,
Sivas University of Science and Technology, Sivas, Turkey

CONTENTS

18.1 Introduction 323
18.2 Flexible electrodes 324
    18.2.1 Flexible anodes 324
        18.2.1.1 Carbon materials 324
        18.2.1.2 Mxenes 331
    18.2.2 Flexible cathodes 331
18.3 Flexible electrolytes 333
18.4 Battery structures 335
18.5 Fabrication of FLIBs 338
18.6 Conclusion 338
References 340

18.1 INTRODUCTION

Energy is an indispensable need in our daily life. Especially in mobile applications, energy storage is carried out with batteries. Batteries are devices that store electrical energy as chemical energy and return it as electrical energy when desired. The energy requirement of portable electronic devices has made it necessary to develop special types of batteries. These batteries must meet certain features, such as flexibility, lightweight, and small volume coverage to be used in such devices. Among these battery types, the most widely used ones are lithium-based batteries due to their high energy/low weight ratio.

With the rise of new technologies, such as wearable and flexible electronics, there is also high demand for compatible and flexible energy storage devices. During the last decade, studies focused on the flexibility of Li-ion batteries (LIBs), where various forms such as film, fiber, textile, and other different geometries have been reported to enable bendable, twistable, foldable, and stretchable properties. Many different fabrication techniques were implemented in these reported results, such as CVD, ultrasonication, nanofabrication, sputtering, printing, and spray-painting. Nevertheless, these techniques are still not feasible for bulk industrial productions. A flexible battery requirement may differ for various applications. For instance, a typical mobile phone of ~3 mm thickness is exposed to a ~5% nominal strain in a bent condition [1]. To avoid possible damage of battery components and thermal runaways, unconventional designs that allow bending and twisting of the battery with no energy losses and fractures are required. Another important aspect of the flexible Li-ion battery (FLIB) is its power output during deformation. Accordingly, the overpotential change should be less than 100 mV, and the cell impedance should not double during the charging/discharging [1].

18.2 FLEXIBLE ELECTRODES

18.2.1 FLEXIBLE ANODES

18.2.1.1 Carbon materials

Traditional LIBs with metal-foiled electrodes have disadvantages of lack of bendability, stretchability, and portability. Two approaches are reported in the literature to solve the flexibility issue of electrodes (i) using conductive and flexible materials together with active materials, and (ii) applying a non-conductive material having excellent mechanical properties [2]. Each method has both advantages and disadvantages in a LIB in terms of its electrochemical properties, such as energy and power capacity, rate-capability, and cycle life. Traditional LIBs usually contain carbon-based anode materials, which have the ability of intercalating and deintercalating of lithium ions (Li^+) between the carbon layers. Inorganic metal oxides of Fe, Mn, Ni, Co, Mo, Sn, Sb, etc., may also be utilized, along with the conductive carbon. However, due to the brittle nature of the traditional materials used, it is not possible to employ them as a flexible anode material. Therefore, bendable conductive support materials (carbon nanotubes, graphene, carbon nanofiber, carbon cloth, etc.) are used together with nonconductive highly elastomeric flexible materials, such as cellulose, polymer, and paper, to constitute the anode in FLIBs [3].

Carbon nanotubes (CNTs) are rolled graphene sheet(s) structurally and due to the delocalized π-bonds and continuous sp^2 hybridization, they exhibit excellent conductivity (~103 S/cm) and high surface area (~1600 m^2/g) with a reversible capacity between 300 mAh/g and 600 mAh/g [4]. CNTs demonstrate high electrochemical and mechanical performances in FLIBs. Among CNTs, multi-wall carbon nanotubes (MWCNTs) display better charge/discharge performances compared to those of single-wall carbon nanotubes (SWCNTs) [5]. CNTs may also be used as free-standing anodes without the need for copper or aluminum current collectors, conductive additives, and binders [2].

Bensalah et al. introduced MWCNT-Si synthesized by physical vapor deposition followed by acid-treatment, and this process demonstrated 2150 mAh/g after 100 cycles [6]. Another 3D free-standing binderless electrode fabrication was introduced by Elizabeth et al., where Sb_2S_3 rods were first synthesized by a solvothermal process in ethylene glycol media and used for the preparation of Sb_2Sn_3/CNT composites through a vacuum-filtration method to give a capacity of 450 mAh/g after 100 cycles [7]. Several prelithiation methods have been explored to prevent the capacity loss of anode materials, among which Sugiawati et al. reported an electropolymerization of p-sulfonated poly(allylphenylether) (SPAPE) and direct pre-lithiation treatment on CNT as a flexible anode for Li-ion micro batteries (Figure 18.1a–e). A lightweight, free-standing, binderless, and flexible SPAPE/CNT with electrochemical pre-lithiation showed an excellent capacity of 508 mAh/g at 10C, even after 500 cycles (Figure 18.1f) [8]. In conclusion, CNT-based cathodes are mostly reported in two types structurally; (i) CNTs with active materials embedded to inner space (inner implementation), and (ii) active materials attached to an outer surface of CNTs (outer deposition) [3]. In the case of outer deposition, since the electrolyte has no direct contact with the CNT, a poor electrochemical performance causing an unstable solid-electrolyte interface (SEI) is observed. Moreover, inner implementation prevents active material volume expansion. Thus, systems with inner implementation are considered better than those of outer deposition. Other CNT-based flexible electrode structures are also summarized in Table 18.1 at the end of this section.

Graphene has received significant interest due to its unique thermal, mechanochemical, electrical, and optical properties since its first discovery in 2004 [9]. In the last decade, potential applications of graphene grew in the fields of nanoelectronics, sensors, capacitors, composites, photovoltaics, and batteries. Owing to its outstanding stability, mechanical flexibility, and high specific surface area, and electrical

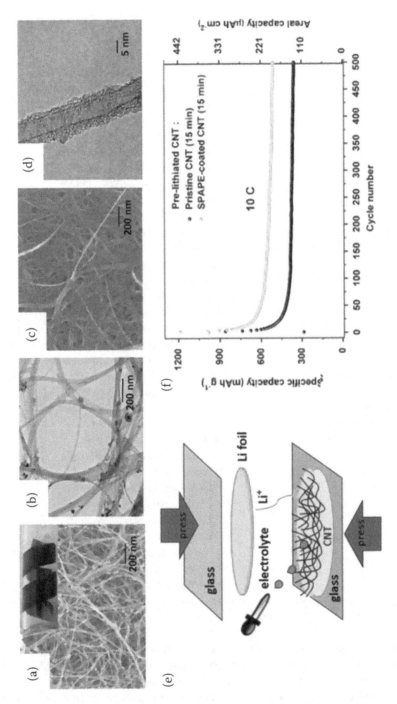

FIGURE 18.1 (a) an SEM image of pristine carbon nanotube (CNT) with an inset photograph of a flexible CNT, (b) a TEM image of pristine CNT, (c) an SEM image of CNT coated with a polymer, p-sulfonated poly(allyl phenyl ether (SPAPE), (d) TEM image of SPAPE-CNT, (e) an illustration of the direct pre-lithiation method, and (f) Cycle life graph of the prepared CNT electrodes. (Figure 18.1 adapted with permission from ref [8]. Copyright The Authors, some rights reserved; exclusive licensee MDPI. Distributed under a Creative Commons Attribution License 4.0 (CC BY), https://creativecommons.org/licenses/by/4.0/).

TABLE 18.1

Various Carbon-Based Materials Utilized in Glexible Anode Material in Flexible Li-ion Batteries

Carbon-based materials	Electrochemical performance	Ref.
MWCNT-Si	2150 mAh/g – 100 cycles – 420 mA/g	[6]
Co-Co$_3$O$_4$/CNT	505 mAh/g – 100 cycles – 500mA/g	[33]
SnO$_2$/CNT	2030 mAh/cm^2 – 20 cycles – 250 mA/g	[37]
SnSb-CNTs@NCNFs	451 mAh/g – 1000 cycles – 2000 mA/g	[38]
Sb$_2$S$_3$/CNT	443 mAh/g – 100 cycles – 200 mA/g	[7]
Pre-lithiated polymer-coated CNT	463 mAh/g – 500 cycles – 1.2 mA/cm^2	[8]
GN/NC/Si	1251 mAh/g – 100 cycles – 100mA/g	[11]
fNiO/GP	359 mAh/g – 600 cycles – 1000mA/g	[14]
SnO$_2$ QDs@GF	320 mAh/g – 1000 cycles – 5000mA/g	[16]
Hierarchical SnS$_2$ nanoplates decorated on graphene	1987.4 mAh/g – 150 cycles – 300mA/g	[39]
Germanium/CNFs	1420 mAh/g – 100 cycles – 244 mA/g	[40]
Sn/CNFs	350 mAh/g – 200 cycles – 200 mA/g	[41]
NiSb@NCNFs	510 mAh/g – 2000 cycles – 2000 mA/g	[25]
SnS/CNFs	349 mAh – 500 cycles – 200 mA/g	[42]
ECNFs	715 mAh/g – 80 cycles – 50 mA/g	[43]
Si-CNF composite	580 mAh/g – 100 cycles – 1000 mA/g	[23]
LiCoO$_2$/CNF	412 mAh/g – 100 cycles – 100 mA/g	[44]
CoO/graphene/CNFs	690 mAh/g – 352 cycles – 500 mA/g	[45]
Fe$_3$O$_4$/CNFs	754 mAh/g – 100 cycles – 100 mA/g	[46]
MnO/CNFs	987 mAh/g – 50 cycles – 100 mA/g	[47]
Fe$_2$O$_3$/carbon cloth	1350 mAh/g – 100 cycles – 100 mA/g	[48]
ZnO@TiO$_2$ nanorod arrays on carbon cloth	1000 mAh/g – 250 cycles – 1000 mA/g	[26]
NiO/carbon cloth	3.08 mAh/cm^2 – 300 cycles – 0.25 mA/cm^2	[49]
MoS$_2$ nanoflake array/carbon cloth	3 mAh/cm^2 – 30 cycles – 0.15 mA/cm^2	[28]
TiO$_2$@titanium nitride nanowires/carbon cloth	202 mAh/g – 650 cycles – 3350 mA/g	[50]
SnO$_2$@TiO$_2$ double-shell nanotubes/carbon cloth	779 mAh/g – 100 cycles – 780 mA/g	[51]
VO$_2$(B)/carbon cloth	138 mAh/g – 100 cycles – 100 mA/g	[52]
FeS@C/carbon cloth	420 mAh/g – 100 cycles – 91.35 mA/g	[53]
FeP@C/carbon fabric	718 mAh/g – 670 cycles – 500 mA/g	[30]
NiCo$_2$O$_4$/V$_2$O$_5$ /carbon fabric	364.2 Wh/g – 100 cycles – 240 W/kg	[29]
TiO$_2$/carbon textile	192 mAh/g – 2000 cycles – 680 mA/g	[31]
Nanostructured graphite paper	322 mAh/g – 100 cycles – 100 mA/g	[32]
Co$_9$O$_8$ /carbon foam	615 mAh/g – 450 cycles – 500 mA/g	[36]
N-doped carbon foam@rGO-TiO$_2$	156 mAh/g – 3000 cycles – 10C	[34]
SnS/C/carbon foam	747 mAh/g – 100 cycles – 50 mA/g	[35]

conductivity, graphene has already been exploited in FLIBs. A wide range of graphene-containing flexible anode composite materials has already been investigated, including Si, SiO, Mn$_3$O$_4$, Co$_3$O$_4$, Fe$_2$O$_3$, and TiO$_2$ [3–10]. Despite the features of graphene mentioned above, it still suffers from low theoretical capacity and rapid capacity loss in cycles. Therefore, graphene has been used as a backbone conductive material in most cases.

Although silicon as an anode material has a very high theoretical capacity, its volume expansion is ~300%, deteriorating the cycle life. Zhou et al. reported the fabrication of silicon-based anode together with graphene sheets to construct a self-supported, flexible, free-standing composite anode [11]. Graphene blended with nanocellulose acted as both an electronic conductor and a current collector. Electrochemical results showed a reversible capacity of 1251 mAh/g after 100 cycles at 100 mA/g and 405 mAh/g at 6.4 A/g [11].

Sang et al. introduced a highly flexible anode that was composed of silicon oxycarbide ceramic (SiOC) fibers cloth with a quasi-3D graphene framework. Enhanced Li$^+$ ion mobility due to the porous and rough structure of the SiOC fibers combined with the high electron network in the flexible electrode owing to the 3D interconnected graphene enabled a long cycle life, achieving also a high capacity of 924 mAh/g at 0.1 A/g. In addition, it displayed a better rate performance of 330 mAh/g at 2.0 A/g and very good cycling stability (~686 mAh/g at 0.5 A/g) after 500 cycles [12]. Murata et al. reported a flexible anode consisting of thick multilayer graphene (G), nickel catalyst, and polyimide polymer layer giving no significant capacity fading during cycling [13]. Graphene was also utilized as a conductive material together with NiO micro-flower (fNiO) nanoparticles (NPs) to construct a 3D porous hybrid paper (Figure 18.2a) [14]. fNiO NPs were dispersed in a solution of graphene oxide (GO), which was thermally reduced to graphene at 400°C under Ar atmosphere. This hybrid fNiO/G paper electrode demonstrated a good cycling capability of 355 mA/g after 600 cycles (Figure 18.2b) [14]. Deposition of SnO$_2$ quantum dots (QDs) on graphene framework (GF) was achieved by a hydrothermal method, and the resulting SnO$_2$/GF exhibited an excellent electrochemical performance compared to those of sole SnO$_2$ QDs (initial: 1494 mAh/g; after 100 cycles: 885 mAh/g) and reduced graphene oxide (rGO), with a capacity of 1260 mAh/g after 100 cycles at 0.1 A/g (Figure 18.2c–f) [15]. It also showed a long cycle performance of 320 mAh/g at 5 A/g after 1000 cycles [16]. These excellent electrochemical properties are due to the homogeneous dispersion of SnO$_2$ QDs on conductive GF, which in return gives a decrease of large volume change and a shortened path length for the lithium-ion diffusion, providing an increased electron transport. Other graphene-based flexible electrode structures are summarized in Table 18.1.

Similar to carbon nanotubes and graphene, carbon nanofibers (CNFs) are also lightweight and display remarkable mechanical properties, good electrical conductivity, and high surface area. Because of these features, CNFs are employed as an anode material in FLIBs. A wide variety of CNFs are used together with previously known anode active materials, some of which are Li$_4$Ti$_5$O$_{12}$, MnO, Fe$_3$O$_4$, TiO$_2$, Ge, Si [3–10]. Nano-Fe$_3$O$_4$ anchored on 3D CNT and aerogels (Fe$_3$O$_4$@BC-CNF) produced a highly flexible structure without using any current collector, binder, nor conductive additives, providing a free-standing flexible anode [17]. This anode was stable and gave a good rate-capability (~755 mAh/g up to 80 cycles). Bismuth has a five-fold higher volumetric capacity (~3800 mAh/cm^3) compared to graphite and also has a low potential hysteresis of 0.11 V [18]. Therefore, Bismuth NPs were also used in a cucumber-shaped 3D cross-linked CNT network by electrospinning and carbonization method, where the resulting free-standing flexible electrode delivered a reversible capacity of 322 mAh/g after 500 cycles at 0.25 A/g [19]. TEMPO-oxidized cellulose nanofibrils as a binder and carbon fiber (CF) as current collectors were combined with Li$_4$Ti$_5$O$_{12}$ LTO to form LTO/CF. The rate-capability tests conducted before and after bending showed only small capacity retention. It was also noted that after 4000 repeated bending tests, the electrochemical performance remained almost stable [20]. Thirugnanam et al. reported another study where TiO$_2$ synthesized with N-doped helical CNF as a hybrid electrode material showed a specific surface area of 295 m^2/g and a highly reversible capacity of 316 mAh/g and 244 mAh/g at 75 mA/g and 186 mA/g, respectively, after 100 cycles [21].

In addition to metal oxide-based flexible anode active compounds, there are also sole metal/CNF structures [22]. Germanium (Ge) nanowires together with a thin amorphous carbon layer were grown on 3D interconnected CNFs as a self-supported flexible electrode, showing 1520 mAh/g after 100 cycles at 0.1C. In addition, a long cycle with 840 mAh/g after 200 cycles at 2C was observed. The remarkable electrochemical performance and stable electrical conductivity observed were most likely due to the synergistic effect of carbon, Ge nanowires, and 3D structured CNF framework that prevents the volume expansion of embedded Ge NPs [22]. Another metal/CNF hybrid structure is Si NP/CNF self-standing

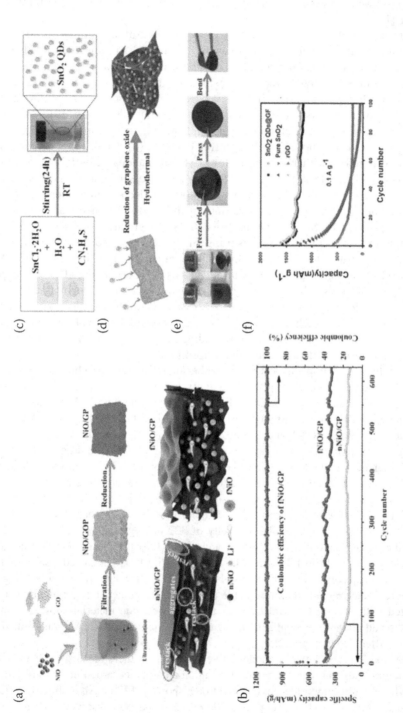

FIGURE 18.2 (a) A scheme of the fabrication of the anode based on graphene with NiO NPs (NiO/GP), (b) Cycle life graph of the nano- and microflower-NiO/GP based anodes at 1 A/g. (a–b: Adapted with permission from ref. [14], Copyright (2021), Elsevier). (c) An illustration of the synthesis of SnO$_2$ quantum dots (QDs), (d) An illustration of the anode material prepared by the hydrothermal synthesis method, (e) Photographs of an aqueous solution of SnO$_2$ QDs@GO, SnO$_2$ QDs@GF gel, and SnO$_2$ QDs@GF monoliths, (f) Cycle life graph of SnO$_2$ QDs@GF, SnO$_2$, and rGO at 100 mA/g. (c–f: Adapted with permission from ref. [16], Copyright (2020), American Chemical Society).

anode material without any binder nor a current collector. The electrode demonstrated a high capacity of 580 mAh/g after 100 cycles at 1 A/g, and a high-rate capacity of 242 mAh/g at 6C. The mean diameter of CNFs was ~230 nm, with a high surface area improved kinetics of Li^+ diffusion and electron transportation. Moreover, the electrolyte had no direct access to the encapsulated Si NPs in the CNFs [23]. Antimony with a high theoretical capacity (~660 mAh/g) is an important metal for LIBs [24]. Chen et al. used antimony, however, to provide flexibility and prevent the disadvantageous properties of antimony, such as poor cycling stability and structural deterioration, a composite of nitrogen-doped CNF (NCNF) with the intermetallic compound, NiSb was prepared [25]. Due to the highly conductive network of CNF and the inhibition of aggregation of NiSb, the resulting composite, NiSb@NCNF, delivered a high specific capacity (~720 mAh/g at 100 mA/g) and a very good cycle life (510 mAh/g after 2000 cycles at 2000 mA/g) with a 98% capacity fading (Figure 18.3). Other CNF-based flexible electrode structures are summarized in Table 18.1.

Providing flexibility is an important issue when combining electrode materials. One of the explored materials, carbon cloth, has high corrosion resistance and tensile strength, as well as good mechanical flexibility and conductivity [3]. What is more, ease of commercial availability makes the carbon cloth an ideal candidate. Various numbers of anode materials on carbon cloth were employed to construct flexible anode materials, such as Fe_2O_3, NiO, $ZnO@TiO_2$ nanorods, MoS_2 nanoflakes, $TiNb_2O_7$ [3–10].

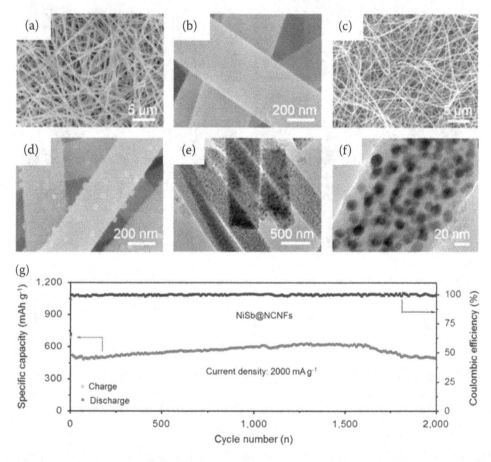

FIGURE 18.3 (a–b) SEM images of the nanofabric prior to annealing process, (c–d) SEM and (e–f) TEM images of the NiSb@NCNFs, and (g) Cycle life and coulombic efficiency graph of the anode based on NiSb@NCNF at 2000 mA/g. (Adapted with permission from ref. [25], Copyright (2019), Royal Society of Chemistry).

ZnO@TiO$_2$ core/shell nanorod grown on carbon cloth improved the electronic conductivity of TiO$_2$ while preventing the ZnO volume expansion during charge/discharge cycles [26]. Compared with pure ZnO with a theoretical capacity of 987 mAh/g, high capacity and high rate capability properties were obtained within 1000 mAh/g for 250 cycles at 1 A/g. NiO nanosheets mesoporous in nature were grown on a 3D-carbon cloth as a flexible anode for LIBs with no binder added, and their electrochemical characterization indicated a high reversible capacity and a good rate capability of 893 mAh/g after 120 cycles at 100 mA/g and 758 mAh/g at 700 mA/g after 150 cycles (Figure 18.4). The woven morphology of the unloaded and loaded carbon cloth with NiO nanosheets can be seen in Figure 18.4a [27].

Another carbon cloth study was reported by Yu et al. by synthesizing a one-step hydrothermal synthesis of 3D MoS$_2$ nanoflakes grown directly on the carbon cloth from MoO$_3$. The electrochemical performance was found to be 3–3.5 mAh/cm^2 at 0.15 mA/cm^2, with the contribution of the carbon cloth by providing high stability and conductivity (Figure 18.5). MoS$_2$ nanoflakes act as reaction sites for

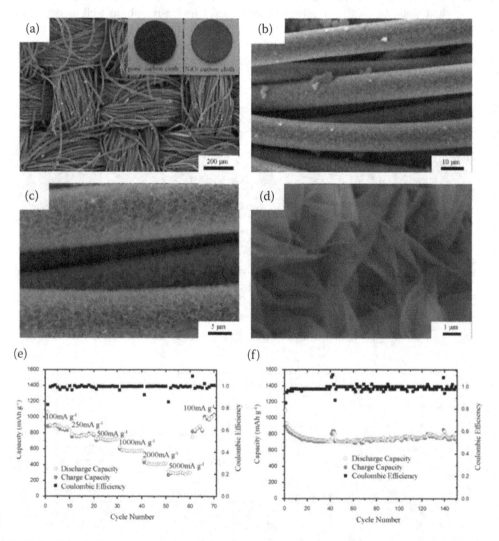

FIGURE 18.4 (a–d) SEM images of the anode based on a NiO-carbon cloth composite. The first image includes the photographs of an unloaded carbon cloth and one loaded with NiO nanosheets. (e) Rate-capability graph of the anode made of NiO nanosheets-carbon cloth composite. (f) Cycle life and coulombic efficiency graph of the corresponding electrode at 700 mA/g. (Adapted with permission from ref. [27], Copyright (2014), Springer Nature).

Li insertion/deintercalation and increase the overall capacity while the high conductivity of the carbon cloth provides stability to the electrode [28]. Carbon fabric [29,30], carbon textile [31], expanded graphite paper [32], and carbon foam [33–36] have the potential to be used as flexible anode materials due to their lightness and excellent flexible properties (Table 18.1; Figure 18.5).

18.2.1.2 Mxenes

Transitional metal carbides, nitrides, and carbonitrides having a two-dimensional structure are a class of inorganic compounds called MXenes [54]. MXenes combine metallic conductivity and hydrophilicity, offering a good electronic structure and a high specific surface area. Therefore, Mxenes are good candidates with their compatible electrochemical properties for supercapacitors and battery applications. Recently, MXenes have directed a new route to build free-standing electrodes without any binder or conductive agents. Thus, this property makes MXenes popular in flexible Li-ion batteries. MXene layers are found to be promising for intercalation/de-intercalation reactions occurred in LIBs for charge/discharge cycles [54]. In a recent study, a 2D layered Ti_2C etched by HF from Ti_2AlC showed five times higher electrochemical performance with a capacity of 225 mAh/g at C/25 rate and a high rate capability of 80 mAh/g at 3C after 120 cycles compared to that of pristine Ti_2AlC. It also gave a 70 mAh/g capacity after 200 cycles at 10C [55]. Later, the same group employed a porous 2D $Ti_3C_2T_x$ MXenes combined with CNT to fabricate a flexible anode with an electrochemical capacity of 1250 mAh/g at 0.32 A/g and 330 mAh/g at 3.2 A. Pristine $Ti_3C_2T_x$ films gave only 35 mAh/g while as-produced porous p-$Ti_3C_2T_x$ was 110 mAh/s. With 10 wt % of CNT addition, the first-cycle capacity of $Ti_3C_2T_x$ MXenes/CNTs was found to be 790 mAh/g (Figure 18.6). These satisfactory results were most likely due to the porous hybrid structure of the anode with a high surface area that improved the accessibility of the lithium ions to active electrochemical sites [56].

MXenes are very promising anode materials for FLIBs but progress still needs to be made. As MXene has the ability to rapid charge/discharge at high C-rates and the capability to store a high amount of energy per unit volume or mass, they are very strong candidates for flexible Li-ion anodes.

18.2.2 FLEXIBLE CATHODES

Cathode materials are oxides of transition metals that are oxidized to a higher valence by lithium migration. When transition metals turn into oxides, their crystal structures acquire a stable morphology over a wide range, in which lithium ions should have the opportunity to settle and separate. While all lithium ions leave the structure during charging, lithium ions settle back into the structure in discharge. In the process of incorporating lithium ions into the structure, electrons in the anode reduce the transition metal ions to a lower valence at the cathode and the process is completed. These two processes are constrained by the time required for lithium ions to diffuse into the electrolyte and reach the electrodes. This time (activity rate) determines the maximum current that can be drawn from the battery. Lithium-ion transfer occurs in the electrolyte at the electrolyte-electrode interface. Therefore, the cathode performance is directly dependent on the microstructure, morphology, and electrochemical properties of the electrode. In other words, the effectiveness of the cathode materials determines the performance in LIBs. For this reason, most of the studies focusing on increasing the performance of the LIBs concentrate on the cathode component.

A wide range of cathode materials has been used in LIBs, such as $LiCoO_2$, $LiMn_2O_4$, $LiFePO_4$, V_xO_y, and the integration of these cathode materials into the flexible conducting materials, mostly of carbon in nature (carbon nanofibers, graphene, carbon nanotubes (CNTs), ultrathin graphite foam, carbon cloth, etc.) [3–10]. Addition of metals and/or organic polymers is also an effective way toward the cathode performance and stability. Since flexible substrates that need to be used in integration with cathode active materials are unstable at elevated temperatures (>700°C), studies on this subject are quite challenging. Wang et al. developed electrodeposition and solvothermal lithiation strategy for the fabrication of a $LiMn_2O_4$/CNT paper-based flexible cathode [57]. For this purpose, MnO_2/CNT was synthesized first by

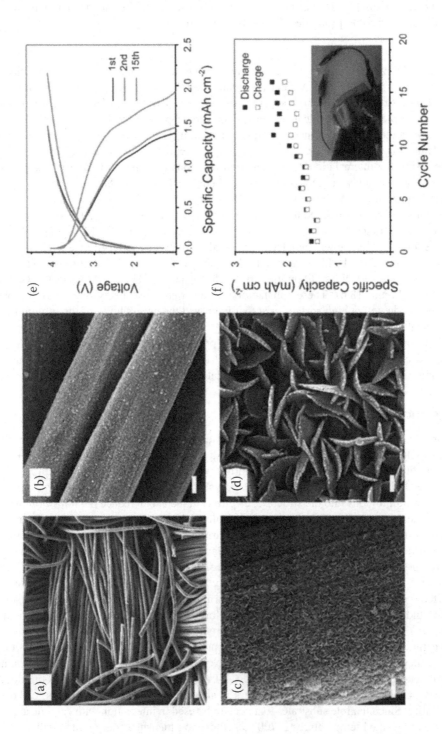

FIGURE 18.5 SEM images of MoS$_2$ nanoflake-carbon cloth composite with the scale bars of (a) 50 mm, (b) 3 mm, (c) 1 mm and (d) 100 nm, (e) capacity graph of full FLIB with different cycles, (f) Cycle life of the full FLIB with an inset photo of a LED lit of the full FLIB in bent condition. Adapted with permission from ref. [28]. Copyright The Authors, some rights reserved; exclusive licensee The Royal Society of Chemistry. Distributed under a Creative Commons Attribution-Non Commercial-ShareAlike 4.0 International License https://creativecommons.org/licenses/by-nc-sa/4.0/).

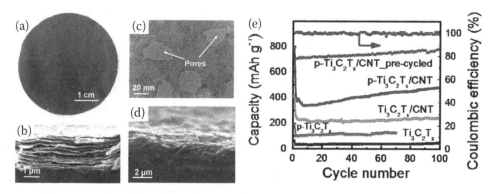

FIGURE 18.6 (a) A photo of a free-standing p-$Ti_3C_2T_x$ film, (b) A cross-sectional SEM image of the same p-$Ti_3C_2T_x$ film, (c) HRTEM image of a chemically etched p-$Ti_3C_2T_x$ flake, (d) A cross-sectional SEM image of a p-$Ti_3C_2T_x$/CNT composite, and (e) Cycle life and coulombic efficiency graph of the corresponding materials. (Adapted with permission from ref. [56], Copyright (2016), Wiley-VCH Verlag GmbH & Co. KGaA).

electrodeposition and then solvothermal lithiation was performed to obtain $LiMn_2O_4$/CNT. This flexible cathode remained 60% of its initial capacity after 4000 cycles.

Xia et al. synthesized a porous $LiMn_2O_4$ nanowall/carbon cloth composite with a hydrothermal lithiation strategy, which in return gave a discharge capacity of 127 mAh/g after 200 cycles [58]. This flexible cathode with the carbon cloth as the current collector material was achieved by coating Mn_3O_4 nanowall arrays onto a carbon cloth followed by cathodic deposition and lithiation in an autoclave at 240°C. Fan and Shen reported a fabrication process including CVD, solvothermal synthesis, and electrodeposition in order to form a flexible V_2O_5 nanobelt array/polymer core/3D graphene foam, which gave a highly stable capacity (~260 mAh/g), even after 1000 cycles at 1.5A/g C-rate [59]. Wang et al. reported potassium vanadate (KVO) nanowire on titanium (Ti) fabric to obtain a flexible cathode [60]. KVO was hydrothermally deposited on the Ti fabric substrate, and the resulting flexible cathode displayed 270 mAh/g after 300 cycles at 100 mA/g. Polyurethane as a polymer matrix was also employed to achieve a flexible $LiFePO_4$-based cathode where $LiFePO_4$ and Super P carbon particles were uniformly distributed into the porous polyurethane matrix by the phase-separation method [61]. With this technique, fabricated $LiFePO_4$/Super P/polyurethane flexible cathode remained 99% of its initial capacity, even after 100 cycles at 170 mA/g. An ultrasonication and co-deposition technique was applied to obtain a binder-free flexible cathode based on a $LiCoO_2$/CNT composite, where CNT acted as both a structural framework and a conductive additive. This structure demonstrated highly flexibility and conductivity with an outstanding performance such as 151 mAh/g at 0.1C and 137 mAh/g at 2C after 50 cycles (Figure 18.7). Whatismore, unlike traditional $LiCoO_2$/Super P cathodes, binder-free $LiCoO_2$/CNT cathode increased its specific mass capacity by 21% and its specific volume capacity by 64% [62].

Fundamentally, most of the commercial inorganic active materials, as well as the ones at the research phase, suffer from low conductivity even if they demonstrate high capacity. In addition, flexible carbon-based substrates used, along with these active materials, have good conductivity but low capacity. Therefore, optimization of these two properties is essential and still poses a challenge for a successful flexible electrode.

18.3 FLEXIBLE ELECTROLYTES

Electrolytes in LIBs act as an electronic insulator allowing the diffusion of Li^+ ions between the electrodes. Traditionally, organic-based electrolytes are used for LIBs due to the sensitivity of the Li^+ ion to water and the narrow electrochemical window of water during operation. As a result of intense research in the 1960–1970s, the first commercial LIBs using organic (propylene carbonate) and inorganic (thionyl

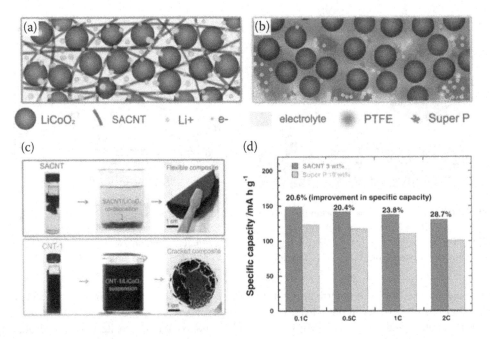

FIGURE 18.7 (a) An illustration of the super aligned carbon nanotube (SACNT) based cathode with LiCoO$_2$ and no binder, (b) An illustration of the cathode with LiCoO$_2$-Super P and the binder, PTFE, (c) Images of the properties of the two cathode materials; SACNT and CNT-1. (d) Rate-capability of the electrodes. (Adapted with permission from ref. [62], Copyright (2012), Wiley-VCH Verlag GmbH & Co. KGaA).

chloride and sulfur dioxide) electrolytes were introduced [63]. In the following years, many obstacles such as cycle life and safety were encountered in lithium rechargeable batteries. Nowadays, cycle life and safety concerns originate from the electrolyte and continue, and thus, intensive studies focus on the development of state-of-art solid-state electrolytes (SSE).

An ideal electrolyte should have the following features:

1. Inert to all of the cell components during operation,
2. Safe with a high flash point (T$_f$), nontoxic, and cheap,
3. Its dielectric constant should be high to dissolve the salts sufficiently,
4. Its viscosity should be low for facile ion transport,
5. High boiling point for operation at a wide temperature range,
6. It must also be safe with a high flash point in addition to being nontoxic and economical [63].

As a component in LIBs, electrolytes transport lithium ions between electrodes for charging and discharging operations to provide ionic conductive media. Also, it is very important to choose an appropriate electrolyte since it plays a significant role in the formation of the solid-electrolyte interface (SEI) layer. Liquid-type electrolytes seem to be more appropriate candidates than solid-state electrolytes for FLIBs due to the fragile nature of the solid-state and the contact problems when the battery is bent and/or twisted. Widely used liquid electrolytes in LIBs are ethylene carbonate (EC) or propylene carbonate (PC), incorporated with linear carbonates, namely, diethyl carbonate (DEC), dimethyl carbonate (DMC) or ethyl methyl carbonate (EMC) [63]. Some ionic liquids are also reported such as N-methoxyethyl-N-methylpyrrolidinium bis(trifluoromethanesulfonyl)-imide, N-butyl-N-ethylpyrrolidinium bis(trifluoromethanesulfonyl)imide, N-butyl-N-methylpyrrolidinium bis(fluorosulfonyl)imide, ethyl methyl sulfone, ethyl vinyl sulfone, and tetramethyl sulfone [64,65].

Solid polymer electrolytes contain an appropriate polymer, lithium salt and a solvent for use as a plasticizer. In solid electrolyte lithium batteries, poly(vinylidene fluoride) (PVDF), poly(vinylidene fluoride)-cohexafluoropropylene (PVDF-HFP), poly(ethylene oxide) (PEO), poly(acrylonitrile) (PAN), poly(acrylonitrile)/poly(vinyl acetate) (PAN-PVA) mixtures, poly(oxyethylene) with tri(oxy-ethylene) side chains and 2-(2-methoxyethoxy) ethyl glycidyl ether (MEEGE) are intensively used [66]. On the other hand, for FLIBs, organic electrolytes may be replaced with aqueous Li-ion electrolytes due to their compatibility with the flexible structure, enhanced safety, and environmentally friendly nature. However, an electrochemical window for aqueous electrolytes is rather narrow with 2.5–3.8 V vs Li/Li$^+$, limiting the range of possible cathode and anode active materials for flexible aqueous electrolytes. Dong et al. recently published the usage of aqueous electrolyte together with lithium-rich spinel $Li_{1.1}Mn_2O_4$ cathode and carbon-coated $LiTi_2(PO_4)_3$ anode [67]. To our best knowledge, spinel lithium manganese oxide (LMO) is the only reported cathode active material that can be used with aqueous electrolytes.

$LiTi_2(PO_4)_3$ [67], polyimide/CNT [68], LiV_3O_8 coated with polypyrrole [69] are the anode active materials used with flexible aqueous electrolytes. Furthermore, a flexible aqueous gel polymer electrolyte called "water-in-salt" has been introduced with symmetrical $LiVPO_4F$ used both as anode and cathode [70]. Another aqueous polyethyleneglycol (PEG)-containing acrylate gel electrolyte operating in the range of 3.7–4.2 V exhibited convenient impedance characteristics with a conductivity of ≥0.3 mS/cm [71]. Recently, Zhou et al. reported a flexible aqueous electrolyte with $LiTi_2(PO_4)_3$-graphite felt anode (LTP/GF) and spinel LMO cathode demonstrating 93% capacity retention after 500 cycles and 72% after 3000 cycles. It was also noted that a high stability rate of 96% was kept after 1000 cycles under dynamic bending conditions (Figure 18.8) [72].

All electrolytes must contain lithium salts to ensure their ionic mobility and conductivity in LIBs. Most common salts used are lithium perchlorate ($LiClO_4$), lithium tetrafluoroborate ($LiBF_4$), lithium hexafluoroarsenate ($LiAsF_6$), lithium bis(trifluoromethanesulfonyl)imide (LiTFSI) and lithium hexafluorophosphate ($LiPF_6$) [63].

18.4 BATTERY STRUCTURES

FLIBs can operate like traditional batteries while they at the same time display mechanical deformations (bending, twisting, stretching, and folding). FLIBs are expected to show the same electrochemical performances, such as capacity, cycle life, and rate capability under mechanical deformations in charge/discharge cycles. However, these mechanical deformations come along with contact problems between anode and cathode, electrolyte leakage, and capacity retention with cracks formed by the active material.

Fang et al. classified a variety of battery structures based on their fabrication strategy and flexibility as a thin film, fiber, wavy, paper folding, island connection, and bamboo slip structures [73]. Cylindrical cells and recently developed prismatic and pouch-type cells are general structures of LIBs. Cylindrical cell, the most commonly used structure in a commercial LIB, has an outer protective layer that will not allow any kind of external deformation. Therefore, new structure designs are essential for FLIBs. In a thin film structure, electrodes and electrolytes are vertically stacked within traditional Al-plastic films or flexible polymers. Koo et al. reported a thin film structure by $LiCoO_2$ cathode and lithium metal anode with PDMS encapsulation layer (Figure 18.9a) [74]. This thin-film structure offers a simple fabrication method by reducing the amount of packing material, as well as the thickness of the battery, and thus, improves volumetric and gravimetric energy density. Fiber-type structures are based on twisted fiber electrodes (anode and cathode) that are tightly connected with each other (Figure 18.9b). Ren et al. developed this type of fiber structure within $Li_4Ti_5O_{12}$ anode and $LiMn_2O_4$ cathode [75]. Another type is wavy structure, providing highly stretchability in FLIBs, employing CNT-based composites together with LTO anode and NMC cathode exhibiting a good electrochemical performance after 2000 tensile cycles by 150% strain in different axes (Figure 18.9c) [76].

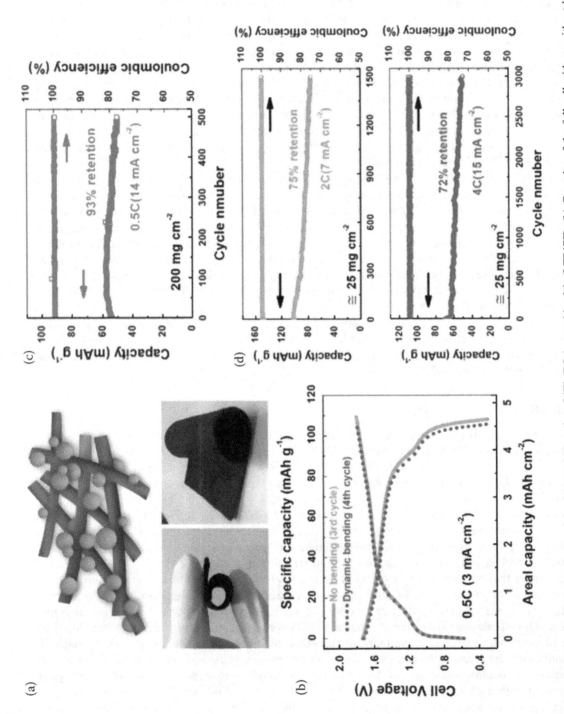

FIGURE 18.8 (a) Structure and the photos of the felectrode based on LiTi$_2$(PO$_4$)$_3$-graphite felt (LTP/GF). (b) Capacity of the full cell with or without bending conditions. Specific capacity versus cycle number versus coulombic efficiency percentage graphs of the electrode with (c) 200 mg/cm^2 and (d) 25 mg/cm^2 loading of LiTi$_2$(PO$_4$)$_3$. (Adapted with permission from ref. [72], Copyright (2021), Elsevier).

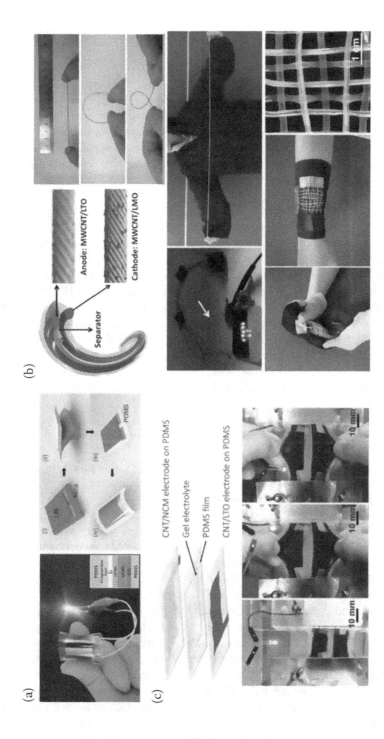

FIGURE 18.9 (a) Left: An image of a thin film type FLIB under bent condition, right: step-by-step layers of the thin film type FLIB. (Adapted with permission from ref. [74]. Copyright (2012) American Chemical Society). (b) An illustration of wire-type FLIB and some photos representing its flexibility. (Adapted from ref. [75] with permission from John Wiley & Sons, Copyright 2014 Wiley-VCH Verlag GmbH & Co. KGaA). (c) Top: An illustration of a FLIB with a wavy structure and bottom: photos displaying the battery under different strain conditions. (Adapted with permission from ref. [76], Copyright (2018), Royal Society of Chemistry).

A paper-folding strategy has been applied by Liu et al., providing mass-energy and volume-energy density with several parallel cells having great mechanical strength and electrochemical performance (Figure 18.10a–b) [77]. An island connection structure type LIB consisting of $LiCoO_2$ and $Li_4Ti_5O_{12}$ as cathode and anode, respectively, gave reversible strain levels of stretchability of 300% with a capacity 1.1 mAh/cm^2 and a little capacity loss after 20 cycles [78]. Wang later named this type of FLIB due to its island-shaped units and wires over electrodes (Figure 18.10c–e).

18.5   FABRICATION OF FLIBS

Fabrication of the traditional batteries is mostly based on the blade-casting method. In addition to the blade-casting method, there are various processes for lab-scale production, such as ultrasonication, CVD, sputtering, nanofabrication, printing, and spray-painting for the production of FLIBs. Among the electrode fabrication techniques, printing and spray painting take great attention since they allow quick electrode coating with a combination of different active materials in a cost-effective and easy to scale-up way.

A novel 3D printing method for the electrode fabrication of FLIBs was performed to construct $LiFePO_4$ and $Li_4Ti_5O_{12}$ as a cathode and an anode, respectively [79]. Ethylene glycol (EG) and water were used for the preparation of the active material inks by dispersing them in a concentration of 60% $LiFePO_4$ and 57% $Li_4Ti_5O_{12}$ by weight. After the printing process, the electrode structures were annealed at 600^0C under Ar. This full cell demonstrated 9.7 J/m^2 energy density and 2.7 mW/cm^2 power density [79]. Park et al. studied with printed $LiCoO_2$ cathode together with PVdF-HFP gel electrolyte and printed $Li_4Ti_5O_{12}$ anode. Airbrush spraying was used for the preparation of the electrodes. This full cell was cycled between 1.5V to 2.7V at a rate of C/8 [80]. Zhao et al., for the first time, demonstrated an inkjet-printed half cell with an initial capacity of 813 mAh/g between 0.05–1.2V at 33 $\mu A/cm^2$ current density. The anode active material was prepared by dispersion of SnO_2 nanoparticles in a mixture of distilled water, ethanol, diethylene glycol, triethanolamine, and 2-propanol. This suspension was mixed with acetylene black and carboxymethyl cellulose (CMC), which were used as a conductive agent and a binder, respectively [81].

18.6   CONCLUSION

Flexibility in energy storage devices is an important issue because FLIBs are needed for use in roll-up displays, solar panels, wearable electronics, and smart cards to present a stable energy output under mechanical deformation. Unlike traditional LIBs, flexible batteries can be bendable, twistable, and stretchable by delivering the same energy and power compared to those of the existing batteries. The advantages of flexibility is lightweight, stretchability, bendability, and portability to offer end-users a compatible one-size-fits-all technology solution. LIB components are stacked together by connecting the electrodes and sealing the battery to avoid electrolyte leakage. Mechanical deformations may cause the delamination of the electrodes, and the breaking off the active materials, which in return may cause the touch of the current collectors, resulting in a short-circuit. To prevent these deformations, novel materials, substrates, components, and new system designs should be applied. FLIB needs a flexible and highly conductive current collector and a strong and highly adhesive binder. In addition, novel flexible active materials with high surface area and high electrochemical activity should be used in the electrodes. To prevent electrolyte leakage, a highly flexible and stable protective case is also vital. Flexible busbars are needed to tightly connect the electrodes in FLIBs. Apart from the necessary advancements on the cell components, an investigation on the fundamental cell degradation mechanism to fully understand the effect of the flexibility on the cells is also crucial. Although liquid electrolytes are currently more compatible than both polymer-based gel and solid-state electrolytes for FLIBs, electrolyte leakage is a major issue. These types of electrolytes usually have a narrow electrochemical window with low ionic

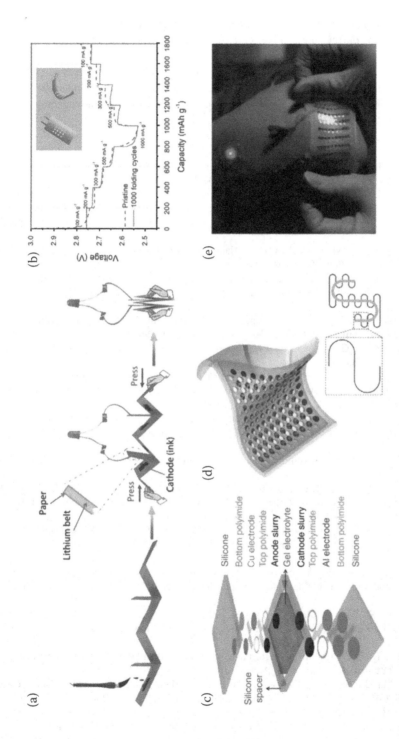

FIGURE 18.10 (a) An illustration of a paper-folding Li–O$_2$ battery pack, (b) Rate performance of the Li–O$_2$ battery with and without folding conditions. (a–b: Adapted with permission from ref. [77], Copyright (2015), Wiley-VCH Verlag GmbH & Co. KGaA). (c) Illustrated structure of a island-shaped FLIB, (d) Illustrations of top: bent and stretched state of a island-shaped FLIB completed device and bottom: serpentine structured interconnects, (e) An image of a functioning island-shaped FLIB on the arm. (c–e: Adapted with permission from ref. [78], Copyright (2013), Springer Nature).

conductivity and poor kinetic properties. Therefore, these types of electrolytes need to be further optimized for better electrochemical performance in FLIBs.

New large-scale fabrication methods should be developed for the commercialization of FLIBs. Printing, spray-painting, vapor deposition, and wetting methods for their production are still not applicable for serial production. Furthermore, novel cell packaging materials durable under deformation conditions need to be developed. As new structures become available, further application of FLIBs in our daily life will be more applicable. Progress is certainly needed to exceed the limitations of FLIBs and to solve the flexibility issues of traditional LIBs. We highly believe that mass production and wide usage of FLIBs on a variety of applications will be feasible in the near future with these advancements.

REFERENCES

1. Qian, G., Liao, X., Zhu, Y., Pan, F., Chen, X., Yang, Y. Designing Flexible Lithium-Ion Batteries by Structural Engineering. *ACS Energy Lett.* 2019, *4* (3), 690–701.
2. Wang, Z., Zhang, W., Li, X., Gao, L. Recent Progress in Flexible Energy Storage Materials for Lithium-Ion Batteries and Electrochemical Capacitors: A Review. *J. Mater. Res.* 2016, *31* (12), 1648–1664.
3. Tao, T., Lu, S., Chen, Y. A Review of Advanced Flexible Lithium-Ion Batteries. *Adv. Mater. Technol.* 2018, *3* (9), 1700375.
4. Zhu, S., Sheng, J., Chen, Y., Ni, J., Li, Y. Carbon Nanotubes for Flexible Batteries: Recent Progress and Future Perspective. *Natl. Sci. Rev.* 2021, *8* (5), nwaa261.
5. Chew, S. Y., Ng, S. H., Wang, J., Novák, P., Krumeich, F., Chou, S. L., Chen, J., Liu, H. K. Flexible Free-Standing Carbon Nanotube Films for Model Lithium-Ion Batteries. *Carbon N. Y.* 2009, *47* (13), 2976–2983.
6. Bensalah, N., Kamand, F. Z., Zaghou, M., Dawoud, H. D., Tahtamouni, T. Al. Silicon Nanofilms as Anode Materials for Flexible Lithium Ion Batteries. *Thin Solid Films* 2019, *690*, 137516.
7. Elizabeth, I., Singh, B. P., Gopukumar, S. Electrochemical Performance of Sb2S3/CNT Free-Standing Flexible Anode for Li-Ion Batteries. *J. Mater. Sci.* 2019, *54* (9), 7110–7118.
8. Sugiawati, V. A., Vacandio, F., Yitzhack, N., Ein-Eli, Y., Djenizian, T. Direct Pre-Lithiation of Electropolymerized Carbon Nanotubes for Enhanced Cycling Performance of Flexible Li-Ion Micro-Batteries. *Polymers* 2020, 12 (2). doi:10.3390/polym12020406
9. Novoselov, K. S., Geim, A. K., Morozov, S. V., Jiang, D., Zhang, Y., Dubonos, S. V., Grigorieva, I. V., Firsov, A. A. Electric Field Effect in Atomically Thin Carbon Films. *Science (80-.).* 2004, *306* (5696), 666–669.
10. Zhou, G., Li, F., Cheng, H.-M. Progress in Flexible Lithium Batteries and Future Prospects. *Energy Environ. Sci.* 2014, *7* (4), 1307–1338.
11. Zhou, X., Liu, Y., Du, C., Ren, Y., Li, X., Zuo, P., Yin, G., Ma, Y., Cheng, X., Gao, Y. Free-Standing Sandwich-Type Graphene/Nanocellulose/Silicon Laminar Anode for Flexible Rechargeable Lithium Ion Batteries. *ACS Appl. Mater. Interfaces* 2018, *10* (35), 29638–29646.
12. Sang, Z., Yan, X., Wen, L., Su, D., Zhao, Z., Liu, Y., Ji, H., Liang, J., Dou, S. X. A Graphene-Modified Flexible SiOC Ceramic Cloth for High-Performance Lithium Storage. *Energy Storage Mater.* 2020, *25*, 876–884.
13. Murata, H., Nakajima, Y., Kado, Y., Saitoh, N., Yoshizawa, N., Suemasu, T., Toko, K. Multilayer Graphene Battery Anodes on Plastic Sheets for Flexible Electronics. *ACS Appl. Energy Mater.* 2020, *3* (9), 8410–8414.
14. Fu, J., Kang, W., Guo, X., Wen, H., Zeng, T., Yuan, R., Zhang, C. 3D Hierarchically Porous NiO/Graphene Hybrid Paper Anode for Long-Life and High Rate Cycling Flexible Li-Ion Batteries. *J. Energy Chem.* 2020, *47*, 172–179.
15. Cevher, O., Akbulut, H. Electrochemical Performance of SnO_2 and SnO_2/MWCNT/Graphene Composite Anodes for Li-Ion Batteries. *Acta Phys. Pol. A* 2017, *131*, 204–206.
16. Gao, L., Wu, G., Ma, J., Jiang, T., Chang, B., Huang, Y., Han, S. SnO2 Quantum Dots@Graphene Framework as a High-Performance Flexible Anode Electrode for Lithium-Ion Batteries. *ACS Appl. Mater. Interfaces* 2020, *12* (11), 12982–12989.
17. Wan, Y., Yang, Z., Xiong, G., Guo, R., Liu, Z., Luo, H. Anchoring Fe3O4 Nanoparticles on Three-Dimensional Carbon Nanofibers toward Flexible High-Performance Anodes for Lithium-Ion Batteries. *J. Power Sources* 2015, *294*, 414–419.

18. Hong, W., Ge, P., Jiang, Y., Yang, L., Tian, Y., Zou, G., Cao, X., Hou, H., Ji, X. Yolk–Shell-Structured Bismuth@N-Doped Carbon Anode for Lithium-Ion Battery with High Volumetric Capacity. *ACS Appl. Mater. Interfaces* 2019, *11* (11), 10829–10840.

19. Shen, K., Zhang, Z., Wang, S., Ru, Q., Zhao, L., Sun, L., Hou, X., Chen, F. Cucumber-Shaped Construction Combining Bismuth Nanoparticles with Carbon Nanofiber Networks as a Binder-Free and Freestanding Anode for Li-Ion Batteries. *Energy & Fuels* 2020, *34* (7), 8987–8992.

20. Lu, H., Hagberg, J., Lindbergh, G., Cornell, A. Li4Ti5O12 Flexible, Lightweight Electrodes Based on Cellulose Nanofibrils as Binder and Carbon Fibers as Current Collectors for Li-Ion Batteries. *Nano Energy* 2017, *39*, 140–150.

21. Thirugnanam, L., Palanisamy, M., Kaveri, S., Ramaprabhu, S., Pol, V. G., Dutta, M. TiO2 Nanoparticle Embedded Nitrogen Doped Electrospun Helical Carbon Nanofiber-Carbon Nanotube Hybrid Anode for Lithium-Ion Batteries. *Int. J. Hydrogen Energy* 2021, *46* (2), 2464–2478.

22. Li, W., Li, M., Yang, Z., Xu, J., Zhong, X., Wang, J., Zeng, L., Liu, X., Jiang, Y., Wei, X., Gu, L., Yu, Y. Carbon-Coated Germanium Nanowires on Carbon Nanofibers as Self-Supported Electrodes for Flexible Lithium-Ion Batteries. *Small* 2015, *11* (23), 2762–2767.

23. Ghanooni Ahmadabadi, V., Shirvanimoghaddam, K., Kerr, R., Showkath, N., Naebe, M. Structure-Rate Performance Relationship in Si Nanoparticles-Carbon Nanofiber Composite as Flexible Anode for Lithium-Ion Batteries. *Electrochim. Acta* 2020, *330*, 135232.

24. He, J., Wei, Y., Zhai, T., Li, H. Antimony-Based Materials as Promising Anodes for Rechargeable Lithium-Ion and Sodium-Ion Batteries. *Mater. Chem. Front.* 2018, *2* (3), 437–455.

25. Chen, R., Xue, X., Lu, J., Chen, T., Hu, Y., Ma, L., Zhu, G., Jin, Z. The Dealloying–Lithiation/Delithiation–Realloying Mechanism of a Breithauptite (NiSb) Nanocrystal Embedded Nanofabric Anode for Flexible Li-Ion Batteries. *Nanoscale* 2019, *11* (18), 8803–8811.

26. Wang, L., Gu, X., Zhao, L., Wang, B., Jia, C., Xu, J., Zhao, Y., Zhang, J. ZnO@TiO2 Heterostructure Arrays/Carbon Cloth by Charge Redistribution Enhances Performance in Flexible Anode for Li Ion Batteries. *Electrochim. Acta* 2019, *295*, 107–112.

27. Long, H., Shi, T., Hu, H., Jiang, S., Xi, S., Tang, Z. Growth of Hierarchal Mesoporous NiO Nanosheets on Carbon Cloth as Binder-Free Anodes for High-Performance Flexible Lithium-Ion Batteries. *Sci. Rep.* 2014, *4* (1), 7413.

28. Yu, H., Zhu, C., Zhang, K., Chen, Y., Li, C., Gao, P., Yang, P., Ouyang, Q. Three-Dimensional Hierarchical MoS2 Nanoflake Array/Carbon Cloth as High-Performance Flexible Lithium-Ion Battery Anodes. *J. Mater. Chem. A* 2014, *2* (13), 4551–4557.

29. Son, J.-M., Oh, S., Bae, S.-H., Nam, S., Oh, I.-K. A Pair of NiCo2O4 and V2O5 Nanowires Directly Grown on Carbon Fabric for Highly Bendable Lithium-Ion Batteries. *Adv. Energy Mater.* 2019, *9* (18), 1900477.

30. Xu, X., Liu, J., Liu, Z., Wang, Z., Hu, R., Liu, J., Ouyang, L., Zhu, M. FeP@C Nanotube Arrays Grown on Carbon Fabric as a Low Potential and Freestanding Anode for High-Performance Li-Ion Batteries. *Small* 2018, *14* (30), 1800793.

31. Xia, Y., Xiong, W.-S., Jiang, Y., Zhou, S.-Y., Hu, C.-L., He, R.-X., Sang, H.-Q., Chen, B., Liu, Y., Zhao, X.-Z. Controllable In-Situ Growth of 3D Villose TiO2 Architectures on Carbon Textiles as Flexible Anode for Advanced Lithium-Ion Batteries. *Mater. Lett.* 2018, *229*, 122–125.

32. Son, D.-K., Kim, J., Raj, M. R., Lee, G. Elucidating the Structural Redox Behaviors of Nanostructured Expanded Graphite Anodes toward Fast-Charging and High-Performance Lithium-Ion Batteries. *Carbon N. Y.* 2021, *175*, 187–201.

33. Liu, W., Fu, Y., Li, Y., Chen, S., Song, Y., Wang, L. Three-Dimensional Carbon Foam Surrounded by Carbon Nanotubes and Co-Co3O4 Nanoparticles for Stable Lithium-Ion Batteries. *Compos. Part B Eng.* 2019, *163*, 464–470.

34. Zhang, X., Wang, B., Yuan, W., Wu, J., Liu, H., Wu, H., Zhang, Y. Reduced Graphene Oxide Modified N-Doped Carbon Foam Supporting TiO2 Nanoparticles as Flexible Electrode for High-Performance Li/Na Ion Batteries. *Electrochim. Acta* 2019, *311*, 141–149.

35. Yang, H.-R., Kim, J.-H. Synthesis and Electrochemical Properties of Carbon Foam/SnS/C Composite for Flexible Lithium Ion Batteries. *ECS Meet. Abstr.* 2020, *MA2020-02* (2), 418.

36. Zhang, P., Tian, R., Cao, M., Feng, Y., Yao, J. Embedding Co9S8 Nanoparticles into Porous Carbon Foam with High Flexibility and Enhanced Lithium Ion Storage. *J. Electroanal. Chem.* 2020, *863*, 114062.

37. Abnavi, A., Faramarzi, M., Sanaee, Z., Ghasemi, S. SnO2 Nanowires on Carbon Nanotube Film as a High Performance Anode Material for Flexible Li-Ion Batteries. *J. Nanostructures* 2018, *8* (3), 288–293.

38. Chen, R., Xue, X., Hu, Y., Kong, W., Lin, H., Chen, T., Jin, Z. Intermetallic SnSb Nanodots Embedded in Carbon Nanotubes Reinforced Nanofabric Electrodes with High Reversibility and Rate Capability for Flexible Li-Ion Batteries. *Nanoscale* 2019, *11* (28), 13282–13288.

39. Wang, M., Huang, Y., Zhu, Y., Wu, X., Zhang, N., Zhang, H. Binder-Free Flower-like SnS2 Nanoplates Decorated on the Graphene as a Flexible Anode for High-Performance Lithium-Ion Batteries. *J. Alloys Compd.* 2019, *774*, 601–609.

40. Li, W., Yang, Z., Cheng, J., Zhong, X., Gu, L., Yu, Y. Germanium Nanoparticles Encapsulated in Flexible Carbon Nanofibers as Self-Supported Electrodes for High Performance Lithium-Ion Batteries. *Nanoscale* 2014, *6* (9), 4532–4537.

41. Wang, J., Song, W.-L., Wang, Z., Fan, L.-Z., Zhang, Y. Facile Fabrication of Binder-Free Metallic Tin Nanoparticle/Carbon Nanofiber Hybrid Electrodes for Lithium-Ion Batteries. *Electrochim. Acta* 2015, *153*, 468–475.

42. Xia, J., Liu, L., Jamil, S., Xie, J., Yan, H., Yuan, Y., Zhang, Y., Nie, S., Pan, J., Wang, X., Cao, G. Free-Standing SnS/C Nanofiber Anodes for Ultralong Cycle-Life Lithium-Ion Batteries and Sodium-Ion Batteries. *Energy Storage Mater.* 2019, *17*, 1–11.

43. Ma, X., Smirnova, A. L., Fong, H. Flexible Lignin-Derived Carbon Nanofiber Substrates Functionalized with Iron (III) Oxide Nanoparticles as Lithium-Ion Battery Anodes. *Mater. Sci. Eng. B* 2019, *241*, 100–104.

44. Shen, X., Cao, Z., Chen, M., Zhang, J., Chen, D. A Novel Flexible Full-Cell Lithium Ion Battery Based on Electrospun Carbon Nanofibers Through a Simple Plastic Package. *Nanoscale Res. Lett.* 2018, *13* (1), 367.

45. Zhang, M., Yan, F., Tang, X., Li, Q., Wang, T., Cao, G. Flexible CoO–Graphene–Carbon Nanofiber Mats as Binder-Free Anodes for Lithium-Ion Batteries with Superior Rate Capacity and Cyclic Stability. *J. Mater. Chem. A* 2014, *2* (16), 5890–5897.

46. Wan, Y., Yang, Z., Xiong, G., Luo, H. A General Strategy of Decorating 3D Carbon Nanofiber Aerogels Derived from Bacterial Cellulose with Nano-Fe3O4 for High-Performance Flexible and Binder-Free Lithium-Ion Battery Anodes. *J. Mater. Chem. A* 2015, *3* (30), 15386–15393.

47. Zhao, X., Du, Y., Jin, L., Yang, Y., Wu, S., Li, W., Yu, Y., Zhu, Y., Zhang, Q. Membranes of MnO Beading in Carbon Nanofibers as Flexible Anodes for High-Performance Lithium-Ion Batteries. *Sci. Rep.* 2015, *5* (1), 14146.

48. Narsimulu, D., Nagaraju, G., Sekhar, S. C., Ramulu, B., Yu, J. S. Designed Lamination of Binder-Free Flexible Iron Oxide/Carbon Cloth as High Capacity and Stable Anode Material for Lithium-Ion Batteries. *Appl. Surf. Sci.* 2019, *497*, 143795.

49. Chen, S., Tao, R., Tu, J., Guo, P., Yang, G., Wang, W., Liang, J., Lu, S.-Y. High Performance Flexible Lithium-Ion Battery Electrodes: Ion Exchange Assisted Fabrication of Carbon Coated Nickel Oxide Nanosheet Arrays on Carbon Cloth. *Adv. Funct. Mater.* 2021, *n/a* (n/a), 2101199.

50. Balogun, M.-S., Li, C., Zeng, Y., Yu, M., Wu, Q., Wu, M., Lu, X., Tong, Y. Titanium Dioxide@titanium Nitride Nanowires on Carbon Cloth with Remarkable Rate Capability for Flexible Lithium-Ion Batteries. *J. Power Sources* 2014, *272*, 946–953.

51. Zhang, H., Ren, W., Cheng, C. Three-Dimensional SnO2@TiO2 Double-Shell Nanotubes on Carbon Cloth as a Flexible Anode for Lithium-Ion Batteries. *Nanotechnology* 2015, *26* (27), 274002.

52. Li, S., Liu, G., Liu, J., Lu, Y., Yang, Q., Yang, L.-Y., Yang, H.-R., Liu, S., Lei, M., Han, M. Carbon Fiber Cloth@VO2 (B): Excellent Binder-Free Flexible Electrodes with Ultrahigh Mass-Loading. *J. Mater. Chem. A* 2016, *4* (17), 6426–6432.

53. Wei, X., Li, W., Shi, J., Gu, L., Yu, Y. FeS@C on Carbon Cloth as Flexible Electrode for Both Lithium and Sodium Storage. *ACS Appl. Mater. Interfaces* 2015, *7* (50), 27804–27809.

54. Anasori, B., Xie, Y., Beidaghi, M., Lu, J., Hosler, B. C., Hultman, L., Kent, P. R. C., Gogotsi, Y., Barsoum, M. W. Two-Dimensional, Ordered, Double Transition Metals Carbides (MXenes). *ACS Nano* 2015, *9* (10), 9507–9516.

55. Naguib, M., Come, J., Dyatkin, B., Presser, V., Taberna, P.-L., Simon, P., Barsoum, M. W., Gogotsi, Y. MXene: A Promising Transition Metal Carbide Anode for Lithium-Ion Batteries. *Electrochem. commun.* 2012, *16* (1), 61–64.

56. Ren, C. E., Zhao, M.-Q., Makaryan, T., Halim, J., Boota, M., Kota, S., Anasori, B., Barsoum, M. W., Gogotsi, Y. Porous Two-Dimensional Transition Metal Carbide (MXene) Flakes for High-Performance Li-Ion Storage. *ChemElectroChem* 2016, *3* (5), 689–693.

57. Wang, J., Zhang, L., Zhou, Q., Wu, W., Zhu, C., Liu, Z., Chang, S., Pu, J., Zhang, H. Ultra-Flexible Lithium Ion Batteries Fabricated by Electrodeposition and Solvothermal Synthesis. *Electrochim. Acta* 2017, *237*, 119–126.

58. Xia, H., Xia, Q., Lin, B., Zhu, J., Seo, J. K., Meng, Y. S. Self-Standing Porous LiMn2O4 Nanowall Arrays as Promising Cathodes for Advanced 3D Microbatteries and Flexible Lithium-Ion Batteries. *Nano Energy* 2016, *22*, 475–482.

59. Chao, D., Xia, X., Liu, J., Fan, Z., Ng, C. F., Lin, J., Zhang, H., Shen, Z. X., Fan, H. J. A V2O5/Conductive-Polymer Core/Shell Nanobelt Array on Three-Dimensional Graphite Foam: A High-Rate, Ultrastable, and Freestanding Cathode for Lithium-Ion Batteries. *Adv. Mater.* 2014, *26* (33), 5794–5800.

60. Wang, C., Cao, Y., Luo, Z., Li, G., Xu, W., Xiong, C., He, G., Wang, Y., Li, S., Liu, H., Fang, D. Flexible Potassium Vanadate Nanowires on Ti Fabric as a Binder-Free Cathode for High-Performance Advanced Lithium-Ion Battery. *Chem. Eng. J.* 2017, *307*, 382–388.

61. Bao, J.-J., Zou, B.-K., Cheng, Q., Huang, Y.-P., Wu, F., Xu, G.-W., Chen, C.-H. Flexible and Free-Standing LiFePO4/TPU/SP Cathode Membrane Prepared via Phase Separation Process for Lithium Ion Batteries. *J. Memb. Sci.* 2017, *541*, 633–640.

62. Luo, S., Wang, K., Wang, J., Jiang, K., Li, Q., Fan, S. Binder-Free LiCoO2/Carbon Nanotube Cathodes for High-Performance Lithium Ion Batteries. *Adv. Mater.* 2012, *24* (17), 2294–2298.

63. Xu, K. Nonaqueous Liquid Electrolytes for Lithium-Based Rechargeable Batteries. *Chem. Rev.* 2004, *104* (10), 4303–4418.

64. Abouimrane, A., Belharouak, I., Amine, K. Sulfone-Based Electrolytes for High-Voltage Li-Ion Batteries. *Electrochem. Commun.* 2009, *11* (5), 1073–1076.

65. Lewandowski, A., Świderska-Mocek, A. Ionic Liquids as Electrolytes for Li-Ion Batteries—An Overview of Electrochemical Studies. *J. Power Sources* 2009, *194* (2), 601–609.

66. Foreman, E., Zakri, W., Hossein Sanatimoghaddam, M., Modjtahedi, A., Pathak, S., Kashkooli, A. G., Garafolo, N. G., Farhad, S. A. Review of Inactive Materials and Components of Flexible Lithium-Ion Batteries. *Adv. Sustain. Syst.* 2017, *1* (11), 1700061.

67. Dong, X., Chen, L., Su, X., Wang, Y., Xia, Y. Flexible Aqueous Lithium-Ion Battery with High Safety and Large Volumetric Energy Density. *Angew. Chemie Int. Ed.* 2016, *55* (26), 7474–7477.

68. Zhang, Y., Wang, Y., Wang, L., Lo, C.-M., Zhao, Y., Jiao, Y., Zheng, G., Peng, H. A Fiber-Shaped Aqueous Lithium Ion Battery with High Power Density. *J. Mater. Chem. A* 2016, *4* (23), 9002–9008.

69. Liu, Z., Li, H., Zhu, M., Huang, Y., Tang, Z., Pei, Z., Wang, Z., Shi, Z., Liu, J., Huang, Y., Zhi, C. Towards Wearable Electronic Devices: A Quasi-Solid-State Aqueous Lithium-Ion Battery with Outstanding Stability, Flexibility, Safety and Breathability. *Nano Energy* 2018, *44*, 164–173.

70. Yang, C., Ji, X., Fan, X., Gao, T., Suo, L., Wang, F., Sun, W., Chen, J., Chen, L., Han, F., Miao, L., Xu, K., Gerasopoulos, K., Wang, C. Flexible Aqueous Li-Ion Battery with High Energy and Power Densities. *Adv. Mater.* 2017, *29* (44), 1701972.

71. Cresce, A., Eidson, N., Schroeder, M., Ma, L., Howarth, Y., Yang, C., Ho, J., Dillon, R., Ding, M., Bassett, A., Stanzione, J., Tom, R., Soundappan, T., Wang, C., Xu, K. Gel Electrolyte for a 4V Flexible Aqueous Lithium-Ion Battery. *J. Power Sources* 2020, *469*, 228378.

72. Zhou, Y., Wang, Z., Lu, Y.-C. Flexible Aqueous Lithium-Ion Batteries with Ultrahigh Areal Capacity and Long Cycle Life. *Mater. Today Energy* 2021, *19*, 100570.

73. Fang, Z., Wang, J., Wu, H., Li, Q., Fan, S., Wang, J. Progress and Challenges of Flexible Lithium Ion Batteries. *J. Power Sources* 2020, *454*, 227932.

74. Koo, M., Park, K.-I., Lee, S. H., Suh, M., Jeon, D. Y., Choi, J. W., Kang, K., Lee, K. J. Bendable Inorganic Thin-Film Battery for Fully Flexible Electronic Systems. *Nano Lett.* 2012, *12* (9), 4810–4816.

75. Ren, J., Zhang, Y., Bai, W., Chen, X., Zhang, Z., Fang, X., Weng, W., Wang, Y., Peng, H. Elastic and Wearable Wire-Shaped Lithium-Ion Battery with High Electrochemical Performance. *Angew. Chemie Int. Ed.* 2014, *53* (30), 7864–7869.

76. Yu, Y., Luo, Y., Wu, H., Jiang, K., Li, Q., Fan, S., Li, J., Wang, J. Ultrastretchable Carbon Nanotube Composite Electrodes for Flexible Lithium-Ion Batteries. *Nanoscale* 2018, *10* (42), 19972–19978.

77. Liu, Q.-C., Li, L., Xu, J.-J., Chang, Z.-W., Xu, D., Yin, Y.-B., Yang, X.-Y., Liu, T., Jiang, Y.-S., Yan, J.-M., Zhang, X.-B. Flexible and Foldable Li–O2 Battery Based on Paper-Ink Cathode. *Adv. Mater.* 2015, *27* (48), 8095–8101.

78. Xu, S., Zhang, Y., Cho, J., Lee, J., Huang, X., Jia, L., Fan, J. A., Su, Y., Su, J., Zhang, H., Cheng, H., Lu, B., Yu, C., Chuang, C., Kim, T., Song, T., Shigeta, K., Kang, S., Dagdeviren, C., Petrov, I., Braun, P. V., Huang, Y., Paik, U., Rogers, J. A. Stretchable Batteries with Self-Similar Serpentine Interconnects and Integrated Wireless Recharging Systems. *Nat. Commun.* 2013, *4* (1), 1543.

79. Sun, K., Wei, T.-S., Ahn, B. Y., Seo, J. Y., Dillon, S. J., Lewis, J. A. 3D Printing of Interdigitated Li-Ion Microbattery Architectures. *Adv. Mater.* 2013, *25* (33), 4539–4543.

80. Park, M.-S., Hyun, S.-H., Nam, S.-C., Cho, S. B. Performance Evaluation of Printed LiCoO2 Cathodes with PVDF-HFP Gel Electrolyte for Lithium Ion Microbatteries. *Electrochim. Acta* 2008, *53* (17), 5523–5527.

81. Zhao, Y., Zhou, Q., Liu, L., Xu, J., Yan, M., Jiang, Z. A Novel and Facile Route of Ink-Jet Printing to Thin Film SnO2 Anode for Rechargeable Lithium Ion Batteries. *Electrochim. Acta* 2006, *51* (13), 2639–2645.

19 Flexible Na-Ion Batteries

Jun Mei

School of Chemistry and Physics, Queensland University of Technology, Brisbane, Queensland, Australia

Centre for Materials Science, Queensland University of Technology, Brisbane, Queensland, Australia

CONTENTS

19.1 Introduction...345
19.2 Flexible Na-ion batteries..345
    19.2.1 Configurations...345
    19.2.2 Electrolytes 346
    19.2.3 Electrode materials 350
    19.2.4 Separators 353
19.3 Conclusion..353
References...353

19.1 INTRODUCTION

The development of flexible energy storage systems (ESSs) can meet the ever-growing demands of flexible electronics in various application fields, such as bendable phones and roll-up displays [1–5]. In spite of much progress on flexible lithium-based ESSs, the limited lithium resource is a major challenge for large-scale applications [6,7]. Developing low-cost and high-performance flexible ESSs remains a crucial research topic [8]. Flexible Na-based ESSs are regarded as highly promising candidates due to the high earth-abundant and global distribution of sodium resources [9–11]. Moreover, the working mechanisms for Na-based ESSs are similar to lithium-based counterparts [12–14]. The electrochemical performance of flexible Na-based ESSs largely depends on cell configurations and components, including electrodes, electrolytes, and separators [15]. In contrast to the small radius of Li-ion (0.76 Å), Na-ion has a much larger radius (1.02 Å), which will pose negative effects on the ion-intercalation kinetics, resulting in slow ion diffusion and obvious volume change over cycling [8,16–19]. Therefore, exploring suitable electrode hosts and ion transport mechanisms are urgently needed for Na-based ESSs, particularly for flexible Na-ion batteries.

In these common flexible Na-ion batteries, nonaqueous liquid electrolytes with high volatile and flammable properties are often used, which may lead to serious safety issues, such as fire and explosion, and thus hinder the grid-level application of Na-ion batteries [20,21]. To address this issue, flexible solid-state Na-ion batteries offer a promising solution [22–26]. Herein, the recent major progress on flexible Na-ion batteries based on the cell configurations, electrolytes, and electrode materials, as well as separators, are summarized to offer us some useful guidance for developing next-generation smart ESSs.

19.2 FLEXIBLE NA-ION BATTERIES

19.2.1 Configurations

Flexible electronics, which always feature bendable, rollable, rugged, or foldable features, are one of the promising smart technologies for the next-generation portable and wearable devices [31]. The most commonly configured for flexible Na-ion batteries is the two-dimensional (2D) sandwiched structure. Figure 19.1a shows the typical 2D flexible Na-ion battery mode composed of a P2/P3-$Na_{0.7}CoO_2$ nanosheet array as cathode and a hard carbon on carbon cloth as the anode. This assembled full cell delivered a reversible capacity of 120.5 mAh g^{-1} at a rate of 0.1C [27]. Besides the asymmetric configuration, a symmetric flexible Na-ion battery constructed with carbon-coated $Na_3V_2(PO_4)_3$/reduced graphene oxide as both cathode and anode could maintain stable electrochemical performance under various bending conditions (Figure 19.1b) [28]. In addition to the full cells, flexible half-cell Na-ion batteries have been designed by using a metallic Na foil as the counter electrode. The designed one-dimensional (1D) tube-type flexible Na-ion battery, as depicted in Figure 19.1c, was mainly composed of an inner hollow tube intertwined by copper wire as a support, a metallic Na foil, a flexible Na-ion separator, a synthesized cathode (Prussian blue/graphene on Ni-coated cotton textile, PB@GO@NCT), and a shrinkable tube on the outermost [29].

Based on the electrolyte systems, this can be divided into aqueous and nonaqueous Na-ion batteries. Guo et al. developed flexible belt-shaped (Figure 19.1d) and fiber-shaped (Figure 19.1e) aqueous Na-ion batteries by using a $Na_{0.44}MnO_2$-based cathode and a carbon-coated $NaTi_2(PO_4)_3$-based anode [30]. For the belt-shaped aqueous Na-ion battery, a porous polyacrylonitrile (PAN) separator treated by a Na-ion-containing aqueous Na_2SO_4 electrolyte was used. The resultant battery delivered a maximum volumetric energy density of 23.8 mWh cm^{-3} together with a power density of 3.8 W cm^{-3}. For the fiber-shaped aqueous Na-ion battery, an aligned carbon nanotube was employed for modifying electrode materials to promote mechanical flexibility. The assembled fiber-shaped battery presented an energy density of 25.7 mWh cm^{-3} at 0.054 W cm^{-3}. Furthermore, these flexible Na-ion batteries could still work well when saline (0.9 wt % NaCl) or a cell-culture medium (DMEM) was utilized as electrolytes [30].

19.2.2 ELECTROLYTES

The widely used combustible organic liquid electrolytes in Na-ion batteries always lead to serious safety issues, particularly for large-scale applications for the electrical vehicle. To address this problem, the design of solid-state batteries by using nonflammable solid electrolytes is one of the potential solutions. The commonly reported solid-state electrolytes include Na-ion conductive ceramics, polymers, and ceramic-polymer hybrids. Ceramic solid electrolytes (e.g., NASICON, Na_3PS_4, and $Na_3Zr_2Si_2PO_{12}$) possess a high ion conductivity, a wide electrochemical window, and a good thermal tolerance without obvious leakage or pollution; however, their electrochemical performance is not satisfying due to a large interfacial electrode-electrolyte resistance, and the brittle property of the ceramic materials greatly limits their use for flexible batteries [32,33]. As for solid polymers, such as poly(vinylidene fluoridehexafluoropropylene) (PVDF-HFP) and poly(ethylene oxide) (PEO), they possess obvious advantages on mechanical flexibility [34,35]. Unfortunately, most polymer electrolytes possess a low ionic conductivity and a weak resistivity toward undesired oxidation reactions. Solid ceramic-polymer composites can well combine the advantages of ceramics and polymers in terms of ionic conductivity and mechanical flexibility [36]. However, the solid-solid electrode-electrolyte interfacial resistance issue should be further optimized. Therefore, exploring solid electrolytes with desired mechanical properties and low interfacial resistance remains a major challenge for developing flexible and rechargeable Na-ion batteries [37,38].

As early as 2015, Kim et al. reported a NASICON ($Na_3Zr_2Si_2PO_{12}$)-based electrolyte for safe solid-state Na-ion battery [39]. As illustrated in Figure 19.2a, there are three possible Na-ion transport mechanisms, namely, ion hopping between ceramic particles, cross direction through the ceramic-liquid interfaces, and plasticizer ion transport in a liquid electrolyte shell formed on ceramic particles, within the obtained hybrid solid electrolyte. The resultant solid-state electrolyte delivered a high ambient conductivity of 3.6×10^{-4} S cm^{-1} as well as an electrochemical window as wide as 5.0 V. By using this hybrid electrolyte, the assembled solid-state batteries exhibited a specific capacity of 330 mAhg^{-1} for a Na/hard carbon half-cell and 131 mAhg^{-1} for a Na/NaFePO$_4$ half-cell on discharge in the first cycle.

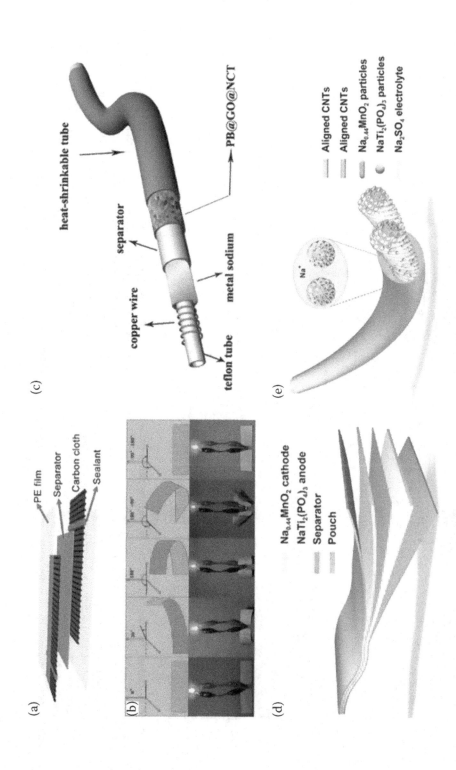

FIGURE 19.1 (a) Schematic illustration of the battery configuration of a flexible full Na-ion battery by using a P2/P3-Na$_{0.7}$CoO$_2$ cathode and a hard carbon anode. Reproduced with permission from ref. [27]. Copyright (2020) Elsevier. (b) Schematic illustration of the bending angels of a flexible symmetric Na-ion full cell and the corresponding photographs for lighting a LED light under various bending states. Reproduced with permission from ref. [28]. Copyright (2017) Royal Society of Chemistry. (c) Schematic illustration for the battery configuration of the tube-type flexible Na-ion batteries. Reproduced with permission from ref. [29]. Copyright (2017) Wiley-VCH. (d, e) Schematic illustration of the configuration of flexible (d) belt-shaped and (e) fibre-shaped aqueous Na-ion batteries by using a Na$_{0.44}$MnO$_2$-based cathode and a nano-sized carbon-coated NaTi$_2$(PO$_4$)$_3$-based anode. Reproduced with permission from ref. [30]. Copyright (2017) Elsevier.

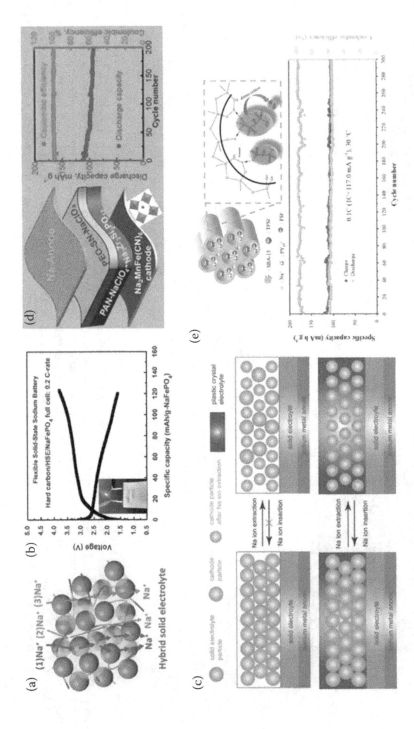

FIGURE 19.2 (a) Schematic illustration of the Na-ion conducting pathways within hybrid solid-state Na-ion battery, and the inset shows the photograph of the assembled pouch-type full cell. Reproduced with permission from ref. [39]. Copyright (2015) The authors under a CC BY-NC 3.0. (b) Initial charge/discharge curve of a flexible solid-state sodium battery with a laminated composite electrolyte and the electrochemical cycling stability of the full cell comprised of metallic sodium anode and a Prussian blue cathode. Reproduced with permission from ref. [40]. Copyright (2017) Wiley-VCH. (d) Schematic illustration of structural configuration of an all-solid-state sodium battery with a laminated composite electrolyte and the electrochemical cycling stability of the full cell comprised of metallic sodium anode and a Prussian blue cathode. Reproduced with permission from ref. [41]. Copyright (2020) Wiley-VCH. (e) Schematic illustration of the configuration and molecular structure of the Na-ionic liquid (IL)@SBA-15 electrolyte and electrochemical cycling performance at 0.1 C of a full quasi-solid-state Na-ion cell with a $Na_3V_2(PO_4)_3$ cathode and a metallic sodium anode at room temperature. Reproduced with permission from ref. [42]. Copyright (2020) American Chemical Society.

Furthermore, a pouch-type flexible battery by using the hybrid sandwiched between $NaFePO_4$ cathode and hard carbon anode, exhibited an initial Coulombic efficiency of 98% (Figure 19.2b), a discharge capacity of about 120 mAhg^{-1}, together with 96% of the capacity retention ratio after 200 cycles, and good flexibility and bendability [39].

Another crucial issue is that the undesired solid-solid interfacial resistance increases upon charge/discharge cycles, which is primarily due to the three-dimensional (3D) volume change of the solid electrode particles. One of the effective solutions is to introduce a plastic-crystal electrolyte interphase between electrolytes and electrodes [40]. As illustrated in Figure 19.2c, the common method by adding solid electrolyte particles can only provide a small fraction of electrode materials access to the electrolyte ions, particularly for the thick electrodes. In contrast, the introduced plastic-crystal electrolytes can effectively penetrate the whole electrode to access most solid electrode particles. More importantly, the reversible deformation with the volume change over cycling could be easily achieved by using plastic-crystal electrolytes, which are quite similar to a liquid electrolyte. By comparison with solid-state Na-ion batteries with $Na_3Zr_2(Si_2PO_4)$ particles, the assembled Na-ion cell exhibited a significantly improved electrochemical performance without obvious irreversible capacity loss on cycling, which manifested a 98% of discharge capacity retention after 100 cycles [40].

Recently, a flexible composite solid electrolyte has been reported by combining poly(ether-acrylate) with the $Na_3Zr_2Si_2PO_{12}$/PVDF-HFP porous skeleton, and the resultant multilayered polymer/inorganic electrolyte exhibited a measured Na-ion transference number of 0.63 and a calculated ion conductivity over 10^{-4} S cm^{-1} at the temperature of 60°C [43]. Meanwhile, the hybrid electrolyte can reduce interfacial ion transfer resistance and also inhibit the dendritic growth during the plating/stripping of sodium. As a result, the full cell by coupling with a $Na_3V_2(PO_4)_3$ (NVP) cathode with a metallic sodium anode possessed a specific capacity of 85 mAhg^{-1} after 100 discharge/charge cycles, and over 90% of capacity retention ratio was achieved by using a $Na_{2/3}Ni_{1/3}Mn_{1/3}Ti_{1/3}O_2$ (NTMO) cathode at 60°C [43]. Besides, Yu et al. reported a ion conductive polymer/ceramic polymer electrolyte for all-solid-state batteries operated in room temperature [41]. In this design, as shown in Figure 19.2d, at the sodium anode side, the PEO matrix in the presence of succinonitrile (SN) was utilized to enhance Na-ion conductivity. At the Prussian blue cathode side, the poly(acrylonitrile) (PAN) polymer matrix coupled with a NASICON-type ($Na_3Zr_2Si_2PO_{12}$) electrolyte was applied for enhancing Na-ion conductivity and meanwhile preventing the sodium dendrite from destroying the sandwiched electrolyte membrane. This double-layer electrolyte endowed a room-temperature conductivity of 1.36×10^{-4} S cm^{-1}, and presented a stable electrochemical stability window of 4.8 V. When the laminated solid electrolyte was incorporated into an all-solid-state Na-ion batteries with a metallic sodium negative electrode and a Prussian blue-type $Na_2MnFe(CN)_6$ positive electrode, the cell delivered remarkably stable cycling at room temperature with ~83.3% capacity retained after 200 discharge/charge cycles, together with a stable coulombic efficiency of 98.5–99.9% over cycles (Figure 19.2d) [41].

Ionogels, which hold the superior advantages on mechanical strength, structural stability, and ionic conductivity, are promising electrolytes for safe electrochemical systems [44,45]. This type of quasi-solid electrolyte can be made by the immobilization of nonvolatile salt-containing ionic liquids in thermal-resistant solid-state hosts. Gao et al. synthesized an ionogel electrolyte by using an SBA-15 matrix and a Na-containing ionic liquid in the presence of PVDF-HFP binder, which presented an optimal conductivity of 2.48×10^{-3} S cm^{-1} at a temperature of 30 °C, and an electrochemical window of 0–4.8 V [42]. It was speculated that the active Si-OH groups were generated in bridged silanols, accompanied by the formation of hydrogen bonds with fluorine in the as-obtained Na-ionic liquid (IL)@ SBA-15 electrolyte (Figure 19.2e). Further spectroscopic characterizations revealed the hydrogen bonding was beneficial to the immobilization of anions and the dissociation of Na-containing salts. After paired with a $Na_3V_2(PO_4)_3$ cathode and a metallic anode, the quasi-solid-state battery delivered a specific capacity of 110.7 mAhg^{-1} on discharge in the first cycle at a temperature of 30°C with a capacity retention ratio of 92% after 300 discharge/charge cycles (Figure 19.2e).

19.2.3 ELECTRODE MATERIALS

As for the cathode, one of the widely used cathode materials in hybrid sodium-ion capacitors is the commercial-activated carbon; however, its capacitance is very limited, particularly at high rates. Xu et al. designed a flexible electrode by anchoring N-doped carbon array on carbon fiber cloth built from 2D metal-organic frameworks (MOFs), followed by the deposition of electroactive VO_2 and $Na_3V_2(PO_4)_3$ as free-standing electrodes [50]. This design could suppress the undesired aggregation of electrode materials and accommodate the structural changes over cycling. As a result, the VO_2-based electrode possessed a pseudocapacitive Na-ion storage characteristic, and the $Na_3V_2(PO_4)_3$ cathode with battery-type behaviors contributed to improving rate capability. Finally, a quasi-solid-state hybrid Na-ion capacitor was assembled by utilizing VO_2- and $Na_3V_2(PO_4)_3$-based electrodes as the anode and the cathode, respectively, in the presence of a Na-ion conducting polymer electrolyte, which delivered a gravimetric energy density of 161 $Whkg^{-1}$ (volumetric density of 8.83 $mWhcm^{-3}$) and a gravimetric power density of 24 $kWkg^{-1}$ (volumetric density of 1.32 $Whcm^{-3}$) [50]. Besides, it has been verified that reconfiguring these electrode materials in 2D geometry into three-dimensional (3D) architectures can effectively improve battery performance. As shown in Figure 19.3a, a flexible Na-ion hybrid device was constructed by an intercalation-type sodium vanadium fluorophosphate $(Na_3(VO)_2(PO_4)_2F)$ positive electrode and a VO_2 negative electrode, and the assembled device presented a maximum energy density of 215 $Whkg^{-1}$ and a maximum power density of 5200 Wkg^{-1} (Figure 19.3b) [46]. Recently, Yao et al. prepared an all-solid-state sodium battery by using a carbon-coated $Na_3V_2(PO_4)_3$ hybrid as the cathode, and metallic sodium as the anode in the presence of a solvent-free solid polymer electrolyte. The soft-pack Na-ion batteries presented a cycling life of 740 cycles with 95% of the reversible capacity retained. Furthermore, the bendable battery delivered a specific capacity of 106 $mAhg^{-1}$ under repeated bending for 535 cycles (Figure 19.3c) [47].

Flexible aqueous rechargeable Na-ion batteries possess the merits of low cost and high security without the use of organic electrolytes. Figure 19.3d shows the synthesis of microcube-like $K_2Zn_3(Fe(CN)_6)_2 \cdot 9H_2O$ on carbon cloth as a free-standing cathode, which exhibited a capacity of 0.76 $mAhcm^{-2}$ at the density of 0.5 $mAcm^{-2}$ and capacity retention of 57.9% at a high density of 20 $mAcm^{-2}$ [48]. By paired with $NaTi_2(PO_4)_3$ on carbon cloth as the anode, the quasi-solid-state flexible aqueous rechargeable Na-ion batteries delivered an output voltage plateau of 1.6 V and an energy density of 0.92 $mWhcm^{-2}$. Particularly, the assembled device maintained 90.3% of the initial specific capacity after 3,000 bending cycles [48]. Then, He et al. reported a hollow-structure $NaTi_2(PO_4)_3$ in N-doped carbon nanofiber (HNTP@PNC) binder-free anode for flexible aqueous Na-ion batteries (Figure 19.3e) [49]. Theoretical calculations concluded that the combination of $NaTi_2(PO_4)_3$ and N-doped carbon coating enhanced the electronic conductivity and meanwhile promoted the Na-ion diffusion kinetics. By coupling with a potassium zinc hexacyanoferrate cathode, the constructed quasi-solid-state aqueous rechargeable Na-ion battery exhibited a volumetric capacity of 24.5 $mAhcm^{-3}$ as well as an energy density of 39.2 $mWhcm^{-3}$ [49].

Owing to the high theoretical ion storage capacity and large natural abundance, metal oxides, such as TiO_2 [54–56], $Na_2Ti_3O_7$ [57,58], and Fe_3O_4 [59], have been explored as promising anode candidates for flexible Na-ion batteries [60–63]. As demonstrated in Figure 19.4a, a flexible hybrid Na-ion capacitor was assembled with a $Na_2Ti_2O_{5-x}$ arrays anode and a reduced graphene oxide/activated carbon (rGO/AC) cathode, and the resultant device presented a voltage output from 1.0 to 3.8 V (Figure 19.4b), together with an energy density of 70 $Whkg^{-1}$ at a power density of 240 $W kg^{-1}$, and a maximum volumetric energy density of 15.6 WhL^{-1} and a maximum power density of 120 WL^{-1} based on the whole package (Figure 19.4c) [51]. Metal sulfides are another important family as potential anode material for Na-ion batteries. Various metal sulfides, such as Co_9S_8 [64], MoS_2 [65,66], NiS_2 [67], Fe_xS_y [52,68], WS_2 [69], Bi_2S_3 [70], Sb_2S_3 [71], have been reported for flexible Na-ion batteries. Figure 19.4d shows a hierarchical 3D carbon-networks/Fe_7S_8/graphene electrode for a flexible Na-ion battery, which presented a capacity of 1.42 $mAhcm^{-2}$ at the density of 0.3 $mAcm^{-2}$, and an energy density of 144

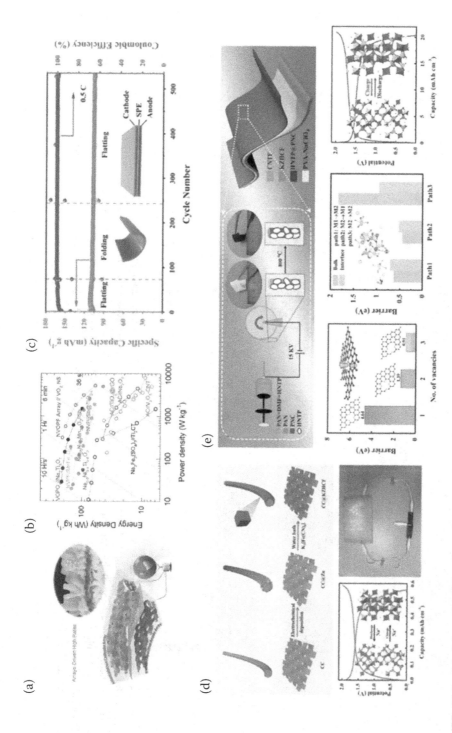

FIGURE 19.3 (a) Schematic illustration of the configuration of a flexible Na-ion hybrid device constructed by a sodium vanadium fluorophosphate ($Na_3(VO)_2(PO_4)_2F$) positive electrode and a VO_2 negative electrode, and (b) Ragone plot comparing the hybrid device with energy storage systems. Reproduced with permission from ref. [46]. Copyright (2018) Wiley-VCH. (c) Cycling stability and Coulombic efficiency of an all-solid-state sodium battery by using a carbon coated $Na_3V_2(PO_4)_3$ composite as the cathode and a metallic sodium as the anode. Reproduced with permission from ref. [47]. Copyright (2020) Wiley-VCH. (d) Schematic illustration of the synthesis of microcube-like $K_2Zn_3(Fe(CN)_6)_2·9H_2O$ on carbon cloth (CC@KZHCF), and charge/discharge plots of the assembled quasi-solid-state flexible aqueous rechargeable Na-ion batteries with a photograph of a lighted LED. Reproduced with permission from ref. [48]. Copyright (2019) Wiley-VCH. (e) Schematic illustration of the synthesis of hollow-structure $NaTi_2(PO_4)_3$ in the porous N-doped carbon nanofiber (HNTP@PNC) binder-free anode for flexible aqueous rechargeable Na-ion batteries, and comparison on the diffusion barriers of Na-ion through or along different surfaces or interfaces, and charge/discharge curves of the assembled flexible aqueous rechargeable Na-ion batteries. Reproduced with permission from ref. [49]. Copyright (2021) Elsevier.

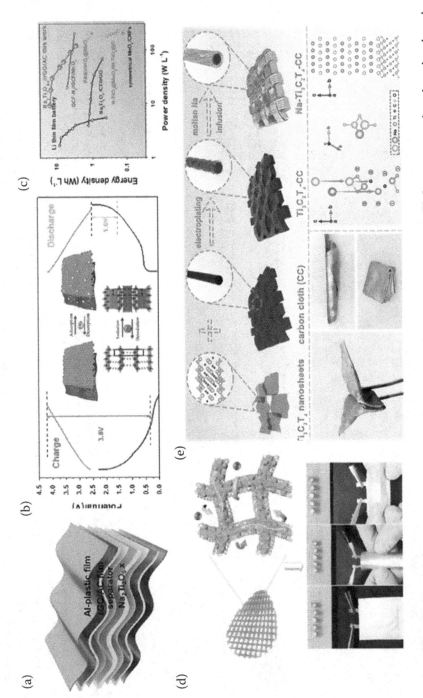

FIGURE 19.4 (a) Schematic illustration of the configuration of a flexible hybrid Na-ion capacitor with a $Na_2Ti_2O_{5-x}$ arrays anode and a reduced graphene oxide/activated carbon (rGO/AC) cathode with (b) Voltage range and (c) Ragone plots. Reproduced with permission from ref. [51]. Copyright (2017) American Chemical Society. (d) Schematic illustration of the structure of a hierarchical 3D carbon-networks/Fe_7S_8/graphene electrode for flexible Na-ion batteries and the photographs of the lighted LED bulbs powered by the pouch-type full cells. Reproduced with permission from ref. [52]. Copyright (2019) Wiley. (e) Schematic illustration of the synthesis of $Ti_3C_2T_x$-coated carbon cloth with sodium, and its foldable and bendable demonstration. Reproduced with permission from ref. [53]. Copyright (2020) American Chemical Society.

Whkg^{-1} [52]. Besides, metal-containing selenides (e.g., VSe$_2$), phosphides (e.g., FeP and CoP$_4$), and MXenes have been studied for flexible Na-ion batteries [53,72–76]. For example, a foldable and bendable Ti$_3$C$_2$T$_x$-coated carbon cloth with modification by metallic sodium via a thermal infusion treatment, was employed as a metal anode for flexible Na-ion batteries by coupling with a Na$_3$V$_2$(PO$_4$)$_3$ cathode, which exhibited stable electrochemical cycling performance under different bending states (Figure 19.4e) [53]. In addition, carbonaceous materials, such as graphene [77–79], and carbon fiber or cloth [80–82], have been considered as the promising anode materials [83,84], which is due to their favorable operating voltage and good cycle stability for flexible Na-ion batteries.

19.2.4 SEPARATORS

For electrochemical storage devices, an ideal separator should possess attractive electrolyte uptake properties and good chemical, mechanical, and thermal stability, as well as a suitable production cost [85]. Several reported separators for Na-ion batteries include glass fiber paper, polyolefin microporous separators, and nonwoven organic polymers [86,87]. Glass fiber paper based on inorganic materials possesses good thermal stability and rich porous structure; however, the insufficient mechanical strength and the high production costs are two major drawbacks. Polyolefin separators exhibit good mechanical performance and chemical stability; however, the weak thermal stability and the undesired wettability to electrolyte hinder their application for Na-ion batteries. Nonwoven organic polymers, such as PVDF and PVDF-HFP, manifest good mechanical strength, electrochemical stability, and high porosity at reasonable costs; however, improving the inherent unsatisfactory thermal stability of PVDF separator remains a major challenge.

19.3 CONCLUSION

Flexible batteries based on Na-ion are promising for future applications in portable electronics and even large-scale grid-level storage, which are primarily due to the sodium resource abundance and rich intercalation chemistry. However, the overall performance is far away from practical applications. Specifically, (i) many reported cell configurations are only at the conceptual stage, and the optimal configuration is still difficult to confirm; (ii) most electrode materials for flexible Na-ion batteries are the same as these traditional Na-ion batteries, and the mechanical requirements for flexible batteries are primarily achieved by substrate optimization; (iii) the development of solid electrolytes are still in very early stages, and most reported ones could be only used as quasi-solid-state batteries; (iv) the research on understanding the storage mechanisms, and tracking the electrochemical reactions within the flexible Na-ion batteries under practical operation conditions are lacking. Although some major challenges to improving electrochemical performance exist, it is expected that the flexible Na-ion batteries would be a potential candidate for wearable ESSs. In the future, more work should be done to explore the design on cell configurations, the optimization on electrode materials, and the accurate match on electrolytes to achieve good mechanical strength and satisfying battery performance.

REFERENCES

1. H.G. Wang, W. Li, D.P. Liu, X.L. Feng, J. Wang, X.Y. Yang, X.B. Zhang, Y. Zhu, Y. Zhang, Flexible Electrodes for Sodium-Ion Batteries: Recent Progress and Perspectives, *Adv. Mater.* 29 (2017) 1703012.
2. H. Li, X. Zhang, Z. Zhao, Z. Hu, X. Liu, G. Yu, Flexible Sodium-ion Based Energy Storage Devices: Recent Progress and Challenges, *Energy Storage Mater.* 26 (2020) 83–104.
3. Q. Zhai, F. Xiang, F. Cheng, Y. Sun, X. Yang, W. Lu, L. Dai, Recent Advances in Flexible/stretchable Batteries and Integrated Devices, *Energy Storage Mater.* 33 (2020) 116–138.
4. B. Yao, J. Zhang, T. Kou, Y. Song, T. Liu, Y. Li, Paper-Based Electrodes for Flexible Energy Storage Devices, *Adv. Sci.* 4 (2017) 1700107.
5. C. Zhao, Y. Lu, L. Chen, Y. Hu, Flexible Na Batteries, *InfoMat.* 2 (2020) 126–138.

6. C. Zhou, S. Bag, V. Thangadurai, Engineering Materials for Progressive All-Solid-State Na Batteries, *ACS Energy Lett.* 3 (2018) 2181–2198.

7. Z. Xi, X. Zhang, Y. Ma, C. Zhou, J. Yang, Y. Wu, X. Li, Y. Luo, D. Chen, Recent Progress in Flexible Fibrous Batteries, *ChemElectroChem.* 5 (2018) 3127–3137.

8. Y. Liu, Z. Sun, K. Tan, Di.K. Denis, J. Sun, L. Liang, L. Hou, C. Yuan, Recent Progress in Flexible Non-lithium Based Rechargeable Batteries, *J. Mater. Chem. A.* 7 (2019) 4353–4382.

9. W. Luo, F. Shen, C. Bommier, H. Zhu, X. Ji, L. Hu, Na-ion Battery Anodes: Materials and Electrochemistry, *Acc. Chem. Res.* 49 (2016) 231–240.

10. J.-Y. Hwang, S.-T. Myung, Y.-K. Sun, Sodium-ion Batteries: Present and Future, *Chem. Soc. Rev.* 46 (2017) 3529–3614.

11. S. Dong, P. Nie, X. Zhang, Flexible Sodium Ion Batteries: From Materials to Devices, in: C.Zhi and L. Dai (Ed.), *Flexible Energy Conversion and Storage Devices*, Wiley-VCH, 2018: pp. 97–125.

12. T. Jin, Q. Han, L. Jiao, Binder-Free Electrodes for Advanced Sodium-Ion Batteries, *Adv. Mater.* 32 (2020) 1806304.

13. P.K. Nayak, L. Yang, W. Brehm, P. Adelhelm, From Lithium-Ion to Sodium-Ion Batteries: Advantages, Challenges, and Surprises, *Angew. Chem. Int. Ed.* 57 (2018) 102–120.

14. J. Mei, T. Liao, Z. Sun, 2D/2D Heterostructures: Rational Design for Advanced Batteries and Electrocatalysis, *Energy Environ. Mater.* (2021) 10.1002/EEM2.12184.

15. M.I. Jamesh, Recent Advances on Flexible Electrodes for Na-ion Batteries and Li–S Batteries, *J. Energy Chem.* 32 (2019) 15–44.

16. T. Zhang, L. Yang, X. Yan, X. Ding, Recent Advances of Cellulose-Based Materials and Their Promising Application in Sodium-Ion Batteries and Capacitors, *Small.* 14 (2018) 1802444.

17. J. Mei, T. Liao, J. Liang, Y. Qiao, S.X. Dou, Z. Sun, Toward Promising Cathode Catalysts for Nonlithium Metal-oxygen Batteries, *Adv. Energy Mater.* 10 (2020) 1901997.

18. J. Mei, T. Liao, G.A. Ayoko, Z. Sun, Two-Dimensional Bismuth Oxide Heterostructured Nanosheets for Lithium- and Sodium-Ion Storages, *ACS Appl. Mater. Interfaces.* 11 (2019) 28205–28212.

19. J. Mei, J. Wang, H. Gu, Y. Du, H. Wang, Y. Yamauchi, T. Liao, Z. Sun, Z. Yin, Nano Polymorphism-Enabled Redox Electrodes for Rechargeable Batteries, *Adv. Mater.* 33 (2021) 2004920.

20. A. Banerjee, K.H. Park, J.W. Heo, Y.J. Nam, C.K. Moon, S.M. Oh, S.-T. Hong, Y.S. Jung, Na_3SbS_4: A Solution Processable Sodium Superionic Conductor for All-Solid-State Sodium-Ion Batteries, *Angew. Chem.* 128 (2016) 9786–9790.

21. Y. Lu, L. Li, Q. Zhang, Z. Niu, J. Chen, Electrolyte and Interface Engineering for Solid-State Sodium Batteries, *Joule.* 2 (2018) 1747–1770.

22. T. Wang, J. Mei, J. Liu, T. Liao, Maximizing Ionic Transport of $Li_{1+x}Al_xTi_{2-x}P_3O_{12}$ Electrolytes for All-solid-state Lithium-ion Storage: A Theoretical Study, *J. Mater. Sci. Technol.* 73 (2021) 45–51.

23. Y. Wang, S. Song, C. Xu, N. Hu, J. Molenda, L. Lu, Development of Solid-state Electrolytes for Sodium-ion Battery-A Short Review, *Nano Mater. Sci.* 1 (2019) 91–100.

24. W. Hou, X. Guo, X. Shen, K. Amine, H. Yu, J. Lu, Solid Electrolytes and Interfaces in All-solid-state Sodium Batteries: Progress and Perspective, *Nano Energy.* 52 (2018) 279–291.

25. C. Zhao, L. Liu, X. Qi, Y. Lu, F. Wu, J. Zhao, Y. Yu, Y.S. Hu, L. Chen, Solid-State Sodium Batteries, *Adv. Energy Mater.* 8 (2018) 1703012.

26. K. Mishra, N. Yadav, S.A. Hashmi, Recent Progress in Electrode and Electrolyte Materials for Flexible Sodium-ion Batteries, *J. Mater. Chem. A.* 8 (2020) 22507–22543.

27. L. Xue, X. Shi, B. Lin, Q. Guo, Y. Zhao, H. Xia, Self-standing P2/P3 Heterostructured $Na_{0.7}CoO_2$ Nanosheet Arrays as 3D Cathodes for Flexible Sodium-ion Batteries, *J. Power Sources.* 457 (2020) 228059.

28. W. Wang, Q. Xu, H. Liu, Y. Wang, Y. Xia, A Flexible Symmetric Sodium Full Cell Constructed Using the Bipolar Material $Na_3V_2(PO_4)_3$, *J. Mater. Chem. A.* 5 (2017) 8440–8450.

29. Y.H. Zhu, S. Yuan, D. Bao, Y. Bin Yin, H.X. Zhong, X.B. Zhang, J.M. Yan, Q. Jiang, Decorating Waste Cloth via Industrial Wastewater for Tube-Type Flexible and Wearable Sodium-Ion Batteries, *Adv. Mater.* 29 (2017) 1603719.

30. Z. Guo, Y. Zhao, Y. Ding, X. Dong, L. Chen, J. Cao, C. Wang, Y. Xia, H. Peng, Y. Wang, Multi-functional Flexible Aqueous Sodium-Ion Batteries with High Safety, *Chem.* 3 (2017) 348–362.

31. Y.H. Zhu, X.Y. Yang, T. Liu, X.B. Zhang, Flexible 1D Batteries: Recent Progress and Prospects, *Adv. Mater.* 32 (2020) 1901961.

32. T. Lan, C.L. Tsai, F. Tietz, X.K. Wei, M. Heggen, R.E. Dunin-Borkowski, R. Wang, Y. Xiao, Q. Ma, O. Guillon, Room-temperature All-solid-state Sodium Batteries with Robust Ceramic Interface Between Rigid Electrolyte and Electrode Materials, *Nano Energy*. 65 (2019) 104040.

33. H. Wan, J.P. Mwizerwa, F. Han, W. Weng, J. Yang, C. Wang, X. Yao, Grain-boundary-resistance-less $Na_3SbS_{4-x}Se_x$ Solid Electrolytes for All-solid-state Sodium Batteries, *Nano Energy*. 66 (2019) 104109.

34. W. Niu, L. Chen, Y. Liu, L.Z. Fan, All-Solid-State Sodium Batteries Enabled by Flexible Composite Electrolytes and Plastic-crystal Interphase, *Chem. Eng. J.* 384 (2020) 123233.

35. G. Chen, K. Zhang, Y. Liu, L. Ye, Y. Gao, W. Lin, H. Xu, X. Wang, Y. Bai, C. Wu, Flame-retardant Gel Polymer Electrolyte and Interface for Quasi-solid-state Sodium Ion Batteries, *Chem. Eng. J.* 401 (2020) 126065.

36. Z. Zhang, K. Xu, X. Rong, Y.S. Hu, H. Li, X. Huang, L. Chen, $Na_{3.4}Zr_{1.8}Mg_{0.2}Si_2PO_{12}$ filled Poly(ethylene oxide)/$Na(CF_3SO_2)_2N$ as Flexible Composite Polymer Electrolyte for Solid-state Sodium Batteries, *J. Power Sources*. 372 (2017) 270–275.

37. S. Lou, F. Zhang, C. Fu, M. Chen, Y. Ma, G. Yin, J. Wang, Interface Issues and Challenges in All-Solid-State Batteries: Lithium, Sodium, and Beyond, *Adv. Mater.* 33 (2021) 2000721.

38. H. Dai, Y. Chen, W. Xu, Z. Hu, J. Gu, X. Wei, F. Xie, W. Zhang, W. Wei, R. Guo, G. Zhang, A Review of Modification Methods of Solid Electrolytes for All-Solid-State Sodium-Ion *Batteries, Energy Technol.* 9 (2021) 2000682.

39. J.K. Kim, Y.J. Lim, H. Kim, G.B. Cho, Y. Kim, A Hybrid Solid Electrolyte for Flexible Solid-state Sodium Batteries, *Energy Environ. Sci.* 8 (2015) 3589–3596.

40. H. Gao, L. Xue, S. Xin, K. Park, J.B. Goodenough, A Plastic-Crystal Electrolyte Interphase for All-Solid-State Sodium Batteries, *Angew. Chem.* 129 (2017) 5633–5637.

41. X. Yu, L. Xue, J.B. Goodenough, A. Manthiram, Ambient-Temperature All-Solid-State Sodium Batteries with a Laminated Composite Electrolyte, *Adv. Funct. Mater.* 31 (2021) 2002144.

42. Y. Gao, G. Chen, X. Wang, H. Yang, Z. Wang, W. Lin, H. Xu, Y. Bai, C. Wu, $PY_{13}FSI$-Infiltrated SBA-15 as Nonflammable and High Ion-Conductive Ionogel Electrolytes for Quasi-Solid-State Sodium-Ion Batteries, *ACS Appl. Mater. Interfaces*. 12 (2020) 22981–22991.

43. W. Ling, N. Fu, J. Yue, X.X. Zeng, Q. Ma, Q. Deng, Y. Xiao, L.J. Wan, Y.G. Guo, X.W. Wu, A Flexible Solid Electrolyte with Multilayer Structure for Sodium Metal Batteries, *Adv. Energy Mater.* 10 (2020) 1903966.

44. N. Chen, H. Zhang, L. Li, R. Chen, S. Guo, Ionogel Electrolytes for High-Performance Lithium Batteries: A Review, *Adv. Energy Mater.* 8 (2018) 1702675.

45. H. Sun, G. Zhu, X. Xu, M. Liao, Y.Y. Li, M. Angell, M. Gu, Y. Zhu, W.H. Hung, J. Li, Y. Kuang, Y. Meng, M.C. Lin, H. Peng, H. Dai, A Safe and Non-flammable Sodium Metal Battery Based on An Ionic Liquid Electrolyte, *Nat. Commun.* 10 (2019) 3302.

46. D. Chao, C.H.M. Lai, P. Liang, Q. Wei, Y.S. Wang, C.R. Zhu, G. Deng, V.V.T. Doan-Nguyen, J. Lin, L. Mai, H.J. Fan, B. Dunn, Z.X. Shen, Sodium Vanadium Fluorophosphates (NVOPF) Array Cathode Designed for High-Rate Full Sodium Ion Storage Device, *Adv. Energy Mater.* 8 (2018) 1800058.

47. Y. Yao, Z. Wei, H. Wang, H. Huang, Y. Jiang, X. Wu, X. Yao, Z.S. Wu, Y. Yu, Toward High Energy Density All Solid-State Sodium Batteries with Excellent Flexibility, *Adv. Energy Mater.* 10 (2020) 1903698.

48. B. He, P. Man, Q. Zhang, C. Wang, Z. Zhou, C. Li, L. Wei, Y. Yao, Conversion Synthesis of Self-Standing Potassium Zinc Hexacyanoferrate Arrays as Cathodes for High-Voltage Flexible Aqueous Rechargeable Sodium-Ion Batteries, *Small*. 15 (2019) 1905115.

49. B. He, K. Yin, W. Gong, Y. Xiong, Q. Zhang, J. Yang, Z. Wang, Z. Wang, M. Chen, P. Man, P. Coquet, Y. Yao, L. Sun, L. Wei, $NaTi_2(PO_4)_3$ Hollow Nanoparticles Encapsulated in Carbon Nanofibers as Novel Anodes for Flexible Aqueous Rechargeable Sodium-ion Batteries, *Nano Energy*. 82 (2021) 105764.

50. D. Xu, D. Chao, H. Wang, Y. Gong, R. Wang, B. He, X. Hu, H.J. Fan, Flexible Quasi-Solid-State Sodium-Ion Capacitors Developed Using 2D Metal-Organic-Framework Array as Reactor, *Adv. Energy Mater.* 8 (2018) 1702769.

51. L.F. Que, F. Da Yu, K.W. He, Z.B. Wang, D.M. Gu, Robust and Conductive $Na_2Ti_2O_{5-x}$ Nanowire Arrays for High-Performance Flexible Sodium-Ion Capacitor, *Chem. Mater.* 29 (2017) 9133–9141.

52. W. Chen, X. Zhang, L. Mi, C. Liu, J. Zhang, S. Cui, X. Feng, Y. Cao, C. Shen, High-Performance Flexible Freestanding Anode with Hierarchical 3D Carbon-Networks/Fe_7S_8/Graphene for Applicable Sodium-Ion Batteries, *Adv. Mater.* 31 (2019) 1806664.

53. Y. Fang, R. Lian, H. Li, Y. Zhang, Z. Gong, K. Zhu, K. Ye, J. Yan, G. Wang, Y. Gao, Y. Wei, D. Cao, Induction of Planar Sodium Growth on MXene ($Ti_3C_2T_x$)-Modified Carbon Cloth Hosts for Flexible Sodium Metal Anodes, *ACS Nano.* 14 (2020) 8744–8753.

54. S. Liu, Z. Luo, G. Tian, M. Zhu, Z. Cai, A. Pan, S. Liang, TiO_2 Nanorods Grown on Carbon Fiber Cloth as Binder-free Electrode for Sodium-ion Batteries and Flexible Sodium-ion Capacitors, *J. Power Sources.* 363 (2017) 284–290.

55. X. Zhang, B. Wang, W. Yuan, J. Wu, H. Liu, H. Wu, Y. Zhang, Reduced Graphene Oxide Modified N-doped Carbon Foam Supporting TiO_2 Nanoparticles as Flexible Electrode for High-performance Li/Na Ion Batteries, *Electrochim. Acta.* 311 (2019) 141–149.

56. C. Wang, J. Zhang, X. Wang, C. Lin, X.S. Zhao, Hollow Rutile Cuboid Arrays Grown on Carbon Fiber Cloth as a Flexible Electrode for Sodium-Ion Batteries, *Adv. Funct. Mater.* 30 (2020) 2002629.

57. H. Li, L. Peng, Y. Zhu, X. Zhang, G. Yu, Achieving High-Energy-High-Power Density in a Flexible Quasi-Solid-State Sodium Ion Capacitor, *Nano Lett.* 16 (2016) 5938–5943.

58. S. Dong, L. Shen, H. Li, G. Pang, H. Dou, X. Zhang, Flexible Sodium-Ion Pseudocapacitors Based on 3D $Na_2Ti_3O_7$ Nanosheet Arrays/Carbon Textiles Anodes, *Adv. Funct. Mater.* 26 (2016) 3703–3710.

59. Q. Zhao, Z. Xia, T. Qian, X. Rong, M. Zhang, Y. Dong, J. Chen, H. Ning, Z. Li, H. Hu, M. Wu, PVP-assisted Synthesis of Ultrafine Transition Metal Oxides Encapsulated in Nitrogen-doped Carbon Nanofibers as Robust and Flexible Anodes for Sodium-ion Batteries, *Carbon.* 174 (2021) 325–334.

60. J. Mei, T. Liao, L. Kou, Z. Sun, Two-dimensional Metal Oxide Nanomaterials for Next-generation Rechargeable Batteries, *Adv. Mater.* 29 (2017) 1700176.

61. J. Mei, T. Liao, Z. Sun, Two-dimensional Metal Oxide Nanosheets for Rechargeable Batteries, *J. Energy Chem.* 27 (2018) 117–127.

62. Y. Wang, Y. Zheng, J. Zhao, Y. Li, Flexible Fiber-shaped Lithium and Sodium-ion Batteries with Exclusive Ion Transport Channels and Superior Pseudocapacitive Charge Storage, *J. Mater. Chem. A.* 8 (2020) 11155–11164.

63. J. Mei, T. Liao, G.A.G.A. Ayoko, J. Bell, Z. Sun, Cobalt Oxide-based Nanoarchitectures for Electrochemical Energy Applications, *Prog. Mater. Sci.* 103 (2019) 596–677.

64. X. Ma, X. Xiong, P. Zou, W. Liu, F. Wang, L. Liang, Y. Liu, C. Yuan, Z. Lin, General and Scalable Fabrication of Core–Shell Metal Sulfides@C Anchored on 3D N-Doped Foam toward Flexible Sodium Ion Batteries, *Small.* 15 (2019) 1903259.

65. H. Wang, H. Jiang, Y. Hu, P. Saha, Q. Cheng, C. Li, Interface-engineered MoS_2/C Nanosheet Heterostructure Arrays for Ultra-stable Sodium-ion Batteries, *Chem. Eng. Sci.* 174 (2017) 104–111.

66. B. Lu, J. Liu, R. Hu, H. Wang, J. Liu, M. Zhu, C@MoS_2@PPy Sandwich-like Nanotube Arrays as An Ultrastable and High-rate Flexible Anode for Li/Na-ion Batteries, *Energy Storage Mater.* 14 (2018) 118–128.

67. Q. Chen, S. Sun, T. Zhai, M. Yang, X. Zhao, H. Xia, Yolk–Shell NiS_2 Nanoparticle-Embedded Carbon Fibers for Flexible Fiber-Shaped Sodium Battery, *Adv. Energy Mater.* 8 (2018) 1800054.

68. Y. Liu, Y. Fang, Z. Zhao, C. Yuan, X.W. (David) Lou, A Ternary Fe_{1-x}S@Porous Carbon Nanowires/Reduced Graphene Oxide Hybrid Film Electrode with Superior Volumetric and Gravimetric Capacities for Flexible Sodium Ion Batteries, *Adv. Energy Mater.* 9 (2019) 1803052.

69. Y. Wang, D. Kong, S. Huang, Y. Shi, M. Ding, Y. Von Lim, T. Xu, F. Chen, X. Li, H.Y. Yang, 3D Carbon Foam-Supported WS_2 Nanosheets for Cable-Shaped Flexible Sodium Ion *Batteries, J. Mater. Chem. A.* 6 (2018) 10813–10824.

70. C. Lu, Z. Li, L. Yu, L. Zhang, Z. Xia, T. Jiang, W. Yin, S. Dou, Z. Liu, J. Sun, Nanostructured Bi_2S_3 Encapsulated Within Three-dimensional N-doped Graphene as Active and Flexible Anodes for Sodium-ion Batteries, *Nano Res.* 11 (2018) 4614–4626.

71. S. Liu, Z. Cai, J. Zhou, M. Zhu, A. Pan, S. Liang, High-performance Sodium-ion Batteries and Flexible Sodium-ion Capacitors Based on Sb_2X_3 (X = O, S)/carbon Fiber Cloth, *J. Mater. Chem. A.* 5 (2017) 9169–9176.

72. S. Shi, Z. Li, L. Shen, X. Yin, Y. Liu, G. Chang, J. Wang, S. Xu, J. Zhang, Y. Zhao, Electrospun Free-standing FeP@NPC Film for Flexible Sodium Ion Batteries with Remarkable Cycling Stability, *Energy Storage Mater.* 29 (2020) 78–83.

73. D. Sun, X. Zhu, B. Luo, Y. Zhang, Y. Tang, H. Wang, L. Wang, New Binder-Free Metal Phosphide–Carbon Felt Composite Anodes for Sodium-Ion Battery, *Adv. Energy Mater.* 8 (2018) 1801197.

74. Y. Wu, W. Zhong, W. Tang, L. Zhang, H. Chen, Q. Li, M. Xu, S. juan Bao, Flexible Electrode Constructed by Encapsulating Ultrafine VSe_2 in Carbon Fiber for Quasi-solid-state Sodium Ion Batteries, *J. Power Sources.* 470 (2020) 228438.

75. Y. Wang, Y. Zheng, J. Zhao, Y. Li, Assembling Free-standing and Aligned Tungstate/MXene Fiber for Flexible Lithium and Sodium-ion Batteries with Efficient Pseudocapacitive Energy Storage, *Energy Storage Mater.* 33 (2020) 82–87.

76. J. Mei, G.A. Ayoko, C. Hu, J.M. Bell, Z. Sun, Two-dimensional Fluorine-free Mesoporous Mo_2C MXene via UV-induced Selective Etching of Mo_2Ga_2C for Energy Storage, *Sustain. Mater. Technol.* 25 (2020) e00156.

77. H. An, Y. Li, Y. Gao, C. Cao, J. Han, Y. Feng, W. Feng, Free-standing Fluorine and Nitrogen co-doped Graphene Paper as A High-performance Electrode for Flexible Sodium-ion Batteries, *Carbon.* 116 (2017) 338–346.

78. G. Zhou, Y.E. Miao, Z. Wei, L.L. Mo, F. Lai, Y. Wu, J. Ma, T. Liu, Bioinspired Micro/Nanofluidic Ion Transport Channels for Organic Cathodes in High-Rate and Ultrastable Lithium/Sodium-Ion Batteries, *Adv. Funct. Mater.* 28 (2018) 1804629.

79. X. Guo, S. Zheng, G. Zhang, X. Xiao, X. Li, Y. Xu, H. Xue, H. Pang, Nanostructured Graphene-based Materials for Flexible Energy Storage, *Energy Storage Mater.* 9 (2017) 150–169.

80. Y. Wang, N. Xiao, Z. Wang, Y. Tang, H. Li, M. Yu, C. Liu, Y. Zhou, J. Qiu, Ultrastable and High-capacity Carbon Nanofiber Anode Derived from Pitch/polyacrylonitrile Hybrid for Flexible Sodium-ion Batteries, *Carbon.* 135 (2018) 187–194.

81. R. Chen, Z. Cheng, Y. Hu, L. Jiang, P. Pan, J. Mao, C. Ni, Discarded Clothing Acrylic Yarns: Low-cost Raw Materials for Deformable C Nanofibers Applied to Flexible Sodium-ion Batteries, *Electrochim. Acta.* 359 (2020) 136988.

82. J.Z. Guo, Z.Y. Gu, X.X. Zhao, M.Y. Wang, X. Yang, Y. Yang, W.H. Li, X.L. Wu, Flexible Na/K-Ion Full Batteries from the Renewable Cotton Cloth–Derived Stable, Low-Cost, and Binder-Free Anode and Cathode, *Adv. Energy Mater.* 9 (2019) 1902056.

83. Q. Ni, Y. Bai, Y. Li, L. Ling, L. Li, G. Chen, Z. Wang, H. Ren, F. Wu, C. Wu, 3D Electronic Channels Wrapped Large-Sized $Na_3V_2(PO_4)_3$ as Flexible Electrode for Sodium-Ion Batteries, *Small.* 14 (2018) 1702864.

84. Z. Wu, Y. Wang, X. Liu, C. Lv, Y. Li, D. Wei, Z. Liu, Carbon-Nanomaterial-Based Flexible Batteries for Wearable Electronics, *Adv. Mater.* 31 (2019) 1800716.

85. L. Coustan, J.M. Tarascon, C. Laberty-Robert, Thin Fiber-Based Separators for High-Rate Sodium Ion Batteries, *ACS Appl. Energy Mater.* 2 (2019) 8369–8375.

86. W. Chen, L. Zhang, C. Liu, X. Feng, J. Zhang, L. Guan, L. Mi, S. Cui, Electrospun Flexible Cellulose Acetate-Based Separators for Sodium-Ion Batteries with Ultralong Cycle Stability and Excellent Wettability: The Role of Interface Chemical Groups, *ACS Appl. Mater. Interfaces.* 10 (2018) 23883–23890.

87. X. Ma, F. Qiao, M. Qian, Y. Ye, X. Cao, Y. Wei, N. Li, M. Sha, Z. Zi, J. Dai, Facile Fabrication of Flexible Electrodes with Poly(vinylidene fluoride)/Si_3N_4 Composite Separator Prepared by Electrospinning for Sodium-ion Batteries, *Scr. Mater.* 190 (2021) 153–157.

20 Flexible Batteries Based on K-ion

Yu Liu

Institute of New Energy on Chemical Storage and Power Sources,
Yancheng Teachers University, Yancheng, Jiangsu, People's Republic of China

Department of Chemical Engineering and Waterloo Institute for
Nanotechnology, University of Waterloo, 200 University Avenue West,
Waterloo, Ontario, Canada

Yuzhen Sun, Zhiyuan Zhao, Xiaowei Yang, and Rong Xing
Institute of New Energy on Chemical Storage and Power Sources,
Yancheng Teachers University, Yancheng, Jiangsu, People's Republic of China

CONTENTS

20.1 Introduction.. 359
20.2 Working principle.. 360
20.3 Influence of electrolytes and solid electrolyte interphase in K-ion based batteries 361
    20.3.1 Thermodynamic understanding of electrolyte reduction............................ 361
    20.3.2 Comparison of K-ion SEI and Li-/Na-ion SEI 362
    20.3.3 Effects of electrolyte selection on SEI 363
    20.3.4 Mechanical stability of SEI 363
20.4 Anode materials for K-ion based flexible batteries 364
    20.4.1 Carbon materials.. 364
    20.4.2 Phosphorus compounds.. 365
    20.4.3 Titanium-based compounds 365
    20.4.4 Alloying-type compounds 367
    20.4.5 Organic compounds.. 368
20.5 Cathode materials for K-ion based flexible batteries....................................... 368
    20.5.1 Hexacyanometallate... 368
    20.5.2 Layered metal oxides 370
    20.5.3 Polyanionic compounds 370
    20.5.4 Organic materials 371
20.6 Summary and future outlooks.. 371
References... 372

20.1 INTRODUCTION

In 2004, the design of potassium batteries by using Prussian blue cathode was introduced by Ali Eftekhari [1]. After one year, a patent related to the KPF_6 electrolyte was submitted for the potassium battery in 2005 [2]. In 2007, Starsway Electronics company in China developed the first portable media player driven by a potassium battery [3]. Later, battery research based on K-ion has persisted for many years owing to the growing popularity of Li and Na-ion battery technologies and the safety problems of potassium. Until 2015, a greatly increased number of scientific publications could be indexed, numbering

DOI: 10.1201/9781003186755-20

more than 50. In recent years, the number of publications on potassium batteries grew to approximately 400 [3,4]. All these contributions are markedly promoting the investigation of PIBs.

With the fast development of science and technology, wearable electronics have gained increasing attention in recent years because of their flexible properties and application in many fields. For instance, wearable electronics could be placed on a body or integrated into clothes, which is enriching our daily lives and becoming more significant every day [5]. Apart from their miniature, flexible, lightweight, stretchable, and foldable properties, these electronics can be also regarded as "intelligent" apparatuses because they can be interconnected by direct data communication or cloud interactions through the wireless network [6]. Nonetheless, the lack of suitable energy storage systems inhibits the wide applications of these wearable smart apparatuses.

Compared to traditional LIBs, PIBs provide many advantages and have obtained increasing concern in recent years. The main advantages are shown as follows [3]: (i) Potassium has a relatively high content in the earth's crust (1.5 wt.%). (ii) PIBs possess the low reduction potential (−2.93 V vs. standard hydrogen electrode (SHE)), which is less than that (−2.73 V vs. SHE) of sodium-ion batteries (SIBs) and near to that (−3.04 V vs. SHE) of LIBs. In addition, the theoretical and practical calculations showed that the K^+ deposition potential could be approximately −2.88 V vs. SHE in organic solvents (e.g., propylene carbonate (PC)) compared with Na^+/Na (−2.56 V vs. SHE) and Li^+/Li (−2.79 V vs. SHE), indicating that PIBs have a wider potential window and high energy density. (iii) The Lewis acidity of K-ion is much weaker, resulting in smaller solvated ions compared with sodium and lithium ions. Moreover, ion migration through the electrolyte/electrode interface could be enhanced by the small desolvation energy of potassium ions. In other words, although the atomic radius (1.38 Å) of K is larger than that of Na (0.97 Å) and Li (0.68 Å), the Stokes' radius (3.6 Å) of K^+ is less than that of Na^+ (4.6 Å) and Li^+ (4.8 Å) in PC electrolytes, demonstrating that K^+ possesses the best ion conductivity and mobility. Additionally, ab initio molecular dynamics simulations (MDS) suggested that the diffusion coefficient of K^+ is very high, which is approximately three times greater than that of lithium ions. (iv) Graphite can be intercalated/deintercalated by potassium and lithium ions reversibly but failed to accommodate sodium ions. For example, graphite can display a specific capacity of 279 mAh/g in PIBs by forming KC_8. However, the SIBs can only show 35 milliampere hours per gram by generating NaC_{64} [4]. (v) Similar to Na, potassium cannot react with aluminum used as a current collector to generate alloy at low voltage. Therefore, the copper current collector of the negative electrode could be substituted by the aluminum foil; therefore, the battery cost could be reduced markedly [3].

In this chapter, we aim to summarize the current development of flexible PIBs, including working principles, solid electrolyte interphase (SEI), electrode materials, fabrication methods, electrochemical performance, and applications. At the same time, the prospects and challenges in the practical application of PIBs are also discussed. We anticipate that this chapter would provide a direction for developing flexible PIBs with high electrochemical performance.

20.2 WORKING PRINCIPLE

Like lithium- and sodium-ions batteries, PIBs obey a "rocking chair" mechanism. To be specific, the K^+ ions migrate from the cathode to the anode in the charging process. During the discharge process, the K^+ ions transport reversely from the anode to the cathode. The corresponding schematic illustration of the working principle is shown in Figure 20.1 [7]. It shows different components of PIBs, which take responsibility for the battery working during cycling. The cathode and anode are isolated by a separator, named porous membrane; meanwhile, the suitable electrolyte is selected to soak the membrane, which is to ensure ionic migration and resist short circuit between the anode and cathode. In the charging process, the potassium ions are extracted from the cathode host by releasing the electrons to the external circuit, and then they migrate to the negative electrode through the electrolyte and separator. Once the potassium ions arrive at the negative electrode, the external electrons will be consumed. Therefore, the energy from the external power source is charged in the battery by the way of chemical energy. In the discharge

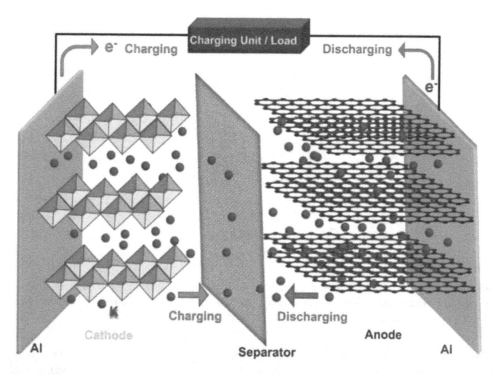

FIGURE 20.1 Schematic illustration of working principle of PIBs. Reprinted with permission from ref. [7]. Copyright 2020, Wiley.

process, the behaviors get reversed. Briefly, the potassium ions extract from the negative electrode and travel to the positive electrode through the electrolyte and separator in the period of discharge, the electrons move to the outside circuit from the negative electrode. Most importantly, the flexible batteries based on K-ion have the same working principle as this conventional electrochemical storage mechanism.

20.3 INFLUENCE OF ELECTROLYTES AND SOLID ELECTROLYTE INTERPHASE IN K-ION BASED BATTERIES

The solid electrolyte interphase (SEI) has been a crucial issue at the beginning of the appearance of PIBs, which affects the self-discharge behavior, cycling stability, Coulombic efficiency, and even safety of PIBs significantly. The SEI is generated from the electrolyte and the anode during the battery's functioning, which is regarded as a significant interface in PIBs [8]. Current studies on K-ion SEI are primarily concentrated on its morphology, composition, and structure, which the performance of the battery is probably related to. In this section, electrolyte reduction-related thermodynamic behavior in PIBs is covered first. Subsequently, we discuss the difference of K-ion SEI by comparing it with Li-ion and Na-ion interfaces. Lastly, the interfacial reaction mechanisms of K-ion SEI are explored, followed by focusing on the mechanical stability of SEI.

20.3.1 THERMODYNAMIC UNDERSTANDING OF ELECTROLYTE REDUCTION

The SEI was considered as an important part of organic-based electrolyte rechargeable batteries; after that, it was formed on negative electrodes and reported in the 1970s [9]. Figure 20.2 demonstrates the energy states of electrolytes and electrodes. The driving force of the SEI formation and thermodynamic

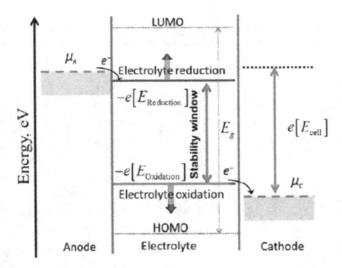

FIGURE 20.2 Schematic illustration for energy states of the electrolyte and the electrodes, resulting in the formation of the SEI and cathode electrolyte interphase. μ_A and μ_C are energy levels of anode and cathode, respectively. Reprinted with permission from ref. [10]. Copyright 2018, Royal Society of Chemistry.

behavior of electrolytes on the negative electrode can be determined by the different energy values between the lowest unoccupied molecular orbitals (LUMO) of electrolytes and the Fermi level of the anode. If $\mu_A > E_{LUMO}$, electrons can be transferred from the anode to the electrolyte, causing the spontaneous formation of SEI [10]. For instance, the potassium can be stored in anodes with the potential region of 0~1 V, like graphite (0.1 V), metallic potassium (0 V), red phosphorus (0.7 V), and antimony (0.8 V). In these cases, the LUMO energies of organic-based electrolytes are smaller than Fermi levels [10]. As a result, the electrolyte reduction on these anode surfaces is thermodynamically driven.

Additionally, reduction stability of the anion/solvent is strongly affected by the property of cation-anion or cation-solvent interaction because the coordination number of the solvent around the cation and the solvation structure is influenced. Because of the electron pair's donation to cation, LUMO levels of anions or solvents decrease when they coordinate with cations [11]. Therefore, ion-pair formation and solvation can accelerate the decomposition of electrolytes. In conclusion, the SEI generation and electrolyte stability can be affected by the positively charged ions in the electrolyte.

20.3.2 COMPARISON OF K-ION SEI AND LI-/NA-ION SEI

Based on the literature studies, the potassium, sodium, and lithium battery chemistries are much dissimilar. The unique potassium chemistries may result in the different structure, morphology, mechanical stability, and composition of K-ion SEI. Consequently, it is hard to obtain the chemical and physical properties of K-ion SEI based on the existed lithium and sodium SEI. Herein, a few comparative investigations of lithium, sodium, and potassium-ion SEI are listed.

According to previous findings, the SEI generated on the electrode surfaces in propylene carbonate (PC), including K and Li electrolytes ($KClO_4$ and $LiClO_4$), were stable. However, the SEI generated in the $NaClO_4$ electrolyte was destroyed easily [12]. This fact possibly comes from different Lewis's acidity of alkali ions ($Li^+ > Na^+ > K^+$). Generally, coulombic interaction is deserved to be noted between strong Lewis's acid and anions or solvent because the corresponding electrolytes are more soluble. Therefore, $KClO_4$ in PC is more stable than $NaClO_4$ [12]. However, Li salts demonstrate an abnormal phenomenon because of the small ionic size of Li. The strong bonds in lithium carbonates, hydroxides, and oxides restrict their solubility in a polar solvent. Consequently, the stability of lithium formed cross-linked or polymerized SEI is more pronounced, as compared to sodium [13]. Considering the much weaker Lewis's acidity of potassium ions, the stability of sodium ion SEI is worse than that of potassium ion.

Additionally, the K-ions SEI possessed different chemical composition compared with Li-ions SEI. The amount of KF was small in the K-ions SEI, yet too much LiF was in the Li-ions SEI. However, the amount of -SO$_x$F species was largest in the K system. The presence of K$_x$S$_y$ and S^0, which were seen in the K system, implied that negatively charged ions decomposition is notable [14]. These featured SEI compositions were observed in 1M KFSI/PC dilute electrolyte, while they were only reported in the highly concentrated bis(trifluoromethanesulfonyl)amide anion (TFSI$^-$) or bis(fluorosulfonyl)amide anion (FSI$^-$)-based electrolytes in lithium and sodium systems. This implies a dissimilar principle for K-ion SEI formation.

In addition, the diversity of SEI for different alkali metal ions was identified by the electrochemical test, which is a powerful and convenient method for analyzing SEI. Generally, the SEI is regarded as an Ohmic resistor, even though it has complex chemical composition. It was also reported that the overpotential of symmetric K, Na, and Li batteries in EC/DMC solvent with electrolyte (e.g., KPF$_6$, NaPF$_6$, LiPF$_6$) shows a nonlinear curve [15]. Comparing with control (no current), the offsets are 227, 152, and 32 mV at a current density of 0.7 mA cm^{-2} for K, Na, Li, whereas among the current densities of 0.7~28 mA cm^{-2} the overpotential varied from 495 to 628 mV, respectively. At low current densities, the increased overpotential for K and Na over Li is ascribed to the poor ionic conductivity of their SEI layers [15]. An increase of the resistance values from 45 to 244 and 2771 Ohm for Li, Na, and K was revealed, respectively. This suggests that the drawback that PIBs have, and the electrochemical performances of SEI in PIBs, need to be improved urgently [15].

20.3.3 EFFECTS OF ELECTROLYTE SELECTION ON SEI

Electron blocking and ion transport are two primary features of SEI. The former inhibits the electrolyte reduction, yet the latter influences on rate performance and reversibility of batteries. One of the main reasons for battery failure is the augmented polarization because of a thickened SEI.

The SEI structure, composition, coverage, and thickness on the electrode surface decide the efficiency of ion transport. The SEI properties greatly rely on the electrolyte composition because the SEI is generated at the interface of electrolyte/anode through an electrochemical or chemical reaction.

Graphite is also widely used as an anode for PIBs because of the similar structure of potassium and lithium intercalated graphite. This implies that commercial technologies and knowledge for graphite anode in lithium-ion batteries also can be utilized for PIBs. At the same time, the accumulated experiences from Li-ion SEI also provide some valuable references for the electrolyte selection in PIBs. Generally, three categories for the positively charged ion intercalated graphite are listed as below [16]: (1) the solvent is easy to reduce, which induces graphite exfoliation; (2) the solvent is stable, so the positively charged ion with solvation shell can co-intercalate reversibly; (3) the solvent is not unstable, yet a stable protective layer is formed by the decomposed products, which cover graphite edges; therefore, the cation can intercalate reversibly without solvation shell.

PC belongs to the first case, which made premature Li$^+$ fail to intercalate into graphite. This problem was tackled until the third case was clarified; it was replaced with ethylene carbonate-based electrolyte [16]. Therefore, electrolyte selection is very important to generate protective SEI, which inhibits the co-intercalation of solvent molecules and graphite exfoliation.

20.3.4 MECHANICAL STABILITY OF SEI

As we may know, the volume of electrode materials could be changed during the electrochemical reaction process. Due to the bigger potassium ions, the volume change of electrode material is approximately 60% at charge state. The potassium ions are approximately 13% larger than lithium ions [17].

However, the volume expansion of conversion-type materials (e.g., SnS_2 and CoS) and alloy-type materials (e.g., Sn, Sb, and P) is much larger during potassium storage compared with intercalation-type materials [17]. The negative electrodes could be destroyed and side reactions, including electrolyte reduction and SEI generation, could be triggered because of the large volume change, which would result in capacity decreasing. Because passivation reactions and volume expansion would cause SEI layer breaking, a flexible and mechanically robust SEI layer is needed to tolerate volume expansion during the electrochemical reaction.

In general, the viscosity, solubility, and ionic conductivity of potassium salts are balanced by adjusting electrolyte concentration in PIBs, which is normally smaller than 1.5 M [18]. However, poor electrochemical properties are induced by the interfacial reaction on the potassium electrode surface because of such diluted electrolytes. As mentioned earlier, in addition to the potassium salts and solvents chemistries, electrolyte concentration is crucial for SEI generation. Like lithium/sodium-ion batteries, the stability of potassium anode can be achieved by increasing electrolyte concentration. This is because the concentrated electrolytes will decrease the amount of free solvent molecules and increase the cation-anion or cation-solvent interaction. This solvation state variability shifts the LUMO situation from solvent to solutes [19]. Consequently, at low potential, negatively charged ions are reduced before solvent, leading to an anion-derived SEI formation.

As we know, dendrite growth inhibits the development of metal negative electrodes for Li/Na ion batteries. The high reactivity of metallic K anode is similar to lithium and sodium anodes. The potassium electrodeposition leads to unstable SEI formation by reacting with electrolytes. The volume variability of potassium electrodes further results in cracking of fragile SEI, thereby newly exposed potassium triggers subsequent deposition and induces dendrite formation. The continuous SEI cracking and reformation leads to low coulombic efficiency and high impedance [20]. All these problems are crucial for the stability of PIBs.

As we discussed above, designing a stable and robust SEI is an effective strategy to tackle these problems. However, the natural SEI generated on the metallic K surface is still not so stable; the dendrites can be seen universally. Due to the high reactivity of metallic K, introducing an artificial barrier at the interface of anode/electrolyte is a promising way to stabilize the interphase. This artificial barrier is named the artificial SEI for metal anode [21]. In fact, an artificial SEI for potassium anode should meet some conditions: (1) flexible and mechanically robust to keep its integrity from volume expansion and inhibit dendrite formation; (2) lower the charge transfer resistance; (3) regulate ion migration with ion-flux distribution; (4) technologically scalable and feasible to be introduced on the anode surface.

Although the development of man-made SEI is in the initial stage, undoubtedly, the man-made SEI method is supplementary to optimize electrolytes. This method is a potential strategy to address the serious dendrite growth because the artificial SEI is a potential strategy to address the dendrite growth because the artificial SEI possesses tunable composition, controllable structure with obvious mechanical property, and unique functions. In the future, more investigations on artificial SEI are needed, along with a further understanding of the potassium deposition mechanism.

20.4   ANODE MATERIALS FOR K-ION BASED FLEXIBLE BATTERIES

20.4.1   CARBON MATERIALS

Many carbon materials, including graphite and derivatives, carbon nanotubes (CNTs), doped carbons, hard carbons, carbon fibers (CFs), and mesoporous carbons, have been used as anode materials for PIBs. Graphite is widely utilized as negative electrode material in the lithium-ion battery, while it can accommodate K^+ ions till to generate KC_8 with a theoretical capacity of 279 mAh/g. The electrochemical behavior of K^+ intercalating in graphite is in a sequence of $KC_{36} \rightarrow KC_{24} \rightarrow KC_8$ [22]. Soft carbons fabricated by pyrolyzing aromatic organic molecules demonstrate higher voltage when compared with graphite; they are unsuitable for use as electrode materials, but they show better cycling and rate performances.

Among other carbonaceous materials, CNTs appear as promising anode materials based on the interconnected conductive network they form. N-doped CNTs fabricated by chemical vapor deposition (CVD) demonstrated a discharge capacity of 236 mAh/g after 100 cycles at 20 mA/g, while multiwall CNT is not able to be intercalated by K^+ ions reversibly [23]. N-doped CNTs fabricated by pyrolyzing metal-organic framework exhibited a capacity of 255 mAh/g after 300 cycles at 50 mA/g and a superior rate performance with 100 mAh/g at 2 A/g in 0.8 M KPF_6 EC/DEC electrolyte [24]. CFs have been widely studied. For instance, CFs synthesized by electrospinning demonstrated good cycling performance over 1200 cycles with 210 mAh/g in 0.8 M KPF_6 EC/DEC electrolyte at 200 mA/g [25]. The performance of hard carbons fabricated by pyrolysis of sucrose was also studied in PIBs. Good rate performance and capacity retention are gained for potassium after 100 cycles at C/10 rate with 216 mAh/g in 0.8 M KPF_6 EC/DEC electrolyte [26].

In conclusion, a large number of carbonaceous materials have been investigated as anode for PIBs. Graphite can store potassium ions reversibly at low potential, but it shows a relatively poor rate performance. Porous carbons (e.g., CNTs and CFs) are studied by adsorbing/desorbing K^+ in the porous structure with high specific surface area, demonstrating better rate performance, capacity retention, and higher working potential than graphite. Therefore, the synthesized materials with tailored structures combing both advantages of porous carbons and graphite might be a good candidate for anode of flexible PIBs.

20.4.2 PHOSPHORUS COMPOUNDS

Phosphorus is regarded as a promising anode material thanks to resource abundance and a large theoretical capacity of 2594 mAh/g with the generation of K_3P. The phase diagram of P-K shows below alloys, such as KP, KP_{15}, K_3P, K_3P_7, K_3P_{11}, K_4P_3, and K_4P_6, which might be seen during the potassiation process of phosphorus [27]. However, phosphorus has both a substantial volume change and low electrical conductivity when compared with all alloying materials. The utilization of nanostructure in carbon porous frameworks and intermetallic compounds is probably a way to address those issues. The first investigation of a C/Sn_4P_3 composite showed 307 mAh/g after 50 cycles at 50 mA/g, much better than Sn/C and P/C anodes [28]. Based on Sn_4P_3 research method, the GeP_5 obtained from ball milling storing potassium ions was observed, which displayed 495 mAh/g after 50 cycles at 50 mA/g [29].

Additionally, a flexible and freestanding PIB made of a bilayer-copper phosphide/copper nanowires (CuP_2/Cu NWs) anode and perylene–3,4,9,10-tetracarboxylic dianhydride/carbon nanotubes (PTCDA@CNTs) cathode with excellent cycling and rate performances was described in Figure 20.3 [30]. The cathode and anode exhibited excellent rate performances with 113 mAh/g at 5250 mA/g. Moreover, the coin-typed full battery showed 80% retention at 400 mA/g after 842 cycles and a high capacity of 117.3 mAh/g at 12000 mA/g. Regarding stability and flexibility, pouch battery characterized in form of bending state displayed stable open-circuit voltage during 5000 cycles. These superior performances show the chance to develop wearable PIBs with high safety and energy density under ultrafast reaction rates.

20.4.3 TITANIUM-BASED COMPOUNDS

In addition to carbonaceous and phosphorus-containing compositions, most inorganic intercalation-type materials were studied as anodes in PIBs, like Ti^{3+}/Ti^{4+} redox pairs. Typical investigations on the performances of titanium phosphates, carbides, and oxides will be discussed in the following text.

The titanium oxides, including $K_2Ti_4O_9$, $K_2Ti_6O_{13}$, and $K_2Ti_8O_{17}$, have been investigated as anodes. For instance, micrometric-sized $K_2Ti_4O_9$ was fabricated through the solid-state method and demonstrated 97 mAh/g at 30 mA/g in 1M KPF_6 EC/PC electrolyte [31]. $K_2Ti_6O_{13}$ nanorods in micro

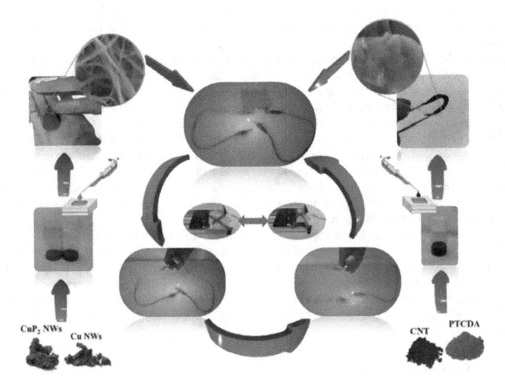

FIGURE 20.3 Schematically illustration of the preparation and advantages of flexible and freestanding bilayer CuP_2/Cu NWs and PTCDA/CNTs mesh electrodes. Reprinted with permission from ref. [30]. Copyright 2020, Elsevier.

frameworks were fabricated via the hydrothermal method under alkaline conditions. The first capacity was approximately 300 mAh/g at 50 mA/g in 0.8M KPF_6 with PC solvent containing 5 vol.% FEC additives. The capacity was still approximately 90 mAh/g, even though a significant loss happened in the following cycling [32]. In a similar method, nanostructured $K_2Ti_8O_{17}$ was also fabricated by the hydrothermal method. The obtained acanthosphere-like $K_2Ti_8O_{17}$ nanorods exhibited a specific capacity of over 180 mAh/g in 0.8M KPF_6 EC/DEC electrolyte at 20 mA/g. This material kept a discharge capacity over 115 mAh/g at low current density within 50 cycles, despite an irreversible capacity in the initial cycle [33]. Low coulombic efficiency was observed in oxides, which can be ascribed to the important trapping of potassium ions into the layered structure or to SEI formation.

The electrochemical behavior of K^+ ions in MXene compounds was also investigated. These early two-dimensional transition metal carbides provide inter-slab space allowing fast ionic diffusion. According to theoretical calculation, the Ti_3C_2 delivered a discharge capacity of 192 mAh/g [34]. $Ti_3C_2T_x$ as anode displayed an initial capacity 260 mAh/g. However, the capacity faded to 45 mAh/g at 120th cycle. Three-dimensional alkalized porous Ti_3C_2 nanoribbons with a larger inter-slab space were prepared from Ti_3C_2 in aqueous KOH, which delivered 42 mAh/g at 200 mA/g after 500 cycles [35].

Potassium-containing polyanionic compounds such as $KTi_2(PO_4)_3$ have also been studied to work as anodes. For example, $KTi_2(PO_4)_3$ nano cubic prepared via hydrothermal method was reported. In addition, carbon-coated $KTi_2(PO_4)_3$ was also synthesized by utilizing cane sugar as a precursor [36]. The $KTi_2(PO_4)_3$ showed that the capacity rapidly fades to 75 mAh/g at C/2 rate in EC/DEC solvent with 0.8M KPF_6. However, the electrolyte can effectively infiltrate in carbon-coated $KTi_2(PO_4)_3$ nanomaterials fabricated via electrospray because of their large amount of porosity. Therefore, the nanocomposites demonstrated a large capacity of 293 mAh/g at 20 mA/g and exhibited good rate performance with 133 mAh/g at 1 A/g.

20.4.4 ALLOYING-TYPE COMPOUNDS

The insertion-type materials often suffer from structural damage because of repeated insertion of potassium ions during long periods of charge/discharge processes. In contrast, the alloying-type materials do not have such a problem because the original material is thoroughly reshaped, and new species are generated along with phase transformations during the electrochemical process.

Among the alloying elements, bismuth can form K_3Bi with potassium, displaying a theoretical capacity of 385 mAh/g. No matter what the nature of the bismuth anode and the electrolyte, a potential plateau at about 0.35 V can be seen in initial discharging implying two-phase transition. However, in the range of 0.5–1.3 V, there are three platforms in the following steps, corresponding to independent three reactions. It was reported that monoglyme molecules were adsorbed on the bismuth surface, resulting in better performance and a stable SEI [37]. Therefore, a capacity of about 320 mAh/g after 300 cycles at 800 mA/g was obtained in a DME-containing electrolyte. A Bi/rGO nanomaterial prepared at room temperature showed 290 mAh/g at 0.05 A/g after 50 cycles [38].

Similar to bismuth, antimony is very interesting due to its large theoretical capacity of 660 mAh/g with the generation of K_3Sb and low working potential versus K^+/K. For instance, a C/Sb nanomaterial synthesized through ball milling exhibited 600 mAh/g at 0.035 A/g in EC/PC solvents with 1 M KPF_6 [39]. Furthermore, an interconnecting porous carbon nanofiber loaded with 4 nm Sb nanocrystalline (Sb@C PNFs) were designed as flexible PIBs negative electrode (Figure 20.4) [40]. In this case, Sb nanocrystalline was confined in N-doping porous carbon nanofiber, which served as a three-dimensional conductive substrate. Because of efficient electrolyte transportation passages, shortened K^+ migration distance, and sufficient room to tolerate volume change, the Sb@C PNFS flexible anode showed superior potassium ions accommodation capability with capacities of 399.7 mAh/g at 100 mA/g and 208.1 mAh/g even at 5.0 A/g, respectively. Moreover, the capacity of 264.0 mAh/g can be kept at 2000 mA/g after 500 cycles.

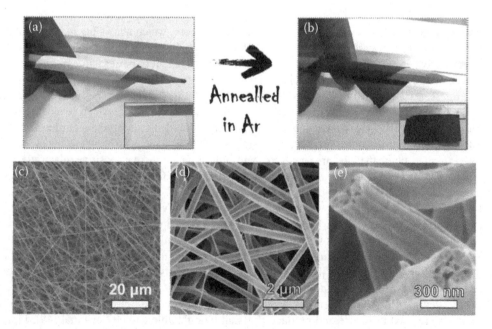

FIGURE 20.4 Digital photos of the (a) Sb@C PNFs precursor and (b) Sb@C PNFs twisted around a pencil and in the plane (the inset of (a) and (b)); (c–e) SEM images of the as-prepared Sb@C PNFs. Reprinted with permission from ref. [40]. Copyright 2020, Wiley.

20.4.5 ORGANIC COMPOUNDS

Organic compounds are becoming popular as anode since they not only are environmentally friendly and low cost but also have good cycling. In the case of K^+ ions storage, more free space and lower barrier energy for large ions intercalation are achieved through Van der Waals forces among organic molecules. Most reported negative electrode materials were modified from the terephthalate structure. For example, the dipotassium terephthalate ($K_2C_8H_4O_4$, named K_2TP) demonstrated a capacity of 180 mAh/g at C/20 after 100 cycles with 1M KFSI in EC/DMC electrolyte [41]. Interestingly, K_2TP was first prepared from potassiation of terephthalic acid. Subsequently, two conjugated carboxylate groups provide the activation sites for redox reactions. Because of low reduction potential (~0.6 V), K_2TP is an appropriate anode material.

Other carboxylates-based aromatic organic compounds were also studied. It was expected that intermolecular electron percolation would be increased by larger π conjugation. As for 4,4'-E-stilbenedicarboxylate potassium (K_2SBDC) and 1,1'-biphenyl–4,4'-dicarboxylate potassium (K_2BPDC), the capacities were less than that of K_2TP, but they are still promising with 143 mAh/g at 0.2 A/g and 100 mAh/g at 1 A/g for K_2BPDC [42]. Among the organic materials, a series of oxocarbon salts were also synthesized and studied as anodes. It was found that $K_2C_5O_5$ and $K_2C_6O_6$ can be intercalated/deintercalated by K^+ ions fast. For instance, the $K_2C_6O_6$ delivered a capacity of 212 mAh/g under 0.2C, whereas remained 164 mAh/g under 10C with 1.25 M KPF_6 in DME solvent [43].

Finally, it is worth mentioning an anode based on vitamin K [44]. Actually, a quinone structure exists in this molecule, making it active electrochemically. For example, a composite of graphene nanotubes and commercially available vitamin K displayed an initial capacity of 300 mAh/g at 0.1 A/g, which maintained rather a large capacity of 222 mAh/g for 100 cycles and 165 mAh/g even at a current density of 1 A/g.

20.5 CATHODE MATERIALS FOR K-ION BASED FLEXIBLE BATTERIES

20.5.1 HEXACYANOMETALLATE

Hexacyanoferrates (HCF) possess an open metal-ligand framework, which allows big ions intercalation and migration, making hexacyanoferrates favorable hosts. Additionally, it can offer high discharge capacity because two equivalents of alkali ions can be stored in each formula unit of hexacyanoferrates. Consequently, hexacyanoferrates can be selected as a cathode in PIBs. The typical structure is $KM[Fe(CN)_6]_y$ (nH_2O), where M represents Fe and Mn [45]. In 1986, it was reported that potassium hexacyanoferrates can accommodate potassium along with valence changes of Fe^{3+}/Fe^{2+} or M^{3+}/M^{2+}, which founded a theoretical basis for developing PIBs [45]. As we knew, Prussian blue is the delegate in potassium hexacyanoferrates due to its superior electrochemical properties and simple synthesis.

Many Prussian blue analogs are prepared and investigated as the positive electrode in PIBs. According to the first-principal calculations, it was concluded that CuHCF is much better than FeHCF as a cathode material in PIBs. In addition, the $K_{1.89}Mn[Fe(CN)_6]_{0.92}\cdot0.75H_2O$ compound, having a theoretical capacity of 156 mAh/g, showed two plateaus around 3.6 V [46]. Furthermore, a novel Prussian white, $K_{0.22}Fe[Fe(CN)_6]_{0.805}\cdot4.01H_2O$ nanoparticles, showed a capacity of 73.2 mAh/g at 0.05 A/g and exhibited a large working voltage of 3.1–3.4 V [47]. The $K_{1.81}Ni[Fe(CN)_6]_{0.97}\cdot0.086H_2O$ microporous material showed an ultralong lifespan over 8000 cycles at 0.2 A/g thanks to its large specific surface area [48]. Besides, Prussian green ($Fe_{III}Fe_{III}(CN)6$) also exhibited a superior ability to accommodate potassium ions as the positive electrode in PIBs.

Recently, flexible PIBs have been fabricated through a photographic printing technique. To be specific, photos were used to induce and produce Prussian blue flexible cathode material [49]. Figure 20.5 shows the appearance and flexible property of the PIBs, which consisted of graphite on a copper substrate as a negative electrode, Prussian blue as cathode, and 0.8M KPF_6 as an electrolyte in EC/DEC.

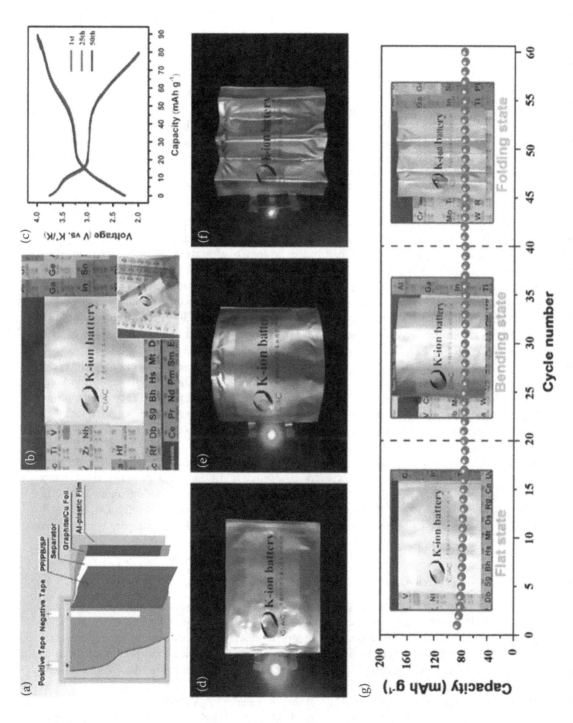

FIGURE 20.5 (a–g) Structure and bending properties of the flexible KIBs (cathode: Super P sprayed photographic printed prussian blue; anode: graphite; electrolyte: 0.8 M KPF_6 in EC/DEC (1:1)). Reprinted with permission from ref. [49]. Copyright 2018, Elsevier.

This flexible battery exhibited a large energy density of 232 watts hour per kilogram and superior flexible performance. Figure 20.5c exhibits two voltage platforms at 3.28/3.54V and 3.95/4.13V versus K^+/K, which are redox reactions of Fe^{3+}/Fe^{2+} linked by N with high-spin and Fe^{3+}/Fe^{2+} connected by C with low-spin, respectively. This material demonstrated 69% of initial capacity after 500 cycles at 200 mA/g [49].

20.5.2 LAYERED METAL OXIDES

A common formula of A_xMO_2 can be used to express layered metal oxides, where M represents Ni, Mn, Co, Fe; A stands for K, Na, Li. Layered metal oxides have several merits, such as low cost, a stable structure, and large theoretical capacity, which are confirmed to be suitable for cathodes in PIBs. For instance, the layered P3-type $K_{0.5}MnO_2$ showed a discharge capacity of 100 mAh/g between 1.5 and 3.9 V [50]. Additionally, the layered P3-$K_{0.5}MnO_2$ hollow microspheres prepared via a two-step self-templating method showing good retention of 89% at 200 mA/g after 400 cycles due to its unique hollow interior structure [51]. Moreover, based on effective AlF_3 surface modification, the $K_{1.39}Mn_3O_6$ microspheres demonstrated a large discharge capacity of 110 mAh/g at 0.01 A/g [52]. Besides, the birnessite $K_{0.77}MnO_2 \cdot 0.23H_2O$ nanosheet demonstrated high retention of 80.5% and a superior rate capability with 77 mAh/g after 1000 cycles at 1000 mA/g [53].

In addition, vanadate, chromium, and cobalt-based layered oxides were utilized as the positive electrode in PIBs. For instance, the pristine $K_{0.8}CrO_2$ first transformed to P'3-$K_{0.48}CrO_2$ after the first charge and eventually formed a new O'3-$K_{0.9}CrO_2$ phase in the following discharge. The existing P'3-phases were known to be helpful in improving the potassium ions' transport rate and cycling performance. Consequently, this material displayed good retention of 99% for 300 cycles under 1 C [54]. Furthermore, the $K_{0.5}V_2O_5$ belts showed a capacity of 90 mAh/g at 0.01 A/g and good retention of 81% for 250 cycles under 0.1 A/g thanks to its large interlayer distance of 9.505 Å [55].

Recently, binary and ternary metal-layered oxides have also been suggested as cathodes in PIBs. For instance, a new Fe/Mn-based layered oxide nanowire showed an initial capacity of 178 mAh/g. Further, the assembled PIBs combined with soft carbon as anode, demonstrating both good cycling and rate performances [56]. Besides, P2-type $K_{0.65}Fe_{0.5}Mn_{0.5}O_2$ microspheres displayed a large discharge capacity of 151 mAh/g under 0.02 A/g. Even at 0.1 A/g, the capacity was up to 103 mAh/g [57]. Another potential P3-$K_{0.54}[Co_{0.5}Mn_{0.5}]O_2$ material exhibited a large capacity of 78 mAh/g under 0.5 A/g and outstanding rate performance because it had quicker potassium ion transportation when compared with other layered materials. Additionally, a new layered ternary material $K_{0.67}Ni_{0.17}Co_{0.17}Mn_{0.66}O_2$ displayed a discharge capacity of 76.5 mAh/g under 0.02 A/g with a working voltage of 3.1V [58]. Moreover, it was observed that P2-type $K_{2/3}Ni_{1/3}Co_{1/3}Te_{1/3}O_2$ with honeycomb framework exhibited a larger voltage of 4 V versus K^+/K compared with other reported layered oxides.

20.5.3 POLYANIONIC COMPOUNDS

The polyanionic compounds with a universal formula of A $M_x[(XO_4)]_y$, where X represents nonmetal elements like S and P; M refers to transition metals such as V, Mn, Ti, Fe, which demonstrate some superior properties, including stable structure, diverse structure, and high thermal stability. For instance, three-layered $KVOPO_4$ fabricated via the hydrothermal method with various morphologies, such as nanosheets, bulk, and flower-like microspheres, were reported. The experimental results showed that the $KVOPO_4$ nanosheets had better cycling and rate performances, as well as higher capacity, because of its low volume change of approximately 9.4% [59]. Another material, $K_3V_2(PO_4)_2F_3$, was prepared by exchanging sodium ions with an electrochemical method, which exhibited a capacity of 100 mAh/g and a working voltage of 3.7V. More importantly, the assembled battery using graphite as a negative electrode also demonstrated good cycling performance [60].

20.5.4 ORGANIC MATERIALS

As negative electrode materials, organic materials have been applied as a cathode in PIBs as well, thanks to its advantages of recyclability, low cost, and environmental friendliness. Because Van der Waals force existed in organic compounds instead of covalent/ionic bonding, the interlayer space in organic materials is usually bigger than that in polyanionic materials and layered metal oxides. Also, it was reported that low solubility in organic-based electrolytes of organic compounds generated by aromatic π-π stacking was achieved. Additionally, to increase organic compounds, potentially strong electron-withdrawing groups are adopted to graft onto aromatic rings. This leads to significant deficiency of electrons on rings, thereby reducing the energy barrier of LUMO.

Benefiting from electrochemical deposition techniques, organic materials display discharge capacities of 120–220 mAh/g, which is higher than other organic materials in capacities. Take tetrahydroxyquinone (o-$Na_2C_6H_2O_6$), for example; it showed a high specific capacity of 168.1 and 26.7 mAh/g under 0.025 and 5 A/g, respectively [61]. Additionally, the poly(N-phenyl-5,10-dihydrophenazine) as cathode displayed a large capacity of 162 mAh/g under 0.2 A/g, superior rate capability with 120 mAh/g under 10 A/g, and a long lifespan of approximately 2000 cycles under 2000 mA/g [62]. It was also reported that the copper-tetracyanoquinodimethane with a three-electron redox reaction delivered a large capacity of 125 mAh/g under 1000 mA/g [63]. Furthermore, the poly (pentacenetetrone sulfide) provided a high discharge capacity of 260 mAh/g under 0.1 A/g and exhibited 3000 long cycles with 190 mAh/g at 5000 mA/g [64]. Interestingly enough, a new organic material [N,N'-bis(2-anthraquinone)]-perylene–3,4,9,10-tetracarboxydiimide displayed a high capacity of 133 mAh/g under ultra-large current density of 20 A/g [65].

20.6 SUMMARY AND FUTURE OUTLOOKS

PIBs are inexpensive alternatives to LIBs for future energy storage, which cannot be managed by LIBs because of expensive costs and restricted lithium resources. Although some promising progress has been achieved on electrolytes and electrodes for PIBs, flexible PIBs can affect the market of energy storage only if important breakthroughs are obtained. The widespread investigation of potassium ion chemistry is still necessary to improve energy density, coulombic efficiency, rate performance, and safety. Meanwhile, our knowledge of potassium ion SEI formation, component, and structure are insufficient. In the partial section, we exhaustively concluded the important findings, current understanding, and recent progress of the K-ion SEI, and we truly expect to encourage study on clarifying property of potassium ion SEI and then designing its functions.

As for the main challenges of various anodes, no matter in conventional or flexible PIBs, some strategies are suggested to address the following issues. Due to the big atomic radius of potassium, the different anodes normally undergo serious volume expansion. Nanostructure materials, such as carbonaceous materials, offer more chances to buffer the volume change. Also, the buffer is widely applied in alloy-type anodes. Meanwhile, designing a robust SEI layer is an effective strategy to enhance negative electrode stability. Another issue related to interfacial side reactions can be tackled by optimizing electrolytes. Moreover, this method can be operated conveniently. But the severe challenge is the surface issue of potassium metal due to its high reactivity, which may trigger serious electrolyte degradation. Recently, ionic liquids provide novel chances for developing the good performance of PIBs, considering their electrochemical and chemical stabilities and aluminum anti-corrosion features. In addition, artificial SEI strategy may fundamentally tackle the unstable SEI generation challenge. Consequently, much more attention to this field should be paid in the future.

In conclusion, organic materials, layered metal oxides, hexacyanoferrates, and polyanionic compounds are introduced in PIBs. Among the cathodes, the hexacyanoferrates display better electrochemical performances, such as specific capacity. However, they show lower volumetric energy than layered metal oxides owing to their low density. In addition, the main disadvantages of the layered oxides are low working voltage, high sensitivity toward water, high voltage polarization, and poor specific capacity. Therefore, the main challenge to building novel materials for PIBs is to tackle the large strain caused by the accommodation of the big potassium ion. It is well known that inorganic compounds possess rigorous crystal structure, whereas short-range ordered and amorphous compounds can demonstrate good electrochemical performance. Although these are the advantages of organic compounds, the low operating voltage, poor electronic conductivity, and bulky material size prevent their practical application. Consequently, it is worth undertaking deep studies to explore the natures of organic materials because of their versatile properties.

Besides, several challenges for developing the PIBs are listed: (1) more investigations are needed on the kinetic activity of K-ions; (2) the suitable composition and ratio of the additive in the electrolyte will greatly support the formation of a stable SEI, which enhances the improvment of the battery's electrochemical performances. In this consideration, more studies are needed; (3) the main safety issue is from thermal runaway of the battery because of poor heat dissipation. Moreover, metallic K possesses stronger activity and a lower melting point than lithium. Therefore, the safety problem is challenging and complex in PIBs; it needs all-around thinking from various directions, such as battery materials, electrolytes, and devices.

As mentioned above, although the development of PIBs can draw on many mature experiences from LIBs and SIBs, it is still in its infancy, and many technological and scientific challenges must be overcome, not to mention flexible ones.

REFERENCES

1. Eftekhari A., (2004) Potassium secondary cell based on Prussian blue cathode. *J. Power Sources* 126: 221–228.
2. Li W., Oyama Y., (2005) US20080286649A1.
3. Rajagopalan R., Tang Y., Ji X., Jia C., Wang H., (2020) Advancements and challenges in potassium ion batteries: a comprehensive review. *Adv. Funct. Mater.* 30: 1909486.
4. Jian Z., Luo W., Ji X., (2015) Carbon electrodes for K-ion batteries. *J. Am. Chem. Soc.* 137: 11566–11569.
5. Huang X., Cai X., Xu D., Chen W., Wang S., Zhou W., Meng Y., Fang Y., Yu X., (2018) Hierarchical Fe_2O_3@ CNF fabric decorated with MoS_2 nanosheets as a robust anode for flexible lithium-ion batteries exhibiting ultrahigh areal capacity. *J. Mater. Chem. A* 6: 16890–16899.
6. Xiang P., Chen X., Xiao B., Wang Z., (2019) Highly flexible hydrogen boride monolayer as potassium-ion battery anodes for wearable electronics. *ACS Appl. Mater. Interfaces* 11: 8115–8125.
7. Rajagopalan R., Tang Y., Ji X., Jia C., Wang H., (2020) Advancements and challenges in potassium ion batteries: a comprehensive review. *Adv. Funct. Mater.* 30: 1909486.
8. Heiskanen S.K., Kim J., Lucht B.L., (2019) Generation and evolution of the solid electrolyte interphase of lithium-ion batteries. *Joule* 3: 2322–2333.
9. Dey A., (1970) Film formation on lithium anode in propylene carbonate. *J. Electrochem. Soc.* 117: C248.
10. Peljo P., Girault H.H., (2018) Electrochemical potential window of battery electrolytes: the HOMO–LUMO misconception. *Energy Environ. Sci.* 11: 2306–2309.
11. Chen X., Li H.R., Shen X., Zhang Q., (2018) The origin of the reduced reductive stability of ion-solvent complexes on alkali and alkaline earth metal anodes. *Angew Chem. Int. Ed.* 57: 16643–16647.
12. Moshkovich M., Gofer Y., Aurbach D., (2001) Investigation of the electrochemical windows of aprotic alkali metal (Li, Na, K) salt solutions. *J. Electrochem. Soc.* 148, E155–E167.
13. Mogensen R., Brandell D., Younesi R., (2016) Solubility of the solid electrolyte interphase (SEI) in sodium ion batteries. *ACS Energy Lett.* 1: 1173–1178.
14. Hosaka T., Kubota K., Hameed A.S., Komaba S., (2020) Research development on K-ion batteries. *Chem. Rev.* 120: 6358–6466.
15. Hess M., (2017) Non-linearity of the solid-electrolyte-interphase overpotential. *Electrochimica Acta* 244: 69–76.
16. Winter M., Barnett B., Xu K., (2018) Before Li ion batteries. *Chem. Rev.* 118: 11433–11456.
17. Li Y., Lu Y., Adelhelm P., Titirici M.M., Hu Y.S., (2019) Intercalation chemistry of graphite: alkali metal ions and beyond. *Chem. Soc. Rev.* 48: 4655–4687.

18. Hosaka T., Kubota K., Hameed A.S., Komaba S., (2020) Research development on K-ion batteries. *Chem. Rev.* 120: 6358–6466.
19. Yamada Y., Furukawa K., Sodeyama K., Kikuchi K., Yaegashi M., Tateyama Y., Yamada A., (2014) Unusual stability of acetonitrile-based superconcentrated electrolytes for fast-charging lithium-ion batteries. *J. Am. Chem. Soc.* 136: 5039–5046.
20. Liu P., Wang Y., Gu Q., Nanda J., Watt J., Mitlin D., (2020) Dendrite-free potassium metal anodes in a carbonate electrolyte. *Adv. Mater.* 32: e1906735.
21. Li T., Zhang X.Q., Shi P., Zhang Q., (2019) Fluorinated solid-electrolyte interphase in highvoltage lithium metal batteries. *Joule* 3: 2647–2661.
22. Jian Z., Luo W., Ji X., (2015) Carbon electrodes for K-ion batteries. *J. Am. Chem. Soc.* 137: 11566–11569.
23. Zhao X., Tang Y., Ni C., Wang J., Star A., Xu Y., (2018) Free-standing nitrogen-doped cup-stacked carbon nanotube mats for potassium-ion battery anodes. *ACS Appl. Energy Mater.* 1: 1703–1707.
24. Xiong P., Zhao X., Xu Y., (2018) Nitrogen-doped carbon nanotubes derived from metal-organic frameworks for potassium-ion battery anodes. *ChemSusChem* 11: 202–208.
25. Zhao X., Xiong P., Meng J., Liang Y., Wang J., Xu Y., (2017) High rate and long cycle life porous carbon nanofiber paper anodes for potassium-ion batteries. *J. Mater. Chem. A* 5: 19237–19244.
26. Jian Z., Xing Z., Bommier C., Li Z., Ji X., (2016) Hard carbon microspheres: potassium-ion anode versus sodium-ion anode. *Adv. Energy Mater.* 6: 1501874.
27. Sangster J. M., (2010) K-P (Potassium-Phosphorus) system. *J. Phase Equilibria Diffus.* 31: 68–72.
28. Zhang W., Mao J., Li S., Chen Z., Guo Z., (2017) Phosphorus-based alloy materials for advanced potassium-ion battery anode. *J. Am. Chem. Soc.* 139: 3316–3319.
29. Zhang W., Wu Z., Zhang J., Liu G., Yang N. H., Liu R. S., (2018) Unraveling the effect of salt chemistry on long-durability high-phosphorus concentration anode for potassium ion batteries. *Nano Energy* 53: 967–974.
30. Huang S. B., Hsieh Y. Y., Chen K. T., Tuan H. Y., (2021) Flexible nanostructured potassium-ion batteries. *Chem. Eng. J.* 416: 127697.
31. Kishore B. G. V., Munichandraiah N., (2016) $K_2Ti_4O_9$: a promising anode material for potassium ion batteries. *J. Electrochem. Soc.* 163: A2551–A2554.
32. Dong S., Li Z., Xing Z., Wu X., Ji X., Zhang X., (2018) Novel potassium ion hybrid capacitor based on an anode of $K_2Ti_6O_{13}$ microscaffolds. *ACS Appl. Mater. Interfaces* 10: 15542–15547.
33. Han J., Xu M., Niu Y., Li G. N., Wang M., Zhang Y., (2016) Exploration of $K_2Ti_8O_{17}$ as an anode material for potassium-ion batteries. *Chem. Commun.* 52: 11274–11276.
34. Xie Y., Dall'Agnese Y., Naguib M., Gogotsi Y., Barsoum M. W., Zhuang H. L., (2014) Prediction and characterization of MXene nanosheet anodes for non-lithium-ion batteries. *ACS Nano* 8: 9606–9615.
35. Lian P., Dong Y., Wu Z.S., Zheng S., Wang X., Wang S., (2017) Alkalized Ti_3C_2 MXene nanoribbons with expanded interlayer spacing for high-capacity sodium and potassium ion batteries. *Nano Energy* 40: 1–8.
36. Han J., Niu Y., Bao S. J., Yu Y. N., Lu S. Y., Xu M., (2016) Nanocubic $KTi_2(PO_4)_3$ electrodes for potassium-ion batteries. *Chem. Commun.* 52: 11661–11664.
37. Lei K., Wang C., Liu L., Luo Y., Mu C., Li F., (2018) A porous network of bismuth used as the anode material for high-energy-density potassium ion batteries. *Angew. Chem. Int. Ed.* 57: 4687–4691.
38. Zhang Q., Mao J., Pang W. K., Zheng T., Sencadas V., Chen Y., (2018) Boosting the potassium storage performance of alloy-based anode materials via electrolyte salt chemistry. *Adv. Energy Mater.* 2018: 1703288.
39. McCulloch W. D., Ren X., Yu M., Huang Z., Wu Y., (2015) Potassium-ion oxygen battery based on a high capacity antimony anode. *ACS Appl. Mater. Interfaces* 7: 26158–26166.
40. Cao K. Z., Liu H., Jia Y., Zhang Z., Jiang Y., Liu X., Huang K., Jiao L., (2020) Flexible antimony@carbon integrated anode for high-performance potassium-ion battery. *Adv. Mater. Technol.* 5: 2000199.
41. Deng Q., Pei J., Fan C., Ma J., Cao B., Li C., (2017) Potassium salts of para-aromatic dicarboxylates as the highly efficient organic anodes for low-cost K-ion batteries. *Nano Energy* 33: 350–355.
42. Li C., Deng Q., Tan H., Wang C., Fan C., Pei J., (2017) Para-conjugated dicarboxylates with extended aromatic skeletons as the highly advanced organic anodes for K-ion battery. *ACS Appl. Mater. Interfaces* 9: 27414–27420.
43. Zhao Q., Wang J., Lu Y., Li Y., Liang G., Chen J., (2016) Oxocarbon salts for fast rechargeable batteries. *Angew. Chemie Int. Ed.* 55: 12528–12532.
44. Xue Q., Li D., Huang Y., Zhang X., Ye Y., Fan E., (2018) Vitamin K as a high-performance organic anode material for rechargeable potassium ion batteries. *J. Mater. Chem. A* 6: 12559–12564.

45. Wessells C. D., Huggins R. A., Cui Y., (2011) Copper hexacyanoferrate battery electrodes with long cycle life and high power. *Nat. Commun.* 2: 550.

46. Xue L., Li Y., Gao H., Zhou W., Lü X., Kaveevivitchai W., Manthiram A., Goodenough J. B., (2017) Low-cost high-energy potassium cathode. *J. Am. Chem. Soc.* 139: 2164–2167.

47. Zhang C., Xu Y., Zhou M., Liang L., Dong H., Wu M., Yang Y., Lei Y., (2017) Potassium Prussian blue nanoparticles: a low-cost cathode material for potassium-ion batteries. *Adv. Funct. Mater.* 27: 1604307.

48. Chong S., Wu Y., Guo S., Liu Y., Cao G., (2019) Potassium nickel hexacyanoferrate as cathode for high voltage and ultralong life potassium-ion batteries. *Energy Storage Mater.* 22: 120–127.

49. Zhu Y.H., Yang X., Bao D., Bie X.F., Sun T., Wang S., Jiang Y.S., Zhang X.B., Yan J.M., Jiang Q., (2018) High-energy-density flexible potassium-ion battery based on patterned electrodes. *Joule* 2: 736–746.

50. Kim H., Seo D. H., Kim J. C., Bo S. H., Liu L., Shi T., Ceder G., (2017) Investigation of potassium storage in layered P3-type $K_{0.5}MnO_2$ cathode. *Adv. Mater.* 29: 1702480.

51. Peng B., Li Y., Gao J., Zhang F., Li J., Zhang G., (2019) High energy K-ion batteries based on P3-type $K_{0.5}MnO_2$ hollow submicrosphere cathode. *J. Power Sources* 437: 226913.

52. Zhao S., Yan K., Munroe P., Sun B., Wang G., (2019) Construction of hierarchical $K_{1.39}Mn_3O_6$ spheres via AlF3 coating for high-performance potassium-ion batteries. *Adv. Energy Mater.* 9: 1803757.

53. Lin B., Zhu X., Fang L., Liu X., Li S., Zhai T., Xue L., Guo Q., Xu J., Xia H., (2019) Birnessite nanosheet arrays with high K content as a high-capacity and ultrastable cathode for K-ion batteries. *Adv. Mater.* 31: 1900060.

54. Naveen N., Han S. C., Singh S. P., Ahn D., Sohn K.S., Pyo M., (2019) Highly stable P3-$K_{0.8}CrO_2$ cathode with limited dimensional changes for potassium ion batteries. *J. Power Sources* 430: 137–144.

55. Deng L., Niu X., Ma G., Yang Z., Zeng L., Zhu Y., Guo L., (2018) Layered potassium vanadate $K_{0.5}V_2O_5$ as a cathode material for nonaqueous potassium ion batteries. *Adv. Funct. Mater.* 28: 1800670.

56. Wang X., Xu X., Niu C., Meng J., Huang M., Liu X., Liu Z., Mai L., (2016) Earth abundant Fe/Mn-based layered oxide interconnected nanowires for advanced K-ion full batteries. *Nano Lett.* 17: 544–550.

57. Deng T., Fan X., Chen J., Chen L., Luo C., Zhou X., Yang J., Zheng S., Wang C., (2018) Layered P2-type $K_{0.65}Fe_{0.5}Mn_{0.5}O_2$ microspheres as superior cathode for high-energy potassium-ion batteries. *Adv. Funct. Mater.* 28: 1800219.

58. Liu C., Luo S., Huang H., Wang Z., Hao A., Zhai Y., Wang Z., (2017) $K_{0.67}Ni_{0.17}Co_{0.17}Mn_{0.66}O_2$: a cathode material for potassium-ion battery. *Electrochem. Commun.* 82: 150–154.

59. Liao J., Hu Q., Che B., Ding X., Chen F., Chen C., (2019) Competing with other polyanionic cathode materials for potassium-ion batteries via fine structure design: new layered $KVOPO_4$ with a tailored particle morphology. *J. Mater. Chem. A* 7: 15244–15251.

60. Lin X., Huang J., Tan H., Huang J., Zhang B., (2019) $K_3V_2(PO_4)_2F_3$ as a robust cathode for potassium-ion batteries. *Energy Storage Mater.* 16: 97–101.

61. Chen L., Liu S., Wang Y., Liu W., Dong Y., Kuang Q., Zhao Y., (2019) Ortho-di-sodium salts of tetrahydroxyquinone as a novel electrode for lithium-ion and potassium-ion batteries. *Electrochim. Acta* 294: 46–52.

62. Obrezkov F. A., Ramezankhani V., Zhidkov I., Traven V. F., Kurmaev E. Z., Stevenson K. J., Troshin P. A., (2019) High-energy and high-power-density potassium ion batteries using dihydrophenazine-based polymer as active cathode material. *J Phys. Chem. Lett.* 10: 5440–5445.

63. Ma J., Zhou E., Fan C., Wu B., Li C., Lu Z.H., Li J., (2018) Endowing CuTCNQ with a new role: a high-capacity cathode for K-ion batteries. *Chem. Commun.* 54: 5578–5581.

64. Tang M., Wu Y., Chen Y., Jiang C., Zhu S., Zhuo S., Wang C., (2019) An organic cathode with high capacities for fast-charge potassium-ion batteries. *J. Mater. Chem. A* 7: 486–492.

65. Hu Y., Tang W., Yu Q., Wang X., Liu W., Hu J., Fan C., (2020) Novel insoluble organic cathodes for advanced organic K-ion batteries. *Adv. Funct. Mater.* 30: 2000675.

21 Flexible Batteries Based on Zn-Ion

Haobo Dong and Ivan P. Parkin
Christopher Ingold Laboratory, Department of Chemistry,
University College London, United Kingdom

Guanjie He
Christopher Ingold Laboratory, Department of Chemistry,
University College London, United Kingdom

School of Chemistry, University of Lincoln, Brayford Pool, Lincoln, United
Kingdom

CONTENTS

21.1   Introduction.. 375
21.2   Zinc-ion batteries and mechanisms 377
21.3   Flexible zinc-ion batteries.. 378
    21.3.1   Polymer electrolytes.. 378
        21.3.1.1   PEO and derivatives.. 379
        21.3.1.2   PVA and derivatives.. 379
        21.3.1.3   PAM and derivatives... 381
    21.3.2   Functionalities.. 381
    21.3.3   Flexible device constructions (electrodes).. 385
21.4   Current challenges and perspectives.. 388
    21.4.1   Voltage issue 388
    21.4.2   Structural enhancement 391
    21.4.3   Multifunctionalities.. 392
21.5   Conclusions... 392
Reference 393

21.1   INTRODUCTION

In this digital age, the development of electronics has increased demand for various types of energy storage devices emphasizing different applications. Flexible batteries, the ones combining structural and electrochemical properties, have attracted the most attention in the fields of healthcare, flexible displays, and smart clothes that help build up the Internet of Things (IoT). As an efficient and flexible energy storage device, Li-ion batteries (LIBs) have been widely used in the fields of consumer electronics, electric vehicles, and grid-scale energy storage. However, the production cost of LIBs is relatively high, Li-reserves are limited, and most of them use toxic and flammable organic electrolytes. There are potential safety hazards and environmental pollution associated with LIBs. These problems seriously hinder their large-scale application for grid-energy storage. Therefore, researchers have started to seek new energy storage systems that can replace LIBs and have the characteristics of low cost, high energy, high safety, and environmental benignity.

DOI: 10.1201/9781003186755-21

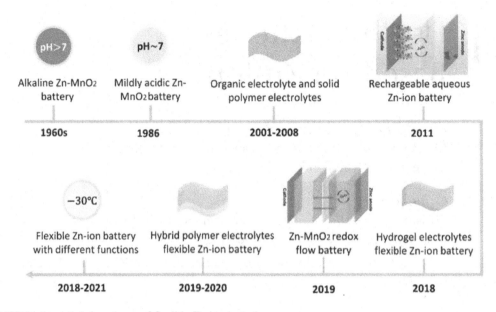

FIGURE 21.1 A brief roadmap of flexible Zn-ion batteries.

Flexible zinc-ion batteries (FZIBs) are regarded as a promising energy source in healthcare applications owing to their biocompatibility, high stretchability, high safety, and cost-efficiency. As shown in Figure 21.1, zinc-ion batteries (ZIBs) have been developed since the 1960s when alkaline-based Zn-MnO_2 were first commercialized [1]. However, the low reversibility induced by the formation of ZnO and Zn dendrites resulted in great capacity fade and low cycling ability after recharging. Until 2012, rechargeable ZIBs with mildly acidic aqueous electrolytes had been developed [2]. Aqueous electrolytes exhibit high ionic conductivity, high safety, and low cost; hence, they have been widely used in secondary battery systems, such as Na^+, K^+, and multivalent metal ions, including Mg^{2+}, Al^{3+}, Ca^{2+}, Zn^{2+}, etc. Despite the fact that the earth is abundant in Na and K elements and they are inexpensive, their energy densities are low, and their ion radii are relatively large. They have higher requirements on the structure of positive and negative host materials. Currently, Na and K-ion batteries need more efforts to satisfy the requirement of commercial applications. Metallic zinc, on the other hand, has attracted broad interests by its high specific capacity, low oxidation-reduction potential (−0.763 V relative to the standard hydrogen electrode (SHE)), high hydrogen evolution overpotential, abundant reserves, and nontoxicity. Compared with LIBs, ZIBs exhibit a high theoretical volumetric capacity (5855 mAh cm^{-3}) and gravimetric capacity (820 mAh g^{-1}). Meanwhile, the facile fabrication process and abundant raw materials improve the cost efficiency ($25/kWh in the lab) of AZIBs [3], which is beneficial for FZIBs to be applied in different scenarios. Contributing to the exploitation of aqueous ZIBs is the tremendous development of FZIBs since 2018 [4]. Hydrogel electrolytes were initially utilized in FZIBs rather than solid polymer electrolytes offering high ionic conductivities of 10^{-1} to 10^{-2} S cm^{-1}. As new candidates in flexible batteries, extensive studies have been exploited in recent decades, with the emphasis on enhancing ionic conductivities of polymer electrolytes for FZIBs. Besides the performance improvement, polymer electrolytes exhibit additional functionalities, such as flexibility, biocompatibility, self-healing ability, anti-freezing, and cooling-recovery abilities.

This chapter summarizes FZIBs in different functionalities and gives a guide to current research progress. The fundamental background of ZIBs and multiple cathode materials will be briefly introduced, followed by a detailed description of the use of polymer electrolytes. Challenges and perspectives will be discussed in the end, with the emphasis on the interface interaction between polymer electrolytes and electrodes, as well as structural enhancement strategies.

21.2 ZINC-ION BATTERIES AND MECHANISMS

Similar to LIBs, rechargeable ZIBs are "rocking chair" batteries, where during the discharging process, the zinc anode loses two electrons and oxidizes to Zn^{2+} with a SHE potential of -0.76 V, while the cathode restores two electrons accompanied by the Zn^{2+} intercalation. The charging is exactly the opposite process of the discharging, in which Zn^{2+} is released from the cathode to the anode. Energy storage mechanisms are amended extensively when mild acid aqueous electrolytes are substituted from organic electrolytes. As hydrogel electrolytes further improved FZIBs with potential application in implantable devices, the mechanism for aqueous ZIBs will be illustrated in detail.

To date, there are four categories of cathode materials for ZIBs, which are manganese-based materials, vanadium-based materials, Prussian blue analogues, and organic materials. In general, there are three charge storage mechanisms: (1) Zn^{2+} intercalation/deintercalation; (2) Zn^{2+}/H^+ co-insertion; (3) chemical conversion reaction. Chemical conversion reactions normally exist in MnO_2 where the average valence states of Mn varies from +4 to +3 oxidation state during the discharge, forming MnOOH [5]. Even for vanadium oxides, a self-charging process happens where V varies from +3 to +5, recovering to its charged states through spontaneous oxidation reaction with dissolved oxygen in the electrolyte [6]. Apart from this common mechanism, coordination/release reactions are exhibited in organic cathodes, such as quinone-based compounds, where Zn^{2+} are stored at the active sites of the electronegative oxygen atoms [7,8]. Typically, charge carriers used in aqueous solutions are $ZnSO_4$, $Zn(CF_3SO_3)_2$, $ZnCl_2$, and $Zn(TFSI)_2$. Zn^{2+} will be surrounded by six water molecules forming the solvation sheath $[Zn(OH_2)_6]^{2+}$. For nonaqueous electrolytes, the solvation sheath forming with the solvent, such as acetonitrile, requires higher energy to de-solvate at the interface. During discharging, Zn^{2+} de-solvates from the sheath and inserts into the cathode associated with an additional activation energy at the cathode-electrolyte interface (CEI). In terms of the zinc anode, metallic zinc suffers from various reactions in aqueous electrolytes. At the anode-electrolyte interface (AEI), because of the uneven distribution of the current density, zinc plating/stripping will result the formation of zinc dendrites. Then, the plated Zn prefers depositing on the existing nucleus resulting in "dead zinc" and causing a short circuit. As displayed in the Pourbaix diagram (Figure 21.2a), the dissolved zinc complexes are pH dependent, for which by-products $Zn(OH)_4^{2-}$ and $HZnO_2^{2-}$ are likely formed in the alkaline-based electrolyte, resulting in an irreversible reaction of ZnO ($Zn(OH)_4^{2-} \leftrightarrow ZnO + 2H_2O + 2OH^-$). However, under mildly acidic solution, the only restriction is to avoid an operating voltage window beyond the region for hydrogen evolution reaction (HER) on the anode and oxygen evolution reaction (OER) on the cathode. Owing to the nonuniform distribution of current densities at the anode, dendrites will generate for both alkaline and acidic ZIBs. Zn metal corrosion results in the formation of insulating layers on the anode interface, inhibiting the reversible Zn^{2+} stripping/deposition and eventually promoting capacity fade. The dissolution of positive structural components in aqueous electrolytes and the formation of side-reaction products can also cause battery capacity decay and poor cycling stability. In addition, the low conductivity and poor rate performance of vanadium and manganese-based materials have severely restricted their development. In typical $Zn-MnO_2$ batteries (Figure 21.2b), Mn^{4+} ions transform to Mn^{3+} during the discharging process forming $Zn-Mn_2O_4$, as shown in Eq. 21.1. MnOOH is also formed because of the H^+ ion intercalation. As shown in Figure 21.2c, a phase change occurs for $\alpha-MnO_2$ cathode materials, where a reversible formation of Zn-birnessite structure happens during the discharging process from tunnel structures.

$$Zn^{2+} + 2e^- + 2MnO_2 \leftrightarrow ZnMn_2O_4 \qquad (21.1)$$

$$MnO_2 + H_2O + e^- \leftrightarrow MnOOH + OH^- \qquad (21.2)$$

Focusing on the flexibility of ZIBs, researchers have made some progress on flexible electrodes and polymer electrolytes to construct different structures of FZIBs. Among these components, extensive

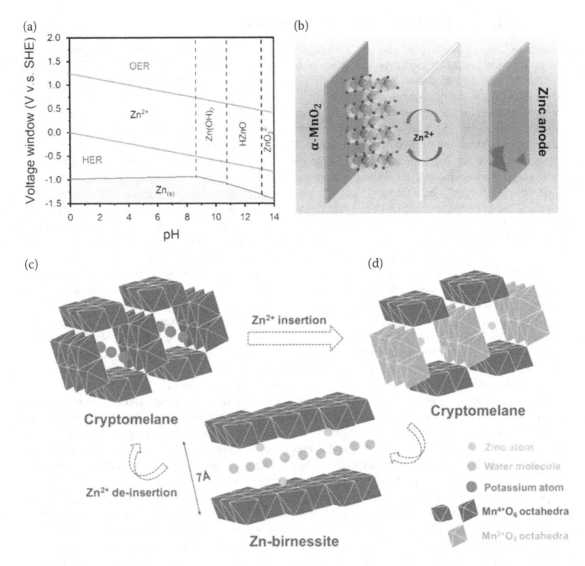

FIGURE 21.2 (a) Pourbaix diagram of Zn in aqueous electrolyte. (b) Schematic diagram of Zn-MnO$_2$ battery for energy storage mechanism in ZIBs. (c) Phase transfer of α-MnO$_2$ in ZIBs. (Song et al., 2018) Reproduced by permission from Hong Jin Fan, *Recent Advances in Zn-Ion Batteries* (*Advanced Functional Materials*). Copyright (2018) WILEY-VCH Verlag GmbH & Co. KGaA, Weinheim.

efforts have been applied in polymer electrolytes that offer flexibility and additional functionalities. This compromise allows for greater flexibility and other functions while maintaining competitible electro-chemical performance. Polymer electrolytes, functionalities, and flexible device configurations for FZIB are discussed in the following sections.

21.3 FLEXIBLE ZINC-ION BATTERIES

21.3.1 POLYMER ELECTROLYTES

Polymer electrolytes are currently classified into three types: solid polymer electrolytes (SPEs), hydrogel polymer electrolytes (HPEs), and hybrid polymer electrolytes (HPEs) (HBPEs). Generally, segmental motions in the host polymer are believed to be the basis of the ionic transfer mechanism in polymer electrolytes. Zn^{2+} ions are located at the active site where there are free electron donors, such as active oxygen atoms and nitrogen atoms of the polymer host. As shown in Figure 21.3b, Zn^{2+} ions connected with two oxygen atoms will hop to the neighboring active site, transmitting in the electrolyte by segmental motions. Segmental motions are the relaxation process of the polymer activating from the lowest energy state to an excited state in all dimensions, including backbone rotation and segment internal motion. Hence, the ion transportation in the polymer electrolyte is highly relevant with intra- and intermolecular reactions in the polymer framework. Inspired by flexible LIBs, organic SPEs have been developed for two decades by simply complexing zinc salts into organic frameworks, such as poly(ethylene oxide) (PEO), poly(vinylidene fluoride) (PVDF), and polyacrylonitrile (PAN). However, the low solubility and the strong intermolecular reactions of certain conducting salts, such as $ZnSO_4$ and $Zn(CF_3SO_3)_2$, restrict the ionic conductivities to the range of $10^{-6} \sim 10^{-4}$ S cm^{-1}. Until the breakthrough of rechargeable aqueous ZIBs in 2012 by Kang and coworkers [2], [9], the application of aqueous electrolytes facilitates the development of quasi-solid electrolytes by using hydrogels as the electrolyte host. Polyacrylamide (PAM), poly(vinyl alcohol) (PVA), and polysaccharides are widely utilized as the host for HPEs. Due to the large water content in the host, the ionic conductivity improved to the magnitude of $10^{-3} \sim 10^{-1}$ S cm^{-1}. However, the drawback of weak mechanical properties of HPEs leads to the development of hybrid electrolytes. HBPEs are composites combining at least two polymer frameworks with enhancing bonding strength or other functionalities. Currently, HBPEs are designed based on three polymer electrolyte frameworks; these are PEO, PVA and PAM derivatives. As shown in the radar diagram (Figure 21.3a), among three polymer electrolyte categories, SPEs exhibit the best mechanical performance, while HPEs and HBPEs are preponderant in ionic conductivity and multifunctionalities. As the presence of water in HPEs and HBPEs, side reactions from HER and dendrite formation result in less inhibition of zinc dendrites on the anode. The details of each category are illustrated in the following content, especially the hybrid electrolytes.

21.3.1.1 PEO and derivatives

For SPEs, PEO is a promising candidate not only for LIBs but also for ZIBs. However, the ionic conductivity for pristine PEO polymer electrolyte is only 10^{-6} S cm^{-1}; even at 100 °C, there is only a three-fold increase to 10^{-3} S cm^{-1}. Although different zinc salts have been investigated, including $ZnCl_2$, $Zn(CF_3SO_3)_2$, $Zn(BF_4)_2$, and $ZnBr_2$, there is seemingly little difference in the ionic conductivity at room temperature. Plasticizers like propylene carbonate (PC) and ceramic nanofillers, such as Al_2O_3, SiO_2 and TiO_2, have been utilized as additives, thus expanding the free volume for ionic transportation reaching ionic conductivities of 10^{-4} S cm^{-1} at room temperature. Recently, with the application of 1-ethyl-3-methyl-imidazolium tetrafluoroborate ([EMIM]BF$_4$) ionic liquids mixed with 2 M $Zn(BF_4)_2$, 10^{-3} S cm^{-1} were attained [10,11]. The limited improvement in ionic conductivity for SPEs further accelerates the development of hybrid electrolytes with the aim of additional functionalities. Apart from tensile-strength enhancement, one PEO derivative, poly(ethylene oxide)-poly(propylene oxide)-poly(ethylene oxide) (F77), is a thermo-reversible polymer exhibiting a sol-gel reaction. Cui et al. [12] reported a pluronic hydrogel electrolyte (PHE) with a cooling-recovery ability for which the hybrid hydrogel electrolyte was a gel at room temperature but became liquid at −5 °C. The conductivity was 6.33×10^{-3} S cm^{-1} at 25°C. At low temperature, F77 aggregates into micelles hydrating the PEO shell, although, as the temperature increases to 25°C, hydrophobic interactions induce the dehydration of PEO and PPO interfaces, eventually causing hard-sphere crystallisation. Once there is a crack between electrodes and HBPE under an external force, the electrode-electrolyte interface can be refreshed upon a cooling process. Integrating with LiMnO$_2$ cathode and Zn anode, the recovered capacity healed to an efficiency of 98% compared to its initial state.

21.3.1.2 PVA and derivatives

In addition to PEO, PVA-based polymer electrolytes are widely used in both alkaline and acidic ZIBs. They benefit from hydroxyl groups attached to the backbones, and PVA hydrogel electrolytes

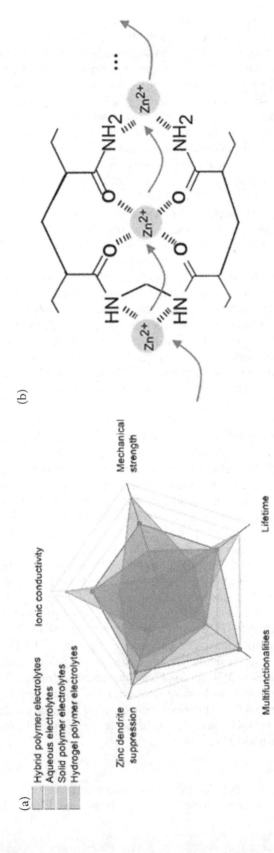

FIGURE 21.3 (a) Comparison of polymer electrolytes in a radar diagram. (b) An example of zinc ion transportation in a PAM-based hydrogel electrolyte. (Dong, Li, Guo, et al. 2021) Reproduced by permission from Haobo Dong, *Insights on Flexible Zinc-Ion Batteries from Lab Research to Commercialization* (*Advanced Materials*). Copyright (2021) The Authors. Advanced Materials published by Wiley-VCH GmbH.

exhibit a superior aqueous electrolyte absorbing ability. Zeng [13] invented a PVA-based HPE accompanied by $ZnCl_2$ and $MnSO_4$. The hydrogel electrolyte delivers a high ionic conductivity of 1.26×10^{-2} S cm^{-1}, and 77% capacity fades under 100% coulombic efficiency (CE) within 300 cycles after integrating with the cathode MnO_2@PEDOT (poly(3,4-ethylenedioxythiophene)). At the same time, for alkaline FZIBs, KOH with conducting zinc triflate ($Zn(CF_3SO_3)_2$) was applied [14]. Altering the composition mass ratio to 40:35:25 (PVA: KOH: zinc triflate), the fabricated HPE shows an ionic conductivity of 2.548×10^{-2} S cm^{-1}. After integrating into a Zn/AgO battery, only 3% capacity decay was attained over 20 cycles. Owing to the high electrochemical performance for pristine PVA HPE and stable mechanical properties, a PVA derivative with an anti-freezing ability was developed by complexing glycerol into the polymer host. The formation of strong hydrogen-bonding interactions with water molecules reduced the freezing point to −60°C. Chen et al. [15,16] have synthesized PVA/glycerol HPEs, for which a high ionic conductivity of 10.1 mS cm^{-1} was achieved at a temperature of −35°C.

21.3.1.3 PAM and derivatives

Among all-polymer frameworks for hydrogel electrolytes, PAM is the most promising host for which different derivatives are developed with additional functionalities. The porous structures formed by polymerizing acrylamide monomers and N,N-methylene bis(acrylamide) exhibit excellent water ab-sorption abilities. Zhi and coworkers [17,18] made extensive efforts to exploit HPEs and HBPEs based on PAM. Immersing PAM into the solution of 2M $ZnSO_4$ and 0.2M $CoSO_4$, the ionic conductivity of 0.12 S cm^{-1} was achieved under a swelling ratio of 300%. Although PAM electrolyte exhibits a superior electrochemical performance, the tensile strength is limited by the single network polymer structure, which is only 1.25 MPa. Polysaccharides, biomass hydrogel structures, were added to form the double network (DN) hydrogel electrolytes, which strengthened the structure without alleviating ionic con-ductivities. Literally, pristine polysaccharides, such as carrageen [19], gum-based hydrogel [20,21], alginate [22], and agar [23], were investigated and shown to have ionic conductivities in the range of $10^{-3} \sim 10^{-2}$ S cm^{-1}. In terms of protein-based hydrogels, gelatin [24] and collagen [25] were exploited as HPEs. Generating DN hydrogel electrolytes are essential to enhance the intermolecular bonding, hence strengthening the structure. Zhang [23] has investigated the gelatin/PAM dual-network gel as-sociated with robust mechanical properties, for which the tensile modulus and stress increase to 84 kPa and 0.268 MPa, respectively, compared to the ones for PAM 60 kPa and 0.178 MPa. Although me-chanical properties are ameliorated, the limited improvement has little benefit for practical applications. Zhi has pioneered the development of a PAM derivative grafted with gelatin chain on the poly-acrylonitrile (PAN) membrane. With the additional host structure, the as-prepared HBPE enhances the tensile strength to 7.76 MPa compared with the initial 1.25 MPa for the PAM hydrogel electrolyte. The electrolyte also delivers a high ionic conductivity of 1.76×10^{-2} S cm^{-1} when accompanied by a $ZnSO_4$ and $MnSO_4$ complex. Assembled with MnO_2 and the Zn anode in a sandwich structure, this FZIB exhibits a high specific capacity of 306 mAh g^{-1} and a capacity retention of 97%, even under extreme conditions. Therefore, a novel separator design, such as PAN membrane, is the most efficient strategy to improve the load-bearing ability. As the most promising polymer framework, PAM derivatives were developed with different functionalities by combining with SPEs and organic additives. Details are demonstrated in Section 3.2.

21.3.2 Functionalities

Multifunctionalities are noteworthy characteristics for polymer electrolytes in FZIBs. Apart from the flexibility enhancement, other properties in HBPEs were developed with the emphasis on anti-freezing ability, thermo-responsibility, self-healing ability, and zinc dendrite prevention. Various functions are beneficial for FZIBs to apply in realistic scenarios. Anti-freezing ability is an essential ability to operate at low temperature. Owing to the development of hydrogel electrolytes in HPEs, binary solutions, such as ethylene glycol and glycerol, are low-cost additives offering anti-freezing ability. Similar to a coolant, as shown in Figure 21.4a, ethylene glycol (EG) lowers the freezing point of the hydrogel by strong hydrogen bonding. Zhi [18] has fabricated a flexible anti-freezing aqueous ZIB working at −20°C comprised of PAM with low molecular vicinal alcohol, EG. To anchor the EG onto the polymer chains, isophorone diisocyanate (IPDI) as monomers and dimethylol propionic acid (DMPA) as chain extenders were utilized, ensuring the hydroxyl groups are bonded covalently with the isocyanate groups. By laminating with α-MnO$_2$ cathodes, this FZIB exhibits a high specific capacity of 226 mAh g^{-1} (0.2 A g^{-1}) at −20°C and maintains 82% of its initial capacity at ambient temperature (Figure 21.4b). Glycerol in PVA, as mentioned previously, also provides a lower freezing point, resulting in a stable electrical output at −35°C.

Thermal responsiveness enables the materials to recover from cracking at the interface, as indicated in the section of PEO and derivatives, but it can also enable them to perform as a smart temperature-controllable switch. Zhi [26] and Niu [27] have developed thermo-sensitive FZIBs based on polymer electrolyte poly(N-isopropyl acrylamide) (PNIPAM). This sol-gel electrolyte exhibits a porous structure that is in the solid-state at 70°C. The gelation point of the polymer is 50°C. Therefore, as shown in Figure 21.5a, at room temperature, the porous structure will swallow the aqueous electrolyte exhibiting hydrophilic behaviors; however, once the temperature is greater than the volume phase transition temperature, the sudden formation of intramolecular hydrogen bonds in PNIPAM will increase the internal resistance ten-fold to 160 MΩ, inhibiting the transmission of Zn^{2+} ions. Hence, the FZIB switches off at high temperature, avoiding an explosion. Apart from PAM derivatives, PEO derivative F77 is also a thermal-sensitive polymer. Coating F77 on PAM, the HBPE shows similar sol-gel behavior, while different from former HBPEs; only the surface of the polymer became the inhibitor at high temperature. Integrating with the CoFe(CN)$_6$ cathode, the FZIB exhibits a specific capacity of 142.34 mAh g^{-1} (1 A g^{-1}) and a high capacity retention of 93.4% over 2000 cycles (Figure 21.5b).

Prior to the application of hydrogel electrolytes, self-healing ZIBs become feasible due to the reversible formation of hydrogen bonds in both polymer electrolytes and cathode substrates. Both PAM and PVA hydrogels exhibit self-healing ability. Huang et al. [28] have developed a self-healing, integrated, all-in-one FZIB based on PVA/Zn(CF$_3$SO$_3$)$_2$. A scratch gradually disappeared after cutting, indicating the autonomous self-healing property. Integrating with polymer cathode polyaniline (PANI), the FZIB exhibited a constant capacity of 80 mAh g^{-1} at 0.5 A g^{-1} under the cutting-self healing process. Wang [29] developed PAM-based self-healing ZIB, for which carboxylated-polyurethane (CPU) was used as the electrode substrates offering self-healing ability for the electrodes. After four cuts, the battery still attained 81.2% capacity.

Apart from the functionalities, polymer electrolytes provide cathode stabilization and Zn anode dendrite prevention properties. In terms of cathode stabilization, pre-intercalated metal ions will be released from the cathode and dissolved into the electrolyte, resulting in capacity decay. Dong et al. [22] have developed a cathode-stabilized HBPE, where alginate/PAM electrolyte was applied with the sodium pre-intercalated manganese oxide Na$_{0.65}$Mn$_2$O$_4$·1.31H$_2$O. The dissociated Na$^+$ in the electrolyte from the biomass sodium alginate naturally stabilizes the cathode by offering additional Na$^+$ ions. The as-fabricated FZIB delivers a high capacity of 305 mAh g^{-1} at 0.1 A g^{-1}, 10% higher than the ZIBs in an aqueous electrolyte. Moreover, this HBPE exhibits capacity retention of 96% within 1000 cycles at 2 A g^{-1}, 4% greater than aqueous ZIBs. Regarding the zinc anode, the presence of polymer electrolytes physically suppresses zinc dendrite formation compared with aqueous electrolytes. As demonstrated in section 2, Zn metal anodes suffer from dendrites and severe side reactions, including HER and passivation in both alkaline and mildly acidic solutions, thus restricting the shelf life of ZIBs. The application of SPEs offers a dendrite-free property on the anode owing to the hydrogen-free polymer electrolytes. Chen [30] developed the dendrite-free SPE utilizing poly(vinylidene fluoride-co-hexafluoropropylene) filled by the poly(methyl acrylate) grafted MXenes, which showed an ionic conductivity of 2.69 × 10^{-4} S cm^{-1} at room temperature. This HBPE not only exhibits a stable Zn anode stripping/plating mechanism, but it also expands the operating working temperature from −35°C to 100°C. Although there

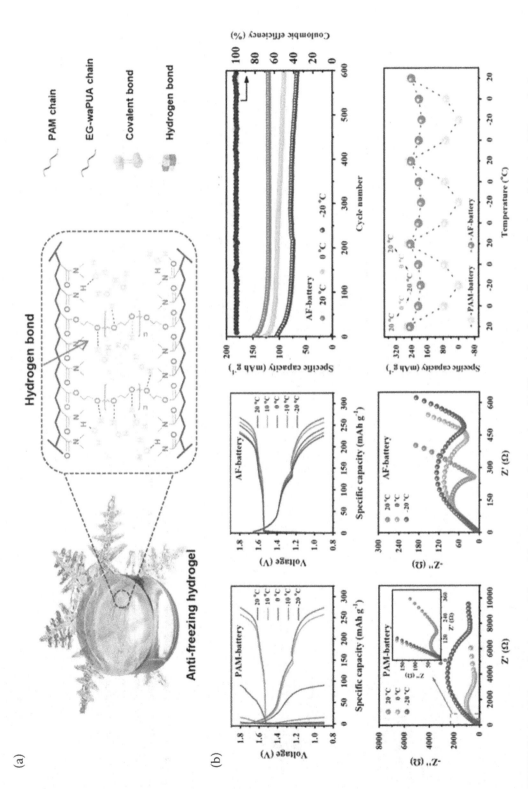

FIGURE 21.4 (a) Internal structure of the anti-freezing hydrogel electrolyte. (b) Electrochemical performance for the anti-freezing ZIBs at the low temperature [18]. Reproduced by permission from Chunyi Zhi, *A flexible rechargeable aqueous zinc manganese-dioxide battery working at −20°C* (*Energy and Environmental Science*), Copyright (2019) The Royal Society of Chemistry 2019.

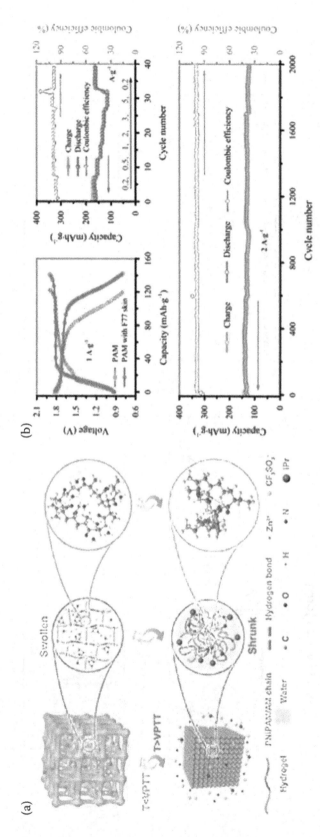

FIGURE 21.5 (a) Working mechanism of the thermal-gated PNIPAM hydrogel electrolyte (VPTT: volume phase transition temperature) [27]. Reproduced by permission from Zhiqing Niu, *Thermal-Gated Polymer Electrolytes for Smart Zinc-Ion Batteries (Angewandte Chemie International Edition)*, Copyright (2020) Wiley-VCH Verlag GmbH & Co. KGaA, Weinheim. (b) Charging/discharging rate and cycling performances for the thermo-responsive polymer electrolytes [18]. Reproduced by permission from Chunyi Zhi, *Achieving High-Voltage and High-Capacity Aqueous Rechargeable Zinc Ion Battery by Incorporating Two-Species Redox Reaction (Advanced Energy Materials)*, Copyright (2019) WILEY-VCH Verlag GmbH & Co. KGaA, Weinheim.

are still minor side effects for HPEs because of the water content, HPEs could inhibit the zinc dendrite by providing a uniformly deposited Zn anode utilizing a polyzwitterion electrolyte. The polyzwitterion hydrogel electrolytes reported by Mo [31] were composited from cellulose nanofibrils and poly[2-(methacryloyloxy)ethyl] diethy-(3-sulopropyl) (PMAEDS) through physical entanglement and intertwining. There is an equal amount of anionic and cationic counterions in the zwitterionic chains, i.e., Zn^{2+} and SO_4^{2-} ions. These positive/negative ions form an immigration channel (Figure 21.6c) and homogenize the ion distribution of Zn platting, resulting in longer cycling stability. The as-prepared HBPE maintained a constant voltage-time profile within the symmetric cell test for over 200 h at 0.5 mAg cm^{-2}; however, the other cells based on aqueous and pristine PAM HPE only keep the constant voltage for 100 hours (see Figure 21.6a). Luo [31] also indicated that poly [2-(methacryloyloxy) ethyl] dimethyl-(3-sulfopropyl) (PSBMA), with an ionic conductivity of 32.0 mS cm^{-1}, exhibits a superior long shelf life up to 3500 hours under symmetric cell evaluation. Besides these summarized functionalities, others, such as shape memory, medical-grade biocompatibility, and pH sensitivity, could be further exploited.

21.3.3 FLEXIBLE DEVICE CONSTRUCTIONS (ELECTRODES)

FZIBs, as opposed to conventional batteries, necessitate flexibility for all components. Substrates for electrodes determine that the flexible device configurations can be either sandwich laminates or cable-type structures. As shown in Figure 21.7a, the sandwich structure consists of laminates of electrodes immersed in the polymer electrolyte; the cable-type structure consists of two derivatives, yarn-based and fiber-based FZIBs (Figure 21.7b and 21.7c, respectively). The former cable-type configuration is based on two helix yarns coated with anode and cathode materials, respectively, while the latter is the lamination of electrodes layer-by-layer coated on a single fiber substrate. Regarding the sandwich structure, a two-dimensional (2D) architecture, carbon nanotube papers, and carbon clothes are normally utilized as the substrates. As for cable-type FZIBs in the one-dimensional (1D), wired shape, carbon nanotube (CNT), yarns are applied. For the Zn anode, pristine Zn foils and Zn wires are initially utilized; however, considering the adhesive bonding between polymer electrolytes and anodes under the deformation, flexible Zn anodes are normally deposited on carbon-based substrates. For hydrogel electrolyte, adhesive bonding enforces the anode-electrolyte interface, which is directly related to the contact surface. As adhesive bonding is formed by mechanical interlocking, covalent bonding, and electrostatic forces, the larger the contact area, the greater the adhesive forces. Compared with a woven carbon-based substrate, delamination for Zn foil could happen at a small curvature radius; hence, pristine Zn metal anode substrates should be substituted by the substrate with a large surface area, avoiding the poor interfacial bonding.

Fabrication strategies for FZIBs are still being developed in the lab. In the sandwich structure, a simple layer-by-layer assembly is commonly used. The as-fabricated device has the advantage of having a high areal energy density. Integrating with PAM-g-gelation hydrogel electrolyte and α-MnO_2 cathode, the device exhibits an areal energy density of 6.18 mWh cm^{-2}. For the two-helix cable type, MnO_2 paste was roll-dip coated on the CNT fibers, while the anode was fabricated by electrodepositing Zn metal onto the fiber. Finally, electrode yarns were produced by twisting together eight fibers. After being wound on an elastic fiber, PAM hydrogel electrolyte was coated on and between these electrodes. A silicone tube was applied as the outside sealing, providing extra stretchability. The structure enabled the entire device to exhibit an advantage in volumetric energy density of 53.8 mWh cm^{-3} compared to the sandwich FZIBs. In terms of the coaxial-fiber based FZIBs, components including Zn nanosheet arrays, cellulose-based CMC-$ZnSO_4$ electrolyte, and Zn hexacyanoferrate are directly coated on the CNT fiber by a roll-electrodeposition process. The entire device also possesses a high volumetric energy density of 195.39 mWh cm^{-3}. Distinguished from the sandwich FZIBs, cable-type FZIBs have the advantages of a higher volumetric energy density and better deformation robustness. Dunn [32] indicates that for a small flexible IoT device, areal energy densities are vital metrics to increase construction integrity. Energy

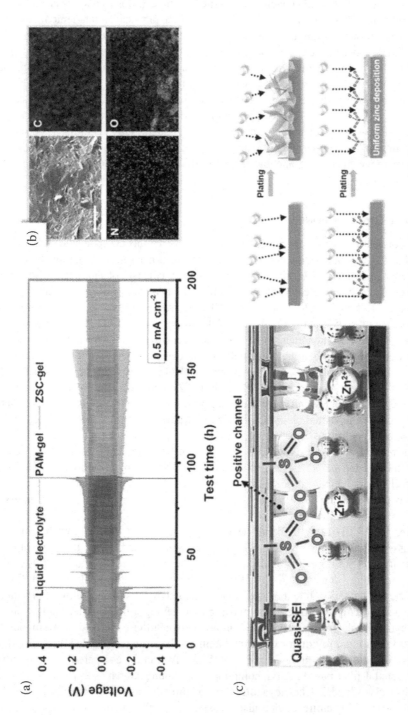

FIGURE 21.6 (a) Comparison of voltage–time profiles reflecting the cycling zinc stripping/plating performances of Zn//Zn symmetric cells based on various electrolytes. (b) SEM image and EDS element maps of the anode after 15 electrochemical cycles. Scale bar: 10 μm. (c) Schematic of the formation of the immigration channel between electrolyte and Zn anode [11]. Reproduced by permission from Chunyi Zhi, *Zwitterionic Sulfobetaine Hydrogel Electrolyte Building Separated Positive/Negative Ion Migration Channels for Aqueous Zn-MnO₂ Batteries with Superior Rate Capabilities (Advanced Energy Materials)*, Copyright (2020) WILEY-VCH Verlag GmbH & Co. KGaA, Weinheim.

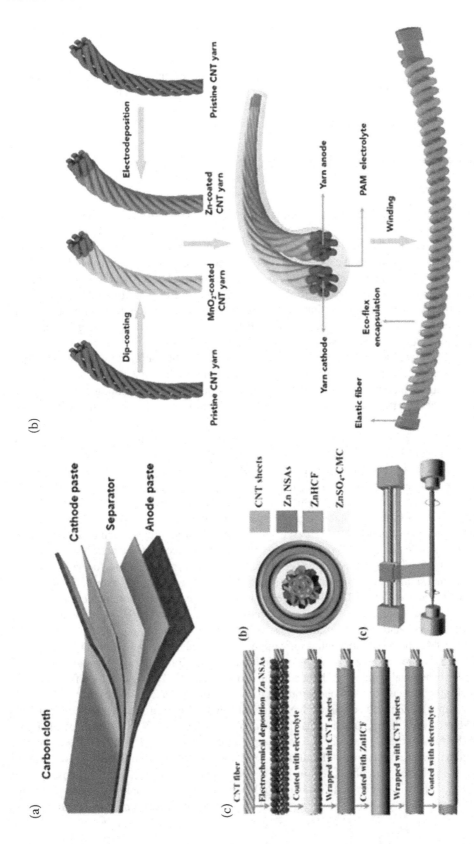

FIGURE 21.7 (a) Sandwich structure. (b) Two helix yarns cable-type structure.(Li, Liu, et al. 2018) Reproduced by permission from Hongfei Li, *Waterproof and Tailorable Elastic Rechargeable Yarn Zinc Ion Batteries by a Cross-Linked Polyacrylamide Electrolyte (ACS Nano)*. Copyright (2018) American Chemical Society. (c) Coaxial cable-type structure.(Q. Zhang et al. 2019) Reproduced by permission from Qichong Zhang, *Flexible and High-Voltage Coaxial-Fiber Aqueous Rechargeable Zinc-Ion Battery (Nano Letters)*. Copyright (2019) American Chemical Society.

density for the entire device could be simply enhanced by stacking more single FZIBs lamination in parallel. Cable-type FZIBs have potential in the garment industry, offering a highly safe, biocompatible, and smart cloth. Current configurations are still in their early stage; hence, more detailed fabrication and packing strategies should be evaluated for further commercialization. Further, packing materials such as poly(ethylene terephthalate) (PET) are regarded as suitable substrates to encapsulate FZIBs. With the application of FZIBs, employing flexible and three-dimensional (3D) architecture is also an approach to further increase the energy density compared to current 2D laminated and 1D cable types.

21.4 CURRENT CHALLENGES AND PERSPECTIVES

21.4.1 Voltage issue

For FZIBs, achieving a high operational voltage is an insurmountable challenge. The water content in the hydrogel electrolytes limits the operating voltage range for HER at the anode and OER at the cathode in hydrogel-based FZIBs. As a flexible device, it is necessary to maintain a high voltage and a stable energy supply; thus, the voltage window must be expanded. The voltage window is not only related to CEI and AEI, with the energy levels of the highest occupied molecular orbital (HOMO) and lowest unoccupied molecular orbital (LUMO), but is also related to the redox potentials of the electrolyte.[33] When only considering LUMO-HOMO energies at the interfaces, Na-ion batteries could be used in aqueous solutions. While restricting by the HER at −4.02 eV and OER at −5.25 eV of water band gaps, the thermodynamic potential window of water is only 1.23 V at pH 7. Although the solid-solid interfaces between hydrogel electrolytes and electrodes are believed to expand the voltage range due to higher interfacial activation energy, hydrogel electrolytes still suffer from unstable side effects, resulting in the same operational voltage window as aqueous ZIBs. Therefore, instead of determining the LUMO and HOMO energy levels at the interface, redox reactions in the electrolyte are the determining factor. The redox potential range is located within the LUMO-HOMO energy levels, providing the actual stable window (as shown in Figure 21.8a). Extensive efforts were applied in expanding the electrolyte stability window. Generally, most of FZIBs are operated below 1.9 V, while the application of "water-in-salt" or "polymer-in-salt" strategies have slightly increased the working voltage range up to 2.2 V. Water-in-salt solutions, as the name suggests, are highly concentrated electrolytes aiming to suppress water hydrolysis by increasing the salt concentration. In the highly concentrated electrolytes, fewer water molecules are surrounded by Zn^{2+} ions resulting in stable reactions at AEI rather than the HER and zinc dendrite formations. As a promising strategy, researchers have developed large-potential-range ZIBs utilizing different conducting salts. Wang et al.[164] applied the ultra-highly concentrated electrolyte of 20 M/kg of LiTFSI and 1 M/kg $Zn(TFSI)_2$. The strong $(Zn-TFSI)^+$ interactions impede the formation of $(Zn-(H_2O)_6)^{2+}$, resulting in a stable electrochemical performance rather than generating hydroxyl ions (OH^-). Similarly, 30M $ZnCl_2$ reported by Ji [34] also deliver a wider voltage range. A three-electrode test was performed, where Ti foil was applied as the working electrode, and activated carbon (AC) and Ag/AgCl were applied as the counter and reference electrodes, respectively. By increasing the concentration of $ZnCl_2$ from 5 to 30M, the high concentration suppresses the reaction between Zn^{2+} and OH^-. Meanwhile, there is an increase for Zn plating potential and a reduction in HER potential at the AEI; hence, a wider electrochemical window is induced from 1.6 V to 2.3 V. The coulombic efficiency also increases along with the increase of salt concentration, for which there is a 20% increase from 73.2% in 5M $ZnCl_2$ to 95.4% in 30M. Zhi and coworkers have utilized a hydrogel electrolyte immersed into highly concentrated 4M $Zn(OTf)_2$, offering a stable and high voltage window of 0.7–2.0 V for $Zn/CoFe(CN)_6$ batteries. Due to the full interaction between the conducting salts and the polymer matrix, the proportion of amorphous regions has been maximized. In addition, positive ion transmission is realized, not only by the movement of the polymer amorphous region, but also by the unique ion channel composed of ion clusters. Although the water-in-salt approach is beneficial for both Zn dendrite prevention and expansion of the voltage range, the highly acidic electrolyte and the high cost of salts promote the exploration of

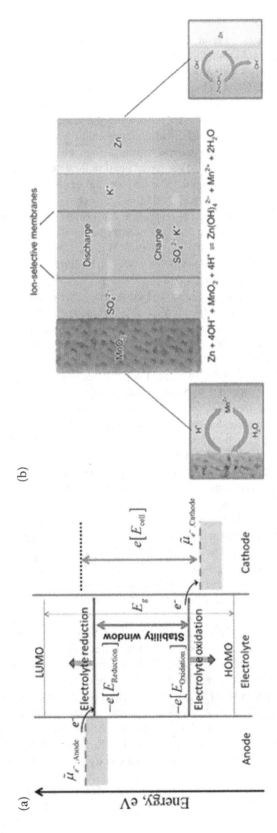

FIGURE 21.8 (a) Stable voltage window limits for the electrolyte stability, and the energy levels of HOMO and LUMO. (Peljo and Girault 2018) Reproduced by permission from Pekka Peljo, *Electrochemical potential window of battery electrolytes: The HOMO-LUMO misconception (Energy and Environmental Science).* Copyright (2018) The Royal Society of Chemistry. (b) Working mechanisms of the decoupled Zn–MnO₂ battery [37]. Reproduced by permission from Wenbin Hu, *Decoupling electrolytes towards stable and high- energy rechargeable aqueous zinc–manganese dioxide batteries (Nature Energy).* Copyright (2020) The Author(s). under exclusive licence to Springer Nature Limited.

additives. Sodium dodecyl sulfate (SDS) is an anionic surfactant used as a detergent in engine degreasing and domestic cleaning. It is a synthetic organic compound containing a 12-carbon tail attached to a sulfate group. The special structure combines hydrophilicity from the hydrocarbon and lipophilicity from the polar sulphate. Adding into the $ZnSO_4$ aqueous solution, a wider window was obtained from 2.0 to 2.5 V accompanying with $Na_2MnFe(CN)_6$ cathode [35]. SDS exhibits a positive influence for both anodes and cathodes, suppressing HER and OER. At the AEI, the onset HER potential was decreased from −0.70 V to −1.15 V (vs. SHE); whereas at the CEI, the onset potential of OER was increased from 1.1 V to 1.4 V (vs. SHE). SDS was adsorbed on CEI with its hydrophilic group facing the electrode and its hydrophobic group facing the electrolyte, hence presenting a high energy barrier for water transmission. Inspired by this organic additive, owing to the presence of polar groups, succinic acid, dodecyl-trimethylammonium bromide, and tartaric acid could also be potential candidates to expand the voltage window. Highly concentrated electrolyte and organic additives further promote the application of the combined strategy. Ma[11] has developed a HBPE based on polymer poly(vinylidene fluoride-hexafluoropropylene) (PVDF-HFP). Dissolving a highly concentrated $Zn(BF_4)_2$ and ionic liquid 1-ethyl-3-methyl-imidazolium tetrafluoroborate ([EMIM]BF_4) mixture, FZIBs achieved a high voltage up to 2.05 V with the cobalt hexacyanoferrate cathode (CoHCF). Organic additive [EMIM]BF_4 extends the water stability window from −0.05 ~ 1.7 V (vs. Zn/Zn^{2+}) to −0.75 ~ 2.8 V (vs. Zn/Zn^{2+}).

Evaluating the electrolytes by suppressing the HER and OER certainly expands the working stability window; however, the increment is restricted to 2.2 V. To further increase the voltage range, hybrid redox flow and decoupled zinc–manganese dioxide batteries have been developed. Qiao [36] constructed the first redox flow batteries using a two-electron redox electrolysis reaction of Mn^{4+}/Mn^{2+}. In comparison with conventional ZIBs, only one electron is transferred during discharge, forming MnOOH along with the by-product $[Zn(OH)_4]SO_4$. After adjusting the pH value with H_2SO_4, a highly reversible Mn deposition/dissolution reaction facilitated a high output voltage of 1.95 V. The proposed Mn^{4+}/Mn^{2+} ZIBs are revolutionary, allowing the investigation of high voltage and high energy density ZIBs. A concept of decoupled ZIBs has been investigated where decoupled ZIBs contain two electrolyte reservoirs, the acidic electrolyte in the cathode side and alkaline-based electrolyte in the zinc anode side. Hu et al. [37] have proposed a decoupled ZIB with two ion-selective membranes. As displayed in Figure 21.8b, KOH and $MnSO_4$ chambers are separated by cation-exchange and anion-exchange membranes, respectively. A buffer layer with K_2SO_4 is located in between for the ion exchange. As shown in the following reactions, manganese deposition/dissolution reactions occur on the cathode, where Mn^{4+} is reduced to Mn^{2+} during discharging; $SO_4^{2−}$ ions can diffuse through the anion-exchange membrane to the buffer region. As for the anode chamber, $Zn(OH)_4^{2−}$ ions are formed by releasing electrons to the cathode. K^+ ions can diffuse through the cation-exchange membrane to the buffer region at the same time. During the charging process, $SO_4^{2−}$ and K^+ stored in the buffer chamber will diffuse back to their original chamber and enable the reactions of two half cells. Notably, the acidic and alkaline hybrid cell exhibits an open-circuit voltage of 2.83 V in contrast to 1.5 V for conventional Zn/MnO_2 ZIBs.

Despite having a high voltage window, the high cost and short lifetime of selective membranes stymie commercialization. Yadav et al. [38] have developed a membrane-free redox flow battery using polymer electrolytes. Poly(acrylic acid) (PAA) was utilized with KOH generating the gelled KOH electrolyte for Zn anode. HPE, PAA/KOH at the anode side, enables the diffusion of cation and anions in the electrolyte without neutralization. This redox cell exhibits a high open-circuit potential voltage around 2.8 V and a reversible charge/discharge capacity of 20~100% of its theoretical capacity (308 mAh g^{-1}). Inspired by this, a novel FZIB can achieve a high voltage by laminating two polymer electrolyte layers with acidic and alkaline electrolytes, respectively. Apart from PAA/KOH gelled electrolyte, $MnSO_4$ and H_2SO_4 could be stored in the PAM structure. An investigation of the electrolyte-electrolyte interface between PAM and PAA should be exploited.

Anode:

$$Zn + 4OH^- \leftrightarrow Zn(OH)_4^{2-} + 2e^- \ (E^o = -1.199 \text{ V } vs. \text{ SHE})$$ (21.3)

Cathode:

$$MnO_2 + 4H^+ + 2e^- \leftrightarrow Mn^{2+} + 2H_2O \ (E^o = 1.224 \text{ V } vs. \text{ SHE})$$ (21.4)

21.4.2 STRUCTURAL ENHANCEMENT

As a flexible battery, the mechanical load-bearing property is a crucial index determining whether the FZIB is feasible for commercialization. Until now, most FZIBs exhibit a stable energy output under external forces, such as bending, twisting, hammering, and cutting; however, few mechanical properties of the fabricated FZIBs are quantitatively reported. In the laminated structure, once the intraplanar or interplanar stress exceeds the tolerance limit, delamination of components results in the disconnection of the battery. Under the bending mode, the most frequent scenario, a convex strain of a single layer can be determined as the functions below, in which b, the distance from the outer convex to the middle neutral surface (Figure 21.9b), is simplified with the average of the thickness of polymer film itself and the electrode substrate. d and R represent the thickness and bending radius, respectively. For a multilayer structure with n layers, a complicated calculation with plane strain and Poisson's ratio is implied to determine the bending curvature [39]. The equation is advised to be simplified for the experimental use by three parameters: L (end-to-end length); R (bending radius); θ (bending angle), as shown in Figure 21.9a. L is the distance between two ends, which indicates the bending state irrelevant to the shape of the deformation. θ describes the movement of the end regarding the bending angle. Bending radius R directly relates to bending strain can be measured with a given-radius cylinder. These parameters are suggested to be specified for future FZIBs.

$$b = \frac{1}{2}(d_f + d_s)$$ (21.5)

$$\varepsilon = \frac{b}{R} = \frac{d_f + d_s}{2R}$$ (21.6)

Theoretically, hydrogel-based HBPEs exhibit fewer tensile strengths compared with SPEs. It is a fixed mindset that there is a trade-off design in flexible energy storage batteries between mechanical and

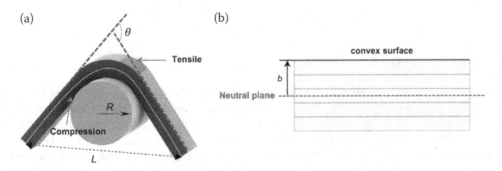

FIGURE 21.9 (a) A FZIB under the bending mode with three describing parameters: L, θ, and R. (b) Schematic diagram of the neutral plane in a multilayer flexible energy device.

electrochemical performances. Focusing on polymer electrolytes, a stronger inter and intramolecular bonding in the polymer structure could definitely enhance the structural stiffness, whereas the strong interactions impede the transportation of Zn^{2+} ions through intermolecular chains in the polymer electrolyte. This trapped mindset, however, would not be true for FZIBs. For a flexible battery, it is essential to evaluate the structural properties considering the integrated structure. A novel separator design, as mentioned before, is a promising strategy. Separators carrying the majority external loading are necessary components in FZIBs, despite the application of polymer electrolytes to combine the separator function. By grafting polymer electrolytes on a membrane host such as PAN or PET film, the tensile stress can reach 7 MPa, as illustrated in section 3.1.3. Other woven-glass fabrics, for example, have the potential to be applied as the separator generating a fiber-reinforced plastic. Carbon clothes, widely used electrode substrates, can provide a carbon fiber reinforced structure once an epoxy-based electrolyte is synthesized. Packing materials are other components that are frequently neglected in FZIBs. Encapsulating flexible devices either through sandwich or cable types wrapping materials can provide extra load-bearing properties under deformation.

21.4.3 MULTIFUNCTIONALITIES

In flexible LIBs, the application of polymer electrolytes enables the use of lithium metals as the anode, eventually increasing the energy density. Unlike LIBs, there is a limited enhancement for the integrated capacity with the development of polymer electrolytes; hence, multifunctionalities are characteristics of FZIBs. Polymer electrolytes with an emphasis on thermal sensitivity, anti-freezing ability, and dendrite prevention have been developed in the past five years; however, there is no study to quantify the multifunctionality for FZIBs. Herein, we propose a multifunctional coefficient (Eq. 21.7) to quantify the multifunctionalities. As shown in the equation, η represents the coefficient, where m and e are abbreviations of multifunctionality and electrochemical performance. i refers to various functionalities other than charge storage. Comparing the additional functionality performed in FZIBs to the average magnitude required in certain application scenarios, multifunctional coefficient can be amended. This coefficient is also a ranking factor conforming to the FZIBs under the same test criteria. Considering this factor in wearable energy storage batteries, as an example, bifunctional FZIBs consisting of electrochemical and mechanical properties can be evaluated in Eq. 21.8, where specific capacities and tensile strengths are selected as determinants for electrical and mechanical properties, respectively [40]. If η_m is close to 2, the entire device has the potential for industrialization.

$$\eta_m = \eta_e + \eta_i = \frac{C_{FZIBs}}{C_{AVG}} + \left(\frac{f_{FZIBs}}{f_{AVG}}\right)_i \qquad (21.7)$$

$$\eta_m = \frac{C_{FZIBs}}{C_{AVG}} + \frac{\sigma_{FZIBs}}{\sigma_{AVG}} = \frac{C_{FZIBs}}{5.79} + \frac{\sigma_{FZIBs}}{11.58} \qquad (21.8)$$

21.5 CONCLUSIONS

The advantages of FZIBs, including high safety, low cost, high energy, and environmental friendliness, mean they are good prospects in the field of wearable and implant devices. Among many cathode materials, vanadium and manganese-based compounds have excellent Zn-ion storage properties, and they have attracted widespread attention in recent years. There is a tremendous improvement for electrochemical properties in FZIBs for which the ionic conductivities of polymer electrolytes and specific capacity can reach the same magnitude for aqueous ZIBs. Extensive efforts have been applied in polymer electrolytes, including functionalities and the prevention of side effects. Zinc dendrite formation is inhibited for all types of polymer electrolytes, including SPEs, HPEs, and HBPEs. Moreover, functionalities such as self-healing, thermo-responsibility, and anti-freezing ability will help accelerate them into the market. However, aiming for realistic applications, it is essential for FZIBs to reach a high operational voltage. With the understanding of electrode-electrolyte interfacial reactions, strategies for achieving a wider stable window are discussed in detail. Manipulating manganese deposition/dissolution reactions, polymer electrolytes even offer a benefit in the low-expense membrane-free decoupled ZIBs. Further investigations of interfacial mechanisms at CEI and AEI are required to better understand energy storage mechanisms. In this chapter, functions and the multifunctional index of FZIBs are well demonstrated. Although FZIBs are still at an early stage, it is likely they will transform into consumer devices. Under a specific scenario, electrochemical properties are influenced by external deformation. It is not, however, a contradiction to increase energy densities while also improving structural performance, such as through novel separator design. In this diverse era, a solution for the energy storage sector based solely on LIBs cannot stand alone; other strategies will enter the market for various application scenarios. The low cost and biocompatibility will undoubtedly provide opportunities for FZIBs to satisfy market requirements.

REFERENCE

1. Reddy, D. Linden and T. B. 2002. *Linden's Handbook of Batteries*, 4th Edition. Handbook of Batteries.
2. Xu, Chengjun, Baohua Li, Hongda Du, and Feiyu Kang (2012) Energetic Zinc Ion Chemistry: The Rechargeable Zinc Ion Battery. *Angewandte Chemie – International Edition* 51 (4): 933–935.
3. Kundu, Dipan, Brian D. Adams, Victor Duffort, Shahrzad Hosseini Vajargah, and Linda F. Nazar (2016) A High-Capacity and Long-Life Aqueous Rechargeable Zinc Battery Using a Metal Oxide Intercalation Cathode. *Nature Energy* 1 (10): 16119.
4. Li, Hongfei, Cuiping Han, Yan Huang, Yang Huang, Minshen Zhu, Zengxia Pei, Qi Xue, et al (2018) An Extremely Safe and Wearable Solid-State Zinc Ion Battery Based on a Hierarchical Structured Polymer Electrolyte. *Energy and Environmental Science* 11 (4): 941–951.
5. Pan, Huilin, Yuyan Shao, Pengfei Yan, Yingwen Cheng, Kee Sung Han, Zimin Nie, Chongmin Wang (2016) Reversible Aqueous Zinc/Manganese Oxide Energy Storage from Conversion Reactions. *Nature Energy* 1 (April): 16039.
6. Zhang, Yan, Fang Wan, Shuo Huang, Shuai Wang, Zhiqiang Niu, and Jun Chen (2020) A Chemically Self-Charging Aqueous Zinc-Ion Battery. *Nature Communications* 11 (1): 1–10.
7. Zhao, Qing, Weiwei Huang, Zhiqiang Luo, Luojia Liu, Yong Lu, Yixin Li, Lin Li, Jinyan Hu, Hua Ma, and Jun Chen (2018) High-Capacity Aqueous Zinc Batteries Using Sustainable Quinone Electrodes. *Science Advances* 4 (3): 1761.
8. Kundu, Dipan, Pascal Oberholzer, Christos Glaros, Assil Bouzid, Elena Tervoort, Alfredo Pasquarello, and Markus Niederberger (2018) Organic Cathode for Aqueous Zn-Ion Batteries: Taming a Unique Phase Evolution toward Stable Electrochemical Cycling. *Chemistry of Materials* 30 (11): 3874–3881.
9. Wei, Chunguang, Chengjun Xu, Baohua Li, Hongda Du, and Feiyu Kang (2012) Preparation and Characterization of Manganese Dioxides with Nano-Sized Tunnel Structures for Zinc Ion Storage. *Journal of Physics and Chemistry of Solids* 73 (12): 1487–1491.
10. Candhadai Murali, Sai Prasanna, and Austin Suthanthiraraj Samuel (2019) Zinc Ion Conducting Blended Polymer Electrolytes Based on Room Temperature Ionic Liquid and Ceramic Filler. *Journal of Applied Polymer Science* 136 (24): 47654.
11. Ma, Longtao, Shengmei Chen, Na Li, Zhuoxin Liu, Zijie Tang, Juan Antonio Zapien, Shimou Chen, Jun Fan, and Chunyi Zhi (2020) Hydrogen-Free and Dendrite-Free All-Solid-State Zn-Ion Batteries. *Advanced Materials (Deerfield Beach, Fla.)* 32 (14): 1908121.
12. Zhao, Jingwen, Keval K. Sonigara, Jiajia Li, Jian Zhang, Bingbing Chen, Jianjun Zhang, Saurabh S. Soni, Xinhong Zhou, Guanglei Cui, and Liquan Chen (2017) A Smart Flexible Zinc Battery with Cooling Recovery Ability. *Angewandte Chemie – International Edition* 56 (27): 7871–7875.
13. Zeng, Yinxiang, Xiyue Zhang, Yue Meng, Minghao Yu, Jianan Yi, Yiqiang Wu, Xihong Lu, and Yexiang Tong (2017) Achieving Ultrahigh Energy Density and Long Durability in a Flexible Rechargeable Quasi-Solid-State Zn–MnO2 Battery. *Advanced Materials* 29 (26): 1700274.
14. Vatsalarani, J., N. Kalaiselvi, and R. Karthikeyan (2009) Effect of Mixed Cations in Synergizing the

Performance Characteristics of PVA-Based Polymer Electrolytes for Novel Category Zn/AgO Polymer Batteries-a Preliminary Study. *Ionics* 15 (1): 97–105.

15. Chen, Minfeng, Weijun Zhou, Anran Wang, Aixiang Huang, Jizhang Chen, Junling Xu, and Ching Ping Wong (2020). Anti-Freezing Flexible Aqueous Zn-MnO2 Batteries Working at −35 °c Enabled by a Borax-Crosslinked Polyvinyl Alcohol/Glycerol Gel Electrolyte. *Journal of Materials Chemistry* A 8 (14): 6828–6841.

16. Zhou, Weijun, Jizhang Chen, Minfeng Chen, Anran Wang, Aixiang Huang, Xinwu Xu, Junling Xu, and Ching Ping Wong (2020) An Environmentally Adaptive Quasi-Solid-State Zinc-Ion Battery Based on Magnesium Vanadate Hydrate with Commercial-Level Mass Loading and Anti-Freezing Gel Electrolyte. *Journal of Materials Chemistry* A 8 (17): 8397–8409.

17. Li, Hongfei, Zhuoxin Liu, Guojin Liang, Yang Huang, Yan Huang, Minshen Zhu, Zengxia Pei (2018) Waterproof and Tailorable Elastic Rechargeable Yarn Zinc Ion Batteries by a Cross-Linked Polyacrylamide Electrolyte. *ACS Nano* 12 (4): 3140–3148.

18. Mo, Funian, Guojin Liang, Qiangqiang Meng, Zhuoxin Liu, Hongfei Li, Jun Fan, and Chunyi Zhi (2019) A Flexible Rechargeable Aqueous Zinc Manganese-Dioxide Battery Working at −20 °c. *Energy and Environmental Science* 12 (2): 706–715.

19. Huang, Yuan, Jiuwei Liu, Jiyan Zhang, Shunyu Jin, Yixiang Jiang, Shengdong Zhang, Zigang Li, Chunyi Zhi, Guoqing Du, and Hang Zhou (2019) Flexible Quasi-Solid-State Zinc Ion Batteries Enabled by Highly Conductive Carrageenan Bio-Polymer Electrolyte. *RSC Advances* 9 (29): 16313–16319.

20. Huang, Yuan, Jiyan Zhang, Jiuwei Liu, Zixuan Li, Shunyu Jin, Zigang Li, Shengdong Zhang, and Hang Zhou (2019) Flexible and Stable Quasi-Solid-State Zinc Ion Battery with Conductive Guar Gum Electrolyte. *Materials Today Energy* 14: 100349.

21. Zhang, Silan, Nengsheng Yu, Sha Zeng, Susheng Zhou, Minghai Chen, Jiangtao Di, and Qingwen Li (2018) An Adaptive and Stable Bio-Electrolyte for Rechargeable Zn-Ion Batteries. *Journal of Materials Chemistry* A 6 (26): 12237–12243.

22. Dong, Haobo, Jianwei Li, Siyu Zhao, Yiding Jiao, Jintao Chen, Yeshu Tan, Dan J.L. Brett, Guanjie He, and Ivan P. Parkin (2021) Investigation of a Biomass Hydrogel Electrolyte Naturally Stabilizing Cathodes for Zinc-Ion Batteries. *ACS Applied Materials and Interfaces* 13 (1): 745–754.

23. Yan, Xiaoqiang, Qiang Chen, Lin Zhu, Hong Chen, Dandan Wei, Feng Chen, Ziqing Tang, Jia Yang, and Jie Zheng (2017) High Strength and Self-Healable Gelatin/Polyacrylamide Double Network Hydrogels. *Journal of Materials Chemistry* B 5 (37): 7683–7691.

24. Han, Qi, Xiaowei Chi, Shuming Zhang, Yunzhao Liu, Biao Zhou, Jianhua Yang, and Yu Liu (2018) Durable, Flexible Self-Standing Hydrogel Electrolytes Enabling High-Safety Rechargeable Solid-State Zinc Metal Batteries. *Journal of Materials Chemistry* A 6 (45): 23046–23054.

25. Kong, Weiqing, Tian Li, Chaoji Chen, Gegu Chen, Alexandra H. Brozena, Dapeng Liu, Yang Liu (2019) Strong, Water-Stable Ionic Cable from Bio-Hydrogel. *Chemistry of Materials* 31 (22): 9288–9294.

26. Mo, Funian, Hongfei Li, Zengxia Pei, Guojin Liang, Longtao Ma, Qi Yang, Donghong Wang, Yan Huang, and Chunyi Zhi (2018) A Smart Safe Rechargeable Zinc Ion Battery Based on Sol-Gel Transition Electrolytes. *Science Bulletin* 63 (16): 1077–1086.

27. Zhu, Jiacai, Minjie Yao, Shuo Huang, Jinlei Tian, and Zhiqiang Niu (2020) Thermal-Gated Polymer Electrolytes for Smart Zinc-Ion Batteries. *Angewandte Chemie International Edition* 59 (38): 16480–16484.

28. Huang, Shuo, Fang Wan, Songshan Bi, Jiacai Zhu, Zhiqiang Niu, and Jun Chen (2019) A Self-Healing Integrated All-in-One Zinc-Ion Battery. *Angewandte Chemie – International Edition* 58 (13): 4313–4317.

29. Wang, Donghong, Lufeng Wang, Guojin Liang, Hongfei Li, Zhuoxin Liu, Zijie Tang, Jianbo Liang, and Chunyi Zhi (2019) A Superior δ-MnO 2 Cathode and a Self-Healing Zn-δ-MnO 2 Battery. *ACS Nano* 13 (9): 10643–10652.

30. Chen, Ze, Xinliang Li, Donghong Wang, Qi Yang, Longtao Ma, Zhaodong Huang, Guojing Liang, et al (2021) Environmental Science Grafted MXene/Polymer Electrolyte for High Shelf Life at Low/High Temperatures. *Energy and Enviormental Sience* 14: 3492–3501.

31. Leng, Kaitong, Guojie Li, Jingjing Guo, Xinyue Zhang, Aoxuan Wang, Xingjiang Liu, and Jiayan Luo (2020) A Safe Polyzwitterionic Hydrogel Electrolyte for Long-Life Quasi-Solid State Zinc Metal Batteries. *Advanced Functional Materials* 30 (23): 2001317.

32. Hur, Janet I, Leland C. Smith, and Bruce Dunn (2018) High Areal Energy Density 3D Lithium-Ion Microbatteries. *Joule* 2 (6): 1187–1201.

33. Peljo, Pekka, and Hubert H. Girault (2018) Electrochemical Potential Window of Battery Electrolytes: The HOMO-LUMO Misconception. *Energy and Environmental Science* 11 (9): 2306–2309.

34. Zhang, Chong, John Holoubek, Xianyong Wu, Aigerim Daniyar, Liangdong Zhu, Cheng Chen, Daniel P.

Leonard (2018) A ZnCl2 Water-in-Salt Electrolyte for a Reversible Zn Metal Anode. *Chemical Communications* 54 (100): 14097–14099.

35. Hou, Zhiguo, Xueqian Zhang, Xiaona Li, Yongchun Zhu, Jianwen Liang, and Yitai Qian (2017) Surfactant Widens the Electrochemical Window of an Aqueous Electrolyte for Better Rechargeable Aqueous Sodium/Zinc Battery. *Journal of Materials Chemistry* A 5 (2): 730–738.

36. Chao, Dongliang, Wanhai Zhou, Chao Ye, Qinghua Zhang, Yungui Chen, Lin Gu, Kenneth Davey, and Shi Zhang Qiao (2019) An Electrolytic Zn–MnO2 Battery for High-Voltage and Scalable Energy Storage. *Angewandte Chemie – International Edition* 58 (23): 7823–7828.

37. Zhong, Cheng, Bin Liu, Jia Ding, Xiaorui Liu, Yuwei Zhong, Yuan Li, Changbin Sun, et al (2020) Decoupling Electrolytes towards Stable and High-Energy Rechargeable Aqueous Zinc–Manganese Dioxide Batteries. *Nature Energy* 5: 440–449.

38. Yadav, Gautam G., Damon Turney, Jinchao Huang, Xia Wei, and Sanjoy Banerjee (2019) Breaking the 2 v Barrier in Aqueous Zinc Chemistry: Creating 2.45 and 2.8 v MnO2-Zn Aqueous Batteries. *ACS Energy Letters* 4 (9): 2144–2146.

39. Mao, Lijuan, Qinghai Meng, Aziz Ahmad, and Zhixiang Wei (2017) Mechanical Analyses and Structural Design Requirements for Flexible Energy Storage Devices. *Advanced Energy Materials* 7 (23): 1700535.

40. Dong, Haobo, Jianwei Li, Jian Guo, Feili Lai, Fangjia Zhao, Yiding Jiao, Dan J L Brett, Tianxi Liu, Guanjie He, and Ivan P. Parkin (2021) Insights on Flexible Zinc-Ion Batteries from Lab Research to Commercialization. *Advanced Materials* 33 (1800): 2007548.

22 Fabrication Techniques for Wearable Batteries

Ifra Marriam, Hiran Chathuranga, and Cheng Yan
School of Mechanical, Medical and Process Engineering,
Faculty of Engineering, Queensland University of Technology,
Brisbane, Queensland, Australia

Mike Tebyetekerwa
School of Chemical Engineering, The University of Queensland, St Lucia,
Brisbane, Queensland, Australia

Shengyuan Yang
State Key Laboratory for Modification of Chemical Fibers and Polymer Materials,
College of Materials Science and Engineering, Donghua University, North Renmin
Road, Shanghai, People's Republic of China

CONTENTS

22.1 Introduction...398
22.2 Wearable batteries 398
22.3 Electrode fabrication approaches 401
    22.3.1 Substrate-enabled techniques 402
        22.3.1.1 Chemical vapor deposition...402
        22.3.1.2 Hydrothermal deposition 402
        22.3.1.3 Electrochemical deposition...404
        22.3.1.4 Electrospinning and electrospraying 404
        22.3.1.5 Solution dip coating 404
        22.3.1.6 Spray painting...406
        22.3.1.7 Biscrolling...406
    22.3.2 Substrateless techniques 406
        22.3.2.1 Electrospinning 406
        22.3.2.2 Wet spinning...408
        22.3.2.3 Melt spinning 408
        22.3.2.4 3D extrusion printing 408
22.4 Structures and approaches for unification of electrodes 408
    22.4.1 Winding 409
    22.4.2 Twisting 409
    22.4.3 Coaxial assembly...411
22.5 Integration of fiber electrodes/batteries into textiles for wearable applications 411
    22.5.1 Weaving..411
    22.5.2 Knitting..411
22.6 Conclusion...413
References...413

DOI: 10.1201/9781003186755-22

22.1 INTRODUCTION

The ease that comes with using miniaturized electronics is related to their portability since they are easy to carry from one place to another, aiding movements in the existing complex human-environment filled with unpredictable events. For example, these portable devices can help in communication, tracking, direction, monitoring, and many other advanced functionalities. For this reason, innovations in existing electronic products are driving electronics to miniaturized smart wearable electronics, which are predicted to significantly profit the future electronics and textile industries.[1] It is apparent that miniaturization and integration of the technology into wearable smart textiles will readily make everything handy and wireless, whereas, simultaneously, the need for powering and charging these new fancy electronics will become complex and challenging. An innovative wearable yarn/fiber battery for energy storage can overcome this problem.

The first image that comes to mind with the term "battery" is something that is heavy, rigid, made of metal, and dangerous. Indeed, these qualities were true until the recent innovations in flexible and wearable electronics. With the rapid research progress in wearable electronics, numerous flexible and wearable energy devices, such as batteries, supercapacitors, solar cells, fuel cells, and triboelectric generators have been developed. For the case of batteries, those based on Li-ion, Na-ion, Zn-ion, and metal-air, as discussed in previous chapters, are considered the possible leading power sources for wearable electronics.[2–4] The main research goals of the next-generation wearable batteries are to obtain lightweight, flexible, portable, and user-friendly devices with high-performance, shape-conformability, and excellent mechanical durability (Figure 22.1).[1] To make wearable batteries, the battery counterparts should be similar to conventional textile fibers and yarns in terms of their flexibility, breathability, stretchability, and safety for easy integration into the garments for wearability.

22.2 WEARABLE BATTERIES

Just like traditional batteries, wearable fiber batteries also are made to consist of five main parts: two electrodes (the anode and the cathode), an electrolyte, a separator, and an outer layer. Using Li-ion battery (LiB) as an example, the charge-discharge mechanism of LiB goes as follows. During charging, the cathode is oxidized, and Li^+ ions transfer from the cathode to the anode. Conversely, during discharging (during use), Li-ions intercalate into the cathodes via the electrolytes, thus realizing energy conversion and storage within the batteries.[5,6] This charge-discharge process can be explained with help of the "rocking-chair" mechanism, which involves the back and forth swapping of Li-ions between the two electrodes, as shown in Figure 22.2a. The suitable electrolyte (i.e., solid, liquid, or gel type) allows the lithium ions to transfer from one electrode to another, and the separator acts as a barrier to avoid direct contact of the two electrodes. The current collectors can also be used to collect charges and facilitate the charge transfer process, depending on the electrode materials.[5]

Conventional solid-state electrochemical batteries employ planar metal substrates that are very heavy, rigid, not bendable, unwashable, and user-hostile.[9–11] Here, fiber batteries are a solution (Figure 22.2b). Wearable fiber batteries are meant to be in contact with human skin, which means they should be soft, light in weight, pliable, stretchable, flexible, breathable, washable, durable, and deformable.[1] To shape batteries according to these properties, several key parameters should be carefully considered.[1] For example, (1) the traditional rigid electrode materials and current collectors should be made soft, flexible, strong, and pliable in one dimension and be capable of withstanding twisting, bending, and even stretching effectively, (2) the conventional liquid electrolyte should be replaced with solid or gel safe electrolytes, (3) the various carcinogenic components in traditional batteries should be substituted to safe and nontoxic materials and, (4) finally, the individual yarn electrodes should be strong enough to withstand the tradition textile processes, such as weaving, knitting,

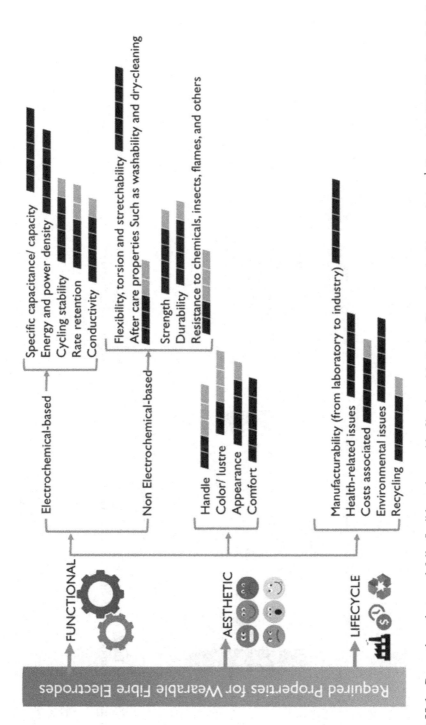

FIGURE 22.1 Research goals toward fully flexible and wearable fiber-based batteries. Reproduced with permission.[1] Copyright 2019 The Royal Society of Chemistry.

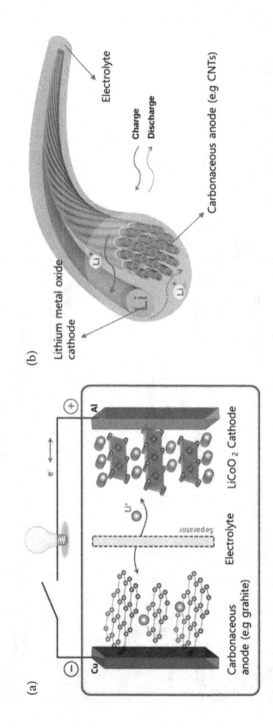

FIGURE 22.2 (a) The working mechanism of LiBs. Reprinted with permission.[7] Copyright 2013 American Chemical Society. (b) Fiber battery showing the arrangement of electrodes and electrolyte in a one-dimensional fiber structure. Adapted with permission.[8] Copyright 2013 WILEY-VCH Verlag GmbH & Co. KGaA, Weinheim.

pressing, mild wet processing, packaging, etc. The next section of fabrication techniques discusses enabling the utilization of various active materials into wearable yarn electrodes, covering various techniques capable of obtaining fiber electrodes that are soft, flexible, strong, and pliable in one-dimension fiber structure. Furthermore, the assembling of fiber batteries into the textiles is evaluated, along with their key fundamentals and textile design structures.

22.3   ELECTRODE FABRICATION APPROACHES

There are various techniques to fabricate fiber-shaped wearable batteries which can be broadly categorized into two (Figure 22.3); (1) substrate-enabled and (2) substrateless techniques. The substrate-enabled fabrication techniques involve the utilization of any off-the-shelf fibers (conductive or insulating) as a substrate to enable deposition of active materials that are later utilized as yarn anodes or cathodes. The common substrate-enabled fabrication approaches are deposition and coating techniques, such as dip coating, chemical vapor deposition, hydrothermal deposition, electrochemical deposition, electrospraying and electrospinning and spray printing. In the case of the substrateless techniques, the different electrodes are made using bottom-up approaches where the starting materials are the electrode active materials. Some key typical examples of these substrateless techniques include wet spinning, melt spinning, electrospinning, 3D printing, biscrolling, and filtration.

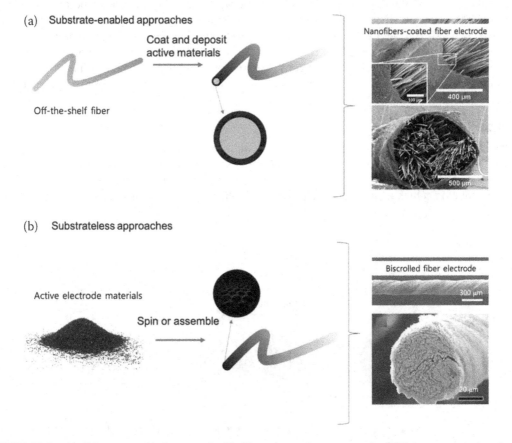

FIGURE 22.3   (a) Substrate-enabled approach. (b) The substratelesss approach. SEM images reproduced with permission.[12,13] (a) Copyright 2018 American Chemical Society. (b) Copyright 2018 The Authors.

22.3.1 Substrate-enabled techniques

These are deposition and coating techniques in which the active electrode materials are deposited on substrate yarns. These fabrication techniques are facile, though, for excellent performance in terms of robust interfacial bonding, multiple coating steps might be required. Coatings can readily deliver a uniform and continuous layer if done with care. The utilized yarn substrates can be the strong and existing natural/synthetic textile fibers or yarns as the core material which are then coated with conductive materials to form an electrode. In addition, coating and deposition techniques will enable electrodes with various textile properties, depending on the substrate fiber selected. For example, these properties of substrate fibers can range from metallic and conducting, to stretchable, polymeric, and insulating.

22.3.1.1 Chemical vapor deposition

Chemical vapor deposition (CVD) is a material-processing and deposition technique that involves introducing solid materials, typically under vacuum-controlled environments of pressure and temperature. CVD as a process typically involves reacting various chemicals together, which are then deposited on target substrates, all within the CVD chamber. For fiber electrodes, the fiber substrate is inserted into a vacuum in the presence of volatile materials. The gases react with the active sites and deposit a uniform coating on the fiber surface. The by-products generated by this process are eliminated using gas flow in a closed chamber.[14] The desired deposition materials and thickness can be achieved by controlling the temperature, pressure, type, and ratio of the gases. This process is popular for the integration of carbon-based materials with TMOs, and the synergistic effect between them can enhance the performance of the electrodes. The abundant defects and surface-active sites on the carbon fibers due to the interconnected conductive network can help in achieving superior electrochemical and mechanical performance. A heated concentric tube reactor design for roll-to-roll CVD of 2D materials on flexible fiber substrates (as presented in Figure 22.4) allows the substrate to pass through the helical path continuously and wrap around the tube. The circular tubes and the annular reactor used in this technique can be operated at any pressure, as required for various materials.[15]

22.3.1.2 Hydrothermal deposition

Similar to the CVD method, the hydrothermal deposition technique decorates various nano or macro-materials onto the yarn substrates under high temperatures and pressure in a vacuum to prepare flexible electrodes (Figure 22.5a).[18] In this technique, a supersaturated solution is produced by dissolving metal ions in the solution at high temperature and pressure. The crystals of active materials are then grown onto

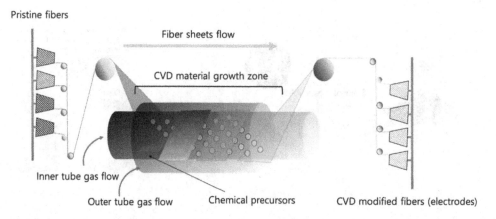

FIGURE 22.4 CVD process.

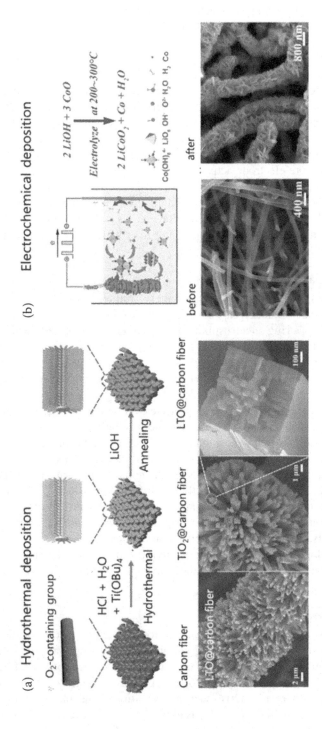

FIGURE 22.5 (a) Hydrothermal deposition process. Reproduced with permission.[16] Copyright 2019 WILEY-VCH Verlag GmbH & Co. KGaA, Weinheim. (b) Electrochemical deposition process. Reproduced with permission.[17] Copyright 2017 The Authors, some rights reserved; exclusive licensee American Association for the Advancement of Science. Distributed under a Creative Commons Attribution NonCommercial License 4.0 (CC BY-NC).

the surface of the fiber substrate.[19] The growth of the precursor can be improved on the surface by adjusting the pH level of the solution. The resultant deposited materials then undergo thermal treatment to achieve the desired hardness of the materials. Usually in the case of thermal treatment alone, the active materials tend to get aggregated, creating a nonuniform coating, but with hydrothermal deposition, a uniform, monodispersed, and controllable layer is formed.[20]

22.3.1.3 Electrochemical deposition

An electrochemical deposition involves adding a thin coherent active material coating on the target substrate with a help of current in a solution with two electrodes, as shown in Figure 22.5b. Various electrochemical cell parameters can be tuned to adjust the final properties of the active material coating on the fiber surface. The parameters include current density, voltage, solution type, additives, substrate surface, temperature, and many others. Unlike the coating and deposition approaches, which involve the use of a slurry technology, sometimes at high temperatures, the electrochemical deposition technique involves deposition of pure electrode active materials in a wet but concentrated environment. It is capable of giving strong interactive material or active material to substrate connections capable of limiting the detachment of active materials during bending.[21] Electrochemical deposition enables the synthesis of active materials on the fiber substrate with various morphologies, such as size, shape, porosity, and many others; these are the key to improving fiber battery performance in terms of their energy and power densities.

22.3.1.4 Electrospinning and electrospraying

Electrospinning and electrospraying are among the widely studied fabrication techniques in materials processing.[22] The two methods are capable of giving nanostructured, high surface area, and electroactive nanomaterials. During electrospinning, nanofibers are formed, while with electrospraying, nanoparticles are formed. For electrospinning, fibers of a few nanometers to micrometers in size are electrospun under applied high voltage (in kV)[12,23] from a functionalized viscous polymer and/or composite solution (spinning dope). For electrospraying, functional nanoparticles in dilute solutions are utilized. Both approaches are capable of depositing active materials with a range of morphologies, such as nanofibers, nanotubes, nanoparticles, and nanospheres. For substrate-enabled approaches, both conductive and insulating yarns can be utilized, as shown in Figure 22.6a. For nonconducting yarn substrates, a dummy conducting collector can be placed in the vicinity; the nonconducting yarn substrate can be placed between the spinning nozzle and the collector conducting plate because it is a fiber formation zone. Also, the conducting yarn can be placed stationary or in a rotary motion on which the nanofibers are then uniformly deposited.[28–30] This technique has been commercially used to fabricate composite nanofibers and nanoparticles for flexible fiber batteries.[31] There are numerous advantages of electrospinning. For example, it has versatility for spinning polymer materials, inorganic, and hybrid composites with ease; also, one can easily functionalize the spinning dope before or after fiber spinning and increase the electroactive ability of the resultant yarn. In addition, it can deposit nanofibers on various types of yarns, and it is easy to set up at a low cost.[22]

22.3.1.5 Solution dip coating

The dip coating is one of the oldest coating techniques to form an extra layer on the desired substrate. Dip coating, as the name describes, is carried out by dipping the substrate into the solution, which is similar to the textile pad-dyeing process (Figure 22.6b). In the case of electrodes for fiber batteries, the precursor solution is prepared by required aqueous or sol-gel active materials, and the substrate can be a flexible fiber.[32,33] Various active materials, such as transition metal oxides, transition metal dichalcogenides, and carbon-based materials, can be used for electrodes because of their good interfacial properties.[34] For the dip-coating process, the fiber substrate is first cleaned using wet processes to remove any surface impurities that can cause nonuniform coating.[35] Uneven coating can also be caused

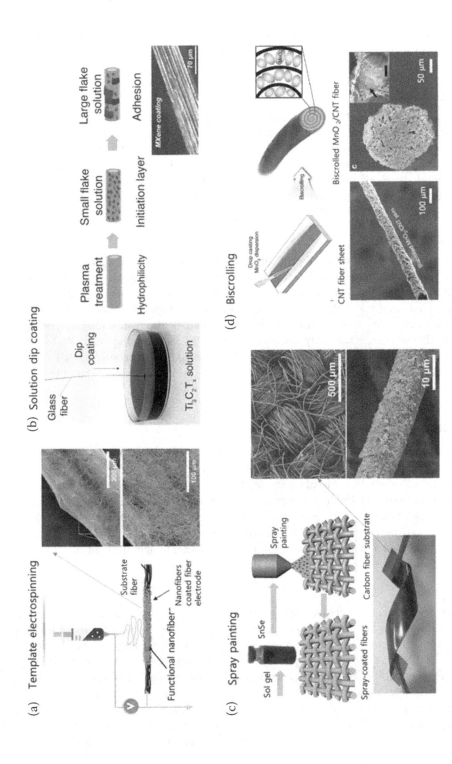

FIGURE 22.6 (a) Template electrospinning. Reproduced with permission.[12,24] Copyright 2018 American Chemical Society and Copyright 2018 The Royal Society of Chemistry. (b) Solution dip coating. Reproduced with permission.[25] Copyright 2020 by the authors. Licensee MDPI, Basel, Switzerland. This article is an open access article distributed under the terms and conditions of the Creative Commons Attribution (CC BY) license. (c) Spray coating. Reproduced with permission.[26] Copyright 2014 WILEY-VCH Verlag GmbH & Co. KGaA, Weinheim. (d) Biscrolling. Reproduced with Permission.[27] Copyright 2016, The Author(s).

by the agglomeration in the solution; thus, it is essential for the precursor to disperse well into the solvent. A single dip might not result in a proper layer formation. For this situation, the dipping process contains several dipping cycles to allow a uniform layer of the active materials onto the fibers. For mass production, the continuous fiber electrodes can be coated by several immersions into the solution with the help of padders or rollers.[32] The drying process takes place after the dipping and results in a fiber electrode.

22.3.1.6   Spray painting

Spray painting is a very simple technique to coat active materials onto the substrate via spraying using manual or automatic procedures. Here, the sol-gel solution with active materials is prepared; it is then spray painted onto the textile substrates under high pressure (Figure 22.6c). Using this technique, a fine spray-coated layer, in a variety of structures, can be formed onto the fiber substrate.[36] Uneven coating might occur with multiple layers of spraying.[37] Therefore, as a precaution, individual continuous substrate fiber to be coated can be held between two opposite rotating rollers, a technique that allows a uniform coating all over the yarn cross-section.

22.3.1.7   Biscrolling

To fabricate biscrolled yarns, a thin or flexible textile substrate (the host) is utilized, onto which a uniform layer of guest material is introduced. Here, both the host and guest can be electroactive electrode materials. The formed double-layer guest/host assembly is then twisted to form a biscrolled yarn (Figure 22.6d). Depending upon end constraints and the symmetry of applied stresses, twist insertion results in various biscrolled yarns with a variety of properties ranging from tuneable morphologies, rigidity, and porosity. In batteries, electrically conductive electrodes are very important. In this vein, conducting sheets of CNTs can be readily used. In scenarios where insulating yarns such as PET are to be employed, the metallization or dip-coating process may be chosen prior to deposition and the biscrolling step.

22.3.2   SUBSTRATELESS TECHNIQUES

These entail all the bottom-up approaches, which form fiber electrodes directly from the constituent active materials. These fabrication techniques enable high-capacity, fully flexible electrodes with high mechanical strength and no delamination issues because the techniques do not rely on substrates, as in the case of coating and deposition techniques summarized above. The substrateless techniques are mainly spinning techniques. Generally, spinning involves processing materials into a continuous length of fiber/yarn. Spinning can be carried out in various methods. To convert the solution containing electrode materials into the continuous fiber, the extrusion of the solution to fiber is mainly carried out by electrospinning, wet spinning, melt spinning, or 3D printing.

22.3.2.1   Electrospinning

The method for using the electrospinning technique is mainly the same as discussed above. However, in this approach, long and continuous electrodes in a single 1D fiber form, without any further coating or deposition, are formed. Nanofiber yarn electrospinning employs needle/needleless electrospinning or both, as shown in Figure 22.7a. Employing needleless electrospinning is known to offer higher productivity as compared to needle electrospining,[43] although it is complex to manipulate due to the elevated voltages and involved multiple jets. To fabricate continuous nanofiber yarns from needless electrospinning, a disc trough for nanofiber generation is used, onto which a high voltage is applied. Then a rotating intermediate ring for collecting the formed nanofibers is added as an attachment capable of collecting the nanofibers yarns onto a collector moving bobbin.

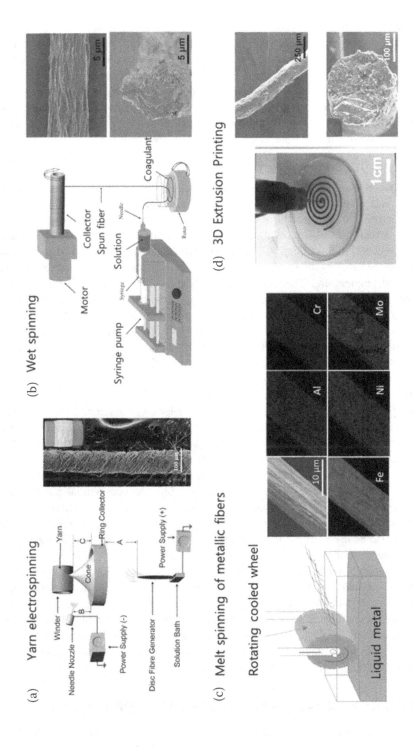

FIGURE 22.7 (a) Yarn electrospinning. Reproduced with permission.[38] Copyright 2015 The Royal Society of Chemistry. (b) Wet-spinning process. Reproduced with permission.[39,40] Left: Copyright 2017 Elsevier Ltd. Right: Copyright 2015 The Authors. (c) Metal fiber melt spinning (Left: internet source: Bekaert). Copyright 2021. Bekaert.[41] Right: Reproduced with permission.[22] Copyright 2020 Elsevier Inc. (d) 3D extrusion printing technique. Reproduced with permission.[42] Copyright 2017 WILEY-VCH Verlag GmbH & Co. KGaA, Weinheim.

22.3.2.2 Wet spinning

The wet-spinning technique is a very promising method of manufacturing fiber electrodes for wearable electronics. It allows the spinning of a variety of existing nanomaterials into filaments. It can maintain the properties and structures of electrode materials.[44] It is a scalable process that can be attenuated to be manual or automatic. The wet-spinning process starts with the preparation of the solution that contains well-dispersed electrode materials. Then, the solution is extruded via the nozzle, and the required fiber electrodes are spun into a coagulation bath (Figure 22.7b). This convenient technique offers another advantage of controlling the fiber diameter, structure, and design with the change in nozzle size and variation in the viscosity of the solution. After the actual spinning is done, the resultant fiber electrode is usually washed and dried after few minutes, when the solidification takes place.[45] This technique is also promising for the large-scale production of fiber electrodes for wearable batteries.

22.3.2.3 Melt spinning

The melt-spinning technique is widely used to produce nanocrystalline structures, which can be polymers, metals, or both. For wearable fiber electrodes, the required electrode materials can be mixed with polymers to allow well-spun fibers. This technique allows the insertion of polymer granulates or chips through a hopper, which is then melted in a heating zone. The polymer granulates or chips must be homogeneously melted for good processability.[46] The pressurized melted solution is pushed by a melt pump and extruded through the holes of the spinneret. The spun fibers are dried in a quenching chamber with adjustable airflow and temperature. Then the dried fibers pass through drawing rollers for the extension and parallelization of macromolecules and crystallites. In the case of metal-based materials, this technique can form very thin ribbons or sheets of a particular atomic structure.[41] However, the manufacturing technique operates at very high temperatures to melt the materials, and it requires the operators to address the related safety concerns. Here, the semi-continuous fibers are formed on rapid solidification of liquid metal with the help of a rotating cooled wheel (Figure 22.7c).

22.3.2.4 3D extrusion printing

The conventional printing technique can be used to build a layer-by-layer assembly, but it utilizes a substrate on which the active materials are printed, as discussed earlier. Not only does the substrate add more weight, but it is difficult to maintain the quality of the process and to promise strong interfacial contact and mechanical stability during bending or folding. Hence, 3D printing is used to enhance the interfacial properties and make the device flexible by creating all-in-one fibers. This technique was introduced around three decades ago, and it has rapidly emerged into the commercial manufacturing world courtesy of its ease and versatility. 3D printing can automatically produce accurate shape and design, with uniform thickness, by providing the required input data, which makes it reliable and mechanically robust.[47] It can also be used to produce a variety of structures and fiber designs for wearable electrodes (Figure 22.7c). Here, electrode active materials are prepared as ink sol-gel formulation in the syringe barrels. The formulation is pushed by a fluid dispenser with a controlled flow rate and passed through the nozzles of the desired structure. The fiber electrodes are then printed directly into the coagulation bath/support platforms in which the fibers solidify. The air is blown at the end to dry the fibers completely. To design a complete fiber battery, this same method can be repeated to print both electrodes, i.e., anode and cathode, separately. After printing, the gel-coated electrodes can be twisted together; the result is weavable or knittable into the textile fabric.[48] This technique reduces the production and material cost by cutting material wastage.

22.4 STRUCTURES AND APPROACHES FOR UNIFICATION OF ELECTRODES

For a complete fiber battery, the individual fiber electrodes, i.e., anode and cathode, fabricated by the above-discussed techniques, need to further be combined into various structures with the introduction of

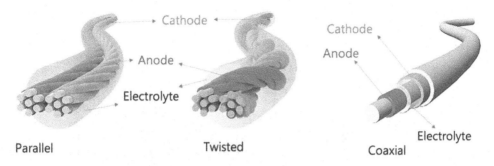

FIGURE 22.8 Structures of the fiber electrode integration to full fiber battery. Reproduced with permission.[49] Copyright 2019 WILEY-VCH Verlag GmbH & Co. KGaA, Weinheim.

the electrolyte. The common structures of the fiber battery integration are coaxial, parallel, and twisted structures, as shown in Figure 22.8. Techniques such as coating, winding, twisting, and rolling can be used to obtain the named structures. The approach to be used in unification will rely upon the structure required, physical and mechanical properties, together with the techniques used to fabricate the single electrode and the final end-use application.

22.4.1 WINDING

The winding technique is used to fabricate fiber batteries with a wrapped or coaxial structure. In this technique, the structure is simply attained by winding one fiber electrode onto another one coated with an electrolyte and separator (Figure 22.9a). This structure can also be formed by wrapping a sheet of electrode materials around the fiber electrode or by scrolling the sheet of electrode materials to form a wrapped fiber.[52] In this technique, the core fiber electrode is fixed (stretched in some cases) from both ends with stationary jaws on which another fiber electrode is wound at a certain angle. The helical turns of the wound fiber make the battery stretchable. An elastic textile fiber can also be introduced on which the two electrodes are wrapped together, distanced by a separator.[53] This type of fiber battery can allow better electrolyte penetration and facilitate ion diffusion and transport. However, if the winding turns are narrow and the diameter is too small, then the device can cause a short circuit; that is why it is recommended to use some blank spaces between the two fiber electrodes to act as a separator.[50]

22.4.2 TWISTING

The principle of twisting comes from the conventional ring-spinning method, which is utilized in the yarn manufacturing process. Using the same concept, a fiber battery with a twisted structure can be formed by either twisting the two fiber electrodes with each other at a certain twist angle or twisting both fiber electrodes individually with a conventional twisting method, and then wrapping them around one another (Figure 22.9b).[42] To twist the fiber electrodes, one electrode is held with jaws from both sides, and the jaws are connected to a rotating motor. Another electrode is linked to the first electrode from one side and to a linear stage from the other side. With the rotation of the motor, the fiber electrode also rotates in a clockwise direction, and the other fiber starts to move in a counter-clockwise direction; hence, a twist is inserted.[8,54] Furthermore, two fiber electrodes coated with an electrolyte can also be placed parallel to each other forming a parallel structure without the need to form a twist. A spacer wire can be used as a separator, if required. But if the twist is needed, then the parallel fibers can be twisted with a spring-like structure to make

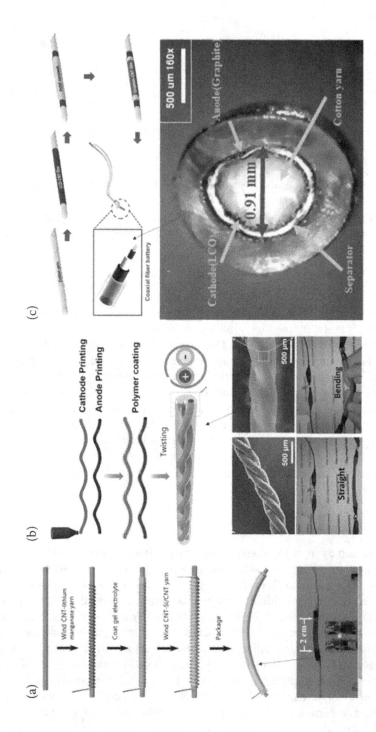

FIGURE 22.9 (a) Winding integration. Reproduced with permission.[50] Copyright 2014 American Chemical Society. (b) Twisting integration. Reproduced with permission.[42] Copyright 2017 WILEY-VCH Verlag GmbH & Co. KGaA, Weinheim. (c) Coating integration LiB. Reproduced with permission.[51] Copyright 2019 Elsevier Ltd.

the resulting fiber more stretchable.[55] The twisted structures are strong, thin, lightweight, and compatible for weaving and knitting for wearable applications.

22.4.3 COAXIAL ASSEMBLY

Coaxial structures have two major sections, i.e., the core and the shell. Here, the inner fiber electrode acts as core and an outer layer as shell, sandwiching a separator (if available) (Figure 22.9c). This structure can be obtained from the coating and deposition techniques at various stages of preparation, as discussed earlier. The fabrication method involves the preparation of the core material, i.e., fiber electrode, and sequential coating of a separator and an electrolyte. Another electrode layer followed by another layer of an electrolyte can also be coated to prepare a complete battery. The coaxial technique also suffers from issues like unrepairability, if the core electrode should fail. Therefore, care should be made to ensure robust layers with reliable mechanical properties during assembly.

22.5 INTEGRATION OF FIBER ELECTRODES/BATTERIES INTO TEXTILES FOR WEARABLE APPLICATIONS

A fully prepared one-dimensional fiber battery, or individual electrodes fabricated by the techniques discussed above, can also further be integrated into the textile fabrics. A fiber battery in a twisted, wrapped, coaxial, or parallel structure can be transferred into the fabric by traditional fabric manufacturing techniques of weaving or knitting. In the same way, fiber electrodes can also be integrated, although not in a straightforward approach.

22.5.1 WEAVING

The process of weaving involves interlacing the two sets of yarns perpendicular to each other at a right angle to create a woven fabric. It produces one of the most common and sturdy structures worn in daily life. The fiber battery can be woven directly into the fabric structure in any direction (Figure 22.10a).[56] However, in the case of a fiber electrode, the weaving process can be carried out by interlacing the anode fiber in a warp direction (vertical axis) and the cathode fiber in a weft direction (horizontal axis), spaced by an intersecting yarn acting as a separator.[57,58] During the weaving process, the fibers pass through several machine parts and suffer several mechanical stresses, including friction, abrasion, and tension. Therefore, the fiber electrodes/battery should be mechanically and chemically stable to withstand all the deformations during this process. Once the fiber electrodes or batteries are successfully integrated into the fabric, the resultant woven structure is usually lightweight, flexible, breathable, strong, and can be worn as a normal textile fabric.

22.5.2 KNITTING

The knitting process requires only one continuous fiber, as opposed to weaving, where wefts and warps are needed for successful interlacing into a fabric. During knitting, a long continuous fiber battery is interlocked into a knitted fabric with consecutive loops in the course and wale direction. The knitted structure makes the fabric highly stretchable, warm, bulky, and adaptable to body movements. This process requires the fiber battery to be highly flexible, fine, long, and mechanically strong for the loop formation, which means not all the fiber electrodes can be knitted effectively.[59] The process is cost-effective and does not require extra chemical preparation if compared to weaving, as fewer mechanical processes are involved. But still, the fibers are subjected to mechanical deformation during the loop formation due to the fast-moving needles, which can cause some cracks in the structure. Therefore, the fibers prepared as a battery should be fabricated by a suitable technique to withstand the stress during knitting.

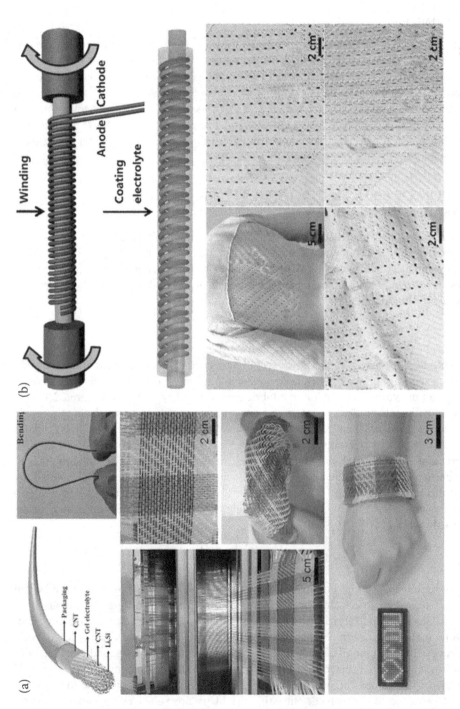

FIGURE 22.10 (a) Weaving and (b) Knitting of fiber batteries into textiles for wearable applications. (a) Reproduced with permission.[56] Copyright 2017 Wiley-VCH Verlag GmbH & Co. KGaA, Weinheim. (b) Reproduced with permission.[53] Copyright 2014 The Royal Society of Chemistry.

22.6 CONCLUSION

With the growth of portable electronics, consumers are being driven to wanting their devices beyond portability, which means devices need to be inconspicuous and to be powered in the smartest way possible. One of the possible approaches is flexible textile and wearable fiber batteries. Therefore, the next generation of rechargeable batteries should be flexible, portable, and inconspicuous, and yet capable of delivering the required power, as traditional batteries do. Hence, traditional textile manufacturing techniques, together with the chemical processing techniques, as discussed in this chapter, should be married to design fiber-based wearable batteries that can be easily integrated into the textile fabric or clothing and still ensure reliable electrochemical performance.

REFERENCES

1. Tebyetekerwa, M., Marriam, I., Xu, Z., Yang, S. Y., Zhang, H., Zabihi, F., Jose, R., Peng, S. J., Zhu, M. F., Ramakrishna, S., Critical insight: challenges and requirements of fibre electrodes for wearable electrochemical energy storage. *Energy & Environmental Science* 2019, *12* (7), 2148–2160.
2. Kang, B., Ceder, G., Battery materials for ultrafast charging and discharging. *Nature* 2009, *458* (7235), 190–193.
3. Kang, K., Meng, Y. S., Bréger, J., Grey, C. P., Ceder, G., Electrodes with high power and high capacity for rechargeable lithium batteries. *Science* 2006, *311* (5763), 977.
4. Zhai, Q., Xiang, F., Cheng, F., Sun, Y., Yang, X., Lu, W., Dai, L., Recent advances in flexible/stretchable batteries and integrated devices. *Energy Storage Materials* 2020, *33*, 116–138.
5. Ji, L., Lin, Z., Alcoutlabi, M., Zhang, X., Recent developments in nanostructured anode materials for rechargeable lithium-ion batteries. *Energy & Environmental Science* 2011, *4* (8), 2682–2699.
6. Armand, M., Tarascon, J. M., Building better batteries. *Nature* 2008, *451* (7179), 652–657.
7. Goodenough, J. B., Park, K.-S., The Li-ion rechargeable battery: a perspective. *Journal of the American Chemical Society* 2013, *135* (4), 1167–1176.
8. Lin, H., Weng, W., Ren, J., Qiu, L., Zhang, Z., Chen, P., Chen, X., Deng, J., Wang, Y., Peng, H., Twisted aligned carbon nanotube/silicon composite fiber anode for flexible wire-shaped lithium-ion battery. *Advanced Materials* 2014, *26* (8), 1217–1222.
9. Lui, G., Li, G., Wang, X., Jiang, G., Lin, E., Fowler, M., Yu, A., Chen, Z., Flexible, three-dimensional ordered macroporous TiO2 electrode with enhanced electrode–electrolyte interaction in high-power Li-ion batteries. *Nano Energy* 2016, *24*, 72–77.
10. Zhou, G., Li, F., Cheng, H.-M., Progress in flexible lithium batteries and future prospects. *Energy & Environmental Science* 2014, *7* (4), 1307–1338.
11. Gaikwad, A. M., Whiting, G. L., Steingart, D. A., Arias, A. C., Highly flexible, printed alkaline batteries based on mesh-embedded electrodes. *Advanced Materials* 2011, *23* (29), 3251–3255.
12. Tebyetekerwa, M., Xu, Z., Li, W., Wang, X., Marriam, I., Peng, S., Ramakrishna, S., Yang, S., Zhu, M., Surface self-assembly of functional electroactive nanofibers on textile yarns as a facile approach toward super flexible energy storage. *ACS Applied Energy Materials* 2018, *1* (2), 377–386.
13. Lee, J. M., Choi, C., Kim, J. H., de Andrade, M. J., Baughman, R. H., Kim, S. J., Biscrolled carbon nanotube yarn structured silver-zinc battery. *Scientific Reports* 2018, *8* (1), 1–8.
14. Zhang, J., Wang, F., Shenoy, V. B., Tang, M., Lou, J., Towards controlled synthesis of 2D crystals by chemical vapor deposition (CVD). *Materials Today* 2020, *40*, 132–139.
15. Polsen, E. S., McNerny, D. Q., Viswanath, B., Pattinson, S. W., John Hart, A., High-speed roll-to-roll manufacturing of graphene using a concentric tube CVD reactor. *Scientific Reports* 2015, *5* (1), 10257.
16. Wang, C., Wang, X., Lin, C., Zhao, X. S., Lithium titanate cuboid arrays grown on carbon fiber cloth for high-rate flexible lithium-ion batteries. *Small* 2019, *15* (42), 1902183.
17. Zhang, H., Ning, H., Busbee, J., Shen, Z., Kiggins, C., Hua, Y., Eaves, J., Davis, J., Shi, T., Shao, Y.-T., Electroplating lithium transition metal oxides. *Science Advances* 2017, *3* (5), e1602427.
18. Zeng, Y., Zhang, X., Qin, R., Liu, X., Fang, P., Zheng, D., Tong, Y., Lu, X., Dendrite-free zinc deposition induced by multifunctional CNT frameworks for stable flexible Zn-Ion batteries. *Advanced Materials* 2019, *31* (36), 1903675.

19. Meng, W., Chen, W., Zhao, L., Huang, Y., Zhu, M., Huang, Y., Fu, Y., Geng, F., Yu, J., Chen, X., Porous Fe3O4/carbon composite electrode material prepared from metal-organic framework template and effect of temperature on its capacitance. *Nano Energy* 2014, *8*, 133–140.

20. Wang, C., Wang, X., Lin, C., Zhao, X. S., Lithium titanate cuboid arrays grown on carbon fiber cloth for high-rate flexible lithium-ion batteries. *Small* 2019, *15* (42), 1902183.

21. Pu, J., Shen, Z., Zhong, C., Zhou, Q., Liu, J., Zhu, J., Zhang, H., Electrodeposition technologies for Li-based batteries: new frontiers of energy storage. *Advanced Materials* 2020, *32* (27), 1903808.

22. Tebyetekerwa, M., Ramakrishna, S., What is next for electrospinning? *Matter* 2020, *2* (2), 279–283.

23. Shao, W., Tebyetekerwa, M., Marriam, I., Li, W., Wu, Y., Peng, S., Ramakrishna, S., Yang, S., Zhu, M., Polyester@ MXene nanofibers-based yarn electrodes. *Journal of Power Sources* 2018, *396*, 683–690.

24. Marriam, I., Wang, X. P., Tebyetekerwa, M., Chen, G. Y., Zabihi, F., Pionteck, J., Peng, S. J., Ramakrishna, S., Yang, S. Y., Zhu, M. F., A bottom-up approach to design wearable and stretchable smart fibers with organic vapor sensing behaviors and energy storage properties. *Journal of Materials Chemistry A* 2018, *6* (28), 13633–13643.

25. Hatter, C. B., Sarycheva, A., Levitt, A., Anasori, B., Nataraj, L., Gogotsi, Y., Electrically conductive MXene-coated glass fibers for damage monitoring in fiber-reinforced composites. *C* 2020, *6* (4), 64.

26. Wang, X., Liu, B., Xiang, Q., Wang, Q., Hou, X., Chen, D., Shen, G., Spray-painted binder-free SnSe electrodes for high-performance energy-storage devices. *ChemSusChem* 2014, *7* (1), 308–313.

27. Choi, C., Kim, K. M., Kim, K. J., Lepró, X., Spinks, G. M., Baughman, R. H., Kim, S. J., Improvement of system capacitance via weavable superelastic biscrolled yarn supercapacitors. *Nature Communications* 2016, *7* (1), 1–8.

28. Ramakrishna, S., *An introduction to electrospinning and nanofibers*. World Scientific: 2005.

29. Thavasi, V., Singh, G., Ramakrishna, S., Electrospun nanofibers in energy and environmental applications. *Energy & Environmental Science* 2008, *1* (2), 205–221.

30. Marriam, I., Wang, X., Tebyetekerwa, M., Chen, G., Zabihi, F., Pionteck, J., Peng, S., Ramakrishna, S., Yang, S., Zhu, M., A bottom-up approach to design wearable and stretchable smart fibers with organic vapor sensing behaviors and energy storage properties. *Journal of Materials Chemistry A* 2018, *6* (28), 13633–13643.

31. Jung, J.-W., Lee, C.-L., Yu, S., Kim, I.-D., Electrospun nanofibers as a platform for advanced secondary batteries: a comprehensive review. *Journal of Materials Chemistry A* 2016, *4* (3), 703–750.

32. Brinker, C. J., Dip coating. In *Chemical solution deposition of functional oxide thin films*, Springer: 2013; pp 233–261.

33. Zhu, H.-W., Ge, J., Peng, Y.-C., Zhao, H.-Y., Shi, L.-A., Yu, S.-H., Dip-coating processed sponge-based electrodes for stretchable Zn-MnO2 batteries. *Nano Research* 2018, *11* (3), 1554–1562.

34. Park, S.-K., Seong, C.-Y., Piao, Y., A Simple Dip-coating Approach for Preparation of Three-dimensional multilayered graphene-metal oxides hybrid nanostructures as high performance lithium-ion battery electrodes. *Electrochimica Acta* 2015, *176*, 1182–1190.

35. Marriam, I., Xu, F., Tebyetekerwa, M., Gao, Y., Liu, W., Liu, X., Qiu, Y., Synergistic effect of CNT films impregnated with CNT modified epoxy solution towards boosted interfacial bonding and functional properties of the composites. *Compos Part a-Appl S* 2018, *110*, 1–10.

36. Lee, S. H., Huang, C., Grant, P. S., High energy lithium ion capacitors using hybrid cathodes comprising electrical double layer and intercalation host multi-layers. *Energy Storage Materials* 2020, *33*, 408–415.

37. Praveen, S., Santhoshkumar, P., Joe, Y. C., Senthil, C., Lee, C. W., 3D-printed architecture of Li-ion batteries and its applications to smart wearable electronic devices. *Applied Materials Today* 2020, *20*, 100688.

38. Shuakat, M. N., Lin, T., Highly-twisted, continuous nanofibre yarns prepared by a hybrid needle-needleless electrospinning technique. *Rsc Adv* 2015, *5* (43), 33930–33937.

39. Wang, X., Meng, S., Tebyetekerwa, M., Li, Y., Pionteck, J., Sun, B., Qin, Z., Zhu, M., Highly sensitive and stretchable piezoresistive strain sensor based on conductive poly (styrene-butadiene-styrene)/few layer graphene composite fiber. *Compos Part a-Appl S* 2018, *105*, 291–299.

40. Xu, Z., Gao, C., Graphene fiber: a new trend in carbon fibers. *Materials Today* 2015, *18* (9), 480–492.

41. Lescanne, Y. WHITEPAPER – An introduction to Metal Fiber Technology 2017. https://www.bekaert.com/en/product-catalog/content/Metal-fibers/replacement-of-glass-fiber-media-by-metal-fiber-media (accessed 8May 2021).

42. Wang, Y., Chen, C., Xie, H., Gao, T., Yao, Y., Pastel, G., Han, X., Li, Y., Zhao, J., Fu, K., 3D-printed all-fiber li-ion battery toward wearable energy storage. *Advanced Functional Materials* 2017, *27* (43), 1703140.

43. Yarin, A., Zussman, E., Upward needleless electrospinning of multiple nanofibers. *Polymer* 2004, *45* (9), 2977–2980.

44. Jalili, R., Razal, J. M., Innis, P. C., Wallace, G. G., One-Step Wet-Spinning Process of Poly(3,4-ethylenedioxythiophene):Poly(styrenesulfonate) Fibers and the Origin of Higher Electrical Conductivity. *Advanced Functional Materials* 2011, *21* (17), 3363–3370.

45. Garcia-Torres, J., Roberts, A. J., Slade, R. C. T., Crean, C., One-step wet-spinning process of CB/CNT/MnO2 nanotubes hybrid flexible fibres as electrodes for wearable supercapacitors. *Electrochimica Acta* 2019, *296*, 481–490.

46. Zhang, Y.-h., Liu, Z.-c., Li, B.-w., Ma, Z.-h., Guo, S.-h., Wang, X.-l., Structure and electrochemical performances of Mg2Ni1−xMnx (x=0–0.4) electrode alloys prepared by melt spinning. *Electrochimica Acta* 2010, *56* (1), 427–434.

47. Lipson, H., Kurman, M., *Fabricated: The new world of 3D printing*. John Wiley & Sons: 2013.

48. Wang, Y., Chen, C., Xie, H., Gao, T., Yao, Y., Pastel, G., Han, X., Li, Y., Zhao, J., Fu, K., Hu, L., 3D-printed all-fiber Li-Ion battery toward wearable energy storage. *Advanced Functional Materials* 2017, *27* (43), 1703140.

49. Mo, F., Liang, G., Huang, Z., Li, H., Wang, D., Zhi, C., An overview of fiber-shaped batteries with a focus on multifunctionality, scalability, and technical difficulties. *Advanced Materials* 2020, *32* (5), 1902151.

50. Weng, W., Sun, Q., Zhang, Y., Lin, H., Ren, J., Lu, X., Wang, M., Peng, H., Winding aligned carbon nanotube composite yarns into coaxial fiber full batteries with high performances. *Nano Letters* 2014, *14* (6), 3432–3438.

51. Song, H., Jeon, S.-Y., Jeong, Y., Fabrication of a coaxial high performance fiber lithium-ion battery supported by a cotton yarn electrolyte reservoir. *Carbon* 2019, *147*, 441–450.

52. Liu, L., Zhu, M., Huang, S., Lu, X., Zhang, L., Li, Y., Wang, S., Liu, L., Weng, Q., Schmidt, O. G., Artificial electrode interfaces enable stable operation of freestanding anodes for high-performance flexible lithium ion batteries. *J Mater Chem A* 2019, *7* (23), 14097–14107.

53. Zhang, Y., Bai, W., Ren, J., Weng, W., Lin, H., Zhang, Z., Peng, H., Super-stretchy lithium-ion battery based on carbon nanotube fiber. *J Mater Chem A* 2014, *2* (29), 11054–11059.

54. Sun, H., Zhang, Y., Zhang, J., Sun, X., Peng, H., Energy harvesting and storage in 1D devices. *Nature Reviews Materials* 2017, *2* (6), 1–12.

55. Fu, Y., Cai, X., Wu, H., Lv, Z., Hou, S., Peng, M., Yu, X., Zou, D., Fiber supercapacitors utilizing pen ink for flexible/wearable energy storage. *Advanced Materials* 2012, *24* (42), 5713–5718.

56. Zhang, Y., Jiao, Y., Lu, L., Wang, L., Chen, T., Peng, H., An ultraflexible silicon–oxygen battery fiber with high energy density. *Angewandte Chemie International Edition* 2017, *56* (44), 13741–13746.

57. Zhao, Z., Yan, C., Liu, Z., Fu, X., Peng, L.-M., Hu, Y., Zheng, Z., Machine-washable textile triboelectric nanogenerators for effective human respiratory monitoring through loom weaving of metallic yarns. *Advanced Materials* 2016, *28* (46), 10267–10274.

58. Huang, Y., Hu, H., Huang, Y., Zhu, M., Meng, W., Liu, C., Pei, Z., Hao, C., Wang, Z., Zhi, C., From industrially weavable and knittable highly conductive yarns to large wearable energy storage textiles. *ACS Nano* 2015, *9* (5), 4766–4775.

59. Kwon, Y. H., Woo, S.-W., Jung, H.-R., Yu, H. K., Kim, K., Oh, B. H., Ahn, S., Lee, S.-Y., Song, S.-W., Cho, J., Shin, H.-C., Kim, J. Y., Cable-type flexible lithium ion battery based on hollow multi-helix electrodes. *Advanced Materials* 2012, *24* (38), 5192–5197.

23 Carbon-Based Advanced Flexible Supercapacitors

Anuj Kumar
Nano-Technology Research Laboratory, Department of Chemistry,
GLA University, Mathura, Uttar Pradesh, India

Ram K. Gupta
Department of Chemistry, Kansas Polymer Research Center,
Pittsburg State University, Pittsburg, Kansas, USA

CONTENTS

23.1 Introduction...417
23.2 Carbon-based materials for FSCs 419
    23.2.1 Graphene...420
    23.2.2 Carbon nanotubes 423
    23.2.3 Bio-based carbon 423
23.3 Synthesis of carbon-based materials...423
23.4 Mechanisms of energy storage in supercapacitors..425
    23.4.1 Electrochemical double-layer capacitors 426
    23.4.2 Pseudo-capacitors 428
    23.4.3 Hybrid capacitors 429
23.5 Carbon-based flexible supercapacitors..430
    23.5.1 Graphene-based flexible supercapacitors..430
    23.5.2 CNT-based flexible supercapacitors 432
    23.5.3 Bio-based FSCs 432
23.6 Conclusion..434
References...436

23.1 INTRODUCTION

Nowadays, because energy demand is increasing, efficient energy storage is one of the most serious challenges for researchers working in the field of energy sectors. Waste is also increasing with the demand. With the powerful devices and technologies working on the sole mechanism of energy (in many forms), it has become essential to not just prevent energy wastage, but also to store energy for future innovations. In this regard, energy storage devices play a crucial role in helping the researchers and developers of electronic devices to store energy in the most advanced and convenient ways possible. These storage methods can be classified as long-term, such as battery storage, compressed air energy storage, hydrogen storage, and as short-term, such as supercapacitors, flywheel, and inductor systems, as illustrated in Figure 23.1. Considering the many desirable properties of advanced energy storage systems, one of the prominent types is electrochemical energy storage devices. These devices exhibit high energy power densities, longer discharge time, high power range, and high cycle efficiency, all of which are needed to successfully store energy for longer durations [2]. Due to these unique properties, these devices can be used for efficient energy management and applications in advanced electronic systems,

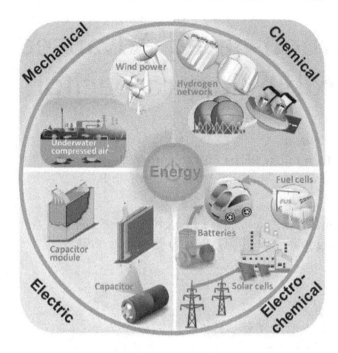

FIGURE 23.1   For stationary uses and classification of several types of energy storage technologies. Adapted with permission from ref. [1]. Copyright 2018, The Authors. Distributed under a Creative Commons Attribution License 4.0 (CC BY).

including flexible energy storage devices. They can store power ranging from a few watts to many megawatts [3].

There are two main ways to store energy when it comes to energy storage equipment and processes: batteries and supercapacitors. These two have their unique mechanisms and potential to store energy and enhance the power of the devices. The major differences between batteries and supercapacitors are their power and energy capacities, along with operating life. For instance, batteries are best for showing high energy densities with durability performance, while supercapacitors project more power density, which is enough for a shorter yet robust use. Similarly, batteries possess a superior breakdown voltage, whereas supercapacitors are more like portable devices, having a thriving power capacity that can run stronger devices. Apart from these differences, batteries and supercapacitors also differ in longevity and life span. In this sense, batteries may tire out soon since they rely on charge/discharge chemical reactions in the anodes and cathodes. On the other hand, supercapacitors do not rely on these chemical reactions; hence, they have longer durability. Batteries also have a longer charging time, but they also last longer. On the contrary, supercapacitors charge faster but they have a slightly higher discharge rate. Finally, there is a difference in cost between the two devices. Batteries are more expensive due to their anode and cathode materials, while supercapacitors are cost-friendly, which is why they are favored by manufacturers and innovators. Due to this accessibility, they have applications in advanced systems like electronic transportation and hybrid technologies.

In supercapacitor technology, flexible supercapacitors (FSCs) are the most advanced systems of energy storage in electronic devices. FSCs are lightweight and have high energy storage capacity. As a result of the increasing power density, good charge/discharge rate, and extended life duration, they are regarded promising devices for innovative applications. Also, their flexible nature makes them applicable to many advanced wearable devices. Due to these desirable aspects, FSCs are gaining considerable commercial attention recently. This is because flexible electronics have a huge market

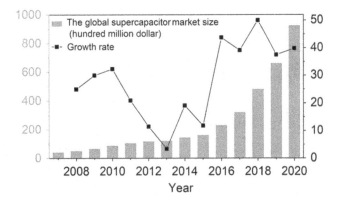

FIGURE 23.2 The global supercapacitor market size and the corresponding growth rate. Adapted with permission from ref. [6]. Copyright 2019, The Authors. Distributed under a Creative Commons Attribution License 4.0 (CC BY).

application, especially, in consumer goods. On the contrary, the conventional supercapacitors are not very convenient for lightweight devices; hence, FSCs make things easier by adjusting to any shape, size, weight, and design diversity of devices. Some of the prominent mechanical advantages of flexible supercapacitors are small size, bendability, high-temperature tolerance, and easily processbaility. They are used in many advanced technologies like electric vehicles, hybrid cars, industrial power systems, memory back-ups, portable devices like mobile phones, wearable electronics, etc. [4]. Moreover, the unique capability of FSCs to endure high-temperature makes them stand out among other energy storage devices. According to a study, FSCs can endure up to 200°C when using reduced graphene oxide (rGO) and a clay composite electrolyte [5]. Hence, with these advantages, FSCs have become more prominent in use and demand. Their demand and growth in development are increasing in the market, as seen in Figure 23.2.

The ferrites, metal oxides, transition metal sulfides, conducting polymers, and carbon-based materials are commonly used as electrode materials in supercapacitors. These materials show high performance when incorporated in the supercapacitors as electrode materials. They have high specific capacities and high electrode penetration rates [7]. In particular, carbonaceous materials are favored as electrode materials and supercapacitors for their high specific capacitance and high specific surface area. Carbonaceous materials for supercapacitors have various manifestations, such as activated carbon (AC), carbide-derived carbon (CDC), carbon fiber cloth, carbon wires, carbon aerogel, carbon nanotubes (CNTs), and graphite (graphene), as shown in Figure 23.3. All these materials are critical and applicable in almost all advanced electronic systems. These materials exhibit excellent mechanical and chemical properties, which increase the performance of supercapacitors, such as high capacitance, high electrical conductivity, increased chemical and electrical stability, and low costs [9]. These can be symmetric or asymmetric, porous, or solid-state materials. They are physically and mechanically efficient and possess longer durability than other materials. Due to these properties, carbonaceous electrode materials are favored for use in supercapacitors, which are installed in advanced electronic systems.

23.2 CARBON-BASED MATERIALS FOR FSCS

Due to high electrical and mechanical properties and rich redox chemistry, nano carbons like CNTs and graphene sheets, in particular, have been utilized to fabricate FSCs. This study intends to give a detailed review of a wide range of nanocarbon-based stretchable, transparent, and integrated FSCs, as shown in Figure 23.4 [18–21].

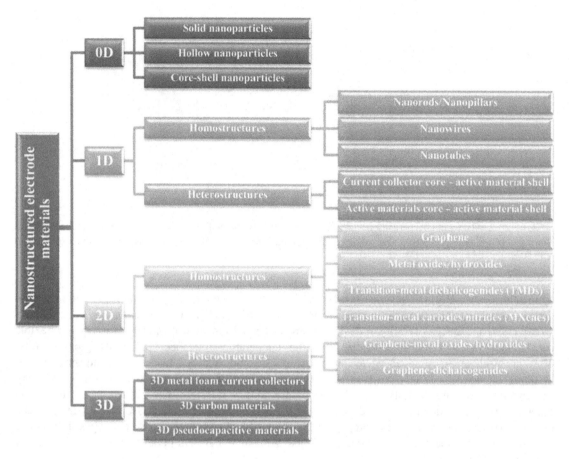

FIGURE 23.3   A schematic representation of nanostructured electrode materials summarizing their key aspects from 0 to 3D. Adapted with permission from ref. [8]. Copyright 2014, The Authors. Distributed under a Creative Commons Attribution License 4.0 (CC BY).

23.2.1 GRAPHENE

Graphene is a 2D regular honeycomb lattice structure made up of monolayers of the hexatomic ring of carbon and belongs to a vast family of 2D carbon nanostructures [22,23]. Due to its remarkable features in terms of excellent thermal and electrical conductivity, mechanical behavior, and large specific surface area (SSA), it has been created as a novel but a competitive candidate for employment in energy storage tools for electrochemical processes [24,25]. Using graphene as electrode material in supercapacitors, productive research has so far confirmed its potential application. Capacitive products of graphene depend solely on the number of graphene layers [26], which decide the active SSA of prepared electrodes for the electrochemical reactions. Conceptually, the parallel layers might offer huge pathways by which electrolyte ions can be smoothly transported along with small diffusion resistance on the surface of each graphene layer. Graphene-based electrodes provide an enhanced energy density due to their super conductivity. Moreover, due to the strong cohesiveness of the Vander Waals as well as π-π stacking forces, graphene layers have a strong tendency toward re-stacking of layers, resulting in the reduction of accessible SSA for ionic/ electronic transportation [27,28].

It is therefore vital to discover the strategies to minimize such adverse impacts. Until now, several methods have been studied to avoid the aggregation of graphene layers to increase the SSA,

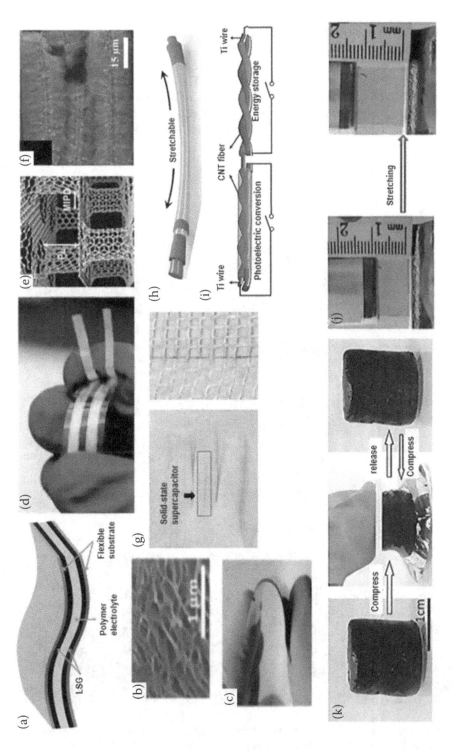

FIGURE 23.4 Illustration of different nano-carbons based bendable solid-state supercapacitors having different shapes and morphologies: graphene-based supercapacitor (a–c). Reprinted with permission [10]. Copyright 2021, AAAS. (d) Supercapacitor fabricated using ultrathin graphene. Reprinted from ref. [11] with Copyright 2021, ACS publishing group. (e and f) Nano-carbons-based 3D-architectured supercapacitors. Reprinted with permission [12]. Copyright 2021, ACS publishing group. (g) Supercapacitor comprising of wearable fiber. Reprinted with permission [13]. Copyright 2021, Nature Publishing Group. (h) flexible fiber-shaped supercapacitor. Reproduced with permission [14]. Copyright 2021, John Wiley and Sons publishing group. (i) A wire with interconnected conversion and storage systems. Reproduced with permission [15]. Copyrights 2021, John Wiley and Sons publishing group. (j) Supercapacitor comprising of transparent and flexible graphene. Reproduced with permission [16]. Copyrights 2021, ACS publishing group. (k) Compressible supercapacitor comprising of graphene-based 3D composite. Reproduced with permission [17]. Copyright 2021, John Wiley and Sons publishing group.

minimize the inner resistance, and allow the electrolyte disbursement in the supercapacitors [27]. As Liu et al., [28] reported a laser-based strategy for the fabrication of Gr/CNT based flexible micro-supercapacitors. In this strategy, to avoid the re-stacking of graphene-layers, CNTs acted as spacers to generate an accessible surface for the ionic movement. On the other hand, using an in situ microwave irradiation strategy, Oh and coworkers [29] created a graphene-coated and Co_3O_4-inserted hybrid (Figure 23.5a). In addition, graphene layers may serve as effective conductive routes between active materials and current collectors. The electrochemical performance was excellent due to the integration of functional Co_3O_4 nanoparticles onto the conductive graphene layers. Similarly, graphene nanosheets vertically oriented with Co_3O_4 nanoparticles supported by carbon fabric were effectively produced with outstanding electrochemical performance (Figure 23.5b–d) [27].

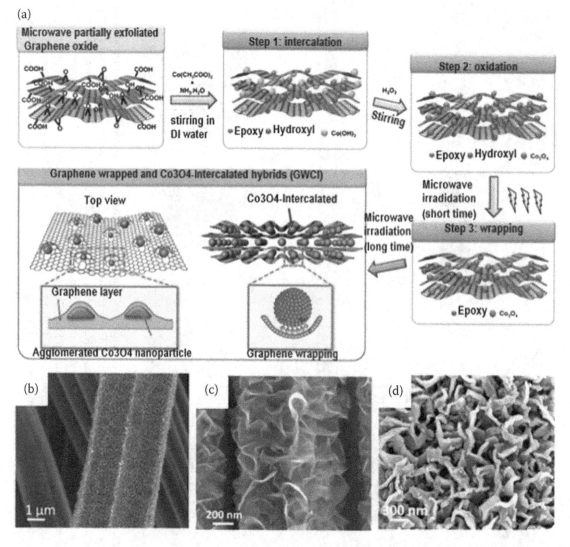

FIGURE 23.5 (a) A three-step procedure for creating a GWCI hybrid is depicted schematically. (b) SEM pictures of the VAGN at low resolution. (c) VAGN high-resolution SEM pictures, and (d) Co3O4-H high-resolution SEM images. Reprinted with permission [27,29,30]. Copyright 2021, ACS and Elsevier publishing group.

23.2.2 CARBON NANOTUBES

The CNTs are 1D carbon nanostructures made up of pure sp^2-bonded hexagonal carbon grids folded into a tubular structure [31,32]. CNTs are being extensively researched as electrodes or active materials in supercapacitors due to their excellent structural stability, robust conductivity, and mechanical properties. These characteristics authorize them to be favorable contenders for the modification and inclusion of different active materials [33]. Huge attempts have been devoted to manipulating CNTs for the development of high-functionality CNT-based active electrode materials. An effective technique is to spread CNTs through dry or wet spinning operations into subsequent fibers. Chen et al. [34] produced a hierarchically structured carbon fiber that included oriented and interconnected single-walled carbon nanotubes (SWNTs). Two parallel fiber electrodes were made of all-solid-state micro-supercapacitors with exceptional volumetric power (6.3 mWh/cm^3); this is because the reduced graphene oxide (rGO) was capable of providing a highly selective ion adsorption surface area. Moreover, the strong conductive strength of SWNTs can minimize the interlayer as well as contact resistance. The incorporation of functional O/N-heteroatoms has meanwhile boosted the surface weighting resilience of the hybrid fibers. Peng et al. [35] reported a hybrid fiber of graphene/CNTs via the addition of graphene layers to neighboring CNTs (Figure 23.6a). The graphene layers acted as "bridges" for the cargo transfer due to the strong π-π stacking force between both the graphene layers and CNTs, reducing the interaction barriers. Thus, the new composite fiber gained the requisite electrical and electric performance at the same time. Luo et al. [36] recently prepared an ultrathin zinc sulfide (ZnS)@CNT by coating hierarchical ultrathin ZnS nanosheets onto multiwall carbon nanotubes (MWCNTs), as illustrated in Figure 23.6b. Furthermore, Zhao et al. [39] described a novel strategy for fabricating ZnCo$_2$O$_4$/ZnO@MWCNT hybrids. A hydrothermal process was used to make ZnCo$_2$O$_4$/ZnO (Zn$_1$Co$_2$) coated on Ni foam, which was then used to coat MWCNTs with Zn$_1$Co$_2$ at room temperature using a dipping-drying technique (Figure 23.6c).

23.2.3 BIO-BASED CARBON

Carbon-based biomass-derived active materials have now been recognized as a viable choice because of the lower cost, diversity in morphologies, mechanical, and compositional behavior of such materials. Renewable precursors are of particular importance because they may offer a wide variety of chemical functional groups (–COOH, –NH$_2$, and –OH), which are necessary for the creation of functional carbon nanomaterials [40–43]. Several functional molecules, like cellulose, chitosan, starch, tannins, and lignins, can be derived more quickly from the biomass components than crude biomass, like shells and strips, leaves, stems, and wood. Due to high regularity and minimal self-association, biomass-determined antecedents can be utilized related to customary design coordinating strategies like aqueous carbonization, delicate and severe templated plans, and fast pyrolysis, as well as spinning.

Activated carbons have become the most widely utilized carbon material in supercapacitors to date because of their simple assembling methodology, high electrical conductivity, and plentiful porosity. However, the great mechanical strength was just accomplished after the presentation of more contemporary materials, like CNTs, conductive polymers, and graphene, which dominated the field of FSCs with a capacitance value of up to 763 F/g and more than 1,000 cycles with an 85.5% retention force (Table 23.1).

23.3 SYNTHESIS OF CARBON-BASED MATERIALS

Carbon materials such as graphene, CNTs, carbon spheres, carbon ribbons, carbon sponges, carbon nanosheets, carbon nano-cage, etc., have been produced employing several strategies, including

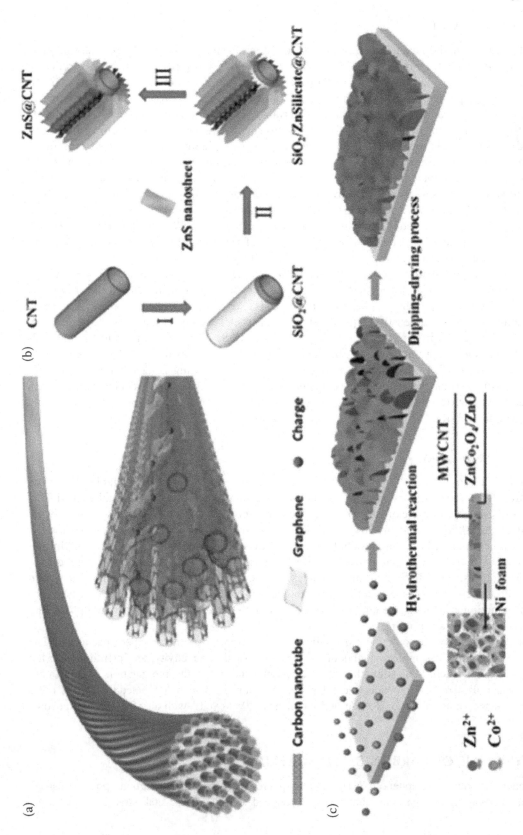

FIGURE 23.6 (a) Schematic depiction of the graphene and/or CNT composite fibre construction. (b) Diagrammatic depicting the multistep conversion method used to make ZnS@CNT composites. (c) Schematic depiction of the manufacturing procedures for a layered Zn1Co2@MWCNT composite electrode. Reprinted with permission [36–38]. Copyrights 2021, Royal society of chemistry publishing group.

TABLE 23.1

Performances of FSCs utilizing the biomass and modern non-biomass based electrodes

Raw material	Specific capacitance	Retention force (%)	Surface area ($m^2\ g^{-1}$)	Ref.
Flexible materials derived from biomass				
Cherry blossom petals	154 F/g (at 10 mV/s)	92.3 over 10,000 cycles	509	[44]
Jute fibers	51 F/g (at 1 mV/s)	100 over 5,000 cycles	1769	[45]
Bamboo fibers	1.52 F/cm^2 (at 5 mV/s)	100 over 5,000 cycles	1120	[46]
Ficus religiosa leaves	3.4 F/g (at 1 mA/g)	88 over 4,000 cycles	157	[47]
Bacterial cellulose	0.289 F/cm^2 (at 0.1 mA/cm^2)	66.7 over 100 cycles	490	[48]

hydrothermal carbonization, pyrolysis, chemical vapor reaction, ball milling, etc. [49–52]. Pyrolysis is a thermochemical process in which materials are degraded at high temperatures in an inert environment like Ar or N_2, as shown in Figure 23.7. Carbon precursors can decompose at high temperatures into tenuous volatile compounds and exit the oven, leaving carbon residue and potentially inorganic cinders in the utensil. Pores in carbon are formed when the carbon porous structure is combined with the escape of volatile molecules. Pyrolysis is a common and easy way to make a variety of useful carbon compounds [54,55].

The graphene, graphene-based nanoribbons, CNTs, and other high-quality carbon compounds are commonly produced via chemical vapor deposition (CVD) [56,57]. To make the necessary deposition, carbon raw materials are poured over the surface of transition metals or dielectric substrates and are allowed to react, as illustrated in Figure 23.7. In addition to the carbon precursor, the temperature of deposition, carrier gas flow rate, chamber pressure, and other variables have a significant influence on the physicochemical behavior of the generated nanocarbons [58]. To transform carbohydrates, organic compounds, or biomass into carbon materials, hydrothermal carbonization (HTC) mimics the organic phenomenon of coal creation. HTC reactions take place in an aqueous media at temperatures exceeding 100°C and under elevated pressure (Figure 23.7) [59]. HTC can be divided into two categories based on the varied working temperatures; (i) HTC at high temperatures such as 300–800°C, which permits the production of carbon nanomaterials with a high degree of graphitization, like CNTs or activated carbon, among other things; and (ii) HTC at minimal temperatures (less than 300°C) [60].

Further, ball milling was invented to make samples smaller and more controllable. This strategy is often used to make nanostructured materials to alter the reaction rate of grinded particles, to start chemical reactions, and to change the state of materials, including amorphization, polymorphic transformations [61,62]. This method is an appropriate strategy to offer low installation and grinding medium cost. The mill's balls (steel or ceramic) cascade with the rotation of the grinding jars (Figure 23.7), and particle mixes are subjected to huge energy impacts by the mill's balls (steel/ceramic).

23.4 MECHANISMS OF ENERGY STORAGE IN SUPERCAPACITORS

Energy storage mechanisms for supercapacitors can be explained based on the following three capacitive behaviors: (i) at an electrode interface, electrochemical double-layer capacitors (EDLCs) consume authentic electrical power, (ii) high-pressure and high-reversion surface redox phenomena create pseudo-capacitance (PC), and (iii) hybrid capacitors involve the above-mentioned phenomena.

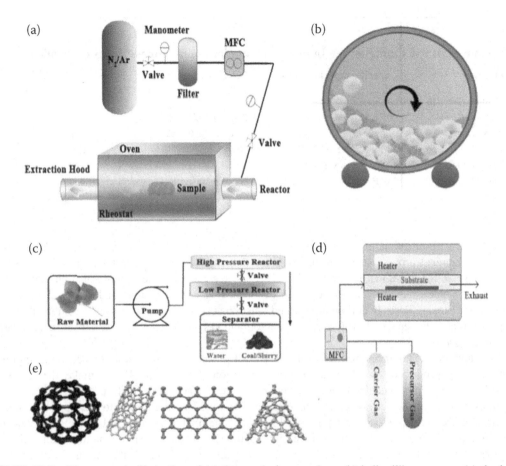

FIGURE 23.7 Diagrammatic illustration of (a) the pyrolysis procedure; (b) ball-milling process; (c) the hydro-thermal process; (d) the chemical vapor deposition process; (e) fullerene, carbon nanotube, graphene nanosheets, and carbon nanocone are examples of carbon-based nanomaterials, from left to right. Reprinted with permission [53]. Copyrights 2021, Elsevier publishing group.

23.4.1 Electrochemical Double-Layer Capacitors

EDLCs offer significant energy density as compared to traditional capacitors because of their larger surface area and small load separation distance. EDLCs use electrolytes like KOH, H_2SO_4, or Na_2CO_3 between electrodes instead of the dielectric material used by traditional capacitors. Various physical factors, including the shape of these two charging plates [63], restrict the performance of conventional capacitors, most notably, of the insufficient storage load. Helmholtz first conceptualized and formulated the theory of EDLCs via investigations into the opposing charges of colloidal particle surfaces in the 19th century, as shown in Figure 23.8a. The proposed concept suggested that the adverse charges are laid on the electrode/electrolyte contact point and divided by atomic distance, similar to the traditional double-plate capacitors. The Helmholtz model was changed into the diffuse layer model by Gouy and Chapman (Figure 23.8b) [64,65]. With this distance, the capacity for two distinct charging arrays should rise reciprocally.

In general, EDLCs are supposed to act in the same capacitation way as capacitors on the parallel plate [66]. Thus, as they approach the electrode surface, a huge capacitance with point charges would increase. Both zones of particle distribution have an interior area known as a compact layer and a diffuse layer (Figure 23.8c), which must be recognized. Stern created a model that summed up the Helmholtz and

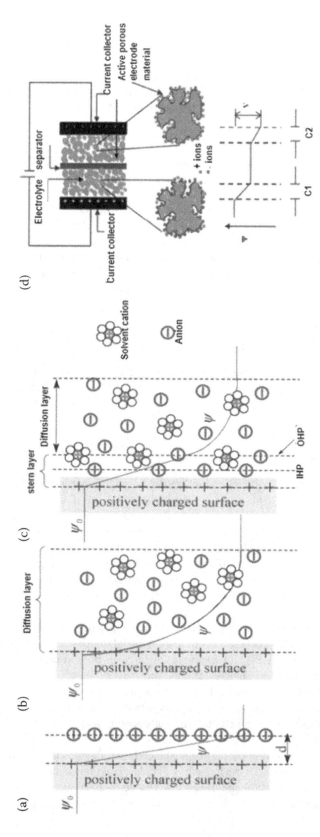

FIGURE 23.8 The Helmholtz model (a), the Gouy–Chapman model (b), and the Stern model (c), respectively shown, are three versions for the electrical double layer at a positively charged surface. (d) An EDLC composed of porous electrode materials is depicted in this diagram. Reprinted with permission [63]. Copyrights 2021, The Royal Society of Chemistry.

Gouy-Chapman models. In the Stern layer, the ions (which are usually hydrated) are a good adsorbent for the conductors. Additionally, the compact layer adsorbs particular adsorbing ions, as well as non-specific counter ions. The Helmholtz internal and exterior planes distinguish the two types of adsorbed ions. According to the Gouy-Chapman model, the diffuse layer area is explainable because the kinetic energy of the counter ions leads to a diffuse dual-thickening layer. The EDLC capacity (C_{dl}) is compatible with the Stern capability (CH), and the diffused region capacity (C_{diff}). C_{dl} can be written mathematically as:

$$\frac{1}{Cdl} = \frac{1}{H} + \frac{1}{Cdiff}$$ (23.1)

The EDLC's electrical conduction on the planar surface of the electrode is controlled throughout the electrode and electrolyte by the electrical field, as well as chemical affinities. Factors including the affinity of the tortuous mass transfer route, pores' surfaces, solution-related electrical phenomena, and solution-related pores' surface play a crucial role in particle transport in the limited system. The mechanism for EDLCs employing porous carbon conduction materials is shown in Figure 23.8d schematically by de Todes. EDLCs can be estimated with the specific capacitance:

$$C = \frac{\varepsilon r \varepsilon_0 \times A}{d}$$ (23.2)

The dielectric electrolyte constant, the permittivity of vacuum, the specific surface area of the electrode obtainable to ionic elements of the electrolyte, and the efficiency EDL (Debye length) correspondingly describe the daily dielectric constant of the electrolyte. The close link between the specified capability (C) and the available capability (A) is relevant in Eq. (23.2).

23.4.2 PSEUDO-CAPACITORS

The mechanism of pseudo-capacitors is dependent on the Faradaic charge technique, which can be reversible or irreversible (transport of electrons throughout the interface of an electrode by the redox reaction of a chemical species). During chemical redox reactions, the reversible process generates no new chemical entities, whereas the irreversible Faradaic process produces new species. Pseudo-capacitive materials, in particular, exhibit the same response as double-layer capacitors. The pseudo-capacitor is fundamentally reliant on the size of particles and morphologies. Extrinsic pseudo-capacitors can only be accessed via nanostructured surfaces. The same material does not display pseudo-capacitance farther into the bulk due to phase changes throughout the ion storage procedure [67].

Electroanalytical studies, on either side, give a method for distinguishing between pseudo-capacitance and battery-type materials. Redox processes of transition metal oxides and reversible doping and de-doping in conducting polymers (CPs), for example, distinguish pseudo-capacitive electrodes. The most common materials used to form such electrodes are carbon, metal oxides, and CPs [68,69]. The capacitance of a pseudo-capacitor can be 10–100 times more than that of an EDLC. The power performance of a pseudo-capacitor, on the contrary, is worse than that of an EDLC because the slower Faradaic mechanism shrinks and swells throughout charging and discharging, resulting in poor life cyclic and mechanical strength.

The majority of the electric charge in pseudo-capacitance-based devices is delivered to the bulk or surface next to the solid electrode surfaces. As a result, a Faradaic reaction occurs between the electrolyte and the solid material, reflecting the voltage-dependent charge-transfer mechanism. As a result, the pseudo-capacitors are voltage dependent. Surface adsorption of electrolyte ions, redox reactions involving electrolyte ions, and doping/de-doping of pseudo-capacitance materials in the electrode are all electrochemical processes that are frequently used in the production of ultra-capacitors that use pseudo-capacitors. Because the first two processes rely heavily on surface mechanics, the area of the electrode materials significantly affects performance. The fourth approach, which uses pseudo-capacitive materials in the electrode, is a bulk procedure; thus, capacitance is unaffected by surface area. For the distribution and collection of electric current, the electrode materials must have maximal electronic conductivity in all circumstances.

23.4.3 HYBRID CAPACITORS

Supercapacitors are categorized as either EDLCs or pseudo-capacitors and are a combination of capacitors and batteries. The primary distinction between the EDLCs-based charge storage methods is that these processes do not involve chemical reactions on the electrode surface, but charge accumulates at the electrode/electrolyte interface. Besides, pseudo-capacitors possess high capacitance due to redox reactions that occur at the electrode surface. Two elements determine the energy density of supercapacitors: electrode capacitance and cell voltage. Develop nanosized and porous electrode materials with increased energy density as a viable way of enhancing their capacitance. Another technique is to construct asymmetric/hybrid supercapacitors with two types of electrodes that may use the potential gap between the two types of electrodes to increase total cell voltage. For example, positive and negative electrodes are constructed of pseudo-capacitors and EDLC, respectively.

The possibility of transition metal chalcogenides (TMCs), such as nitrides, metal oxides, carbides, and sulfides, as supercapacitors has just lately been studied. They can give greater specific capacitances than carbon electrodes. Binitha et al. [70] utilized an electro-spinning technique to create porous-Fe_2O_3 fiber and nanograin structures with electrochemical characteristics in 1 M LiOH but capacitances of 256 F/g and 102 F/g, respectively. Co-based oxides/hydroxides, on either side, have attracted people's curiosity because of their potential usage in supercapacitors. The capacitance of supercapacitors constructed with porous $Co(OH)_2$ was 609 F/g [71]. The use of a suitable electrode material with a large surface area that facilitates electron and ion mobility is required for the creation of a battery-like supercapacitor. Several strategic developments have been proposed in this regard, including (1) increasing electrode conductivity [72]; (2) using nano-architectures for electrodes to reduce diffusional pathways while achieving higher surface areas [73]; (3) evolving 3D materials to accelerate ionic diffusion in multiple directions (i.e., porous structures) [74]; and (4) lowering diffusional activation.

To meet this requirement, researchers have looked at a wide range of carbon-based materials with sp^2 and sp^3 atomic orbital hybridizations, varied allotropic forms (e.g., diamond, fullerenes, graphene, nanotubes, and so on), and various dimensionalities (0D–3D) [75–78]. Organic binders, such as polytetrafluoroethylene, were employed as pressed or coated electrode supports and coupled with carbon-based materials for the majority of the supercapacitors tested in these procedures. Due to the poor properties of the binders, electronic transport inside electrodes and ionic species diffusion on electrode surfaces are frequently hampered (e.g., low conductivity and relatively poor stability). Similarly, manufacturing battery-like supercapacitors require a new synthesis of binder-free carbon electrodes.

The choice of electrolyte is another key challenge in the construction of battery-like supercapacitors. EDLCs typically utilize inert electrolytes, whereas pseudocapacitors prefer redox species electrolytes covered with metal oxides and polymers on conductive substrates. Modifying the electrolytes by adding soluble redox species is another effective method for producing pseudocapacitors-based supercapacitors. The number of redox species employed in this novel method is an important factor in determining the charge storage capacity, which can be readily controlled. The strategy's impact on supercapacitors' overall performance, on the other hand, remains uncertain.

23.5 CARBON-BASED FLEXIBLE SUPERCAPACITORS

Two of the most popular carbon allotropes utilized in the storage of electrochemical energy are CNTs and graphene. High-performance FSCs with planar topologies have been created using CNT, graphene, and hybrid electrodes [79,80].

23.5.1 GRAPHENE-BASED FLEXIBLE SUPERCAPACITORS

Generally, CNTs with a huge SSA, strong mechanical and electrical characteristics have been used as a building block in numerous graphene-based electrode materials for supercapacitors [81–88]. Various methods for the production of graphene have been studied, including chemical graphene oxide (GO) reduction, CVD, and ball-milling strategies. The resulting graphene-based materials were widely employed as electrolytes/electrodes in supercapacitors. The capacity of 200 F/g for one single electrode from two-electrode supercapacitors employing high surface graphene-electrode (3100 m^2/g) generated by chemical activation was particularly interesting. Alongside the development of supercapacitors based on graphene with electrolytes, there have been numerous fascinating reports on the existence of graphene-based flexible electrodes for all-solid-state supercapacitors. For example, El-Kady and other coworkers have utilized a conventional optical light scribe DVD drive to reduce GO layers in a laser tool to make electrodes of graphite for supercapacitors, which do not require either a binder or a current gatherer (Figure 23.9a–f). High power (20 W/cm^3, 20 times greater than the activated carbon analog) and energy density (1.36 mWh/cm^3, more than two times the active carbon analog) were achieved using flexible, all-solidarity supercarriers, with outstanding stability even while bending from 0 to 180 degree (Figure 23.9g–i).

Usually, the restacking of graphene layers can minimize the SSA of graphene, resulting in poor ionic transportation/diffusion through them. As a result, much lower capacitance (80–118 F/g) was observed as compared to the corresponding theoretical values for the most solid-state supercapacitors based on free-standing graphene materials. To address such issues, 3D porous graphene networks, like graphene-aerogels and hydrogels, were explored as suitable electrode materials for electrochemical energy storage devices [89,90]. The freezing of graphene oxide (GO) dispersed solution or direct chemical vapor deposition of graphene on nickel foam can produce a 3D porous graphene nanocomposite. In both situations, the restacking of graphene layers was prohibited and the transport of electrolytes through the graphene network improved considerably. Therefore, the electrochemical performance of 3D graphene hydrogels toward 3-electrode supercapacitor systems was significantly improved with a specific capability of up to 220 F/g for one single electrode in an aqueous solution of 5 M KOH.

Zhao et al. [17] synthesized Gr-foams and their composite molds (Figure 23.9e–h), which can sustain large-form deformations under manual compaction (e.g., 50% strain) in the process of hydrothermal reducing of aqueous GO dispersions and return to the original form within 10 years of stress release with no structural fatigue. They also found that the CV curves for supercapacitors based on these 3D graphene foams remained unchanged after 1,000 compression-dis-compression cycles with 50% compressed strain (Figure 23.9i). Edge-functional graphite sheets (EFGs) containing various rim groups (e.g., –H, –N, –Br, –Cl, –I, –COOH, –SO$_3$H) have been used in recent attempts [91–95]. EFGs' unique structure, with numerous active sites near the basal plane's edge, and perfect conjugation (conductivity), should result in high-performance, high-speed capabilities. The presence of solvent-processable EFGs allowed for the production of large-area graphene using various solution techniques for processing and self-assembling, which may be utilized as electrodes in a variety of energy-related and other devices, such as fuel cells and ex-substantial supercapacitors. It is also possible to obtain EFGs that have been processed in a solution.

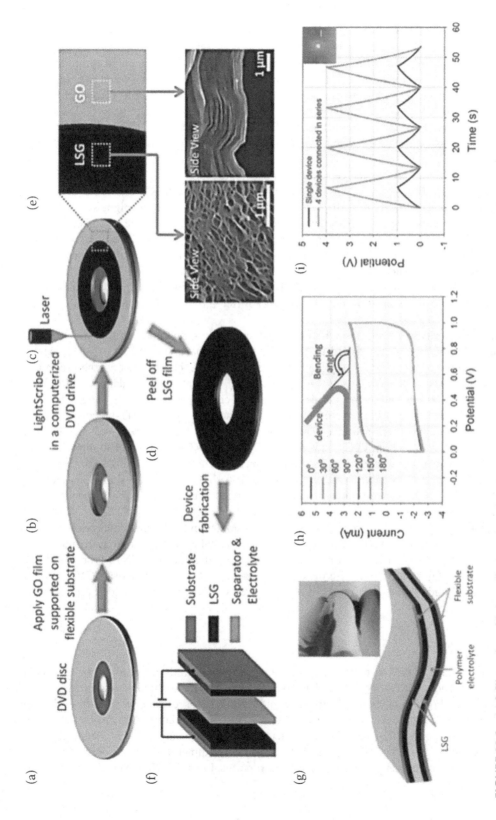

FIGURE 23.9 (a–f) Illustration of laser-scribed graphene-incorporated supercapacitors fabrication. (e) The scheme indicating the shift of golden brown color to black color with the reduction of GO film into laser-scribed graphene. The cross-sectional SEM pictures, indicating a low power application for mobile-laser stacked GO sheets into a Ni-film. (g) The crystallized electrolyte can act as both the electrolyte and the separator in the all-solid-state LSG–EC, as shown in the schematic picture. The inset is a digital image that demonstrates the device's versatility. (h) CV curves received when the device was bent at various angles at a scan rate of 1000 mV s1. (i) Charge–discharge curves for four devices linked in series, driven by the light of an LED (the inset image). Portion of ref. [10] has been reprinted. Copyright 2021, AAAS.

23.5.2 CNT-BASED FLEXIBLE SUPERCAPACITORS

CNTs are frequently utilized as electrode materials in FSCs with both liquid and gel electrolytes [79,96]. To this end, the CNTs can be brushed or spray covered directly on either the present electrode or the current collector, or onto flexible conductive substrates, like electrode materials, onto flexible nonconductive substrates (such as plastically fine, cellulose paper, and office paper) [97–100]. In this respect, Kaempgen and his colleagues reported thin-film printable super-capacities that are both electrodes and charging collectors utilizing one-wall spray-coated CNTs on polyethylene phthalate (PET) films (Figure 23.10a). The gel electrolyte (PVA/H_3PO_4), which combines the separator and the electrolyte into a single layer, was used to make the device entirely printable.

The CNT electrodes and gel electrolytes were sanded, and the thin/tiny superstrometers were used together (Figure 23.10b). Figures 23.10c–d illustrate the curves of the CV and GCD to acquire a high specific capacitance. In addition to plastic films, other low-cost lightweight substrates on which CNTs have been coated (e.g., bureau paper, bacterial nano-cellulose) have also been utilized as electrodes in supercapacitors [96,100]. Kang et al., for example, deposited CNTs with a vacuum modification procedure on a bacterial nano-cellulose substratum, and the resulting papers demonstrated excellent re-existence, broad specific surface area, and strong chemical stability. All-solid-state supercapacitors were assembled with an outstanding rating of 46.9 F/g at a rate of scanning 100 mV/s with a lower capacity loss of 0.5% after 5,000 load discharge cycles at an outstanding density of 10 A/g.

To rectify the performance of CNT-based EDLCs, CNT composites incorporating transitional metal oxides or conductive polymers were utilized to boost the pseudo-capacity of FSCs [102–105]. Meng et al., in particular, reported the use of paper-like polyaniline (PANI)-CNT networks for the use in all-solid-state supercapacitors as the electrodes (Figure 23.10e–g). The electrode materials had a high specific capacitance of 350 F/g, and the overall device had a specific capacitance of 31.4 F/g, with only an 8.1% decrease in specific capacitance after 1,000 charge-discharge cycles. CNT-based electrodes, distributed randomly, exhibited high resistance and complicated pores (micro-pores), causing sluggish transit of ions. In contrast to random CNT networks, CNT's (VA-CNT's) with vertical alignment can generate a well-aligned porous structure that has well-defined intertube space to allow more efficient loading/saving of electrolyte-accessible surfaces [37,38,106,106,107]. Conceptual simulation and expertise have shown that supercapacitors based on VA-CNTs display greater capacity. In addition, via appropriate procedures (e.g., plasma etching) [108,109], the top end cap of VA-CNTs may be opened appropriately, enabling the electrolyte access for extra charge storage in the normally inaccessible inner chamber of the VA-CNT. Recent research has shown, as expected, that a three-electrode system that uses liquid electrolytes can provide a better VA-rate CNT's capability than random CNTs [106,110–113]. Specifically, in 1 M H_2SO_4 for the VA-CNT array electrode produced by template-assisted CVD showed 440 F/g for VA-CNT ionic liquid electrolytes, a template-free CVD technique provided a high capacitance of 365 F/g.

23.5.3 BIO-BASED FSCS

Several publications have looked at biomass as a supercapacitor electrode material [114]. Chemical activation in the presence of KOH or K_2CO_3 has been shown to have the highest gravimetric capacitances (around 300 F/g for aqueous electrolytes and 150 F/g for organic electrolytes) due to their adjustable pore size distribution and extremely huge surface areas (above 3000 m^2/g) [115–117]. The energy density of carbons that have been physically operative by air, CO_2, or steam are considerable, reaching 300 F/g in 0.5 M H_2SO_4 [118,119]. Eventually, numerous biomass-derived pseudo-capacitive carbons with high energy density have been reported [46,119]. However, the method for

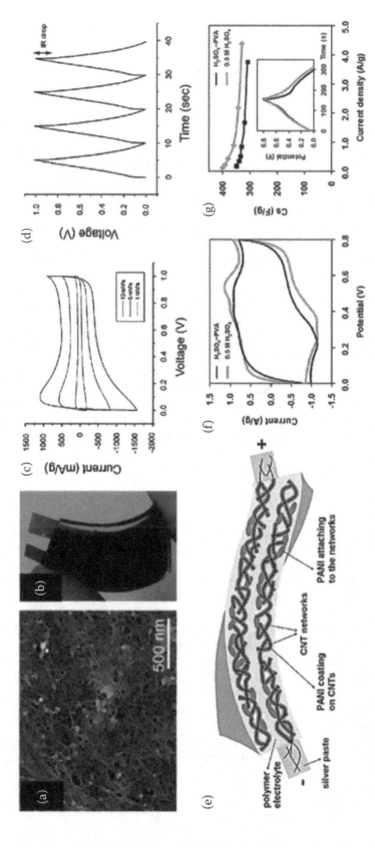

FIGURE 23.10 (a) When the SWCNT networks were deposited, they were observed using a scanning electron microscope. (b) Fabrication of thin-film SCs utilizing the SWCNT films coated on PET-electrodes and PVA/H3PO4-based polymer electrolyte and separator. The CV (c) and GCD (d) curves of thin-film SCs. Reproduced with permission [97]. Copyright 2021, ACS publishing group. (e) Illustration of fabrication of PANI/CNTs hybrid electrode. (f and g) CVs recorded in H$_2$SO$_4$–PVA gel electrolyte and 0.5 M H$_2$SO$_4$ aqueous solution at 5 mV/s (f) and discharge abilities (g) of flexible PANI/CNT composite thin-film electrodes. The inset in (g) depicts one cycle of galvanostatic charge–discharge, recorded in H$_2$SO$_4$–PVA gel electrolyte and an aqueous solution of 0.5 M H$_2$SO$_4$. Reproduced with permission [101]. Copyright 2021, ACS publishing group.

getting pseudo-capacitive functions from heteroatoms, such as nitrogen, oxygen, or phosphorous, is still unknown.

The number of reports on flexible solid-state devices is relatively limited. Flexible electrodes made entirely of renewable precursors have been made using a variety of approaches. To make fiber-based FSCs, Selvan et al. [47] utilized dead focus leaves as the active material and PVA/H_3PO_4 as the gel electrolyte. The electrodes were created by dipping a stainless steel wire in the active ingredient slurry and then coating it with the electrolyte. The BET surface area of 157 m^2/g is 10 times lower than typical activated carbon for FSCs, despite the micro-mesopore structure, and additional information concerning the material's outstanding performance is unavailable. In a paper-based FSC system, Ogale et al. [120] used laser scribing, a new simple and highly scalable approach to produce the interdigitated electrodes. They utilized mushrooms that had been carbonized hydrothermally and chemically activated in the presence of KOH, resulting in a nano-carbon material having a surface area of 2604 m^2/g.

In another work, a slurry including the active component, binder, and conductive additive was sprayed on an indium tin oxide current collector on a PET substrate. Using a PVA/H_2SO_4 gel-electrolyte, the volumetric energy and power density of 1.8 mWh/cm^3@ 0.05 mA/cm^2 and 720 mW/cm^3@ 1 mA/cm^2, respectively. Yu et al. [121] reported a biomass precursor-based FSC for the first time in 2017. It was a free-standing film about petal blossoms. The petal's thin thickness (10–20 μm) allows for flexibility, while the wrinkled surface is perfect for ion storage. They measured capacitances of 154 F/g in a solid-state flexible device with a surface area of 509 m^2/g in a typical PVA/H_3PO_4 electrolyte. In 2016, Zhi et al. [48] developed the world's first sustainable polymer-based FSCs. The processed bacterial cellulose with silica template synthesis and self-activation, pure cotton activated with KOH, cotton activated with NH_3, cotton activated with NH_3 were also reported [122–124]. Other studies have used electro-spun pure lignin nanofibers activated in air [48] and sisal hemp leaves [125]. All of these components were evaluated as free-standing materials in acidic and basic aqueous environments with or without current collectors. Figure 23.11 shows the micro and nanostructured carbon materials produced from biomass, as well as bio-inspired nanostructures.

23.6 CONCLUSION

Due to the extended cycle stability and high power density, electro-chemical supercapacitors are one of the best new energy storage technologies. They are flexible, elastic, and wearable, as compared to their contemporaries, unlike their traditional equivalents, and they can be assembled with the help of highly active and stable nanomaterials. Usually, the exceptional chemical and electrochemical features of carbon-based nanomaterials, such as CNTs and graphene, attracted the concerned scientific community to utilize them as active electrode materials for the development of supercapacitors. These nano-carbons have been used in different forms, such as planar and fiber-like along with programmable topology, including aligned, pillared, and foam. Flexible supercapacitors have also been shown to work with other energy devices in planar and wire self-powered systems, such as solar cells and nanogenerators. The material and/or device qualities nevertheless need to be changed in many situations to enhance the performance of the devices. The flexibility of carbon nanomaterials for the production of FSCs with unique characteristics for a wide range of specialized applications has been clearly shown in recent progress in the area. There have been substantial advances in published research during the past few years. More research has to be done, though, if additional study is carried out in this promising field, soft supercapacitors might become one of the powerful and efficient technologies for energy storage that affect every part of modern life.

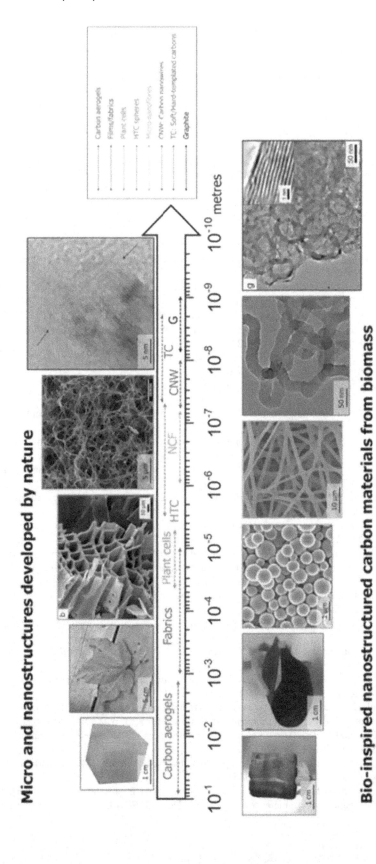

FIGURE 23.11 Carbon materials with micro and nanostructures made from biomass and bio-inspired nanostructures. Reproduced with permission [46]. Copyright 2021, Springer nature publishing group.

REFERENCES

1. Liu, J., et al., Advanced energy storage devices: Basic principles, analytical methods, and rational materials design. *Advanced Science*, 2018. **5**(1): p. 1700322.
2. Behabtu, H.A., et al., A review of energy storage technologies' application potentials in renewable energy sources grid integration. *Sustainability*, 2020. **12**(24): p. 10511.
3. Sufyan, M., et al., Sizing and applications of battery energy storage technologies in smart grid system: A review. *Journal of Renewable and Sustainable Energy*, 2019. **11**(1): p. 014105.
4. Wang, X., et al., Flexible energy-storage devices: Design consideration and recent progress. *Advanced materials*, 2014. **26**(28): p. 4763–4782.
5. Kim, D.W., S.M. Jung, and H.Y. Jung, A super-thermostable, flexible supercapacitor for ultralight and high performance devices. *Journal of Materials Chemistry A*, 2020. **8**(2): p. 532–542.
6. Huang, S., et al., Challenges and opportunities for supercapacitors. *APL Materials*, 2019. **7**(10): p. 100901.
7. Maksoud, M.A., et al., Advanced materials and technologies for supercapacitors used in energy conversion and storage: A review. *Environmental Chemistry Letters*, 2021. **19**(1): p. 375–439.
8. Yu, Z., et al., Supercapacitor electrode materials: Nanostructures from 0 to 3 dimensions. *Energy & Environmental Science*, 2015. **8**(3): p. 702–730.
9. Sharma, K., A. Arora, and S.K. Tripathi, Review of supercapacitors: Materials and devices. *Journal of Energy Storage*, 2019. **21**: p. 801–825.
10. El-Kady, M.F., et al., Laser scribing of high-performance and flexible graphene-based electrochemical capacitors. *Science*, 2012. **335**(6074): p. 1326–1330.
11. Yoo, J.J., et al., Ultrathin planar graphene supercapacitors. *Nano letters*, 2011. **11**(4): p. 1423–1427.
12. Du, F., et al., Preparation of tunable 3D pillared carbon nanotube–graphene networks for high-performance capacitance. *Chemistry of Materials*, 2011. **23**(21): p. 4810–4816.
13. Lee, J.A., et al., Ultrafast charge and discharge biscrolled yarn supercapacitors for textiles and micro-devices. *Nature communications*, 2013. **4**(1): p. 1–8.
14. Yang, Z., et al., A highly stretchable, fiber-shaped supercapacitor. *Angewandte Chemie*, 2013. **125**(50): p. 13695–13699.
15. Chen, T., et al., An integrated "energy wire" for both photoelectric conversion and energy storage. *Angewandte Chemie International Edition*, 2012. **51**(48): p. 11977–11980.
16. Chen, T., et al., Transparent and stretchable high-performance supercapacitors based on wrinkled graphene electrodes. *ACS Nano*, 2014. **8**(1): p. 1039–1046.
17. Zhao, Y., et al., Highly compression-tolerant supercapacitor based on polypyrrole-mediated graphene foam electrodes. *Advanced Materials*, 2013. **25**(4): p. 591–595.
18. Wu, Z.S., et al., Graphene-based in-plane micro-supercapacitors with high power and energy densities. *Nature Communications*, 2013. **4**(1): p. 1–8.
19. Bae, J., et al., Fiber supercapacitors made of nanowire-fiber hybrid structures for wearable/flexible energy storage. *Angewandte Chemie International Edition*, 2011. **50**(7): p. 1683–1687.
20. Gu, T. and B. Wei, High-performance all-solid-state asymmetric stretchable supercapacitors based on wrinkled MnO 2/CNT and Fe 2 O 3/CNT macrofilms. *Journal of Materials Chemistry A*, 2016. **4**(31): p. 12289–12295.
21. Wee, G., et al., Printable photo-supercapacitor using single-walled carbon nanotubes. *Energy & Environmental Science* 2011. **4**: p. 413.
22. Raccichini, R., et al., The role of graphene for electrochemical energy storage. *Nature Materials*, 2015. **14**(3): p. 271–279.
23. Shao, Y., et al., Graphene-based materials for flexible supercapacitors. *Chemical Society Reviews*, 2015. **44**(11): p. 3639–3665.
24. Cong, H.-P., et al., Flexible graphene–polyaniline composite paper for high-performance supercapacitor. *Energy & Environmental Science*, 2013. **6**(4): p. 1185–1191.
25. Kim, H.-K., et al., Dual coexisting interconnected graphene nanostructures for high performance super-capacitor applications. *Energy & Environmental Science*, 2016. **9**(7): p. 2249–2256.
26. Jiang, H., P.S. Lee, and C. Li, 3D carbon based nanostructures for advanced supercapacitors. *Energy & Environmental Science*, 2013. **6**(1): p. 41–53.
27. Liao, Q., et al., All-solid-state symmetric supercapacitor based on Co3O4 nanoparticles on vertically aligned graphene. *ACS nano*, 2015. **9**(5): p. 5310–5317.
28. Wen, F., et al., Enhanced laser scribed flexible graphene-based micro-supercapacitor performance with reduction of carbon nanotubes diameter. *Carbon*, 2014. **75**: p. 236–243.

29. Kumar, R., et al., Graphene-wrapped and cobalt oxide-intercalated hybrid for extremely durable supercapacitor with ultrahigh energy and power densities. *Carbon*, 2014. **79**: p. 192–202.

30. Zhang, Y., et al., Morphology effect of vertical graphene on the high performance of supercapacitor electrode. *ACS applied materials & interfaces*, 2016. **8**(11): p. 7363–7369.

31. Seo, D.H., et al., Synergistic fusion of vertical graphene nanosheets and carbon nanotubes for high-performance supercapacitor electrodes. *ChemSusChem*, 2014. **7**(8): p. 2317–2324.

32. Lv, T., et al., Highly stretchable supercapacitors based on aligned carbon nanotube/molybdenum disulfide composites. *Angewandte Chemie International Edition*, 2016. **55**(32): p. 9191–9195.

33. Wu, J., et al., Three-dimensional hierarchical interwoven nitrogen-doped carbon nanotubes/CoxNi1-x-layered double hydroxides ultrathin nanosheets for high-performance supercapacitors. *Electrochimica Acta*, 2016. **203**: p. 21–29.

34. Yu, D., et al., Scalable synthesis of hierarchically structured carbon nanotube–graphene fibres for capacitive energy storage. *Nature Nanotechnology*, 2014. **9**(7): p. 555–562.

35. Sun, H., et al., Novel graphene/carbon nanotube composite fibers for efficient wire-shaped miniature energy devices. *Advanced Materials*, 2014. **26**(18): p. 2868–2873.

36. Hou, X., et al., Ultrathin ZnS nanosheet/carbon nanotube hybrid electrode for high-performance flexible all-solid-state supercapacitor. *Nano Res*, 2017. **10**(8): p. 2570–2583.

37. Cai, Z., et al., Flexible, weavable and efficient microsupercapacitor wires based on polyaniline composite fibers incorporated with aligned carbon nanotubes. *Journal of Materials Chemistry A*, 2013. **1**(2): p. 258–261.

38. Ghosh, A., et al., TLM-PSD model for optimization of energy and power density of vertically aligned carbon nanotube supercapacitor. *Scientific Reports*, 2013. **3**(1): p. 1–10.

39. Sun, J., et al., Superior performance of ZnCo 2 O 4/ZnO@ multiwall carbon nanotubes with laminated shape assembled as highly practical all-solid-state asymmetric supercapacitors. *Journal of Materials Chemistry A*, 2017. **5**(20): p. 9815–9823.

40. Wang, Z., et al., Cellulose-based supercapacitors: material and performance considerations. Advanced Energy *Materials*, 2017. **7**(18): p. 1700130.

41. Wu, K., et al., Three-dimensional flexible carbon electrode for symmetrical supercapacitors. *Materials Letters*, 2016. **185**: p. 193–196.

42. De, S., et al., Biomass-derived porous carbon materials: synthesis and catalytic applications. *ChemCatChem*, 2015. **7**(11): p. 1608–1629.

43. Zhu, H., et al., Wood-derived materials for green electronics, biological devices, and energy applications. *Chemical Reviews*, 2016. **116**(16): p. 9305–9374.

44. Yu, X., et al., Soft and wrinkled carbon membranes derived from petals for flexible supercapacitors. *Scientific Reports*, 2017. **7**(1): p. 1–8.

45. Zequine, C., et al., High-performance flexible supercapacitors obtained via recycled jute: bio-waste to energy storage approach. *Scientific Reports*, 2017. **7**(1): p. 1–12.

46. Zequine, C., et al., High per formance and flexible supercapacitors based on carbonized bamboo fibers for wide temperature applications. *Scientific Reports*, 2016. **6**(1): p. 1–10.

47. Senthilkumar, S. and R.K. Selvan, Flexible fiber supercapacitor using biowaste-derived porous carbon. *ChemElectroChem*, 2015. **2**(8): p. 1111–1116.

48. Wang, X., et al., All-biomaterial supercapacitor derived from bacterial cellulose. *Nanoscale*, 2016. **8**(17): p. 9146–9150.

49. Fan, M., et al., Recent progress in 2D or 3D N-doped graphene synthesis and the characterizations, properties, and modulations of N species. *Journal of Materials Science*, 2016. **51**(23): p. 10323–10349.

50. Vineesh, T.V., et al., Bifunctional electrocatalytic activity of boron-doped graphene derived from boron carbide. *Advanced Energy Materials*, 2015. **5**(17): p. 1500658.

51. Alatalo, S.-M., et al., Soy protein directed hydrothermal synthesis of porous carbon aerogels for electrocatalytic oxygen reduction. *Carbon*, 2016. **96**: p. 622–630.

52. Yu, X., et al., Strategies to curb structural changes of lithium/transition metal oxide cathode materials & the changes' effects on thermal & cycling stability. *Chinese Physics B*, 2015. **25**(1): p. 018205.

53. Wang, J., et al., Carbon-based electrocatalysts for sustainable energy applications. *Progress in Materials Science*, 2021. **116**: p. 100717.

54. Gao, J., et al., High performance of N, P co-doped metal-free carbon catalyst derived from ionic liquid for oxygen reduction reaction. *Journal of Solid State Electrochemistry*, 2018. **22**(2): p. 519–525.

55. Chen, S., et al., Ionic liquid-assisted synthesis of N/S-double doped graphene microwires for oxygen evolution and Zn–air batteries. *Energy Storage Materials*, 2015. **1**: p. 17–24.

56. Yan, Z., Z. Peng, and J.M. Tour, Chemical vapor deposition of graphene single crystals. *Accounts of Chemical Research*, 2014. **47**(4): p. 1327–1337.

57. Zhao, D., et al., A hierarchical phosphorus nanobarbed nanowire hybrid: its structure and electrochemical properties. *Nano Letters*, 2017. **17**(6): p. 3376–3382.

58. Chen, K., et al., Scalable chemical-vapour-deposition growth of three-dimensional graphene materials towards energy-related applications. *Chemical Society Reviews*, 2018. **47**(9): p. 3018–3036.

59. Rahman, M.M., et al., Fabrication of 4-aminophenol sensor based on hydrothermally prepared ZnO/Yb 2 O 3 nanosheets. *New Journal of Chemistry*, 2017. **41**(17): p. 9159–9169.

60. Hu, B., et al., Engineering carbon materials from the hydrothermal carbonization process of biomass. *Advanced Materials*, 2010. **22**(7): p. 813–828.

61. Wang, H., et al., Ball-milling synthesis of Co 2 P nanoparticles encapsulated in nitrogen doped hollow carbon rods as efficient electrocatalysts. *Journal of Materials Chemistry A*, 2017. **5**(33): p. 17563–17569.

62. Xing, T., et al., Ball milling: a green mechanochemical approach for synthesis of nitrogen doped carbon nanoparticles. *Nanoscale*, 2013. **5**(17): p. 7970–7976.

63. Zhang, L.L. and X. Zhao, Carbon-based materials as supercapacitor electrodes. *Chemical Society Reviews*, 2009. **38**(9): p. 2520–2531.

64. Chapman, D.L., LI. A contribution to the theory of electrocapillarity. *The London, Edinburgh, and Dublin Philosophical Magazine and Journal of Science*, 1913. **25**(148): p. 475–481.

65. Gouy, M., Sur la constitution de la charge électrique à la surface d'un électrolyte. *Journal of Theoretical and Applied Physics*, 1910. **9**(1): p. 457–468.

66. Frackowiak, E., Carbon materials for supercapacitor application. *Physical Chemistry Chemical Physics*, 2007. **9**(15): p. 1774–1785.

67. Augustyn, V., P. Simon, and B. Dunn, Pseudocapacitive oxide materials for high-rate electrochemical energy storage. *Energy & Environmental Science*, 2014. **7**(5): p. 1597–1614.

68. Miller, J.R. and P. Simon, Electrochemical capacitors for energy management. *Science Magazine*, 2008. **321**(5889): p. 651–652.

69. Naoi, K. and P. Simon, New materials and new configurations for advanced electrochemical capacitors. *The Electrochemical Society Interface*, 2008. **17**(1): p. 34.

70. Binitha, G., et al., Electrospun α-Fe 2 O 3 nanostructures for supercapacitor applications. *Journal of Materials Chemistry A*, 2013. **1**(38): p. 11698–11704.

71. Chou, S.-L., et al., Electrochemical deposition of porous Co (OH) 2 nanoflake films on stainless steel mesh for flexible supercapacitors. *Journal of The Electrochemical Society*, 2008. **155**(12): p. A926.

72. Ates, M., M. El-Kady, and R.B. Kaner, Three-dimensional design and fabrication of reduced graphene oxide/polyaniline composite hydrogel electrodes for high performance electrochemical supercapacitors. *Nanotechnology*, 2018. **29**(17): p. 175402.

73. Xu, J., et al., Fabrication of hierarchical MnMoO4· H2O@ MnO2 core-shell nanosheet arrays on nickel foam as an advanced electrode for asymmetric supercapacitors. *Chemical Engineering Journal*, 2018. **334**: p. 1466–1476.

74. Yang, B.-j., et al., Three-dimensional porous biocarbon wrapped by graphene and polypyrrole composite as electrode materials for supercapacitor. *Journal of Materials Science: Materials in Electronics*, 2018. **29**(3): p. 2568–2572.

75. Yao, L., et al., Scalable 2D hierarchical porous carbon nanosheets for flexible supercapacitors with ultrahigh energy density. *Advanced Materials*, 2018. **30**(11): p. 1706054.

76. Song, D., et al., Freestanding two-dimensional Ni (OH) 2 thin sheets assembled by 3D nanoflake array as basic building units for supercapacitor electrode materials. *Journal of Colloid and Interface Science*, 2018. **509**: p. 163–170.

77. Guan, Y., et al., Core/shell nanorods of MnO2/carbon embedded with Ag nanoparticles as high-performance electrode materials for supercapacitors. *Chemical Engineering Journal*, 2018. **331**: p. 23–30.

78. Qi, D. and X. Chen, Flexible Supercapacitors Based on Two-Dimensional Materials. *Flexible and Stretchable Medical Devices*, 2018. p.161–197

79. Park, S., M. Vosguerichian, and Z. Bao, A review of fabrication and applications of carbon nanotube film-based flexible electronics. *Nanoscale*, 2013. **5**(5): p. 1727–1752.

80. Chen, T. and L. Dai, Flexible supercapacitors based on carbon nanomaterials. *Journal of Materials Chemistry A*, 2014. **2**(28): p. 10756–10775.

81. Novoselov, K.S., et al., A roadmap for graphene. *Nature*, 2012. **490**(7419): p. 192–200.
82. Geim, A.K. and K.S. Novoselov, *The rise of graphene, in Nanoscience and technology: a collection of reviews from nature journals*. 2010, World Scientific. p. 11–19.
83. Dai, L., Functionalization of graphene for efficient energy conversion and storage. *Accounts of Chemical Research*, 2013. **46**(1): p. 31–42.
84. Hou, J., et al., Graphene-based electrochemical energy conversion and storage: fuel cells, supercapacitors and lithium ion batteries. *Physical Chemistry Chemical Physics*, 2011. **13**(34): p. 15384–15402.
85. Liu, S., S. Sun, and X.-Z. You, Inorganic nanostructured materials for high performance electrochemical supercapacitors. *Nanoscale*, 2014. **6**(4): p. 2037–2045.
86. Alvi, F., et al., Graphene–polyethylenedioxythiophene conducting polymer nanocomposite based super-capacitor. *Electrochimica Acta*, 2011. **56**(25): p. 9406–9412.
87. Xu, C., et al., Graphene-based electrodes for electrochemical energy storage. *Energy & Environmental Science*, 2013. **6**(5): p. 1388–1414.
88. Parvez, K., et al., Exfoliation of graphite into graphene in aqueous solutions of inorganic salts. *Journal of the American Chemical Society*, 2014. **136**(16): p. 6083–6091.
89. Xu, Z., et al., ACS Nano 2012, 6, 7103; d) Z. Chen, W. Ren, L. Gao, B. Liu, S. Pei, H.-M. Cheng. *Nature Materials*, 2011. **10**: p. 424.
90. Xu, Y., et al., Self-assembled graphene hydrogel via a one-step hydrothermal process. *ACS Nano*, 2010. **4**(7): p. 4324–4330.
91. Jeon, I.-Y., et al., Edge-carboxylated graphene nanosheets via ball milling. *Proceedings of the National Academy of Sciences*, 2012. **109**(15): p. 5588–5593.
92. Jeon, I.-Y., et al., Large-scale production of edge-selectively functionalized graphene nanoplatelets via ball milling and their use as metal-free electrocatalysts for oxygen reduction reaction. *Journal of the American Chemical Society*, 2013. **135**(4): p. 1386–1393.
93. Jeon, I.Y., et al., Edge-selectively sulfurized graphene nanoplatelets as efficient metal-free electro-catalysts for oxygen reduction reaction: the electron spin effect. *Advanced Materials*, 2013. **25**(42): p. 6138–6145.
94. Jeon, I.-Y., et al., Direct nitrogen fixation at the edges of graphene nanoplatelets as efficient electrocatalysts for energy conversion. *Scientific Reports*, 2013. **3**(1): p. 1–7.
95. Jeon, I.-Y., et al., Facile, scalable synthesis of edge-halogenated graphene nanoplatelets as efficient metal-free eletrocatalysts for oxygen reduction reaction. *Scientific Reports*, 2013. **3**(1): p. 1–7.
96. Dai, L., et al., Carbon nanomaterials for advanced energy conversion and storage. *Small*, 2012. **8**(8): p. 1130–1166.
97. Kaempgen, M., et al., Printable thin film supercapacitors using single-walled carbon nanotubes. *Nano Letters*, 2009. **9**(5): p. 1872–1876.
98. Hu, M., Pasta, FL Mantia, L. Cui, S. Jeong, HD Deshazer, JW Choi, S. M. Han, and Y. Cui. *Nano Lett*, 2010. **10**: p. 708.
99. Kang, Y.J., et al., All-solid-state flexible supercapacitors based on papers coated with carbon nanotubes and ionic-liquid-based gel electrolytes. *Nanotechnology*, 2012. **23**(6): p. 065401.
100. Kang, Y., et al., ACS Nano, 2012, 6, 6400–6406. *This article is licensed under a Creative Commons Attribution*. **3**.
101. Meng, C., et al., Highly flexible and all-solid-state paperlike polymer supercapacitors. *Nano Letters*, 2010. **10**(10): p. 4025–4031.
102. Hu, L., et al., Symmetrical MnO2–carbon nanotube–textile nanostructures for wearable pseudocapacitors with high mass loading. *ACS Nano*, 2011. **5**(11): p. 8904–8913.
103. Yu, G., et al., Hybrid nanostructured materials for high-performance electrochemical capacitors. *Nano Energy*, 2013. **2**(2): p. 213–234.
104. Nyholm, L., et al., Toward flexible polymer and paper-based energy storage devices. *Advanced Materials*, 2011. **23**(33): p. 3751–3769.
105. Wang, K., et al., Flexible supercapacitors based on cloth-supported electrodes of conducting polymer na-nowire array/SWCNT composites. *Journal of Materials Chemistry*, 2011. **21**(41): p. 16373–16378.
106. Ye, J.S., et al., Preparation and characterization of aligned carbon nanotube–ruthenium oxide nano-composites for supercapacitors. *Small*, 2005. **1**(5): p. 560–565.
107. Dai, L., et al., Aligned nanotubes. *ChemPhysChem*, 2003. **4**(11): p. 1150–1169.
108. Lu, W., et al., High performance electrochemical capacitors from aligned carbon nanotube electrodes and ionic liquid electrolytes. *Journal of Power Sources*, 2009. **189**(2): p. 1270–1277.

109. Huang, S. and L. Dai, Plasma etching for purification and controlled opening of aligned carbon nanotubes. *The Journal of Physical Chemistry B*, 2002. **106**(14): p. 3543–3545.

110. Hoefer, M. and P. Bandaru, Determination and enhancement of the capacitance contributions in carbon nanotube based electrode systems. *Applied Physics Letters*, 2009. **95**(18): p. 183108.

111. Zhang, H., et al., Growth of manganese oxide nanoflowers on vertically-aligned carbon nanotube arrays for high-rate electrochemical capacitive energy storage. *Nano Letters*, 2008. **8**(9): p. 2664–2668.

112. Chen, Q.-L., et al., Fabrication and electrochemical properties of carbon nanotube array electrode for supercapacitors. *Electrochimica Acta*, 2004. **49**(24): p. 4157–4161.

113. Lu, W., et al., Superior capacitive performance of aligned carbon nanotubes in ionic liquids. *ECS Transactions*, 2008. **6**(25): p. 257.

114. Gao, Z., et al., Biomass-derived renewable carbon materials for electrochemical energy storage. *Materials Research Letters*, 2017. **5**(2): p. 69–88.

115. Fuertes, A.B. and M. Sevilla, Superior capacitive performance of hydrochar-based porous carbons in aqueous electrolytes. *ChemSusChem*, 2015. **8**(6): p. 1049–1057.

116. Zhao, L., et al., Nitrogen-containing hydrothermal carbons with superior performance in supercapacitors. *Advanced Materials*, 2010. **22**(45): p. 5202–5206.

117. Sevilla, M. and A.B. Fuertes, A green approach to high-performance supercapacitor electrodes: the chemical activation of hydrochar with potassium bicarbonate. *ChemSusChem*, 2016. **9**(14): p. 1880–1888.

118. Lu, H. and X. Zhao, Biomass-derived carbon electrode materials for supercapacitors. *Sustainable Energy & Fuels*, 2017. **1**(6): p. 1265–1281.

119. Enock, T.K., et al., Status of biomass derived carbon materials for supercapacitor application. *International Journal of Electrochemistry*, 2017. **2017**.

120. Yadav, P., et al., Highly stable laser-scribed flexible planar microsupercapacitor using mushroom derived carbon electrodes. *Advanced Materials Interfaces*, 2016. **3**(11): p. 1600057.

121. Anothumakkool, B., et al., High-performance flexible solid-state supercapacitor with an extended nanoregime interface through in situ polymer electrolyte generation. *ACS Applied Materials & Interfaces*, 2016. **8**(2): p. 1233–1241.

122. Cheng, P., et al., Biomass-derived carbon fiber aerogel as a binder-free electrode for high-rate supercapacitors. *The Journal of Physical Chemistry C*, 2016. **120**(4): p. 2079–2086.

123. Li, L., et al., Nitrogen-doped carbonized cotton for highly flexible supercapacitors. *Carbon*, 2016. **105**: p. 260–267.

124. Hao, X., et al., Bacterial-cellulose-derived interconnected meso-microporous carbon nanofiber networks as binder-free electrodes for high-performance supercapacitors. *Journal of Power Sources*, 2017. **352**: p. 34–41.

125. Li, Y., et al., A top-down approach for fabricating free-standing bio-carbon supercapacitor electrodes with a hierarchical structure. *Scientific Reports*, 2015. **5**(1): p. 1–10.

24 2D Materials for Flexible Supercapacitors

Wei Ni

State Key Laboratory of Vanadium and Titanium Resources Comprehensive Utilization, ANSTEEL Research Institute of Vanadium & Titanium (Iron & Steel), Chengdu, People's Republic of China

Material Corrosion and Protection Key Laboratory of Sichuan Province, Sichuan University of Science and Engineering, Zigong, People's Republic of China

Vanadium and Titanium Resource Comprehensive Utilization Key Laboratory of Sichuan Province, Panzhihua University, Panzhihua, People's Republic of China

Ling-Ying Shi

College of Polymer Science and Engineering, Sichuan University, Chengdu, People's Republic of China

CONTENTS

24.1   Introduction... 441
24.2   Classification, methods, and merits 442
    24.2.1   Graphene and its carbonaceous analogs 443
    24.2.1   TMCs and TMDs 443
    24.2.3   MXenes.. 448
    24.2.4   2D MOFs 449
    24.2.5   Other 2D/layered materials 450
    24.2.5   Hybrid 2D nanostructures 451
    24.2.6   Configurations of FSCs 454
24.3   Challenges and prospects 455
References.. 456

24.1   INTRODUCTION

Flexible and wearable electronic devices (e.g., flexible displays, sensors, and health monitors) are attracting significantly increasing interest, and the accompanied flexible, lightweight energy storage and conversion systems are highly desirable, of which the safe and fast energy storage device flexible supercapacitors (FSCs) with high power density, excellent mechanical compliance, and cycle stability are of great research and application interest [1–5]. Among various FSCs electrode materials, 2D and layered materials are ideal electrode materials due to their great variety and many unique physical and chemical advantages, such as flexible 2D structures, expanded surface area, high specific capacitance, improved electronic/ionic conductivity, and high surface tunability to form synergistic hybrids/composites [6–10]. These various classes of 2D or layered materials mainly include graphene [11–16] and carbonaceous analogues [17], transition metal oxides (TMOs) [18], hydroxides (TMHs; or layered double hydroxides, LDHs), chalcogenides (TMCs) and dichalcogenides (TMDs) [19,20], MXenes [21,22], black phosphorus (BP, or phosphorene) [23], as well as conducting polymers [24,25], 2D metal-organic frameworks (MOFs) or covalent organic frameworks (COFs) [26]. There are some excellent reviews on this topic [2,6,9,14], and here we focus on the recent two to three years of significant achievements, plus some of our previous interesting works, to highlight the advantages and give new insight into this research area.

24.2 CLASSIFICATION, METHODS, AND MERITS

Flexible SCs may be classified into several categories by the configuration, i.e., 1D fiber SCs (parallel, or twisted), 2D planar SCs (in-plane, thin or sandwich-like), micro-supercapacitors (MSCs i.e., in-plane, interdigital or 3D-printing configuration), and other novel 3D stretchable architectures (e.g., honeycomb-lantern-like structure).

Electrode manufacturing techniques are of great importance for FSCs (Figure 24.1). These techniques include deposition (e.g., magnetron sputtering), (vacuum) filtration, wet-spinning, roll-to-roll, printing, and laser scribing, etc.; however, the scalability and cost-effectiveness should be considered [6]. For example, 3D printing is an emerging and effective technique for the fabrication of MSCs, typically all-solid-state MSCs [27]. This versatile technique endows the MSCs with diverse structure and configuration, not only useful for the design and construction of patterns but also controllable in

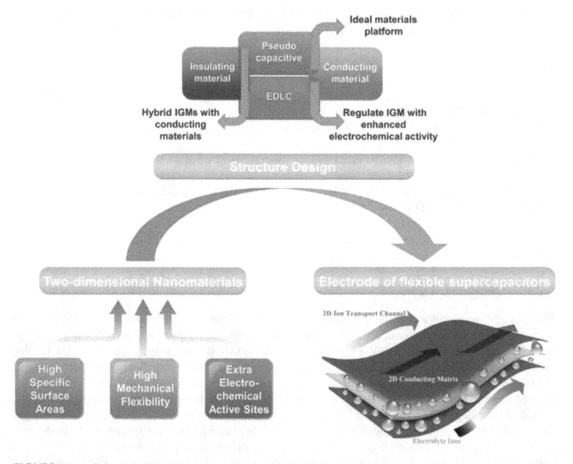

FIGURE 24.1 Schematic illustration of the structure design of high-performance flexible supercapacitors (FSCs) from emerging 2D nanomaterials, mainly 2D inorganic graphene-like materials (IGMs), as well as some other 2D or layered materials. Reprinted with permission from ref. [2]. Copyright (2014) The Royal Society of Chemistry.

the electrode thickness, which is beneficial to the tuning of the loading active materials and related energy/power densities thereof. Ultra-short pulsed laser processing possesses many advantages, e.g., significantly low thermal effect surrounding the focal point, high precision, as well as in situ laser reduction, thus a promising tool for manufacturing and surface process of graphene-based MSCs [28,29]. Screen printing is also a versatile technique that can be used for high-quality and scalable FSCs or MSCs, as well as integrated circuit paths or conductive tracks [30,31].

24.2.1 Graphene and its carbonaceous analogs

As a star 2D material with exceptional mechanical, thermal, and electrical properties, high specific surface area, and electrochemical stability, graphene has attracted great attention since its emergence in 2004 [13,14,32]. As a 2D, monolayered, and flexible building block, the single-atom-thick carbon flake that can be assembled into macroscopically ordered structures is suitable for flexible and/ or wearable 2D planar supercapacitors (film-shaped FSCs) [13] or wet-spun fiber supercapacitors (fiber-shaped FSCs) [11,14].

Due to the difficulty in melt processing of graphene materials, fluid assembly is the major method to scalably produce flexible graphene-based energy-storage devices; thus, the highly soluble or dispersive graphene derivatives, especially graphene oxide (GO) with large lateral size, are practically applicable based on the rich chemistry for chemical or physical functionalization [11]. The as-prepared graphene fiber (GFs) or graphene papers (GPs) can have good overall performance, i.e., mechanically flexible and strong, electrically conductive, chemically resistive, and/or electrochemically active, which would be advantageous for smart multifunctional applications including flexible supercapacitors (Figure 24.2) [11,33]. Among the various physical and chemical reduction methods of GO, laser reduction in recent years has attracted significant interest due to its rapid, highly flexible, and chemical-free route that allows direct writing on a diverse solid substrate with sub-micrometer feature size [29]. The laser-reduced graphene (LRG) or laser-scribed graphene (LSG) based FSCs with superior conductivity, smaller feature size, and more flexible 2D/3D electrode design would provide tremendous opportunities in energy-related flexible device applications [29,34]. These graphene-based planar FSCs usually possess outstanding EDLC performances with tunable voltages and superior cycling stability (up to 10,000 cycles) for commercial application, although a flexible substrate is generally required.

Although the graphene-based FSCs are of great electrochemical stability and thus for ultralong cycle life and decayless high-voltage output in a tandem configuration, the pure graphene supercapacitors are founded on an EDLCs mechanism (i.e., capacitive mechanism), the incorporation of some battery-type or pseudocapacitive materials such as 2D or layered transition-metal oxides/dichalcogenides, electrochemically active polymers [35], would further enhance their capacitance; furthermore, the combination of a capacitor-type electrode and a battery-type electrode to construct next-generation hybrid supercapacitors (e.g., lithium-ion capacitors, LICs, or called lithium-ion hybrid supercapacitors, LIHSs) with simultaneous high-power and high-energy features can be effectively released [36,37]. The quasi-graphene carbons such as 2D hierarchical porous carbon nanosheets (PCNS) or heteroatom-doped analogs may have comparably high-volumetric-capacitance in aqueous or gel-electrolyte FSCs, although their mechanical strength and conductivity are not that remarkable [17].

24.2.1 TMCs and TMDs

2D transition-metal chalcogenides (TMCs), especially dichalcogenides (TMDs), are recently emerging 2D/layered materials with specific advantages, such as superior electronic properties than their oxide analogs. These 2D TMDs that have been investigated or exploded for potential FSCs include VS_2 nanosheets [38], WS_2 nanosheets, SnSe nanosheets, $SnSe_2$ nanodisks [39], VSe_2 nanosheets [40], metallic hydrogenated-Cu_2WS_4 nanosheet film, and WTe_2 nanosheets [41]. Due to the enhanced electronic

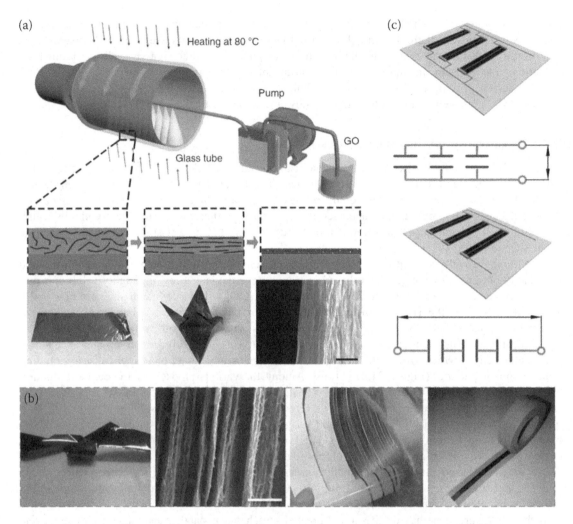

FIGURE 24.2 (a) Schematic diagram of the preparation of highly aligned and compact GO film by a continuous centrifugal casting (CCC) method, and the as-obtained flexible rGO film and its cross-sectional SEM image (scale bar = 1 μm). (b) Morphology and structure of the obtained 2D rGO/SWCNT hybrid film for flexible tape SCs: (from left to right) folded rGO/SWCNT hybrid film; cross-sectional SEM image of rGO/SWCNT film (scale bar = 1 μm); continuous long rGO/SWCNT hybrid ribbon via film cutting; tape SC based on the hybrid ribbon. (c) Schematic and equivalent circuit of the three tape SCs connected in parallel (upper) and in series (bottom). Adapted from ref. [33]. Open Access under CC BY 4.0 (2018) Springer Nature.

properties and unique mechanical flexibility, the synthesis of 2D TMCs are of great importance. For example, vanadium disulfide (VS_2), as a typical V-containing layered chalcogenide, has been utilized as high-performance active materials in in-plane FSCs. Xie and coworkers have developed few-layered metallic VS_2 ultrathin nanosheets (~2.5 nm, 4–5 single layers of conducting S–V–S stacked by weak interlayer interactions of Van der Waals) via exfoliating the NH_3-intercalated $VS_2 \cdot NH_3$; these high-conductivity 2D materials endow the FSCs with superior specific capacitance (4,760 μF cm^{-2} in a 150 nm planar configuration) and high cycle stability (no obvious degradation after 1,000 charge-discharge cycles) (Figure 24.3) [38]. The indispensable requirements for high-performance in-plane SCs, e.g., high-efficiency permeable channels, high-conductivity scaffolds, and high-specific-area electrode materials, are met with for these TMDs.

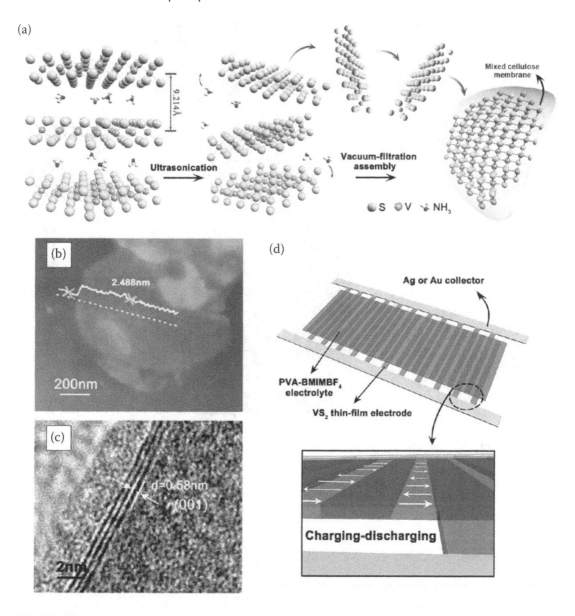

FIGURE 24.3 (a) Schematic illustration of the preparation of VS_2 nanosheet film via liquid-exfoliation and vacuum-filtration assembly; (b and c) AFM and HRTEM images of the as-synthesized few-layered VS_2 nanosheets; (d) schematic illustration of the planar interdigital FSC with facilitated ion migration pathways. Adapted with permission from ref. [38], Copyright (2011) American Chemical Society.

By precursor optimization and phase-controlled synthesis, 2D selenide nanostructures such as SnSe nanosheets and $SnSe_2$ nanodisks can be obtained (Figure 24.4a–d). The resulting TMC/TMD nanostructures as electrode materials showed typically (phase-dependent) pseudocapacitive features, i.e., the orthorhombic-phase SnSe nanosheets deliver a specific capacitance of 228 F g^{-1} and hexagonal-phase $SnSe_2$ nanodisks of 168 F g^{-1}, respectively, at 0.5 A g^{-1}, although the specific capacitance of mixed-phase nanodisks (31 F g^{-1} for $SnSe–SnSe_2$) is much lower than those of pure phase ones. When fabricated in an all-solid-state FSCs device (polymer gel PVA/H_2SO_4 serving as both solid electrolyte and

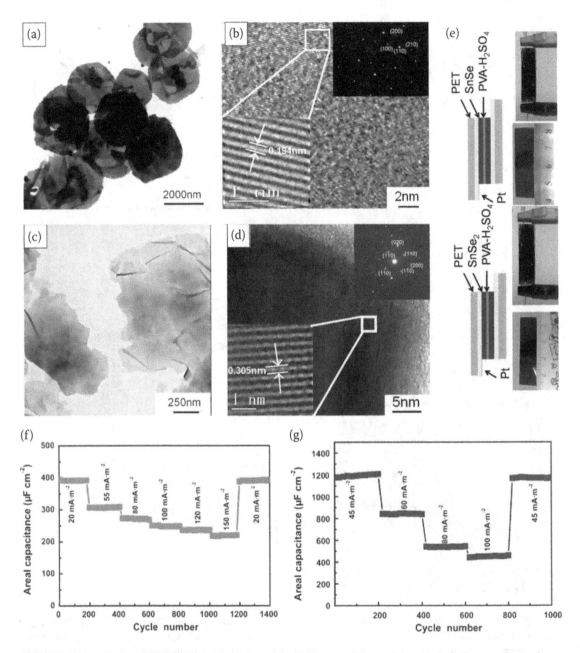

FIGURE 24.4 TEM and HRTEM images of (a and b) $SnSe_2$ nanosisks and (c and d) SnSe nanosheets (inset: SAED patterns and zoomed-in lattice fringes); (e) schematic illustration and photos and (f and g) cycling/rate performances of the corresponding flexible symmetric all-solid-state SCs based on the as-synthesized 2D selenides (left: $SnSe_2$ ND-SSC, right: SnSe NS-SSC). Adapted with permission from ref. [39]. Copyright (2014) American Chemical Society.

separator, Pt layer deposited PET substrates coated with active materials as electrodes), the two 2D selenide nanostructures demonstrated superior areal capacitances, high cycling stabilities, as well as good flexibilities and desirable mechanical strength. Moreover, the $SnSe_2$ nanodisk, however, shows much better rate capability than of SnSe nanosheet in the FSCs devices; thus, it can serve as a kind of promising active material for high-performance all-solid-state FSCs (Figure 24.4e and f) [39].

Compared to the semiconductive S- or Se-based TMDs, metallic Te-based 2D TMDs may have much better electric conductivity and thus enhanced performance in FSCs. Via controllable liquid-phase exfoliation semimetal few-layered 1Td WTe_2 nanosheets (1.3–5.3 nm, 2–7 layers) can be synthesized and then assembled into air-stable films for all-solid-state FSCs. As expected, the PET-supported FSCs can deliver high mass and volumetric capacitances of 221 F g^{-1} and 74 F cm^{-3}, respectively, as well as excellent volumetric power and energy densities of 83.6 W cm^{-3} and 0.01 W h cm^{-3}, respectively. These indexes are superior to that of typical thin-film batteries/electrolytic capacitors (Figure 24.5). In addition, the WTe_2-based solid-state FSCs show superior cycle stability (~91% capacitance retention after 5,500 cycles) and excellent mechanical flexibility [41].

MoS_2 is an intensively studied layer-structured material, which can also be exfoliated into single-layer nanosheets. To further enhance its electric conductivity, 2D graphene (e.g., reduced graphene oxide) can be incorporated and/or embedded into carbon nanotubes (or carbon fibers) for solid-state, flexible-fiber

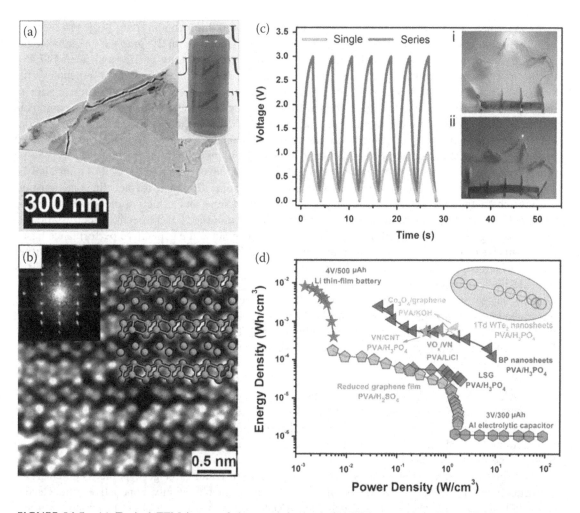

FIGURE 24.5 (a) Typical TEM image of the as-synthesized 1Td WTe_2 nanosheet (inset: digital photo of its translucent solution). (b) Atomic resolution Z-contrast STEM image of the 1Td WTe_2 nanosheet (inset: corresponding FFT pattern). (c) GCD curve of the FSC in series with 1Td WTe_2 nanosheet film electrodes to light red and blue LEDs, and (d) the corresponding Regone plot (volumetric energy density *vs.* power density) of the as-fabricated device compared with other typical devices, commercial and reported. Adapted with permission from ref. [41]. Copyright (2017) WILEY-VCH Verlag GmbH & Co. KGaA, Weinheim.

asymmetric SCs (or hybrid paper-based planar FSCs) to maintain high energy density without compromising rate stability [42,43].

24.2.3 MXENES

Due to its excellent electrochemical properties, post-graphene 2D material MXene has become a popular material in the field of energy storage in recent years, shortly after its emergence in 2011, particularly with flexible fast rechargeable devices [21,22,44,45]. These clay-inspired 2D materials have tunable element composition and bandgap, high charge-carrier mobility, metallic nature (excellent electric conductivity), high surface area, hydrophilicity, and favorable mechanical properties, conferring their superior performance and significant impact in the burgeoning field of FSCs [22,45]. However, for MXenes, they have specific challenges to be addressed, i.e., the oxidation of titanium, in addition to the many similar challenges, such as restacking and re-crushing that graphene and their analogs encounter [44].

In addition to the pure but flexible strong MXene films with high conductivity [46], MXenes can be further combined with conductive polymers, 1D carbon nanotubes, or 2D graphene to enhance the conductivity and to fully utilize their surface area to alleviate restacking of 2D nanosheets [47,48]. Furthermore, by the introduction of electrolyte-mediated factors (or other functional molecules), the synergistic layers of MXene and graphene are controllably assembled in a compact structure with an interlayer filled with electrolyte (or molecules) for better electrolyte accessibility. And the composite films can directly serve as self-supporting electrodes for FSCs without additional binders or conductive agents, showing an enhanced volumetric specific capacity, energy/power density compared to pure graphene electrode or porous composite electrode by conventional methods [47,49]. Some other components, such as bacterial cellulose (BC) or cellulose nanofibrils (CNFs), can also be incorporated to enhance the mechanical properties of MXene films and to tune the ion transport channel characteristics (for high areal mass loading of MXene and effective electrode structural design) [50,51]. Furthermore, MXenes can be chemically modified via large-size/electrostatic terminations (e.g., $-SO_4$) or doped by nitrogen or sulfur to enhance the capacitive performance based on an enlarged interlayer spacing (or ionic channel), facilitated electrolyte ion infiltration, further destroyed surface atomic structure, and the replacing heteroatoms (or additional bonding), as well as improved conductivity.

Yarn-shaped SCs (YSCs) comprising bi-scrolled MXene and CNTs (or rGO) can be prepared and integrated into fabrics, which is a promising way for the application of FSCs due to easy integration and the high mechanical durability and cycling stability [52,53]. The YSC electrode could trap up to ~98wt% MXene within the CNT yarn scrolls, and when configured into an asymmetric SC, it could reach a maximum power density of 5428 mW cm^{-3} (at 11.9 mW h cm^{-3}) and a maximum energy density of 61.6 mW h cm^{-3} (at 358 mW cm^{-3}) (Figure 24.6) [54]. Similar YSCs such as MXene-coated silver-plated nylon, MXene-coated carbon fiber (or carbon cloth, cotton textiles), wet-spun rGO/MXene yarns (or textile) SCs are also developed; these knittable composite fibers introduce a new family of wearable fiber supercapacitors [52]. Furthermore, through the effects of electrolyte mediation (e.g., H_2SO_4 preincorporation) and MXene size (favoring larger size in general), the performances of these fiber-shaped SCs may be further enhanced.

3D printing via robocasting or direct ink writing (DIW) of colloidal or gel-type ink is an emerging extrusion-based method for 3D MSCs. By using these printable, highly conductive, and electrochemically active materials based inks including MXenes can be used to construct patterned electrodes of FSCs or MSCs. As a typical example, the highly concentrated additive-free MXene ink can be used for scalable fabrication of all-state-solid MSCs with diverse architectures and electrode thickness on varied flexible substrates, including paper and polymer film. These as-prepared MSCs demonstrate exceptional high areal capacitance (up to 1035 mF cm^{-2}) and energy densities (51.7 μW h cm^{-2}) (Figure 24.7) [27].

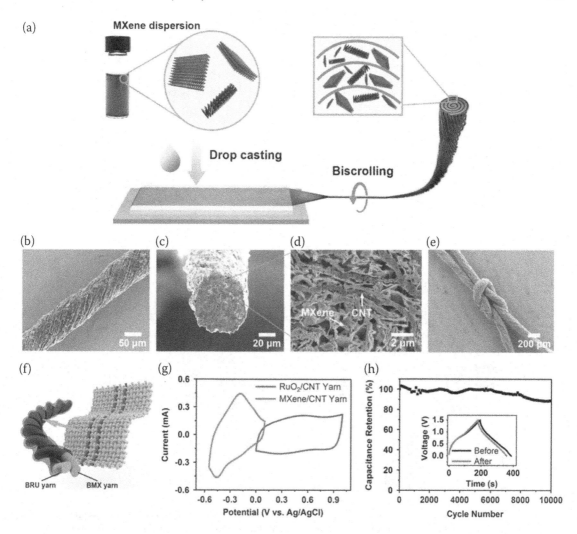

FIGURE 24.6 (a) Schematic illustration of the fabrication of biscrolled MXene/CNT (BMX) yarns. (b–d) SEM images of the surface, cross-sectional, and zoomed-in section of the BMX yarn (containing 97.4 wt% MXene). (e) SEM image of a BMX yarn tied into a reef knot (containing 90 wt% MXene). (f) Schematic illustration of the textile asymmetric YSC (aYSC) prototype (BRU: RuO₂/CNT), and (g and h) the related CVs and cycling performance (inset: GCD curves). Adapted with permission from ref. [54]. Copyright (2018) WILEY-VCH Verlag GmbH & Co. KGaA, Weinheim.

24.2.4 2D MOFs

2D conjugated MOFs (2D *c*-MOFs) are a novel generation of layer-stacked MOFs with strong in-plane interaction but weak interlayer Van der Waals interaction. Owing to the intrinsic conductivity, well-defined active sites, tunable porosity, and versatile structures, it is a rising candidate for FSCs and MSCs [55]. And the 2D MOF nanosheets show significantly enhanced electrochemical performances compared to that of bulk MOFs (for an interdigital MSC based on 2D ultrathin Cu-porphyrin MOF, see Figure 24.8) [56]. However, additional 2D conductive agents such as graphene [55], MXene [57], or substrates (e.g., 2D conductive MOF assembled on CNFs [58]) are still needed for the high-performance device fabrication, and the cycling stability improvement is required for future practical application.

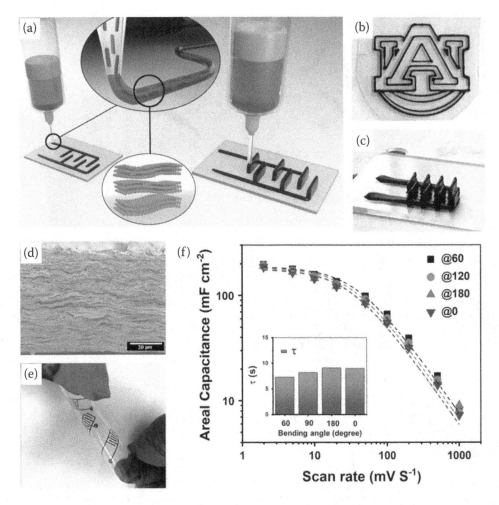

FIGURE 24.7 (a) Schematic illustration of the 3D printing of interdigital MSCs with varied electrode height via increased MXene ink layers. (b and c) The digital photos of the as-printed logo and MSC-10 (10 printed layers, 75 μm). (d) Cross-sectional SEM image of MSC-10 electrode. (e and f) Digital photo showing the flexibility of the all-MXene MSCF-1 (flexible MSC-1) device and its areal capacitance *vs.* scan rate at various bending angles (inset: an increase in time constant, τ, reflecting a sluggish charge/discharge process). Adapted with permission from ref. [27]. Copyright (2019) American Chemical Society.

Their derivatives, such as porous carbon nanosheets (or nanoporous graphene films, planar or hierarchical) and the transition-metal compounds/composites thereof, e.g., $Ni(OH)_2$–MnO_2/C, $Ni(OH)_2$@ZnCoS-NSs/CC, VN@ZnCoS-NSs/CC, $Zn_{0.76}Co_{0.24}$S/$NiCo_2S_4$/CC, are also of great use in FSCs.

24.2.5 OTHER 2D/LAYERED MATERIALS

Besides the abovementioned 2D/layered materials, some 2D structured metals, such as Zn nanosheets can also be developed into hierarchically nanostructured electrodes for flexible Zn-ion supercapacitors (ZIC), these fiber-type or flat-type aqueous FSCs usually have excellent pseudocapacitive performances with high energy/power densities. Although the output voltage is not as high as that of commercial Li-ion batteries/capacitors based on organic electrolytes, its advantages regarding safety and environmental benignity endow it with promising practical application in wearable energy-storage devices. For

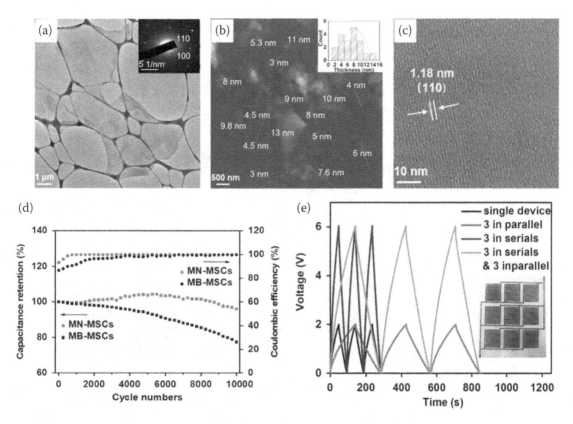

FIGURE 24.8 (a–c) TEM, AFM, and HRTEM images of the 2D ultrathin MOF CoTCPP-PZ nanosheets (insets: SAED pattern and thickness distribution diagram; TCPP = 5,10,15,20-tetra(carboxyphenyl) porphyrin, PZ = pyrazine). (d) Cycling stability and coulombic efficiency of the MSCs (MN-MSCs: MOF-nanosheets-based MSCs, MB-MSCs: MOF-bulk-based MSCs). (e) The integrated MN-MSCs (in series/parallel). Adapted with permission from ref. [56]. Copyright (2020) Elsevier B.V.

example, the full-cell fiber-type ZIC based on 2D Zn structures anchored on carbon fibers (2D-Zn@CF) as anode and activated carbon on carbon fibers (AC@CF) as cathode, can deliver a high energy density of 12–25 μWh cm^{-2} at a varied power density of 3000–50 μW cm^{-2}, along with the superb stability up to 10,000 cycles (almost no capacity decay) (Figure 24.9) [59]. Of course, in principle, all kinds of material may be processed into 2D materials, and these flexible and high-surface-area nanomaterials can be incorporated into electrodes of FSCs; some may be also of great research interest, e.g., 2D ultrathin $NiCo_2O_4$ nanosheet arrays in situ grown on Ni foam, 2D ultrathin nickel-cobalt phosphate nanosheets (Ni/Co = 4:5, thickness ~5 nm), porous Cu_xO nanosheets arrays on 3D metallized cellulose fiber membrane, anisotropic boron–carbon hetero-nanosheets (derived from 2D g-C_3N_4 and boron nanosheet), and metallic layered GeP_5, etc.

24.2.5 Hybrid 2D Nanostructures

The integration of two or more 2D/layered materials may endow the FSCs with both high capacity and high stability. For example, although the pseudocapacitive transition-metal oxides (TMOs) are an important kind of electrochemically active material, the combined conductivity and hydrophilicity for the binder/filler need to be improved for wider application. The synergistic hybrid of molecularly stacked MnO_2/Ti_3C_2 (MXene) with individual advantages is combined to solve these drawbacks. It is noteworthy

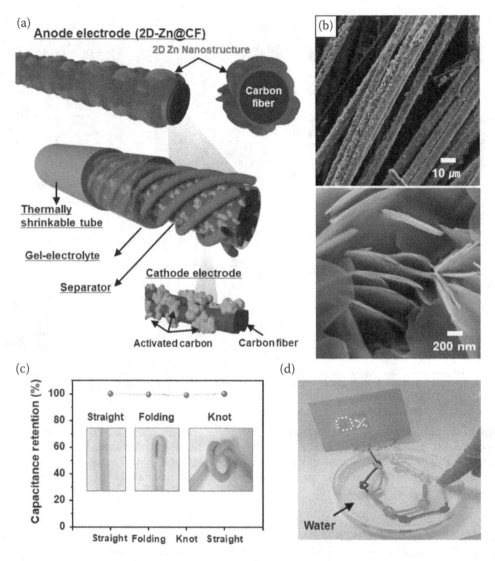

FIGURE 24.9 (a) Schematic illustration of the flexible Zn-ion supercapacitor (ZIC, 2D-Zn@CF//AC@CF, ZnSO$_4$-PVA gel electrolyte). (b) SEM images of electroplated 2D Zn@CF. (c) Capacitance retention of the ZIC under different bending states. (d) Fiber-type ZIC in series-parallel connection to light red LED arrays. Adapted with permission from ref. [59]. Copyright (2019) WILEY-VCH Verlag GmbH & Co. KGaA, Weinheim.

to mention the preparation process through scalable but simple mixing and filtration, which, however, shows top-class electrochemical performances for the symmetric planar FSCs, e.g., the maximum energy density of 8.3 W h kg^{-1} at 0.22 kW kg^{-1} and maximum power density of 2.38 kW kg^{-1} at 3.3 W h kg^{-1} (Figure 24.10) [60]. Hierarchical PPy@MnO$_2$-NSs@rGO@conductive metal yarns (PPy: polypyrrole, NSs: nanosheets), which are industrially wearable and knittable, can be developed into large-scale wearable textile supercapacitors [61].

Pseudocapacitive conducting polymer nanosheets with huge redox-active sites and rapid electron/ion transport between lamellar permeable spaces are another novel candidate for FSCs electrodes [24,25]. When incorporated with capacitive graphene nanosheets or functional TMOs nanowires for densely packed or tunable heterostructure films, for example, they demonstrate superior volumetric

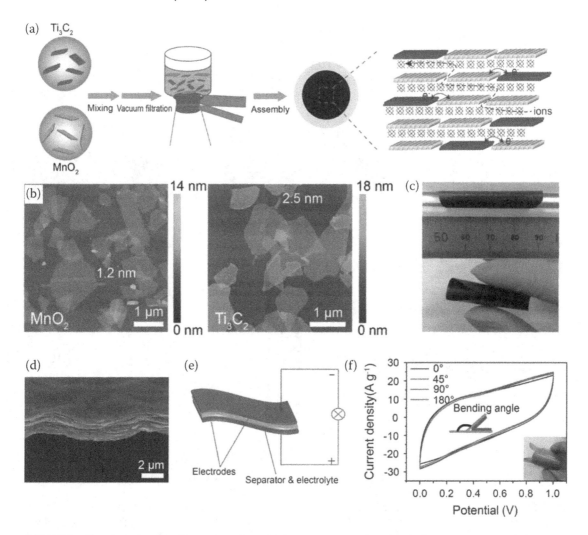

FIGURE 24.10 (a) Schematic illustration of the fabrication process of 2D MnO_2/Ti_3C_2 hybrid film via vacuum-assisted filtration. (b) AFM images of the delaminated 2D MnO_2 and Ti_3C_2 nanosheets, respectively. (c) Digital photo of the as-prepared flexible MnO_2/Ti_3C_2 hybrid film. (d) Cross-sectional SEM image of the hybrid film. (e) Schematic diagram of the flexible sandwich-structured SSC and its CV curves at varied bending angles. Adapted with permission from ref. [60]. Copyright (2017) WILEY-VCH Verlag GmbH & Co. KGaA, Weinheim.

capacitance and high rate capability, as well as mechanical robustness owing to the integrated synergistic effect [24,25]. As an example, the polythiophene nanosheet/graphene stacked-layer heterostructure film, serving as additive- and binder-free electrode supported on PET substrate for all-solid-state planar FSCs, can deliver a volumetric capacitance of 375 F cm^{-3} and areal capacitance of 3.9 mF cm^{-2} (much higher than those of single graphene-based FSCs), as well as high energy and power densities of 13 mWh cm^{-3} and 776 W cm^{-3}, respectively. Moreover, the MSCs with heterostructure electrode covered by PVA/H_2SO_4 gel electrolyte can be operated at a superior high rate of up to 1 kV s^{-1} and can provide an ultrahigh rate capability, i.e., the volumetric capacitance of 123 F cm^{-3} and areal capacitance of 1.3 mF cm^{-2} at 100 V s^{-1} (Figure 24.11) [24].

Some other examples are also listed, as follows, for further information, e.g., CNTs enhanced flexible MXene film electrode, single-layer MnO_2 nanosheet/rGO free-standing binder-free film or planar

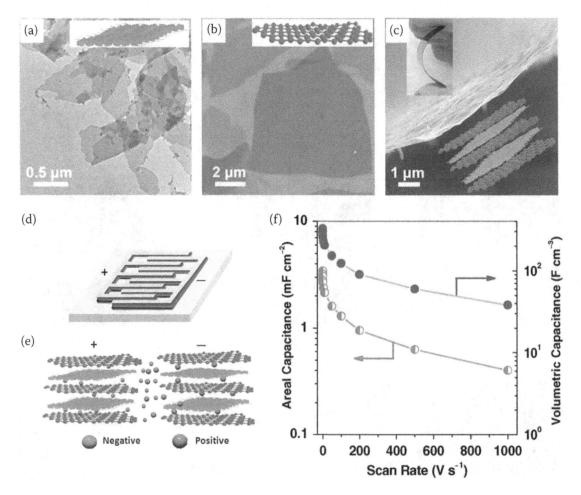

FIGURE 24.11 (a and b) TEM images of the thiophene (TP) and exfoliated graphene (EG) nanosheets, respectively (inset: schematic structures). (c) SEM of the TP/EG heterostructure film (insets: PET substrate supported film and the schematic heterostructure). (d and e) Schematic illustrations of the planar TP/EG-MSC and its side view of charging state. (f) The areal/volumetric capacitances of the TP/EG-MSCs as a function of scan rate. Adapted with permission from ref. [24]. Copyright (2016) WILEY-VCH Verlag GmbH & Co. KGaA, Weinheim.

interdigital electrode, 3D graphene/MnO_2 composite network electrode, MnO_x or Fe_2O_3 decorated MXene film electrode, solution-processable MXene/conducting polymer composite film electrode, in situ grown MWCNTs/MXene on carbon cloth electrode (MWCNTs as interlayer pillars in MXene sheets), LDH nanosheets/rGO lay-by-layer (LBL) film electrode, ultrathin $VOPO_4$ nanosheet/graphene film electrode, SnS_2 nanoflakes@PANi-NF@GF (PANi-NF = polyaniline nanofiber, GF = graphene foam), etc., these hybrid 2D electrodes efficiently help to enhance the overall performance of FSCs [2].

24.2.6 CONFIGURATIONS OF FSCs

As aforementioned, these FSCs can be used as a power source for many portable smart devices. Here we give specific attention to planar micro-supercapacitors (MSCs) as an example. Planar MSCs can be regarded as a kind of specific supercapacitor that is usually featured with flexible but miniaturized energy-storage components (e.g., on-chip energy storage devices) [62,63]. They demonstrate excellent electrochemical performance and attractive flexible design revealed from considerable experimental and theoretical investigations; however, as a promising candidate for real-world application, their intrinsically low energy density had to be overcome [62]. As such, electrode materials, microfabrication methods, precise structure/dimension control, substrates, and electrochemical properties should be optimized [62,63].

All the abovementioned 2D/layered materials could be fabricated into planar MSCs, laser machining, microstamping, mask/spray coating, inkjet printer, or 3D printing process is usually needed, and these typical examples include graphene-based MSCs [64], MXene-based MSCs [27,65], ink-jet printed MnO_2-nanosheet based MSCs [18,66], quasi-2D ultrathin MnO_2/graphene nanosheets based MSCs [67], 2D polythiophene nanosheets/graphene pseudocapacitor [24], and rGO//MXene interdigitated asymmetric flexible MSCs. Most of these are planar interdigital electrode-based all-solid-state configuration, and flexible substrates such as PET are required.

24.3 CHALLENGES AND PROSPECTS

Structural collapse or aggregation during repeated ion insertion/extraction is one of the major drawbacks that reduce the reversible capacitance and cycling lifetime of 2D materials; for layered materials, some samples own relatively small interlayer distances and strong coulombic interactions between host lattice anions and working ions, which will cause slow ion diffusion [6,7]. To optimize the structural integrity, ion transmission kinetics, and enhance the electrical conductivity of these 2D/layered materials, many strategies have been developed to overcome these issues, such as interlayer engineering of layered materials [7] and design of hybrid and hierarchical 3D structures [6]. In addition, the accurate control of size/thickness, improvement of chemical/electrochemical stability is also critical challenges [6].

For the FSCs, the flexible hybrid supercapacitors combining the advantages of capacitive cathode and a battery-type anode is an ideal and effective alternative for EDLC-based FSCs; however, the balance or matching of electrode kinetics and charge-storage capacity between the two electrodes should be considered and optimized. Fortunately, the 2D inorganic materials, i.e., graphene analogs, with rich electrochemical active sites have addressed these drawbacks, and further optimization on the lifetime could be conducted. The fabrication of coplanar asymmetric FSCs or MSCs with both electrodes comprised of 2D/layered components is a promising strategy for the achievement of high-voltage output and high energy/power densities.

The mechanical properties (including flexibility, toughness, strength, and stretchability, etc.), cycle stability, and compatibility with the application conditions should be comprehensively evaluated. The mechanical properties of the FSCs could be enhanced by synergistic effect from interfacial interactions (biomimics or "learning from nature") [32] and combination with other polymers/metal substrates (e.g., nylon yarn, fabrics, commercial carbon fibers, carbon cloth, metal wires, glass fibers, nickel foam, metal mesh, polymer films).

All-solid-state flexible supercapacitors are a trend in the next generation for real practical application in wearable energy storage devices. These all-solid-state systems are more safe and robust than traditional liquid electrolyte-based ones; however, some drawbacks, such as lower ionic conductivity of the electrolytes, should be further improved for high-power conditions. Furthermore, nowadays used flexible solid-state SCs are usually PVA/strong-electrolyte (e.g., H_2SO_4, H_3PO_4, KOH, KNO_3, LiCl, Na_2SO_4) based gel electrolytes, which may be harmful to human skin; thus, more skin-friendly and eco-friendly electrolytes are urgently wanted, and the ionic liquid gel electrolytes may be a kind of promising candidate. Self-healing FSCs or MSCs are also promising development trends in specific practical applications.

Although some FSCs show high performances, the cost-effectiveness and scalability (or scale-up) of the device-manufacturing process should be evaluated for eventual practical application [6]. The massive, low-cost production of the 2D materials with demanded quality, a critical precondition for commercial uses, should be addressed. Many of these challenges have been overcome and are showing inspiring opportunities for industrial applications [68].

ABBREVIATIONS

AFM	atomic force microscopy
CC	carbon cloth
CF	carbon fiber
CNFs	carbon nanofibers
CV	cyclic Voltammetry
(MW)CNTs	(multi-walled) carbon nanotubes
EC(s)	electrochemical capacitor(s)
EDLCs	electrochemical double layer capacitors
FFT	fast Fourier transform
GCD	galvanostatic charging-discharging
(r)GO	(reduced) graphene oxide
HRTEM	high-resolution TEM
LSG	laser-scribed graphene
MSCs	micro-supercapacitors
NDs	nanodisks
NSs	nanosheets
PET	poly(ethylene terephthalate)
PVA	polyvinyl alcohol
SAED	selected area electron diffraction
SC(s)	supercapacitor(s)
SSC	symmetric SC
STEM	scanning transmission electron microscope
TEM	transmission electron microscopy
YSCs	Yarn-shaped supercapacitors
XRD	X-ray diffraction.

REFERENCES

1. Wang F, Wu X, Yuan X, Liu Z, Zhang Y, Fu L, et al. Latest advances in supercapacitors: from new electrode materials to novel device designs. *Chem Soc Rev* 2017;46(22):6816–6854.
2. Peng X, Peng L, Wu C, Xie Y. Two dimensional nanomaterials for flexible supercapacitors. *Chem Soc Rev* 2014;43(10):3303–3323.
3. Cao XY, Halder A, Tang YY, Hou CY, Wang HZ, Duus JO, et al. Engineering two-dimensional layered nanomaterials for wearable biomedical sensors and power devices. *Mater Chem Front* 2018;2(11): 1944–1986.
4. Xue Q, Sun JF, Huang Y, Zhu MS, Pei ZX, Li HF, et al. Recent progress on flexible and wearable super-capacitors. *Small* 2017;13(45):1701827.
5. Gao L. Flexible device applications of 2D semiconductors. *Small* 2017;13(35):1603994.
6. Mendoza-Sanchez B, Gogotsi Y. Synthesis of two-dimensional materials for capacitive energy storage. *Adv Mater* 2016;28(29):6104–6135.
7. Zhang Y, Ang EH, Yang Y, Ye M, Du W, Li CC. Interlayer chemistry of layered electrode materials in energy storage devices. *Adv Funct Mater* 2021;31(4):2007358.
8. Gu TH, Kwon NH, Lee KG, Jin XY, Hwang SJ. 2D inorganic nanosheets as versatile building blocks for hybrid electrode materials for supercapacitor. *Coord Chem Rev* 2020;421:213439.
9. Han Y, Ge Y, Chao YF, Wang CY, Wallace GG. Recent progress in 2D materials for flexible super-capacitors. *J Energy Chem* 2018;27(1):57–72.
10. Guo YP, Wei YQ, Li HQ, Zhai TY. Layer structured materials for advanced energy storage and conversion. *Small* 2017;13(45):1701649.
11. Xu Z, Gao C. Graphene in macroscopic order: liquid crystals and wet-spun fibers. *Acc Chem Res* 2014; 47(4):1267–1276.

12. Huo PP, Zhao P, Wang Y, Liu B, Yin GC, Dong MD. A Roadmap for achieving sustainable energy conversion and storage: graphene-based composites used both as an electrocatalyst for oxygen reduction reactions and an electrode material for a supercapacitor. *Energies* 2018;11(1):167.

13. Dong YF, Wu ZS, Ren WC, Cheng HM, Bao XH. Graphene: a promising 2D material for electrochemical energy storage. *Sci Bull* 2017;62(10):724–740.

14. Huang L, Santiago D, Loyselle P, Dai LM. Graphene-based nanomaterials for flexible and wearable supercapacitors. *Small* 2018;14(43):1800879.

15. Peng L, Zhu Y, Li H, Yu G. Chemically integrated inorganic-graphene two-dimensional hybrid materials for flexible energy storage devices. *Small* 2016;12(45):6183–6199.

16. Chen T, Dai LM. Flexible supercapacitors based on carbon nanomaterials. *J Mater Chem A* 2014;2(28): 10756–10775.

17. Ni W, Shi L. Layer-structured carbonaceous materials for advanced Li-ion and Na-ion batteries: beyond graphene. *J Vac Sci Technol A* 2019;37(4):040803.

18. Ten Elshof JE, Wang Y. Advances in ink-jet printing of MnO_2-nanosheet based pseudocapacitors. *Small Methods* 2019;3(8):1800318.

19. Guan ML, Wang QW, Zhang X, Bao J, Gong XZ, Liu YW. Two-dimensional transition metal oxide and hydroxide-based hierarchical architectures for advanced supercapacitor materials. *Front Chem* 2020;8:390.

20. Lin LX, Lei W, Zhang SW, Liu YQ, Wallace GG, Chen J. Two-dimensional transition metal dichalcogenides in supercapacitors and secondary batteries. *Energy Storage Mater* 2019;19:408–423.

21. Aslam MK, Niu Y, Xu M. MXenes for non-lithium-ion (Na, K, Ca, Mg, and Al) batteries and supercapacitors. *Adv Energy Mater* 2021;11(2):2000681.

22. Wang H, Wu Y, Yuan XZ, Zeng GM, Zhou J, Wang X, et al. Clay-inspired MXene-based electrochemical devices and photo-electrocatalyst: state-of-the-art progresses and challenges. *Adv Mater* 2018;30(12): 1704561.

23. Wu YP, Yuan W, Xu M, Bai SG, Chen Y, Tang ZH, et al. Two-dimensional black phosphorus: properties, fabrication and application for flexible supercapacitors. *Chem Eng J* 2021;412:128744.

24. Wu ZS, Zheng YJ, Zheng SH, Wang S, Sun CL, Parvez K, et al. Stacked-layer heterostructure films of 2D thiophene nanosheets and graphene for high-rate all-solid-state pseudocapacitors with enhanced volumetric capacitance. *Adv Mater* 2017;29(3):1602960.

25. Zhang J, Fan X, Meng X, Zhou J, Wang M, Chen S, et al. Ice-templated large-scale preparation of two-dimensional sheets of conjugated polymers: thickness-independent flexible supercapacitance. *ACS Nano* 2021;15:8870–8882.

26. Ni W, Shi L. Metal-organic-framework composites as proficient cathodes for supercapacitor applications. *Mater Res Found* 2019;58:177–238.

27. Orangi J, Hamade F, Davis VA, Beidaghi M. 3D printing of additive-free 2D $Ti_3C_2T_x$ (MXene) ink for fabrication of micro-supercapacitors with ultra-high energy densities. *ACS Nano* 2020;14(1):640–650.

28. Yu YC, Bai S, Wang ST, Hu AM. Ultra-short pulsed laser manufacturing and surface processing of microdevices. *Engineering* 2018;4(6):779–786.

29. Wan ZF, Streed EW, Lobino M, Wang SJ, Sang RT, Cole IS, et al. Laser-reduced graphene: synthesis, properties, and applications. *Advanced Materials Technologies* 2018;3(4):1700315.

30. Azadmanjiri J, Thuniki NR, Guzzetta F, Sofer Z. Liquid metals-assisted synthesis of scalable 2D nanomaterials: prospective sediment inks for screen-printed energy storage applications. *Adv Funct Mater* 2021;31(17):2010320.

31. Abdolhosseinzadeh S, Schneider R, Verma A, Heier J, Nuesch F, Zhang CF. Turning trash into treasure: additive free MXene sediment inks for screen-printed micro-supercapacitors. *Adv Mater* 2020;32(17): 2000716.

32. Gong SS, Ni H, Jiang L, Cheng QF. Learning from nature: constructing high performance graphene-based nanocomposites. *Mater Today* 2017;20(4):210–219.

33. Zhong J, Sun W, Wei QW, Qian XT, Cheng HM, Ren WC. Efficient and scalable synthesis of highly aligned and compact two-dimensional nanosheet films with record performances. *Nat Commun* 2018;9:3484.

34. El-Kady MF, Strong V, Dubin S, Kaner RB. Laser scribing of high-performance and flexible graphene-based electrochemical capacitors. *Science* 2012;335(6074):1326–1330.

35. Hong XD, Fu JW, Liu Y, Li SG, Wang XL, Dong W, et al. Recent progress on graphene/polyaniline composites for high-performance supercapacitors. *Materials* 2019;12(9):1451.

36. Su F, Hou XC, Qin JQ, Wu ZS. Recent advances and challenges of two-dimensional materials for high-energy and high-power lithium-ion capacitors. *Batteries Supercaps* 2020;3(1):10–29.

37. Lang JW, Zhang X, Liu B, Wang RT, Chen JT, Yan XB. The roles of graphene in advanced Li-ion hybrid supercapacitors. *J Energy Chem* 2018;27(1):43–56.

38. Feng J, Sun X, Wu C, Peng L, Lin C, Hu S, et al. Metallic few-layered VS_2 ultrathin nanosheets: high two-dimensional conductivity for in-plane supercapacitors. *J Am Chem Soc* 2011;133(44): 17832–17838.

39. Zhang C, Yin H, Han M, Dai Z, Pang H, Zheng Y, et al. Two-dimensional tin selenide nanostructures for flexible all-solid-state supercapacitors. *ACS Nano* 2014;8(4):3761–3770.

40. Wang C, Wu X, Ma Y, Mu G, Li Y, Luo C, et al. Metallic few-layered VSe_2 nanosheets: high two-dimensional conductivity for flexible in-plane solid-state supercapacitors. *J Mater Chem A* 2018;6(18): 8299–8306.

41. Yu P, Fu W, Zeng Q, Lin J, Yan C, Lai Z, et al. Controllable synthesis of atomically thin type-II Weyl semimetal WTe_2 nanosheets: an advanced electrode material for all-solid-state flexible supercapacitors. *Adv Mater* 2017;29(34):1701909.

42. Sun G, Zhang X, Lin R, Yang J, Zhang H, Chen P. Hybrid fibers made of molybdenum disulfide, reduced graphene oxide, and multi-walled carbon nanotubes for solid-state, flexible, asymmetric supercapacitors. *Angew Chem Int Ed* 2015;54(15):4651–4656.

43. Jeon H, Jeong JM, Kang HG, Kim HJ, Park J, Kim DH, et al. Scalable water-based production of highly conductive 2D nanosheets with ultrahigh volumetric capacitance and rate capability. *Adv Energy Mater* 2018;8(18):1800227.

44. Zang XB, Wang JL, Qin YJ, Wang T, He CP, Shao QG, et al. Enhancing capacitance performance of $Ti_3C_2T_x$ MXene as electrode materials of supercapacitor: from controlled preparation to composite structure construction. *Nano-Micro Lett* 2020;12(1):77.

45. Ma C, Ma M-G, Si C, Ji X-X, Wan P. Flexible MXene-based composites for wearable devices. *Adv Funct Mater* 2021;31:2009524.

46. Zhang JZ, Kong N, Uzun S, Levitt A, Seyedin S, Lynch PA, et al. Scalable manufacturing of free-standing, strong $Ti_3C_2T_x$ MXene films with outstanding conductivity. *Adv Mater* 2020;32(23):2001093.

47. Zhang M, Cao J, Wang Y, Song J, Jiang T, Zhang Y, et al. Electrolyte-mediated dense integration of graphene-MXene films for high volumetric capacitance flexible supercapacitors. *Nano Res* 2021;14(3): 699–706.

48. Xu S, Wei G, Li J, Han W, Gogotsi Y. Flexible MXene–graphene electrodes with high volumetric capacitance for integrated co-cathode energy conversion/storage devices. *J Mater Chem A* 2017;5(33): 17442–17451.

49. Tian WQ, Vahid Mohammadi A, Wang Z, Ouyang LQ, Beidaghi M, Hamedi MM. Layer-by-layer self-assembly of pillared two-dimensional multilayers. *Nat Commun* 2019;10:2558.

50. Wang YM, Wang X, Li XL, Bai Y, Xiao HH, Liu Y, et al. Engineering 3D ion transport channels for flexible MXene films with superior capacitive performance. *Adv Funct Mater* 2019;29(14):1900326.

51. Tian WQ, Vahid Mohammadi A, Reid MS, Wang Z, Ouyang LQ, Erlandsson J, et al. Multifunctional nanocomposites with high strength and capacitance using 2D MXene and 1D nanocellulose. *Adv Mater* 2019;31(41):1902977.

52. Seyedin S, Yanza ERS, Razal JM. Knittable energy storing fiber with high volumetric performance made from predominantly MXene nanosheets. *J Mater Chem A* 2017;5(46):24076–24082.

53. Yang Q, Xu Z, Fang B, Huang T, Cai S, Chen H, et al. MXene/graphene hybrid fibers for high performance flexible supercapacitors. *J Mater Chem A* 2017;5(42):22113–22119.

54. Wang Z, Qin S, Seyedin S, Zhang J, Wang J, Levitt A, et al. High-Performance biscrolled MXene/carbon nanotube yarn supercapacitors. *Small* 2018;14(37):1802225.

55. Wang MC, Shi HH, Zhang PP, Liao ZQ, Wang M, Zhong HX, et al. Phthalocyanine-based 2D conjugated metal-organic framework nanosheets for high-performance micro-supercapacitors. *Adv Funct Mater* 2020; 30(30):2002664.

56. Dai FN, Wang XK, Zheng SH, Sun JP, Huang ZD, Xu B, et al. Toward high-performance and flexible all-solid-state micro-supercapacitors: MOF bulk vs. MOF nanosheets. *Chem Eng J* 2021;413:127520.

57. Zhao WW, Peng JL, Wang WK, Jin BB, Chen TT, Liu SJ, et al. Interlayer hydrogen-bonded metal porphyrin frameworks/MXene hybrid film with high capacitance for flexible all-solid-state supercapacitors. *Small* 2019;15(18):1901351.

58. Zhao SH, Wu HH, Li YL, Li Q, Zhou JJ, Yu XB, et al. Core-shell assembly of carbon nanofibers and a 2D conductive metal-organic framework as a flexible free-standing membrane for high-performance supercapacitors. *Inorg Chem Front* 2019;6(7):1824–1830.

59. An GH, Hong J, Pak S, Cho Y, Lee S, Hou B, et al. 2D metal Zn nanostructure electrodes for high-performance Zn ion supercapacitors. *Adv Energy Mater* 2020;10(3):1902981.
60. Liu W, Wang Z, Su Y, Li Q, Zhao Z, Geng F. Molecularly stacking manganese dioxide/titanium carbide sheets to produce highly flexible and conductive film electrodes with improved pseudocapacitive performances. *Adv Energy Mater* 2017;7(22):1602834.
61. Huang Y, Hu H, Huang Y, Zhu M, Meng W, Liu C, et al. From industrially weavable and knittable highly conductive yarns to large wearable energy storage textiles. *ACS Nano* 2015;9(5):4766–4775.
62. Zhang JH, Zhang GX, Zhou T, Sun SH. Recent developments of planar micro-supercapacitors: fabrication, properties, and applications. *Adv Funct Mater* 2020;30(19):1910000.
63. Zhang P, Wang F, Yu M, Zhuang X, Feng X. Two-dimensional materials for miniaturized energy storage devices: from individual devices to smart integrated systems. *Chem Soc Rev* 2018;47(19):7426–7451.
64. Zhang GF, Han YY, Shao CX, Chen N, Sun GQ, Jin XT, et al. Processing and manufacturing of graphene-based microsupercapacitors. *Mater Chem Front* 2018;2(10):1750–1764.
65. Qin L, Tao Q, Liu X, Fahlman M, Halim J, Persson POÅ, et al. Polymer-MXene composite films formed by MXene-facilitated electrochemical polymerization for flexible solid-state microsupercapacitors. *Nano Energy* 2019;60:734–742.
66. Wang Y, Zhang YZ, Gao YQ, Sheng G, Ten Elshof JE. Defect engineering of MnO_2 nanosheets by substitutional doping for printable solid-state micro-supercapacitors. *Nano Energy* 2020;68:104306.
67. Peng L, Peng X, Liu B, Wu C, Xie Y, Yu G. Ultrathin two-dimensional MnO_2/graphene hybrid nanostructures for high-performance, flexible planar supercapacitors. *Nano Lett* 2013;13(5):2151–2157.
68. Yang LS, Chen WJ, Yu QM, Liu BL. Mass production of two-dimensional materials beyond graphene and their applications. *Nano Res* 2021; 14:1583–1597.

25 Flexible Supercapacitors Based on Metal Oxides

Haradhan Kolya and Chun-Won Kang
Department of Housing Environmental Design, and Research Institute of
Human Ecology, College of Human Ecology, Jeonbuk National University,
Jeonju, Jeonbuk, Republic of Korea

Sarbaranjan Paria, Subhadip Mondal, and Changwoon Nah
Department of Polymer-Nano Science and Technology, Jeonbuk National
University, Jeonju, Jeonbuk, Republic of Korea

CONTENTS

25.1 Introduction.. 461
25.2 Key characteristics of flexible supercapacitors 463
25.3 Equations of super-capacitance measurement 463
25.4 Metal oxide based-flexible supercapacitors.. 464
    25.4.1 Drawbacks of MOs for supercapacitors 464
    25.4.2 How to overcome the drawbacks?.. 464
    25.4.3 Various electrolytes in supercapacitors 465
    25.4.4 Various metal oxides in supercapacitors.. 466
        25.4.4.1 Ruthenium oxide-based supercapacitors........................... 466
        25.4.4.2 Manganese-oxide-based supercapacitors........................... 467
        25.4.4.3 Other metal oxide-based supercapacitors.......................... 467
    25.4.4 Transparent supercapacitors 474
25.5 Conclusions.. 476
Acknowledgments.. 477
References... 477

25.1 INTRODUCTION

Nowadays, the shortage of nonrenewable energy supply and pollution have led to a significant advance in the strategy and expansion of energy storage devices (ESD). Two important factors, safety and flexibility for energy storage systems, improve energy delivery and have a quick charging and discharging rate, energy density (ED), and a long lifetime [1]. The devices used to store energy are known as supercapacitors (SCs). SCs are made up of two parallel-plate electrodes, an ion-permeable separator, and an electrolyte solution. They have adequate energy and power densities for moderate to high-power applications. They are storage devices that fall somewhere between ordinary capacitors and batteries [2]. SCs are distinct from batteries and fuel cells by their long charging/discharging cycles and broad operating temperature [3]. The energy storage mechanism (ESM) in SCs mostly depends on charge accumulation or redox reactions. Based on the ESM, SCs are classified into three categories, such as EDLCs (electrical double-layer capacitors), PCs (pseudocapacitors), and HCs (hybrid capacitors) [4].

EDLCs are ultracapacitors that are made up of conducting polymer electrodes or porous electrodes that are immersed in an electrolyte solution. Graphene-activated carbon, active carbon fiber, and carbon nanotubes (CNTs) are examples of extremely porous carbonaceous materials with high specific surface area (SSA) [5]. Carbonaceous materials, which are a popular form of electrode material in EDLCs, might be made from biomass. Carbonaceous materials exhibit high conductivity and mechanical durability; therefore, these electrode materials have become attractive for various uses. The storage time is very short, i.e., 30–60 seconds [4]. It functions as reversible ion adsorption on the electrode/electrolyte interfaces to physically disperse charges. The mechanism relies purely on the physical adsorption of ions and hence does not require any chemical reactions [6]. In detail, when a voltage is supplied to the EDLC, electrons, pass across the external loop from the negative electrode to the positive electrode, accumulating positive (+ve) and negative (-ve) charges on the corresponding electrodes. The electrolyte solution's cations attract toward the negative electrode, whereas anions tend toward the positive electrodes [7]. Furthermore, because the potential decrease is restricted to a narrow region (0.1–10 nm), EDLCs have a greater energy density (ED) than traditional capacitors. However, the capacitance is restricted by the charging mechanism in the energy storage process, resulting in a high power density (PD) but a relatively low capacitance [8].

PCs can store energy chemically due to redox reactions that occur concerning electrode and electrolytes. In detail, the ESM in PCs involves a fast and reversible Faraday reaction at or near the active material's surface. This is like the charging and discharging process in batteries; however, it does not change the phase of electrode materials [9]. MOs and conductive polymers (CPs) are the common electrode products used in such redox reactions [10]. The PCs combine two types of pseudo-capacitance: intercalation and redox pseudo-capacitance [11]. The electron transfer occurs in the redox pseudo-capacitance process when the ions in the electrolyte solution get adsorbed on the electrode's surface. However, CPs have less cycle stability due to frequent swelling and shrinking of the polymeric backbone throughout the doping/de-doping method [12]. To address the issues with EDLCs and PCs, several efforts have been made to develop HCs that combine the benefits of both EDLCs and PCs. HCs have a higher specific capacitance compared to EDLCs and PCs [13].

HCs have been described as the third form of SCs. In HCs or hybrid supercapacitors (HSCs), there are mainly two asymmetrical electrode arrangements [14]. The HSCs are not the equivalent of asymmetric supercapacitors [15]. A capacitive electrode (CE) and a battery-type Faradaic electrode are found in HSCs. ASCs contain two CEs, which possibly will be EDLCs or PCs and one EDLC/one PCs [16]. Earliest MOs, ruthenium dioxide (RuO_2), and manganese dioxide (MnO_2) have been showcased to have redox PCs characteristics [17–19]. The electrochemical activities of redox PCs exhibit almost similar properties to EDLCs electrodes. It also has galvanic charge-discharge (GCD) curves with linear and triangular shapes, and cyclic voltammetry (CV) curves with rectangular shapes [20]. PCs materials (e.g., V_2O_5, Nb_2O_5, TiO_2, WO_3, and MoO_3) can store charge [21]. However, the crucial factor for energy-efficient applications is the blend of innovative materials, which provides a much higher surface-to-volume ratio [22]. In rapid Faraday redox reactions, MOs can transmit multiple electrons to increase discharge time and therefore enhance ED. It's because there are numerous oxidation states present in MOs [23]. In addition, transition metal hydroxides (MHOs), sulfides, and selenides have received attention in this regard [24]. In modern electronics, hierarchical nanoarchitectures with significant void space, unusual porous networks, and multiple active sites have piqued interest for preparing advanced SCs electrodes [25]. Metal hydroxides or derivatives of copper (Cu), nickel (Ni), cobalt (Co), and cadmium (Cd) compounds are commonly used as charge storage materials in batteries [26,27]. MOs, MHOs, and LDH (layered double hydroxides) have all been studied for SCs until now. However, researchers have concentrated on synthetic methods, nanostructure construction, material optimization, and composite electrode design patterns [28,29]. Besides, micro-supercapacitors (MSCs) with in-plane electrode configurations have also piqued the attention of researchers working on flexible electronic devices. Because they are easier to incorporate on chips or thin devices, MSCs are typically used when high energy and extended cycle stability are required, and they are well-matched with more conventional micro-batteries despite their low ED [30].

The relevance of MOs in the making of flexible SCs is the key attention of this chapter. It describes in detail the recent advances in MOs-based flexible SCs. The fabrication and the electrochemical performances of different electrodes are also explored to provide a broad view of the topic. Finally, the limitations, prospects, and future scope for improving the capacitive efficiency of MOs-based flexible SCs electrodes are addressed.

25.2 KEY CHARACTERISTICS OF FLEXIBLE SUPERCAPACITORS

Specific capacitance (C) (normalized by electrode mass, volume, or area), ED, PD, rate capability (retained capacitance under high current loading), and cycle stability are some of the most important key metrics of SCs. To improve the energy and PD of a SC, the C and operating voltage window should be increased, whereas the equivalent series resistance should indeed be reduced. The maximum operating voltage window for EDLC Scs is mostly determined by the electrolyte used, which is restricted by the electrolyte's stability [31]. The following characteristics are required of ideal electrode materials:

1. Low cost, flexibility, high cycle stability, good electrochemical characteristic (s), and more typical mechanical properties
2. Controlled porosity that impacts the C and effective use of time
3. C depends on SSA
4. Higher electrical conductivity, significant for calculating rate and PD
5. More electroactive sites, crucial for pseudo-capacitance
6. Excellent thermo-chemical stability

The key properties of several kinds of SCs are provided in Table 25.1.

25.3 EQUATIONS OF SUPER-CAPACITANCE MEASUREMENT

The equations that are commonly used to determine capacitance are below:

$$\text{(i) Specific capacitance: } C = \frac{Q}{V} \tag{25.1}$$

TABLE 25.1

Characteristics of Various Supercapacitors

SCs types	Electrode materials	Charge storage mechanism	Merits/limitations
EDLCs	Carbon materials for both electrodes.	Charge absorption/desorption at the electrode-electrolyte interface.	Good cycling performance and high PD; low ED and C.
PCs	Metal oxides, conductive polymers and composites of pseudocapacitive materials with carbon materials.	Reversible surface Faradic redox processes.	High capacitance and high ED; poor cycling performance and low working voltage.
EDLC// PCs- ASCs	Carbon materials for the anode and pseudocapacitive materials for the cathode.	One electrode with redox reactions and EDL absorption/desorption for another electrode without Faradic process.	High PD, ED and good cycling ability.

(ii) Electrochemical double – layer capacitance: $EDLCs = \dfrac{\varepsilon_r \; \varepsilon_0}{d} A$ (25.2)

(iii) Pseudo – capacitance: $PCs = \dfrac{n \times F}{M \times V}$ (25.3)

(iv) Energy density: $ED = \dfrac{CV^2}{2}$ (25.4)

(v) Power density: $PD = \dfrac{V^2}{4R}$ (25.5)

Where C: specific capacitance, F: Faraday constant, d: the effective thickness of EDL, A: SSA of the electrode, M: molar mass of MOs, n: mean of electrons participated in the redox reaction, Q: charge of electrode/unit mass, R: equivalent resistance, ε_r: relative permittivity of the medium in EDLCs, ε_0: permittivity of vacuum, and V: operating voltage window.

25.4   METAL OXIDE BASED-FLEXIBLE SUPERCAPACITORS

Metal oxides have been identified as potential materials for use as electrodes in ESDs due to their high SSA, flexible components and surface morphologies, and high C (theoretical). MOs play a key role in the electrodes of electrochemical SCs, and by modifying and regulating their defects and surface/interfaces on a nanoscale, they can significantly increase the capacitance. The C of MOs electrodes is greater than that of carbonaceous electrodes. Many studies reported that only MOs electrodes with a low current potential can produce significant C and high ED [31]. Metal oxides have a higher ED than ordinary carbon electrode materials and have superior electrochemical stability to polymer materials. Depositing metal oxide/hydroxide particles on electrodes to improve their C is a good concept. The MOs/hydroxide might be ruthenium oxide (RuO_2), manganese oxide (MnO_2), cobalt oxide (Co_3O_4), nickel oxide (NiO), tin oxide (SnO_2), zinc oxide (ZnO), titanium oxide (TiO_2), ferric oxide (Fe_3O_4) or mixed metals ions, proven to play a crucial role in improving electrode capacitance through rapid faradaic pseudo-capacitance effects.

25.4.1   DRAWBACKS OF MOS FOR SUPERCAPACITORS

The following are the drawbacks of MOs in supercapacitors applications.

a. Most MOs, besides RuO_2, have extremely poor conductivity.
b. MOs' high resistivity raises at electrode's sheet and charges the transfer process, which is responsible for significant energy loss at higher current density. The energy densities and low rate capability (RC) restricts the end-use.
c. Charge-discharge cycles cause electrode cracking in pure MOs, giving in poor long-term stability.
d. MOs are challenging to modify in terms of development SSA, pore dispersion, and porosity.

25.4.2   HOW TO OVERCOME THE DRAWBACKS?

It becomes feasible to develop an HSCs electrode that combines the benefits of both carbon and MOs while minimizing the drawbacks of each. Carbon nanostructures functioned as charge-transfer channels and physically supported MOs in carbonaceous–MOs HSCs [31]. Carbonaceous materials have good

TABLE 25.2

Comparison of Three Electrochemical Electrodes

Carbon materials	MOs materials	MOs–carbon composite
1. High SSA	1. Low SSA	1. SSA is used to regulate various carbon support.
2. Pore size distribution changes depending on how the surface is modified.	2. Changes in pore size distribution are difficult to attain.	2. Various tailoring options based on carbon support Changes in pore size distribution
3. Low specific capacitance	3. High specific capacitance	3. High C
4. High conductivity	4. Low conductivity	4. Tunable conductivity
5. High rate capabilities	5. Low rate capabilities	5. Good rate capabilities
6. Good stability	6. Poor stability	6. Good stability
7. Low cost	7. High cost	7. Moderate cost

electrical conductivity, which increases ED and RC at high charge/discharge currents. MOs' electro-activities contribute to the high C and ED of carbonaceous–MOs HSCs electrodes [31]. The effectiveness of carbon, MOs, and MOs–carbon composite electrodes are compared, as shown in Table 25.2.

25.4.3 VARIOUS ELECTROLYTES IN SUPERCAPACITORS

Electrolytes are also an important factor for the performance of SCs. The breakdown voltage of SCs is controlled by the decomposition potential range of electrolytes, which restricts their energy and power densities. A high operating voltage might result in ED and PD. Furthermore, electrolyte ionic conductivity is a crucial characteristic related to ED. High electrochemical stability, temperature coefficient, low viscosity, high ionic concentrations (ICs), low toxicity, relatively low cost, and low volatility are all considerations in the choosing of electrolytes. Three electrolytes, such as aqueous, organic, and ionic liquid, are majorly used in SCs [32]. Electrolytes have their unique set of benefits and drawbacks, which are detailed in Table 25.3.

TABLE 25.3

Benefits and Difficulties of Electrolytes

Electrolytes	Examples	Advantages	Disadvantages
Aqueous	i. Acidic solutions (H_2SO_4) (H_3PO_4) ii. Alkaline solutions (KOH) iii. Neutral solutions (Li_2SO_4, Li Cl, $LiClO_4$ and Na_2SO_4)	• Improved capacitance and rate performance • Provide ICs • Low resistance • Low cost	• Flammability • Health issues in practice • Restricting the ED • Thermodynamic decomposition of water • Narrow potential window
Organic electrolytes	Acetonitrile (CH_3CN) Propylene carbonate ($C_4H_6O_3$)	• Broad potential window • Higher ED and PD • More expensive	
Water-in-salt	$LiTFSILiTFSI/MnO_2$	• Provide a broad potential window • Resolve the issue of organic electrolyte flammability	• Expensive

25.4.4 Various metal oxides in supercapacitors

25.4.4.1 Ruthenium oxide-based supercapacitors

It was the first candidate in SCs application due to good stability, best conductivity, more numbers of oxidation states, high potential window, and extended cycle stability. But, high cost in production, rare-earth element, agglomeration effects, and barriers issues restrict its commercial usage. Therefore, RuO_2-based nanocomposites have been extensively investigated to reduce material costs while improving electrochemical performance. Generally, researchers have used a strategy of mixing RuO_2 with other low-cost carbonaceous compounds to reduce cost while maintaining high capacitance [33]. A schematic representation of RuO_2 with other low-cost materials was shown in Figure 25.1.

Researchers reported several techniques of fabrication for the synthesis of both hydrous and anhydrous nano-forms of RuO_2 to get high C with broad potential windows, which leads to higher ED and PD. Various synthetic techniques for preparing RuO_2 nanostructures for energy storage purposes are represented schematically, as shown in Figure 25.2.

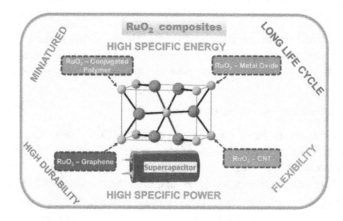

FIGURE 25.1 A schematic of RuO_2 with other low-cost materials. Adapted with permission from ref. [33]. Copyright (2019) John Wiley and Sons.

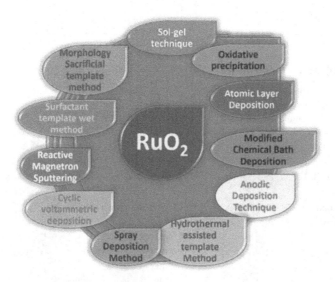

FIGURE 25.2 Schematic of the various synthesis process of RuO_2 nanostructures. Adapted with permission from ref. [33]. Copyright (2019) John Wiley and Sons.

The charge-discharge mechanism associated with the proton-coupled-electrons transfer process from mixed-conducting hydrous RuO_2, shown below.

$$RuO_2 + xH^+ = RuO_{2-x,} OH_x (O < x < 2) \tag{25.6}$$

It is clear from the above equation that the oxide surface can be charged in two ways: by using a proper electrode potential and by modifying the pH of the medium. The stability of higher oxidation states of interfacial Ru atoms in an alkaline solution are due to increased oxygen evolution potential [33]. Besides, RuO_2 based micro-supercapacitors are developed by laser irradiating a spin-coated bilayered tetra-chloroauric acid-cellulose acetate/RuO_2 film, resulting in adherent Au/RuO2 electrodes with a distinct pillar morphology. The as-prepared microdevices provide 27 mF cm^2/540 F cm^3 in 1 (M) H_2SO_4 and retain 80% of their original capacitance after 10,000 cycles, as shown in Figure 25.3 [34].

Furthermore, the carbon component in RuO_2/carbon composites primarily functions for two purposes. First, carbon might help prevent RuO_2 particles from aggregating together and lead to a more uniform RuO_2 dispersion, as shown in Figure 25.4. Second, adding carbon to the electrode/electrolyte interfaces makes it easier to transport ions and electrons. Furthermore, as compared to single RuO_2, the conductivity is improved. There are still some issues to be handled, such as the complicated preparation process, lower electroactive species-usage efficiency, and nonuniform nanoparticle dispersion in the carbon matrix.

25.4.4.2   Manganese-oxide-based supercapacitors

MnO_2-based composites electrodes have attracted more attention recently because of their high PD and extended cycle life for energy-related applications. But, its performance and usage were limited due to its poor electrical conductivity. There are typically two methods to increase the electrochemical performance of MnO_2-based SCs. The first method is to make SCs with carbon materials as anodes and nanostructure MnO_2 with a large SSA as cathodes; this method has been intensively explored as promising for ASCs. Another method to improve the electrochemical properties of MnO_2-based SCs is to combine them with carbon or CPs, to increase conductivity and power capability [36]. The advances of MnO_2 based SCs are presented in Table 25.4.

25.4.4.3   Other metal oxide-based supercapacitors

Zinc oxide (ZnO) is widely employed in gas sensors, solar cells, piezo-electric nanogenerators, photodetectors, nanolasers, short-wavelength light-emitting devices, and electroluminescence devices; it is chosen because of excellent optical, electrical properties, such as wide-band-gap (3.37 eV) semiconductor material [10]. ZnO is a good candidate for SCs applications as it has a high ED. It's a brilliant idea to combine one-dimensional and two-dimensional ZnO structures with other one-dimensional/two-dimensional materials to form nanocomposites. It increases the electrochemical characteristics of ZnO in composite materials. ZnO-based materials have recently developed as a viable candidate for advanced SCs. Studies have evaluated a composite of ZnO with a variety of materials, as shown in Figure 25.5 [37].

In 2019, a two-step approach has been used to develop an extraordinary flexible electrode (FE) consisting of porous carbonized cotton, ZnO nanoparticles, and CuS microspheres. The porous carbonized cotton/ZnO/CuS composite electrode exhibited promising C (1830 mF cm^{-2} at 2 mA cm^{-2}), a high rate performance, and excellent cycle stability 14.8% loss after 5000 cycles.

Titanium oxide: TiO_2 is a useful metal oxide for SCs because of its facile synthesis, high ED, low cost, nontoxicity, environmental friendliness, and high purity. TiO_2 is an n-type semiconductor that is utilized in solar cells, hydrogen generation and ESDs, SCs, and batteries, and plenty of other things. Incorporating inorganic nanoparticles with organic conducting polymers to produce hybrid nanocomposite materials appeared to be a promising way to develop electrochemical ESDs. Because of its

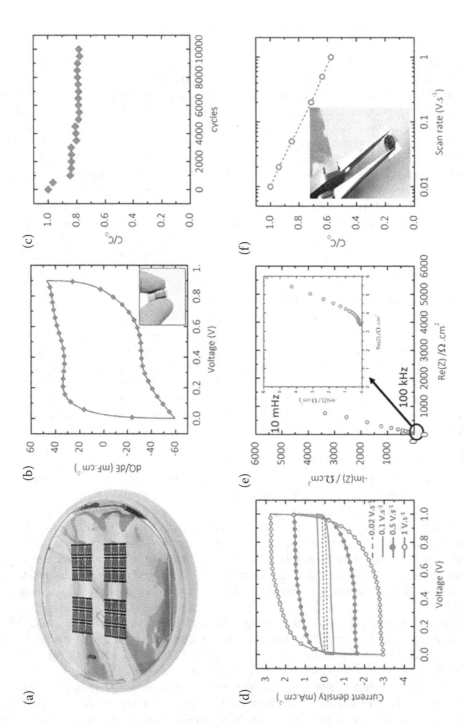

FIGURE 25.3 Laser-writing of flexible RuO_2/Au micro-supercapacitors (a) Integration of 36 micro-supercapacitors on polyimide sheet, (b) CV of a RuO_2/Au flexible micro-supercapacitor recorded at 5 mV s^{-1} in 1 (M) H_2SO_4 and (c) the corresponding capacitance retention of the device over 10,000 cycles, (d) Cyclic voltammogram of a RuO_2/Au flexible micro-supercapacitors (power design) at various scan rate, (e) the corresponding Nyquist plot, and (f) change of the areal capacitance with the scan rate. Adapted with permission from ref. [34]. Copyright (2019) John Wiley and Sons.

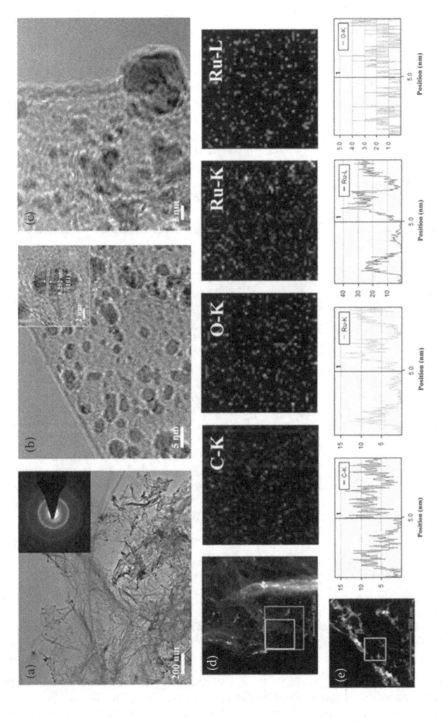

FIGURE 25.4 (a) TEM image of RuO$_2$@C/RGO. Inset is the corresponding SAED pattern. (b and c) HRTEM images of RuO$_2$@C/RGO. Inset is a lattice-resolved high-resolution TEM image. (d) TEM image and corresponding EDX mapping of C, O and Ru atoms. (e) Cross-sectional TEM image and EDX line scan profiles of C, O and Ru atoms. The red short line in (e) denotes the path examined in the EDX line scan of panels. Adapted with permission from ref. [35]. Copyright (2013) Royal Society of Chemistry.

TABLE 25.4

Advances of MnO$_2$-Based Supercapacitors

Supercapacitors	Voltage window (V)	The ED	Retention (%)	Cycles
PPy–MnO$_2$–CC	1.8	8.67 mWh cm^{-3}	98.6	1000
CNTs@DNA-MnO$_2$//CNTs@DNA	1.5	11.6 Wh kg^{-1}	91.6	10 000
MnO$_2$	1.8	17.0 Wh kg^{-1}	94.0	23 000
MnO$_2$	2.0	21.0 Wh kg^{-1}	88.0	1000
CuO@ MnO$_2$//MEGO	1.8	22.1 Wh kg^{-1}	101.5	10 000
MnO$_2$/CNT	2.0	23.2 Wh kg^{-1}	95.0	5000
MnO$_2$-graphene	2.0	25.2 Wh kg^{-1}	96.0	500
Mn^{2+}/Mn^{3+}	2.0	27.7 Wh kg^{-1}	92.9	10000
p-BC@ MnO$_2$//p-BC	2.0	32.91 Wh kg^{-1}	95.4	2000
MnO$_2$/PANI@cellulose//rGO@cellulose	1.7	41.5 Wh kg^{-1}	78.0	3000
MnO$_2$//FMCNTs	2.0	47.4 Wh kg^{-1}	90.0	1000
MnO$_2$-graphene	2.0	97.2 Wh kg^{-1}	97.0	10 000

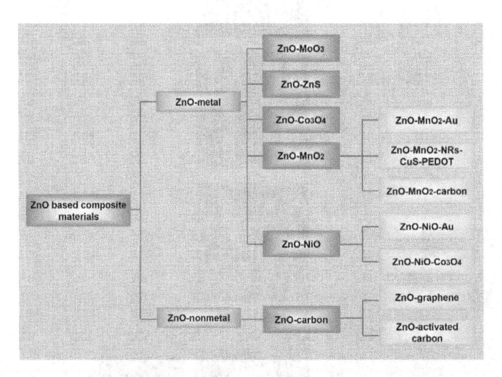

FIGURE 25.5 Chart of ZnO based composite materials for advanced supercapacitors applications. Adapted with permission from ref. [37]. Copyright (2018) John Wiley and Sons.

high stabilities in air and water, high conductivity in its oxidized/protonated state, acid-base char-acteristics, electrochromic behavior, and electrochemical capacitance, polyaniline (PANI) has attracted the scientific community's interest among the various polymers. It has been estimated that the capaci-tance of PANI/CNT/Si2.8 (77.18 F/g), PANI/CNT/GP (136.84 F/g), and PANI/CNT/TiO$_2$ (477.1 F/g) capacitor depends on the materials used. The high 477.1 F/g for the PANI/CNT/TiO$_2$ capacitor sheet reflects the high capacitance of transition metal oxides like TiO_2. Besides, chemically reduced graphene oxide (rGO) has been combined with TiO_2 nanobelts (NBs) and nanoparticles (NPs) to generate nanocomposites that are employed as SCs electrodes. rGO–TiO_2 composites have a greater C than monolithic rGO, TiO_2 NPs, or NBs, as shown in Figure 25.6 [38].

On the other hand, incorporating the Ag/TiO_2 nanocomposite into the SC anode provided exceptional capacitance and ED values of 1193.2 Fg^{-1} and 238.4 Wh kg^{-1}, respectively. The SC device made with Ag/TiO_2 nanocomposite had a stable output voltage of 0.49 V, but when the Ag/N-doped TiO_2 nanocomposite was added to the SCs, the capacitance, and energy densities were reduced to 208.3 2 Fg^{-1} and 41.6 Wh kg^{-1} (compared to the device made with Ag/TiO_2 nanocomposite), and the output voltage was increased to 1.91 V [39].

Nickel/cobalt oxide: Nickel oxide (NiO) is naturally available, less toxic, and has high C (theoretical) (3750 Fg^{-1}) that's considered a viable alternative for SCs electrodes. Nickel oxide and Cobalt oxide-based electrodes are considered pseudocapacitive materials instead of battery types. Equation (25.7) can be used to represent the redox processes that occur in alkaline electrolytes [40].

$$NiO + OH^- \rightarrow NiOOH^- + e^- \tag{25.7}$$

Besides, Cobalt oxide (Co_3O_4) is also an appealing candidate because of its well-regulated surface morphology, high charge storage capability, and strong redox activity [41]. Pure Co_3O_4 always has weak conductivity, which limits its applicability. Incorporating Co_3O_4 into carbonaceous materials is an important way to boost conductivity, and minimizing aggregation is the most effective way to overcome this disadvantage. When compared to pure RGO and Co_3O_4, the electrochemical measurements revealed that the 3D Co_3O_4–RGO aerogel had a higher capacitance and better cycle stability. Mixed MOs like $NiCo_2O_4$, $MnMoO_4$, $CoMoO_4$, $NiMoO_4$, and others, have attracted a lot of attention due to rapid redox reaction and highly conductive and high capacitance value compared to unitary MOs like NiO and Co_3O_4. In addition, onion-like $CuCo_2O_4$ hollow spheres were manufactured using a bimetal-organic framework derived technique and shown excellent electrochemical properties (ED of 48.75 Wh kg^{-1} and ~92% capacity retention over 10 000 cycles).

On the other hand, ferric oxide (Fe_2O_3) is one of the best choices because of its natural availability, low cost, environmental friendliness, and high theoretical C (1005 mAh g^{-1}). However, the Fe_2O_3 particles' weak conductivity and substantial agglomeration during charge-discharge operations might be a problem, resulting in fast capacity loss [42]. A layered society technique was developed in 2013 for the manufacture of a freestanding flexible nanocomposite of nanoporous graphene sheets with iron oxide (NGP–Fe_3O_4), which was enhanced by carbon nanofiber [43]. Besides, the general capacity of various MOs is influenced by intercalation pseudocapacitance. Molybdenum oxide (MoO_3) has the highest theoretical C (1256 Fg^{-1}) among other intercalation pseudocapacitive materials due to its 2D layered structure and varied oxidation states of α-MoO_3 [44]. There is a lot of interest these days in the use of hollow structures morphology for energy storage because of the amazing surface features, such as high aspect ratio, short charge transport lengths, abundant electrochemical active sites accessible to the electrolyte, and well-defined interior space [45]. Therefore, hierarchical molybdenum dioxide@nitrogen-doped carbon (MoO_2@NC) and copper-cobalt sulfide ($CuCo_2S_4$) tubular nanostructures have recently been grown on flexible carbon fibers for usage as electrode materials in SCs. It had an ultrahigh ED of 65.1 W h kg^{-1} at a PD of 800 W kg^{-1}, a good ED of 27.6 W h kg^{-1} at an ultrahigh PD of 12.8 kW kg^{-1}, good electrochemical cycling stability (90.6% retention after 5000 cycles), and superior mechanical flexibility, with 92.2% retention of the initial capacitance after 2000 bending cycles [46]. In aqueous electrolytes, typical potential windows of various electrode materials are shown in Figure 25.7 [47] and performances of various MOs-based supercapacitors are shown in Table 25.5.

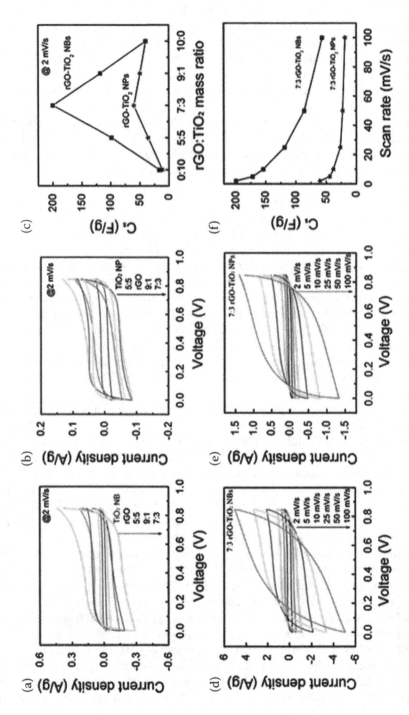

FIGURE 25.6 CV curves of (a) rGO–TiO$_2$ NBs and (b) rGO–TiO$_2$ NPs with different rGO:TiO$_2$ ratios at a scan rate of 2 mV s^{-1}; (c) C of rGO–TiO$_2$ NBs and rGO–TiO2 NPs as a function of rGO:TiO$_2$ ratio at a scan rate of 2 mV s^{-1}; CV curves at various scan rates of (d) rGO–TiO$_2$ NBs and (e) rGO–TiO$_2$ NPs with a 7:3 rGO:TiO$_2$ ratio at various scan rates; and (f) C of rGO–TiO$_2$ NBs and rGO–TiO$_2$ NPs with a 7:3 rGO:TiO$_2$ ratio as a function of scan rate. Adapted with permission from ref. [38]. Copyright (2012) Royal Society of Chemistry.

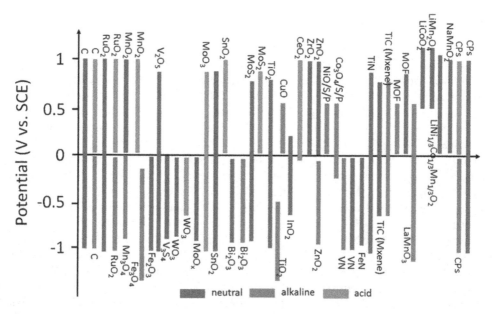

FIGURE 25.7 Typical working potential windows of various electrode materials in aqueous electrolytes. Adapted with permission from ref. [47]. Copyright (2017) Royal Society of Chemistry.

TABLE 25.5
Recent Progress of Various MOs-Based Supercapacitors

Supercapacitors	Voltage Window (V)	The ED	Retention (%)	Cycles
CoNi–LDH	1.5	86.4 W h kg^{-1}	77.0	3000
Ni(OH)$_2$/CNTs	1.7	139 W h kg^{-1}	78.0	2000
Ni wire/Co$_3$O$_4$@ MnO$_2$//CF/graphene	1.5	4.34 μWh cm^{-2}	82.0	1000
CuCo$_2$O$_4$/CuO	1.6	33 W h kg^{-1}	83.0	5000
Al$_2$O$_3$-doped Co$_3$O$_4$/graphene//AC	1.65	40.1 Wh kg^{-1}	84.2	5000
NiMoO$_4$//AC	1.7	60.9 Wh kg^{-1}	85.7	10 000
CoWO$_4$/Co$_3$O$_4$	1.5	57.8 Wh kg^{-1}	85.9	5000
Li$_3$VO$_4$/CNF composite//graphene	3.8	110 Wh kg^{-1}	86.0	2400
ZrO$_2$/CNTs	2.2	65 W h kg^{-1}	89.9	3000
Co$_3$O$_4$//AC	1.5	24.9 Wh kg^{-1}	90.0	5000
Co$_3$S$_4$//AC	1.6	60.1 Wh kg^{-1}	90 .0	5000
FeCo$_2$O$_4$@PPy//AC	1.6	68.8 Wh kg^{-1}	91.0	5000
ZnO/SnO$_2$@NF//AC@NF	1.6	47.28 Wh kg^{-1}	92.8	4000
3D Co$_3$O$_4$-RGO//HPC	1.5	40.65 Wh kg^{-1}	92.9	2000
NiCo$_2$O$_4$//RGO	1.4	23.9 Wh kg^{-1}	93.5	10 000
rGO/V$_2$O$_5$//rGO	1.6	75.9 Wh kg^{-1}	94.0	3000
NiO//rGO	1.7	39.9 Wh kg^{-1}	95.0	3000
Ni(OH)$_2$@3D Ni	1.3	21.8 Wh kg^{-1}	96.0	3000
ZnO@TiO$_2$@Ni(OH)$_2$	1.6	52.2 W h kg^{-1}	96.6	5000
VO$_2$@Ni NWs	1.6	27.3 W h kg^{-1}	116.6	20000

25.4.4 Transparent supercapacitors

In recent years, there has been an increase in the usage of flexible transparent supercapacitors (FTSCs). Therefore, flexible and optically transparent technologies are becoming more common in lightweight smart electronics. Advances in FTSCs power sources with great energy storage activity and high optical transparency are urgently needed. In the last decade, FTSCs have made huge progress because of their ease of use, fast charge/discharge, high ED/PD, and long-term cycle stability. All-solid-state FTSCs involve four main flexible and transparent parts, such as current collector, electrode material, substrate, and solid-state gel electrolyte (Figure 25.8).

The ESM of FTSCs is the same as traditional SCs, such as EDLC and PCs. Typical PCs materials for FTSCs undergo a redox reaction during charge/discharge, including MOs, metal sulfides, metal carbides, metal hydroxides (e.g., RuO_2, MnO_2, MoS_2, CuS, Ti_3C_2, and $Ni(OH)_2$). The use of conductive polymers, such as polypyrrole and PANI as electrode materials in FTSCs, has also been reported. Besides, metal-based flexible conductive electrodes (FTCEs) could well be prepared by coating metal nanoparticles (MNPs) on the surface of substrates like polyethylene terephthalate (PET), polyurethane acrylate, polyethylene, cyclic olefin copolymer, and so on. Nickel, copper, silver, and gold have been used broadly for making FTCEs. Their pristine utilization in FTSCs was restricted by their low chemical stability and expensive [48].

MOs have been studied as representative PCs materials for SCs. But, getting high optical transmittance from opaque metal oxides remains difficult. Furthermore, the metal oxide coatings are so stiff and brittle that mechanical flexibility, such as folding, bending, and twisting, is difficult. MOs-based FTCEs have been developed to address these issues by depositing MOs, such as ITO, Co_3O_4, MnO_2, RuO_2 and MoO_3, on the substrates' surface using advanced instruments.

Indium tin oxide (ITO) is a transparent material with good optoelectronic properties. It has a long history of usage as a flexible device substrate. However, because of its high ceramic content, it fractures at a low strain of 2–3%, restricting its use as a standalone FTCE. ITO film was deposited onto a flexible transparent PET substrate for usage as FTCEs in the lab or the marketplace. Recently, ITO/PET has indeed been broadly applied as a current collector to load active materials along with Co_3O_4, ITO, $NiO@MnO_2$, Fe_2O_3, RuO_2, and MnO_2 into FTCEs for FTSCs. The optical transmittance of the mesoporous and laminated ITO NPs film could be changed by adjusting the mass loading of ITO NPs by 700–800 nm. The electrodes of Co_3O_4/ITO/PET in FTSCs were shown to have better bending stability and retain 100% of their capacitance after 20000 cycles. On the other hand, direct coating of MnO_2 nanoparticles cladding around metals network (e.g., Au and Ni) is a popular technique for fabricating $MnO_2@$metal networks FTCEs with outstanding optoelectronic properties, superior electrochemical performance, and mechanical flexibility. Because of the strong interactions between MnO_2 and metal NPs located in or the core-shell structure, increased electron transit from MnO_2 to metal networks was enhanced [48]. Interconnected MnO_2 sheets' SSA and redox-active sites were enhanced, allowing ion/electron transport to be expedited. High transparency can be achieved by selective electrodeposition of MnO_2 on the surface of metal networks. Another important method for fabricating transparent MnO_2 film for FTSCs is to pattern ordered MnO_2 arrays. The array interspace provided excellent transparency by allowing natural light to pass and increased device stability by removing internal stress during the bending process, as shown in Figure 25.9. Moreover, RuO_2 is a famous PCs electrode material due to its high theoretical C and high electrical conductivity. The deposition of RuO_2 NPs on a substrate surface is a low-cost technique for generating solid transparent electrodes for FTSCs. For example, RuO_2 deposition on PET substrate produces the required transparent electrode for FTSCs. The high cost, unavailability, and agglomeration effect restrict its extensive use. Nowadays, CO_3O_4 nanostructures, NPs, nanobelts, and nanosheets are useful redox material for FTSCs due to their high SSA, more active sites, easily synthesized by hydrothermal, sol-gel method or electrochemical methods, and greater cycle stability. For example, Co_3O_4 FTCEs were initially developed by dropping Co_3O_4 dispersion over ITO/PET, resulting in a 58 % T550 nm. The as-prepared FTSCs had a somewhat lower T550 nm of 51%, a PCs of 177 F g1, 100% capacitance retention after 20000 cycles. It also exhibits good mechanical flexibility at various bending angles.

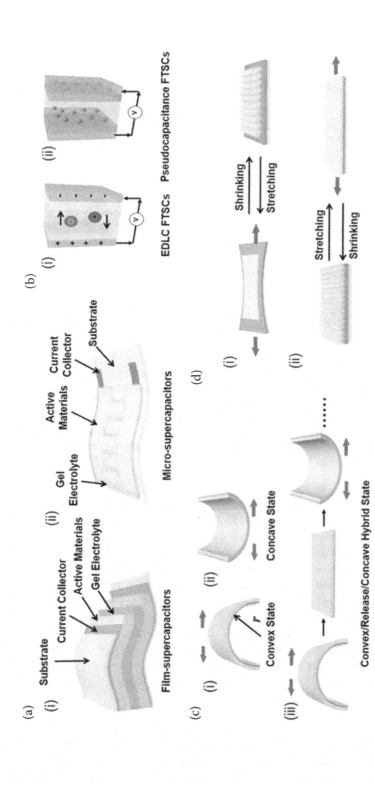

FIGURE 25.8 (a) Schematic diagram of the device configuration of FTSCs. (i) The in-plane structure of micro-supercapacitors. (ii) The sandwiched structure of film-supercapacitors. (b) Schematic illustration of two mechanisms of symmetric FTSCs. (i) EDLC FTSCs. (ii) Pseudocapacitance FTSCs. The green panel represents the current collector; the yellow panel represents EDLC materials; the purple panel represents pseudocapacitance materials and blue represents gel hydrogel. The red, blue, and green balls represent cation, anion, and electron, respectively. (c) Schematic illustration of three categories of bending deformations of FTCEs under a curvature radius (r). (i) Convex state. (ii) Concave state. (iii) Convex/release/concave hybrid state. (d) Schematic illustration of different stretching states of FTSCs. (i) Out-of-planar structure. (ii) In-plane structure. Adapted with permission from ref. [48]. Copyright (2020) Wiley-VCH GmbH.

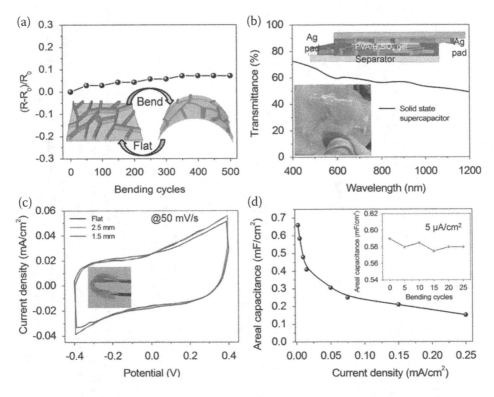

FIGURE 25.9 (a) Relative change in resistance of Au/MnO_2 (60 min) electrode while bent to 2.5 mm bending radius. (b) Transmittance spectrum of cross-assembled solid-state supercapacitor device. Schematic of electrode alignment (top) and the photograph of the device (bottom) are shown in the inset. (c) CV curves of $Au–MnO_2$ (60 min) solid-state supercapacitor with PVA/H_2SO_4 gel electrolyte in flat and while bent to 2.5 and 1.5 mm radii. Inset shows the photograph of a flexible supercapacitor device is bent geometry. (d) Areal capacitance as a function of current densities and inset is a change in areal capacitance as a function of bending cycles. Adapted with permission from ref. [49]. Copyright (2017) WILEY-VCH Verlag GmbH & Co. KGaA, Weinheim.

In terms of capacitance and stretchability, FTCEs of graphene films supported on PET, MnO_2 nanosheets on ITO-PET substrates, or single-walled CNT on polydimethylsiloxane and PANI substrates have proven to be highly promising for transparent electronics. However, the electrical conductivity of plastic substrates is low that restricts their uses. Recently, paper-based flexible SCs are appealing options for electronic displays in portable devices because they are lightweight, flexible, transparent, and easy to manufacture (e.g., digital cameras and laptops, and mobile phones). Therefore, both free-standing electrodes, such as CNT-paper composites and PC deposition on recyclable paper substrates, have now been explored extensively. Flexible sponge substrates with hierarchical macroporous and network-free morphology provide high liquid absorption, greater SSA, continuous coating, and improved interaction between electrodes and electrolytes [25]. In summary, flexible SCs are extremely appealing due to their high power density; even so, the main issues in the technology are their low energy density and high production cost.

25.5   CONCLUSIONS

Even though the usages of flexible SCs and FTSCs are rapidly expanding, there are still several problems to overcome and solutions to consider to continue the improvement of high-performance and cost-effective flexible devices. There are various studies available on successful attempts to create composite electrodes with the successful fabrication of FEs using a free-standing composite film without the insertion of any binders or additives. The existence of a ternary component, such as metal oxides, which were defined to help in sustaining a 3D framework of graphene, gave a maximum SSA for the infusion of electrolyte ions and enhanced the capacitance gained, was easily recognized. Electrolytes based on nontoxic and/or low-toxic materials with outstanding mechanical performance, strong ionic conductivity, exceptional stability, and low cost are particularly desirable for manufacturing flexible SCs. For next-generation flexible SCs, researchers should concentrate on less hazardous electrode materials.

ACKNOWLEDGMENTS

Professor Kang is thankful to the Basic Science Research Program through the National Research Foundation (NRF) of Korea funded by the Ministry of Education (NRF-2019R1I1A3A02059471) and was supported under the framework of an international cooperation program managed by the NRF of Korea (NRF-2020K2A9A2A08000181).

REFERENCES

1. Z. Pan, M. Liu, J. Yang, Y. Qiu, W. Li, Y. Xu, X. Zhang, Y. Zhang, High electroactive material loading on a carbon nanotube@ 3D graphene aerogel for high-performance flexible all-solid-state asymmetric supercapacitors, *Adv. Funct. Mater.* 27 (2017) 1701122.
2. B.E. Conway, *Electrochemical supercapacitors: scientific fundamentals and technological applications*, Springer Science & Business Media, 2013.
3. H. El Brouji, O. Briat, J.-M. Vinassa, H. Henry, E. Woirgard, Analysis of the dynamic behavior changes of supercapacitors during calendar life test under several voltages and temperatures conditions, *Microelectron. Reliab.* 49 (2009) 1391–1397.
4. J. Yan, Q. Wang, T. Wei, Z. Fan, Recent advances in design and fabrication of electrochemical supercapacitors with high energy densities, *Adv. Energy Mater.* 4 (2014) 1300816.
5. C. Choi, J.A. Lee, A.Y. Choi, Y.T. Kim, X. Lepró, M.D. Lima, R.H. Baughman, S.J. Kim, Flexible supercapacitor made of carbon nanotube yarn with internal pores, *Adv. Mater.* 26 (2014) 2059–2065.
6. X. Yang, C. Cheng, Y. Wang, L. Qiu, D. Li, Liquid-mediated dense integration of graphene materials for compact capacitive energy storage, *Science (80-.)* 341 (2013) 534–537.
7. A.S.F.M. Asnawi, M.H. Hamsan, S.B. Aziz, M.F.Z. Kadir, J. Matmin, Y.M. Yusof, Impregnation of [Emim] Br ionic liquid as plasticizer in biopolymer electrolytes for EDLC application, *Electrochim. Acta.* 375 (2021) 137923.
8. R. Liu, A. Zhou, X. Zhang, J. Mu, H. Che, Y. Wang, T.-T. Wang, Z. Zhang, Z. Kou, Fundamentals, advances and challenges of transition metal compounds-based supercapacitors, *Chem. Eng. J.* 412 (2021) 128611.
9. J. Kang, S. Zhang, Z. Zhang, Three-dimensional binder-free nanoarchitectures for advanced pseudocapacitors, *Adv. Mater.* 29 (2017) 1700515.
10. S.A. Delbari, L.S. Ghadimi, R. Hadi, S. Farhoudian, M. Nedaei, A. Babapoor, A. Sabahi Namini, Q. Van Le, M. Shokouhimehr, M. Shahedi Asl, M. Mohammadi, Transition metal oxide-based electrode materials for flexible supercapacitors: A review, *J. Alloys Compd.* 857 (2021) 158281.
11. V. Augustyn, P. Simon, B. Dunn, Pseudocapacitive oxide materials for high-rate electrochemical energy storage, *Energy Environ. Sci.* 7 (2014) 1597–1614.
12. M.S. Halper, J.C. Ellenbogen, *Supercapacitors: a brief overview*, MITRE Corp. McLean, Virginia, USA. (2006) 1–34.
13. A. Muzaffar, M.B. Ahamed, K. Deshmukh, J. Thirumalai, A review on recent advances in hybrid supercapacitors: design, fabrication and applications, *Renew. Sustain. Energy Rev.* 101 (2019) 123–145.
14. H. Wang, W. Ye, Y. Yang, Y. Zhong, Y. Hu, Zn-ion hybrid supercapacitors: Achievements, challenges and future perspectives, *Nano Energy.* 85 (2021) 105942.
15. W. Zuo, R. Li, C. Zhou, Y. Li, J. Xia, J. Liu, Battery-supercapacitor hybrid devices: recent progress and future prospects, *Adv. Sci.* 4 (2017) 1600539.
16. N. Choudhary, C. Li, J. Moore, N. Nagaiah, L. Zhai, Y. Jung, J. Thomas, Asymmetric supercapacitor electrodes and devices, *Adv. Mater.* 29 (2017) 1605336.

17. S. Trasatti, G. Buzzanca, Ruthenium dioxide: a new interesting electrode material. Solid state structure and electrochemical behaviour, *J. Electroanal. Chem. Interfacial Electrochem.* 29 (1971) A1–A5.

18. M. Toupin, T. Brousse, D. Bélanger, Charge storage mechanism of MnO2 electrode used in aqueous electrochemical capacitor, *Chem. Mater.* 16 (2004) 3184–3190.

19. Z.-H. Huang, Y. Song, D.-Y. Feng, Z. Sun, X. Sun, X.-X. Liu, High mass loading MnO_2 with hierarchical nanostructures for supercapacitors, *ACS Nano.* 12 (2018) 3557–3567.

20. N.R. Chodankar, H.D. Pham, A.K. Nanjundan, J.F.S. Fernando, K. Jayaramulu, D. Golberg, Y. Han, D.P. Dubal, True meaning of pseudocapacitors and their performance metrics: asymmetric versus hybrid supercapacitors, *Small.* 16 (2020) 2002806.

21. M. Liu, C. Yan, Y. Zhang, Fabrication of Nb_2O_5 nanosheets for high-rate lithium ion storage applications, *Sci. Rep.* 5 (2015) 1–6.

22. E. Frackowiak, F. Beguin, Carbon materials for the electrochemical storage of energy in capacitors, *Carbon N. Y.* 39 (2001) 937–950.

23. C. Choi, D.S. Ashby, D.M. Butts, R.H. DeBlock, Q. Wei, J. Lau, B. Dunn, Achieving high energy density and high power density with pseudocapacitive materials, *Nat. Rev. Mater.* 5 (2020) 5–19.

24. X. Rui, H. Tan, Q. Yan, Nanostructured metal sulfides for energy storage, *Nanoscale.* 6 (2014) 9889–9924.

25. S. Palchoudhury, K. Ramasamy, R.K. Gupta, A. Gupta, Flexible Supercapacitors: a materials perspective, *Front. Mater.* 5 (2019) 83.

26. P. Simon, Y. Gogotsi, B. Dunn, Where do batteries end and supercapacitors begin?, *Science (80–)* 343 (2014) 1210–1211.

27. H. Liu, X. Liu, S. Wang, H.-K. Liu, L. Li, Transition metal based battery-type electrodes in hybrid supercapacitors: a review, *Energy Storage Mater* 28 (2020) 122–145.

28. V. Sharma, S.J. Kim, N.H. Kim, J.H. Lee, All-solid-state asymmetric supercapacitor with MWCNT-based hollow $NiCo_2O_4$ positive electrode and porous Cu_2WS_4 negative electrode, *Chem. Eng. J.* 415 (2021) 128188.

29. A.K. Das, U.N. Pan, V. Sharma, N.H. Kim, J.H. Lee, Nanostructured $CeO_2/NiV–LDH$ composite for energy storage in asymmetric supercapacitor and as methanol oxidation electrocatalyst, *Chem. Eng. J.* (2020) 128019.

30. M.M. Korah, T. Nori, S. Tongay, M.D. Green, Materials for optical, magnetic and electronic devices, *J. Mater. Chem.* 100 (2020) 10479–10489.

31. M. Zhi, C. Xiang, J. Li, M. Li, N. Wu, Nanostructured carbon–metal oxide composite electrodes for supercapacitors: a review, *Nanoscale.* 5 (2013) 72–88.

32. C. An, Y. Zhang, H. Guo, Y. Wang, Metal oxide-based supercapacitors: progress and prospectives, *Nanoscale Adv* 1 (2019) 4644–4658.

33. D. Majumdar, T. Maiyalagan, Z. Jiang, Recent progress in ruthenium oxide-based composites for supercapacitor applications, *ChemElectroChem.* 6 (2019) 4343–4372.

34. K. Brousse, S. Pinaud, S. Nguyen, P.-F. Fazzini, R. Makarem, C. Josse, Y. Thimont, B. Chaudret, P.-L. Taberna, M. Respaud, P. Simon, Facile and scalable preparation of ruthenium oxide-based flexible microsupercapacitors, *Adv. Energy Mater.* 10 (2020) 1903136.

35. B. Shen, X. Zhang, R. Guo, J. Lang, J. Chen, X. Yan, Carbon encapsulated RuO_2 nano-dots anchoring on graphene as an electrode for asymmetric supercapacitors with ultralong cycle life in an ionic liquid electrolyte, *J. Mater. Chem. A.* 4 (2016) 8180–8189.

36. L. Wang, Y. Ouyang, X. Jiao, X. Xia, W. Lei, Q. Hao, Polyaniline-assisted growth of MnO_2 ultrathin nanosheets on graphene and porous graphene for asymmetric supercapacitor with enhanced energy density, *Chem. Eng. J.* 334 (2018) 1–9.

37. Y. Wang, X. Xiao, H. Xue, H. Pang, Zinc oxide based composite materials for advanced supercapacitors, *ChemistrySelect.* 3 (2018) 550–565.

38. C. Xiang, M. Li, M. Zhi, A. Manivannan, N. Wu, Reduced graphene oxide/titanium dioxide composites for supercapacitor electrodes: shape and coupling effects, *J. Mater. Chem.* 22 (2012) 19161–19167.

39. R. Mendoza, V. Rodriguez-Gonzalez, A.I. Oliva, A.I. Mtz-Enriquez, J. Oliva, Stabilizing the output voltage of flexible graphene supercapacitors by adding porous Ag/N-doped TiO_2 nanocomposites on their anodes, *Mater. Chem. Phys.* 255 (2020) 123602.

40. F. Luan, G. Wang, Y. Ling, X. Lu, H. Wang, Y. Tong, X.-X. Liu, Y. Li, High energy density asymmetric supercapacitors with a nickel oxide nanoflake cathode and a 3D reduced graphene oxide anode, *Nanoscale.* 5 (2013) 7984–7990.

41. Y. Jiang, C. He, S. Qiu, J. Zhang, X. Wang, Y. Yang, Scalable mechanochemical coupling of homogeneous Co_3O_4 nanocrystals onto in-situ exfoliated graphene sheets for asymmetric supercapacitors, *Chem. Eng. J.* 397 (2020) 125503.

42. X. Zhu, Y. Zhu, S. Murali, M.D. Stoller, R.S. Ruoff, Nanostructured reduced graphene oxide/Fe$_2$O$_3$ composite as a high-performance anode material for lithium ion batteries, *ACS Nano.* 5 (2011) 3333–3338.

43. X. Huang, B. Sun, S. Chen, G. Wang, Self-assembling synthesis of free-standing nanoporous graphene–transition-metal oxide flexible electrodes for high-performance lithium-ion batteries and supercapacitors, *Chem. – An Asian J.* 9 (2014) 206–211.

44. B. Yao, L. Huang, J. Zhang, X. Gao, J. Wu, Y. Cheng, X. Xiao, B. Wang, Y. Li, J. Zhou, Flexible transparent molybdenum trioxide nanopaper for energy storage, *Adv. Mater.* 28 (2016) 6353–6358.

45. Y.M. Chen, X.Y. Yu, Z. Li, U. Paik, X.W. (David) Lou, Hierarchical MoS$_2$ tubular structures internally wired by carbon nanotubes as a highly stable anode material for lithium-ion batteries, *Sci. Adv.* 2 (2016) e1600021.

46. S. Liu, Y. Yin, K.S. Hui, K.N. Hui, S.C. Lee, S.C. Jun, High-performance flexible quasi-solid-state supercapacitors realized by molybdenum dioxide@nitrogen-doped carbon and copper cobalt sulfide tubular nanostructures, *Adv. Sci.* 5 (2018) 1800733.

47. J. Sun, C. Wu, X. Sun, H. Hu, C. Zhi, L. Hou, C. Yuan, Recent progresses in high-energy-density all pseudocapacitive-electrode-materials-based asymmetric supercapacitors, *J. Mater. Chem. A.* 5 (2017) 9443–9464.

48. W. Zhao, M. Jiang, W. Wang, S. Liu, W. Huang, Q. Zhao, Flexible transparent supercapacitors: materials and devices, *Adv. Funct. Mater.* 31 (2021) 2009136.

49. S. Kiruthika, C. Sow, G.U. Kulkarni, Transparent and flexible supercapacitors with networked electrodes, *Small.* 13 (2017) 1701906.

26 Recent Advances in Transition Metal Chalcogenides for Flexible Supercapacitors

Somoprova Halder, Souhardya Bera, and Subhasis Roy
Department of Chemical Engineering, University of Calcutta, Kolkata, India

CONTENTS

26.1 Introduction.. 481
26.2 Substrates for flexible supercapacitors 483
26.3 Metal chalcogenides and their categories.. 484
    26.3.1 Unary-metal chalcogenides.. 485
    26.3.2 Binary or ternary metal chalcogenide 487
26.4 How does the electrode system work?.. 490
26.5 Recent developments in the metal chalcogenide-based Flexible Supercapacitors.... 493
26.6 Conclusion 495
Acknowledgment 496
References... 496

26.1 INTRODUCTION

With developing economies replanning their growth strategies and large-scale consumption of fossil fuels still abundant, environmental destruction, ecological crisis, and energy wastage have become severe restrictions for creating a sustainable society. There is an immediate requirement to explore and implement clean and renewable sources to alleviate this overdependence on fossil fuels. This has already garnered considerable interest, only now that one requires supplementing the systems' energy conversion and storage strategies, which are low-cost and have an easy fabrication. Energy obtained from fluctuating renewable sources must be stored in the electrochemical form to be properly utilized and supplied [1]. The main pioneers in this field of storing generated charges are supercapacitors (SCs) and batteries. In this aspect, Flexible SCs (FSCs) are highly attractive for their inherently high-power density, fast charging/discharging capability, longer operation time, and of course, mechanical stability.

Figure 26.1a and b show the two different supercapacitor models: conventional and flexible. FSCs indeed take an upper hand over conventional SCs since they are bendable and can be twisted in any manner without any significant loss of performance. This can be noticed from the different bending angles an FSC is subjected to, as shown in its corresponding cyclic voltammetry (CV) curves in Figure 26.1c and d, where it is evident that it performs invariantly well for all the conformations. The transition metal chalcogenides (TMCs) form an attractive class of electrode material for FSCs for their exceptionally high surface area to host redox reactions. They contain different metal ions, which allow us to modulate its bandgap as a function of the number of layers that result in variable catalytic activity with boundary defects. Weak Van der Waals interaction between layers allows for mechanical exfoliation of most TMCs to be dissociated into nanosheets [4]. Figure 26.1e and f show SEM images of a copper and

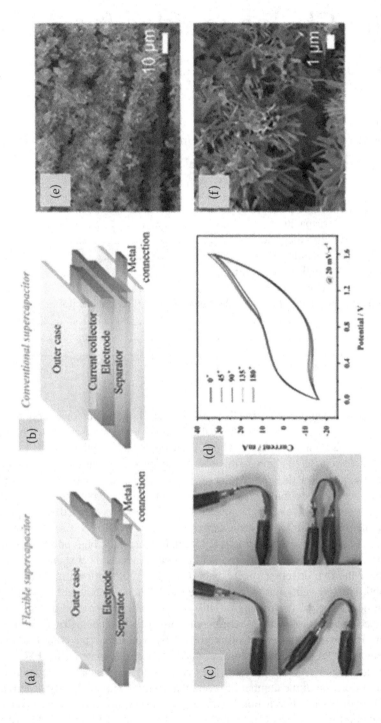

FIGURE 26.1 (a–b) Representation of a (a) flexible supercapacitor; (b) conventional supercapacitor [2]. Reprinted with permission from ref. [2]. Copyright (2019) *Frontiers in Materials*; (c–d) Images of the fabricated asymmetric device captured at different bending states and the corresponding CV curves shown respectively [Reprinted with permission from ref. [3]. Copyright 2020 *Frontiers in Materials*]; (e–f) SEM images for CuCo$_2$S$_4$/CC at 8 hours at different magnifications [3]. [Reprinted with permission from ref. [3]. Copyright 2020 *Frontiers in Materials*].

cobalt-based composite sulfide for application as supercapacitor electrodes. The immense amount of surface area provided by these nanostructures makes them efficient electrodes, allowing the exchange of electrolyte solution and ions.

Reversible faradic reactions in the pseudo-capacitors lead to capacitor-like behavior, much like a battery where electrochemical phenomena are the basis of charge storage. These charge transfer reactions are caused by the reversible mass transfer of the electrolyte ions at the interfacial region, where ion adsorption at the surface of the active material is succeeded by formation of an intermediate complex or intercalation within the layers of the electrode material. The layer-like lattice structure of TMCs allows for easy intercalation processes, which leads to its acceptance as efficient charge storage materials. Such electrochemical storage pathways, when combined with conventional capacitors that store charge by forming an electrical double layer, have the potential of resulting in high efficiencies. Intercalation capacitance, as a parameter, is capable of providing excellent capacitance with high-rate capability in comparison to conventional double-layer capacitance, and additionally, pseudo-capacitance leads to higher charge storage through redox reaction. Asymmetric supercapacitors (ASCs) can reach a high operating potential window (OPW) of 2.2 V, rather than the limited range of 1.1 V for the SSCs, ASCs remain more prevalent over their symmetric counterparts [5].

26.2 SUBSTRATES FOR FLEXIBLE SUPERCAPACITORS

Carbon cloths are the most pioneering material in use when it comes to flexible supercapacitors. They consist of woven carbon fibers in a diameter range of 5–10 μm [1]. Carbon cloths have found extreme popularity in research and commercial applications of supercapacitors (both symmetric and asymmetric), solar cells, batteries, and catalysis. The high surface area and requisite conductivity prove to be advantageous in addition to its excellent flexibility, and they take the upper hand over the more commonly used Ni foam substrates. Figure 26.2a shows a highly magnified SEM image of a single carbon fiber. Several such carbon fibers are woven together in a regular criss-cross fashion in a carbon cloth. High conductivity and a large surface area allow for the diffusion of ions, decreasing the resistance to mass transfer and adsorption. Where flexibility is a parameter, it needs to be ascertained whether fibers are mechanically strong, robust, and damage-resistant under twisting or bending of carbon cloth substrates. To,this test, carbon cloth fulfills all the parameters. In addition to the above, they are cheap and environment friendly [8]. Their good biocompatibility makes their application in biomedical applications very interesting and promising. A simple question arises: "Is it so superior that there's no drawback?" Indeed, there are certain demerits. Characteristics like conductivity and mechanical strength are superior in other metallic counterparts, and undeniably, they become a better option. Other than carbon cloths, flexible inter-digital micro-supercapacitors (MSCs) are highly efficient candidates to be used directly as an energy component in wearable and portable electric devices [9–11].

Modification of the as-purchased carbon cloth is essential to activate the surface and prepare it for the next fabrication steps. Carbon, being highly hydrophobic, repels an aqueous solution; that may be a disadvantage while working with such media during the growth of nanostructures on pristine carbon cloth. In addition, it is always much better to use the substrate for direct growth of the desired material than to deposit binder-based slurries that drastically cover up a lot of accessible surface area of the cloth and form disordered lumps that have no proper alignment of nanostructures. Surface activation involves etching through chemical treatment, oxidation, and functionality with appropriate chemical groups or doping, making them much more active electrochemically [1]. Figure 26.2b demonstrates a typical SEM image of a carbon fiber modified by an electro-etching process. Wang et al. have demonstrated that there can be a 1733-fold increase of areal capacitance with much enhanced specific surface areas in comparison to that of pristine carbon fibers (CFs) when fabricated using the electrochemical H_2SO_4-HNO_3 activation methodology [12]. Doping of hetero-conductivity is accomplished by creating the positive charge centers in carbon materials [13]. Coated with a thin metallic layer deploying the electro-less deposition route, pristine twisted CFs exhibit an exceptional

FIGURE 26.2 (a) SEM image of the carbon fiber cloth. (b) Carbon fiber cloth after electro-etching. Grooves on the carbon fiber are due to electro-etching, and the surface area is enlarged. (c) SEM image of electro-etched carbon fiber cloth after PANI coating [6]. (Reprinted (adapted) with permission from Cheng Q, Tang J, Ma J, Zhang H, Shinya N, Qin L-C (2011) Polyaniline-Coated Electro-Etched Carbon Fiber Cloth Electrodes for Supercapacitors. J Phys Chem C. 115:23584–23590 Copyright (2011) American Chemical Society); (d) Photo of flexible super-capacitor composed of poly(3,4-ethylenedioxythiophene) and graphene oxide composite films on flexible graphite-poly (ethylene terephthalate) substrate along with a setup of the supercapacitor demonstration with a solar cell and electrochromic display [7]. (Reprinted (adapted) with permission from Lehtimaki S, Suominen M, Damlin P, Tuukkanen S, Kvarnstron C, Lupo D (2015) Preparation of Supercapacitors on Flexible Substrates with Electrodeposited PEDOT/Graphene Composites. ACS Appl Mater Interfaces. 7:22137–22147, Copyright (2015) American Chemical Society).

conductivity – an improvement of nearly 300% with much superior mechanical strength [14]. Fibers may also be coated with a conductive polymer like PANI where the surface of the carbon cloth would become rugged, as we can see in Figure 26.2c. A set-up of an electrical circuit using a solar cell along with a flexible supercapacitor has been demonstrated by Lehtimatki et al. It has been represented in Figure 26.2d.

Precursors with high carbon content, such as viscous rayon jute fibers, can be used to directly obtain chemically treated or activated carbon cloths (ACCs) through a carbonization process. The most important part of a fabrication that comes after surface activation is the synthesis of the electrode material, which may be deposited on CFs directly, applying the hydrothermal (HT) or the electrochemical deposition (ELD) route.

26.3 METAL CHALCOGENIDES AND THEIR CATEGORIES

As mentioned above, TMCs have played an essential role in the evolution of research on supercapacitors. In this regard, metal chalcogenides can be classified into two basic types: unary and binary or ternary-metal-based chalcogenides. Though unary metal chalcogenides are more popular to work with, due to their simplicity of synthesis with a lot of available literature, binary and ternary metal chalcogenides somewhat surpass the former in terms of efficiency and behavior. This has opened up new possibilities of research in this domain, and the study of such composite systems is indeed interesting.

26.3.1 UNARY-METAL CHALCOGENIDES

Unary-metal chalcogenides (UMCs) usually consist of substances like MoS_2, CoS, VS_2, MnS_2, NiS, $InSe$, etc., which are sulfides, selenides, or tellurides of only one metal component. These are conventional metal chalcogenides having produced very satisfactory results owing to their electrochemically active nature in redox reactions. Wide varieties of synthetic procedures allow for creating a wide variety of morphology of nanostructures on the carbon cloth, which in turn effectively helps the electrolyte to maximize the area of contact. It also gives more scope for catalytic reactions to take place preferentially at the exposed facets.

Hydrothermal or solvothermal methods are currently the most common for the synthesis of TMCs because it is simple and highly flexible in terms of the selected synthetic pathway. Two-step hydrothermal reactions have been practiced by many research groups in which sulfide or chalcogenide formation occurs in two steps, the first stage being the hydrothermal reaction of precursors to form an oxy-hydride or double hydroxide, followed by sulfurization in the second step. Various other methods like electrodepositing, chemical bath deposition (CBD), and chemical vapor transport have been used. Electrodeposition allows very precise control over the morphology of nanostructures, and it can be vastly exploited by tuning several parameters, like precursors, medium, voltage sweep rates, current applied, etc., and is, by far, a very convenient method to work with. CBD, although employed regularly as a thin-film deposition technique for metal sulfides and selenides, has shown to yield lower uniformity and with much less control on the morphology of nanostructures. It is one of the simplest techniques to synthesis nanoparticles on a substrate; however, precise study on monitoring of the deposition parameter is still somewhat unexploited [15]. Layered III–IV metal chalcogenides, such as InSe attracts increasing attention for their high carrier mobility, which exceeds $103\ cm^2V^{-1}s^{-1}$ and $104\ cm^2V^{-1}s^{-1}$ at room and liquid-helium temperatures [16,17]. InSe nanosheets have been fabricated using the chemical vapor transport method, followed by liquid exfoliation by Mudd et al. A one-step chemical expanded graphite, followed by an ultrasonic exfoliation, was used to prepare graphene nanosheets [4]. Subramani et al. had carried out the synthesis of CoS-nanoparticles in two steps, deploying the development of cobalt hexacyanoferrate (CoHCF) complex followed by solvothermal conversion into crystalline CoS nanoparticles [18]. Pandit et al. showed the synthesis of hexagon-like VS_2 nanostructures on the hexagonal matrix of multiwalled carbon nanotubes (MWCNTs) by employing the simple and cost-effective successive ionic layer adsorption and reaction (SILAR) [19]. Generally, SILAR as a process does not prove to be effective while depositing sulfides on planar substrates; however, due to the structural and conductive support of MWCNTs, 'dip and dry' processed MWCNTs served as the core, with VS_2 nanostructures growing on them like a shell. MWCNTs have a mesoporous structure due to the web-like nano-network, which proves favorable for anchoring VS_2 via the SILAR method. The process has been illustrated in Figure 26.3a. Once VS_2 nanoparticles have been grafted on MWCNTs, the VS_2/MWCNT composites can produce relatively porous and active nanostructures, promoting fast electrochemical reactions during charging-discharging. The SEM images in Figure 26.3b depict the coverage of VS_2 on the carbon fibers, giving them a rugged morphology. The figure next to it, Figure 26.3c, is a photograph of the fabricated electrode held in a bending position. This SILAR process is much easier and faster than any of the other ones and can be a great choice for simple, cost-effective fabrication. However, the creation of uniform or hierarchical structures is difficult.

The dimensions could also be an important factor to achieve the optimal contact area during the choice between two different materials to simplify the process of charge transfer. To this end, Congpu et al. proposed the superiority of graphene, a two-dimensional material that can achieve the highest possible contact area of two-dimensional InSe nanosheets [16]. They debated that materials like carbon nanotubes may not be as efficient as a pair compared to graphene; that's because the former does not correspond well with the 2D morphology of chalcogenides, and the contact area becomes small, consequently unfavorable for the electron transfer. Due to the high congruence of InSe nanosheets with

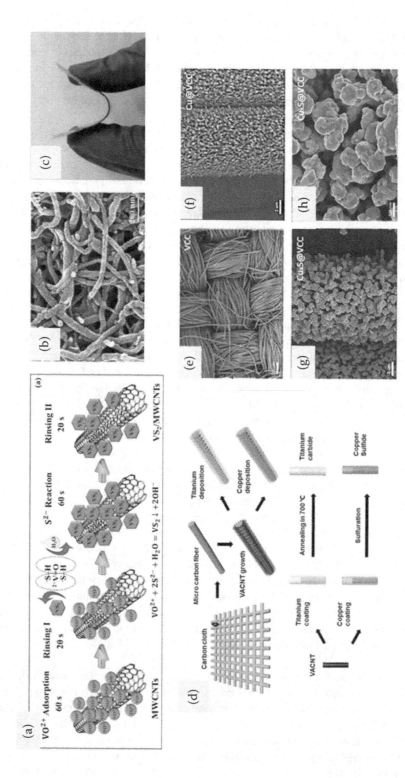

FIGURE 26.3 (a) Schematic illustrating the deposition of hexagonal VS2 on MWCNTs (b) FESEM image of VS2-CNWT [19] (c) Flexible device fabricated [19]. (Reprinted with permission from ref. [19]. Pandit B, Karade SS, Sankapal BR (2017) Hexagonal VS₂ Anchored MWCNTs: First Approach to Design Flexible Solid-State Symmetric Supercapacitor Device. ACS Appl Mater Interfaces. 9:44880-44891 Copyright (2017) American Chemical Society; (d) Schematics of the fabrication processes of the cathode and anode electrode with TiC@VCC as anode and Cu$_x$S@VCC as cathode. (e) SEM images of VCC (f) SEM images of Cu@VCC electrode before functionalization (g–h) SEM images of Cu$_x$S@VCC electrode after functionalization [20]. (Reprinted with permission from ref. [20]. Copyright 2019 *Frontiers in Chemistry*).

graphene, 93.6% of specific capacitance could be retained, even after 5000 and 10000 cycles, along with sufficient mechanical flexibility (retaining capacity at 98.2% after 2000 cycles under bending angle of 90°). Let's note an interesting fact here: when two MSCs are connected in parallel, the current density is seen to be twice at a voltage window of 0–1.4 V, and the voltage of two (three) MSCs in series increased from 1.4 to 2.8 V (4.2V). In addition, the charge/discharge time of two MSCs in parallel was twice that of a single MSC, and the potential operating window of two (three) MSCs devices in series was two (three) times 1.4 V.

Leimeng et al. proposed the growth of a well-distributed vertical-aligned carbon nanotube (VACNT) array on a piece of flexible carbon cloth as current collectors by using a plasma-enhanced chemical vapor deposition (PECVD) process, followed by deposition of copper through a magnetron sputtering system [20]. A second electrode was fabricated in a similar process where the growth of VACNTs was followed by the deposition of titanium. The copper was converted to sulfide in a thermal chemical vapor deposition furnace. Figure 26.3d shows the fabrication process of the initial carbon cloth to the final copper sulfide and titanium carbide electrodes. The SEM images of the VACNTs, Cu@VACNTs, and Cu_xS@VANCTs are provided in Figures 26.3e–h.

Thus, starting from fundamental methods like chemical bath deposition, one might choose complicated and extensive procedures like chemical vapor deposition for the fabrication of electrodes. However, the hydrothermal method remains the most employed and most economical procedure for developing FSC electrodes. Table 26.1, provided in one of the following sections of this chapter, summarizes the compounds and their synthesis route that have been reported by some groups, along with their respective performances. The better the control over the synthesis, the better shall the morphology of the nanostructures be, which would consequently lead to a better specific capacitance.

26.3.2 BINARY OR TERNARY METAL CHALCOGENIDE

Chalcogenides composed of two or multiple metal components are referred to as binary or ternary chalcogenides. The synergistic interaction between two or metal components results in higher performance of multi-metal composites over single-metal composite. Binary metal sulfides usually have better electrochemical conductivity and activity than oxides [30]. Researchers have therefore developed many binaries, such as $CoNi_2S_4$ [31], $CuCo_2S_4$ [32], $MnCo_2S_4$ [33], $SnNi_2S_4$ [34], $ZnCo_2S_4$ [35], Ni_4AlS_2 [36], Cu_4SnS_4 [37], $FeCo_2S_4$ [38], Ni-Mn sulfides [39] and so forth. In the following paragraphs, we will discuss how these composite chalcogenides were synthesized and put to the application by various groups.

Gao and his group developed a nickel-coated carbon-fiber-based electrode followed by conductive ink deposition and growth of nickel-cobalt double hydroxide. They employed an electroless pathway to build the metallic surface on the carbon fibers, followed by dip coating in conductive ink. The pen ink/nickel/ CF was further coated with Ni-Co double hydroxide deploying an electrodeposition route. The fabrication method is demonstrated in Figure 26.4a. The application of pen-ink film enables the conductive graphite carbon nanoparticles to form a porous morphology and coarse surface, significantly increasing the loading mass of electrochemically active materials and offering a robust adhesion for Ni-Co double hydroxides to enhance the cycling ability. Jun et al. have reported the synthesis of hierarchical ZnO@H-Ni@ Al-Co-S as an electrode material on carbon cloth. In the first step, they had grown ZnO nanorods through a hydrothermal approach, followed by electro-deposition of a layer of Ni. The ZnO when washed off with ammonia solution led to the formation of Ni nanotubes. These were coated with Al-doped cobalt sulfide nanosheets via the electro-deposition method. The process is graphically represented in Figure 26.4b. The nanorod arrays of ZnO can be observed from the SEM image in Figure 26.4c. Figure 26.4d–f shows Al-doped CoS nanosheets grown on the Ni nanotube at different magnifications. These nanosheets form an interconnected mesh-like structure with a lot of perforations, providing a high amount of active surface area. Kim et al. have grown a novel crumpled quaternary sulfur-doped nickel-cobalt selenide nano-architecture on carbon cloth (S-(NiCo) Se/CC) through a facile one-step

TABLE 26.1

Unary Metal Chalcogenides that have been Worked upon so Far, their Synthesis Route, and Performance Parameters

Electrode	Preparation methodology	Capacitance $(F.g^{-1})$	Current density $(A.g^{-1})$	Research comments	Ref.
α-MnSe	Hydrothermal (Slurry @ Stainless Steel)	96.76	-	Phenomenal cyclic stability (103.40%) after 2000 cycles	[21]
$Cu_{1.4}S$	Hydrothermal (Direct CC)	485	0.25	Capacity retention of 80.2% after 2000 cycles	[22]
Ni_3S_2-based CCNS-RGO	Hydrothermal (Slurry @ Ni foam)	860.1	5	Capacity retention at 98.6% after 500 cycles	[23]
2-NiS/CFs	Hydrothermal (Direct CC)	534.8	1	Capacity retention of 86% after 2000 cycles	[24]
CoS_2NWs	Hydrothermal (Direct CC)	828.2	5	A loss of 0-2.5% capacity after 4250 cycles	[25]
Ni_9S_8/O-MoS_2nano-composite	Hydrothermal (Direct CC)	907	2	Capacity retention at 85.7% after 1200 cycles	[26]
MoS_2hierarchical nanospheres	Hydrothermal (Slurry @ CC)	368	-	Capacity reduction of 3.5% after 5000 cycles	[27]
$CoSe_2$nanoarray	Hydrothermal selenization (Slurry @ Ni foam)	759.5	-	Capacity retention of 94.5% after 5000 cycles	[28]
CoS	Solvothermal (Slurry @ graphite foil)	310	5	95% capacity retention after 5000 cycles	[18]
VS_2/ MWCNTs	SILAR (CC)	830	-	95.9% stability over 10000 cycles	[19]
$CoTe_2$	Hydrothermal (Slurry @	360	5	Retains 52% at 2500 cycles	[29]
$CoSe_2$	Ti foil)	951		Retains 42% at 2500 cycles	

Abbreviations: CC: Carbon Cloth; CCNS-RGO: Carbon Coated Nickel Sulfide-Reduced Graphene Oxide; NWs: Nano Wires; MWCNTs: Multi-Walled Carbon Nanotubes

solvothermal process. The procedure is schematically shown in Figure 26.4g in a stepwise manner. A co-precipitation process in the hydrothermal autoclave leads to the formation of nanosheet-like structures over time, which grows in size until continuous crumpled S-(NiCo)Se nanosheet arrays are formed on carbon cloth. Thermal annealing is an important post-synthesis step that increases the grain size of the metals and improves the crystallinity of the formed metal structures, resulting in excellent electrical and electrochemical characteristics. The SEM images reproduced in Figures 26.4h–j show an excellently interconnected porous network of joined nanosheets.

Copper antimony sulfides (Cu_3SbS_4) nanowires were synthesized directly via a microwave irradiation method operating in the pulse mode as a binder-free positive electrode for SCs. The use of Cu_2MoS_4 nanoparticles, processed via facile one-pot hydrothermal route, was reported as a counter electrode [5]. Hu et al. obtained hydrogenated-Cu_2WS_4 by lithium intercalation-assisted exfoliation strategy, where an intermediate precursor of $Li_XCu_2WS_4$ was obtained by treating bulk Cu_2WS_4 with n-butyl lithium io-dide. The precursor led to a lithium intercalation process, which resulted in deteriorated interaction between the layers of $CuWS_4$ lattice. The intermediate precursor formation itself is an essential para-meter for the exfoliation and successive hydrogen doping process, which on successive treatment takes part in a hydrolysis reaction ($Li_XCu_2WS_4 + H_2O \rightarrow H_2\uparrow + LiOH + H_YCu_2WS_4$). Parts of hydrogen would

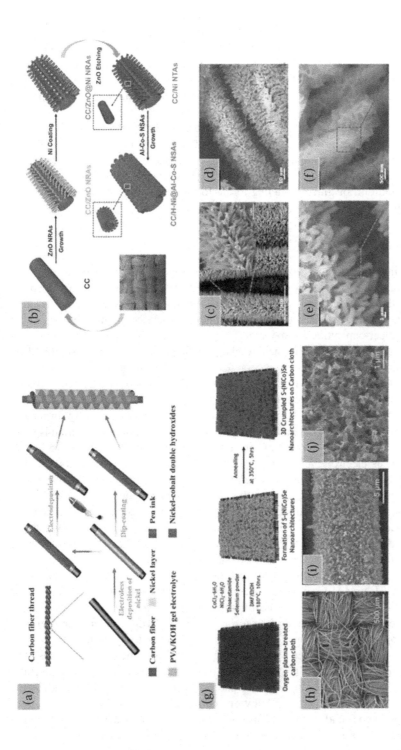

FIGURE 26.4 (a) Schematic illustration of fabrication for the flexible fiber-type solid-state asymmetric fiber supercapacitor device [14]. (Reprinted (adapted) with permission from Gao L, Surjadi JU, Cao K, Zhang H, Li P, Xu S, Jiang C, Song J, Sun D, Lu Y (2017) Flexible Fiber-Shaped Supercapacitor Based on Nickel-Cobalt Double Hydroxide and Pen Ink Electrodes on Metallized carbon Fiber. ACS Appl Mater Interfaces. 9:5409–5418 Copyright (2017) American Chemical Society; (b) Schematic illustration of the fabrication of a hierarchical core-branch CC/H-Ni@Al-Co-S nanosheet electrode; (c) SEM image of CC/ZnO nanorod arrays [40] (d–f) Low- and high-magnification SEM images of the CC/H-Ni@Al-Co-S nanosheet electrode [40]. (Reprinted (adapted) with permission from Huang J, Wei J, Xiao Y, Xu S, Xiao Y, Wang Y, Tan L, Yuan K, Chen Y (2018) When Al-doped Cobalt Sulfide Nanosheets Meet Nickel Nanotube Arrays: A Highly Efficient and Stable Cathode for Asymmetric Supercapacitors. ACS Nano. 12:3030–3041. Copyright (2018) American Chemical Society. (g) Schematic Representations of synthesis steps of 3D Crumpled S-(NiCo)Se Nanoarchitectures on a Carbon Cloth Electrode; (h–j) High-resolution SEM images of S-(NiCo)Se/CC. with different resolutions of S-(NiCo)Se [41]. (Reprinted (adapted) with permission from Kim J, Tabassian R, Nguyen VH, Umrao S, Oh I-K (2019) Crumbled Quaternary Nanoarchitecture of Sulfur-Doped Nickel Cobalt Selenide Directly Grown on Carbon Cloth for Making Stronger Ionic Soft Actuators. ACS Appl Mater Interfaces. 11:40451–40460 Copyright (2019) American Chemical Society).

get incorporated into the exfoliated Cu_2WS_4 structure to create ultrathin hydrogenated nanosheets [42]. This process of hydrogenation proved to enhance the conductivity of pristine $CuWS_4$ remarkably. Ramasamy and his group reported an easy and simple colloidal method of synthesizing $CuSbSe_xS_{2-x}$ ($0 \leq x \leq 2$) meso-crystals with belt-like morphology by substituting S by Se. The results of the X-ray diffraction show that the substitution of sulfur for selenium in $CuSbS_2$ allows the width of the inter-layer gap to be fine-tuned [43]. Following two-step synthesis through solvothermal methods, $CoTe_2$ and $CoSe_2$ nanostructures have been directly cultivated through an anion-exchange reaction between pre-synthesized $Co(OH)_2$ hexagonal nanosheets and chalcogen (tellurium and selenium) ions in the hydrothermal environment [29].

Chenxu et al carried out doping of the carbon-skeleton by nitrogen for enhanced conductivity. This N-doped C-skeleton was used as a base for synthesis of $CoSe_2$ by the calcination of the Co-MOF and selenysation with Se powder in an argon atmosphere [44]. Zheng et al. prepared nickel-copper sulfide-loaded carbon for SC electrode with an excellent electrochemical performance by a hydrothermal approach. It was found that the addition of glucose leading to the formation of carbon nanoparticles has better electrochemical performance than single $Ni_{0.9}Cu_{0.1}S$ [45].

26.4 HOW DOES THE ELECTRODE SYSTEM WORK?

We have already mentioned in the earlier portions that supercapacitors generally employ two main mechanisms during their operation: first, capacitive, which arises from the formation of an electrical double layer on the surface of the electrode, and second, from an intercalation mechanism, where small ions are trapped between the layered crystal lattices, forming an intermediate compound through redox reactions. If we consider the super-capacitive behavior of TiS_2, as shown in Figure 26.5a, we might notice that 85% of the capacitance arises from capacitive charge storage while the rest is offered by diffusion of Li^+ ions to the electrolyte in TiS_2 to form a complex Li_xTiS_2. This reaction takes place with a reduction in the oxidation state of Ti^{4+}, the reason for which such behavior is ascribed to redox reactions. The reduction in oxidation state and the lithium-ion intercalation leads to an increase in lattice dimension, slightly expanding the volume of the TiS_2 [46]. Transition metal oxides generally form these redox complexes by forming an oxide-hydroxide state; while diffusion-based cation intercalation is more common in the case of transition metal chalcogenides, as reported by Libo et al. in their work with nickel-cobalt double hydroxides as electrode material (Figure 26.5b). However, the formation of complexes with OH^- has also been explored in literature, which takes place if we use an electrolyte with a rather large-sized cation. The interlayer distance between the layers in metal chalcogenides depends on the type and amount of the anionic substituent. Jun et al. in their work had fabricated a core–branch CC/H-Ni@Al-Co-S nanosheet-based electrode. They explain that the high efficiency arises from Al doping, which improves the conductivity and electrochemical activity of cobalt sulfide, resulting in optimized cobalt sulfide nanosheets. Besides, the direct growth of Ni nanotube ensures an efficient conduction channel and reduces resistance. The mechanism of charge flow is shown in Figure 26.5c. The hierarchical nanoarchitecture of CC/H-Ni@Al-Co-S allows electrons to be directly and rapidly transferred from one material to the other [40]. Cobalt selenide nanostructures are grown on carbon cloth by Chenxu et al., which suggest the conversion of the selenide into a reversible oxy-hydroxy compound through XRD analysis, with nitrogen doping of the carbon cloth ensuring better conductivity. Figure 26.5d gives a diagrammatic representation of $CoSe_2$ embedded on CC and the movement of the electrolyte ions [44].

Karthik et al. have demonstrated that in $CuSbS_2$, the gap between the layers can be systematically varied by changing the sulfur to selenium ratio. Figure 26.5e and f show these compounds and adjust the interlayer distance between them. By tuning the gap between the layers, it is possible to host various intercalated atoms or ions. They have found that the diffusion barrier is the major factor that determines the rate of charge and discharge processes in these compounds, and, thereby, the specific capacitance implying the dominance of redox reactions [43]. The redox reactions in copper-based ternary chalcogenide materials studied for lithium-ion batteries indicate the occurrence of a combination displacement/

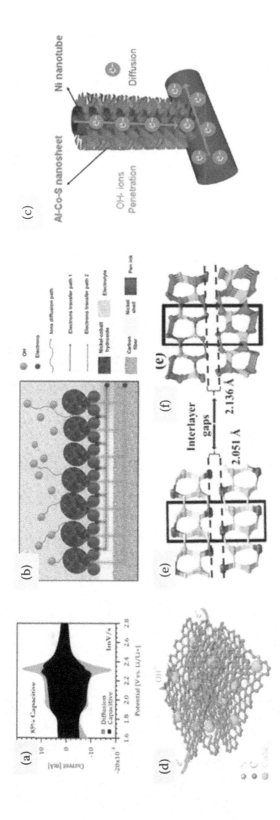

FIGURE 26.5 (a) Capacitive and diffusion-controlled contributions to charge storage in 100 nm 2D-TiS$_2$ cycled in Li$^+$ electrolyte at 1 mV s^{-1}; Capacitive contributions are in black [46]. [Reprinted with permission from Muller GA, Cook JB, Lim H-S, Tolbert SH, Dunn B (2015) High performance Pseudocapacitor Based on 2D Layered Metal Chalcogenide Nanocrystals. Nano Lett. 15:1911–1917 Copyright (2015) American Chemical Society](b) Schematic illustration of the charge storage advantages of Ni-Co DHs/Pen ink/Nickel/CF [14]. (Reprinted with permission from Pandit B, Karade SS, Sankapal BR (2017) Hexagonal VS$_2$ Anchored MWCNTs: First Approach to Design Flexible Solid-State Symmetric Supercapacitor Device. ACS Appl Mater Interfaces. 9:44880–44891) Copyright (2017) American Chemical Society; (c) Schematic illustration displaying the merits of the core–branch CC/H-Ni@Al-Co-S electrode for energy storage [40]. [Reprinted with permission from Huang J, Wei J, Xiao Y, Xu Y, Xiao Y, Wang Y, Tan L, Yuan K, Chen Y (2018) When Al-doped Cobalt Sulfide Nanosheets Meet Nickel Nanotube Arrays: A Highly Efficient and Stable Cathode for Asymmetric Supercapacitors. ACS Nano. 12:3030–3041 Copyright (2018) American Chemical Society; (d) schematic illustration of CoSe$_2$/NC nanosheets with excellent electrochemical performance [44]. [Reprinted with permission from Miao C, Xiao X, Gong Y, Zhu K, Cheng K, Ye K, Yan J, Cao D, Wang G, Xu P (2020) Facile Synthesis of Metal-Organic Framework-Derived CoSe$_2$ Nanoparticles Embedded in the N-Doped Carbon Nanosheet Array and Application for Supercapacitors. ACS Appl Mater Interfaces. 12:9365–9375 Copyright (2020) American Chemical Society] (e–f) showing the van der Waals gap and distance between two quadruple layers [43]. [Reprinted with permission from Ramasamy K, Gupta RK, Palchoudhury S, Ivanov S, Gupta A (2015) Layer-Structured Copper Antimony Chalcogenides (CuSbSe$_x$S$_{2-x}$): Stable Electrode Materials for Supercapacitors. Chem Mater. 27:379–386 Copyright (2015) American Chemical Society].

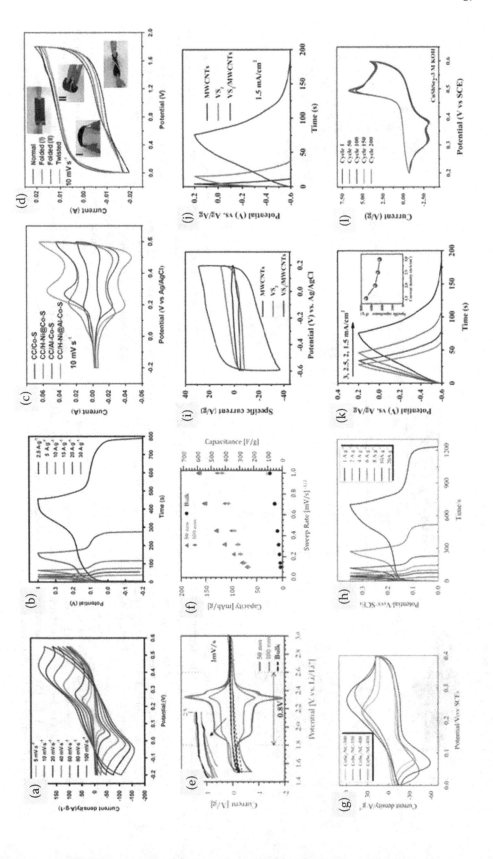

intercalation (CDI) mechanism rather than the traditional intercalation pathway. In this CDI mechanism, the formation of the intercalation complex is followed by the complete reduction of one of the metal components. For instance, the CDI mechanism goes by a reaction sequence like $CuSbS_2 \rightarrow M_xCu_{1-x}SbS_2 \rightarrow MSbS_2 + Cu$, leaving the electrodes somewhat distorted, and the electrode gets somewhat damaged after repeated cycles [43]. It has been observed that that the conductive nature of layered ternary chalcogenide Cu_2WS_4 can be switched from semiconducting to metallic by hydrogen incorporation, accompanied by 1010 times increase in electrical conductivity at the room temperature itself, and it can be maintained quite stably over high temperatures [42]. Hydrogen incorporation strongly affects the electronic structure of materials by providing external electrons.

26.5 RECENT DEVELOPMENTS IN THE METAL CHALCOGENIDE-BASED FLEXIBLE SUPERCAPACITORS

Evaluation of the electrochemical performance of different supercapacitor electrodes or materials requires characterizations like galvanic charge/discharge tests and cyclic voltammetry. These analyses happen to be the most fundamental in determining the performance. While cyclic voltammetry gives information about the extent of capacitive or pseudocapacitive behavior, galvanic charge-discharge tests plot the charge and discharge current with time for different scanning rates.

Figure 26.6a and b shows the CV and GCD curves of the Ni-Co DHs/Pen ink/Nickel/CF, CV curves at different scan rates ranged from 5 to 100 $mV \cdot s^{-1}$, the redox peaks in the CV arising from the different redox reactions that take place by conversion of double hydroxides to oxy-hydroxides or oxides. With all

---

FIGURE 26.6 Electrochemical performance of Ni-Co DHs/Pen ink/Nickel/CF in 6 M KOH electrolyte in a three-electrode system. (a) CV curves at different scan rates from 5 to 100 $mV.s^{-1}$. (b) GCD profiles at different current densities ranging from 2.5 to 30 $A \cdot g^{-1}$ [14]. [Reprinted with permission from Gao L, Surjadi JU, Cao K, Zhang H, Li P, Xu S, Jiang C, Song J, Sun D, Lu Y (2017) Flexible Fiber-Shaped Supercapacitor Based on Nickel-Cobalt Double Hydroxide and Pen Ink Electrodes on Metallized carbon Fiber. ACS Appl Mater Interfaces. 9:5409–5418 Copyright (2017) American Chemical Society]; (c) CV curves of CC/Co-S, CC/H-Ni@Co-S, CC/Al-Co-S, and CC/H-Ni@Al-Co-S electrodes at a scan rate of 10 $mV\ s^{-1}$. (d) CV curves of the flexible ASC under different bending conditions. The insets of (d) show the photographic images of the device under different bending states [40]. [Reprinted with permission from Huang J, Wei J, Xiao Y, Xu Y, Xiao Y, Wang Y, Tan L, Yuan K, Chen Y (2018) When Al-doped Cobalt Sulfide Nanosheets Meet Nickel Nanotube Arrays: A Highly Efficient and Stable Cathode for Asymmetric Supercapacitors. ACS Nano. 12:3030–3041 Copyright (2018) American Chemical Society; (e) CVs (10 cycles) at 1 $mV\ s^{-1}$ in a Li^+ electrolyte for 2D-TiS_2 nanocrystals, comparing 50 nm (red) and 100 nm (green) diameters to B-TiS_2 particles (black). (f) Kinetic analysis of 2D-TiS_2 and B-TiS_2 cycled in a Li^+ electrolyte at various sweep rates from 1–100 $mV \cdot s^{-1}$ [46]. [Reprinted with permission from Muller GA, Cook JB, Lim H-S, Tolbert SH, Dunn B (2015) High performance Pseudocapacitor Based on 2D Layered Metal Chalcogenide Nanocrystals. Nano Lett. 15:1911–1917 Copyright (2015) American Chemical Society (g) Comparison of CV curves of $CoSe_2$/NC-NF (h) GCD curves of $CoSe_2$/NC-NF [44]. [Reprinted with permission from Miao C, Xiao X, Gong Y, Zhu K, Cheng K, Ye K, Yan J, Cao D, Wang G, Xu P (2020) Facile Synthesis of Metal-Organic Framework-Derived $CoSe_2$ Nanoparticles Embedded in the N-Doped Carbon Nanosheet Array and Application for Supercapacitors. ACS Appl Mater Interfaces. 12:9365–9375 Copyright (2020) American Chemical Society; (i) CV curves for MWCNTs, VS_2 and VS_2/MWCNTs samples at scan rate of 100 mV/s, (j) GCD curves of MWCNTs, VS_2 and VS_2/MWCNTs at specific current of 1.5 mA/cm^2, (k) GCD curves at different current densities ranging from 1.5 to 3 mA/cm^2, inset shows specific capacitance at a function of current density [19]. [Reprinted with permission from Pandit B, Karade SS, Sankapal BR (2017) Hexagonal VS_2 Anchored MWCNTs: First Approach to Design Flexible Solid-State Symmetric Supercapacitor Device. ACS Appl Mater Interfaces. 9:44880–44891 Copyright (2017) American Chemical Society; (l) Galvanostatic charge–discharge characteristics of CV curves of CuSbSe2 for different CV cycles [43]. [Reprinted with permission from Ramasamy K, Gupta RK, Palchoudhury S, Ivanov S, Gupta A (2015) Layer-Structured Copper Antimony Chalcogenides ($CuSbSe_xS_{2-x}$): Stable Electrode Materials for Supercapacitors. Chem Mater. 27:379–386 Copyright (2015) American Chemical Society].

curves showing similar shapes, the current density increases with the growing scan rate. Even at a scan rate of 100 mV·s^{-1}, the CV redox peaks still can be seen, suggesting the structure is favorable to fast redox reactions. Near symmetric GCD curves in Figure 26.6b indicate high charge-discharge coulombic efficiency and low polarization. Figure 26.6c depicts the CV of the core–branch CC/H-Ni@Al-Co-S electrode. The pair of redox peaks observed in all the samples can be attributed to the faradaic reactions of cobalt sulfide in KOH solution. The CC/H-Ni@Al-Co-S exhibits a larger CV curve area and redox peak intensity compared to CC/Co-S and CC/Al-Co-S, implying a significantly improved specific capacitance and faster redox reaction kinetics processes due to Al doping [40]. The brilliant performance exhibited by the flexible supercapacitor electrode can also be perceived from the unaffected nature and trajectory of the CV curve, as shown in Figure 26.6d. The insets show photographs of the carbon-cloth-based electrode when bent or twisted through different angles.

Cyclic voltammograms at 1 mV s^{-1} for 2D-TiS$_2$ nanocrystals of different thicknesses (50 nm and 100 nm) compared to bulk-TiS2 particles are shown in Figure 26.6e [46]. While the peaks in the CV can be traced to the (Li/Li$^+$) redox, we observe that 2D-TiS$_2$ nanocrystals exhibit much higher levels of charge storage compared to the B-TiS$_2$ material. Their group had carried out a characterization to understand the kinetics of the electrode processes given in Figure 26.6f, where the capacity of the TiS$_2$ materials was plotted against the square root of the inverse sweep rate (i.e., $v^{-1/2}$) from 1 to 100 mV s^{-1}. It was observed from the linear dependence of the plot that for the bulk-TiS$_2$ materials, the kinetics are characterized by diffusion-controlled behavior, while for 2D-TiS$_2$, both linear and nonlinear regions are present, accounting for both capacitive and diffusion processes [46]. The effect of calcination temperature of the carbon cloth on capacitive properties was examined by Chenxu et al. through CV studies. Figure 26.6g exhibits cyclic voltammograms of CoSe$_2$/NC-NF sintered at 300, 350, 400, and 450 °C, respectively. The presence of redox peaks in all the curves, however, indicates that the capacitance arises from pseudo-capacitance behavior; largely, though, the energy storage efficiency is the highest at 400°C. So, the sample calcinated at 400°C was taken for further GCD analysis, as shown in Figure 26.6h. The GCD curves, symmetric and nonlinear as they are, imply high reversibility and the pseudo-capacitive processes taking place.

Cu$_{1.4}$S coated carbon cloth was used as an electrode by Xin's group. An analysis of their cyclic-voltammetry curves leads them to decipher the two possible redox reactions, the first one being Cu/Cu$^+$ conversion and the redox reaction between CuS and CuSOH. As expected, they have observed the position of peaks to extend with scan rates increased, stemming from the faradic energy storage process. Their proposition is backed by the nature of the GCD, which is quite distorted from symmetry, making pseudo-capacitive processes in the electrode very evident. While specific capacitance decreases with increasing current density, 66% of the initial capacitance is maintained, even under a high current density as 10 A g^{-1}. Further, due to copper sulfides' adhesive solid force to the carbon cloth, the specific capacitance retains most of the initial capacitance under 160° bending (98.3%) and 100 times folded (96.8%).

Figure 26.6i compares the CV performance of three different materials: MWCNTs, VS$_2$, and MWCNTs/VS$_2$ [19]. As stated in one of the previous sections, a rectangular CV curve demonstrates capacitive behavior through the EDLC formation process. In either of the CV curves VS$_2$ and VS$_2$/MWCNT, the CV plots of VS$_2$/MWCNTs manifest a distorted, rectangular shape without distinct redox peaks. Such a distortion of the CV curve from the ideal rectangular shape and the lack of distinct redox peaks implies a reversible redox reaction related to electrolyte intercalation/de-intercalation of electrolyte ions into VS$_2$. Coming to the GCD curves, similar to what's shown in Figure 26.6j, the charge and discharge regions are not perfectly linear, which demonstrates the reversible redox mechanism related to intercalation/de-intercalation of electrolyte ions onto VS$_2$. The GCD variation of VS$_2$/MWCNTs at different current densities, as depicted in Figure 26.6k, shows a drop in the specific capacitance at greater current densities [19]. Cyclic voltammetry studies, as we have mentioned, are used to study the stability of the supercapcitor electrode under changing number of cycles or with different bending angles. A highly efficient flexible electrode would show nearly no deviation in its graphical plot under increasing number of cycles. Such an observation can be noted in Figure 26.6l, where the stability of a $CuSbSe_2$ electrode has been assessed by CV studies.

Theerthagiri et al. has already comprehensively mentioned advances in transition metal chalcogenides for advanced supercapacitor systems. Unary metal chalcogenides have been extensively discussed with a focus on Ni and Co sulfides and selenide compounds [47]. But compared to sulfides, selenides have often gone unreported and unstudied. Certain breakthrough developments on unary metal chalcogenide electrode-based FSCs have been reported below in Table 26.1.

It is a known idea that doping a metal atom into MoS_2 and FeS_2 has effectively tuned their morphology to attain higher electrochemical performance [48,49]. Therefore, Ni-doping on Mo-S and Fe-S proved to be highly effective. Processed through scalable in situ hydrothermal processes, and followed by effective anion-exchange methodology for sulfurization, hierarchical Ni-Mo-S and Ni-Fe-S nanosheets (NSs) were prepared [40]. Ni^+ ions helped achieve higher electrical conductivity, larger surface area, and exclusive porous network. Both Ni-Mo-S and Ni-Fe-S NSs exhibited ultrahigh specific capacities of 312 and 246 mAh.g^{-1} at 1 mA.cm^{-2}. The compositional tenability of the Mo-S and Fe-S system with Ni helped achieve exceptional rate capabilities (78.85% and 78.46% capacity retention at 50 mA.cm^{-2}) and optimal cycling stabilities an important role in fabricating all-solid-state ASC without incorporating any polymer binders or conductive additives.

Occasionally, copper tins sulfides (CTSs) have been extensively investigated as potential electrode material for abundant presence on earth, environment friendliness, low cost, and high theoretical capacitance. However, they suffer from poor electrical conductivity and large volume change, especially during the redox reaction, leading to inefficient electrochemical performance. Zhou et al. investigated incorporating carbon quantum dots (CQDs) in CTSs to expect better performances in terms of rate capability and cycling stability [50]. The electrode developed from CTSs@CQDs exhibited very high specific capacitance reaching 856 F.g^{-1} at 2mV.s^{-1} and a high-rate capability of 474 F.g^{-1} at even 50 mV.s^{-1}, which is considerably higher than electrodes developed only out of CTS. In addition to the superior rate capability and cycling stability, the CQDs helped improve electrical conductivity and increased the volume change during a redox reaction.

$NiCo_2S_4$ (NCS) and nickel-based material had been studied as a promising electrode for pseudocapacitive electrode systems for its enhanced electrochemical activity, but it lagged in charge transfer rate development; increasing the electrical conductivity of a material is critical to improving the energy density and power density of electrode material, and thus, reduced graphene oxide (rGO) can be considered [51–53]. Yu et al. devised a new experimental methodology, wherein in situ growth of NCS nanosheets was made to happen on porous nitrogen-doped rGO (PN-rGO) via a two-step hydrothermal method [54]. The process improved the surface area and structural features, thereby helping to develop the electrical conductivity of the system. Electrochemical tests further hinted at ultra-low charge transfer resistance, excellent specific capacitance, and enhanced rate capability. The PN-rGO/NCS electrode material showed a specific capacitance up to 1687 F.g^{-1}, an excellent rate performance with specific capacitance up to 1470 F.g^{-1} at 10 A.g^{-1} current density and an ultra-low resistance (0.16 Ω). DFT studies supplement the experimental studies with a theoretical basis and make PN-rGO/NCS a promising electrode material. Table 26.2 below elaborates on several binary and ternary metal chalcogenide-based FSC systems that have been studied:

26.6 CONCLUSION

This chapter aimed to highlight transition metal chalcogenide-based electrode materials on flexible substrates for application in supercapacitors, and in the process, we have described the status of the development of flexible supercapacitors based on transition metal chalcogenide electrodes. Research in this field is becoming significant as researchers are trying to design next-generation compact, wearable, and portable electronic systems. While the high flexibility and lightweight nature of these supercapacitors are positive attributes needed for designing, stability remains a concern. Most of these flexible

TABLE 26.2

Binary and Ternary Metal Chalcogenides that have been Worked upon so Far, their Synthesis Route, and Performance Parameters

Electrode	Preparation methodology	Capacitance (F.g^{-1})	Current density (A.g$^{-1)}$	Research comments	Ref.
NiCoS/CC	Solvothermal (Slurry@CC)	128	0.5	Capacity retention of 84% after 3000 cycles	[55]
Ni$_{0.8}$Cu$_{0.2}$S	Hydrothermal (Direct CC)	938.6	1	Capacity retention of 69% after 10000 cycles at 2 A.g^{-1}	[56]
CC/H-Ni@Al-CoS	Template assisted hydrothermal (Direct CC)	2434	1	90.6% retention even after 10000 cycles	[57]
Ni$_{0.6}$Co$_{0.4}$Se$_2$	Solvothermal	1580	1	90% capacity retention at 20000 cycles at 10 A.g^{-1}	[58]
Cu-Sn-S	Hydrothermal (Slurry @ Ni foam)	856	–	–	[50]

devices are fabricated to run with satisfactory efficiency up to 5000 cycles or 10000. This happens due to the repeated charging and discharging processes in the electrode, wearing off or damaging the nanostructures. The chapter has focused extensively on low-cost substrates that have presently found extreme importance in developing FSC systems. While work on conventional unary metal chalcogenides, like copper sulfide, vanadium sulfide, and cobalt sulfide/selenide, finds a place, similarly, the use of binary or mixed metal chalcogenides has also attracted interest over the recent years. A synergy between two different metal components has been reported to generally outperform the conventional unary chalcogenides. Doping in these systems also helps in increasing efficiency. FSC systems based on these compounds have already achieved a high-power density, a critical parameter to characterize FSC electrodes. However, nanostructure engineering is also vital, which needs to be efficiently tuned for exposing the maximum amount of surface area and catalytic sites. One more observation, as a suggestion, might be to grow these nanostructures directly on the flexible substrate instead of using a separate binder, which shields a lot of surface area. The most common synthesis method remains the hydrothermal route, which is easy, economical, and effective. Thus, these latest studies have circumvented low energy density and high production cost, and FSC electrodes suffered from, and have successfully dealt with, a flexible supercapacitor.

ACKNOWLEDGMENT

This work was supported by Science and Engineering Research Board (SERB) grants funded by the Department of Science and Technology (DST) Central, Government of India through Teachers Associateship for Research Excellence (TAR/2018/000195) (Subhasis Roy).

REFERENCES

1. Roy, S., and Botte, G. G. (2018). Perovskite solar cell for photocatalytic water splitting with a TiO2/co-doped hematite electron transport bilayer. *RSC Adv.* 8(10): 5388–5394
2. Palchoudhury S, Ramasamy K, Gupta RK, Gupta A (2019) Flexible supercapacitors: a materials perspective. *Front Mater.* 5:83

3. Xie T, Xu J, Wang J, Ma C, Su L, Dong F, Gong L (2020) Freestanding needle flower structure $CuCo_2S_4$ on carbon cloth for flexible high energy supercapacitors with the gel electrolytes. *Front Chem.* 8:62

4. Mu C, Sun X, Chang Y, Wen F, Nie A, Wang B, Xiang J, Zhai K, Xue T, Liu Z (2021) High-performance flexible all-solid state micro-supercapacitors based on two-dimensional InSe nanosheets. *J. Power Sources.* 482:228987

5. Mariappan VK, Krishnamoorthy K, Pazhamalai P, Sahoo S, Nardekar SS, Kim S-J (2019) Nanostructured ternary metal chalcogenide-based binder-free electrodes for high energy density asymmetric supercapacitors. *Nano Energy.* 57:307–316.

6. Cheng Q, Tang J, Ma J, Zhang H, Shinya N, Qin L-C (2011) Polyaniline-coated electro-etched carbon fiber cloth electrodes for supercapacitors. *J Phys Chem C.* 115:23584–23590.

7. Lehtimaki S, Suominen M, Damlin P, Tuukkanen S, Kvarnstron C, Lupo D (2015) Preparation of super-capacitors on flexible substrates with electrodeposited PEDOT/graphene composites. *ACS Appl Mater Interfaces.* 7:22137–22147.

8. Wu J, Liu W-W, Wu Y-X, Wei T-C, Geng D, Mei J, Liu H, Lau W-M, Liu L-M (2016) Three-dimensional hierarchical interwoven nitrogen-doped carbon nanotubes/Co_xNi_{1-x}-layered double hydroxides ultrathin nanosheets for high-performance supercapacitors. *Electrochim. Acta.* 203:21–29.

9. Gao W, Singh N, Song L, Liu Z, Reddy ALM, Ci L, Vajtai R, Zhang Q, Wei B, Ajayan PM (2011) Direct laser writing of micro-supercapacitors on hydrated graphite oxide films. *Nat Nanotechnol.* 6:496–500

10. Orangi J, Hamade F, Davis VA, Beidaghi M (2020) 3D printing of additive-free 2D Ti_3C_2Tx(MXene) ink for fabrication of micro-supercapacitors with ultra-high energy densities. *ACS Nano.* 14:640–650

11. El-Kady MF, Kaner RB (2013) Scalable fabrication of high-power graphene micro-supercapacitors for flexible and on-chip energy storage. *Nat Commun.* 4:1475

12. Wang Z, Han Y, Zeng Y, Qie Y, Wang Y, Zheng D, Lu X, Tong Y (2016) Activated carbon fiber paper with exceptional capacitive performance as a robust electrode for supercapacitors. *J Mater Chem A.* 4:5828–5833

13. Zhao J, Hongwei L, Lyu Z, Jiang Y, Xie K, Wang X, Wu Q, Yang L, Jin Z, Ma Y, Liu J, Hu Z (2015) Hydrophilic hierarchical nitrogen-doped carbon nanocages for ultrahigh supercapacitiveperformance. *Adv Mater.* 27:3541–3545

14. Gao L, Surjadi JU, Cao K, Zhang H, Li P, Xu S, Jiang C, Song J, Sun D, Lu Y (2017) Flexible fiber-shaped supercapacitor based on nickel-cobalt double hydroxide and pen ink electrodes on metallized carbon fiber. *ACS Appl Mater Interfaces.* 9:5409–5418

15. Abza T, Ampong FK, Hone FG, Nkrumah I, Nkum RK, Boakye F (2017) A new route for the synthesis of CdS thin films from acidic chemical baths. *Int J Thin Films Sci Technol.* 6:67–71

16. Mudd GW, Svatek SA, Ren TH, Patane A, Makarovsky O, Eaves L, Beton PH, Kovalyuk ZD, Lashkarev GV, Kudrynskyi ZR, Dmitriev AL (2013) Tuning the bandgap of exfoliated InSe nanosheets by quantum confinement. *Adv Mater.* 25:5714–5718

17. Jin H, Ji JW, Wan LH, Daai Y, Wei YD, Guo H (2017) Ohmic contact in monolayer InSe-metal interface. *2D Mater.* 4:025116

18. Subramani K, Sudhan N, Divya R, Sathish M (2017) All-solid state asymmetric supercapacitors based on cobalt hexacyanoferrate-derived CoS and activated carbon. *RSC Adv.* 7:6648–6659

19. Pandit B, Karade SS, Sankapal BR (2017) Hexagonal VS_2 anchored MWCNTs: first approach to design flexible solid-state symmetric supercapacitor device. *ACS Appl Mater Interfaces.* 9:44880–44891

20. Sun L, Wang X, Wang Y, Xiao D, Cai W, Jing Y, Wang Y, Hu F, Zhang Q (2019) *In-situ* functionalization of metal electrodes for advanced asymmetric supercapacitors. *Front Chem.* 7:512

21. Sahoo S, Pazhamalai P, Krishnamoorthy K, Kim S-J (2018) Hydrothermally prepared α-MnSe nanoparticles as a new pseudocapacitive electrode material for supercapacitor. *Electrochim Acta.* 268:403–410

22. He X, Mao X, Zhang C, Yang W, Zhou Y, Yang Y, Xu J (2020) Flexible binder-free hierarchical copper sulfide/carbon cloth hybrid supercapacitor electrodes and the application as negative electrodes in asymmetric supercapacitor. *J Mater Sci: Mater Electron.* 31:2145–2152

23. Ma L, Shen X, Ji Z, Wang S, Zhou H, Guoxing Z (2014) Carbon coated nickel sulfide/reduced graphene oxide nanocomposites: facile symsthesis and excellent supercapacitor performance. *Electrochim Acta.* 146:525–532

24. Zhu Y, Xiang D, Ma S, Wang Y, Li H, Jiang Z (2021) Controllable synthesis of nickel sulfide nanosheet/carbon fibers composite and its electrochemical performances. *Int J Electrochem Sci.* 16:210254

25. Ren R, Faber MS, Dziedzic R, Wen Z, Jin S, Mao S, Chen J (2015) Metallic CoS_2 nanowire electrodes for high cycling performance supercapacitors. *Nanotechnology.* 26:494001

26. Li S, Chen T, Wen J, Gu P, Fang G (2017) *In-situ* grown Ni_9S_8 nanorod/O-MoS_2 nanosheet nanocomposite

on carbon cloth as a free binder supercapacitor electrode and hydrogen evolution catalyst. *Nanotechnology.* 28:445407

27. Javed MS, Dai S, Wang M, Guo D, Chen L, Wang X, Hu C, Xi Y (2015) High performance solid state flexible supercapacitor based on molybdenum sulfide nanospheres. *J Power Sources.* 285:63–69

28. Chen T, Li S, Wen J, Gui P, Guo Y, Guan C, Liu J, Fang G (2018) Rational construction of hollow core-branch $CoSe_2$ nanoarrays for high-performance asymmetric supercapacitor and efficient oxygen evolution. *Small.* 14:1700979

29. Bhat KS, Shenoy S, Nagraja HS, Sridharan K (2017) Porous cobalt chalcogenide nanostructures as high performance pseudo-capacitor electrodes. *Electrochim Acta.* 248:188–196

30. Yu X, Zhou T, Cao X, Zhao W, Chang J, Zhu W, Guo W, Du W (2017) Low-cost synthesis and electrochemical characteristics of ternary Cu-Co sulfides for high performance full-cell asymmetric supercapacitors. *Mater Res Bull.* 91:68–76

31. Li Z, Zhao D, Xu C, Ning J, Zhong Y, Zhang Z, Wang Y, Hu Y (2018) Reduced $CoNi_2S_4$ nanosheets with enhanced conductivity for high-performance supercapacitors. *Electrochim Acta.* 278:33–41

32. Zhang M, Annamalai KP, Liu L, Chen T, Gao J, Tao Y (2017) Multiwalled carbon nanotubes-supported $CuCo_2S_4$ as a heterogeneous Fenton-like catalyst with enhanced performance. *RSC Adv.* 7:20724–20731

33. Liu S, Jun SC (2017) Hierarchical manganese cobalt sulfide core-shell nanostructures for high performance asymmetric supercapacitors. *J Power Sources.* 342:629–637

34. Chandrasekharan NI, Muthukumar H, Sekar AD, Pugazhendhi A, Manickam M (2018) High-performance asymmetric supercapacitor from nanostructured tin nickel sulfide ($SnNi_2S_4$) synthesized via microwave-assisted technique. *J Mol Liq.* 266:649–657

35. Cheng C, Zhang X, Wei C, Liu Y, Cui C, Zhang Q, Zhang D (2018) Mesoporous hollow $ZnCo_2S_4$ core-shell nanospheres for high performance supercapacitors. *Ceram int.* 44:17464–17472

36. Feng X, Shi Y, Yang X, Li S, Li Y (2017) Binary Ni-Al sulfide/nickel foam electrodes with hierarchical hyacinth-like structure for supercapacitor performance. *J Alloy Compd.* 705:652–658

37. Lokhande AC, Patil A, Shelke A, Babar PT, Gang MG, Lokhande VC, Dhawale DS, Lokhande CD, Kim JH (2018) Binder-free novel Cu_4SnS_4 electrode for high-performance supercapacitors. *Electrochim Acta.* 284:80–88

38. Deng C, Yang L, Yang C, Shen P, Zhao L, Wang Z, Wang C, Li J, Qian D (2018) Spinel $FeCo_2S_4$ nanoflower arrays grown on Ni foam as novel binder-free electrodes for long-cycle-life supercapacitors. *Appl Surf Sci.* 428:148–153

39. Wan H, Jiang J, Ruan Y, Yu J, Zhang L, Chen H, Miao L, Bie S (2014) Direct Formation of Hedgehog-like Hollow Ni-Mn Oxides and Sulfides for Supercapacitor Electrodes. *Part Part Syst Charact.* 31:857–862

40. Balamurugan J, Li C, Aravindan V, Kim NH, Lee JH (2018) Hierarchical Ni-Mo-S and Ni-Fe-S nanosheets with ultrahigh energy density for flexible all solid state supercapacitors. *Adv Funct Mater.* 28:1803287

41. Kim J, Tabassian R, Nguyen VH, Umrao S, Oh I-K (2019) Crumbled quaternary nanoarchitecture of sulfur-doped nickel cobalt selenide directly grown on carbon cloth for making stronger ionic soft actuators. *ACS Appl Mater Interfaces.* 11:40451–40460

42. Hu X, Shao W, Hang X, Zhang X, Zhu W, Xie Y (2016) Superior electrical conductivity in hydrogenated layered ternary chalcogenide nanosheets for flexible all-solid-state supercapacitors. *Angew Chem Int Ed Engl.* 55:5733–5738

43. Ramasamy K, Gupta RK, Palchoudhury S, Ivanov S, Gupta A (2015) Layer-structured copper antimony chalcogenides ($CuSbSe_xS_{2-x}$): stable electrode meterials for supercapacitors. *Chem Mater.* 27:379–386

44. Miao C, Xiao X, Gong Y, Zhu K, Cheng K, Ye K, Yan J, Cao D, Wang G, Xu P (2020) Facile synthesis of metal-organic framework-derived $CoSe_2$ nanoparticles embedded in the N-doped carbon nanosheet array and application for supercapacitors. *ACS Appl Mater Interfaces.* 12:9365–9375

45. Zheng J, Wang F, Ma J, Zhou K (2019) One-step hydrothermal synthesis of carbon-coated nickel-copper sulfide nanoparticles for high-performance asymmetric supercapacitors. *Eur J Inorg Chem.* 13:1740–1747

46. Muller GA, Cook JB, Lim H-S, Tolbert SH, Dunn B (2015) High performance pseudocapacitor based on 2D layered metal chalcogenide nanocrystals. *Nano Lett.* 15:1911–1917

47. Theerthagiri J, Karuppasamy K, Durai G, Sarwar Rana AuH, Arunachalam P, Sangeetha K, Kuppusami P, Kim H-S (2018) Recent advances in metal chalcogenides (MX; X = S, Se) nanostructures for electrochemical supercapacitor applications: a brief review. *Nanomaterials.* 8:256

48. Merki D, Vrubel H, Rovelli L, Fierro S, Hu X (2012) Fe, Co, and Ni ions promote the catalytic activity of amorphous molybdenum sulfide films for hydrogen evolution. *Chem Sci.* 3:2515–2525

49. Huang S-Y, Sodano D, Leonard T, Luiso S, Fedkiw PS (2017) Cobalt-doped iron sulfide as an electro-catalyst for hydrogen evolution. *J Electrochem Soc.* 164:F276

50. Bi Z, Huang L, Shang C, Wang X, Zhou G (2019) Stable copper tin sulfide nanoflower modified carbon quantum dots for improved supercapacitors. *J Chem.* 2019:6109758

51. Xiong X, Waller G, Ding D, Chen D, Rainwater B, Zhao B, Wang Z, Liu M (2015) Controlled synthesis of $NiCo_2S_4$ nanostructured arrays on carbon fiber paper for high-performance pseudocapacitors. *Nano Energy.* 16:71–80

52. Maitra, S., Sarkar, A., Maitra, Halder S, Roy S and Kargupta K. (2020) Cadmium sulphide sensitized crystal facet tailored nanostructured nickel ferrite @ hematite core-shell ternary heterojunction photoanode for photoelectrochemical water splitting. *MRS Advances.* 5:2585–2593

53. Sudhakar YN, Hemant H, Nitinkumar SS, Poornesh P, Selvakumar M (2016) Green synthesis and elec-trochemical characterization of rGO-CuO nanocomposites for supercapacitors applications. *Ionics.* 23: 1267–1276

54. Hong X, Li J, Zhu G, Xu H, Zhang X, Zhao Y, Zhang J, Yan D, Yu A (2020) Cobalt-nickel sulfide nanosheets modified by nitrogen-doped porous reduced graphene oxide as high-conductivity cathode materials for supercapacitors. *Electrochim Acta.* 362:137156

55. Liu T, Liu J, Zhang L, Cheng B, Yu J (2020) Construction of nickel cobalt-sulfide nanosheet arrays on carbon cloth for performance-enhanced supercapacitor. *J Mater Sci Technol.* 47:113–121

56. Du D, Lan R, Humphreys J, Amari H, Tao S (2018) Preparation of nanoporous nickel-copper sulfide on carbon cloth for high-performance hybrid supercapacitors. *Electrochim Acta.* 273:170–180

57. Huang J, Wei J, Xiao Y, Xu Y, Xiao Y, Wang Y, Tan L, Yuan K, Chen Y (2018) When Al-doped cobalt sulfide nanosheets meet nickel nanotube arrays: a highly efficient and stable cathode for asymmetric supercapacitors. *ACS Nano.* 12:3030–3041

58. Xie S, Gou J, Liu B, Liu C (2019) Nickel-cobalt selenide as high-performance and long-life electrode material for supercapacitor. *J Colloid Interface Sci.* 540:306–314

27 MOFs-Derived Metal Oxides-Based Compounds for Flexible Supercapacitors

Charu Goyal
Department of Chemistry, GLA University, Mathura, India

Anuj Kumar
Department of Chemistry, GLA University, Mathura, India

Ghulam Yasin
Institute for Advanced Study, Shenzhen University, Shenzhen,
 Guangdong, People's Republic of China

Ram K. Gupta
Department of Chemistry, Kansas Polymer Research Center,
Pittsburg State University, Pittsburg, Kansas, USA

CONTENTS

27.1   Introduction.. 501
27.2   MOFs-derived metal oxides for flexible supercapacitors 502
    27.2.1   MOFs-derived binary metal oxides 503
        27.2.1.1   Co-oxides materials 503
        27.2.1.2   Fe-oxides materials.. 503
        27.2.1.3   Ce-oxides materials 505
    27.2.2   MOFs-derived ternary or TMMOs 506
        27.2.2.1   Porous $ZnCo_2O_4$ material................................... 506
        27.2.2.2 $MnCo_2O_4$ materials 508
        27.2.2.3 $Ni_xCo_{3-x}O_4$ material...................................... 510
    27.2.3   MOFs-derived metal oxide/composite... 513
27.3   Conclusions.. 514
References.. 514

27.1   INTRODUCTION

Due to the shortage of fossil fuels, the scheme and expansion of devices for energy storage have grown significantly [1–3]. Flexibility and reliability are needed to significantly enhance renewable energy storage technologies with quick energy supplies, fast charging and discharging, eco-friendly, high energy density, low maintenance costs, long durability, and excellent cycle stability [4–10]. Based on the energy storage procedure, supercapacitors are divided into electrical double-layer condensers (EDLCs), pseudocapacitors (PCs), and hybrid capacitors (HCs) [4–10]. To dynamically distribute charges, EDLCs

DOI: 10.1201/9781003186755-27

employ changeable ion adsorption on the electrode and/or electrolyte interfaces, allowing them to be charged and/or discharged in seconds and over cycles of 100,000 [11–13]. Usually, graphene oxide (GO), graphene (Gr), CNTs (carbon nanotubes), activated carbon (AC), nano-architecture carbon, and aerogels of carbon have been utilized as potential EDLC materials [14,15]; however, these materials exhibited low specific capacitances [16]. Besides, PCs store the energy via redox reactions between the electrode and electrolyte, resulting in higher specific capacitance as compared to EDLCs. Conducting polymers and transition metal oxides were found to be potential materials for PCs. However, in the case of CPs, because of their repetitive enlargement and dwindling, PCs exhibited poor cycling stability [17,18]. After many attempts, recently, HCs have been introduced. combining the benefits of EDLCs as well as PCs, having the prime purpose of improving electrochemical action [17].

Within this framework, energy storage devices, particularly flexible/wearable electronics, are in demand in various sectors. For instance, in-plane micro-supercapacitors (MSCs) electrode outlines have been introduced as powering flexible electronic devices. These devices are easy to integrate onto chips or flexible substrates. MSCs are utilized for high power demand and extended cycle spans, conforming to more ordinary micro-batteries due to their low energy density conventionally [19,20]. To produce MSCs, various techniques have been reported to be used in the microelectronic sector. It is expected that the full potential of SCs in future devices may be realized in an easy, low-cost, flexible, and also eco-friendly manner, allowing SCs with distinctive and appealing designs, such as printed electronics, MSCs, and FSCs [21,22]. Novel printed electronic techniques may now be used to make nonflat and flexible SCs (inkjet printing, graphic design, 3D printing, and many more). In terms of material selection and device setup, printing allows for a greater number of prototypes to be developed and analyzed. Since the study approach may be expedited in this way, important resources are not wasted [23,24].

In this regard, transition metal oxides like Fe_2O_3, MnO_2, ZnO, Co_3O_4, etc., with their composites that have been extensively investigated to be utilized as electrode materials for FSCs due to variable valence state and their significant pseudo-capacity, as well as their significant intrinsic strength [25–28]. Moreover, TMOs can create supercapacitors with significantly higher energy density than typical carbon materials, as well as greater electrochemical strength in comparison to polymer materials. They show electro-chemical faradaic reactions among the ions and material in a suited range, with energy storage like electrostatic carbon. As of now, various chemicals have been modified to improve the wearability of carbonate materials and to improve their capacitive performance. Depositing metal oxide/hydroxide particles on electrodes to improve their specific capacitance is a good concept. The metal oxides, including oxides of manganese, ruthenium, cobalt, and nickel, or oxides of mixed metal perform an important part in improving electrode capacitance via rapid faradaic pseudo-capacitance effects. Enhancing electrode electrical conductivity is a critical aspect in promoting the effect of capacitors of carbon when employing metal [29].

The focus of this chapter is on current advancements in transition metal oxide-based FSCs electrodes: (i) metal oxides, (ii) ternary or mixed-metal oxides, and (iii) metal oxide along with carbon materials. The structural composition, congregation technique, and electrochemical action of these electrodes are also examined to offer a complete picture of the issue. The current developments in the production of transition metal oxide-based FSCs electrodes utilizing several types of TMOs have also been studied. Finally, the limits, prospects, and ways to promote the capacitance performance of existing TMO-based FSCs electrodes are reviewed.

27.2   MOFS-DERIVED METAL OXIDES FOR FLEXIBLE SUPERCAPACITORS

Different research groups have examined transition metal oxides as suitable candidates for SCs usage due to their vast range of oxidation states and redox reactions. A transition metal oxide produced by thermal decomposition of MOF-based nanostructures has an intrinsic surface area that allows ions to easily diffuse through its pores. The capacity of the electrodes to easily diffuse ions is among the most significant characteristics of their electrochemical activity.

27.2.1 MOFS-DERIVED BINARY METAL OXIDES

27.2.1.1 Co-oxides materials

The Co_3O_4 material has a unique nanostructure, extraordinary specific capacitance, and long working life. It also has a spinel crystal structure (space group, cubic Fd3m) and, usually, its octahedral cavity is formed from a cubic cavity. Oxide anions are arranged in a dense lattice. Nano-sized Co_3O_4 materials have attracted a lot of research devotion in recent years since their distinctive nanostructure is reliant on the physicochemical characteristics of many prospective features [30]. Due to the complexity of Co_3O_4 nanostructures, several techniques for regulating their production have been investigated [31]. It is also a cost-effective substitute for RuO_2 due to its high reversibility of theoretical specific capacitance (3,560 F/g) [32].

Zhang et al. published among the earliest instances of Co_3O_4 nanostructures generated from Co(II) MOF precursors [33]. The author projected the formula $[Co(BDC)_2]_n$ for preparing Co(II)-MOF from $CoCl_2.6H_2O$ and p-benzene-dicarboxylic acid (BDC) ligands and used the FT-IR spectroscopy, thermosgravimetric analysis (TGA), and differential scanning calorimetry (DSC) analysis for its characterization. Solid-phase toughening of the Co(II)-MOF at 450°C produced a porous nano/microstructure of Co_3O_4. Further, using electron microscopy and X-ray diffraction (XRD) techniques, the obtained materials were characterized to study the structural features. The current density of 1 A/g and the specific retention capacity of 0.97% after 1000 cycles of continuous charge and discharge in a 6.0 M KOH aqueous solution (compared to a saturated electrode) show their remarkable SCs characteristics.

Further, Lu et al. [34] used a MOF of Co(II) produced using urea, 8-hydroquinoline, and $Co(NO_3)_2$ through a solvothermal technique to create a dendritic porous Co_3O_4 nanostructure comprising of numerous nano-rods (diameter = 15–20 nm and length = 2–3 μm). Calcinating the Co(II) MOF concluded in the development of a black porous nanostructure of Co_3O_4 with the identical shape as the precursor at 450°C for 1 hour in the air. The production of Co_3O_4 was verified by XRD measurements, indicating that Co_3O_4 was formed from the MOF precursor based on Co(II). The symmetric nature of the as-synthesized Co_3O_4 unchanged current galvanostatic charging and/or discharging curves indicated their characteristic high electrochemical capacity. Even at a small density of 0.5 A/g, it maintains its symmetric character. The Co_3O_4 electrodes have a specific capacitance of 207.8 F/g at a current density of 0.5 A/g (far higher than the 77.0 F/g of commercial Co_3O_4 electrodes) and preserve a specific capacitance of 97.5% after 1000 cycles (Figure 27.1). The scanning-electron microscopy (SEM) revealed well-defined nano-sized holes in the Co_3O_4 material, no sorption property was found.

Further, a porous material for the electrode from the Co(II)-based MOF is manufactured using thermolysis and two-stage calcination, as demonstrated by Guo et al. [35]. Single XRD studies of Co(II)-based MOF crystals exhibit that the coordination among the carboxyl ligand azobenzene 3,5,40-tricarboxylic acid (H_3ABTC) and the auxiliary ligand 4,40-bipyridine is no longer. The intermediate product results in a supramolecular 3D multichain structure supported by pi-pi stacking with H-bonding interactions. X-ray diffraction data were used to confirm the purity of the porous crystalline phase of Co_3O_4 and draw the faceted cubic phase. The TEM images show that Co_3O_4 exists in the form of irregular porous aggregates, composed of many NPs, with an average of 10nm size. The surface measurement of Brunauer-Emmett-Teller (BET) showed a surface area of 47.12 m^2A/g, which indicated the porous nature of asymmetric Co_3O_4 materials.

27.2.1.2 Fe-oxides materials

Fe_2O_3 has emerged as one of the best choices because of factors like natural availability, relative inexpense, eco-friendliness, and higher specific capacity (1005 mAh/g). Furthermore, the poor conductive

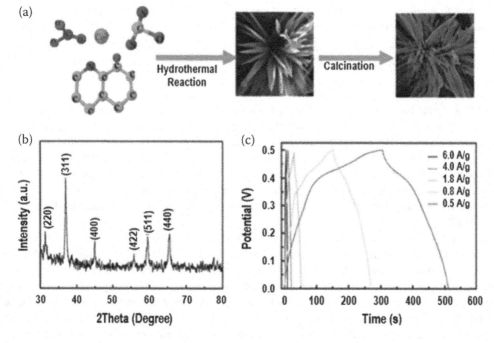

FIGURE 27.1 (a) Schematic synthesis of the Co(II) parent metal and organic frames, and Co_3O_4, (b) X-ray diffraction in powder form dates of the Co_3O_4, (c) charging and unloading curves at a variety of density levels and potentials of Co_3O_4 [34]. Copyright 2021, The Royal Society of Chemistry.

nature of the Fe_2O_3 particles and their significant agglomeration during charge and discharge operations might be a concern, resulting in a rapid loss capacity [36,37]. A layered society approach was devised in 2013 for the production of a free-standing nano-porous graphene sheets flexible nanocomposite with iron oxide ($NGP–Fe_3O_4$) that was aided by nano-fiber of carbon. The loading of Fe_3O_4 nanocrystals using a universal wet-immersion technique was investigated to create not so useful functionalized porous graphene sheets as binder-free anodes for lithium-ion batteries. $NGP-Fe_3O_4$ nanocomposite electrodes offer excellent electrochemical capabilities because of their nano-porous design for reformable lithium storage and supercapacitors: ultra-high specific capability, extended service life, and high rates [38]. Nowadays, there has been a rapid increase in the need for transportable energy storage devices. As a result, an in situ hydrothermal technique for synthesizing three-dimensional iron oxide (Fe_2O_3)/graphene aerogel (GA) hybrid (Fe_2O_3/GA) was developed. The involvement of NH_4OH in Fe^{2+} reduction is outlined in the following mechanism (Eqs. 27.1 and 27.2):

$$GO + Fe^{2+} \rightarrow Fe^{3+} \tag{27.1}$$

$$Fe^{2+} + NH_3.\ H_2O \rightarrow Fe_2O_3 + NH^{4+} \tag{27.2}$$

There are several reasons why the synthetic technique outlined is utilized more commonly to achieve Fe_2O_3/GA hybrid. (i) Its usage as precursors promise to create absolute metal oxide NPs on the surface of the GO sheets; (ii) to utilize a synthesized Fe_3O_4 NPs with certain organic solvents or surfactants such as PVP; and (iii) to substantially load metal oxide NPs onto the hybrids. Similarly, small sections of our Fe_2O_3/GA hybrid electrode were utilized to construct an extremely flexible all-solid-state symmetric SCs device that is suitable for a variety of bending angles and has a high specific capacitance of 440 F/g, with 90% capacitance remaining after cycles of 2200, representing worthy cycling strength. Furthermore, further development of the manufacturing method on the low-cost and simple assemblage of this flexible and frothy device indicates its enhancement as an efficient material for electrodes for practical application in sticky tape-like or suitable electronics [39]. Hematite (α-Fe$_2$O$_3$) is a stable phase of iron oxide found in natural soil and rocks. It is also a suitable substance for applications of supercapacitors, as a capable material in the building of negative electrodes thanks to its large theoretical capacities (3625 F/g), acceptable potential size, richness, thermal strength, corrosion resistance, and nonpoisonous [40,41].

Khatavkar et al. released a paper in 2019 explaining the reposition, by utilizing a simple, cost-effective, and additive and/or binder-free method, of α-Fe$_2$O$_3$ on a tinny, lightweight, and flexible staine-and-state mesh substrate. The supercapacity of α-Fe$_2$O$_3$ thin films deposited with stainless steel mesh are studied in 0.5M Na$_2$SO$_3$ electrolyte. From charge-discharge curves, the good specific capacitance found is 960 F/g at 4 mA/cm^2.

27.2.1.3 Ce-oxides materials

To prepare CeO$_2$/CuO bimetal oxides, two alternative techniques may be used, as illustrated in Figure 27.2: thermal treatment of Ce-MOF with impregnation and direct calcination of Ce(Cu)-MOF [42]. By thermolysis at 650°C, Mahanty et al. [43] produced CeO$_2$ from a Ce-BTC (benzene-1,3,5-tricarboxylate) MOF. The morphological characteristic of the resulting oxide substance was brick-on-tile. Indeed, the fast redox-active interaction between Ce^{3+} and Ce^{4+} and its near potentials have a substantial impact on the CeO$_2$ electrode's supercapacitive performance, as seen by the redox peaks at about 0.23/0.35 V. The electrochemical data of the porous CeO$_2$ nanostructure (galvanostatic charge-discharge scans) revested a certain capacity of 1204 F/g in a current density of 0.2 A/g. It also has exceptional stability, with over 100% capacity retention after 5,000 charge-discharge cycles.

They also illustrated the contribution to supercapacitors of high-performance metal oxide by the additional electron buffer [Fe(CN)$_6$]$^{3-}$/[Fe(CN)$_6$]$^{4-}$ and by the morphological characters on brick-up. In consequence, we concentrated on the significance and application of MOFs for the synthesis of high-quality porous metallic oxides as a sacrificial template for improved performance in this section.

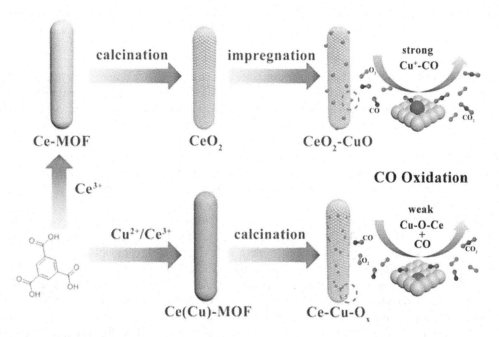

FIGURE 27.2 Thermal action of Ce-MOF to synthesize CeO$_2$/CuO [42].

Thermolysis of MOFs usually has a wide surface area and high conductivity, which are appropriate for supercapacitors. The diffusion of ions has a significant impact on surface faradaic activities, and the pores formed in the metal oxide from MOFs facilitate this diffusion. As a result, developing novel MOFs requires the development of extremely porous metal oxides with appropriate channels of ion transfer and accessible area, particularly for manganese and nickel MOFs, which have yet to be reported. Furthermore, MOFs-derived CeO_2 produced via simple Ce-MOF (Ce-BTC), possessed a regulated nano-to-architecture of 1204 F/g, about double its predicted ability at 0.2 A/g [43]. KOH electrolyte has contributed to the capacitive charge storage of the MOF-derived CeO_2 electrode by intercalation/de-intercalation and surface adsorption/desorption of K^+ ions.

27.2.2 MOFs-DERIVED TERNARY OR TMMOs

The addition of additional metal transition atoms to binary metal oxides is a good approach to enhance cycle stability and capacity. As a result of this concept, mixed transition-metal oxides (TMMOs) were manufactured from bulk crystals to nanoscales [44]. TMMOs (defined as AxB_3xO_4; A or B are Co, Ni, Zn, Mn, Fe, etc.) [45] are very promising materials, typical in a crystal spinel structure, which are stoic or nonstoichiometric in their different possible applications, including multiple ferroics, heat, super conductance, colossal magneto resistivity, etc. [46–49].

As a viable precursor for producing TMMOs for SCs applications, mixed MOFs or hetero-metallic MOFs are examined. The usage of metallo-ligands (porphyrin-based metal ligands), the post-synthetic modification, and direct and partial replacement of the transition metal in a homo-metallic MOF are all typical methods to hetero-metallic MOF by altering the reaction mixture. Caskey and Matzger selectively incorporated a transition metal to a homo-metallic MOF in MOFs using the geometrical preferences of metal ions [50–52]. The hetero-metallic MOF from perylene-3-4,9-10-tetracarboxy-crafty di-anhydrate was synthesized utilizing the "escape by-crafty-scheme" series of spinel mixed-metal oxide MMn_2O_4 (M = Co, Ni, Zn) [53]. $ZnMn_2$-ptcda (perylene-3,4,9,10-tetracarboxylic di-anhydride), a nanoscale hetero-metallic precursor comprising Zn and Mn, was made through the soft chemical assembly of mixed-metal ions and organic ligands on a molecular scale. Mainly, three accounts exist of MOFs being used to make supercapacitors from mixed-metal oxide. This section discusses three TMMOs produced from hetero-metallic MOFs, including $ZnCo_2O_4$ [54], $MnCo_2O_4$ [55], and $Ni_xCo_{3-x}O_4$ [56].

27.2.2.1 Porous $ZnCo_2O_4$ material

Zinc cobaltate ($ZnCo_2O_4$) has the advantages of large reversible capacity, improved cycle stability, and being eco-friendly, and it is one of the most successful TMMOs [57]. $ZnCo_2O_4$ has extensively studied the crystal structure of spinel in Zn and Co at the octahedral position as an efficient SC material [58], and $ZnCo_2O_4$ is an excellent substitute for metal because of its higher conductivity and rich redox reaction caused by the combination of the two metal species [59]. Because of their unique nanostructures based on physicochemical characteristics, nanoscale $ZnCo_2O_4$ has gotten a lot of interest in the scientific community, including as supercapacitors [49]. The researchers have utilized different synthetic techniques, including co-price, sole gels, and synthesis on particular support, to create $ZnCo_2O_4$ at the nanoscale [60–62].

Qiu et. al. [54] prepared a new self-assembled MOF $ZnCo_2O(BTC)_2$ in DMF. The stoichiometric ratio of $ZnCl_2$, $CoCl_2$, $6H_2O$, and BTC in DMF is 1:2:5. It is produced using JUC155 in DMF solvent via standard thermal-assisted strategy. According to the single crystal XRD data, the dissimilar metal MOF JUC155 is formed, having the tetragonal space group. The asymmetric unit is composed of a semi-occupied Zn(II) metal center (occupancy factor), an occupied Co(II) metal molecule, a semi-occupied DMF, oxide (O_2), and lattice molecules (Figure 27.3A). DMF is coordinated with Zn, while BTC and O_2 are combined with the metal center through the metal clustering. Zn(II) has a distorted octahedral coordination geometry. The six oxygen atoms from four independent BTC ligands are coordinated by

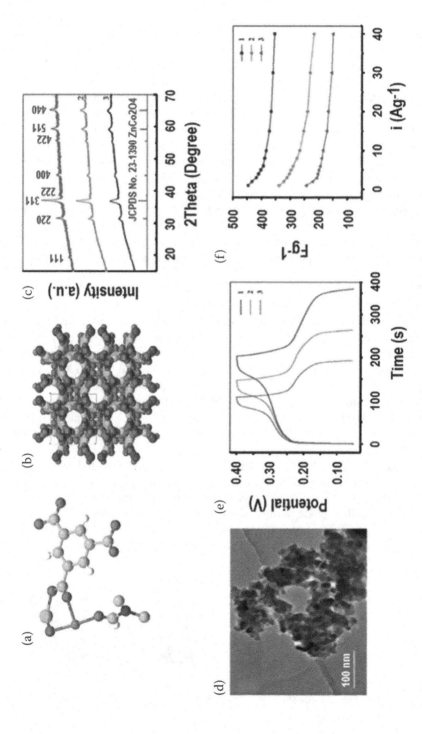

FIGURE 27.3 (a) JUC-155 asymmetric unit. (b) JUC-155's three-dimensional structure as seen adjacent to the "c" axis of crystallography. (c) PXRD plot of synthesized $ZnCo_2O_4$ at different temperatures: 400, 450, and 500 °C (represented by green, red and, black color respectively). (d) picture of nanoparticles of $ZnCo_2O_4$ taken with a transmission electron microscope. (e) Galvanostatic charge/discharge for three electrodes. (f) Capacity derived from the discharge curves of the three electrodes at 1 A/g. The crystallographic coordinates acquired from CSD, 5.31 version, November 2010 were used to produce (a and b). [54] Copyright 2021, © The Partner Organizations 2021.

DMF molecules and the oxygen atoms connecting the core. On the other hand, the Co(II) center has a deformed tetrahedral shape, in which all coordination sites are filled with O_2 and BTCO atoms.

In SBU, a tri-nuclear metal cluster $ZnCoO_2$, the interaction of these ligands with Co(II) and Zn(II) resulted. The creation of 3D porous MOFs on the crystallographic axis "c" with 1D channels is the consequence of the lengthy collaboration between the BTC and such SBUs (Figure 27.3b). To get the optimal temperature to produce a high-quality $ZnCo_2O_4$ electrode material for SCs, the MOF precursor was then calcined to three different temperatures (400, 450, and 500°C). Examination occurred at three different temperatures by PXRD data of those three samples (1, 2, and 3), all of them transitioned from the parent precursor MOF to $ZnCo_2O_4$ pure spinel (Figure 27.3c). The spherical shape of nanoparticles having a diameter of 20 mn or less was shown by TEM imaging of the as-synthesized three different samples 1, 2, and 3 of $ZnCo_2O_2$, and these nanoparticles subsequently aggregated to create a porous structure (Figure 27.3d). The BET surface area dropped (20.4, 45.9 and, 55.0 cm^2/g for $ZnCo_2O_2$ synthesized at 500°C, 450°C, and 400°C, respectively) when the thermolysis temperature was increased, according to the N_2 adsorption-desorption study of all three samples. The as-synthesized three samples of $ZnCo_2O_2$, 1, 2, and 3, were measured using cyclic voltammetry (CV) to show their optimum capacitive behavior. Ion and electron transport may be done at the electrode and electrolyte interface on the high electrode surface area, and sufficient active sites can be provided for the Faraday reaction. Furthermore, the electrodes produced from samples 1, 2, and 3 have specific capacitance values of 451, 330, and 234 F/g at a slow scan rate of 5 mV/s, respectively, and are maintained at 97.9, 96.8, and 94.6% of the original values after 1500 cycles.

27.2.2.2 $MnCo_2O_4$ materials

$MnCo_2O_4$ is one of the successful TMMOs that demonstrated strong pseudo-capacitive behavior due to the synergistic effects of Mn^{2+} and Co^{2+} cations, structural stability, and reversibility of the electrodes. Co^{2+} has a greater oxidation potential, but Mn^{2+} has a higher capacitance and can transport more electrons [63]. In comparison to the binary oxides MnO_2 and Co_3O_4, $MnCo_2O_4$ has a greater electrical conductivity and electrochemical activity. Wang et al. [55] showed that $MnCo_2O_4$ may be synthesized using a dual metal ZIF (Mn-Co-ZIF) [64] as both a precursor and a template. $Mn(NO_3)_2.4H_2O$ and $Co(NO_3)_2.4H_2O$ [with a stoichiometric ratio of 1:2 of Mn(II) and Co(II) salts] were mixed with the 2-methylimidazole linker to produce purple powder of Mn-Co-ZIF, as illustrated in Figure 27.4.

The as-synthesized Mn-Co-ZIF had a PXRD pattern that was similar to ZIF-67 [64], suggesting that they are isomorphous. Mn-Co-ZIF was pyrolyzed to produce a black powder at 400°C. The black powder was identified as $MnCo_2O_4$ by PXRD analysis since all of its diffraction peaks matched the spinel structure, suggesting crystalline phase purity and complete transformation of the MOF precursor into $MnCo_2O_4$ (Figure 27.5a and b). Mn_2, Mn_3, $_{Co2}$, and Co_3 are all present, according to X-ray photoelectron spectroscopic investigations. SEM and TEM were used to examine the nanostructure of the

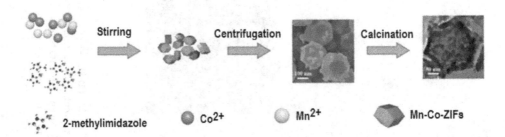

FIGURE 27.4 The numerous stages in the preparation of Mn-Co-zeolitic imidazolate frameworks are depicted in this diagram (ZIFs). Reprinted ref. [55] with permission. Copyright© 2021 Elsevier.

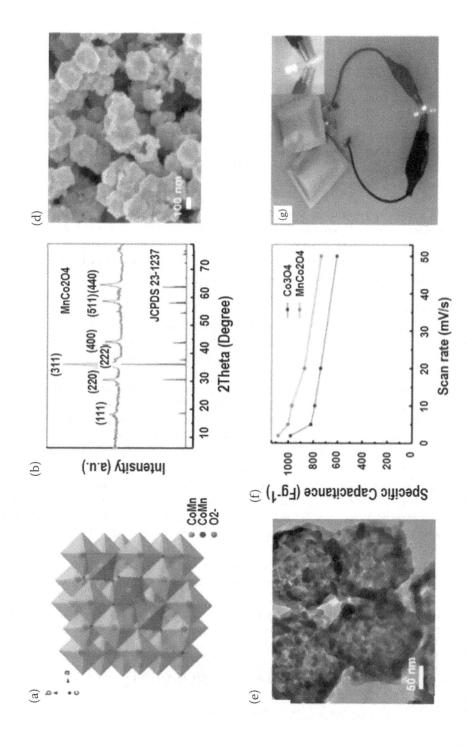

FIGURE 27.5 (a) Crystal structure for MnCo₂O₄. (b) X-ray diffraction pattern for MnCo₂O₄ (experimental (red) and simulated (black)). (c, d) SEM and TEM images MnCo₂O₄, revealing the hollow polyhedral nanocage shape. (e) Plot of current at constant scan rate for both the MnCo₂O₄ and Co₃O₄ electrodes. (f) Fabrication of SCs-based devices. Reprinted from ref. [55] with permission. Copyright© 2021 Elsevier B.V.

MnCo$_2$O$_4$ as-synthesized. The MnCo$_2$O$_4$ had a hollow polyhedral nano-cage shape (average size, 200 nm) in the SEM picture, while the Mn-Co-ZIF precursor's polyhedral structure (average size, 400 nm) was retained without substantial breakdown (Figure 27.5c). The breakdown of the organic component of Mn-Co-ZIF precursor results in a reduction in the size of the nano-cage during the transition to MnCo$_2$O$_4$. On the surface of the nano-cage, the TEM picture indicated the existence of linked nano-particles (5–20 nm), resulting in a porous nanostructure (Figure 27.5d). The ion diffusion channel is made easier by porous materials, which also allow for collaborative electronic transmission. N2 isotherm measurements further supported the porous nature of MnCo$_2$O$_4$ cages. At relative pressures (P/P0) ranging from 0.4 to 1.0, the BET surface area dimension of MnCo$_2$O$_4$ exhibited its usual type IV hysteresis loop, which is representative of a mesoporous structure with a 117 m^2/g surface area.

The authors investigated the electrochemical performance of as-synthesized nano-cage MnCo$_2$O$_4$ to determine the benefit of the porous hollow nanostructure. They also created a Co$_3$O$_4$ nano-cage to compare its electrochemical performance to that of a MnCo$_2$O$_4$ nano-cage. The redox peaks were revealed by examining the CV data of the as-synthesized MnCo$_2$O$_4$ and Co$_3$O$_4$ nano-cages at varied scan speeds ranging from 2 to 50 mV/s. From the CV curves, the average specific capacitance of the MnCo$_2$O$_4$ and Co$_3$O$_4$ nano-cage electrodes was computed and presented at different scan speeds (Figure 27.5e). The electrochemical performance of MnCo$_2$O$_4$ is superior to Co$_3$O$_4$; at a current density of 1 A/g, the highest specific capacitances of MnCo$_2$O$_4$ and Co$_3$O$_4$ electrodes are 1763 and 1209 F/g, respectively. Furthermore, the MnCo$_2$O$_4$ electrode retains 95% of its capacitance after 4500 cycles at 1 A/g, demonstrating its superior cycle stability over Co$_3$O$_4$ (81.2 %).

27.2.2.3 Ni$_x$Co$_{3-x}$O$_4$ material

There is another spinel TMMO, NiCo$_2$O$_4$, which exhibits higher electrical conductivity than NiO or Co$_3$O$_4$ type oxides [65]. As a result of its "stoichiometric ratio"-dependent electrochemical activity, Ni$_x$Co$_{3-x}$O$_4$ with adjustable Ni and Co ratios and mesoporous hierarchical structures attracted researchers [66]. NiCo$_2$O$_4$ in terms of its stoichiometry, Co and Ni have a significant impact on capacitance [67]. However, Ni$_x$Co$_{3-x}$O$_4$ is difficult to synthesize due to the difficulty in accurately regulating the metal ratio (Ni:Co). A new synthesis strategy for the fabrication of Ni$_x$Co$_{3x}$O$_4$ based electrode materials using MOF-74 was described by Qiu et al. [56]. Usually, MOF-74 possesses five isostructural MOFs 74: one with Co, one with Ni, one with a Ni/Co mix, and one with both Ni and Co (in different ratios of Ni:Co 14 1:1, 1:2, 1:4). MOF-74-NiCo1, MOF-74-NiCo$_2$, and MOF-74-NiCo$_4$ were intentionally combined to study the impact of Ni and Co content on the electrochemical properties of these MOFs-based electrode materials in SCs. MOF-74-NiCo1, MOF-74-NiCo$_2$, and MOF-74-NiCo$_4$ have Ni:Co stoichiometric ratios of 1:1, 1:2, and 1:4, respectively, as confirmed by the inductively coupled plasma and energy-dispersive spectrometric investigations. In addition to Co$_3$O$_4$ and NiO, all these MOF-74s were calcined at 400 °C to produce three kinds of Ni$_x$Co$_{3x}$O$_4$ mixed-metal oxide NPs, Ni$_x$Co$_{3-x}$O$_{4-1}$, Ni$_x$Co$_{3-x}$O$_{4-2}$ and Ni$_x$Co$_{3-x}$O$_{4-4}$, respectively (Figure 27.6).

TEM revealed a porous, "coarse" morphology for all these metal oxides, having a pristine crystal phase, as suggested by PXRD investigations (Figure 27.7a). Further, electron diffraction patterns with (220), (311), (400), (422), and (511) planes for Ni$_x$Co3-xO$_{4-1}$ indicated its polycrystalline character (Figure 27.7b). Moreover, a significant rise in low-pressure isotherms was observed in all five oxides, as well as BET surface areas of 64–117 m^2/g, which are extremely high in comparison to other oxides reported.

$$NixCo_{3-x}O_4 + OH^- + H_2O \leftrightarrow xNiOOH + (3 - x)CoOOH + (3 - x)e^-$$

$$CoOOH + OH^- \leftrightarrow CoO_2 + H_2O + e^-$$

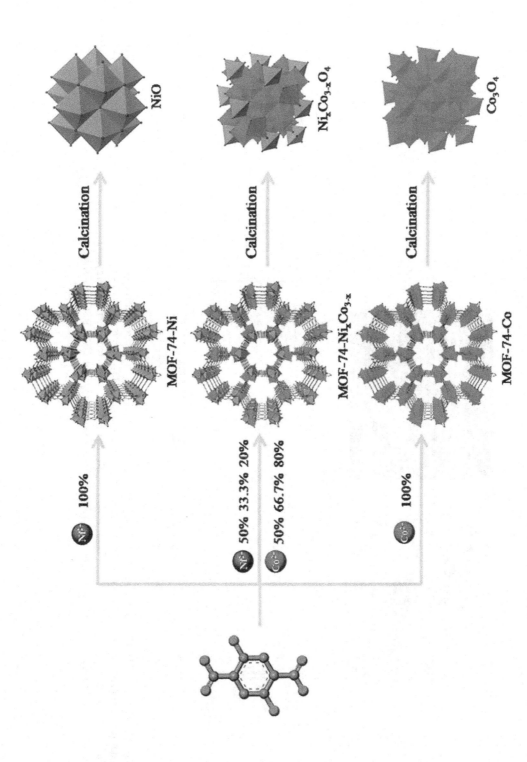

FIGURE 27.6 Schematic fabrication strategy for different MOF-74 crystals [cyan, pink, grey, and red colors are representing the Ni, Co, C, and O]. Reprinted from ref. [56] with permission. Copyright 2021, The Royal Society of Chemistry.

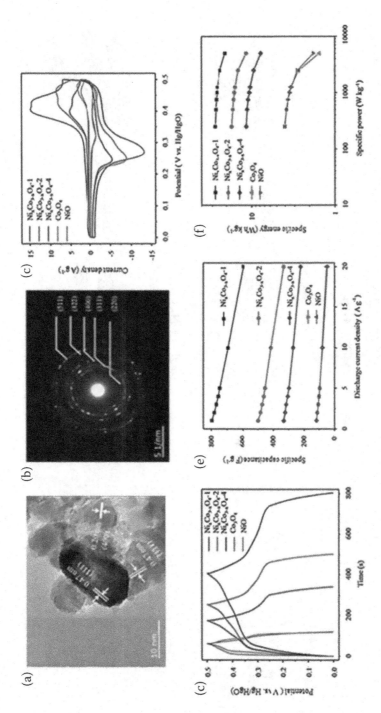

FIGURE 27.7 (a) HR-TEM images of $Ni_xCo_3-_xO_{4-1}$ NPs and (b) selected area in electron diffraction of $Ni_xCo_3-_xO_{4-1}$ NPs. (c) CVs of $Ni_xCo_3-_xO_{4-1}$ and other oxides recorded at a scan rate of 5 mV/s. (d) Charge/discharge curves of $Ni_xCo_3-_xO_{4-1}$ and other oxides @ 1 Ag^{-1} current density. (e) Specific capacitance calculated using discharge curves of $Ni_xCo_3-_xO_{4-1}$ and other oxides. (f) Ragone plots of estimated specific energy and specific power for $Ni_xCo_3-_xO_{4-1}$ and other oxides at various charge–discharge rates. Reprinted from ref. [56] with permission. Copyright 2021, RSC publishing group.

The CV curves indicated that TMMO samples had a six-fold greater capacitance than their binary counterparts Co_3O_4 and NiO, showing that they are superior to binary metal oxides (Figure 27.7c). A specific capacitance of 789 F/g (Figure 27.7d and e) and outstanding cycling stability were also demonstrated for the $Ni_xCo_{3-x}O_{4-1}$ electrode material, which had a Ni:Co 14:1 ratio (specific capacitances remained at 557 F/g after 10,000 charge-discharge cycles). Because $Ni_xCo_{3-x}O_{4-1}$ has a large surface area, ion/electron transfer in the electrode is easier, and, because it contains a large amount of Ni_3, compared to other compounds, its specific capacitance is high. Ni_3-rich surfaces of electrodes are more likely to produce active NiOOH, whereas Ni_3-poor surfaces are more likely to produce active NiOOH (Figure 27.7f). Metallic oxides with a mixed-metal composition have a higher energy density than monometallic oxides. These devices are more energy efficient because of their lower resistance and higher capacitance.

27.2.3 MOFs-DERIVED METAL OXIDE/COMPOSITE

Porous carbons, in particular EDLCs, are the most frequently used electrode materials for SCs due to their high surface area and superior stability [68,69]. Poddar et al. developed the first universal approach to fabricate metal oxide nanoparticles (MO NPs) supported on a porous carbon substrate, displaying remarkable energy storage performance [70]. Most MOF-based material development, however, has focused on producing the MOF-derived metal oxides as the cathode materials. Therefore, some additional carbon materials for the anode fabrication must also be developed. To preserve their form, surface, and flexibility, asymmetric SCs with anode and cathode electrodes nanomaterials must be synthesized from one precursor. If these SCs are employed, synergistic effects with various physicochemical characteristics are produced.

In this way, ZIF-67 was the first MOF to be utilized in an asymmetric SCs with nano-porous carbon and nano-porous cobalt oxides (Co_3O_4), as Yamauchi et al. [71] demonstrated. High temperature treatment in the presence of N_2, the ZIF-67 precursor generated nano-porous carbon, but when this experiment was performed in absence of N_2, organic molecules decomposed, producing nano-porous Co_3O_4. ZIF-67 retains

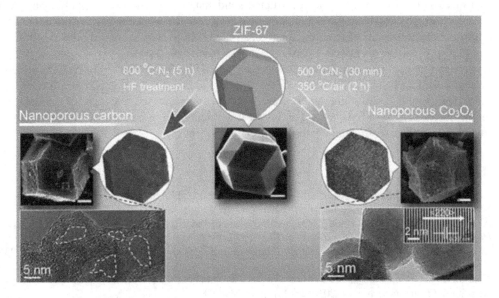

FIGURE 27.8 The schematic images show how ZIF-67 precursor is converted into nanoporous carbon and nanoporous Co_3O_4, with the polyhedral shape of the parent ZIF-67 retained. Reproduced with permission [68]. Copyrights 2021, Elsevier publishing group.

its polyhedral form in both conditions (Figure 27.8). The analytical results suggested that a large number of nano-pores were created on the surface of nano-porous carbon. Nano-porous Co_3O_4 oxide is composed of randomly aggregated granular nanocrystals with an average size of 15–20 nm. These materials possessed 350 and 148 m^2/g surface area, respectively, after being synthesized. Usually, the combination carbon/carbon (~7 Wh/kg) and Co_3O_4/Co_3O_4 (~8 Wh/kg) are their symmetric equivalents, whereas, the asymmetric SCs (Co_3O_4/carbon) has a 36 Wh/kg specific energy and 8000 W/kg power.

For instance, Wang and colleagues [72] created an asymmetric SC using a single 2D-MOF precursor. Generally, Co and 2-methylimidazole were incorporated to formulate the 2D sheet of Co-MOF and grown-up on the carbon surface (CC@Co-MOF). When the PXRD results were refined by Pawley, it was determined that there were two layers to the crystal structure of the Co-MOF. This 2D Co-MOF precursor was thermolyzed to form Co_3O_4 nanosheets (CC@Co_3O_4) or porous carbon nanosheets (CC@NC). Further, CC@Co_3O_4 cathode and CC@NC anode were used in the flexible asymmetric SCs. Comparing the CC@Co_3O_4/CC@NC device to previously reported MOF-based SCs, the Ragone plot highlights the importance of such a unique asymmetric SCs.

27.3 CONCLUSIONS

The creation of TMOs-based FSCs with high performance and low cost has been the subject of numerous efforts, but various barriers must be addressed and solutions must be considered to continue the development of flexible devices. For instance, the manufacturing of integrated 3D core-shell heterostructured with controllable and adjustable morphological characteristics remained a questionable issue. Rather, several accounts of successful efforts to make composite electrodes have been achieved; this was accomplished by conquering manufacturing of flexible devices by employing a free-standing nanocomposite film, free from binders or additives. The viability of this strategy is largely reliant on the composite material's electrical conductivity, which is critical for ensuring the movement of high current and low resistance via electrode interfaces. The second parameter is the addition of graphene to one of the electrode constituents; this is the proper ban of the mechanical breakaway because of the expansion and shrinking of the elements during the continual charging/discharging cycles in the intensive inflow/outflow of electrolyte ions. Most of the constructed solid-state devices were extremely stable, with a continuous cycle range of between 1000 and 5000. It is well recognized that there is a maximum surface area for infusing electrolytes and increasing their capability using ternary components like carbon black, CNTs, or TMOs, which are identified to sustain 3D graphene frameworks.

By this chapter, the readers will get insight into the electric performance of the FSCs described in a wide variety of bending states. In the future, the energy storage-based tools with significant performance will require thin, flexible, lightweight, and roll-up SCs. To summarize, the relevance of transition metal oxides in the manufacturing of FSCs would indirectly generate a fresh research target in the development of flexible energy storage instruments with the ultimate performance.

REFERENCES

1. Xu, X., et al., Extraordinary capacitive deionization performance of highly-ordered mesoporous carbon nano-polyhedra for brackish water desalination. *Environmental Science: Nano*, 2019. **6**(3): p. 981–989.
2. Anasori, B., M.R. Lukatskaya, and Y. Gogotsi, 2D metal carbides and nitrides (MXenes) for energy storage. *Nature Reviews Materials*, 2017. **2**(2): p. 1–17.
3. Zhai, Y., et al., Carbon materials for chemical capacitive energy storage. *Advanced Materials*, 2011. **23**(42): p. 4828–4850.
4. Yan, J., et al., Rapid microwave-assisted synthesis of graphene nanosheet/Co_3O_4 composite for supercapacitors. *Electrochimica Acta*, 2010. **55**(23): p. 6973–6978.
5. Pan, C., H. Gu, and L. Dong, Synthesis and electrochemical performance of polyaniline@ MnO_2/graphene ternary composites for electrochemical supercapacitors. *Journal of Power Sources*, 2016. **303**: p. 175–181.

6. Zhao, Y., et al., Vapor deposition polymerization of aniline on 3D hierarchical porous carbon with enhanced cycling stability as supercapacitor electrode. *Journal of Power Sources*, 2015. **286**: p. 1–9.

7. Yue, L., et al., Improving the rate capability of ultrathin NiCo-LDH nanoflakes and FeOOH nanosheets on surface electrochemically modified graphite fibers for flexible asymmetric supercapacitors. *Journal of Colloid and Interface Science*, 2020. **560**: p. 237–246.

8. Gopalakrishnan, A., N. Vishnu, and S. Badhulika, Cuprous oxide nanocubes decorated reduced graphene oxide nanosheets embedded in chitosan matrix: a versatile electrode material for stable supercapacitor and sensing applications. *Journal of Electroanalytical Chemistry*, 2019. **834**: p. 187–195.

9. Wu, X. and M. Lian, Highly flexible solid-state supercapacitor based on graphene/polypyrrole hydrogel. *Journal of Power Sources*, 2017. **362**: p. 184–191.

10. Kazemi, S.H., et al., Nano-architectured MnO2 electrodeposited on the Cu-decorated nickel foam substrate as supercapacitor electrode with excellent areal capacitance. *Electrochimica Acta*, 2016. **197**: p. 107–116.

11. Yu, H., et al., Facile fabrication and energy storage analysis of graphene/PANI paper electrodes for supercapacitor application. *Electrochimica Acta*, 2017. **253**: p. 239–247.

12. Yan, J., et al., Recent advances in design and fabrication of electrochemical supercapacitors with high energy densities. *Advanced Energy Materials*, 2014. **4**(4): p. 1300816.

13. Chen, J., C. Li, and G. Shi, Graphene materials for electrochemical capacitors. *The Journal of Physical Chemistry Letters*, 2013. **4**(8): p. 1244–1253.

14. Beguin, F. and E. Frackowiak, *Carbons for electrochemical energy storage and conversion systems*. 2009: Crc Press.

15. Peng, C., et al., Carbon nanotube and conducting polymer composites for supercapacitors. *Progress in Natural Science*, 2008. **18**(7): p. 777–788.

16. Augustyn, V., P. Simon, and B. Dunn, Pseudocapacitive oxide materials for high-rate electrochemical energy storage. *Energy & Environmental Science*, 2014. **7**(5): p. 1597–1614.

17. Long, J.W., et al., Asymmetric electrochemical capacitors—Stretching the limits of aqueous electrolytes. *Mrs Bulletin*, 2011. **36**(7): p. 513–522.

18. Halper, M. and J. Ellenbogen, *C-Supercapacitors: A brief overview, The MITRE Corporation, McLean*. 2006, Virginia, Technical report 06-0667.

19. Yue, Y., et al., Highly self-healable 3D microsupercapacitor with MXene–graphene composite aerogel. *Acs Nano*, 2018. **12**(5): p. 4224–4232.

20. Korah, M.M., et al., Materials for optical, magnetic and electronic devices. *Journal of Materials Chemistry*, 2020. **100**: p. 10479–10489.

21. Zhang, Y.-Z., et al., Printed supercapacitors: Materials, printing and applications. *Chemical Society Reviews*, 2019. **48**(12): p. 3229–3264.

22. Zeng, L., et al., Recent progresses of 3D printing technologies for structural energy storage devices. *Materials Today Nano*, 2020: p. 100094.

23. Lv, Z., et al., Honeycomb-lantern-inspired 3D stretchable supercapacitors with enhanced specific areal capacitance. *Advanced Materials*, 2018. **30**(50): p. 1805468.

24. Cheng, T., et al., Inkjet-printed flexible, transparent and aesthetic energy storage devices based on PEDOT: PSS/Ag grid electrodes. *Journal of Materials Chemistry A*, 2016. **4**(36): p. 13754–13763.

25. Jing, C., et al., Optimizing the rate capability of nickel cobalt phosphide nanowires on graphene oxide by the outer/inter-component synergistic effects. *Journal of Materials Chemistry A*, 2020. **8**(4): p. 1697–1708.

26. Jing, C., et al., The pseudocapacitance mechanism of graphene/CoAl LDH and its derivatives: Are all the modifications beneficial? *Journal of Energy Chemistry*, 2021. **52**: p. 218–227.

27. Li, X., et al., Layered double hydroxides toward high-performance supercapacitors. *Journal of Materials Chemistry A*, 2017. **5**(30): p. 15460–15485.

28. Jing, C., B. Dong, and Y. Zhang, Chemical modifications of layered double hydroxides in the supercapacitor. *Energy & Environmental Materials*, 2020. **3**(3): p. 346–379.

29. Wiston, B.R. and M. Ashok, Microwave-assisted synthesis of cobalt-manganese oxide for supercapacitor electrodes. *Materials Science in Semiconductor Processing*, 2019. **103**: p. 104607.

30. Poizot, P., et al., Nano-sized transition-metal oxides as negative-electrode materials for lithium-ion batteries. *Nature*, 2000. **407**(6803): p. 496–499.

31. He, T., et al., Solubility-controlled synthesis of high-quality Co_3O_4 nanocrystals. *Chemistry of materials*, 2005. **17**(15): p. 4023–4030.

32. Kandalkar, S.G., et al., A non-thermal chemical synthesis of hydrophilic and amorphous cobalt oxide films for supercapacitor application. *Applied surface science*, 2007. **253**(8): p. 3952–3956.

33. Zhang, F., et al., Solid-state thermolysis preparation of Co_3O_4 nano/micro superstructures from metal-organic framework for supercapacitors. *Int. J. Electrochem. Sci*, 2011. **6**: p. 2943–2954.

34. Pang, H., et al., Dendrite-like Co 3 O 4 nanostructure and its applications in sensors, supercapacitors and catalysis. *Dalton Transactions*, 2012. **41**(19): p. 5862–5868.

35. Meng, F., et al., Erratum: Porous Co3O4 materials prepared by solid-state thermolysis of a novel Co-MOF crystal and their superior energy storage performances for supercapacitors. *Journal of Materials Chemistry A*, 2013. **1**(48): p. 15554.

36. Wang, B., et al., Quasiemulsion-templated formation of α-Fe_2O_3 hollow spheres with enhanced lithium storage properties. *Journal of the American Chemical Society*, 2011. **133**(43): p. 17146–17148.

37. Zhu, X., et al., Nanostructured reduced graphene oxide/Fe2O3 composite as a high-performance anode material for lithium ion batteries. *ACS Nano*, 2011. **5**(4): p. 3333–3338.

38. Huang, X., et al., Self-assembling synthesis of free-standing nanoporous graphene–transition-metal oxide flexible electrodes for high-performance lithium-ion batteries and supercapacitors. *Chemistry–An Asian Journal*, 2014. **9**(1): p. 206–211.

39. Khattak, A.M., et al., Three dimensional iron oxide/graphene aerogel hybrids as all-solid-state flexible supercapacitor electrodes. *RSC advances*, 2016. **6**(64): p. 58994–59000.

40. Lu, X.-F., et al., α-Fe2O3@ PANI core–shell nanowire arrays as negative electrodes for asymmetric supercapacitors. *ACS Applied Materials & Interfaces*, 2015. **7**(27): p. 14843–14850.

41. Lu, X., G. Li, and Y. Tong, A review of negative electrode materials for electrochemical supercapacitors. *Science China Technological Sciences*, 2015. **58**(11): p. 1799–1808.

42. Guo, Z., et al., CeO_2-CuO bimetal oxides derived from Ce-based MOF and their difference in catalytic activities for CO oxidation. *Materials Chemistry and Physics*, 2019. **226**: p. 338–343.

43. Maiti, S., A. Pramanik, and S. Mahanty, Extraordinarily high pseudocapacitance of metal organic framework derived nanostructured cerium oxide. *Chemical Communications*, 2014. **50**(79): p. 11717–11720.

44. Hu, L., et al., Metaleorganic hybrid interface states of A ferromagnet/organic semiconductor hybrid junction as basis for engineering spin injection in organic spintronics. *Advanced Functional Materials*, 2012. **22**: p. 998–1005.

45. Yuan, C., et al., Mixed transition-metal oxides: Design, synthesis, and energy-related applications. *Angewandte Chemie International Edition*, 2014. **53**(6): p. 1488–1504.

46. Yamasaki, Y., et al., Magnetic reversal of the ferroelectric polarization in a multiferroic spinel oxide. *Physical Review Letters*, 2006. **96**(20): p. 207204.

47. Fujishiro, Y., et al., Synthesis and thermoelectric characterization of polycrystalline $Ni_{1-x} Ca_x Co_2 O_4$ (x = 0–0.05) spinel materials. *Journal of Materials Science: Materials in Electronics*, 2004. **15**(12): p. 769–773.

48. Johnston, D., et al., High temperature superconductivity in the LiTiO ternary system. *Materials Research Bulletin*, 1973. **8**(7): p. 777–784.

49. Akther Hossain, A., et al., Colossal magnetoresistance in spinel type $Zn_{1-x} Ni_x Fe_2 O_4$. *Journal of Applied Physics*, 2004. **96**(2): p. 1273–1275.

50. Lin, Q., et al., New heterometallic zirconium metalloporphyrin frameworks and their heteroatom-activated high-surface-area carbon derivatives. *Journal of the American Chemical Society*, 2015. **137**(6): p. 2235–2238.

51. Das, S., H. Kim, and K. Kim, Metathesis in single crystal: Complete and reversible exchange of metal ions constituting the frameworks of metal– organic frameworks. *Journal of the American Chemical Society*, 2009. **131**(11): p. 3814–3815.

52. Caskey, S.R. and A.J. Matzger, Selective metal substitution for the preparation of heterobimetallic microporous coordination polymers. *Inorganic Chemistry*, 2008. **47**(18): p. 7942–7944.

53. Zhao, J., et al., Spinel $ZnMn_2O_4$ nanoplate assemblies fabricated via "escape-by-crafty-scheme" strategy. *Journal of Materials Chemistry*, 2012. **22**(26): p. 13328–13333.

54. Chen, S., et al., Porous $ZnCo_2O_4$ nanoparticles derived from a new mixed-metal organic framework for supercapacitors. *Inorganic Chemistry Frontiers*, 2015. **2**(2): p. 177–183.

55. Dong, Y., et al., Facile synthesis of hierarchical nanocage $MnCo_2O_4$ for high performance supercapacitor. *Electrochimica Acta*, 2017. **225**: p. 39–46.

56. Chen, S., et al., Rational design and synthesis of $Ni_x Co_{3-x} O_4$ nanoparticles derived from multivariate MOF-74 for supercapacitors. *Journal of Materials Chemistry A*, 2015. **3**(40): p. 20145–20152.

57. Sharma, Y., et al., Nanophase $ZnCo_2O_4$ as a high performance anode material for Li-ion batteries. *Advanced Functional Materials*, 2007. **17**(15): p. 2855–2861.

58. Liu, B., et al., New energy storage option: toward $ZnCo_2O_4$ nanorods/nickel foam architectures for high-performance supercapacitors. *ACS Applied Materials & Interfaces*, 2013. **5**(20): p. 10011–10017.

59. Jiang, J., et al., Recent advances in metal oxide-based electrode architecture design for electrochemical energy storage. *Advanced Materials*, 2012. **24**(38): p. 5166–5180.

60. Qiu, Y., et al., A novel nanostructured spinel $ZnCo_2O_4$ electrode material: morphology conserved transformation from a hexagonal shaped nanodisk precursor and application in lithium ion batteries. *Journal of Materials Chemistry*, 2010. **20**(21): p. 4439–4444.

61. Liu, B., et al., Hierarchical three-dimensional $ZnCo_2O_4$ nanowire arrays/carbon cloth anodes for a novel class of high-performance flexible lithium-ion batteries. *Nano Letters*, 2012. **12**(6): p. 3005–3011.

62. Bao, F., et al., Controlled growth of mesoporous $ZnCo_2O_4$ nanosheet arrays on Ni foam as high-rate electrodes for supercapacitors. *RSC Advances*, 2014. **4**: p. 2393–2397.

63. Xu, Y., et al., Facile synthesis route of porous $MnCo_2O_4$ and $CoMn_2O_4$ nanowires and their excellent electrochemical properties in supercapacitors. *Journal of Materials Chemistry A*, 2014. **2**(39): p. 16480–16488.

64. Banerjee, R., et al., High-throughput synthesis of zeolitic imidazolate frameworks and application to CO2 capture. *Science*, 2008. **319**(5865): p. 939–943.

65. Wei, T.Y., et al., A cost-effective supercapacitor material of ultrahigh specific capacitances: spinel nickel cobaltite aerogels from an epoxide-driven sol–gel process. *Advanced Materials*, 2010. **22**(3): p. 347–351.

66. Wu, H.B., H. Pang, and X.W.D. Lou, Facile synthesis of mesoporous $Ni_{0.3}Co_{2.7}O_4$ hierarchical structures for high-performance supercapacitors. *Energy & Environmental Science*, 2013. **6**(12): p. 3619–3626.

67. Wang, X., et al., High performance porous nickel cobalt oxide nanowires for asymmetric supercapacitor. *Nano Energy*, 2014. **3**: p. 119–126.

68. N.N.,Adarsh , Metal-Organic Frameworks based supercapacitors. In Dubal Pedro, D., & Pedro, G.-R. *Metal Oxides in Supercapacitors*, 2017: Elsevier, p. 165.

69. Salunkhe, R.R., et al., Nanoarchitectures for metal–organic framework-derived nanoporous carbons toward supercapacitor applications. *Accounts of Chemical Research*, 2016. **49**(12): p. 2796–2806.

70. Das, R., et al., Metal and metal oxide nanoparticle synthesis from metal organic frameworks (MOFs): finding the border of metal and metal oxides. *Nanoscale*, 2012. **4**(2): p. 591–599.

71. Salunkhe, R.R., et al., Asymmetric supercapacitors using 3D nanoporous carbon and cobalt oxide electrodes synthesized from a single metal–organic framework. *ACS Nano*, 2015. **9**(6): p. 6288–6296.

72. Guan, C., et al., Cobalt oxide and N-doped carbon nanosheets derived from a single two-dimensional metal–organic framework precursor and their application in flexible asymmetric supercapacitors. *Nanoscale Horizons*, 2017. **2**(2): p. 99–105.

28 Textile-Based Flexible Supercapacitors

Yasin Altin
Department of Polymer Materials Engineering, Bursa Technical University,
Bursa, Turkey

Department of Chemistry, Ordu University, Ordu, Turkey

Ayse Bedeloglu
Department of Polymer Materials Engineering, Bursa Technical University,
Bursa, Turkey

CONTENTS

28.1   Introduction 520
28.2   Classification of supercapacitor based on storage mechanism 520
     28.2.1   Electrical double layer capacitors 520
     28.2.2   Pseudocapacitors 521
     28.2.3   Hybrid supercapacitors 521
28.3   Components of supercapacitors 521
     28.3.1   Electrode 521
     28.3.2   Electrolyte 521
     28.3.3   Separator 522
     28.3.4   Current collector 522
28.4   Textile-based supercapacitors 522
     28.4.1   Fiber-based supercapacitors 523
           28.4.1.1   Parallel fiber structure 523
           28.4.1.2   Twisted fiber structure 523
           28.4.1.3   Coaxial fiber structure 523
     28.4.2   Fabric supercapacitors 524
28.5   Fiber-based electrodes 524
     28.5.1   Conventional fibers based fiber electrodes 526
     28.5.2   Metal yarns/wires/threads-based fiber electrodes 526
     28.5.3   Graphene yarn-based electrodes 526
     28.5.4   CNT yarn-based fiber electrodes 527
     28.5.5   Hybrid fiber-based supercapacitor electrodes 527
28.6   Fabric-based electrodes for supercapacitors 528
     28.6.1   Metal mesh-based electrodes 529
     28.6.2   Carbon fabric-based electrodes 529
     28.6.3   Conventional fabric-based electrodes 529
28.7   Applications 529
28.8   Conclusion and future perspective 533
Reference 535

DOI: 10.1201/9781003186755-28

28.1   INTRODUCTION

The history of textile products almost goes back to the beginning of human history. This adventure, which started with meeting humans' need to dress and cover themselves, has now evolved into smart textiles. Smart textiles are textiles that can sense and respond to changes in the environment and are divided into subclasses according to sense and reactions [1]. Electronic textiles are a subfield of smart textiles, and they are textile products that are capable of adapting electronic circuits to textile products so that they can sense, communicate, and have properties such as data transmitting, heating, radiation, and energy harvesting [2]. Although there are many electronic textile applications, such as textile-based strain sensors, pressure sensors, antennas, and heaters, it is critical to meet the energy requirement for these technologies to be applicable [3].

The energy needs of electronic textiles can be met with externally integrated rigid and bulky energy storage systems, but to increase the performance, comfort, and usability of electronic textiles, energy storage systems must be integrated into flexible and light textile surfaces. Thus, energy-harvesting textiles, such as photovoltaic, triboelectric, piezo-electric, and thermoelectric textiles, can be developed. The fact that instantly generated energy can be stored ensures that the obtained energy can be used when needed. For this reason, it is possible to use energy-harvesting textiles more effectively thanks to the integration of energy-storing textile products.

Textile-based batteries can be counted among the energy-storing textile technologies. However, since the materials used in battery technologies are easily adversely affected by air and humidity, there are difficulties in adapting to textile products. On the other hand, supercapacitors, which fill the gap between batteries and capacitors in terms of energy density and power density, are easier to adapt to textiles. It consists of a supercapacitor electrode, electrolyte, separator, and current collector layers. Electrodes form the most critical component in supercapacitors. The most common classification of supercapacitors is based on the electrode materials used and thus on the energy-storage mechanism. For this purpose, carbon-based materials are used; they include activated carbon, graphene, carbon nanotube, carbon aerogel, carbon and nanofiber, and conductive polymers, such as polyaniline, polypyrrole, polythiophene, PEDOT:PSS, and metal oxides/nitride/sulphites, such as MnO_2, RuO_2, V_2O_5, and Mxenes. The development of textile-based electrodes forms the basis of textile-based supercapacitor studies. Textile-based supercapacitors can be divided into two categories, fiber-based and fabric-based textile supercapacitors.

This section discusses supercapacitor types and components, textile supercapacitors, textile supercapacitor electrodes, textile supercapacitor applications, and future perspectives.

28.2   CLASSIFICATION OF SUPERCAPACITOR BASED ON STORAGE MECHANISM

Various classifications are used for supercapacitors, but the most prominent classification is based on the charge storage mechanisms of supercapacitors. Supercapacitors can be divided into three categories: electrical double-layer capacitors, pseudocapacitors, and hybrid supercapacitors (Figure 28.1).

28.2.1   Electrical double layer capacitors

Electrical double-layer capacitors (EDLCs) are a type of supercapacitor where charges are stored at the electrode-electrolyte interface by electrostatic interactions. In this category of supercapacitors, carbon-based materials are commonly used as electrode materials. Among materials used are: activated carbon, porous carbon, and carbon aerogel with a large surface area, as well as materials such as carbon nanotubes and graphene with a wide specific surface area, excellent mechanical properties, high electrical conductivity, and electrochemical stability.

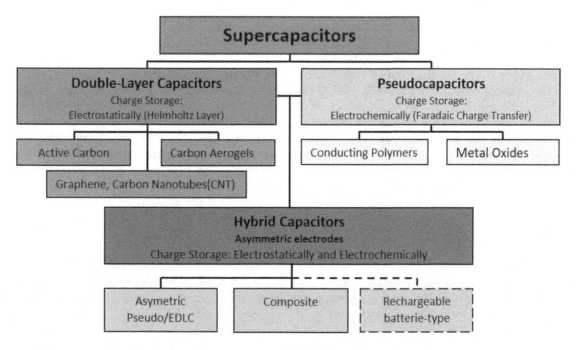

FIGURE 28.1   The classification of supercapacitors based on charge storage mechanism [4].

28.2.2   PSEUDOCAPACITORS

In pseudocapacitors, the charge is stored by electrochemical reactions. Although it has a higher energy density compared to EDCL, it's shorter cycle life constitutes a key disadvantage. Conductive polymers or metal oxides/nitrides/sulfides are used as electrodes.

28.2.3   HYBRID SUPERCAPACITORS

Hybrid supercapacitors are those in which EDLC and pseudocapacitive materials are used together. Two storage mechanisms are used in these capacitors, thus improving supercapacitor performance. In recent years, it is seen that its importance has gradually increased and scientific studies have intensified.

28.3   COMPONENTS OF SUPERCAPACITORS

28.3.1   ELECTRODE

Electrodes constitute the most important component of supercapacitors. The charge-storage mechanism changes according to the electrode material used. The most basic classification of supercapacitors is based on the charge-storage mechanism. Carbon-based materials such as activated carbon, graphene, carbon nanotube, carbon nanofiber, and conductive polymers such as polyaniline, polythiophene, polypyrrole, PEDOT, and metal oxides/nitrides/sulfides such as MnO_2, Vn_2O_5, Co_3O_4, Fe_3O_4 are used as electrode material in supercapacitors [5]. While these materials are used alone in some studies, supercapacitor electrode studies in which these materials are combined in a hybrid way are now coming to the fore.

28.3.2   ELECTROLYTE

The electrolyte is a medium in which a charge is transferred between electrodes. Since it directly affects the working voltage range of the supercapacitor, it is a component that has a very important effect on the performance of supercapacitors. Electrolytes can be liquid or gel. Liquid electrolytes include aqueous electrolytes (H_2SO_4, KOH, KCl aqueous solution), organic electrolytes (acetonitrile and propylene carbonate), and ionic liquids [6]. Although liquid electrolytes are suitable for pouch cell, button cell, and cylindrical cell type supercapacitors, they are not suitable for solid-state supercapacitor applications. For this reason, gel electrolytes are used in solid-state supercapacitors. Among the gel electrolytes, aqueous-based polymeric gel electrolytes such as PVA/H_2SO_4 [7], PVA/H_3PO_4 [8], PVA/HCl [9], PVA/LiCl [10], PVA/Na_2SO_4 [11], $PVA/LiNO_3$ [12], PVA/KOH [13] are the most common. On the other hand, nafion and nonaqueous gel electrolytes such as (poly (ethylene glycol) dimethacrylate (PEGDMA)/propylene carbonate (PC)/tetrabutylammonium tetrafluoroborate ($TEABF_4$)) are also used as electrolytes in solid-state supercapacitors [14]. Another feature of gel electrolytes is that they also act as a separator between the electrodes, so there is no need for conventional separators in device fabrication.

28.3.3 SEPARATOR

Separators are semipermeable membranes that prevent short circuits between electrodes. These materials allow ion transfer between electrodes while preventing supercapacitors from shorting out. Basic properties required in materials used as separators are the ability to electrically insulate, high ion permeability, chemical resistance, wettability, and good mechanical strength [6].

28.3.4 CURRENT COLLECTOR

Current collectors are materials used to collect and transfer charges between electrodes and external charges. It is required when the electrode material does not have sufficient electrical conductivity. Textile-based electrodes such as graphene fiber, CNT fiber, and carbon nanofiber do not need a current collector due to their high electrical conductivity. However, a current collector is needed in supercapacitor applications where conventional textile surfaces (such as cotton fabric, polyester fabric, fiberglass fabric) are used as substrates. In some cases, the surfaces of textile substrates are coated with conductive materials, such as Au, Cu, Ni and converted into conductive textiles, and then electrode materials are applied to it. In some cases, the electrode materials obtained by the modifying process act as current collectors [13,15,16].

28.4 TEXTILE-BASED SUPERCAPACITORS

The flexibility of electronic technologies emerges as an important development for wearable electronics and smart textile applications and paves the way for applications in this field. One of the most critical issues in wearable electronics is the development of lightweight, flexible, and high-performance energy-storage devices to meet the power needs of wearable electronic systems. While textile-based batteries are being developed among wearable energy-storage technologies, textile-based supercapacitors stand out with their superior features, such as flexibility, lightness, environmental friendliness, fast charge/discharge, long life, high energy, and power density [17–23].

The main characteristics of textile materials are that they are thin and flexible materials with sufficient tensile strength and tear resistance. Staple fibers or filaments of natural or synthetic origin form the main component of textiles. These staple fibers or filaments can be spun by applying various twisting techniques, and yarns are obtained by combining these filaments [24]. Yarns are important one-dimensional building blocks that form the basis of textile products. Two-dimensional fabric surfaces are obtained by combining the yarns with methods such as weaving, knitting, or sewing. In addition, nonwoven surfaces are obtained by combining micro/nano-sized fibers with various techniques to form a flat surface from another fabric class. In light of the information mentioned above, supercapacitor textile studies can be examined under two headings, as fiber and fabric-based.

28.4.1 FIBER-BASED SUPERCAPACITORS

Fiber-based supercapacitors are divided into three designs: parallel fiber, twisted fiber, and coaxial fiber structures. These designs are discussed in detail below. Also, in some studies, fiber-based supercapacitors have been encapsulated to protect them from environmental influences.

28.4.1.1 Parallel fiber structure

In this structure, supercapacitors are obtained by combining the fiber-shaped electrodes, coating with gel electrolyte in parallel, or positioning them in parallel on a flexible substrate [25,26]. Figure 28.2 shows a parallel fiber-based supercapacitor schematic structure. The charge-discharge analysis is done by making contact with each of the fiber-based electrodes. The gel electrolyte used provides ion transfer between the electrodes and prevents short circuits.

28.4.1.2 Twisted fiber structure

In another structure, fiber-based electrodes are twisted together to form a single fiber. Fiber electrode surfaces are usually coated with gel electrolyte, and then fiber-based supercapacitors are obtained by twisting the electrodes together. In Figure 28.2, twisted fiber-based supercapacitors are shown schematically.

28.4.1.3 Coaxial fiber structure

In the coaxial fiber supercapacitor design, the supercapacitor structure is produced on a single fiber substrate by covering the layers on top of each other [27]. In some studies, a substrate fiber is used in the core, and then, electrode materials and other layers are applied. In other studies, fiber-based electrodes serve as both a base and an electrode for the upper layers. Zhao et al. developed graphene-based coaxial fiber supercapacitors. For this purpose, first graphene fiber was produced, and then, its surface was coated with PVA gel. The graphene fiber acts as both an electrode and a collecting electrode in this study. It was then covered with GO gel and coagulated in a coagulation bath. The graphene/PVA/graphene coaxial fiber structure was obtained by coagulating the upper GO layer using hydroiodic acid. Finally, the coaxial fiber surface is covered with a PVA gel protective layer. The coaxial structure was kept in 1M H_2SO_4 and then dried to obtain coaxial graphene fiber supercapacitors. This study can be given as a good example of the use of electrode material as a substrate at the same time [28]. The schematic illustration of the fabrication process is shown in Figure 28.3a. In the study by Harrison et al., stainless steel wire was used as a flexible fiber substrate and a current collector. Chinese ink was used as the bottom electrode, PVA/H_3PO_4 as gel electrolyte, active carbon electrode was used as an upper electrode, and silver paste was used as the current collector at the top. The specific capacitance of the

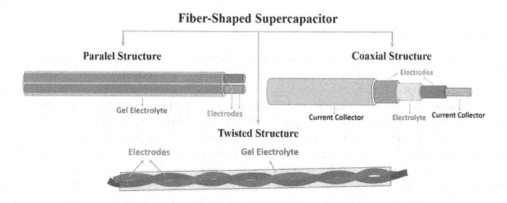

FIGURE 28.2 Fiber-based supercapacitor design.

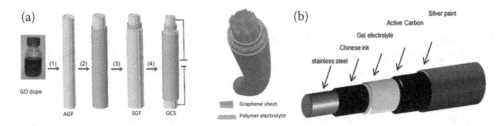

FIGURE 28.3 a) Schematic illustration of (a) the fabrication process and (b) the cross section structure of a GCS with a core fiber and a cylinder sheath as the two electrodes. (1) Wet-spinning and chemical reduction. (2) Dip-coating of polymer electrolyte gel. (3) Dip-coating of GO gel and chemical reduction. (4) Dip-coating of polymer electrolyte gel [28]. Adapted with permission from ref. [28]. Copyright (2015) The Royal Society of Chemistry. (b) Schematic of four coating layers on a 50 mm stainless steel wire [27]. Adapted with permission from ref. [27]. Copyright (2013) The Royal Society of Chemistry.

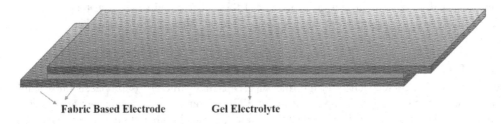

FIGURE 28.4 The illustration of fabric-based supercapacitor.

coaxial fiber supercapacitor was measured as 3.18 mF/cm^2 at 0.1 mF/cm for a 2.5 cm long fiber supercapacitor [27]. The schematic illustration of the fabrication process can be seen in Figure 28.3b.

28.4.2 FABRIC SUPERCAPACITORS

Woven, knitted, or nonwoven fabrics can be used as supercapacitor fabric structures. Fabric supercapacitors can be converted directly into fabric supercapacitors to produce fabric supercapacitors, or fiber supercapacitors can be converted into fabric supercapacitors using techniques such as weaving or knitting. In the most common fabrication technique, fabric supercapacitor electrode surfaces are coated with gel electrolytes and then, by combining them on top of each other, fabric supercapacitors are obtained. Figure 28.4 shows the schematic representation of the fabric supercapacitors.

28.5 FIBER-BASED ELECTRODES

Since fibers are the basic building blocks of textile products, textile-based supercapacitors studies generally focus on fiber-based supercapacitors. In some of the studies, only fiber-based supercapacitors are developed. On the other hand, fabric supercapacitors are obtained by combining fiber-based supercapacitors with techniques such as weaving, knitting, or sewing. The fiber-shaped electrodes, which are expected to have properties such as flexibility, lightness, and mechanical strength, constitute the basic and most important component of fiber-based supercapacitors. In some studies, directly fiber-shaped electrodes such as graphene and CNT yarn are used to obtain supercapacitors, while in some studies, supercapacitors are developed by preparing electrode materials separately and then attaching them to the surface of fiber substrates. Fiber-based electrode shapes are discussed in detail below. Moreover, some fiber-shaped electrodes for supercapacitors are shown in Figure 28.5 [29–31].

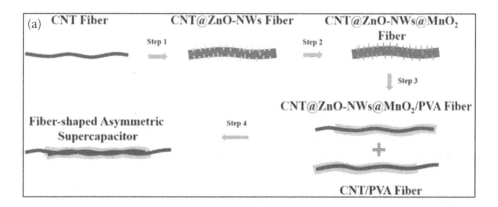

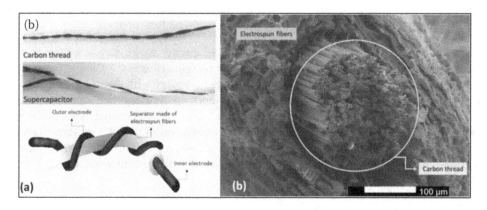

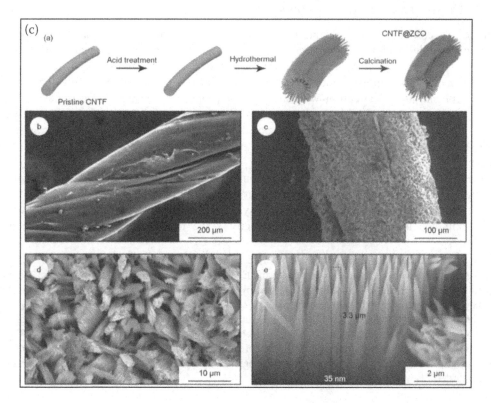

28.5.1 Conventional fibers based fiber electrodes

Conventional textile fibers are also used as substrates in the production of fiber-based super-capacitors. Natural and synthetic polymers, such as cotton, wool, polyester, polyamide, and elastane, which are used as raw materials in textile products, are both highly flexible and insulating materials. Therefore, to obtain fiber-based supercapacitor electrodes, the surfaces of these conventional fibers must be modified with materials that can be used as electrodes in supercapacitors. In the modification of conventional fiber surfaces, materials, such as graphene, CNT, Mxene, polyaniline, polypyrrole, metal oxides, etc., are used. Carbon fiber, a high-performance fiber, stands out with its high electrical and thermal conductivity and high strength properties. Carbon fibers are well suited for use as substrates and current collectors in fiber supercapacitors due to their high electrical conductivity. Therefore, it has been used in many fiber supercapacitor applications [30,32,33].

28.5.2 Metal yarns/wires/threads-based fiber electrodes

As explained above, one of the components of the supercapacitor is the current collector. While the electrode material also functions as a current collector in some structures, a current collector layer is necessary for some other applications. Metal threads come to the fore with their properties, such as easy formability and high electrical conductivity. Fiber-based supercapacitor electrodes are also developed by modifying metal yarn surfaces with various materials. In the literature, materials such as Ni yarn [34], copper wire [25,35], and stainless steel wire [26,27] are used as metal thread/wire/yarn. Nanomaterials such as "graphene and its derivatives, carbon nanotube, and pseudocapacitive materials such as metal oxides/nitrides/sulfides," and conductive polymers are used in the modification of the metal thread surface [34,35].

28.5.3 Graphene yarn-based electrodes

Graphene is a carbon-based nanomaterial formed by sp^2 hybridization of carbon atoms and arranged in a single-layer hexagonal honeycomb structure. Dr. Andre Geim and Dr. Konstantin Novoselov were honored with the Nobel Prize in Physics in 2010 for their work on graphene in 2004. Interest in this important monolayer nanomaterial has increased following these developments. Graphene attracts attention with its superior properties, such as large surface area, high electrical conductivity, and high tensile strength. While it is possible to obtain single-layer graphene using techniques such as chemical vapor deposition and epitaxial growth in graphene production, faster and cheaper graphene derivatives are obtained by using techniques such as chemical exfoliation.

In the first stage of the chemical exfoliation process, graphene oxide (GO) is obtained by oxidizing graphite in strong acid media, and then reduced graphene oxide is obtained by reducing GO with reducing agents, such as hydrazine hydrate, ascorbic acid, hydroiodic acid, etc. This material obtained is a kind of graphene derivative. The most common production method of

FIGURE 28.5   (a) Schematic of the fabrication of the CNT@ZnO-NWs@MnO$_2$ hybrid electrode [29]. Adapted with permission from ref. [29]. Copyright (2016) The Royal Society of Chemistry. (b) Twisted configuration wire-based supercapacitor: a) Photography and schematic of carbon thread and supercapacitor; and b) SEM image of carbon thread coated with electrospun fibres [30]. Adapted with permission from ref. [30]. Copyright (2020) Springer Nature. (c) a) Schematic illustration of the preparation procedure for the hybrid CNT fiber, b) SEM image of the pristine CNT fiber, c–d) The surface morphology of as-prepared hybrid CNT fiber at different magnifications, e) The structure of ZCO nanowire forests at different magnifications [31]. Adapted with permission from ref. [31]. Copyright (2019) Elsevier.

graphene oxide fibers is performed by using a coagulation bath via a wet-spinning method. For this purpose, $CaCl_2$ aqueous solution is the most widely used coagulation bath [36–38]. Reduced graphene oxide fibers (graphene fibers) are obtained by thermal reduction of these fibers in an inert atmosphere at high temperature, or by using chemical reducing agents such as hydrofluoric acid and ascorbic acid. In addition, there are graphene yarn production methods performed with a microfluidic system [39]. In this method, the graphene oxide solution is directed to a thin channel to elongate the graphene plates in the direction of the fiber, then the fibers in the microchannel are dried and reduced at high temperature to obtain graphene fibers [39]. These fibers, which have superior properties such as high electrical conductivity and large surface area, attract attention as a very convenient electrode material for textile supercapacitor applications. For this reason, many studies using graphene fibers as electrodes have entered the literature [40].

28.5.4 CNT YARN-BASED FIBER ELECTRODES

Carbon nanotube is a type of allotrope of carbon, a rolled form of graphene plates. Although it shows similar properties with graphene, it differs from graphene with its high aspect ratio. Carbon nanotubes, which have excellent properties such as very high electrical and thermal conductivity and high aspect ratio, are important nanomaterials used in areas such as supercapacitors, batteries, and electromagnetic shielding. The most widely used production methods can be listed as an arc discharge method, laser ablation method, and chemical vapor deposition (CVD) method [41]. Carbon nanotube fibers, on the other hand, are the microsized fiber form of these nanosized materials. These fibers with high mechanical strength, electrical conductivity, and large surface area, which carry these extraordinary properties of carbon nanotube in nanoscale to macro dimension, find wide application areas for themselves. The methods used in the production of carbon nanotube fiber can be divided into two, as wet spinning and dry spinning techniques. Carbon nanotube fibers are obtained by coagulating carbon nanotube dispersions in a suitable coagulation bath in wet-spinning techniques. On the other hand, CNT yarns are obtained by spinning, twisting/rolling carbon nanotubes produced by the CVD method in dry-spinning techniques [41].

Carbon nanotube threads are used as fiber electrodes in fiber-based supercapacitor applications without the need for a current collector due to their high electrical conductivity. Carbon nanotube fibers can be used as electrodes alone, as well as combined with other carbon-based materials (graphene, activated carbon, etc.), metal oxides, or conductive polymers in fiber-based supercapacitor applications [29].

28.5.5 HYBRID FIBER-BASED SUPERCAPACITOR ELECTRODES

In fiber-based supercapacitor studies, the combination of the materials, which has different charge-storage mechanisms and different properties, increases the performance of the supercapacitors to be produced. Therefore, in almost all fiber-based supercapacitor applications, it is mostly used with more than one material together. Thanks to this synergistic effect, fiber electrode performance is improved. In a study, graphene/CNT and thermoplastic polyurethane wet spun together to produce hybrid fiber, and it was used as an electrode in fiber-based supercapacitor applications [42]. In another study, the solution obtained by mixing graphene and Mxene was dried in a thin capillary tube to obtain hybrid fibers and used as an electrode in the production of fiber-based supercapacitors [43]. In another study, carbon black was added into the graphene solution and then graphene/carbon black fibers were obtained in the coagulation bath. It has been observed that the carbon black additive made to the fibers increases the fiber supercapacitor performance thanks to its effect on fiber morphology. In some cases, the different materials are mixed before the fiber drawing process, while in some cases, there is a modification of the above-mentioned fiber-based electrode surfaces. A summary of fiber-based supercapacitors is provided in Table 28.1.

TABLE 28.1

Summary of Fiber-Based Supercapacitors

	Fiber-based supercapacitor	Electrolyte	Fiber-based supercapacitor type	Capacitance	Scan rate or current density	Ref.
Graphene Fiber	Graphene/carbon black fiber	PVA/ H_3PO_4 gel	parallel	$176.6 \ F/cm^3$	$20 \ mA/cm^3$	[39]
	Graphene/Mxene fiber	–	–	$345.2 \ F/cm^3$	$0.1 \ A/g$	[43]
	Graphene fiber (rGO/SA)	–	–	$93.75 \ mF/cm^2$	$1 \ mA/cm^2$	[36]
	Graphene fiber/Molybdenum disulfide nanosheet	PVA/ H_2SO_4 gel	parallel	$29.1 \ mF/cm^2$	$0.2 \ mA/cm^2$	[38]
Metal Wire/ Fiber/Thread	PEDOT-coated cellulose and SS composite yarns	PVA/ H_2SO_4 gel	parallel	$158.2 \ mF/cm^2$	$1.5 \ mA/cm^2$	[44]
	NiCo-BOH and rGO/CNT@ Ni and Cu coated polyester yarn	PVA/KOH	parallel	$133 \ mF/cm^2$	$5 \ mV/s$	[13]
Conductive Polymer Fiber	PEDOT:PSS fiber	PVA-LiCl gel	parallel	–	–	[10]
CNT Yarn	CNT yarn	PVA/HCl gel	twisted	$1.49 \ F/g$ or $17.0 \ mF/cm_2$	$100 \ mV/s$	[9]
	CNT-chitosan-PEDOT:PSS fibers	1 M PBS	–	$21.4 \ F/g$	$0.5 \ A/g$	[45]
Hybrid Yarns	Graphene/CNT/TPU yarn	PVA/ H_2SO_4 gel	twisted	$14.3 \ F/cm_3$	$10 \ mV/s$	[42]
	CNT fiber@ZCO nanowire forests	PVA/LiCl gel	parallel	$112.67 \ mF/cm_2$	$1 \ mA/cm^2$	[31]
	CNT fiber@ZnO-NW@MnO2	PVA/ H_2SO_4 gel	parallel	$30.03 \ mF/cm_2$	$5 \ mV/s$	[29]
Modified Conventional Fibers	NiCo2S4@Nickel– Cobalt Layered Double Hydroxide Nanotube Arrays on Metallic Cotton Yarns	PVA/KOH	twisted	$37.2 \ mF/cm^2$	$0.4 \ mA/cm^2$	[15]
Modified Carbon Fiber	Polyprole@Carbon thread	Simulated sweat solution		$2.3 \ F/g$	$20 \ mV/s$	[30]
	Carbon fiber/ruthenium dioxides @ graphene based fiber	PVA/ H_3PO_4 gel	parallel	$211.24 \ F/cm^3$	$5 \ mV/s$	[32]
	Surface activated carbon fiber	PVA/ H_3PO_4 gel	twisted	$25 \ mF/cm^2$	$2 \ \mu A/cm^2$	[33]
Activated Carbon Based	Thermally drawable supercapacitor activated carbon based yarn	PVDF/ LiTFSI	parallel	$244 \ mF/cm^2$	$0.5 \ mA/cm^2$	[46]

28.6 FABRIC-BASED ELECTRODES FOR SUPERCAPACITORS

The potential to use textile fabrics directly in flexible supercapacitors is one of the most important developments for wearable electronic applications. It is possible to develop smart textile materials that can store energy by using various textile fabrics as substrates. For this purpose, conductive textile materials such as carbon cloth, metal mesh, etc., which also serve as collector electrodes, can be used, and it is possible to use cotton, polyester, and fabric surfaces, which are used directly in clothes, as a substrate in fabric supercapacitors. The fabrics used in fabric supercapacitors are examined in detail below. Fabric surfaces modified with functional materials are used as fabric electrodes in supercapacitors. Fabric-based solid-state supercapacitors are obtained by applying gel electrolytes between these fabric electrodes. An example of a fabric-shaped supercapacitors process is illustrated in Figure 28.6 [47–50].

28.6.1 METAL MESH-BASED ELECTRODES

Metal meshes are used as substrates in fabric supercapacitor applications with their properties such as high conductivity, flexibility, and strength. For this purpose, metal-based materials such as stainless steel and nickel mesh are generally preferred. At the same time, the surface of these substrates, which serve as collector electrodes, is modified with carbon-based materials, conductive polymers, or metal oxides/nitrides/sulfides to obtain electrode materials for fabric supercapacitors.

28.6.2 CARBON FABRIC-BASED ELECTRODES

Carbon fabrics are important conductive textile materials with properties such as high conductivity, being light in weight, and flexibility. They are much lower density and corrosion-resistant materials compared to metal meshes. These materials attract attention as the most widely used conductive textile material in fabric supercapacitors. Carbon fabric surfaces are modified with such materials as graphene, CNT, polypyrrole, Mxene, V_2O_5/CoMoS$_4$, MnO_2 nano-array, etc. [8,11,12,50,50,50,51]. Modified carbon fabric electrodes are ideal electrode materials for fabric supercapacitors.

28.6.3 CONVENTIONAL FABRIC-BASED ELECTRODES

Conventional fabrics such as cotton, polyester, nylon, elastane, etc., used in the clothing material are frequently preferred as textile-based substrates in fabric-based supercapacitor production. Although they have advantages such as flexibility, low cost, light in weight, and easy availability, they are electrically insulating materials. For this reason, when using conventional textile products, it is important to cover their surfaces with materials that will serve as a current collector. In some cases, the layers used as electrode materials act as a current collector as well [16,52,53], but in some studies, it is seen that conventional textile fabrics are coated with conductive materials before application of electrode materials (Table 28.2) [54,55].

28.7 APPLICATIONS

Flexible textile-based supercapacitors have been developed using the materials mentioned above. These devices can be used by integrating into the clothing surfaces, as well as being integrated into more complicated e-textile systems. Some sample applications using flexible textile-based supercapacitors are mentioned below.

Park et al. developed an all-in-one system that combined textile-based biosensors and textile-based supercapacitors in the same system in their study. Thanks to the strain sensor integrated into the textile, the energy requirement of this system, which can detect fingers, wrists, and elbow joint movements, was provided by the textile-based supercapacitor integrated into the same textile structure. For this purpose, a mixture of multiwalled carbon nanotube (MWCNT) and molybdenum trioxide nanowires (MoO$_3$ NWs) was dispersed in ethanol and coated on the surface of stretchable fabric substrate (nylon (82%)/spandex (18%)) by a spray coating method. The MWCNT and MoO$_3$ NWs coated fabric were used as a strain sensor. On the other hand, the stretchable fabric was coated

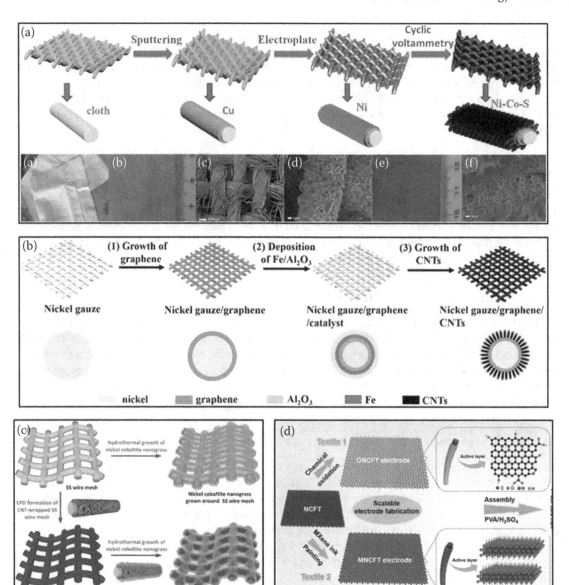

FIGURE 28.6 (a) Schematic illustration of the fabrication of CueNi and its corresponding SEM images [47]. Adapted with permission from ref. [47]. Copyright (2020) Elsevier. (b) Schematic of the fabrication process and structure of textile containing seamlessly connected graphene/CNTs hybrids [48]. Adapted with permission from ref. [48]. Copyright (2020) Elsevier. (c) Schematic illustrating the formation of nickel cobaltite nanograss around SS wire mesh and CNT-wrapped SS wire mesh for high-performance supercapacitors [49]. Adapted with permission from ref. [49]. Copyright (2015) Elsevier. (d) Schematic illustration of the scalable synthesis protocol of the positive O/N-functionalized carbon fiber textile (ONCFT) electrode, negative MXene decorated NCFT (MNCFT) electrode and the assembled MNCFT//ONCFT ASC [50]. Adapted with permission from ref. [50]. Copyright (2020) John Wiley and Sons.

TABLE 28.2

Summary of Fabric-Based Supercapacitors

	Fabric-based supercapacitor	Electrolyte	Device or electrode	Capacitance	Scan rate or current density	Ref.
Metal Mesh	Nickel cobaltite nanograss grown around porous carbon nanotube-wrapped stainless steel wire mesh	1 M KOH	electrode	1223 F/g	1 A/g	[49]
	Graphene/CNT/PANI textile growthed on Ni Mesh	PVA/H_3PO_4	device	164 mF/cm^2	0.2 mA/cm^2	[48]
Carbon Fiber	Polypyrrole/textile polyacrylonitrile-derived activated carbon fiber	2 M H_2SO_4	electrode	302.2 F/g	0.4 A/g	[51]
	O/N functionalized carbon fiber textile/Mxene decorated carbon fiber textile	PVA/H_2SO_4 gel	device	780 mF/cm^2	1 mV/s	[50]
	Mxene coated carbon cloth	PVA-H_3PO_4 gel	device	513 mF/cm^2	2 mV/s	[8]
	3D phosphorus doped graphene foam in carbon cloth to support V_2O_5/CoMoS$_4$ hybrid	PVA/LiNO$_3$	device	113.0 mA h/g	1 A/g	[12]
	Ni-doped MnO_2 nano-array @ carbon cloth	PVA/Na$_2$SO$_4$ gel	device	71.54 mF/cm^2	2 mA/cm^2	[11]
Conventional Fabric	Cellulose textiles/rGO	PVA/LiCl gel	device	1208.4 mF/cm^2	1 mA/cm^2	[16]
	Silver nanoparticles/reduced graphene oxide/cotton fabric	PVA/H_2SO_4 gel	device	353 F/g	0.3 A/g	[56]
	Activated carbon/carbonized asphalt/fiberglass cloth	PVA-H_2SO_4 gel	device	85.4 mF/cm^2	0.5 mA/cm^2	[7]
	Polypyrole/conductive fiberglass cloth	PVA/H_2SO_4 gel	device	549.6 mF/cm^2	0.5 mA/cm^2	[52]
	In-situ deposition of reduced graphene oxide layers on textile surfaces	PVA/H_2SO_4 gel	device	13.3 m F/cm^2	0.1 mA/cm^2	[53]

with PEDOT: PSS as a current collector, and then supercapacitor electrodes were obtained by coating with MWCNT and MoO_3 NWs mixture, as mentioned above. After the developed fabric electrodes were coated with gel electrolyte, the fabric electrodes were overlapped to obtain a supercapacitor structure. A schematic illustration of the fabricated stretchable all-in-one textile system is shown in Figure 28.7 [57].

Wang et al. were combined photovoltaic fiber and fiber supercapacitor in the fiber-shaped photocapacitor they developed. PEDOT: PSS fibers were produced by the wet-spinning method as electrodes. The fiber electrodes were used both in fiber-based supercapacitors and photovoltaic fiber as electrodes. Dye-sensitive solar cell (DSSC) structure has been used for photovoltaic fiber. The total photochemical-electric energy-conversion efficiency (TCE) of the device was measured up to 5.1%. The schematic representation of the produced photocapacitor is shown in Figure 28.8. Thanks to this structure, the energy produced in the presence of light was stored in fiber supercapacitors, and the stored energy was used when needed. Thanks to this combined structure, more effective uses of photovoltaic fibers were developed [10].

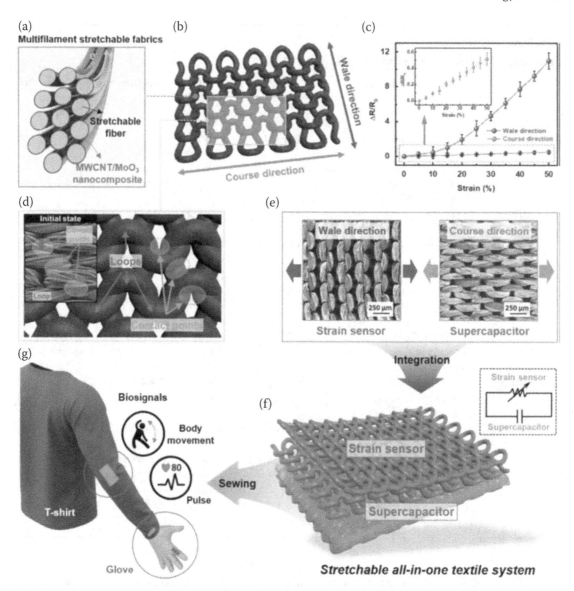

FIGURE 28.7 Schematic illustration of the fabricated stretchable all-in-one textile system. (a) High-magnification illustration of the cross section of the stretchable fiber showing the multifilament structure and MWCNT/MoO$_3$ nanocomposite in the fabric. (b) Schematic diagram of the wale and course directions of a knitted (purl loop) fabric. (c) Relative resistance changes versus strain curves for wale (blue) and course (red) direction stretching. (d) Schematic illustration of contact points between the loops in the knitted fabric structure. Inset is the SEM image of the textile. (e) SEM images showing the fabric with wale (left) and course (right) direction stretching. (f) Integration of the supercapacitor and the strain sensor. (g) Conceptual illustration of the integration system sewn onto a T-shirt and a nylon glove for biosignal monitoring [57]. Adapted with permission from ref. [57]. Copyright (2019) American Chemical Society.

In another important application, Liu et al. developed self-charging power textiles using asymmetric supercapacitor yarns. For this purpose, it was ensured that the waste energy was converted to electrical charge by using fiber-based triboelectric nanogenerators (TENG), and then the energy was stored by fiber-based supercapacitors. First of all, conductive yarns were obtained by coating polyester yarn

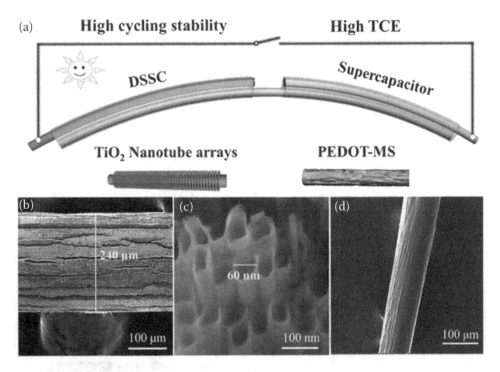

FIGURE 28.8 (a) Schematic illustration of integrated flexible self-powered FPCs for UV photodetector. SEM images of (b and c) TiO$_2$ nanotube array and (d) PEDOT-MS fiber [10]. Adapted with permission from ref. [10]. Copyright (2020) Elsevier.

surfaces with Ni and Cu. rGO/CNT coated yarn via hydrothermal self-assembly method as the negative electrode, and electroplated Ni-Co bimetallic oxyhydroxide (NiCo BOH) coated yarns as positive electrodes were used. Then, these electrodes were coated with PVA/KOH gel electrolyte, fiber-based asymmetric supercapacitors were developed by combining them in parallel. On the other hand, copper-coated polyester yarn surfaces are covered with polydimethylsiloxane. Finally, this yarn (heptadecafluoro-1,1,2,2-tetrahydrodecyl) was functionalized with trichlorosilane to obtain fiber-based TENG. These yarns, which can harvest and store energy, are obtained by weaving into fabric in different forms. The production process and integration of these yarns on the fabric surface can be seen in Figure 28.9 [13].

28.8 CONCLUSION AND FUTURE PERSPECTIVE

Conversion of materials with charge storage features into textile form is a revolutionary development in terms of e-textile technologies. While the energy need of e-textiles is met by external energy-storage devices, meeting this need with textile-based flexible supercapacitors, which stand out with their lightness, flexibility, and comfort, applied directly to the textile surfaces is promising for the future. While the first studies on textile-based supercapacitors were on the development of devices that can store energy in fiber or fabric form, studies have been carried out on systems where these charge-storage systems are used together with other energy and sensor systems. In particular, the ability to store the energy obtained with energy-harvesting textile applications (photovoltaic fiber, triboelectric textile, piezoelectric textile, etc.) with textile-based flexible, light, and comfortable technologies are studies that will improve the effectiveness and applicability of these systems. These energy storage technologies, which will increase the effectiveness of many wearable electronics applications in military, outdoor applications, need to be further developed. In future studies, it is necessary to focus

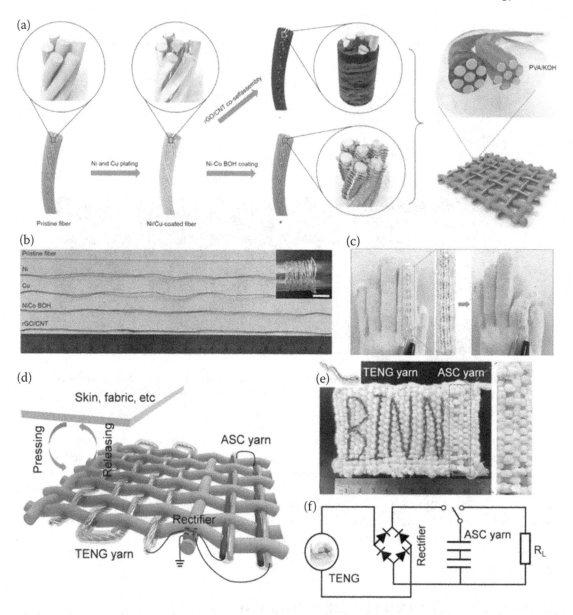

FIGURE 28.9 The fabrication of fiber-shaped electrodes and asymmetric supercapacitor (ASC). (a) Schematic illustration of the fabrication procedure of yarn asymmetric supercapacitor (Y-ASC). (b) The photo of yarns with different coatings (Ni, Cu, Ni-Co BOH, rGO/CNT). (c) Photograph of two Cu-coated yarns being woven in the glove fabric, and a blue LED lighted by a DC source through the conductive fibers, (d) Schematic illustration of the self-charging power textile. The output voltage and current were rectified by a bridge rectifier to charge asymmetric supercapacitor yarns. (e) Photograph of a self-charging power textile by integrating English-letter-shaped (i.e., "BINN") energy-harvesting yarn and ASC yarn in a common fabric. (f) Equivalent circuit of a self-charging system that uses the energy harvesting fabric to charge the ASC to power electronics [13]. Adapted with permission from ref. [13]. Copyright (2019) John Wiley and Sons.

on the production of textile-based supercapacitors at a larger scale and their integration with other wearable electronic systems. Further development of textile-based supercapacitors can pave the way for wearable electronics and e-textile studies.

REFERENCE

1. Koncar, V., (2016). Introduction to smart textiles and their applications, In Smart Text. *Their Appl*, pp: 1–8.
2. Oğuz Gök, M., (2018). Tasarimda ElektronikTekstiller. *International Journal of Social Humanities Sciences Research (JSHSR)*, 5(22): 933–940.
3. Levitt, A., J. Zhang, G. Dion, Y. Gogotsi and J.M. Razal, (2020). MXene-based fibers, yarns, and fabrics for wearable energy storage devices. *Advanced Functional Materials*, 30(47): 2000739.
4. https://en.wikipedia.org/wiki/Supercapacitor. Accessed October 21, 2019.
5. Altin, Yasin, Celik Bedeloglu, A., (2021). Polyacrylonitrile/polyvinyl alcohol-based porous carbon nanofiber electrodes for supercapacitor applications. *International Journal of Energy Research*, 45: 16497–16510.
6. Kim, B.K., S. Sy, A. Yu and J. Zhang, (2015). Electrochemical supercapacitors for energy storage and conversion. *In Handb Clean Energy Syst*, pp: 1–25. 10.1002/9781118991978.hces112.
7. Yang, C., L. Hu, L. Zang, Q. Liu, J. Qiu, J. Yang and X. Qiao, (2020). High-performance all-solid-state supercapacitor based on activated carbon coated fiberglass cloth using asphalt as active binder. *Journal of The Electrochemical Society*, 167(2): 020540.
8. Li, Y., B. Xin, Z. Lu, X. Zhou, Y. Liu and Y. Hu, (2021). Enhancing the supercapacitor performance of flexible MXene/carbon cloth electrodes by oxygen plasma and chemistry modification. *International Journal of Energy Research*, 45(6): 9229–9240.
9. Mun, T.J., S.H. Kim, J.W. Park, J.H. Moon, Y. Jang, C. Huynh, R.H. Baughman and S.J. Kim, (2020). Wearable energy generating and storing textile based on carbon nanotube yarns. *Advanced Functional Materials*, 30(23): 2000411.
10. Wang, Z., J. Cheng, H. Huang and B. Wang, (2020). Flexible self-powered fiber-shaped photocapacitors with ultralong cyclelife and total energy efficiency of 5.1%. *Energy Storage Materials*, 24: 255–264.
11. Zhong, R., M. Xu, N. Fu, R. Liu, A. Zhou, X. Wang and Z. Yang, (2020). A flexible high-performance symmetric quasi-solid supercapacitor based on Ni-doped MnO_2 nano-array @ carbon cloth. *Electrochimica Acta*, 348: 136209.
12. Zhou, X., X. Gao, M. Liu, C. Wang and F. Chu, (2020). Synthesis of 3D phosphorus doped graphene foam in carbon cloth to support V2O5/CoMoS4 hybrid for flexible all-solid-state asymmetry supercapacitors. *Journal of Power Sources*, 453.
13. Liu, M., Z. Cong, X. Pu, W. Guo, T. Liu, M. Li, Y. Zhang, W. Hu and Z.L. Wang, (2019). High-energy asymmetric supercapacitor yarns for self-charging power textiles. *Advanced Functional Materials*, 29(41): 1806298.
14. Chae, J.S., H.N. Kwon, W.S. Yoon and K.C. Roh, (2018). Non-aqueous quasi-solid electrolyte for use in supercapacitors. *Journal of Industrial and Engineering Chemistry*, 59: 192–195.
15. Wang, Y.F., H.T. Wang, S.Y. Yang, Y. Yue and S.W. Bian, (2019). Hierarchical NiCo2S4@nickel-cobalt layered double hydroxide nanotube arrays on metallic cotton yarns for flexible supercapacitors. *ACS Applied Materials and Interfaces*, 11(33): 30384–30390.
16. Chen, R., Y. Yang, Q. Huang, H. Ling, X. Li, J. Ren, K. Zhang, R. Sun and X. Wang, (2020). A multi-functional interface design on cellulose substrate enables high performance flexible all-solid-state super-capacitors. *Energy Storage Materials*, 32: 208–215.
17. Jost, K., G. Dion and Y. Gogotsi, (2014). Textile energy storage in perspective. *Journal of Materials Chemistry A*, 2(28): 10776–10787.
18. Winter, M. and R.J. Brodd, (2004). What are batteries, fuel cells, and supercapacitors? *Chemical Reviews*, 104(10): 4245–4269.
19. Simon, P., Y. Gogotsi and B. Dunn, (2014). Where do batteries end and supercapacitors begin? *Science*, 343(6176): 1210–1211.
20. Jost, K., C.R. Perez, J.K. McDonough, V. Presser, M. Heon, G. Dion and Y. Gogotsi, (2011). Carbon coated textiles for flexible energy storage. *Energy and Environmental Science*, 4(12): 5060–5067.
21. Sun, Y., R.B. Sills, X. Hu, Z.W. Seh, X. Xiao, H. Xu, W. Luo, H. Jin, Y. Xin, T. Li, Z. Zhang, J. Zhou, W. Cai, Y. Huang and Y. Cui, (2015). A bamboo-inspired nanostructure design for flexible, foldable, and twistable energy storage devices. *Nano Letters*, 15(6): 3899–3906.

22. Dong, L., C. Xu, Y. Li, Z.H. Huang, F. Kang, Q.H. Yang and X. Zhao, (2016). Flexible electrodes and supercapacitors for wearable energy storage: A review by category. *Journal of Materials Chemistry A*, 4(13): 4659–4685.

23. Liu, L., Y. Yu, C. Yan, K. Li and Z. Zheng, (2015). Wearable energy-dense and power-dense supercapacitor yarns enabled by scalable graphene-metallic textile composite electrodes. *Nature Communications*, 6: 1–9.

24. Altin, Y. and A. Çelik Bedeloğlu, (2020). Energy storage textile. In ul-Islam, S. & Butol, B. S. *Advances in Functional and Protective Textiles*, pp: 493–529. Woodhead.

25. Purkait, T., G. Singh, D. Kumar, M. Singh and R.S. Dey, (2018). High-performance flexible supercapacitors based on electrochemically tailored three-dimensional reduced graphene oxide networks. *Scientific Reports*, 8(1): 1–13.

26. Senthilkumar, S.T. and R.K. Selvan, (2015). Flexible fiber supercapacitor using biowaste-derived porous carbon. *ChemElectroChem*, 2(8): 1111–1116.

27. Harrison, D., F. Qiu, J. Fyson, Y. Xu, P. Evans and D. Southee, (2013). A coaxial single fibre supercapacitor for energy storage. *Physical Chemistry Chemical Physics*, 15(29): 12215–12219.

28. Zhao, X., B. Zheng, T. Huang and C. Gao, (2015). Graphene-based single fiber supercapacitor with a coaxial structure. *Nanoscale*, 7(21): 9399–9404.

29. Li, Y., X. Yan, X. Zheng, H. Si, M. Li, Y. Liu, Y. Sun, Y. Jiang and Y. Zhang, (2016). Fiber-shaped asymmetric supercapacitors with ultrahigh energy density for flexible/wearable energy storage. *Journal of Materials Chemistry A*, 4(45): 17704–17710.

30. Lima, N., A.C. Baptista, B.M.M. Faustino, S. Taborda, A. Marques and I. Ferreira, (2020). Carbon threads sweat-based supercapacitors for electronic textiles. *Scientific Reports*, 10(1):.

31. Yang, Z., Y. Yang, C. Lu, Y. Zhang, X. Zhang and Y. Liu, (2020). A high energy density fiber-shaped supercapacitor based on zinc-cobalt bimetallic oxide nanowire forests on carbon nanotube fibers. *Carbon*, 159: 687.

32. Li, X., D. Liu, X. Yin, C. Zhang, P. Cheng, H. Guo, W. Song and J. Wang, (2019). Hydrated ruthenium dioxides @ graphene based fiber supercapacitor for wearable electronics. *Journal of Power Sources*, 440: 227143.

33. Lee, J. and G.H. An, (2021). Surface-engineered flexible fibrous supercapacitor electrode for improved electrochemical performance. *Applied Surface Science*, 539: 148290.

34. Zhou, J., N. Chen, Y. Ge, H. Zhu, X. Feng, R. Liu, Y. Ma, L. Wang and W. Hou, (2018). Flexible all-solid-state micro-supercapacitor based on Ni fiber electrode coated with MnO2 and reduced graphene oxide via electrochemical deposition. Science China *Materials*, 61(2): 243–253.

35. Patil, S.J., R.B. Pujari, T.F. Hou and D.W. Lee, (2020). Transition metal sulfide-laminated copper wire for flexible hybrid supercapacitor. *New Journal of Chemistry*, 44(42): 18489–18495.

36. Hu, X., M. Tian, N. Pan, B. Sun, Z. Li, Y. Ma, X. Zhang, S. Zhu, Z. Chen and L. Qu, (2019). Structure-tunable graphene oxide fibers via microfluidic spinning route for multifunctional textile. *Carbon*, 152: 106–113.

37. Tang, M., Y. Wu, J. Yang, H. Wang, T. Lin and Y. Xue, (2021). Graphene/tungsten disulfide core-sheath fibers: High-performance electrodes for flexible all-solid-state fiber-shaped supercapacitors. *Journal of Alloys and Compounds*, 858: 157747.

38. Tang, M., Y. Wu, J. Yang and Y. Xue, (2020). Hierarchical core-shell fibers of graphene fiber/radially-aligned molybdenum disulfide nanosheet arrays for highly efficient energy storage. *Journal of Alloys and Compounds*, 828: 153622.

39. Jia, Y., A. Ahmed, X. Jiang, L. Zhou, Q. Fan and J. Shao, (2020). Microfluidic fabrication of hierarchically porous superconductive carbon black/graphene hybrid fibers for wearable supercapacitor with high specific capacitance. *Electrochimica Acta*, 354: 136731.

40. Hu, Y., H. Cheng, F. Zhao, N. Chen, L. Jiang, Z. Feng and L. Qu, (2014). All-in-one graphene fiber supercapacitor. *Nanoscale*, 6(12): 6448–6451.

41. Lu, Z., R. Raad, F. Safaei, J. Xi, Z. Liu and J. Foroughi, (2019). Carbon nanotube based fiber supercapacitor as wearable energy storage. *Frontiers in Materials*, 6(138):.

42. Wu, G.Q., X.Y. Yang, J.H. Li, N. Sheng, C.Y. Hou, Y.G. Li and H.Z. Wang, (2020). Highly stretchable and conductive hybrid fibers for high-performance fibrous electrodes and all-solid-state supercapacitors. *Chinese Journal of Polymer Science (English Edition)*, 38(5): 531–539.

43. Wang, Z., Y. Chen, M. Yao, J. Dong, Q. Zhang, L. Zhang and X. Zhao, (2020). Facile fabrication of flexible rGO/MXene hybrid fiber-like electrode with high volumetric capacitance. *Journal of Power Sources*, 448: 227398.

44. Du, X., M. Tian, G. Sun, Z. Li, X. Qi, H. Zhao, S. Zhu and L. Qu, (2020). Self-powered and self-sensing energy textile system for flexible wearable applications. *ACS Applied Materials and Interfaces*, 12(50): 55876–55883.

45. Mirabedini, A., Z. Lu, S. Mostafavian and J. Foroughi, (2021). Triaxial carbon nanotube/conducting polymer wet-spun fibers supercapacitors for wearable electronics. *Nanomaterials*, 11(1): 1–16.

46. Khudiyev, T., J.T. Lee, J.R. Cox, E. Argentieri, G. Loke, R. Yuan, G.H. Noel, R. Tatara, Y. Yu, F. Logan, J. Joannopoulos, Y. Shao-Horn and Y. Fink, (2020). 100 m long thermally drawn supercapacitor fibers with applications to 3D printing and textiles. *Advanced Materials*, 32(49).

47. Wang, Z., H. Wang, S. Ji, H. Wang, D.J.L. Brett and R. Wang, (2020). Design and synthesis of tremella-like Ni–Co–S flakes on co-coated cotton textile as high-performance electrode for flexible supercapacitor. *Journal of Alloys and Compounds*, 814: 151789.

48. Liu, K., Y. Yao, T. Lv, H. Li, N. Li, Z. Chen, G. Qian and T. Chen, (2020). Textile-like electrodes of seamless graphene/nanotubes for wearable and stretchable supercapacitors. *Journal of Power Sources*, 446: 227355.

49. Wu, M.S., Z. Bin Zheng, Y.S. Lai and J.J. Jow, (2015). Nickel cobaltite nanograss grown around porous carbon nanotube-wrapped stainless steel wire mesh as a flexible electrode for high-performance super-capacitor application. *Electrochimica Acta*, 182: 31–38.

50. Wang, Y., X. Wang, X. Li, X. Li, Y. Liu, Y. Bai, H. Xiao and G. Yuan, (2021). A high-performance, tailorable, wearable, and foldable solid-state supercapacitor enabled by arranging pseudocapacitive groups and mxene flakes on textile electrode surface. *Advanced Functional Materials*, 31(7): 2008185.

51. Matsushima, J.T., A.C. Rodrigues, J.S. Marcuzzo, A. Cuña and M.R. Baldan, (2020). 3D-interconnected framework binary composite based on polypyrrole/textile polyacrylonitrile-derived activated carbon fiber felt as supercapacitor electrode. *Journal of Materials Science: Materials in Electronics*, 31(13): 10225–10233.

52. Liu, Q., J. Qiu, C. Yang, L. Zang, G. Zhang and E. Sakai, (2020). High-performance textile electrode enhanced by surface modifications of fiberglass cloth with polypyrrole tentacles for flexible supercapacitors. *International Journal of Energy Research*, 44(11): 9166–9176.

53. Stempien, Z., M. Khalid, M. Kozicki, M. Kozanecki, H. Varela, P. Filipczak, R. Pawlak, E. Korzeniewska and E. Sąsiadek, (2019). In-situ deposition of reduced graphene oxide layers on textile surfaces by the reactive inkjet printing technique and their use in supercapacitor applications. *Synthetic Metals*, 256: 116144.

54. Wang, Q., Y. Li, J. Zhao, C. Zhang, X. Luo, J. Shao, M. Zhong, Z. Ye, P. Feng, X. Liu, K. Li and W. Zhao, (2021). A universal strategy for ultra-flexible inorganic all-solid-state supercapacitors. *Journal of Alloys and Compounds*, 852: 156613.

55. Zhang, L., S. Ji, R. Wang, D.J.L. Brett and H. Wang, (2019). Flexible electrode with composite structure for large-scale production. *Journal of Alloys and Compounds*, 810: 151871.

56. Karami, Z., M. Youssefi, K. Raeissi and M. Zhiani, (2021). An efficient textile-based electrode utilizing silver nanoparticles/reduced graphene oxide/cotton fabric composite for high-performance wearable supercapacitors. *Electrochimica Acta*, 368: 137647.

57. Park, H., J.W. Kim, S.Y. Hong, G. Lee, H. Lee, C. Song, K. Keum, Y.R. Jeong, S.W. Jin, D.S. Kim and J.S. Ha, (2019). Dynamically stretchable supercapacitor for powering an integrated biosensor in an all-in-one textile system. *ACS Nano*, 13(9): 10469–10480.

29 Current Development and Challenges in Textile-Based Flexible Supercapacitors

Sunny R. Gurav and Rajendra G. Sonkawade
Radiation and Materials Research Laboratory, Department of Physics,
Shivaji University, Kolhapur, Maharashtra, India

Maqsood R. Waikar
Radiation and Materials Research Laboratory, Department of Physics,
Shivaji University, Kolhapur, Maharashtra, India

Department of Engineering Physics, Padmabhooshan Vasantraodada Patil
Institute of Technology (PVPIT), Sangli (Budhgaon), Maharashtra, India

Akash S. Rasal
Department of Chemical Engineering, National Taiwan University of Science
and Technology, Taipei, Taiwan, Republic of China

Rakesh K. Sonker
Department of Physics, Acharya Narendra Dev College, University of Delhi,
Delhi, India

CONTENTS

29.1 Introduction.. 539
29.2 Classification and properties of textile fabrics... 540
    29.2.1 Classification of textile fabrics 540
        29.2.1.1 Man-made fibers.. 541
        29.2.1.2 Natural fibers 541
    29.2.2 Properties of textile fibers for supercapacitor applications............................. 543
29.3 Textile based flexible supercapacitors (TFSCs)... 544
29.4 Design strategies for TFSCs application 546
29.5 Conclusion and future perspective.. 548
References... 549

29.1 INTRODUCTION

As the world economy continues to grow rapidly, we are faced with an increasingly more urgent ecological emergency brought on by the utilization of nonsustainable energy sources, such as oil and coal. It is important to look into clean energy sources like wind and sun-oriented energy, but the erratic nature of these resources presents a huge demand for efficient and reliable energy-storage devices (ESD) like lithium-ion batteries (LiBs) and supercapacitor (SCs) [1]. In contrast with LiBs, the lifecycle of SCs, particularly in a developing global market, is notable for its long duration, high power density (PD), and low energy density (ED); therefore, SCs are expected to be sought after for a role in future developing markets [2]. There is a wide variety of commercial SCs, each of which must undergo more than one million cycles of testing. Worth noting is that in correlation, LiBs endure around 20% loss after 500-2000 cycle life, contingent on the specific chemistry utilized. Talking about the ED, LiBs have multiple times longer ED than SCs. Excess charging-discharging rate (CD) and a hundred times more current density than LiBs are provided by SCs, with no loss of functionality while giving a hundred times more charging current and lessening inconvenience to the device [3].

As a result of their numerous advantageous characteristics, like higher PD, CD, and cyclic stability, SCs are proven their candidature in ESD. These features make SCs effective in crossover engine vehicles, and they are additionally ideal for different handy electronic devices. There are three main types of SCs. The first type is the electric double layer capacitor (EDLCs), which contain carbon-based electrodes and their subsidiary, the pseudocapacitors (PCs) use metal-oxide or polymer-based electrodes. And third one hybrid capacitors, which include both EDLCs and PCs electrodes [4]. However, the majority of the ESD equipment that is currently in existence is still excessively heavy and bulky. To accept the idea of an elastic ESD with improved ED and PD requires more focus towards device engineering of a flexible ESD with high ED and PD. The amount of weight a device must support is becoming less as future electronic devices become more slender, lighter, and cheaper, requiring more energy to support the same weight. Energizing work has been done over the last several years to produce SCs with the characteristics of a high-performance flexible ESD like flexible supercapacitors (FSCs) [5].

An illustration of the different methodologies used for FSCs is shown; these include fiber-shaped coaxial SCs, carbon nanotubes (CNTs)-based smart textile, wire-shaped, and profoundly stretchable fiber-shaped SCs. While it is true that SCs performance is dependent on their high surface area, it is vital to consider additional characteristics that influence their performance [6]. There are three different types of FSCs, one of which is called a paper-like, fiber-like, and porous three-dimensional structure device [7]. This point of view, in essence, looks to indicate that fibers and fabrics, being considered as expected supercapacitor elements, are both currently under investigation and being utilized to meet their expectations. They can be used to create textile-based supercapacitors (TSCs) while retaining the cost-effectiveness, lightness, and practicality of supercapacitors. Because it has a small volume and inherent flexibility, it can be easily modified into different structures and shapes, making it an excellent base for advanced wearable electronic gadgets and devices that can be produced through well-established textile-based technology [8]. This feature article will explain the three classification schemes for textile fabrics, along with TSCs's design strategy and how textile-based flexible materials can be used for SCs applications. Additionally, this research delivers a new SCs platform that can help advance the performance of portable and wearable electronic appliances in the next generation.

29.2   CLASSIFICATION AND PROPERTIES OF TEXTILE FABRICS

29.2.1   CLASSIFICATION OF TEXTILE FABRICS

Fabric is a flexible material made from interlocking networks of threads or yarns that are created by first twisting, and then lengthwise and repeatedly, coarse fibers into long and curving pieces. The yarns that are being blended are identified by using different abbreviations for weaving, tatting, braiding, and braiding; see in Figure 29.1, as described below [9].

Fibers are divided into two groups by their chemical beginning, such as man-made fibers and natural fibers. Man-made fibers are also referred to as synthetic or manufactured fibers. The classification of textile fibers is as shown in Figure 29.2. This classification is given by the TFPIA (Textile Fiber Products Identification Act), which is discussed in detail [10].

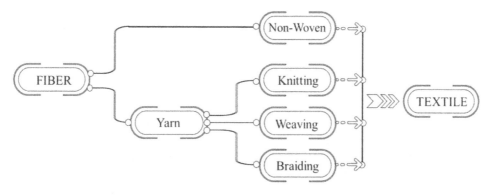

FIGURE 29.1 Various methods of textile formation from fiber.

29.2.1.1 Man-made fibers

A man-made fiber is derived from chemical compounds like synthetic fiber and regenerated fiber. Regenerated fibers are derived from inorganic materials, such as carbon fiber, ceramic fiber, and glass fiber. To supplement the foregoing information, regenerated fibers are composed of various minerals, including graphite, silica, zirconia, alumina, aluminum silicate, boron, boron nitride, boron carbide, silicon carbide, and silicon boride. Concerning regenerated fiber textiles, they are used in composite materials as reinforcements. While traditional fibers are less heat resistant, less rigid, and have a lower melting point; regenerated fibers are more heat resistant, harder, and stronger [11].

In contrast to natural fiber, synthetic fiber is also known as organic fiber, which is then further categorized into synthetic polymer and organic polymer. These polymers are a type of compound, which are described as long, chainlike polymers made up of an extraordinary number of tiny, subatomic weight atoms. Man-made fibers that are commonly created using polymers, such as plastics, surface coatings, rubbers, and glues, have numerous polymers that are the same or similar to mixtures that make up these other things. Polymer fibers have unique properties like electrical conductivity, electro-fluorescence, ultra-high-strength, high thermal stability, flame retardancy, and higher chemical resistance. Polymers can also derive polyethylene terephthalate, cellulose, and polycaprolactam have familiar materials under the business trademarks dacron, rayon, and nylon, respectively, which are fabricated into a variety of nanofiber goods, including cellophane envelope windows, transparent plastic soda bottles, and nano-crystalline fabrics. Most of the commodity fibers (polypropylenes, nylons, and polyesters) have the necessary properties for the application of upholstery and apparel. These materials are often regarded as an important attribute because of their connectivity, strength, resistance to warmth and humidity, and ability to sustain an increase in pressure [12,13]. Because of the diverse chemical structure of polymer fibers, their applications shift broadly, as indicated by the temperature and synthetic conditions.

29.2.1.2 Natural fibers

Natural fibers are chiefly delegated animal fibers, mineral fibers, and vegetable fibers, depending upon the source.

a. Animal fibers

Animal fiber commonly includes fibers made from keratin, such as sheep wool and fiber varieties with a similar purpose, like angora, alpaca, camel, cashmere, and mohair. There are numerous other keratin-based animal fibers in the 'fur' type fibers, and even human hair. Non-keratin-based animal fibers, including cultivated silk and wild silk (Tussah). Creature strands incorporating keratin-based filaments like sheep fleece and filaments of comparable use like angora, alpaca, camel, cashmere, and mohair, all

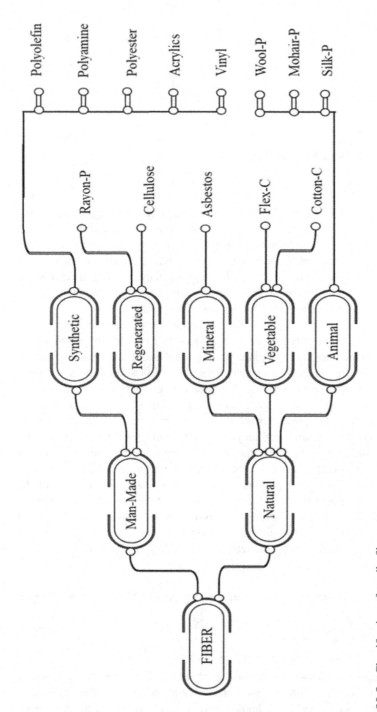

FIGURE 29.2 Classification of textile fibers.

have keratin at their core. Additionally, some other types of keratin-based strands incorporate 'hide' type filaments, and surprisingly, human hair. Non-keratin-based creature filaments incorporate developed silk and wild silk (Tussah) [14].

b. Mineral-based fibers

Mineral-based fibers are classified into basalt, asbestos, and Silexil fibers. Basalt fibers are stable, inert, eco-friendly, nontoxic, and nonreactive reinforcement materials. The composition modification takes into account a wide range of mechanical properties to be revealed, and the higher modulus of elasticity further assists in making this technology suitable for use in composite materials [15].

c. Vegetable fibers

Vegetable fibers, like those found in vegetable matter, are considered nonwood fibers as well (soft and hard) [16]. Five essential sorts of nonwood fibers are given below,

- Grass/reed: miscanthus, switchgrass.
- Straw: corn, wheat, rice, and soy.
- Bast: flax, hemp, ramie, jute, and kenaf.
- Leaf: sisal, pineapple, and abaca.
- Fruit/Seed: coconut, kapok, and coir.

Natural fibers have enormous benefits as compared to man-made fibers because of their plenitude, accessibility, and minimal expense. Natural fibers are introduced to make the composites lighter in contrast with synthetic fibers. The density of glass fiber (2.4 g/cm^3) is higher than the density of other natural fibers (which range from 1.2 to 1.6 g/cm^3), prompting the development of lightweight composites. Also, since interest in the industrial applications of natural fiber-based composites is on the rise, businesses are going to benefit from their bottom line [17].

29.2.2 PROPERTIES OF TEXTILE FIBERS FOR SUPERCAPACITOR APPLICATIONS

The textile has discovered expansive applications in health checking, superior sportswear, wearable displays, and another class of portable gadgets. Due to their essential and integral part in these applications, stretchable, wearable, and lightweight energy sources, such as LiBs and SCs, are in high demand. The perfect wearable energy sources would be constructed into breathable material arrangements that have stretchability (for example, mechanical flexibility) that is applied to the bent surface and can support the wearable's ability to perform during movement [18]. The main property of a fiber or yarn is its tenacity, which is commonly referred to as mechanical strength, and it is defined as a force over linear density, which is essential in breaking a fiber. When it comes to textiles, a single fiber strength of 5 grams per denier (g/d) is generally considered promising. However, some fibers have been found with strengths of 1 g/d that are suitable for a wide variety of textile applications [19]. It is possible to fabricate TSCs by layering a carbon-based material onto the fabric, creating conductive electrodes in the process. Textiles give a porous substrate that is suitable for being used as a scaffold structure for the fabrication of SC electrodes. Also, in case of flexibility, woven textiles that don't contain elastane fibers generally won't extend along the twist direction and will have only a slight amount of stretch in the weft direction. The implication here is that woven fabrics are mechanically stable and promising candidates for the development of stable electrodes [20].

There are many benefits to the great abundance of textile materials (i.e., natural or man-made textiles). Natural and man-made textiles are cheap and have simple changeability. Textile's simple changeability also allows for the facile tunability of structures by weaving, embroidery, and knitting technologies [21].

Woven and knitted textiles can be made using both staple and filament fibers, giving rise to the potential for yarns to be twisted and twisted yarns to be woven and knitted. It is called weaving when a numerous multitude of yarns that run perpendicular to each other intertwine to form a flat woven fabric. Weft knitting is a technique of using a continuous yarn to intertwine the yarn into loops. When compared to weaving, knitted fabrics may only have a single continuous strand of yarn running through them [22]. It is just as conceivable to knit a seamless garment as well as various 3D shapes. One of the attributes of the 3D porous textile-based electrode is its flexibility, and it is possible to place active materials as well as quick ion movements in the electrode [23]. To ensure accuracy, when applying the fibers to more complex processing, such as weaving, the diameter of the fibers should be about 1/100th of their length. The usefulness of synthetic fibers becomes apparent when compared to natural fibers, since synthetic fibers can be customized to the particular size and shape of the orifice opening through which a fibrous material is forced to pass while spinning. Besides, if a yarn is to be formed with the appropriate degree of strength and suppleness, then it is also important that the size and shape must be uniform, and thus coating materials and/or coating techniques have to be excellent when it comes to processing fabrics. When made into yarn, the materials used in combination with the woven fabrics must be able to stick to each other due to the presence of fibers running parallel to the coating. Long-filament fibers, because of their length, can be twisted together to provide added strength, making them a great option for ESD [24].

29.3   TEXTILE BASED FLEXIBLE SUPERCAPACITORs (TFSCs)

As with traditional textiles, these TSCs use textile yarns, fabrics, and fibers to construct their structures. They also have durable and soft mechanical properties that make them excellent SCs candidates. A fibrous structure and 3D textile configuration are given by TSCs in regard to their potential for achieving excellent electrochemical performance and stability as compared to planar thin films [25]. Jie et al. [26] deposited polypyrrole (PPy) and reduced graphene oxide (RGO) was used to prepare TSCs by a thermal reduction of graphene oxide (GO) and a chemical polymerization of pyrrole. The resultant PPy–RGO-fabric held great flexibility and conductivity of the material, with a conductivity of 1.2 S cm^{-1}. The prepared electrode showed high specific capacitance and ED as of 336 Fg^{-1} & 21.1 Whkg^{-1}, respectively, at current density 0.6 mA cm^{-2}. To better integrate and bind the conductive sheets of RGO to the conductive PPy in fabric-based SCs, RGO sheets act as conducting framework under PPy layer and controls the swelling and shrinking of PPy which accelerates the charge transfer leading better electrochemical properties.

The work of Yuanming et al. [27] paved the way for superior, foldable, and tailorable asymmetric solid-state SCs created by one-step scalable chemical oxidation on MXene ink painting of N-doped carbon fiber textile (NCFT) material. Carbon-like electrochemical characteristics (for the employed O/N-functionalized NCFT materials) are incorporated into MXene and MXene-like materials, which posses high pseudocapacitance. With precise control over oxidization time and the placement of MXene on NCFT (MNCFT) and ONCFT electrodes, the active layer of decorated MXene on NCFT (MNCFT) and ONCFT electrodes has a comparable skin structure, in a general sense, reducing the risk of active materials detaching due to mechanical distortion. In other words, the MNCFT/ONCFT device not only meet a high areal ED of 277.3 μWcm^{-2} and 90% retention rate after 30,000 cycles but also reached the goal of 1.6 V working potential window. A novel one-step hydrothermal process has been employed to form highly flexible fiber-shaped graphene hydrogels/MWCNT derived from natural cotton thread by Qianlong et al. [28]. The gel electrolyte PVA-H$_3$PO$_4$ was employed for device fabrication, and it proved to have an excellent capacitance of 97.73 mFcm^{-1} at 2 mVs^{-1} of scan rate with a retention rate of 95.51% over 8000th CD cycles. To fabricate these asymmetric SCs with NiCo$_2$O$_4$@NiCo$_2$O$_4$/ACT gel electrolyte, the group from Zan et al. [29] first made a PVA/KOH gel electrolyte and then fabricated an SCs device out of that material, which features a specific capacitance of 1929 Fg^{-1}, ED of 83.6 Whkg^{-1}, and a PD of 8.4 kW kg^{-1}, which all maintain excellent cyclic stability and provide mechanical robustness. The superior electrochemical performance is thought to be attributed to the novel core/shell nanostructure, which has excellent morphological stability, an extensive active surface area, and a lesser charge-transfer path.

In the study done by Babu et al. [30], PPy-Cotton and PPy-Viscose composites were created using a mild room temperature wet in situ chemical polymerization method. As part of this research, several types of carbohydrate polymers, such as linen (natural cellulosic fiber), cotton, synthetic polymer polyester, and modified cellulosic fiber-viscose rayon fabrics, are composited with PPy to examine the effect on electrochemical capacitance. To be precise, at 5 mVs^{-1}, the particular combination of PPy-Cotton and PPy-Viscose, composed of tensile-based PPy and Cotton, exibited the highest specific capacitance value of 268 and 244 Fg^{-1}, respectively. Energy-harvesting triboelectric nanogenerators (TENG) were built entirely from yarn to create the SCs (Y-ASC) developed by Mengmeng et al. [31]. These Ni/Cu-coated polyester yarns are being used as 1D current collectors in Y-ASC cells and electrodes in TENGs. After undergoing all the processing stages, the Y-ASC electrode showed high areal ED of approximately 78.1 µWcm^{-2}, high PD as of 14 mW cm^{-2}, high cyclic stability of 82.7% retention at 5000th cycles, and excellent flexibility. In a fabrication method developed by Liubing et al. [32], ACFF was used as the basic body material. A carbon nanotube (CNT) or graphene (GN) was then applied as a nanoscale carbon filler on ACFF. The process was simple where both the CNT and GN were simply soaked into the ACFF followed by freeze drying. The prepared CNT/ACFF and GN/ACFF composite textiles showed significantly enhanced areal capacitance, ED, and PD of 3350 mFcm^{-2}, 112 µWhcm^{-2}, and 4155 µWhcm^{-2}, respectively.

In a study led by Lingchang et al. [33], PPy-based nanotube arrays and carbon nano-onions (CNOs) were used to construct a novel wearable SC structure that consists of vertically oriented PPy nanotubes with CNOs growing on textiles by a one-step polymerization method. This hybrid nanostructure uses the CNOs to retain the under-layered PPy film during stretching, with the vertical PPy nanotubes giving a higher charge transfer rate obtained with a superior specific capacitance of 64 Fg^{-1} and a high 99% retention rate after the 5000th cycle. In the study done by Tae et al. [21], MnO$_2$ nanoparticles coated with PPy were applied to CNT textile for SCs. Here, the electrode was prepared by three steps, as shown in Figure 29.3. In the first step, an SWCNTs was coated on textile fiber through dip coating; this was also done in the 2nd step, with the MnO$_2$ nanoparticles deposited onto the CNTs fiber by electroplating technique; and in the third step, the PPy layer was coated on MnO$_2$ nanoparticles through pulse voltage electroplating. The prepared electrode prevents delamination of MnO$_2$ nanoparticles and enhanced 38% of electrochemical performance of 461 Fg^{-1}; also observed an increase in ED, PD, and cyclic stability as of 31.1 Whkg^{-1} and 22.1 Whkg^{-1}, and 93.8% retention rate after 10,000th cycles, respectively.

To fabricate a 3D conductive network on polyester fabrics via the "dip and dry" and electrophoretic deposition method, Li-Na et al. [34] fabricated a 3D conductive network made up of carbon nanotubes (CNTs) and graphene sheets (G) on PETC and created a larger charge transfer rate and a smaller ion-diffusion distance in the electrolyte. When the resultant composite fabric is applied to fabricate the TSC electrodes, it provides an excellent substrate for fabrication. A film of polyaniline (PANI) was applied to the CNT layer, after which a second coat of CNTs was applied to the PANI/CNTs/G/PETC electrode. This layer yielded 791 mFcm^{-2} of areal capacitance at a 1.5 mAcm^{-2} of current density with a 76% of retention rate after 3000 CD cycles. Poonam et al. [35] demonstrated a process that was repeated on numerous occasions that could be replicated by various metal oxide inks that were directly printed on bamboo fabric substrates. The textile-based MnO$_2$–NiCo$_2$O$_4$//rGO ASCs showed excellent areal capacitance, ED, and PD as of 2.12 Fcm^{-2}, 37.8 mWcm^{-3}, and 2,678.4 mWcm^{-3} respectively at 2 mAcm^{-2} of current density. Due to the impressive electrochemical properties of the supercapacitor, it can keep up its electrochemical performance when subjected to a wide range of mechanical deformation conditions, demonstrating its high flexibility and exceptional mechanical strength.

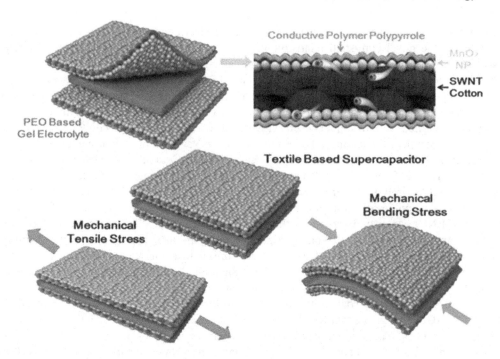

FIGURE 29.3 Schematic representation of device fabrication of PPy, MnO₂ coated TSCs, adapted with permission from ref. [21]. Copyright (2015) American Chemical Society.

29.4 DESIGN STRATEGIES FOR TFSCs application

In smart textile applications, device fabrication of FSCs is a critical issue. The choice of appropriate flexible substrate plays vital role in the device fabrication. Accordingly, various flexible materials such as film, fiber, paper, wipe, texture or other flexible materials have been accounted as a substrate for wearable SCs gadgets. The 2D fabrics are the ideal substrate for all-solid planar wearable SCs because of their exceptional flexibility and elasticity, and because they fit seamlessly into clothing. Despite materials that utilize diverse structural and physical components, as knitted, woven, and nonwoven, or those that utilize various fiber components, like polyester, cotton, and wool, displaying varied performance and producing different electrochemical properties, fabrics consistently showed their unique properties. In contrast to the other structures, the nonwoven fabric produced using only one fiber had a huge macroscopic fibrous structure, and this demonstrated a large amount of visible macroscopic fibrous structure, which made it possible to place more active material atop it [36]. Yongmin et al. [37] described how they developed FSCs, also known as intercalated or electroactive polymers, by forcefully implanting conductive and charge-storage materials into flexible frames that had been very clearly demanded. Due to a specialized ligand-mediated layer-by-layer assembly, the metallic cellulose paper-based SCs electrodes showed exceptional energy storage capability, wherein the contact resistance between adjoining metal or potentially metal oxide nanoparticles is limited by the paper's metallic cellulose surface.

In the study by Huihui et al. [38], the researchers found that traditional sandwich-structured SCs consisted of two substrates, one of which used cotton fabrics and the other nylon lycra, which were woven together to create a web for the interdigital electrode, as shown in Figure 29.4. A single textile-supported SCs shows a higher ultra-flexibility, electrochemical performance, and excellent mechanical stability. In this, the active material and current collector (CC) layers were deposited on PDMS film and silk fabrics via screen printing, respectively, and then a single textile-supported SCs was fabricated through PDMS and silk-based electrode. The designs that resulted were notable. In the experiment of

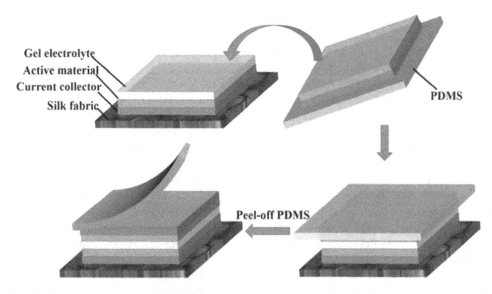

FIGURE 29.4 Schematic representation of the device fabrication of a silk-based wearable SCs, adapted with permission from ref. [38]. Copyright (2016) American Chemical Society.

Tae et al. [39], the authors evaluated the impacts of mechanical straining on CNT-coated textile and discovered a substantial increase in specific capacitance, ED, and PD after pre-straining of the textile. In situ resistance estimation of the textile capacitors found that there was an enhancement in electrochemical performance due to the lower resistance being seen when the capacitor was under tension, also known as tensile straining. Two distinctive CNTs textiles based on polyester and cotton with various mechanical properties are analyzed, where the polyester textile and the cotton textile showed the increment in specific capacitance from 53.6 Fg^{-1} to 85.7 Fg^{-1} and 122.1 Fg^{-1} to 142.0 Fg^{-1} after the textile is pre-strained to 30% and 30% permanent elongation before electrochemical testing, constituting the 37% and 22% increment. The increment in specific capacitance, ED, and PD by the MnO_2 nanoparticles deposited on the substrate of CNTs coated cotton textile via electroplating technique and, afterward, imposing a 30% permanent elongation. Results revealed that the mechanical pre-straining of the textile fibers played a critical part in the enhancement of the SCs performance.

Li et al. [40] studied, the increase in specific capacitance, flexibility, and durability through integrating a thin CC layer into the textile-based CNT SCs. The TSCs were fabricated by using a metal layer as CC on the fabrics of polyethylene terephthalate (PET) through the wet chemical method. The CNT-based SCs showed improvement in electrochemical performance by integrating the addition of metal layer current collector. The electrochemical performance of CNT/Cu/PET and CNT/Au/PET SCs showed a specific capacitance of 4.312×10^{-3} and 3.683×10^{-3} F cm^{-2}, respectively, which is about 60 times greater as compared to the metal collector layer CNT/PET SCs. The ED of a metal layer-based CNT SCs discovered to be nearly 50-fold expanded when contrasted with the CNT/ PET SCs and the PD of metal layer-based CNT SCs two orders of greater magnitude than the CNT/ PET SCs. High flexibility was additionally exhibited in these two metal layered SCs. For the estimation of durability, the CNT/Au/PET SCs showed a stable electrochemical performance at an 89% retention rate over 2500th CD cycles. The CNT/Cu/PET SCs showed relatively less stability with their specific capacitance, which dropped by 12% after 2500th CD cycles. Additionally, polyaniline (PANI) was deposited on the surface of CNT/Au/PET electrode through the electrodeposition technique to improve electrochemical performance. The PANI/CNT/Au/PET SCs showed inherent SCs performance with specific capacitance as of 0.103 Fcm^2 with 30 times higher capacitance as compared to the CNT/Au/ PET SC, and it also exhibited high stability and flexibility.

According to Yuksel et al. [41], the thick layer of MnO_2 on the textile has a high resistance, a loss in capacitance, and dead volume due to electrolyte ions only interacting with the top layer of MnO_2, which does not mean much to the thick layer of MnO_2. This issue demands conquering by implementing a thin layer of MnO_2 on the textile fibers, which should then be used in a composite additive-based conductor application. It has been found that a drastic reduction in capacitance and high resistance are observed with the MnO_2 deposition process when TSCs are applied. In addition, a second problem emerged with bulkier metal oxides. Carboxymethyl cellulose (CMC) or poly-tetrafluoroethylene (PTFE), which are binders, help the electrodes stay together and are useful for enhancing the mechanical stability of the electrodes, but their addition causes a drop in conductivity. Expanding on this, a binder-free conductive additive would prompt both dispersal and improvement in the mechanical properties of MnO_2. It is not possible to use MnO_2 for extra energy storage without causing some other issues, so by wrapping the MnO_2 layer with SWNT on textile fibers, the aforementioned problem was resolved and energy storage capacity is enhanced.

Hao et al. [42] reported TSCs are restricted by their low capacitances of commonly 10^{-3} to 10 F because of the small sizes of electrodes around a few square centimeters. Based on preliminary findings, it appears to be simple to solve this issue by increasing the electrode's diameter to increase the capacitance. However, even with a lot of effort to expand, the specific capacitances diminish to a large extent with the size of the expanding object. The main issue is that the electrodes of TSCs are largely made of carbon-based and polymers-based materials, which have poor electrical conductivities. When the resistance of a large part of an electric circuit increases, this often manifests as a lessened electrochemical performance, especially when using high CD currents. Another way to look at the issue is to consider that because of their high electrical and mechanical properties, metal/graphene hybrid materials are effective as flexible electrodes.

29.5 CONCLUSION AND FUTURE PERSPECTIVE

In this review article, we studied various textile-based materials for supercapacitive properties, such as specific capacitance, ED, PD, and cyclic stability of their materials. Here, we discussed the formation of textiles from fibers, the classification of fibers (i.e., natural and man-made fibers), and textile properties for supercapacitor application. The various design strategies for textile-based flexible supercapacitor are studied. From this review, it is clear that the textile-based substrates exhibit various profits like the large surface area as they posses highly porous structure, mechanical flexibility, and stretchability.

To expand the specific capacitance, flexibility, and durability of TSCs, the use of a thin CC layer on the CNTs is employed. The metallic cellulose paper-based SCs electrodes were superior energy storage performers, as they successfully constrained the contact resistance between the adjoining metal or potentially metal oxide nanoparticles by utilizing a ligand-mediated layer-by-layer assembly. This technology converts the insulating paper into the vastly porous metallic paper with extensive surface areas that can be utilized as CC. When working with the type of material used in the MnO_2 TSCs electrode, one can enhance the electrode's mechanical properties by utilizing the binder-free conductive additive. Through this, the reduction in CNT resistance that resulted from the strain on CNTs impregnated examined can be attributed to the reduction in resistance of TSCs and its enhancement of electrochemical properties such as specific capacitance, ED, and PD of TSCs electrodes by pre-straining of the textile.

Wearable energy gadgets for applications in flexible and stretchable hardware have pulled in expanding consideration and will address a standard direction in present-day gadgets. Flexible hardware like electronic paper, roll-up displays, wearable systems, and stretchable integrated circuits for individual multimedia all need dependable energy supplies with great adaptability flexibility. As a result, to meet this need, it is essential to promote and reinforce an up-to-date and qualified ESD by utilizing current, up-to-date technological advances, which can deliver high ED and PD. Considering the above-mentioned facts, still, the textile-based electrode can be additionally investigated and tried as a promising candidate for supercapacitors. The synergetic effect of EDLC composite with conducting polymers has not been accountedyet, for TSCs application [8]. Research in TSCs is an energizing field introducing reasonable answers for wearable gadgets in a variety of uses. While it's true that a lot of work has yet to be done, it's clear that significant refinement of the fabrication and material testing processes is still required before energy storage can be widely accessible.

REFERENCES

1. A.A. Shaikh, M.R. Waikar, R.G. Sonkawade, Effect of different concentrations of KMnO4 precursor on supercapacitive properties of MnO thin films, *J. Electron. Mater.* 48 (2019) 8116–8128. 10.1007/s11664-019-07648-y.

2. M.R. Waikar, A.A. Shaikh, R.G. Sonkawade, The supercapacitive performance of woollen-like structure of CuO thin films prepared by the chemical method, *Vacuum.* 161 (2019) 168–175. 10.1016/j.vacuum.2018.12.034.

3. M. Horn, B. Gupta, J. Macleod, J. Liu, N. Motta, ScienceDirect Graphene-based supercapacitor electrodes: Addressing challenges in mechanisms and materials, *Curr. Opin. Green Sustain. Chem.* 17 (2019) 42–48. 10.1016/j.cogsc.2019.03.004.

4. M.R. Waikar, A.S. Rasal, N.S. Shinde, S.D. Dhas, A. V. Moholkar, M.D. Shirsat, S.K. Chakarvarti, R.G. Sonkawade, Electrochemical performance of Polyaniline based symmetrical energy storage device, *Mater. Sci. Semicond. Process.* 120 (2020) 105291. 10.1016/j.mssp.2020.105291.

5. G. Shen, S.H. Yu, Y. Xia, X.W. Lou, Themed issue on flexible energy storage and conversion, *J. Mater. Chem. A.* 2 (2014) 10710–10711. 10.1039/c4ta90093f.

6. I. Shown, A. Ganguly, L. Chen, K. Chen, Conducting polymer-based flexible supercapacitor, *Energy Sci. Engineer.* (2014) 2–26. 10.1002/ese3.50.

7. R. Moreno, A. Pinheiro, L. Mário, C. Albuquerque, D.O. Helinando, P. De Oliveira, Wearable supercapacitors based on graphene nanoplatelets/carbon nanotubes / polypyrrole composites on cotton yarns electrodes, *SN Appl. Sci.* (2019). 10.1007/s42452-019-0343-5.

8. A.S. Ghouri, R. Aslam, M.S. Siddiqui, Recent progress in textile-based flexible supercapacitor, 7 (2020) 1–7. 10.3389/fmats.2020.00058.

9. https://en.wikipedia.org/wiki/Textile

10. http://gpktt.weebly.com/classification-of-textile-fibers.html

11. W. Liu, J. Ma, X. Yao, R. Fang, *Inorganic fibers for biomedical engineering applications*, Elsevier Inc., 2019. 10.1016/B978-0-12-818431-8.00001-5.

12. J. Preston, S.A. History, Man-made fibre Chemical composition and molecular structure Linear, branched, and network polymers, (n.d.).

13. G. Bhat, V. Kandagor, *Synthetic polymer fibers and their processing requirements*, Woodhead Publishing Limited, 2014. 10.1533/9780857099174.1.3.

14. C.T. Fibers, *Fibers: Overview*, 2nd ed., Elsevier Ltd., 2013. 10.1016/B978-0-12-382165-2.00088-X.

15. T. Rousakis, *9 - Natural fibre rebar cementitious composites*, Elsevier Ltd, 2017. 10.1016/B978-0-08-100411-1.00009-1.

16. J.K. Pandey, V. Nagarajan, A.K. Mohanty, M. Misra, Commercial potential and competitiveness of natural fiber composites, *Fourteenth*, Elsevier Ltd., 2015. 10.1016/B978-1-78242-373-7.00001-9.

17. Y. Gowda, T. Girijappa, S.M. Rangappa, Natural fibers as sustainable and renewable resource for development of eco-friendly composites: A comprehensive review, 6 (2019) 1–14. 10.3389/fmats.2019.00226.

18. B. Yue, C. Wang, X. Ding, G.G. Wallace, Electrochimica Acta Polypyrrole coated nylon lycra fabric as stretchable electrode for supercapacitor applications, *Electrochim. Acta.* 68 (2012) 18–24. 10.1016/j.electacta.2012.01.109.

19. U. Gulzar, F. De Angelis, R.P. Zaccaria, C. Capiglia, Recent advances in textile based energy storage devices *World J. Text. Eng. Technol.* 2 (2016) 6–15.

20. S. Yong, J. Shi, S. Beeby, Wearable textile power module based on flexible ferroelectret and supercapacitor, *Energy Tech.* 1800938 (2019) 1–8. 10.1002/ente.201800938.

21. T.G. Yun, B. Hwang, D. Kim, S. Hyun, S. Min, J. Han, Polypyrrole, MnO2 coated textile based flexible-stretchable supercapacitor with high electrochemical and mechanical reliability, (2015). 10.1021/acsami.5b01745.

22. K. Jost, Y. Gogotsi, Materials Chemistry A, (2014) 10776–10787. 10.1039/c4ta00203b.
23. M. Barakzehi, M. Montazer, F. Sharif, T. Norby, Electrochimica Acta A textile-based wearable super-capacitor using reduced graphene oxide/polypyrrole composite *Electrochim. Acta.* 305 (2019) 187–196. 10.1016/j.electacta.2019.03.058.
24. J.M. Chem, U. Gulzar, S. Goriparti, E. Miele, T. Li, G. Maidecchi, A. Toma, F. De Angelis, R. Proietti, Next-generation textiles: From embedded supercapacitors to lithium ion batteries, *J. Mater. Chem. A Mater. Energy Sustain.* 4 (2016) 16771–16800. 10.1039/C6TA06437J.
25. Q. Huang, D. Wang, H. Hu, J. Shang, J. Chang, C. Xie, Y. Yang, X. Lepró, R.H. Baughman, Z. Zheng, Additive functionalization and embroidery for manufacturing wearable and washable textile supercapacitors, 1910541 (2020) 1–12. 10.1002/adfm.201910541.
26. J. Xu, D. Wang, Y. Yuan, W. Wei, L. Duan, L. Wang, H. Bao, W. Xu, Polypyrrole/reduced graphene oxide coated fabric electrodes for supercapacitor application, *Org. Electron.* (2015) 1–7. 10.1016/j.orgel.2015.05.037.
27. Y. Wang, X. Wang, X. Li, X. Li, Y. Liu, Y. Bai, H. Xiao, G. Yuan, A High-performance, tailorable, wearable, and foldable solid-state supercapacitor enabled by arranging pseudocapacitive groups and mxene flakes on textile electrode surface, 2008185 (2020) 1–12. 10.1002/adfm.202008185.
28. Q. Zhou, C. Jia, X. Ye, Z. Tang, Z. Wan, A knittable fiber-shaped supercapacitor based on natural cotton thread for wearable electronics, *J. Power Sources.* 327 (2016) 365–373. 10.1016/j.jpowsour.2016.07.048.
29. Z. Gao, N. Song, Y. Zhang, X. Li, Cotton textile enabled, all-solid-state flexible supercapacitors, *RSC Adv.* 5 (2015) 15438–15447. 10.1039/c5ra00028a.
30. K.F. Babu, S.P.S. Subramanian, M.A. Kulandainathan, Functionalisation of fabrics with conducting polymer for tuning capacitance and fabrication of supercapacitor, *Carbohydr. Polym.* 94 (2013) 487–495. 10.1016/j.carbpol.2013.01.021.
31. M. Liu, Z. Cong, X. Pu, W. Guo, T. Liu, M. Li, Y. Zhang, W. Hu, Z.L. Wang, High-energy asymmetric supercapacitor yarns for self-charging power textiles, 1806298 (2019) 1–12. 10.1002/adfm.201806298.
32. A. Manuscript, Review of recent progress in chemical stability of perovskite solar cells, *J. Mater. Chem. A* 3 (2015). 10.1039/C4TA06494A.
33. L. Wang, C. Zhang, X. Jiao, Z. Yuan, Polypyrrole-based hybrid nanostructures grown on textile for wearable supercapacitors, *Nano Res.* 12 (2019).
34. L. Jin, F. Shao, C. Jin, J. Zhang, P. Liu, M. Guo, S. Bian, High-performance textile supercapacitor electrode materials enhanced with three-dimensional carbon nanotubes/graphene conductive network and in situ polymerized polyaniline, *Electrochim. Acta.* (2017). 10.1016/j.electacta.2017.08.035.
35. P. Sundriyal, S. Bhattacharya, Textile-based supercapacitors for flexible and wearable electronic applications, *Sci. Rep.* (2020) 1–15. 10.1038/s41598-020-70182-z.
36. Z. Li, M. Tian, X. Sun, H. Zhao, S. Zhu, Flexible all-solid planar fibrous cellulose nonwoven fabric-based supercapacitor via capillarity-assisted graphene/MnO_2 assembly, *J. Alloys Compd.* 782 (2019) 986–994. 10.1016/j.jallcom.2018.12.254.
37. Y. Ko, M. Kwon, W.K. Bae, B. Lee, S.W. Lee, J. Cho, Metal-like cellulose papers, *Nat. Commun.* 2017 (n.d.) 1–10. 10.1038/s41467-017-00550-3.
38. H. Zhang, Y. Qiao, Z. Lu, Fully-printed ultra-flexible supercapacitor supported by a single-textile substrate, *ACS Appl. Mater. Interfaces* (2016). 10.1021/acsami.6b11172.
39. T. Gwang, M. Oh, L. Hu, S. Hyun, Enhancement of electrochemical performance of textile based super-capacitor using mechanical pre-straining, *J. Power Sources.* 244 (2013) 783–791. 10.1016/j.jpowsour.2013.02.087.
40. V.A. Online, W.C. Li, C.L. Mak, C.W. Kan, C.Y. Hui, RSC Advances Enhancing the capacitive performance of a textile-based CNT supercapacitor, (2014) 64890–64900. 10.1039/c4ra10450a.
41. R. Yuksel, H.E. Unalan, Textile supercapacitors-based on MnO_2/SWNT/conducting polymer ternary composites, (2015) 2042–2052. 10.1002/er.
42. H. Sun, S. Xie, Y. Li, Y. Jiang, X. Sun, B. Wang, Large-area supercapacitor textiles with novel hierarchical conducting structures, *Adv. Mater.* 2016). 10.1002/adma.201602987.

30 Flexible Supercapacitors Based on Nanocomposites

Fang Cheng, Xiaoping Yang, and Wen Lu
College of Chemical Science and Engineering, Yunnan University,
Kunming, China

Institute of Energy Storage Technologies, Yunnan University,
Kunming, China

Liming Dai
School of Chemical Engineering, University of New South Wales,
Sydney, NSW, Australia

CONTENTS

30.1 Introduction 551
30.2 Carbon nanomaterial-incorporated nanocomposites for FSCs 553
    30.2.1 Carbon nanomaterials 553
    30.2.2 Carbon-metal oxide nanocomposites 553
    30.2.3 Carbon-conducting polymer nanocomposites 554
    30.2.4 Carbon-mxene nanocomposites 558
30.3 Device configurations of nanocomposite-based FSCs 560
    30.3.1 One-dimensional fiber-shaped FSCs 560
    30.3.2 Two-dimensional film-shaped FSCs 562
    30.3.3 Three-dimensional structural FSCs 564
30.4 Practical applications of nanocomposite-based FSCs 566
    30.4.1 FSCs for wearable electronic devices 566
    30.4.2 FSCs for flexible electronic devices 568
30.5 Summary and perspectives 568
Acknowledgments 570
References 570

30.1 INTRODUCTION

Over the past decade, a wide range of flexible and wearable electronic devices, such as flexible displays, smartwatches, health monitoring and management, artificial electronic skins, and wearable sensors, have been deeply penetrating our daily lives and will continuously change our lifestyles [1,2]. The rapid development of these emerging electronic devices has attracted widespread attention to the research and development of flexible energy storage devices, including supercapacitors (SCs) and batteries [3–7]. Specifically, having the advantages of high power density, wide operating temperature range, long lifetime, and excellent safety, supercapacitors have been researched for use in flexible fashion, resulting in flexible supercapacitors (FSCs), in the area of this emerging technology.

Ideally, a good FSC would have not only high electrochemical performances (including high energy/power densities and long cycle life) but also excellent mechanical deformabilities (such as

DOI: 10.1201/9781003186755-30

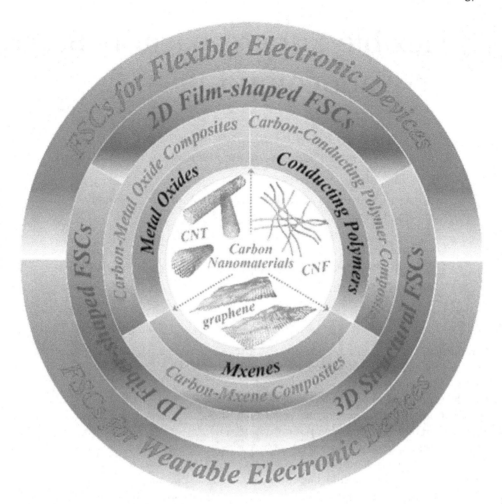

FIGURE 30.1   Schematic outline of major contents discussed in the present chapter.

flexibility, stretchability, bendability, foldability, and twistability), and to which electrode materials, device configurations, and practical applications are pivotal aspects that should be thoroughly considered [4–6,8]. This chapter systematically and comprehensively discusses the recent advances of FSCs in terms of these important aspects, as schematically shown in Figure 30.1. Specifically, the electrode materials, with the focus on nanocomposite-based of FSCs are first elucidated. In particular, owing to their excellent electrical, thermal, and mechanical properties, carbon nanomaterials, including carbon nanotubes (CNTs), carbon nanofibers (CNF), and graphene, have been demonstrated to be promising building components for compositing with pseudocapacitive materials to fabricate nanocomposites. In this regard, the chapter elucidates carbon nanomaterial-incorporated nanocomposites, mainly including carbon-metal oxides, carbon-conducting polymers, and carbon-mxenes, that have been studied for FSCs. This is followed by the discussion about device configurations of nanocomposite-based FSCs, including one-dimensional (1D) fiber-shaped, two-dimensional (2D) film-shaped, and three-dimensional (3D) structural FSCs. Finally, the efforts that have been made to explore the practical applications of nanocomposite-based FSCs are further summarized. Upon conclusion, the current challenges and future opportunities for the research and development of FSCs based on nanocomposites are also highlighted.

30.2 CARBON NANOMATERIAL-INCORPORATED NANOCOMPOSITES FOR FSCS

As for conventional SCs, depending on their difference in energy-storage mechanism [9], FSCs can be divided into two major categories: electrical double-layer capacitors (EDLCs) and pseudo-capacitors. In terms of electrode materials, carbon materials with large specific surface area (SSA) and high electrical conductivity (e.g., carbon nanomaterials [1,4,8]) are usually used for the former, while those possessing good pseudocapacitive properties (e.g., metal oxides [10], conducting polymers [11–13], and mxenes [14]) for the latter. Electrode materials play an important role in determining the overall performances of FSCs [5,7] and should possess both the superior electrochemical and mechanical properties simultaneously [4,5,15]. To this end, considerable efforts have been devoted to design and fabricate the aforementioned materials into flexible electrodes toward high-performance FSCs [8].

30.2.1 CARBON NANOMATERIALS

Carbon nanomaterials have various allotropes, such as 1D CNFs and CNTs [16] and 2D graphene [17] (Figure 30.1). Having excellent electrical, thermal, and mechanical properties, including large SSA, high electrical and thermal conductivity, outstanding flexibility, light in weight, high Young's modulus, and high tensile strength [1,4,18,19], these materials have shown great superiorities in building FSCs through either the direct use as electrodes or as building components for compositing with other electro-active materials to fabricate nanocomposite electrodes [1,4,8,18,20].

Specifically, substantial progress has been made in directly fabricating carbon nanomaterials into flexible electrodes for FSCs [18–20]. For instance, Zang et al. [21] developed a highly stretchable FSC by employing a crumpled graphene paper as the electrode. After being combined with a polyvinyl alcohol (PVA)/H_3PO_4 gel electrolyte, the as-fabricated FSC exhibited an unprecedented stretchability with linear and areal strains reaching 300% and 800%, respectively. Besides graphene, CNTs and CNFs have also been directly fabricated into flexible electrodes for FSCs [18,20,22]. Nevertheless, though marvelous achievements have been realized by the direct use of carbon nanomaterials as electrodes in FSCs, two major challenges remain: 1) the relative low capacitance characteristic of the EDLC mechanism of carbon nanomaterials lowers the energy density of their resultant FSCs [4,5]; 2) carbon nanomaterials usually suffer from serious agglomeration, resulting in their reduced available SSA and hence poor electrolyte infiltration and further decreased capacitance [18,19].

Alternatively, carbon nanomaterials can also be synthesized into various macroscopic morphologies to composite with other electro-active materials to fabricate novel nanocomposites for FSCs. In particular, by integrating the aforementioned advantages of carbon nanomaterials with the high capacitance of pseudocapacitive materials, the resultant nanocomposites exhibit unique morphological, structural, electrochemical, and mechanical characteristics, showing promise as a new class of electrode materials for FSCs [23–25]. To this end, various carbon nanomaterials have been utilized to composite with a range of pseudocapacitive materials, including metal oxides, conducting polymers, and mxenes, to fabricate the so-called carbon nanomaterial-incorporated nanocomposites for FSCs (Figure 30.1), which will be systematically discussed as follows.

30.2.2 CARBON-METAL OXIDE NANOCOMPOSITES

Owing to their high theoretical capacitance originating from the fast reversible redox reactions at the electrode/electrolyte interface [4,26], various metal oxides, such as RuO_2, MnO_2, Fe_3O_4, Fe_2O_3, Co_3O_4, and MoO_3, have drawn significant attention as pseudocapacitive electrode materials for FSCs [10]. Compared to carbon-based materials, however, the inherent brittleness and stiffness of these materials will inevitably cause serious degradation of their electrochemical performances, especially under mechanical deformations [7]. Furthermore, metal oxides suffer from comparatively low electrical conductivity, as well as poor stability and durability during the long-term charge/discharge cycling [26].

These unfavorable properties will inevitably degrade the performances of metal oxides when used for FSCs. To this end, the synergic integration of metal oxides with carbon nanomaterials can enhance not only the conductivity and capacitance but also the mechanical properties of the resultant nanocomposites. This way, numerous carbon-metal oxide nanocomposites, such as carbon-Fe_3O_4 [27], carbon-Fe_2O_3 [28], carbon-Co_3O_4 [29,30], carbon-MnO_2 [31–33], carbon-MoO_3 [34], and carbon-Co_2MnO_4 [35], have been prepared as electrodes for FSCs exhibiting superior performances over those of their pristine counterparts. On the other hand, the enhanced capacitance of carbon-metal oxide nanocomposites over that of pure carbon nanomaterials indicates a good strategy for processing carbon nanomaterials into realistically useful electrodes for FSCs.

During the electrode preparation of such nanocomposites, a design of flexible scaffolds is demonstrated to be simple yet efficient and has been frequently used to fabricate nanocomposite electrodes for FSCs. For example, with this design, Figure 30.2a illustrates the annealing assisted dip-coating method to fabricate a bind-free Fe_3O_4/graphene/carbon fibers (Fe_3O_4/G/CFs) ternary nanocomposite electrode [27]. In the as-synthesized 3D microstructure, ultrafine Fe_3O_4 nanoparticles are evenly and firmly dipcoated on CFs, while graphene layers wrap CFs and bridge adjacent CFs together, leading to a highly efficient conductive network. A symmetric FSC assembled from this nanocomposite electrode with a KOH/PVA gel electrolyte (Figure 30.2b) displayed a good rate capability (56% capacity retention at 10 Ag^{-1} vs 1 A g^{-1}), a considerable energy density (19.2 Wh kg^{-1}), and power density (8.6 kW kg^{-1}), remarkable cycling stability (no capacity decay after 4000 cycles at 1 A g^{-1}), and outstanding flexibility (no performance decay after being folded at various angles).

Among many metal oxides tested, MnO_2 has attracted more attention largely due to its high theoretical capacitance (~1370 F g^{-1}), low cost, environmental friendliness, and abundant resources [10]. Therefore, various approaches have been developed to fabricate carbon-MnO_2 nanocomposite electrodes. Through a stirring hydrothermal method, Lv et al. [36] synthesized an ultralong MnO_2 nanowire/ CNT (MNWs/CNT) composite, followed by sandwiching the as-prepared composite between two thin layers of nanocellulose fibers (NCFs) via vacuum filtration to fabricate a mechanically strengthened MNW-NCF electrode (Figure 30.2c). The nanocomposite electrode thus fabricated was used to assemble a honeycomb-like symmetric FSC with a PVA gel electrolyte, which delivered a high specific capacitance (227.2 mF cm^{-2}) and exhibited excellent stretchability (without capacitance decay after being stretched to 500%, Figure 30.2d) and superior cycling stability and deformation tolerance (with a 95% capacitance retention after 10000 cycles under concurrent bending, folding and twisting, Figure 30.2e). Moreover, the FSC retained nearly 98% of initial capacitance, even after 10000 stretch-release cycles under a 400% tensile strain (Figure 30.2f). In comparison with MnO_2 nanoflowers (MNFs)/CNT-NCF electrodes (Figures 30.2d, e), the robust interconnected fabric-like microstructure of the MNWs/CNT enabled by the intact contact between the ultralong MnO_2 nanowire and CNT should be responsible for these achievements.

Nevertheless, the capacitances of carbon-metal oxide nanocomposites reported so far are still far away from their theoretical values, and the mechanical properties of these materials still need to be further reinforced. Considering the inherently low electrical conductivity and brittleness/stiffness of metal oxides, this indicates a need for better-defined architectures with carbon nanomaterials to address these issues.

30.2.3 CARBON-CONDUCTING POLYMER NANOCOMPOSITES

Conductive polymers (CPs), mainly polyaniline (PANi), polypyrrole (PPy), polythiophene (PTh), and their derivatives, are another typical type of pseudocapacitive materials that have been extensively studied for FSCs, benefiting from their high theoretical capacitance (i.e., 750 Fg^{-1} for PANi and 620 Fg^{-1} for PPy), environmentally friendly nature, ease of synthesis, and especially intrinsically high flexibility and structural diversity (in contrast to metal oxides) [7,11–13]. Since the first highly swelled CP hydrogel electrode used for SCs was reported by Ghosh and Inganäs in 1999 [37], considerable

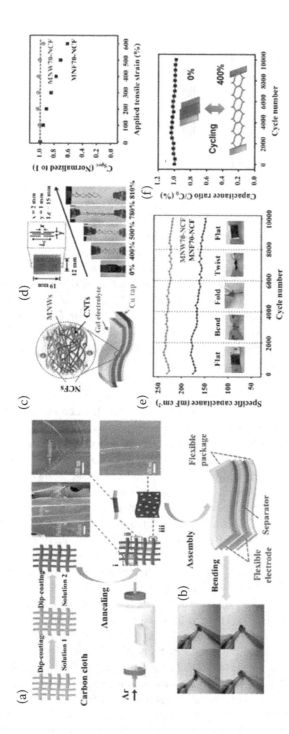

FIGURE 30.2 (a) Schematic illustration of an annealing-assisted dip-coating method to prepare a Fe$_3$O$_4$/G/CFs nanocomposite electrode with SEM images showing the structure of the resultant electrode. (b) Schematic diagram of a FSC fabricated from the Fe$_3$O$_4$/G/CFs nanocomposite electrode. Inset is the FSC being folded at different angles. Adapted with permission from ref. [27] Copyright 2020, American Chemical Society. (c) Schematic illustration of a MNW-NCF based honeycomb-like FSC (side view). (d) Digital images of FSCs and their normalized specific capacitances tested at 1.6 mA cm^{-2} under different strains. (e) Electrochemical performance comparison between the FSCs when cycled at 1.6 mAcm^{-2} under various mechanical deformations with their digital images shown in the inset. (f) Capacitance retention ratio of the MNW-NCF FSC at 1.6 mAcm^{-2} under the recycling tensile strain of 400%. Adapted with permission from ref. [36] Copyright 2018, Wiley-VCH.

efforts have been made to fabricate CPs into FSCs. Unfortunately, CPs based electrodes usually suffer from inferior rate performance determined by their relatively low conductivity [7] and poor cycling stability by their volume change upon doping-dedoping [11,13].

Compositing CPs with carbon nanomaterials has been demonstrated to be an effective strategy to address such drawbacks. Indeed, combing the high capacitance from CPs and the excellent conductivity and stability from carbon nanomaterials, the resultant carbon-CP nanocomposites are endowed with improved rate performance and cycling stability, as well as enhanced capacitance over their original constituent components of CPs and carbons, respectively. In this context, fruitful progress has been achieved to develop a large variety of methods, including in situ growth, vacuum filtration, self-assembly, electro-deposition and coating, and so on, to composite CPs with carbon nanomaterials into various nanocomposite electrodes, such as carbon-PANi [38–40], carbon-PPy [41,42], and carbon-PTh [43,44], for FSCs.

Furthermore, more recently, a novel approach has been proposed to introduce metal oxides into the carbon-CP system to further improve the properties (especially capacitance) for the resultant multi-component composite electrodes. For instance, Wang et al. [39] reported a new type of flexible 3D-activated carbon fiber cloth/CNT/PANi/MnO_2 (ACFC/CNT/PANi/MnO_2) composite textile electrodes through a layer-by-layer construction strategy: PANI, CNT, and MnO_2 were deposited on ACFC in turn through an electropolymerization process, "dipping and drying" method, and in situ chemical reaction, respectively (Figure 30.3a). The FSC assembled with the composite electrodes thus prepared and a PVA/H_2SO_4 gel electrolyte delivered an ultra-high capacitance of 4615 mF cm^{-2} (for a single electrode), along with an enhanced energy and power densities of 413 μWh cm^{-2} and 16120 μW cm^{-2}, respectively. In this unique composite design, the ACFC/CNT hybrid flexible frameworks served as a porous and electrically conductive 3D network for PANi and MnO_2 nanoparticles (providing them with the rapid transmission of electrons and electrolyte ions), together with the synergistic capacitance contributions from both PANi and MnO_2, to significantly enhance the capacitance and mechanical properties for the proposed multi-component composite electrodes.

Apart from PANi, PPy has also drawn tremendous attention for FSC applications because of its high pseudocapacitance, good stability, and ease of synthesis [11,12]. Chen et al. [41] fabricated a freestanding CNFs/PPy/rGO core-double-shell nanocomposite via a combined electrospinning-carbonization-electrochemical deposition-coating process. From this nanocomposite and a PVA/H_3PO_4 gel electrolyte, the as-fabricated FSCs displayed a high specific capacitance of 336.2 F g^{-1} at 2 mV s^{-1} and great flexibility when bent at different angles. However, this composite showed an inferior rate performance, implying its poor contact between the electro-active material and the conductive substrate, and hence indicating the need for a more efficient conductive network. In this sense, as shown in Figure 30.3b, Li et al. [42] firstly painted rGO layers onto $SnCl_2$ modified polyester fibers (M-PEF) to obtain a highly flexible and conductive rGO/M-PEF substrate, followed by polymerizing PPy in situ onto the substrate thus prepared to fabricate a PPy/RGO/M-PEF nanocomposite electrode. The highly conductive rGO/M-PEF scaffold provided PPy with an intact contact and hence a remarkable rate performance for the resultant composite. As a result, the PPy/RGO/M-PEF based FSCs (incorporating a PVA/H_2SO_4 gel electrolyte) exhibited a high areal capacitance (474 mF cm^{-2} at 1 mA cm^{-2}) and great rate capability (350 mF cm^{-2} at 50 mA cm^{-2}), along with an admirable cycling stability and excellent mechanical properties, which should be contributed by the synergistic effects from the high pseudocapacitance of PPy, as well as the high conductivity and flexibility of rGO and M-PEF.

Moreover, poly(3,4-ethylenedioxythiophene) (PEDOT), an important derivative of the PTh family, has been extensively studied for FSCs largely owing to its unique properties of high conductivity, ease of preparation, and well-defined electrochemical characteristics. Most recently, Zeng et al. [44] developed a crosslinking/polymerization strategy to fabricate soft FSCs in an interesting all-hydrogel design (Figure 30.3c). Specifically, both the polyacrylamide/sodium alginate/PEDOT/CNTs (PAM/SA/PEDOT/CNT) electrode and the PAM/SA/Na_2SO_4/potassium ferricyanide/potassium ferrocyanide (PAM/SA/Na_2SO_4/[$K_3Fe(CN)_6$/$K_4Fe(CN)_6$]) electrolyte were made in hydrogels. The all-hydrogel

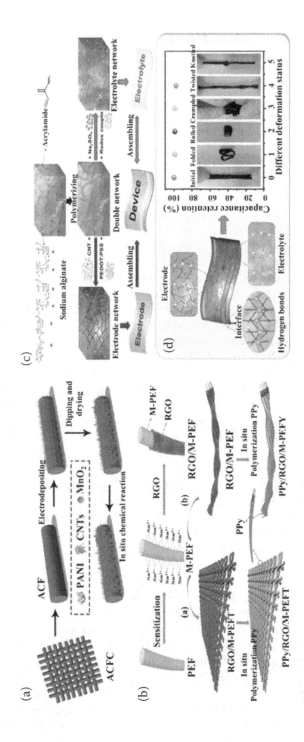

FIGURE 30.3 (a) Schematic illustration of the fabrication of an ACFC/CNT/PANI/MnO₂ nanocomposite electrode. [39] Copyright 2018, American Chemical Society. (b) Schematic illustration of the synthesis route of a) a textile electrode (PPy/RGO/M-PEFT) and b) a yarn electrode (PPy/RGO/M-PEFY) [42]. Copyright 2018, Wiley-VCH. (c) Graphical illustration of the composition and preparation process for an all-hydrogel soft FSC. (d) Schematic illustration of the stable network structure of the FSC and its capacitance retentions under different deformations [44]. Copyright 2020, Elsevier.

FSCs thus fabricated showed a high areal capacitance (232 mF cm^{-2} at 5 mV s^{-1}), high energy density (3.6 μWh cm^{-2}), and long cycle life (over 5000 cycles at 2 mA cm^{-2}). More importantly, benefiting from the 3D cross-linked dual-network structures of both the electrode and electrolyte hydrogels, the FSCs were intrinsically stretchable/compressible, able to maintain their stable energy output even under severe arbitrary deformations, such as folding, rolling, crumpling, twisting, and knotting (Figure 30.3d), thereby indicating their promising prospects for the applications for flexible and wearable electronics.

As can be seen, combining the advantages from CPs and carbon nanomaterials, the resultant nanocomposites have gained remarkable achievements for FSCs. However, it remains a great challenge to further improve the rate capability and cycling stability for these materials to make them practically useful for FSCs.

30.2.4 CARBON-MXENE NANOCOMPOSITES

Compared to metal oxides and CPs discussed above, 2D transition metal carbides and nitrides, known as mxenes, are relatively new members in the pseudocapacitive materials community. Since the discovery of Ti$_3$C$_2$ in 2011 [45], mxenes have received intensive attention as a new class of electrode materials for electrochemical energy storage technologies, including FSCs owing to their extraordinary physical and chemical properties, such as unique 2D layered structure, excellent mechanical strength, metallic conductivity, high surface area, hydrophilic nature, and ability to accommodate intercalants [14,46]. Furthermore, the high theoretical capacity of 615 C g^{-1} of mxenes [47] makes them especially attractive for FSC applications. Nevertheless, mxenes may encounter with a limited practically achievable capacitance due to their restacking of 2D nanosheets (resulting in reduced accessible SSA and hindered electrolyte ion diffusion), as well as an inferior rate capability to their anisotropic nature (resulting in poor interlayer and interparticle conductivity) [48,49].

To overcome these shortcomings and make mxenes useful for FSCs, numerous efforts have been conducted to composite mxenes with carbon nanomaterials, resulting in a large variety of carbon-mxene nanocomposites, including, for example, graphene-mxenes [50,51], CNT-mxenes [25,48,49], and CNF-mxenes [52]. For example, Li *et al.* [50] developed a vacuum-assisted filtration method to synthesize electrochemically exfoliated graphene/mxene (EG/Mxene) nanocomposite film electrodes from graphite and Ti$_3$AlC$_2$ for flexible all-solid-state supercapacitors (ASSSs) (Figure 30.4a). With a PVA/H$_3$PO$_4$ gel electrolyte, the resultant ASSSs deliver an outstanding volumetric capacitance up to 216 F cm^{-3} at 0.1 A cm^{-3} and a good rate performance of 110 F cm^{-3} at 2 A cm^{-3}. Significantly, these flexible ASSSs exhibited excellent cycling stability under severe bending conditions (Figure 30.4b) and sustainability under continuous bending cycles (capacity retaining at 85.2% after 2500 cycles at 1 A cm^{-3}). The unique structure of the EG/Mxene nanocomposite having both EG and mxene as the conductive spacer for each other to prevent the irreversible π-π stacking was demonstrated to be responsible for the distinguished performances achieved.

To further enhance the electrochemical and mechanical properties for carbon-mxene nanocomposites, the scaffold structure well-developed previously has been introduced into the carbon-mxene system. In this regard, Li et al. [48] synthesized flexible mxene/MWCNT/carbon cloth (mxene/MWCNT/CC) nanocomposite electrodes by firstly electrodepositing mxene sheets onto CC, followed by homogeneously growing MWCNT on mxene (Figure 30.4c). The 3D interconnected hierarchical structure of the scaffold established from the multiple carbon components improved the electrochemical and mechanical properties for the resultant composites. Consequently, the FSCs fabricated from the as-prepared nanocomposite electrodes and a PVA/H$_2$SO$_4$ gel electrolyte realized a high areal specific capacitance (994.79 mF cm^{-3} at 1 mA cm^{-2}) and outstanding cycling stability. More admirably, they possessed excellent mechanical deformability, showing a very small resistance increase (~5%), even after 2000 bending cycles at a bending radius of 5 mm. In a similar work, Chen et al. [52] utilized CNF and porous carbon to build the scaffold and from which to fabricate free-standing

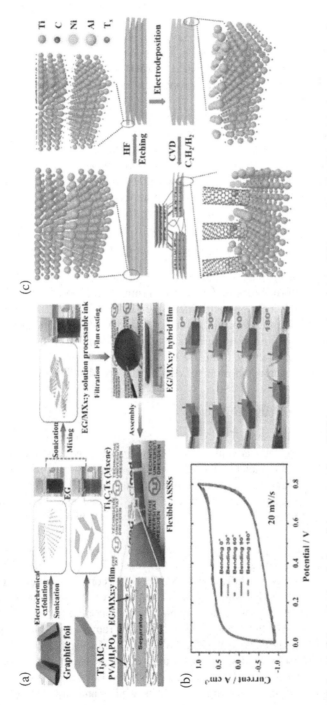

FIGURE 30.4 (a) Schematic description of the preparation of EG/mxene nanocomposites and the EG/mxene-based ASSSs. Digital photographs of a flexible and freestanding EG/MX1:3 film and its resultant ASSS. (b) CV curves and digital photographs of the ASSS bent at different angles between 30°~180° at 20 mV s⁻¹ [50]. Copyright 2017, Wiley-VCH. c. Schematic illustration for the preparation of a Mxene/MWCNT/CC nanocomposite electrode [48]. Copyright 2020, Wiley-VCH.

$Ti_3C_2Tx/CNF/$porous carbon ($Ti_3C_2Tx/CNF/PC$) nanocomposite electrodes via vacuum filtration. The resultant electrodes possessed a high conductivity up to $83.1\,S\,cm^{-1}$. Again, owing to the well-established 3D interconnected hierarchical structure of the nanocomposite, the FSCs assembled from the as-prepared nanocomposite electrodes (with a PVA/KOH gel electrolyte) displayed a high areal capacitance of $143\,mF\,cm^{-2}$ at $0.1\,mA\,cm^{-2}$, excellent rate performance (cyclic voltammetry (CV) curves keeping good rectangularity at $500\,mV\,s^{-1}$), and especially great mechanical deformabilities (CV curves showing little variation under various bending angles). These results indicate the significance of utilizing multiple carbon components in constructing a unique scaffold structure to enhance both the electrochemical and mechanical properties for carbon-mxene nanocomposites to reinforce their capability for FSC applications.

As discussed above, incorporating the unique properties of carbon nanomaterials with those of pseudocapacitive materials, the resultant carbon nanomaterial-incorporated nanocomposites, namely carbon-metal oxides, carbon-conducting polymers, and carbon-mxenes, possess enhanced electrochemical and mechanical properties over their original building components and hence better suitability for FSCs. Furthermore, through proper electrode and device design, these nanocomposites can be fabricated into FSCs with different configurations to meet the application demands for various wearable and flexible electronic devices (Figure 30.1), to be detailed as follows.

30.3   DEVICE CONFIGURATIONS OF NANOCOMPOSITE-BASED FSCS

Structural designing and engineering play a vital role in the determination of both electrochemical performances and mechanical deformability of FSCs [53]. From an appropriate modulation in device configuration, FSCs can be realized in three groups, including 1D fiber-shaped, 2D film-shaped, and 3D structural [53,54], leading to a wide range of opportunities to satisfy the application requirements for varied mechanical deformation circumstances, such as bending, folding, stretching, compressing, and twisting. The FSCs with the configurations from 1D to 3D along with their progress achieved with carbon nanomaterial-incorporated nanocomposites in recent years are elucidated below.

30.3.1   ONE-DIMENSIONAL FIBER-SHAPED FSCs

1D FSCs are characteristic of a fiber-shaped (or wire-like, cable-type) configuration and represent an important group of FSCs owing to their advantages of lightweight and good stretchability and knittability. Various types of carbon nanomaterial-incorporated nanocomposites, including carbon-metal oxides [55], carbon-conducting polymers [56], and carbon-mxenes [25,49], have been fabricated into 1D fiber-shaped FSCs. Based on their variations in electrode structure and fabrication process, 1D fiber-shaped FSCs can be assembled mainly in three configurations, namely parallel, coaxial, and twisted [3].

Specifically, a parallel 1D fiber-shaped FSC is constructed by simply arranging its two electrodes in parallel with each other inside a packaging component. In this regard, Zhang et al. [57] wound a $MnO_2@$PEDOT:PSS@oxidized CNT film ($MnO_2@$PEDOT:PSS@OCNTF) composite fiber as cathode in parallel with a MoS_2 nanosheets@CNTF ($MoS_2@$CNTF) composite fiber as anode, along with a thin layer of LiCl-PVA gel as electrolyte, onto a pre-stretched elastic fiber to fabricate this type of FSC (Figures 30.5a–c). The FSC thus fabricated exhibited a remarkable specific capacitance of $278.6\,mF\,cm^{-2}$ and a superior energy density of $125.37\,\mu Wh\,cm^{-2}$. Mechanically, it showed negligible change in galvanostatic charge/discharge (GCD) curves upon the increase in strain from 0% to 100% (Figure 30.5d) and maintained 92% of its initial capacitance after 3000 stretching cycles under a strain of 100% at $1\,mA\,cm^{-2}$ (Figure 30.5e), demonstrating its great flexibility and stretchability, respectively.

With their simple configuration, parallel 1D fiber-shaped FSCs can be easily scaled up for practical applications. However, the nonuniform electric field formed in the radial direction (across the diameters of the two parallel electrodes) implies a limited effective electrode area. Therefore, arranging the two

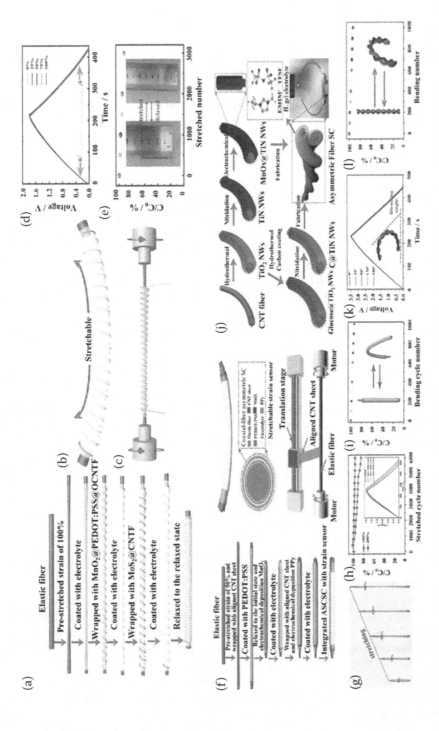

FIGURE 30.5 (a) Schematic illustration of the fabrication of a stretchable parallel 1D fiber-shaped FSC. (b) Structure of the FSC. (c) Wrapping of modified CNT fibers around a pre-stretched elastic fiber to fabricate the FSC. (d) GCD curves of the FSC under the strain from 0% to 100%. (e) Normalized capacitance of the FSC upon stretching cycling at a strain of 100% for 3000 cycles [57]. Copyright 2017, Elsevier. (f) Schematic illustration of the fabrication process of an all-in-one stretchable coaxial 1D ASCSC. (g) Photograph of the ASCSC before and after stretching by 50%, 100%, 150% and 200%. (h) Normalized capacitances of the ASCSC during the stretching cycling at a strain of 100% and 200%, respectively, for 6000 cycles. (i) Normalized capacitance of the ASCSC upon the bending cycling with a bending angle of 90° for 1000 cycles [58]. Copyright 2020, Elsevier. (j) Schematic illustration of the fabrication processes of an all-solid-state twisted 1D AFSC with an ionic liquid incorporated gel-polymer electrolyte. (k) GCD curves of the AFSC measured at 0.5 A cm^{-3} under different bending angles. (l) Normalized capacitance of the AFSC upon bending cycling with a bending angle of 90° for 1000 cycles [59]. Copyright 2019, Wiley-VCH.

electrodes in a more efficient configuration, such as with a coaxial or twisted structure, has been studied, as discussed below.

Unlike the above parallel design, the two electrodes in a coaxial 1D fiber-shaped FSC are arranged in a coaxial configuration, where an internal electrode is wrapped with an external electrode. In this way, the electric field formed between the two electrodes is uniform in both the axial and radial directions, maximizing the effective electrode area and lowering the internal resistance for the device. With this configuration, Pan et al. [58] fabricated a 1D fiber-shaped FSC by using MnO_2/PEDOT: polystyrenesulfonate/CNT (MnO_2/PEDOT:PSS/CNT) and PPy/CNT nanocomposite sheets as the internal (cathode) and external (anode) electrodes, respectively (Figure 30.5f). Benefiting from this unique coaxial architecture and the advanced electrode design, the as-prepared asymmetric stretchable coaxial-fiber supercapacitors (ASCSC) delivered a high stack volumetric energy density of 1.42 mWh cm^{-3} and showed a remarkable flexibility (Figure 32.5g) with high capacitance retention of 85.1% after 6000 stretching cycles under a large strain of 200% (Figure 30.5h) and 93.2% after 1000 bending cycles between 0° to 90° (Figure 30.5i).

Similarly, in a twisted 1D fiber-shaped FSC, the two electrodes are twisted together along with an electrolyte in between, forming a uniform electric field between the two electrodes. Using this configuration, Pan et al. [59] assembled an all-solid-state asymmetric fiber supercapacitor (AFSC) by twisting a MnOx@Titanium Nitride nanowires/CNT (MnOx@TiN NWs@CNT) fiber (as cathode) and a carbon@TiN NWs/CNT (C@TiN NWs@CNT) fiber (as anode) with an ionic liquid incorporated gel polymer electrolyte (Figure 30.5j). The AFSC thus assembled achieved an ultrahigh stack volumetric energy density of 61.2 mWh cm^{-3}, comparable to that of commercial lead-acid batteries (50~90 mWh cm^{-3}). Remarkably, the AFSC showed distinguished flexibility with no evident changes in GCD curves under different bending angles (from 0° to 180°) at 0.5 A cm^{-3} (Figure 30.5k) and maintained high capacitance retention up to 92.7% after 1000 blending cycles at 90° (Figure 30.5l).

As can be seen, based on their differences in electrode structure and fabrication process, 1D fiber-shaped FSCs have been realized in various configurations, showing their potential as power sources for flexible and wearable electronics. However, these FSCs face a common problem, in which their internal resistance would increase upon the increase in length of the fiber electrodes. Furthermore, they suffer from limited total capacity and, hence, the difficulty in necessary packaging (weaving) for practical applications [4]. FSCs with film-shaped electrodes would be able to solve these issues, to be discussed as follows.

30.3.2 TWO-DIMENSIONAL FILM-SHAPED FSCs

Macroscopically, all 2D FSCs have a film-shaped (or paper-like) appearance in common [53]. They have been fabricated from multifarious nanocomposite electrodes in a sandwich structure but with either a symmetric or asymmetric configuration [60–62]. Mechanically, they have been designed to possess various deformabilities, such as bendability, stretchability, and even self-healability, to meet the application demands for varied flexible and wearable electronics.

In a symmetric configuration, as shown in Figure 30.6a, Chen et al. [63] reported a highly robust, stretchable, and real-time omni-healable FSC by sandwiching two PPy-incorporated gold nanoparticle/CNT/poly(acrylamide) (GCP@PPy) hydrogel electrodes with a CNT-free GCP (GP) hydrogel electrolyte. The as-assembled FSC delivered an energy density of 123 μWh cm^{-2}, and very notably, showed a super high-stretching strain up to 800% (Figure 30.6b) without evident performance decay. Also due to the use of hydrogels as both the electrode and electrolyte, the FSC exhibited excellent bendability and significant real-time heal ability during the charge-discharge process, indicating its great potential for next-generation flexible and wearable electronics.

To enhance the output energy for FSCs, an asymmetric configuration has been studied. In such a hybrid design, a pseudocapacitive electrode and an EDLC electrode are used to construct the device,

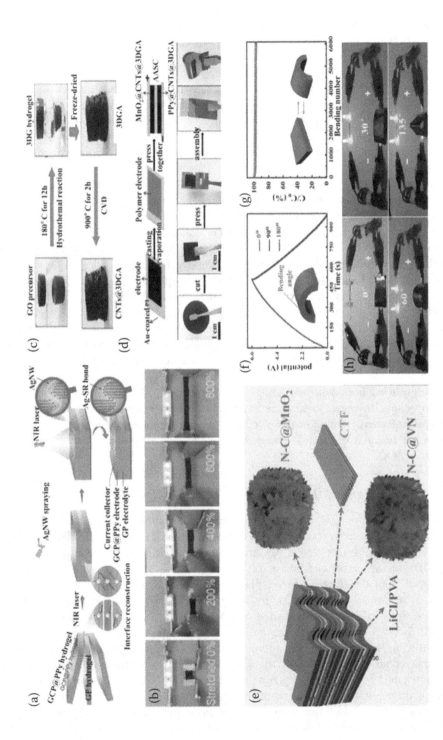

FIGURE 30.6 (a) Schematic illustration of the assembly of a stretchable and self-healable FSC. (b) Optical images show good electrical conductivity of the FSC under stretching at high strains [63]. Copyright 2019, Wiley-VCH. (c) Illustration of the growth procedure of a CNTs@3DGA hybrid. (d) Schematic diagram and photographs of the fabrication process of an AASC from a MnO₂@CNTs@3DGA cathode and a PPy@CNTs@3DGA anode [64]. Copyright 2017, Wiley-VCH. (e) Schematic of an asymmetric all-in-one internal tandem FSC. (f) GCD curves of the FSC measured at 4 mA cm⁻² under different bending angles. (g) Normalized capacitance of the FSC upon bending cycling with a bending angle of 90° for 6000 cycles. (h) Two blue-light LEDs powered by the FSC upon bending from 0 to 135° [65]. Copyright 2020, Elsevier.

resulting in an increased capacitance and an enlarged working voltage and thus improved energy density for the resultant FSC [6]. In this regard, Pan et al. [64] developed an all-solid-state asymmetric super-capacitor (AASC) by incorporating a MnO_2@CNTs@3D graphene aerogel (MnO_2@CNTs@3DGA) electrode as cathode and a PPy@CNTs@3DGA electrode as anode along with a Na_2SO_4/PVA gel as electrolyte (Figure 30.6c, d). In this work, a CNTs@3DGA hybrid prepared by a facile one-step chemical vapor deposition process was used as the 3D porous scaffold for the deposition of PPy to fabricate the anode and for the co-deposition of MnO_2 and PPy to fabricate the cathode, respectively. Due to this asymmetric configuration, the working voltage of the AASC thus fabricated was enlarged to 1.8 V, leading to an improved volumetric energy density up to 3.85 mWh cm^{-3}. Significantly, the flexible nature and excellent integrity of the hybrid electrodes enabled the AASC with perfectly overlapped CV curves, a high capacitance retention of 86.8% after bending at 180° for 500 cycles, and negligible capacitance decay upon bending to different angles, indicating the super flexibility and electrochemical stability of the proposed asymmetric AASC.

One more step further, by integrating multiple asymmetric cells into a tandem structure, the working voltage, and hence energy density, of the resultant FSCs can be significantly boosted. In this context, Zhou et al. [65] developed an all-solid-state internal flexible asymmetric supercapacitor by incorporating a CNT film-supported MnO_2@N-doped carbon skeleton (CNTF/MnO_2@N-C) cathode, a CNTF/N-C@ vanadium nitrogen (CTF/N-C@VN) anode, and a LiCl/PVA gel electrolyte in such a tandem structure, as shown in Figure 30.6e. Consequently, the as-assembled all-in-one FSC exhibited a large working voltage of 6.6 V (versus that of 2.2 V for a single cell), and thus, an exceptionally high energy density up to 118.2 mWh cm^{-3}. Mechanically, the FSC maintained its GCD curves almost unchanged upon the increase of bending angle from 0° to 180° (Figure 30.6f), retained its capacitance at 97.2% even after being bent for 6000 cycles (Figure 30.6g), and showed a stable energy output under various bending angles (Figure 30.6h). These results clearly demonstrate the effectiveness of the combination of the asymmetric configuration at the cell and the tandem structure at the device levels in boosting the energy output for FSCs.

So far, substantial progress has been achieved for 1D fiber-shaped and 2D film-shaped FSCs from a wide range of nanocomposites. However, to enhance their application capabilities, the active material loading, and hence, the overall energy output of these devices, needs to be increased, which has, in turn, triggered the efforts on the research of flexible energy storage devices with higher dimensions [54], resulting in 3D structural FSCs.

30.3.3 THREE-DIMENSIONAL STRUCTURAL FSCS

Compared to 1D and 2D configurations, 3D structural FSCs incorporate more active materials into the device and can omni-directionally accommodate mechanical deformations, making them more suitable toward practical applications. During fabrication, 3D structural FSCs can be readily constructed from low-dimensional (1D, 2D) devices as the building units.

In this regard, Lv et al. [66] developed a novel 3D stretchable FSC with a honeycomb lantern structure. Specifically, as shown in Figure 30.7a, by using a symmetric capacitor composed of a polypyrrole/black-phosphorous oxide electrodeposited on CNT film (PPy/BPO/CNT) nanocomposite electrode and a PVA/H_2SO_4 gel electrolyte as the building unit, the proposed 3D stretchable FSC was constructed in a structure mimicking a honeycomb lantern. As a result, the unique expandable architecture of the honeycomb lantern enabled the FSC with device-thickness-independent ion-transport path and stretchability that allowed the fabrication of FSCs into customizable device thickness for enhancing the energy output and with integrability for fitting with varied shapes of various wearables. Impressively, a 1.0 cm thick FSC was fabricated that displayed a greatly enhanced specific areal capacitance of 7.34 F cm^{-2}, which is about 60 times higher than that of the original 2D SC (120 mF cm^{-2}). Mechanically, the 3D FSC was able to be stretched up to 2400% (Figure 30.7b) and exhibited a device-thickness/shape-independent stretchability (Figure 30.7c).

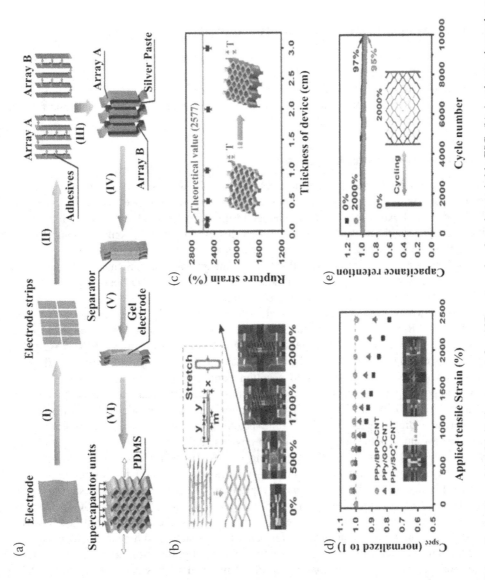

FIGURE 30.7 (a) Schematic drawing of the fabrication process for a 3D rectangular-shaped stretchable FSC. (b) A scheme shows the expandable honeycomb structure and the hexagonal unit cells before and after being stretched of the FSC. Digital images of the FSC under different strain tests. (c) Dependence of rupture strain on device thicknesses (T) of the FSC. (d) Normalized capacitances of FSCs with different PPy-based electrodes tested at 7.8 mA cm^{-2} under different strains. (e) Capacitance retention of the FSC based on PPy/BPO/CNT electrodes tested at 7.8 mA cm^{-2} during the stretching cycling at a strain of 2000% for 10000 cycles [66]. Copyright 2018, Wiley-VCH.

Moreover, it maintained its electrochemical performance under 2400% stretching (Figure 30.7d) and retained its capacitance at 95% even upon stretch-release cycling under 2000% strain for 10000 cycles (Figure 30.7e). These results indicate the superior properties of 3D structural FSCs over their low-dimensional counterparts for the practical applications for flexible and wearables electronics.

Overall, along with the extraordinary progresses in carbon nanomaterial-incorporated nanocomposites, FSCs in multiple dimensions from 1D to 3D have gained considerable achievements in recent years, facilitating their exploration for practical applications, as elaborated in detail below.

30.4 PRACTICAL APPLICATIONS OF NANOCOMPOSITE-BASED FSCS

Although still in their early stage, FSCs have shown great potential for various practical applications for wearable and flexible electronic devices [1,67]. Indeed, many application concepts and even prototypes have been demonstrated with FSCs on their avenues into the market. To this end, the progress that has been made to explore the practical applications of nanocomposite-based FSCs for wearable and flexible electronic devices will be summarized as follows.

30.4.1 FSCS FOR WEARABLE ELECTRONIC DEVICES

Smart wearable electronics have been playing an important role in our daily lives and have received extensive research interests in recent years. Therefore, it is of great significance to provide power for these wearable devices in a flexible manner. FSCs can well fulfill this demand because of their excellent safety, flexibility, and remarkable electrochemical performances. Among various configurations of FSCs, 1D fiber-shaped FSCs are promising flexible power sources for wearable electronics owing to their unique fiber-shaped structures. They can be either knitted/integrated with existing cloths/fabrics or directly woven into textiles without any additional substrates. Furthermore, 1D fiber-shaped FSCs can satisfy the miniaturization requirements of wearable electronics.

In this context, as shown in Figure 30.8a, Wang et al. [25] developed a biscrolling approach to prepare mxene/CNTs (BMX) and RuO_2/CNTs (BRU) nanocomposite yarns and incorporated them as the anode and cathode, respectively, with a PVA-H_2SO_4 gel electrolyte to fabricate yarn-shaped SCs (YSCs). The YSCs thus fabricated were highly flexible and displayed a maximum energy density of 61.6 mWh cm^{-3}. More impressively, when they were woven into a cotton yarn fabric, the resultant "energy textile" prototype was able to power portable electronics under a bending state, for example, the use for a digital watch, as shown in Figure 30.8b, indicating their good application values for wearable electronics.

Apart from wearable digital electronics, real-time health-monitoring systems are another major area where FSCs have found applications for wearable devices [2,53]. In this regard, Park et al. [34] reported a dynamically stretchable fabric SC consisting of an MWCNT/MoO_3 nanocomposite fiber (Figures 30.8c-d) and acetonitrile–propylene carbonate–poly(methyl methacrylate)–lithium perchlorate (ACN–PC–PMMA–$LiClO_4$) gel electrolyte for powering an integrated sensor textile system to detect various bio-signals of different parts of the human body (Figures 30.8e-g). When the FSC and a strain sensor were integrated into a textile via liquid-metal interconnections, the integrated all-in-one textile system showed stable and high electrochemical performances under static and dynamic deformations. It can reliably detect strains derived from the joint movement and wrist pulse, exhibiting a high sensitivity of 46.3 under a strain of 60%, a fast response time of 50 ms, and impressive stability over 10,000 stretching/releasing cycles. These results proved the notable practicability of FSCs in wearable real-time health monitoring systems.

In addition to conventional external power supplies, self-powered wearable devices have received growing attentions in recent years. In such an emerging system, a high-efficiency energy storage device (i.e., FSC) is used to store the energy harvested from sunlight, body motion, or even human biofluids. In this sense, Lv et al. reported a screen-printed textile-based hybrid SC-biofuel cell (SC-BFC) system for on-body testing (Figures 30.8h-n) [68]. Specifically, in this self-powered system, an epidermal BFC

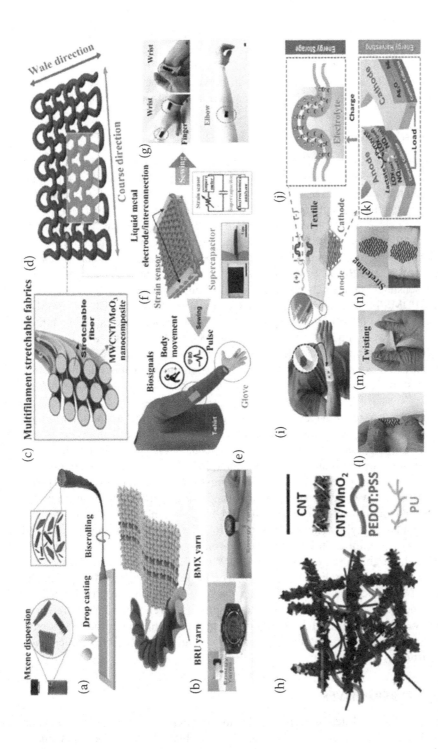

FIGURE 30.8 (a) Schematic of the fabrication process for BMX and BRU yarns. (b) Schematic illustration of an energy textile prototype incorporating YSC devices. Inset shows the energy textile prototype powering a digital watch [25]. Copyright 2018, Wiley-VCH. (c) High-magnification illustration of the cross section of a multifilament stretchable fabric showing the stretchable fibers and MWCNT/MoO₃ nanocomposite in the fabric. (d) Schematic diagram of the wale and course directions of a knitted fabric. (e) Conceptual illustration of an integration system sewn onto a T-shirt and a nylon glove for bio-signal monitoring. (f) Schematic illustration of an all-in-one integrated system fabricated from a strain sensor (blue loop) and a FSC (purl loop). (g) The integrated system is sewn onto a T-shirt and a nylon glove. Detection of bio-signals with the strain sensor is driven by the FSC [34]. Copyright 2019, American Chemical Society. (h) Illustration of the compositions of a MnO₂-CNT/CNT/PEDOT:PSS nanocomposite. (i–k) Schematic illustration of a self-powered textile, in which the energy generated by the BFC from the lactate in sweat is used to charge the FSC, showing the corresponding structures of the BFC (k) and FSC (j). (l–n) Photographs of the textile showing bending, twisting, and 20% stretching, respectively [68]. Copyright 2018, The Royal Society of Chemistry.

(Figure 30.8k) was used to harvest the biochemical energy generated from the wearer's sweat and a CNT/MnO$_2$/PEDOT:PSS nanocomposite-based FSC (Figures 30.8h-j) to store the energy thus harvested. Benefiting from the remarkable electrochemical stability and a high areal energy density (17.5 mWh cm^{-2}) of the FSC, the FSC-BFC hybrid system can deliver a stable energy output over long charging periods, boost the voltage output of the BFC, and exhibit favorable stability under a variety of mechanical deformations (bending, twisting, and stretching, Figures 30.8l-n). These results clearly indicate the great potential of such unique FSC-based hybrid power architecture in providing next-generation self-powered flexible power supplies for wearable electronics and smart textiles.

30.4.2 FSCs FOR FLEXIBLE ELECTRONIC DEVICES

During the past few years, nanocomposite-based FSCs have been exploited to power a variety of flexible electronic devices, including, for example, flexible displays, healthcare monitoring [36,60], skin-attachable sensors [69,70], and transparent biosensors [31]. For example, Park et al. [70] reported a study on the fabrication of pressure/temperature/strain sensors and all-solid-state FSCs using only poly-dimethylsiloxane coated microporous PPy/graphene foam (PDMS/PPy/GF) composite as a common material (Figure 30.9a). In this integrated multifunctional system, the pressure was measured using the thermoelectric voltage induced by the simultaneous increase in temperature caused by a finger touch on the sensor fabricated from the PDMS/PPy/GF composite, while the resistive change of the same composite was used for strain sensing. Interestingly, all these sensing units were powered by the FSCs assembled using the same PDMS/PPy/GF composite as electrodes along with an ACN-PC-PMMA-LiClO$_4$ gel electrolyte. The integrated multifunctional system on a skin-attachable substrate using liquid-metal interconnections functioned well with high durability up to 10,000 cycles of pressure loading. This achievement has been attributed to the well-defined electrochemical performance (with an energy density of 24 μWh cm^{-2}) of the built-in FSCs, importantly indicating a research direction about the integration of nanocomposite-based FSCs into a system to power those flexible electronic devices fabricated from the same composite.

Very recently, flexible and transparent devices have attracted growing interest due to their unique flexibility and transparency [71]. They are also useful in enhancing the aesthetics in civilians and improving the security in military applications [72]. Many attempts have thus been made to fabricate flexible power supplies that are also transparent for these emerging electronics. One example was reported by Yun et al. [31] with a transparent FSC to power a bio-signal detecting system, as shown in Figure 32.9b. Particularly, this was achieved by fabricating a transparent microsupercapcitor (MSC) from a MnO$_2$/CNT nanocomposite electrode and a 1-butyl-3-methylimidazolium bis(tri-fluoromethylsulfonyl)imide/poly(methyl methacrylate) ([BMIM][TFSI]/PMMA) electrolyte and integrating the as-formed MSC with a strain sensor (SS) on a transparent PDMS substrate (Figure 30.9c). The MSC had a high energy density of 1.12 μWh cm^{-2} that could efficiently power the SS to accurately detect both wrist bending (Figure 30.9d) and pulsing (Figure 30.9e). Thanks to the intrinsic flexibility of the electrode, electrolyte, and encapsulations, the proposed MSC-SS system showed excellent mechanical and electrochemical stabilities, even after 2000 stretching cycles with a strain of up to 30%. These results highlight the importance of fabricating FSCs with unique features such as transparency in providing power for emerging flexible electronics such as skin-attachable bioelectronics.

30.5 SUMMARY AND PERSPECTIVES

During the past years, we have witnessed significant progress in the research and development of flexible energy storage devices for flexible and wearable electronic devices, to which FSCs have been playing an important role due to their unique electrochemical and mechanical properties. Furthermore, for FSCs to have both high electrochemical performances and excellent mechanical deformability, three essential

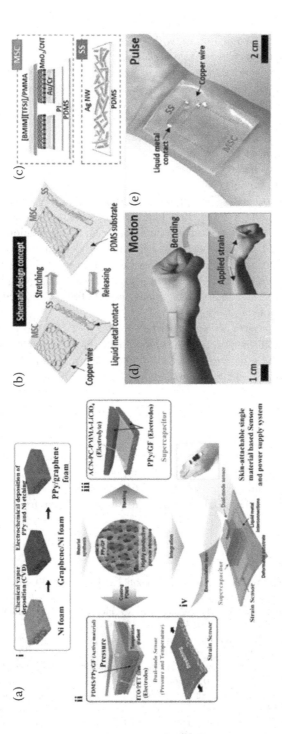

FIGURE 30.9 (a) Schematic illustration of the fabrication of a multi-sensing system powered by an integrated FSC using a single highly conductive porous PPy/GF composite. i) Fabrication of a PPy/GF composite; ii) Fabrication of a dual-mode sensor (pressure and temperature); iii) Fabrication of a FSC; iv) Integration of the dual-mode sensor, a strain sensor, and the FSC. Optical image shows the integrated system attached to hand-skin [70]. Copyright 2018, Wiley-VCH. (b) Schematic illustration of a transparent and stretchable MSC integrated with an Ag nanowire (NW) SS. (c) Cross-sectional view of the MSC and SS. (d) Photograph of wrist bending with the skin-attached MSC-SS system. (e) Photograph of the integrated system attached to the skin of the wrist [31]. Copyright 2020, Elsevier.

aspects of the FSC, including electrode materials, device configurations, and practical applications, should be thoroughly considered.

Specifically, the electrode materials should have both superior electrochemical and mechanical properties to ensure high-performance FSCs. In this regard, combing the excellent electrical, thermal, and mechanical properties of carbon nanomaterials with the high capacitance of pseudocapacitive materials, the resultant carbon nanomaterial-incorporated nanocomposites, including carbon-metal oxides, carbon-conducting polymers, and carbon-mxenes, possess enhanced properties over their original building components and thus show better suitability for FSCs. Nevertheless, for practically useful FSCs, it remains a great challenge to further improve the properties for these composite materials, which in turn indicates a need to continue the research in engineering the composition and morphology to better utilize both the carbon and pseudocapacitive components to synergistically enhance the properties for the composites.

Along with the significant progress in carbon nanomaterial-incorporated nanocomposites, FSCs have been realized in multiple dimensions from 1D fiber-shaped, 2D film-shaped, to 3D structural, resulting in a large variety of opportunities to satisfy the application requirements for varied mechanical deformation circumstances, such as bending, folding, stretching, compressing, and twisting. Although still being in their early stage, FSCs have shown great potential for practical applications and have been integrated into prototype flexible and wearable electronic products. Indeed, the multiple dimensions of FSCs from 1D to 3D provide them with tremendous opportunities to meet the demands for varied deformation scenarios. Specifically, while 1D FSCs are promising for wearable applications, 2D and 3D FSCs have better suitability for large-scale flexible electronics, and 3D FSCs are specifically useful for multi-direction-deformation applications.

Finally, through the significant progresses accomplished in the past and the extensive research under development on the important aspects of nanocomposite electrode materials and device configurations, practically useful FSCs having both well-defined electrochemical performances and excellent mechanical deformatities would be anticipated for the wide applications for flexible and wearable electronic devices.

ACKNOWLEDGMENTS

This work was supported by The Special Significant Science and Technology Program of Yunnan Province (grant number: 2016HE001-2016HE002). Fang Cheng gratefully acknowledges support from the Yunnan Postdoctoral Research Fund (grant number: C615300504006) and Xiaoping Yang acknowledges support from the Yunnan University Innovative Research Fund for Graduate Students (grant number: 2020207).

REFERENCES

1. C. Wang, K. Xia, H. Wang, et al., Advanced Carbon for Flexible and Wearable Electronics, *Adv Mater*, 31 (2019) e1801072.
2. Y. Liu, M. Pharr, G.A. Salvatore, Lab-on-Skin: A Review of Flexible and Stretchable Electronics for Wearable Health Monitoring, *ACS Nano*, 11 (2017) 9614–9635.
3. Y. Zhou, C.H. Wang, W. Lu, et al., Recent Advances in Fiber-Shaped Supercapacitors and Lithium-Ion Batteries, *Adv Mater*, 32 (2020) e1902779.
4. P. Xie, W. Yuan, X. Liu, et al., Advanced Carbon Nanomaterials For State-of-the-Art Flexible Supercapacitors, *Energy Storage Mater*, 36 (2021) 56–76.
5. G. Shao, R. Yu, N. Chen, et al., Stretchable Supercapacitors: From Materials and Structures to Devices, *Small Methods*, 5 (2021) 2000853.
6. C. Cao, Y. Chu, Y. Zhou, et al., Recent Advances in Stretchable Supercapacitors Enabled by Low-Dimensional Nanomaterials, *Small*, 14 (2018) e1803976.

7. T. An, W. Cheng, Recent Progress in Stretchable Supercapacitors, *J Mater Chem A*, 6 (2018) 15478–15494.

8. T. Lv, M. Liu, D. Zhu, et al., Nanocarbon-Based Materials for Flexible All-Solid-State Supercapacitors, *Adv Mater*, 30 (2018) e1705489.

9. Y. Shao, M.F. El-Kady, J. Sun, et al., Design and Mechanisms of Asymmetric Supercapacitors, *Chem Rev*, 118 (2018) 9233–9280.

10. S.A. Delbari, L.S. Ghadimi, R. Hadi, et al., Transition metal oxide-based electrode materials for flexible supercapacitors: A review, *J Alloys Compd*, 857 (2021) 158281.

11. Y. Han, L. Dai, Conducting Polymers for Flexible Supercapacitors, *Macromol Chem Phys*, 220 (2019) 1800355.

12. Y. Wang, Y. Ding, X. Guo, et al., Conductive polymers for stretchable supercapacitors, *Nano Res*, 12 (2019) 1978–1987.

13. C. Zhao, X. Jia, K. Shu, et al., Conducting polymer composites for unconventional solid-state super-capacitors, *J Mater Chem A*, 8 (2020) 4677–4699.

14. S. Nam, J.-N. Kim, S. Oh, et al., $Ti_3C_2T_x$ MXene for wearable energy devices: Supercapacitors and tribo-electric nanogenerators, *APL Mater*, 8 (2020) 110701.

15. D. Qi, X. Chen, Flexible Supercapacitors Based on Two-Dimensional Materials. *Flexible and Stretchable Medical Devices*, John Wiley & Sons, Ltd, (2018).

16. S. Iijima, Helical microtubules of graphitic carbon, *Nature*, 354 (1991) 56–58.

17. K.S. Novoselov, A.K. Geim, S.V. Morozov, et al., Electric field effect in atomically thin carbon films, *Science*, 306 (2004) 666.

18. S. Zhu, J. Ni, Y. Li, Carbon nanotube-based electrodes for flexible supercapacitors, *Nano Res*, 13 (2020) 1825–1841.

19. L. Huang, D. Santiago, P. Loyselle, et al., Graphene-Based Nanomaterials for Flexible and Wearable Supercapacitors, *Small*, 14 (2018) e1800879.

20. D.W. Lawrence, C. Tran, A.T. Mallajoysula, et al., High-energy density nanofiber-based solid-state super-capacitors, *J Mater Chem A*, 4 (2016) 160–166.

21. J. Zang, C. Cao, Y. Feng, et al., Stretchable and high-performance supercapacitors with crumpled graphene papers, *Sci Rep*, 4 (2014) 6492.

22. C. Cao, Y. Zhou, S. Ubnoske, et al., Highly Stretchable Supercapacitors via Crumpled Vertically Aligned Carbon Nanotube Forests, *Adv Energy Mater*, 9 (2019) 1900618.

23. S.-C. Lin, Y.-T. Lu, J.-A. Wang, et al., A flexible supercapacitor consisting of activated carbon nanofiber and carbon nanofiber/potassium-pre-intercalated manganese oxide, *J Power Sources*, 400 (2018) 415–425.

24. X. Guo, N. Bai, Y. Tian, et al., Free-standing reduced graphene oxide/polypyrrole films with enhanced electrochemical performance for flexible supercapacitors, *J Power Sources*, 408 (2018) 51–57.

25. Z. Wang, S. Qin, S. Seyedin, et al., High-Performance Biscrolled MXene/Carbon Nanotube Yarn Supercapacitors, *Small*, 14 (2018) e1802225.

26. P. Kumar Sahoo, C.-A. Tseng, Y.-J. Huang, et al., Carbon-Based Nanocomposite Materials for High-Performance Supercapacitors, Novel Nanomaterials, *IntechOpen* 2021, pp. 1–26.

27. S. Su, L. Lai, R. Li, et al., Annealing-Assisted Dip-Coating Synthesis of Ultrafine Fe_3O_4 Nanoparticles/Graphene on Carbon Cloth for Flexible Quasi-Solid-State Symmetric Supercapacitors, *ACS Appl Energ Mater*, 3 (2020) 9379–9389.

28. T. Gu, B. Wei, High-performance all-solid-state asymmetric stretchable supercapacitors based on wrinkled MnO_2/CNT and Fe_2O_3/CNT macrofilms, *J Mater Chem A*, 4 (2016) 12289–12295.

29. I. Rabani, J. Yoo, H.S. Kim, et al., Highly dispersive Co_3O_4 nanoparticles incorporated into a cellulose nanofiber for a high-performance flexible supercapacitor, *Nanoscale*, 13 (2021) 355–370.

30. Q. Liao, N. Li, S. Jin, et al., All-Solid-State Symmetric Supercapacitor Based on Co_3O_4 Nanoparticles on Vertically Aligned Graphene, *ACS Nano*, 9 (2015) 5310–5317.

31. J. Yun, H. Lee, C. Song, et al., A Fractal-designed stretchable and transparent microsupercapacitor as a Skin-attachable energy storage device, *Chem Eng J*, 387 (2020) 124076.

32. J. Liang, B. Tian, S. Li, et al., All-Printed MnHCF-MnOx-Based High-Performance Flexible Supercapacitors, *Adv Energy Mater*, 10 (2020) 2000022.

33. C. Xiong, M. Li, W. Zhao, et al., Flexible N-Doped reduced graphene oxide/carbon Nanotube-MnO_2 film as a Multifunctional Material for High-Performance supercapacitors, catalysts and sensors, *J Materiomics*, 6 (2020) 523–531.

34. H. Park, J.W. Kim, S.Y. Hong, et al., Dynamically Stretchable Supercapacitor for Powering an Integrated Biosensor in an All-in-One Textile System, *ACS Nano*, 13 (2019) 10469–10480.

35. S. Liu, Y. Yin, D. Ni, et al., New insight into the effect of fluorine doping and oxygen vacancies on electrochemical performance of Co_2MnO_4 for flexible quasi-solid-state asymmetric supercapacitors, *Energy Storage Mater*, 22 (2019) 384–396.

36. Z. Lv, Y. Luo, Y. Tang, et al., Editable Supercapacitors with Customizable Stretchability Based on Mechanically Strengthened Ultralong MnO_2 Nanowire Composite, *Adv Mater*, 30 (2018) 1704531.

37. S. Ghosh, O. Inganäs, Conducting Polymer Hydrogels as 3D Electrodes Applications for Supercapacitors, *Adv Mater*, 11 (1999) 1214.

38. L.-N. Jin, F. Shao, C. Jin, et al., High-performance textile supercapacitor electrode materials enhanced with three-dimensional carbon nanotubes/graphene conductive network and in situ polymerized polyaniline, *Electrochim Acta*, 249 (2017) 387–394.

39. J. Wang, L. Dong, C. Xu, et al., Polymorphous Supercapacitors Constructed from Flexible Three-Dimensional Carbon Network/Polyaniline/MnO_2 Composite Textiles, *ACS Appl Mater Inter*, 10 (2018) 10851–10859.

40. K. Liu, Y. Yao, T. Lv, et al., Textile-like electrodes of seamless graphene/nanotubes for wearable and stretchable supercapacitors, *J Power Sources*, 446 (2020) 227355.

41. L. Chen, L. Chen, Q. Ai, et al., Flexible all-solid-state supercapacitors based on freestanding, binder-free carbon nanofibers@polypyrrole@graphene film, *Chem Eng J*, 334 (2018) 184–190.

42. X. Li, R. Liu, C. Xu, et al., High-Performance Polypyrrole/Graphene/$SnCl_2$ Modified Polyester Textile Electrodes and Yarn Electrodes for Wearable Energy Storage, *Adv Funct Mater*, 28 (2018) 1800064.

43. H.U. Lee, C. Park, J.-H. Jin, et al., A stretchable vertically stacked microsupercapacitor with kirigami-bridged island structure: MnO_2/graphene/Poly(3,4-ethylenedioxythiophene) nanocomposite electrode through pen lithography, *J Power Sources*, 453 (2020) 227898.

44. J. Zeng, L. Dong, W. Sha, et al., Highly stretchable, compressible and arbitrarily deformable all-hydrogel soft supercapacitors, *Chem Eng J*, 383 (2020) 123098.

45. M. Naguib, M. Kurtoglu, V. Presser, et al., Two-dimensional nanocrystals produced by exfoliation of Ti_3AlC_2, *Adv Mater*, 23 (2011) 4248–4253.

46. C. Zhang, Y. Ma, X. Zhang, et al., Two Dimensional Transition Metal Carbides and Nitrides (MXenes): Synthesis, Properties, and Electrochemical Energy Storage Applications, *Energ Environ Mater*, 3 (2020) 29–55.

47. M.R. Lukatskaya, S. Kota, Z. Lin, et al., Ultra-high-rate pseudocapacitive energy storage in two-dimensional transition metal carbides, *Nat Energy*, 2 (2017) 17105.

48. H. Li, R. Chen, M. Ali, et al., In Situ Grown MWCNTs/MXenes Nanocomposites on Carbon Cloth for High-Performance Flexible Supercapacitors, *Adv Funct Mater*, 30 (2020) 2002739.

49. C. Yu, Y. Gong, R. Chen, et al., A Solid-State Fibriform Supercapacitor Boosted by Host-Guest Hybridization between the Carbon Nanotube Scaffold and MXene Nanosheets, *Small*, (2018) e1801203.

50. H. Li, Y. Hou, F. Wang, et al., Flexible All-Solid-State Supercapacitors with High Volumetric Capacitances Boosted by Solution Processable MXene and Electrochemically Exfoliated Graphene, *Adv Energy Mater*, 7 (2017) 1601847.

51. A. Salman, S. Padmajan Sasikala, I.H. Kim, et al., Tungsten nitride-coated graphene fibers for high-performance wearable supercapacitors, *Nanoscale*, 12 (2020) 20239–20249.

52. W. Chen, D. Zhang, K. Yang, et al., Mxene ($Ti_3C_2T_x$)/cellulose nanofiber/porous carbon film as free-standing electrode for ultrathin and flexible supercapacitors, *Chem Eng J*, 413 (2020) 127524.

53. X. Chen, N.S. Villa, Y. Zhuang, et al., Stretchable Supercapacitors as Emergent Energy Storage Units for Health Monitoring Bioelectronics, *Adv Energy Mater*, 10 (2019) 1902769.

54. J. Liang, C. Jiang, W. Wu, Toward fiber-, paper-, and foam-based flexible solid-state supercapacitors: electrode materials and device designs, *Nanoscale*, 11 (2019) 7041–7061.

55. Q. Zhou, X. Chen, F. Su, et al., Sandwich-Structured Transition Metal Oxide/Graphene/Carbon Nanotube Composite Yarn Electrodes for Flexible Two-Ply Yarn Supercapacitors, *Ind Eng Chem Res*, 59 (2020) 5752–5759.

56. S. Wang, N. Liu, J. Su, et al., Highly Stretchable and Self-Healable Supercapacitor with Reduced Graphene Oxide Based Fiber Springs, *ACS Nano*, 11 (2017) 2066–2074.

57. Q. Zhang, J. Sun, Z. Pan, et al., Stretchable fiber-shaped asymmetric supercapacitors with ultrahigh energy density, *Nano Energy*, 39 (2017) 219–228.

58. Z. Pan, J. Yang, L. Li, et al., All-in-one stretchable coaxial-fiber strain sensor integrated with high-performing supercapacitor, *Energy Storage Mater*, 25 (2020) 124–130.

59. Z. Pan, J. Yang, Q. Zhang, et al., All-Solid-State Fiber Supercapacitors with Ultrahigh Volumetric Energy Density and Outstanding Flexibility, *Adv Energy Mater*, 9 (2019) 1802753.

60. G. Lee, J.W. Kim, H. Park, et al., Skin-Like, Dynamically Stretchable, Planar Supercapacitors with Buckled Carbon Nanotube/Mn-Mo Mixed Oxide Electrodes and Air-Stable Organic Electrolyte, *ACS Nano*, 13 (2019) 855–866.

61. Y. Zhou, K. Maleski, B. Anasori, et al., $Ti_3C_2T_x$ MXene-Reduced Graphene Oxide Composite Electrodes for Stretchable Supercapacitors, *ACS Nano*, 14 (2020) 3576–3586.

62. J. Cherusseri, K. Sambath Kumar, D. Pandey, et al., Vertically Aligned Graphene-Carbon Fiber Hybrid Electrodes with Superlong Cycling Stability for Flexible Supercapacitors, *Small*, 15 (2019) e1902606.

63. C.R. Chen, H. Qin, H.P. Cong, et al., A Highly Stretchable and Real-Time Healable Supercapacitor, *Adv Mater*, 31 (2019) e1900573.

64. Z. Pan, M. Liu, J. Yang, et al., High Electroactive Material Loading on a Carbon Nanotube@3D Graphene Aerogel for High-Performance Flexible All-Solid-State Asymmetric Supercapacitors, *Adv Funct Mater*, 27 (2017) 1701122.

65. Z. Zhou, Q. Li, L. Yuan, et al., Achieving ultrahigh-energy-density in flexible and lightweight all-solid-state internal asymmetric tandem 6.6 V all-in-one supercapacitors, *Energy Storage Mater*, 25 (2020) 893–902.

66. Z. Lv, Y. Tang, Z. Zhu, et al., Honeycomb-Lantern-Inspired 3D Stretchable Supercapacitors with Enhanced Specific Areal Capacitance, *Adv Mater*, 30 (2018) 1805468.

67. J. Wen, B. Xu, Y. Gao, et al., Wearable technologies enable high-performance textile supercapacitors with flexible, breathable and wearable characteristics for future energy storage, *Energy Storage Mater*, 37 (2021) 94–122.

68. J. Lv, I. Jeerapan, F. Tehrani, et al., Sweat-based wearable energy harvesting-storage hybrid textile devices, *Energ Environ Sci*, 11 (2018) 3431–3442.

69. H. Park, C. Song, S.W. Jin, et al., High performance flexible micro-supercapacitor for powering a vertically integrated skin-attachable strain sensor on a bio-inspired adhesive, *Nano Energy*, 83 (2021) 105837.

70. H. Park, J.W. Kim, S.Y. Hong, et al., Microporous Polypyrrole-Coated Graphene Foam for High-Performance Multifunctional Sensors and Flexible Supercapacitors, *Adv Funct Mater*, 28 (2018) 1707013.

71. C. Zhang, V. Nicolosi, Graphene and MXene-based transparent conductive electrodes and supercapacitors, *Energy Storage Mater*, 16 (2019) 102–125.

72. T.Q. Trung, L.T. Duy, S. Ramasundaram, et al., Transparent, stretchable, and rapid-response humidity sensor for body-attachable wearable electronics, *Nano Res*, 10 (2017) 2021–2033.

31 Textile-Based Flexible Nanogenerators

Ünsal Ömer Faruk and Çelik Bedeloğlu Ayşe
Bursa Technical University, Polymer Materials Engineering Department,
Bursa, Turkey

Borazan İsmail
Bursa Technical University, Polymer Materials Engineering Department,
Bursa, Turkey

Bartın University, Textile Engineering Department,
Bartın, Turkey

CONTENTS

31.1 Introduction.. 575
31.2 Piezoelectric nanogenerators 577
31.3 Textile-based piezoelectric nanogenerators 579
31.4 Pyroelectric and hybrid nanogenerators 580
31.5 Triboelectric nanogenerators (TENG) 582
31.6 Conclusion 585
References... 585

31.1 INTRODUCTION

Due to the development of technology and the rapid increase in the human population, energy consumption has increased. The increase in energy consumption caused our resources to decrease and nature to be harmed. For these reasons, scientists are researching to efficiently use less harmful and renewable energy sources, such as solar, wind, and biomass energy. Scientists have also begun to discover alternate energy sources around us. The easiest energy to be found in nature is the energy created by the movement of matter, such as mechanical energy. The concept of nanogenerators emerged with the study of mechanical energy. By definition; nanogenerators use a technology that allows the conversion of mechanical or thermal energy into electrical energy.

Nanogenerators are devices that can be integrated into consumer products, such as textiles, electronics, automotive, etc., and convert mechanical and thermal energy to electricity [1]. Wind energy, human motion, vehicle motion, and waste heat in the industry are the most known waste energy sources [2]. Synergistic improvement on smart materials and electronics have also stimulated a dramatic increase in textile or vehicle integrated nanogenerators [3]. Nanogenerators can be classified according to energy type converted and energy conversion principles. According to energy type, converted nanogenerators are two types, mechanical energy-converting nanogenerators and thermal energy-converting nanogenerators. Nanogenerators are also classified according to working principles; triboelectric, piezoelectric, and pyroelectric nanogenerators (Figure 31.1) [5]. However, all nanogenerators work with the displacement current phenomenon [6]. In the displacement current mechanism, partially electrical polarization is necessary. This polarization is provided by mechanically forcing atoms with different

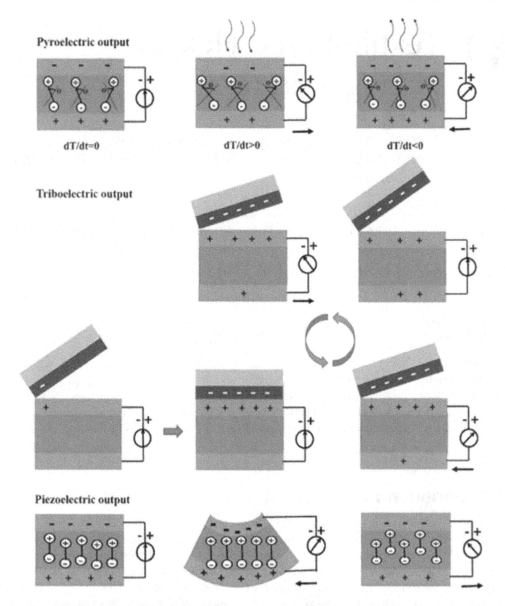

FIGURE 31.1 Schematic illustration of different types of nanogenerators with stimulation mechanisms. Adapted permission from ref. [4]. Copyright (2018) Royal Society of Chemistry.

electronegativity to cumulate different regions of materials, in the piezoelectric effect. Polarization is provided from static electrification, in triboelectric nanogenerators. Tribo-electrification is based on electron transfer from one dielectric material to the other in case of contact with each other. Lastly, in the pyroelectric effect, the atoms forming the matter are polarized by heat energy [7]. Nanogenerators also can be used as sensors. For example; it has been shown that piezoelectric and triboelectric nanogenerators can be used as pressure, vibration, and bending sensors. Pyroelectric nanogenerators can be used as temperature sensors due to their thermo-electric properties [8].

Textile industry is becoming a multidisciplinary work area as the result of developing wearable electronics. Health care sensors, energy-generating textiles, or electric or heat-based therapy

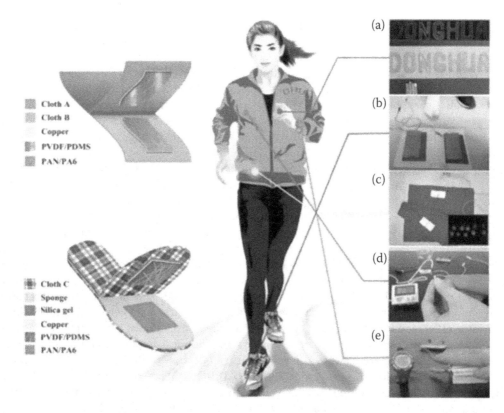

FIGURE 31.2 Applicable nanogenerators for wearables. Adapted permission from ref. [13]. Copyright (2017) Elsevier.

devices could be successfully integrated into textile products. However, some issues, such as washability, cost efficiency, long production times, flexibility, or mechanical stability, need to be solved before commercialization [9]. Joints on daily clothes [10], shoe soles [11], or exterior portions of clothes [12] (to scavenge energy from the environment) are the most commonly used integration areas for nanogenerators (Figure 31.2). Textile-based nanogenerators are not only produced by conventional textile processes like knitting or weaving [14] but they can be also produced by innovative methods like sol-gel [15], electrospinning [16], and different coating methods [17].

31.2 PIEZOELECTRIC NANOGENERATORS

The word piezoelectric is a concept derived from the Latin prefix "piezo" meaning to press. A piezoelectric nonconductive crystal sculpted plate is a phenomenon characterized by the release of reverse-marked charges (+ q and -q) on both sides of the crystal plate at the end of a pressure (tension or compression) applied in a certain direction. The property of piezoelectric materials under the influence of a force is depending on their crystal structure. Under the influence of an electric field, they can change shape in the order of approximately 4% by volume. The piezoelectric property, known to exist since 1665, was first discovered by brothers Pierre and Jacques Curie in 1880. Pierre Curie previously worked on the relationship between pyroelectricity and crystal symmetry. This work forced the brothers to look for electrification caused by pressure only, but the direction in which pressure can be applied and the effect of crystal classes were not explained. It has also been found in many other crystals, such as Rochelle salt. Hankel proposed the name "piezoelectric." Piezoelectric is an inter-action between electrical and mechanical systems. The direct (direct) piezoelectric effect is the electrical polarization generated by mechanical stress. The piezoelectric property is a result of the crystal structure orientation of the material [18].

Piezoelectric effect, basically electric energy generation from a homogeneous material by me-chanical stimuli. In nanogenerators, electrical polarization is necessary for electrical flow. This atomic-electrical polarization is provided by mechanical stress in piezoelectric materials. Piezoelectric crystals have no central symmetrical atom. This situation causes electrical polariza-tion when the mechanical stress forces atoms to approach or become distant from each other [19]. However, the polarization mechanism is different in polymers. A polymer chain having different electronegative side atoms can show a piezoelectric effect based on its conformation (Figure 31.3). For example, β-crystalline phase of polyvinylidene fluoride (PVDF) shows piezoelectric/ferro-electric properties, while α-crystalline phase PVDF has no piezoelectric property [20]. When the piezoelectric material is exposed to mechanical stress, material-forming atoms with different electronegativities accumulate different regions of crystal structure (or different sides of the backbone in polymers). This accumulation causes an electrical charge polarization at different regions of the material. Without integration of electrodes, charge separation forces the electrons to flow from the negative pole to the positive, which is named displacement current. In the case of electrode integration, the resulting system is named piezoelectric nanogenerators. Piezoelectric nanogenerator efficiency largely depends on the surface area of materials. The output power of the nanogenerator increases with surface area (Figure 31.4). The most important property of the ma-terial used for the piezoelectric effect to have is that it is crystals without a center of symmetry. PVDF, PVDF-TrFE, Parylene-C, Polyimide (β-CN) APB/ODPA are some of the important pie-zoelectric materials. Polymer composites have been mostly used for piezoelectric devices in recent studies due to advances in bulk materials, flexibility, ease of application, and biocompatibility [1]. The most used type of piezoelectric material is lead-zirconium-titanium (PZT) piezoceramics. Piezoceramic materials are crystals without a center of symmetry that converts electrical effect into mechanical magnitude and mechanical effect into electrical magnitude. Commonly used piezo-ceramic materials are quartz (SiO_2), $BaTiO_3$, $PbZrO_3$-$PbTiO_3$ alloy (PZT), (Pb, La) (Ti, Zr) O_3 alloy (PLZT), Zinc oxide (ZnO) [18].

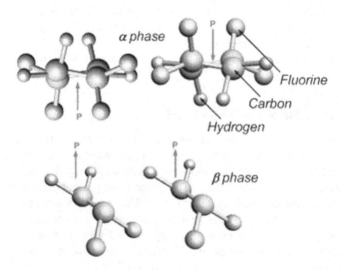

FIGURE 31.3 α and β crystalline phases of PVDF. Adapted permission from ref. [20]. Copyright (2012) Elsevier.

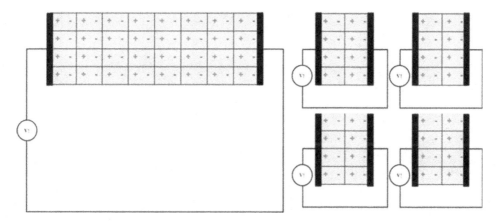

FIGURE 31.4   Surface area effect for piezoelectric materials.

31.3   TEXTILE-BASED PIEZOELECTRIC NANOGENERATORS

Piezoelectric nanogenerators are consist of a sandwiched piezoelectric layer between two electrodes. Usage of piezoelectric polymers or piezoelectric inorganic doped polymers for piezoelectric nanogenerator production brings to devices textile-implantable and comfortable properties. Textile-based piezoelectric nanogenerators can be used as energy harvesters, healthcare sensors with an external energy source, or a self-powered sensor system [21]. Piezoelectric nanogenerators are mostly fabricated using rigid materials like glass, metal foils, inorganic crystals, etc. These types of piezoelectric devices are convenient for use in vibration-based energy conversion. However, in piezoelectric-based smart textiles, sustainable and stable flexibility are required. The nanogenerator's flexibility is required for textiles used in sportswear, military purposes, outdoor wearables, as well as clothes for daily use. Due to these requirements, piezoelectric polymers, piezoelectric-doped elastomers, polymeric base materials instead of glass (sponge, paper, polymeric films, etc.) are commonly used in textile-based nanogenerators [22]. These devices can be based on woven or knitted fabrics, nonwoven fabrics, and/or flexible films. Furthermore, these rigid materials and flexible materials also can be used together for flexible devices. For example, zinc oxide (ZnO) nanowires, are suitable to grow on various surfaces like synthetic or natural textile by a wet chemical (hydrothermal) method. This combined method provides a higher energy output for piezoelectric nanogenerators (Figure 31.5). Another important fact for wearable nanogenerators is electrode design. By improving electrode application methods, various types of conductive materials, like metal nanoparticles, conducting polymers, carbon-based nanoparticles, could be applied to textiles.

Optimization in twisting, weaving, materials selection, and electrode design are some of the parameters that provide high efficiencies to nanogenerators [23]. Nanofibers of PVDF or its derivatives are

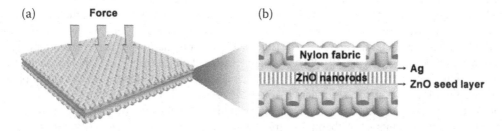

FIGURE 31.5   Schematic illustration of ZnO nanowire-based wearable piezoelectric nanogenerator. Adapted permission from ref. [22]. Copyright (2019) Elsevier.

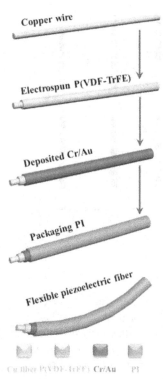

Copper wire

Electrospun P(VDF-TrFE)

Deposited Cr/Au

Packaging PI

Flexible piezoelectric fiber

Cu fiber P(VDF-TrFE) Cr/Au PI

FIGURE 31.6　Core-shell structure of nanofibrous-based piezoelectric textile. Adapted permission from ref. [25]. Copyright (2020) Elsevier.

the most common nonwoven materials that are used in piezoelectric textile due to their high surface area and high beta-crystalline phase formation during the electrospinning process [24]. Piezoelectric nanofibers can be also produced with different additives like carbon nanotubes, inorganic piezoelectrics, graphene and derivatives, and even ionic salts to achieve higher efficiency in nanogenerators. These additives work as charge transfer agents and can also participate in the formation of the crystalline phase. Research has shown that the core-shell nanofibrous structures recovered energy more efficiently due to electrode integration (Figure 31.6) [25]. Many processes, such as sol-gel template method, hydrothermal method, solvent-casting method, electrospinning method, and solid-state techniques can be used to prepare piezoelectric nanostructures. Piezoelectric nanogenerators can be classified into four groups based on their dimensions: zero-dimensional, one-dimensional, two-dimensional, and three-dimensional piezoelectric nanogenerators [26].

31.4　PYROELECTRIC AND HYBRID NANOGENERATORS

Researchers are focusing to recover thermal energy (one of the widely waste energy) by different methods. The pyroelectric effect is one of the two mechanisms to scavenge waste thermal energy. In the pyroelectric effect, thermal energy generates the electrically polarized regions due to the high dipole moment of pyroelectric material (Figure 31.7). Electrically neutral charge couples in anisotropic crystals (for example, H and F atoms on PVDF backbone) change their locations, and polarization sourced voltage difference is observed. Furthermore, thermal expansion can cause further mechanically polarized regions, and this effect is named a secondary pyroelectric effect [27]. The common pyroelectric materials are barium titanate, lead zirconate titanate, and PVDF. Industries, vehicles, solar energy, the human body generates a significant amount of thermal energy, which can be used for these devices. The thermal gradient is very important in the pyroelectric energy conversion mechanism. For example, a pyroelectric nanogenerator will produce more energy in a cold environment than in a hot condition using an equal

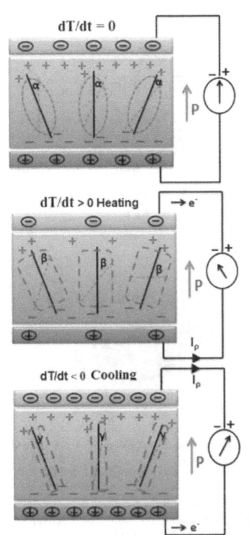

dT/dt = 0

dT/dt > 0 Heating

dT/dt < 0 Cooling

FIGURE 31.7 Schematic illustration of a pyroelectric nano-generator with electric dipoles in heating and cooling states. Adapted permission from ref. [27]. Copyright (2021) Elsevier.

thermal energy source. Different parts of the human body can be used to scavenge waste bio-thermal energy. For example, armpits, chest, dorsal regions, or mouths radiate waste thermal energy. A popular textile-based face mask can be used as a pyroelectric nanogenerator to provide a small voltage and a few microamperes of current by recovering thermal energy of human breath (Figure 31.8) [28]. Pyroelectric energy harvesters can also use a secondary harvester mechanism in hybrid-energy harvesters. Besides piezoelectric properties, working conditions are the secondary motivation for harvester hybridization. For example, hybridization of solar-pyroelectric effects can work efficiently under sunlight. In this system, sunlight energy generates the temperature gradient, and the pyroelectric component can generate electrical energy while the solar cell converts photonic energy [29]. Due to the predominant usage of polymers for piezoelectric nanogenerators, pyroelectric or hybrid nanogenerators are easy to apply to textiles.

PVDF is the most common material used to fabricate pyroelectric nanogenerators, especially in nanofiber form. The electrospinning process is the most known method for nanofiber production. In this method, a high electrical field provides PVDF to gain electroactive β-crystalline phase. Both piezoelectric and pyroelectric effects come with β-crystalline conformation in PVDF polymer.

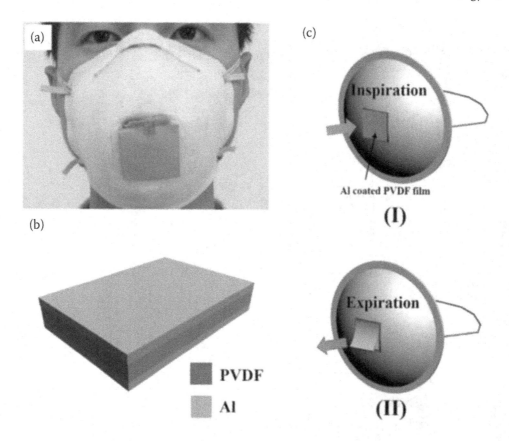

FIGURE 31.8 Photograph and schematic illustration of pyroelectric nanogenerator based face-mask. Adapted permission from ref. [28]. Copyright (2017) Elsevier.

PVDF and its derivatives-based hybridized generators can be efficiently used on heat points of the human body [30]. However, nanogenerator electrodes, which should be highly conductive, and mechanical and thermally stable materials, must be flexible for flexible devices. Graphene and its derivatives, metal particle-filled polymers, conductive polymers, etc., are being used in textile-based nanogenerator electrodes. Pyroelectric nanogenerators, in combination with mechanical energy harvesters (piezoelectric or triboelectric), can provide stretchable devices. Elastomers like thermoplastic polyurethane, polydimethylsiloxane, and different types of rubbers can be used both as covering material and electrode material. Elastomers doped with conducting materials allow the fabrication of highly stretchable electrodes for hybridized nanogenerators [31]. The stretchability of the nanogenerators allows recovering more energy of pyroelectric-piezoelectric textiles by catching more mechanical energy during operation.

31.5 TRIBOELECTRIC NANOGENERATORS (TENG)

Energy consumption is increasing with the development of technology and the population, which decreases the fossil fuel stock. Scientists are looking for alternative energy sources and mechanical energy is one of the ways to secure energy crisis [1]. Triboelectric nanogenerators (TENGs) are devices that convert mechanical energy into electrical energy. In TENG devices, an electrical potential is generated by the means of separated charges on the contact surfaces [32]. When two materials with different polarities are brought into contact, charge transfer occurs due to the triboelectric effect. If these materials are attached to metal electrodes from their noncontacting surfaces, separation of the contact surfaces causes charge accumulation by the effect of electrostatic induction [33]. The TENGs have four working modes: vertical contact-separation mode, lateral sliding mode, single electrode mode, freestanding triboelectric-layer mode. The schematic demonstration is shown in Figure 31.9 [35]. The vertical contact-separation mode was developed in 2012. Energy is produced by the contact and separation of two layers. In this mode, two different dielectric parallel surfaces are connected by separate electrodes from the distant sides of each other. Dielectric films are made perpendicular to the surfaces in contact and separation movements. During contact, opposite electrostatic charges are formed on the surfaces with the difference in electron affinities. When the surfaces are separated, it causes the formation of free charges in the part where the electrodes are attached. With the repetitive contact and separation motion, a current is generated.

In-plane scrolling mode was first introduced in 2013. This mechanism is suitable to generate energy with planar, disc rotation, and cylindrical rotation movements. In-plane sliding motion is a more practical and convenient mode of movement than vertical contact separation motion. The first TENG study in this mode was carried out by moving nylon and polytetrafluoroethylene (PTFE) surfaces in opposite directions to each other. The single-electrode mode was developed in the same year (2013) as the in-plane shift mode. Single electrode mode generally includes a grounded electrode, moving active element, and friction layer. Initially; the active element is made in contact with the friction layer on the back of the electrode. With the electrostatic effect, a positive charge occurs on the active element and a negative charge on the friction surface. While the active part is separated from the friction layer; the electrode is charged with a positive charge as opposed to the charge on the friction surface. This causes a flow of electrons from the electrode to the ground as a result of the potential difference between the earth and the electrode. When the active element is far enough, an electrical balance occurs between the electrode and the friction surface, causing the electric current to stop. Similarly, when the active layer moves back toward the friction surface, the electrons flow in the opposite direction, this time from the ground to the electrode, with the potential difference between the ground and the electrode. Thus, the repetitive contact and separation motion of the earth and active element causes a load current on the external circuit. A single-electrode triboelectric nanogenerator was fabricated using human skin as an active object, a polydimethylsiloxane (PDMS) layer as a friction surface, and indium tin oxide (ITO) as an electrode. The free triboelectric layer mode was first developed in 2014. This mode can be implemented using different configurations. There are two stationary metal electrodes and a freely moving triboelectric layer. The triboelectric layer makes an axial sliding motion on two electrodes located in the same plane. Both electrodes are triboelectrically charged at a varying rate by the sliding layer. The electrostatic induction causes a potential difference in the electrodes, which generates an electric current between the electrodes [36]. The four methods have advantages and disadvantages compared to each other. Each of them can be used for different purposes based on mechanical movement.

The most used materials for triboelectrification are PTFE, PDMS, PI, fluorinated ethylene propylene (FEP) because of their high electron affinity. PTFE is chemically inert and physically strong, PDMS has good stretchable and easy mold casting capability, PI, also known with the trade name of Kapton, is easily processable as elastic thin films and can be stable under high-temperature conditions. Energy generation occurs both in vertical and horizontal directional movement in TEGs. Typical structures of TENGs are plane-shaped, arch-shaped, zig-zag shaped, wavy-shaped, rotor-shaped, disc-shaped, etc. [37]. The materials for TENGs can be divided into four groups: metals, oxides, polymers, and others. The structure of TENGs can be simplified by using metals to function as both TEGs and electrodes. Oxides such as ITO, TiO_2, Al_2O_3, SiO_2, and graphene oxide are widely used materials for these applications [37]. Based on the charge-processing mechanism, materials can be investigated as a charge-trapping layer, charge-collecting layer, and charge-storage layer [32]. Various techniques, such as soft lithography, 3D printing, printing, roll-to-roll producing and textile-like manufacturing, etc., can be used for the fabrication of TENGs [37]. A TENG in single-electrode mode is fabricated with a sandwich structure of silicone rubber and silver-coated glass microspheres that can be stretchable, twistable, and rollable, as seen in Figure 31.10 [38].

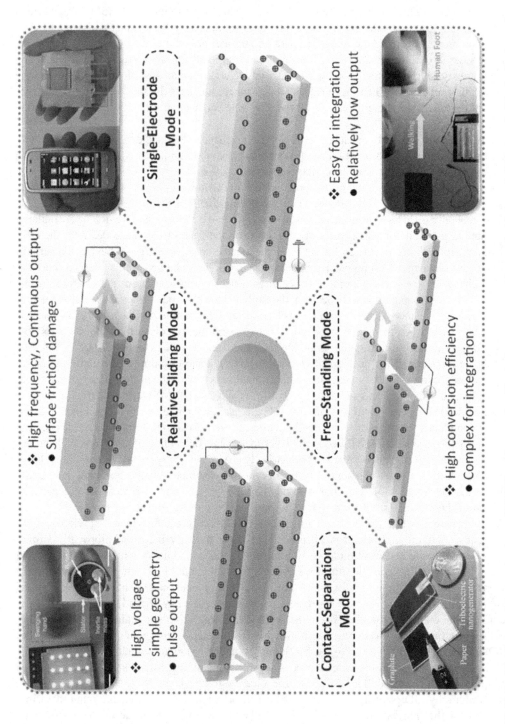

FIGURE 31.9 The four working mechanisms of TENGs with examples. Vertical contact-separation mode (a), lateral sliding mode (b), single electrode mode (c), freestanding triboelectric-layer mode (d). Adapted with permission from ref. [34]. Copyright (2021) Elsevier.

FIGURE 31.10 Images of the TENG with silver-coated glass microspheres by Li et al. Copyright (2021) [38].

31.6 CONCLUSION

Textile-based (flexible) nanogenerators, including piezoelectric, pyroelectric, triboelectric, and hybrid nanogenerators, are described in this chapter. With the advantages of easy production, low cost, lightweight, and abundant material selection, textile-based flexible nanogenerators have become popular for collecting energy for flexible electronics, wearable technologies, biomechanical devices, medical applications, security, and much more. Due to the loss of energy in single energy-conversion systems, hybrid systems are more appealing due to the advantage of combining different energy harvesting methods [39]. Although textile-based nanogenerators are promising for future applications, some problems need to be solved, such as mechanical durability, stable output voltage, compatibility of nanogenerators for the microsystems, selection/synthesizing of materials for multiple nanogenerator systems, development of bio-friendly and ecofriendly materials, etc. [26]. By overcoming these constraints, flexible-energy storage systems can be integrated into a wider range of devices, thus producing renewable energy that does not harm the environment.

REFERENCES

1. F. Greco and V. Mattoli, "Introduction to active smart materials for biomedical applications," pp. 1–27, 2012. doi: 10.1007/978-3-642-28044-3_1.
2. L. Dhakar et al., "Large scale triboelectric nanogenerator and self-powered pressure sensor array using low cost roll-to-roll UV embossing," Scientific Reports, vol. 6, no. February, pp. 1–10, 2016, doi: 10.1038/srep22253.
3. Z. Liu, H. Li, B. Shi, Y. Fan, Z. L. Wang, and Z. Li, "Wearable and implantable triboelectric nanogenerators," Advanced Functional Materials, vol. 29, no. 20, pp. 1–19, 2019, doi: 10.1002/adfm.201808820.
4. W. Wei, J. Gao, J. Yang, J. Wei, and J. Guo, "A NIR light-triggered pyroelectric-dominated generator based on a liquid crystal elastomer composite actuator for photoelectric conversion and self-powered sensing," RSC Advances, vol. 8, no. 71, pp. 40856–40865, Dec. 2018, doi: 10.1039/C8RA08491B.
5. Ö. Faruk Ünsal and A. Çelik Bedeloğlu, "Recent trends in flexible nanogenerators: A review," Material Science Research India, vol. 15, no. 2, pp. 114–130, Aug. 2018, doi: 10.13005/msri/150202.
6. Z. L. Wang, "On Maxwell's displacement current for energy and sensors: the origin of nanogenerators," Materials Today, vol. 20, no. 2. Elsevier B.V., pp. 74–82, Mar. 01, 2017. doi: 10.1016/j.mattod.2016.12.001.
7. S. Wang, Z. L. Wang, and Y. Yang, "A one-structure-based hybridized nanogenerator for scavenging mechanical and thermal energies by triboelectric-piezoelectric-pyroelectric effects," Advanced Materials, vol. 28, no. 15, pp. 2881–2887, 2016, doi: 10.1002/adma.201505684.
8. R. Ghosh et al., "Micro/nanofiber-based noninvasive devices for health monitoring diagnosis and rehabilitation," Applied Physics Reviews, vol. 7, no. 4, 2020, doi: 10.1063/5.0010766.
9. S. Korkmaz and A. Kariper, "Production and applications of flexible/wearable triboelectric nanogenerator (TENGS)," Synthetic Metals, vol. 273, no. November 2020, 2021, doi: 10.1016/j.synthmet.2020.116692.
10. T. Zhou, C. Zhang, C. B. Han, F. R. Fan, W. Tang, and Z. L. Wang, "Woven structured triboelectric nanogenerator for wearable devices," ACS Applied Materials and Interfaces, vol. 6, no. 16, pp. 14695–14701, Aug. 2014, doi: 10.1021/am504110u.

11. T. C. Hou, Y. Yang, H. Zhang, J. Chen, L. J. Chen, and Z. Lin Wang, "Triboelectric nanogenerator built inside shoe insole for harvesting walking energy," *Nano Energy*, vol. 2, no. 5, pp. 856–862, Sep. 2013, doi: 10.1016/j.nanoen.2013.03.001.

12. A. Ahmed *et al.*, "All printable snow-based triboelectric nanogenerator," *Nano Energy*, vol. 60, pp. 17–25, Jun. 2019, doi: 10.1016/j.nanoen.2019.03.032.

13. Z. Li, J. Shen, I. Abdalla, J. Yu, and B. Ding, "Nanofibrous membrane constructed wearable triboelectric nanogenerator for high performance biomechanical energy harvesting," *Nano Energy*, vol. 36, pp. 341–348, Jun. 2017, doi: 10.1016/j.nanoen.2017.04.035.

14. J. Xiong and P. See Lee, "Science and Technology of Advanced Materials ISSN: (Print) (Online) Journal homepage: https://www.tandfonline.com/loi/tsta20 Progress on wearable triboelectric nanogenerators in shapes of fiber, yarn, and textile Progress on wearable triboelectric nanogenerators in shapes of fiber, yarn, and textile," *Science and Technology of Advanced Materials*, vol. 20, no. 1, pp. 837–857, 2019, doi: 10. 1080/14686996.2019.1650396.

15. M. Kim, Y. S. Wu, E. C. Kan, and J. Fan, "Breathable and flexible piezoelectric ZnO@PVDF fibrous nanogenerator for wearable applications," *Polymers*, vol. 10, no. 7, p. 745, Jul. 2018, doi: 10.3390/polym 10070745.

16. Ö. F. Ünsal, Y. Altın, and A. Çelik Bedeloğlu, "Poly(vinylidene fluoride) nanofiber-based piezoelectric nanogenerators using reduced graphene oxide/polyaniline," *Journal of Applied Polymer Science*, vol. 137, no. 13, p. 48517, Apr. 2020, doi: 10.1002/app.48517.

17. A. R. Mule, B. Dudem, H. Patnam, S. A. Graham, and J. S. Yu, "Wearable single-electrode-mode triboelectric nanogenerator via conductive polymer-coated textiles for self-power electronics," *ACS Sustainable Chemistry and Engineering*, vol. 7, no. 19, pp. 16450–16458, Oct. 2019, doi: 10.1021/acssuschemeng. 9b03629.

18. M. A. Ilyas, *Piezoelectric energy harvesting: methods, progress, and challenges.* 2018.

19. N. D. Sharma, R. Maranganti, and P. Sharma, "On the possibility of piezoelectric nanocomposites without using piezoelectric materials," *Journal of the Mechanics and Physics of Solids*, vol. 55, no. 11, pp. 2328–2350, Nov. 2007, doi: 10.1016/j.jmps.2007.03.016.

20. T. Sharma, S. S. Je, B. Gill, and J. X. J. Zhang, "Patterning piezoelectric thin film PVDF-TrFE based pressure sensor for catheter application," *Sensors and Actuators, A: Physical*, vol. 177, pp. 87–92, Apr. 2012, doi: 10.1016/j.sna.2011.08.019.

21. K. Kim and K. S. Yun, "Stretchable power-generating sensor array in textile structure using piezoelectric functional threads with hemispherical dome structures," *International Journal of Precision Engineering and Manufacturing – Green Technology*, vol. 6, no. 4, pp. 699–710, 2019, doi: 10.1007/s40684-019-00127-z.

22. Z. Zhang, Y. Chen, and J. Guo, "ZnO nanorods patterned-textile using a novel hydrothermal method for sandwich structured-piezoelectric nanogenerator for human energy harvesting," *Physica E: Low-Dimensional Systems and Nanostructures*, vol. 105, no. June 2018, pp. 212–218, 2019, doi: 10.1016/ j.physe.2018.09.007.

23. A. Forouzan, M. Yousefzadeh, M. Latifi, and R. Jose, "Effect of geometrical parameters on piezoresponse of nanofibrous wearable piezoelectric nanofabrics under low impact pressure," *Macromolecular Materials and Engineering*, vol. 306, no. 1, pp. 1–10, 2021, doi: 10.1002/mame.202000510.

24. F. Mokhtari, M. Shamshirsaz, M. Latifi, and J. Foroughi, "Nanofibers-based piezoelectric energy harvester for self-powered wearable technologies," *Polymers*, vol. 12, no. 11, pp. 1–15, 2020, doi: 10.3390/ polym12112697.

25. L. Lu, B. Yang, Y. Zhai, and J. Liu, "Electrospinning core-sheath piezoelectric microfibers for self-powered stitchable sensor," *Nano Energy*, vol. 76, no. July, p. 104966, 2020, doi: 10.1016/j.nanoen.2020.104966.

26. A. Chowdhury, "Organic-inorganic hybrid materials for piezoelectric/triboelectric nanogenerator," 2019.

27. S. Korkmaz and A. Kariper, "Pyroelectric nanogenerators (PyNGs) in converting thermal energy into electrical energy: Fundamentals and current status," *Nano Energy*, vol. 84, no. February, 2021, doi: 10.1016/ j.nanoen.2021.105888.

28. H. Xue *et al.*, "A wearable pyroelectric nanogenerator and self-powered breathing sensor," *Nano Energy*, vol. 38, no. May, pp. 147–154, 2017, doi: 10.1016/j.nanoen.2017.05.056.

29. X. Q. Wang *et al.*, "Nanophotonic-engineered photothermal harnessing for waste heat management and pyroelectric generation," *ACS Nano*, vol. 11, no. 10, pp. 10568–10574, 2017, doi: 10.1021/acsnano.7b06025.

30. M. H. You *et al.*, "A self-powered flexible hybrid piezoelectric-pyroelectric nanogenerator based on non-woven nanofiber membranes," *Journal of Materials Chemistry A*, vol. 6, no. 8, pp. 3500–3509, 2018, doi: 10.1039/c7ta10175a.

31. J. H. Lee *et al.*, "Highly stretchable piezoelectric-pyroelectric hybrid nanogenerator," *Advanced Materials*, vol. 26, no. 5, pp. 765–769, 2014, doi: 10.1002/adma.201303570.

32. D. W. Kim, J. H. Lee, J. K. Kim, and U. Jeong, "Material aspects of triboelectric energy generation and sensors," *NPG Asia Materials*, vol. 12, no. 1. Nature Research, pp. 1–17, Dec. 01, 2020. doi: 10.1038/s41427-019-0176-0.

33. N. Deniz Yilmaz, "Nanojeneratörler – Çeyrek Mühendis," 2021. https://www.ceyrekmuhendis.com/nanojeneratorler/ (accessed May12, 2021).

34. X. S. Zhang, M. Han, B. Kim, J. F. Bao, J. Brugger, and H. Zhang, "All-in-one self-powered flexible microsystems based on triboelectric nanogenerators," *Nano Energy*, vol. 47. Elsevier Ltd, pp. 410–426, May 01, 2018. doi: 10.1016/j.nanoen.2018.02.046.

35. W. He *et al.*, "Recent progress of flexible/wearable self-charging power units based on triboelectric nanogenerators," *Nano Energy*, vol. 84. Elsevier Ltd, p. 105880, Jun. 01, 2021. doi: 10.1016/j.nanoen.2021.105880.

36. N. D. Yilmaz, "Triboelektrik nanojeneratörler ile enerji hasadi: teorik köken, çalişma prensibi ve çalişma modlari," *Konya Journal of Engineering Sciences*, vol. 9, no. 1, pp. 232–249, Mar. 2021, doi: 10.36306/konjes.745063.

37. M. Han, X. Zhang, and H. Zhang, *Flexible and stretchable triboelectric nanogenerator devices: Toward self-powered systems*. Wiley, 2019. doi: 10.1002/9783527820153.

38. H. Li *et al.*, "A stretchable triboelectric nanogenerator made of silver-coated glass microspheres for human motion energy harvesting and self-powered sensing applications," *Beilstein Journal of Nanotechnology*, vol. 12, pp. 402–412, May 2021, doi: 10.3762/bjnano.12.32.

39. C. Zhi and L. Dai, *Flexible energy conversion and storage devices*. Wiley-VCH Verlag GmbH & Co. KGaA, 2018. doi: 10.1002/9783527342631.

Index

Abdi, M. A., 43–56, 119–130
Abraham, B. G., 133–151, 138
Abraham, J., 81–95
Acetylene black, 338
Acrylonitrile-butadiene-styrene (ABS), 108
Activated carbons, 423
Additive manufacturing (AM); *See* 3D-printing technologies
Adekunle, A., 166
Adjemian, K. T., 147
Aerosol jet technique-based 3D printing, 113
Altin, Y., 519–535
Aluminum-air batteries, 13; honeycomb-inspired, 13
Aluminum-doped zinc oxide (AZO) thin film, 92
Amorphous silicon, 126
Anctil, A., 313
Anti-freezing hydrogel electrolyte, 381–382, *383*
Antimony selenide (Sb$_2$Se$_3$) solar cell, 252, *253*
Aqueous electrolytes, flexible, 335
Aramid nanofibers (ANFs), 82
Architectures and fundamental designs, 44; flexible batteries, 50–53; flexible solar cells, 45–50; flexible supercapacitors, 53–56
Arora, R., 172
Asymmetric supercapacitors (ASCs), 483, 532–533, *534*
Atomic force microscopy (AFM), 73, *74*
Atomic layer deposition (ALD) method, 17, 214, 275
Ayse, Ç. B., 575–585

Babu, K. F., 545
Bacterial cellulose (BC), 163
Baek, S. W., 300
Ball milling, 425
Balo, F., 307–320
Bandodkar, A. J., 135
Basol, B. M., 245
Batteries, flexible, 5, 7–13, 33–35, *36*, 323; *See also specific type*; advancement in, 53; architecture for, 50–53; lithium-ion (Li-ion) batteries, 7–12, 33, *34*, 51–52; lithium-sulfur (Li-S) batteries, 12; metal-air batteries, 12–13; zinc-ion batteries (ZIBs), *52*, 52–53
Bedeloglu, A., 519–535
Bencherif, H., 43–56, 119–130
Bendib, T., 43–56, 119–130
Bensalah, N., 324
Bera, S., 481–496
Bhargava, K., 261–282
Bicer, E., 323–340
Binary/ternary metal chalcogenide, 487–490, *489*
Binder jetting, 100, *101*
Bio-based carbon, in supercapacitors, 423, **425**
Biobased fibers, for energy devices, 89–94; cellulose-based fibers, 89–92, *91*, *93*; keratin and chitin fiber composites, 93–94

Biological fuel cells (BFCs), 158
Biopolymer-based nanofibers, 82
Biscrolling, 406
Blade-coating method, 269–270, 338
Bragg's Law, 68
Bronger, T., 187
Burschka, J., 312

Cadmium poisoning, 126
Cadmium sulfide (CdS), 127
Cadmium-telluride (CdTe) solar cells, *216*, 216–218, **217**, 228, **263**; *See also* Chalcogenide-based flexible solar cells
Çakar, S., 197–207
Cao, B., 267, 271
Cao, P., 72
Cao, Y. L., 141
Capacitive strain sensor, carbon-based, 3
Carbide-derived carbon (CDC), 54
Carbon-based electrodes, 106–108, *107*, *108*
Carbon-based flexible supercapacitors, 430; *See also* Supercapacitors (SC), flexible; bio-based FSCs, 432, 434, *435*; CNT-based FSCs, 432, *433*; graphene-based FSCs, 430, *431*
Carbon-based materials: as anode materials: for flexible Li-ion batteries, 324–331, **326**; for K-ion batteries, 364–365; for flexible supercapacitors, 2–3, *4*, 87–89, *88*, 419–423, *421*; bio-based carbon, 423, **425**; carbon nanotubes, 423, *424*; graphene, 420, 422, *422*; synthesis of, 423, 425, *426*
Carbon-based nanomaterials, 127
Carbon cloth, **326**, 329–330; MoS$_2$ nanoflake-carbon cloth composite, 330–331, *332*; NiO-carbon cloth composite anode, 330, *330*; in supercapacitors, 483–484, *484*
Carbon-conducting polymer nanocomposites, 554, 556–558
Carbon-metal oxide nanocomposites, 553–554; annealing assisted dip-coating method, 554, *555*; carbon-MnO$_2$ nanocomposite electrodes, 554; electrode preparation of, 554, *555*
Carbon-mxene nanocomposites, 558–560, *559*
Carbon nanofibers (CNFs), 82, *84*, 87–89; as flexible anode for Li-ion batteries, **326**, 327, 329; nitrogen-doped CNF (NCNF), 329
Carbon nanomaterials, 553
Carbon nanotubes (CNTs), 24, 25, 256; as electrode material in supercapacitors, 423, *424*; fibers for fiber electrodes, 527; as flexible anode for Li-ion batteries, 324, *325*, **326**; multi-wall CNTs (MWCNTs), 324; single-wall CNTs (SWCNTs), 324
Carboxymethyl cellulose (CMC), 338
Carlson, D. E., 178

Caskey, S. R., 506
Cathode materials, for flexible Li-ion batteries, 331,
        333; LiCoO$_2$/CNT composite, 333, *334*;
        LiMn$_2$O$_4$/CNT paper-based flexible cathode,
        331, 333; LiMn$_2$O$_4$ nanowall/carbon cloth
        composite, 331, 333
Cavaliere, S., 87
Cellulose-based fibers, 89–92, *91*, *93*
Ceramic nanofibers, 82
Ceramics use, for CIGS solar cells development, *248*,
        248–249
Chalcogenide-based flexible solar cells, 242; CdTe solar
        cells, substrates for: ceramics, 250; metal foils,
        249; polymer, 249, *250*; CIGS solar cells,
        substrates for, 244; ceramic and other materials,
        248–249; metal foils, 245–247, *248*; polyimide,
        244–245, *246*, *247*; CZTS/CZTSSe solar cells:
        metal foils, 250–251, *251*; polymer and other
        materials, 252; UTG, 251; fabrication issues and
        challenges, 252–253, *254*; crack initiation,
        253–254; electrodes issues, 255–256;
        performance degradation under bending,
        254–255; stability and scalability issues, 256;
        substrate choice, 255; future prospects and
        strategies, 256; absorber optimization, 256;
        development of transparent/semitransparent
        solar cells, 257–258; integration with existing
        technologies, 258; machine-learning algorithms,
        257; material database, 257; new chalcogenide
        materials, 256–257; optimizing every
        component, 257; rigorous testing, 257; merits of,
        243–244; progress and development of, *243*;
        properties of metal, ceramic, and plastic
        substrates, comparison of, 242, *243*; Sb$_2$Se$_3$ solar
        cell, 252, *253*; thin-film, 243–244
Chander, S., 227–238
Chapman, D. L., 423
Characterization techniques, 61; atomic force
        microscopy, 73; cyclic voltammetry, 69–70;
        electrochemical impedance spectroscopy, 72–73;
        Fourier transform infrared spectroscopy, 77–78;
        galvanostatic charge-discharge test, 70–72;
        inductively coupled plasma-mass spectroscopy,
        76–77; scanning electron microscopy, 61–65;
        secondary ion mass spectroscopy, 73, *75*, 76;
        transmission electron microscopy, 65–68; x-ray
        diffraction, 68–69
Chathuranga, H., 397–413
Chatterjee, S., 233
Chemical activation process, 162
Chemical bath deposition (CBD), 485, 487
Chemical vapor deposition (CVD), 163, 180, 402,
        402, 425
Chen, C. R., 562
Cheng, F., 551–570
Cheng, M., 110
Chen, L., 556
Chen, M., 381
Chen, R., 329

Chen, W., 558
Chen, Z., 382
Chetty, R., 133–151
Chicken feather waste-based fibers, 93
Chitin fibers, 94
Chu, C. Y., 135
Chung, J., 273
Cindrella, L., 150
Coaxial assembly, 411
Cobalt oxide, as electrode materials in
        supercapacitors, 471
Co-evaporation synthesis method, 247
Conductive polymers (CPs), for FSCs, 554, 556–558
Co$_3$O$_4$ nanostructures, 503, *504*
Copper antimony sulfides nanowires, 488
Copper (Cu) foils, 274
Copper indium gallium diselenide (CIGS), 14, 49,
        126–127, 212–213, 242, **263**;
        See also Chalcogenide-based flexible solar cells
Copper tins sulfides (CTSs), 495
CsPb(I$_{1-x}$Br$_x$)$_3$ solar cells, 219, *221*, 221–223, *222*
Current collectors, metal oxide-based, 149–150
Cyclic voltammetry (CV), 69–70, *70*, *71*, 493–495
Cyclopentasilane (CPS), 187
Cyclopentasiloxane, 187

Dai, L., 551–570
Das, S., 270
Decal transfer, 134
Dendrite growth, 364
Design of flexible energy devices, 30
Dessie, Y., 145
Dey, A. N., 51
Dhanasekaran, P., 133–151
Dip coating, 404–406, *405*
Dipotassium terephthalate, 368
Dipping and drying method, 54
Direct energy deposition (DED), *101*, 102
Direct ink writing (DIW), 100, 106, 448
Direct liquid fuel cells (DLFC), 134, 135, *136*;
        See also Fuel cells
Di Vece, M., 185
Docampo, P., 272
Dong, H., 375–393
Dong, X., 335
Doping metal oxides, 142
Double layer adsorption method, 129
Dual-flow reactor synthesis method, 293, *293*
Dubau, L., 147
Du, H. J., 31
Dunn, B., 385
Du, Z., 298
Dye-sensitized solar cells (DSSCs), 14–16, *17*, 198, 231,
        263; based on metal oxide, 198–200, **201**;
        conversion-efficiency values, 199; fiber-
        structured, 16, *17*; flexible materials and
        fabrication process for, 46–47, *47*; monolithic
        back contact DSSCs, 199; PANI-based
        counter-electrode in, 46; photoanode in, 46;

power conversion efficiency (PCE) of, 46, **47**; structure design and basic concept, 45–46, *46*; studies on, **201**; wire-based, 199, *200*

Edge-functional graphite sheets (EFGs), 430
Eftekhari, A., 359
Ehtesabi, H., 157–168
Elastic-conductive polymers, 26
Electrically conducting polymers, 26
Electric vehicles, 227
Electrochemical deposition, *403*, 404
Electrochemical double-layer capacitors (EDLCs), 2, 35, 426–428, *427*, 462, 520; carbon-based materials in, 2–3, *4*, 462; Gouy–Chapman model, 426, *427*; Helmholtz model, 426, *427*; Stern model, 426–428, *427*; textile, 3, *4*
Electrochemical energy storage devices, 417–417
Electrochemical impedance spectroscopy (EIS), 72–73
Electrodeposition, 134, 485
Electro-etching process, 483
Electrohydrodynamic redox printing (EHD-RP), 113
Electrolytes; *See also* Solid-state electrolytes (SSE); in fuel cells, 147–149; in Li-ion batteries, 333–335; in Na-ion batteries, 346, *348*, 349; in supercapacitors, 465, **465**
Electronics films, 27
Electron transport layer (ETL), 272
Electrophoretic deposition (EPD), 150
Electrospinning, 17, *18*, 404, *405*, 406, *407*
Electrospraying, 404
Electrospun nanofibers, 84, 94–95
Elizabeth, I., 324
El-Kady, M. F., 54, 430
Encapsulation, 130; of FPSCs, 275, **278**
Energy conversion devices, 31, 60–61; flexible solar cells (*See* Solar cells, flexible); nanogenerator, 31; other flexible generators, 32–33; photovoltaic, 31–32
Energy harvesting, 166
Energy storage devices (ESDS), 60, 124–125, 461, 540; *See also specific type*; 1D configuration of: fiber-type, *37*, 37–38; spine-type, *37*, 38; spring-type, *37*, 38; 2D configuration of, *37*, 38; layered sandwich, 39; planar interdigital, 39; self-similar geometries, *37*, 39; wavy structures, *37*, 39; zig-zag like structure, *37*, 39; 3D configuration of: honeycomb architecture, *37*, 39; origami/kirigami, *37*, 39; flexible batteries, 33–35, *36*; flexible supercapacitors (*See* Supercapacitors (SC), flexible)
Environmental effects, of flexible devices, 120–121; designs strategies and processing routes for safe devices, 128; encapsulation, 130; fluor-doped tin oxide substrates, 128–129; improving processing routes, 129; lead-free perovskite, 128; recycling, 129–130; toxic materials replacement, 128–129; flexible materials, 126; amorphous silicon, 126; cadmium, 126; carbon-based nanomaterials, 127; copper indium gallium diselenide, 126–127; lead

halide, 127; tellurium and indium, 127–128; toxic flexible substrate, 128
Ercelik, M., 149
Espinosa, N., 313
Ethylene glycol (EG), 338
Extrinsic self-healing, 30

Fabric-based supercapacitors, 520, 524, 528–529, *530*; *See also* Textile-based supercapacitors (TSCs); carbon fabric-based electrodes, 529; conventional fabric-based electrodes, 529; metal mesh-based electrodes, 529
Face mask, pyroelectric nanogenerator-based, 581, *582*
Facile hydrothermal synthesis, 5
Fadla, M. A., 222
Fang, Z., 335
Fan, Z., 333
Faradaic charge technique, 428
Faruk, Ü. Ö., 575–585
Feng, J., 273
Ferric oxide, as electrode materials in supercapacitors, 471
Fiber-based supercapacitors, 520, 523, *523*, **528**; *See also* Textile-based supercapacitors (TSCs); coaxial fiber structure, 523–524; parallel fiber structure, 523; twisted fiber structure, 523
Fiber-shaped electrodes, for supercapacitors, 524, *525*; carbon nanotube fibers for, 527; conventional fibers for, 526; graphene yarn for, 526–527; hybrid fiber-based electrodes, 527; metal thread/wire/yarn for, 526
Fiber-shaped PSCs, 274
Fiber-type configurations, ESDs, *37*, 37–38
Findik, F., 197–207
Flexible batteries (FBs) *See* Batteries, flexible
Flexible electronic devices, FSCs for, 568, *569*
Flexible electronics, 119–120, *120*; *See also* Flexible energy devices
Flexible energy devices, 1–2, *2*, 24, 44, 59–60, 120; *See also* Energy conversion devices; Energy storage devices (ESDS); batteries, 5, 7–13; characterization, 61; *See also* (Characterization techniques); configuration of, 102; active materials, 102; EES devices, 104–105, *105*; EES electrodes, 102–104, *103*; solid-state electrolytes, 104; nanotechnology for, 44–45; proton-exchange membrane fuel cells, 14; solar cells, 14–18; supercapacitors, 2–5
Flexible fiber supercapacitors, 56
Flexible perovskite solar cells (FPSCs), 262; applications of, *281*, 282; back contact, **277–278**, 280; challenges and future perspectives, 262, 278; environmental stability, 278, 280; large-area fabrication, 280–281; manufacturing cost, 280; mechanical stability, 280; toxicity, 281; charge transport layers, 272–273; electron transport layer (ETL), 272, **276–277**; hole-transport layer (HTL), 273, **277**; development of, 264–265, **266**, **267**; device structure of, 263–264;

flexible substrates, 264; inverted structure (p-i-n), 263, *264*, **267**; normal structure (n-i-p), 263, *264*, **266**; encapsulation of, 275, **278**; techniques for, 275; flexible substrates, 273, **276**; fiber-shaped PSCs, 274; metal substrates, 274; other, 274; polymer (plastic) substrates, 273–274; ultrathin flexible glass, 274; willow glass, 274; laboratory-scale fabrication methods, 265; spin coating, 265, 267–268; thermal evaporation, 268; large scale fabrication methods, 269; blade coating, 269–270; inkjet printing, 269; spray coating, 270–271; lead-free FPSCs, 281; materials for, 271–275, **276–278**; PCE of, 262, 265; perovskite absorber layer, 271–272; recycling of, 275, 278, *279*; transparent conducting layer (TCL), 275, **276**

Flexible supercapacitors (FSCs); *See* Supercapacitors (SC), flexible

Flexible transparent supercapacitors (FTSCs), 474–476, *475*

Flexible zinc-ion batteries (FZIBs); *See* Zinc-ion batteries (ZIBs)

Fluid assembly method, 443

Fluorine-doped tin oxide (FTO), 128–129, 178, 198

Formamidinium lead iodide (FAPbI$_3$), 271

Formamidinium tin iodide (FASnI$_3$), 272

Four-dimensional (4D) printing, 113–114

Fourier transform infrared (FTIR) spectroscopy, 77–78

Fthenakis, V. M., 127

Fuel cells, *32*, 32–33, 134; architecture and materials for, 134–137; for flexible and portable applications, 134; future research perspective, 150–151; material challenges for, 137–138; metal oxides in: bipolar plates and substrates, 149; as catalysts, 143–146; as catalyst supports, 146–147; as co-catalysts, 146; current-collector, 149–150; as electrodes (GDL/MPL), 150; as electrolytes/membranes, 147–149; strategic use of metal oxides for, 138; compositing with carbon/metal-based materials, 142–143; doping, 142; morphological control, 138–139, *139*; oxygen-vacancy control, 141–142; phase structure engineering, *140*, 141

Fused deposition modelling (FDM), 100, *101*

Fu, Y. P., 61

Gallium arsenide solar cells, **263**

Galvanostatic charge-discharge (GCD) method, 70, 72

Gao, Y., 349

Gao, Z., 106, 544

García-Valverde, R., 313

Garnett, E., 180

Gas diffusion layer (GDL), 150

Germanium (Ge) nanowires, 327

Germanium quantum dot (Ge-QD), 9

Ghosh, S., 554

Global warming, 227, 228

Goetzberger, A., 172

Goodman, A. J., 292

Gouy, M., 423

Goyal, C., 501–514

Graphene, 526; as electrode material in supercapacitors, 3, 420, 422, *422*; as flexible anode for Li-ion batteries, 324, 326, *328*

Graphene-based FSCs, 443, *444*

Graphene foam (NGF), 9

Graphene oxide (GO), 26, 443; fibers, 526–527

Gratzel, M., 45

Gribov, B. G., 187

Guo, Z., 503

Gupta, R. K., 1–19, 417–435, 501–514

Gurav, S. R., 539–549

Gwang, T., 547

Halder, S., 481–496

Han, G. S., 274

Hao, M., 286

Harrison, D., 523

He, B., 350

He, D., 99–114

He, G., 375–393

Hematite, 504–505

Heo, J. H., 272

Herz, K., 247

He, W. W., 175

He, X., 25

Hexacyanoferrates (HCF), 368–370, *369*

High cracked-selenium (HC-Se) method, 213

Hole-transport layer (HTL), 273

Hosseinnezhad, M., 211–224

Hot injection method, for QDs synthesis, 292, *293*

Hou, Z., 72

Hsieh, B. J., 142

Hsieh, S. S., 135

Hsu, F. K., 135

Huang, C. F., 135

Huang, S., 382

Huang, S. Y., 147

Huang, T., 84

Huang, X., 95

Hu, L., 92, 286, 300

Hwang, K., 270

Hybrid fiber-based supercapacitor electrodes, 527

Hybrid nanocomposites, 142

Hybrid polymer electrolytes (HBPEs), 378, 379

Hybrid supercapacitors (HCs), 2–3, 462, 521

Hydrogel polymer electrolytes (HPEs), 378, 379

Hydrogen selenide, 127

Hydrothermal carbonization (HTC), 425

Hydrothermal deposition technique, 402–404, *403*

Hydrothermal method, 485

Indium tin oxide (ITO), 25, 178, 198, 223, 474; substrates, 128–129, 255, 264, 280

Indium toxicity, 127–128

Inductively coupled plasma-mass spectroscopy (ICP-MS) technique, 76–77, *77*

Inganäs, O., 554

Ingsel, T., 1–19
Inkjet printing, 269; *See also* Material jetting
Inorganic-based flexible solar cells, 212; CdTe solar cells, *216*, 216–218, **217**; CIGSe-based solar device, *212*, 212–213, *213*; comparison of different types of, **224**; $CsPb(I_{1-x}Br_x)_3$ solar cells, 219, *221*, 221–223, *222*; CZTSSe solar cells, 213–216, *215*; environmental and economic concerns, 223; Sb_2Se_3 solar cells, *218*, 218–219
Inorganic fibers, for flexible energy devices, 86, *86*
Inorganic nanomaterials, for flexible energy devices fabrication, 25; 1D materials, 25; 2D materials, 26
Ion-exchange membrane, 164–165
Ionogels, 349
Ioroi, T., 142
Iron oxide (Fe_2O_3) materials, 503–505
Ismail, B., 575–585
Itagaki, Y., 150

Jani, R., 261–282
Jo, E., 250
Justin, P., 146

Kaempgen, M., 432
Kaltenbrunner, M., 32
Kang, C-W, 461–477
Kang, F., 379
Kang, Y., 432
Karthick, S., 145
Katariya, A., 23–40, 59–79
Kaur, K., 241–258
Keratin and chitin fiber composites, 93–94
Kesterite solar cells, environment-friendly, 256
Ketpang, K., 147
Khang, D. Y., 175
Kim, D., 249
Kim, H. I., 286
Kim, H.-S., 312
Kim, J. K., 346
Kim, J. T., 108
Kim, M., 142
Kim, S. H., 135
Kinayyigit, S., 323–340
K-ion based batteries; *See* Potassium-ion batteries (PIBs)
Knitting process, 411, *412*
Kojima, A., 31
Kolya, H., 461–477
Koo, M., 335
Ko, Y., 546
Kramer, I. J., 286
Kranz, L., 249
Kulbak, M., 312
Kumar, A., 417–435, 501–514
Kumar, M., 241–258
Kumar, M. H., 262, 272
Kumar, P., 285–302

Kumar, S., 285–302
Kuribayashi, I., 92

Lab-on-a-chip sensors, 167
Lacey, S. D., 108
Laser-patterning-based corrugation technique, 176
Laser-reduced graphene (LRG) based FSCs, 443
Laser-scribed graphene (LSG) based FSCs, 443
Laser-scribed graphene films, 54
Layered sandwich-type configuration, 39
Layer transfer technique (LTP), 176
Lead-based perovskite, 31
Lead-free perovskite, 128
Lead toxicity, 127
Lee, J. H., 25, 31
Lee, K. S., 46
Lehtimaki, S., 484
Leijonmarck, S., 92
Lewis, J. A., 110
Liang, K., 312
Li, C., 146
Li, G., 180, 185
Ligand, 295
Li, H., 558
Li-ion flexible batteries (LiBs); *See* Lithium-ion (Li-ion) batteries
Li, K., 219, 252
Li, L., 63, 142, 143
Lin, Y., 179
Li, Q., 139
Liquid electrolytes, 137
Liquid step reduction technique, 92
Lithium-air battery system, 12
Lithium-ion (Li-ion) batteries, 7–12, 33, *34*, 40, 87, 92, 323–324, 540; anodes in: carbon materials, 324–331; MXenes, 331; cathodes in, 333; conversion-type cathodes, 7, *9*; electrolytes in, 333–335; aqueous electrolytes, 335, *336*; liquid-type electrolytes, 334; lithium salts in, 335; solid polymer electrolytes, 335; fabrication of, 338; printing, 338; spray painting, 338; flexibility of, 323; flexible materials for, 51–52; Ge-QD@NG/NGF/PDMS fabrication process, 9, *11*; intercalation cathodes, 7, *9*; island-bridge electrode design, 8, *10*; mechanical deformations in, 335; Miura fold-type battery designs, 9; modern designs of, 125, *125*; one-dimensional electrode designs, 12; structure design and basic concept, 51, *51*; structures, 335–338; cylindrical cell, 335; fiber-type structures, 335; island connection structure type, 338, *339*; paper-folding Li–O_2 battery pack, 338, *339*; thin film structure, 335, *337*; wavy structure, 335, *337*; wire-type FLIB, *337*; thin-film electrode design, 8; wavy electrode design, 9, *10*; working mechanism of, 398, *400*
Lithium-ion capacitors (LICs), 443
Lithium-ion hybrid supercapacitors (LIHSs), 443
Lithium-manganese oxide (Li-MnO_2) batteries, 35, *36*

Lithium-sulfur (Li-S) batteries, 12
Liu, C., 110
Liu, D., 310
Liu, J., 422
Liu, M., 268, 532, 545
Liu, Q.-C., 338
Liu, Y., 359–372
Li, W. C., 547
Li, X., 556
Li, Y., 25, 298
Li, Z., 99–114
Low-pressure chemical vapor deposition technique
        (LPCVD), 178
Luo, J., 385
Lu, W., 551–570
Lv, Z., 554, 564

Ma, F., 298
Mahanty, S., 505
Mahapatra, A., 285–302
Maher, F., 106
Maiyalagan, T., 135
Ma, L., 390
Malinkiewicz, O., 312
Manganese-doped nanofibers, 84, 87
Manganese oxide-based supercapacitors, 467, **470**
Manganese oxides, 143, 145
Manjunatha, K. N., 171–191
Man-made fiber, 541
Marriam, I., 397–413
Marshall, K. P., 312
Ma, S., 273
Mashkour, M., 163, 166
Masuda, T., 187
Material jetting, 100
Materials, for flexible devices, 24–25; inorganic
        nanomaterials, 25; 1D materials, 25; 2D
        materials, 26; organic materials: organometallic
        complexes, 27; polymers, 26; self-healing, 29;
        intrinsic self-healing polymers with reversible
        bond, 29–30; through exhaustion of healing
        agent, 30; strain and strength of, 27; flexible
        substrates and membranes, 27; thickness of
        compound/active layer, 27; wearability
        assessments: softness, 28, 28–29;
        stretchability, 29
Matulionis, I., 249
Matzger, A. J., 506
Mechanical distortion method, 27
Meddour, A., 119–130
Meddour, F., 43–56
Mei, J., 345–353
Melt-spinning technique, 407, 408
Membrane electrode assembly (MEA), 134
Meng, X., 272
Metal-air batteries, flexible, 12–13, 13
Metal chalcogenide-based FSC system; See also Metal
        chalcogenides; binary and ternary, **496**;

developments in, 493–495; electrode system
        working, 490–493, 491
Metal chalcogenides, 484; binary or ternary, 487–490,
        489; unary, 485–487, 486, **488**
Metal-containing selenides, as anode material for Na-ion
        batteries, 353
Metal foil substrate: CdTe solar cells, 249; CIGS solar
        cells, 245–247, 248; CZTS/CZTSSe solar cells,
        250–251, 251
Metallic nanofibers, 86–87
Metallic nanowires (NWs), 25
Metal oxide-based bipolar plates and substrates, 149
Metal oxide-based catalysts, 143–146
Metal oxide-based co-catalysts, 146
Metal oxide based-flexible supercapacitors, 464;
        capacitance of MOs electrodes, 464; cobalt
        oxide-based electrodes, 471; drawbacks of MOs,
        464; minimizing drawbacks of, 464–465;
        electrochemical electrodes, comparison of, **465**;
        manganese oxide-based supercapacitors, 467,
        470; nickel oxide-based electrodes, 471; recent
        progress of, **473**; ruthenium oxide-based
        supercapacitors, **466**, 466–467, 468, 469;
        titanium oxide-based supercapacitors, 467,
        470–471, 472; transparent supercapacitors,
        474–476, 475; working potential windows of
        electrode materials in aqueous electrolytes, 473;
        zinc oxide-based supercapacitors, 467, 470
Metal oxide-based materials, for flexible and portable
        fuel cells; See Fuel cells
Metal oxide-based perovskite solar cells, 203–205, 206,
        235, 235, 237–238; advancement in, 235–237
Metal-oxide nanostructures (MONs), 228–231, **231**, 235
Metal oxides and sulfides, for flexible devices, 3–5, 6
Metal oxides based flexible solar cells, 198, 231–233;
        a-Si:H thin-film solar cells, 205; CuIGSe solar
        cells, 207; dye-sensitized solar cells, 198–200,
        201; with epitaxial Cu_2O absorber layer, 233,
        234; gallium arsenide (GaAs) type solar cells,
        207; $NiO/Cu_2O/ZnO/SnO_2$ heterojunction solar
        cell, 233, 234; organic solar cells, 201–203, 202,
        203; perovskite solar cells, 203–205, 206,
        235–237; transparent window layers in, 232;
        $ZnO-Cu_2O$ flexible solar cells, 232
Metal sulfides, as anode material for Na-ion
        batteries, 350
Methylammonium lead tri-iodide ($MAPbI_3$), 271
Mica, 274
Michael, R., 176
Microbial fuel cells (MFC), 134, 135, 136, 157–158;
        See also Fuel cells; applications, **164**; energy
        harvesting, 166; sensors and portable power
        machines, 167; wastewater treatment, 167; basics
        of, 158–161, 160; construction of flexible MFCs,
        162; electrode materials, 162; bacterial cellulose,
        163; carbonaceous material, 162; graphene sheet,
        163; polypyrrole, 164; fabrication, 166; future
        perspective: anode manipulation, 167; large-scale

uses, 167; membrane-free MFC, 168; instrumental bases, 159; materials used in, *163*; mediator-less, 159; membrane in, 164–165; microorganisms used in, 165, *165*; properties of, **164**; single-compartment, 159, 161, *161*; two-compartment, 159, *161*

Microfibers, 81

Micro-meter devices, 167

Microporous layer (MPL), 150

Micro-supercapacitors (MSCs), 462, 502; 3D printing for fabrication of, 442–443, 449, *450*; inter-digital, 483; planar, 454–455

Microvascular self-healing system, 30

Minami, T., 233

Min, J.-H., 252

Mink, J. E., 163

Mixed transition-metal oxides (TMMOs), for flexible supercapacitors, 506; $MnCo_2O_4$ materials, *508*, 508–510, *509*; $Ni_xCo_{3-x}O_4$ materials, 510–513, *511*, *512*; $ZnCo_2O_4$ materials, 506–508, *507*

Miyasaka, T., 203, 236

Mn-Co-zeolitic imidazolate framework, 508, *508*

Mo, F., 385

MOFs-derived metal oxides, for flexible supercapacitors, 502; binary metal oxides: Ce-oxides materials, *505*, 505–506; co-oxides materials, 503, *504*; Fe-oxides materials, 503–505; MOFs-derived metal oxide/composite, *513*, 513–514; ternary or mixed transition-metal oxides (TMMOs), 506; $MnCo_2O_4$ materials, *508*, 508–510, *509*; $Ni_xCo_{3-x}O_4$ materials, 510–513, *511*, *512*; $ZnCo_2O_4$ materials, 506–508, *507*

Mohanraju, K., 146

Møller, C. K., 235

Molybdenum oxides, 143, 145; as electrode materials in supercapacitors, 471

Molybdenum trioxide nanowires, 521, 529

Mondal, S., 461–477

Mudd, G. W., 485

Multiple exciton generation (MEG), 286, 290, 290–291

Multiwalled carbon nanotubes (MWCNTs), 324, 485, 529

Murata, H., 327

MXenes: as anode materials for flexible Li-ion batteries, 331, *333*; for flexible supercapacitors, 448, *449*, *450*

Nafion composite membranes, 147, *148*

Nafion ionomer, 147

Nah, C., 461–477

Na-ion batteries, flexible; *See* Sodium-ion batteries (SIBs)

Nam, J. K., 312

Nanocomposite-based FSCs, 551–570; applications of, 566; flexible electronic devices, 568, *569*; wearable electronic devices, 566–568, *567*; carbon nanomaterial-incorporated nanocomposites for, 553; carbon-conducting polymer nanocomposites, 554, 556–558; carbon-metal oxide nanocomposites, 553–554; carbon-mxene nanocomposites, 558–560; carbon nanomaterials, 553; device configurations of, 560; 1D fiber-shaped FSCs, 560–562, *561*; 2D film-shaped FSCs, 562–564, *563*; 3D structural FSCs, 564–566, *565*; future perspective, 568, 570

Nanocomposites, for flexible devices, 5, *7*; *See also* Nanocomposite-based FSCs; CNT/Co_3O_4@NiO/GN composites, *7*; PEDOT coated cellulose paper, 5, *8*

Nanofibers, 81–82; in flexible energy devices, 84, *85*; biobased fibers, 89–94; carbon-based fibers, 87–89, *88*; inorganic fibers, 86, *86*; metallic nanofibers, 86–87; synthesis methods and applications of, *83*; types of: aramid fibers, 82; biopolymer fibers, 82; carbon fibers, 82, *84*; ceramic fibers, 82

Nanofiber yarn electrospinning, 406, *407*

Nanogenerators (NGs), 31, *31*, 575; definition of, 575; paper-based triboelectric, 31, *31*, 32; piezoelectric, 577–578; pyroelectric, 576, 580–582, *581*, *582*; textile-based, 577; triboelectric, 576, 582–584, *585*; types of, 575–576, *576*

Nanotechnology, for flexible energy devices, 44–45

Nanowires, 288

Nejand, A., 268, 274, 275

Neopentasilane (NPS), 187, 189, *190*

Nickel oxide, as electrode materials in supercapacitors, 471

Ning, F., 135

Nisika, 241–258

Nitrogen-doped graphene (NG), 9

Niu, Z., 382

Ni, W., 441–456

Noninjection heat-up method, for QDs synthesis, 292–293, *293*

Nonwood fibers, 543

Nozik, A. J., 290

On-chip fuel cells, 137

One-dimensional fiber-shaped FSCS, 560–562, *561*

O'Regan, B., 45

Organic-inorganic hybrid perovskite solar cells (OIHPSCs), 229

Organic solar cells (OSCs), flexible, 31–32, 50, *50*, 175, 198, 231, **263**; based on metal oxide, 201–203, *202*, **203**

Organic waste, 158

Organometallic complexes, 27

Oxygen evolution (OER) reaction, 12, 13

Oxygen reduction reaction (ORR), 12, 13, *13*, 159

Oxygen vacancies, 141–142

Özacar, M., 197–207

Pal, L., 90

Pandey, M., 271

Pandit, B., 485

Pan, Z., 562, 564

Paper-based energy device, flexible, 90, *90*
Paper-based flexible supercapacitors, 476
Paper-based fuel cell (PBFC), 137
Paria, S., 461–477
Parisi, M. L., 313
Park, H., 529, 566, 568
Parkin, I. P., 375–393
Park, Jeong-Il, 274
Park, M.-S., 338
Park, N. G., 231, 236
Paul, S., 171–191
P³awecki, M., 233
PEDOT:PS, 275
Peng, C.-Y., 251
Peng, L., 149
Peng, Z., 299
Perovskite oxide-based flexible ultracapacitor, 5, *6*
Perovskite quantum dots (PQDs), 299–300, *301*
Perovskites as catalysts, 145
Perovskite solar cells (PSCs), 16–17, 31–32, 198, 231, **263**, 308, 310–313; *See also* Flexible perovskite solar cells (FPSCs); Solar cells, flexible; analytic hierarchy process (AHP) analysis, 315–317; characteristics determined, 315, **316**; comparison's scale, 315, **316**; decision matrix, **316**; overall scores, *319*; relative priorities of characteristics, *317*; schematic overview, *319*; scores of alternatives, *318*; conductive oxide substrate, 204; conversion efficiency, 203, 204; efficiency of, 312; elastic, 313–314; electron transporter layer, 204–205; electrospinning process of perovskite fibers, 17, *18*; flexible, 17, *18*; future of, 314–315; hole transporter material, 205; metal oxide-based, 203–205, *206*, 235–237; organic-inorganic halide perovskites, 17, 31; production methods for perovskite, 310, *311*; QDs based, 299–300, *301*; strategic evaluation method, 315; studies on, **204**; titania deposition techniques for, 17; working mechanism, 313–314
Pesce, A., 110
Photoluminescence quantum yield in QDs (PLQY), 294
Photoluminescence (PL) spectroscopy, *294*, 294–295
Photophysical study, 294
Photovoltaic energy, 227, 262; flexible PV technologies, 262, **263**
Photovoltaic solar cells, flexible, 14, 48–50, 172, 308; *See also* Solar cells, flexible; classification of, 172, *174*; hybrid, 175; inorganic, 174; organic, 174–175; silicon-based (*See* Silicon photovoltaic solar cells)
Physical-vapor deposition (PVD), 275
Piezoelectric effect, 578
Piezoelectric nanofibers, 84
Piezoelectric nanogenerators, 93, 577–578; textile-based, 579–580, *580*
Planar interdigital configuration, 39
Plasma-enhanced chemical vapor deposition (PECVD), 126, 180, 275, 487

Poly(acrylic acid) (PAA), 390
Polyaniline (PANI), 470, 554, 556
Polydimethylsiloxane (PDMS), 26
Poly(3,4-ethylenedioxythiophene) (PEDOT), 556
Polyethylene naphthalate (PEN), 198, 264, 273
Polyethylene terephthalate (PET), 198, 264, 273, 286
Polyimide (PI) substrates, 244–245, *246*, 264, 274
Polylactic acid (PLA), 108
Polymer-based electrodes, 108–109, *109*
Polymer electrolyte membrane fuel cells (PEMFC), 134, 135, *136*, 141; *See also* Fuel cells
Poly(N-isopropyl acrylamide) (PNIPAM), 382, *384*
Polypyrrole (PPy), 554, 556
Poly (dimethyl) siloxane (PDMS), 9
Polyvinylidene fluoride (PVDF), 578, 581; nanofibers, 95
Polyzwitterion electrolyte, 385
Potassium-containing polyanionic compounds, 366
Potassium-ion batteries (PIBs), 359–360; advantages of, 360; anode materials for: alloying-type compounds, 367, *367*; carbon materials, 364–365; organic compounds, 368; phosphorus compounds, 365, *366*; titanium-based compounds, 365–366; cathode materials for: hexacyanoferrate, 368–370; layered metal oxides, 370; organic materials, 371; polyanionic compounds, 370; challenges in application of, 371–372; solid electrolyte interphase (SEI), 361; comparison with Li-ion and Na-ion interfaces, 362–363; effects of electrolyte selection on, 363; electrolyte reduction-related thermodynamic behavior, 361–362, *362*; mechanical stability of, 363–364; working principles, 360–361, *361*
Power-conversion efficiency (PCE), 31–32
Pradhan, B., 285–302
Prajapati, U. K., 59–79
Processing routes, improving, 129
Proppe, A. H., 292
Proton-exchange membrane fuel cells (PEMFC), 14; flexible hydrogen generator fabrication process, 14, *15*; prototype flexible, 14, *15*; traditional, 14, *16*
Prussian blue, 368
Pseudocapacitive conducting polymer nanosheets, 452
Pseudocapacitors (PCs), flexible, 2, 3, 35, 428–429, 462, 521; charge-storing capability, 35; intercalation, 462; metal oxides and sulfides in, 3–5, *6*; redox pseudo-capacitance, 462
Pushparaj, V. L., 54, 92
Pyroelectric effect, 580
Pyroelectric nanogenerators, 576, 580–582, *581*, *582*
Pyrolysis, 425

Qiao, S. Z., 390
Qiu, J., 312
Qiu, L., 274
Qiu, Y., 506
Quantum confinement effect, 289

Quantum dots (QDs), 288–289; density of states for, *288*, 289; optoelectronic properties of, 288; quantum confinement effect, 289

Quantum dots-sensitized solar cells (QDSSCs), 45, 231, 298–299; efficiency with different QDs and photo-anode, **49**; flexible materials and fabrication process for, 48; structure design and basic concept, *47*, 47–48

Quantum dots solar cells (QDSCs), 198, **263**, 285–287; development and performance of, **287**; flexible QDs sensitized solar cells, 298–299; perovskite quantum dots (PQDs) solar cells, 299–300, *301*; QDs based heterojunction solar cells, 295–298; QD-silicon hybrid solar cells, 300, 302, *302*; surface engineering, 295, *295*; synthesis of, 292–294, *293*, *294*; theory related to: multiple exciton generation, *290*, 290–291; quantum size effect, *288*, 288–289; ultrafast charge transfer, 291–292

Que, M., 298

Rajeswari, J., 147
Ramasamy, K., 490
Rani, J., 23–40, 59–79
Ranjbar, Z., 211–224
Rao, Li, 272, 275
Rasal, A. S., 539–549
Recycling of FPSCs, 129–130, 275, 278, *279*
Reghunadhan, A., 81–95
Reinhard, P., 245
Remeika, M., 270
Renewable energies, 227, 242, 262
Ren, J., 335
RF-cracked Se technology, 213
Robocasting, 448
Rocking chair mechanism, 360, 377, 398
Roes, A., 313
Roger, J. A., 25
Roll-to-roll (R2R) manufacturing technique, 49, 175, 264, 269
Romeo, A., 250
Ronald, S., 175
Roy, S., 481–496
Ruthenium oxide-based supercapacitors, *466*, 466–467, *468*, *469*
Ruthenium oxides, 146

Salarou, I., 171–191
Salavei, A., 218
Sang, Z., 327
Sarikaya, A., 150
Sari, P. S., 81–95
Sb$_2$Se$_3$ solar cells, *218*, 218–219
Scanning electron microscopy (SEM), 61–65, *62*; lithium cobalt oxide (LCO) and graphite electrodes, 63–65, *64*; lithium-ion batteries, 63, *64*; operating modes, 61; polyaniline (PANI) nanofibers on, 62–63, *63*
Schackmar, F., 269

Schaller, R. D., 291
Screen printing, 134; for flexible supercapacitors, 442–443
Secondary ion mass spectroscopy (SIMS), 73, *75*, 76; dynamic, 76; static, 73, 76; time of flight SIMS (Tof-SIMS), 76
Self-healing mechanism, 29; intrinsic self-healing polymers with reversible bond, 29–30; through exhaustion of healing agent, 30
Selvaganesh, S. V., 133–151
Selvan, R. K., 434
Semiconductor polymers, 26
Seok, S. I., 236, 312
Sequential high vacuum evaporation technique, 213–214
Seung, Y. M., 178
Shang, Y., 25
Shende, P. D., 171–191
Shen, Z. X., 333
Shi, L., 441–456
Shockley–Quessier efficiency, 286
Shockley, W., 174
Silane gas, 126
Silicon: bulk wafer technology, 49; crystalline, 49, 176, 177; as flexible anode for Li-ion batteries, 326–327
Silicon nanopyramid solar cells, 183–185, *186*; ultrathin c-Si flexible solar cell, 185, **187**, *188*
Silicon nanowire (SiNW) solar cells, 180–183, **185**; advantages of, 180; device structure of, 180, *181*; fabrication of, 180–181, *182*; flexible radial tandem junction (RTJ), 181, *184*; high power-to-weight ratio (PTWR), 181
Silicon oxycarbide ceramic (SiOC) fibers, 327
Silicon photovoltaic solar cells, 172, 175, 189–191, 198; flat flexible solar cells, 176; flexible amorphous silicon (a-Si) solar cells, 172, 176–179, *178*, **179**; flexible crystalline silicon solar cells, 176, **177**, *177*; hybrid solar cells, 175; hydrogenated amorphous silicon (a-Si:H) solar cells, 177–178, *178*; inorganic material-based, 174, 175; multi-junction solar cells, 174; nanostructured silicon, 172, 179; silicon ink-based solar cells, 187–189, **191**; silicon nanoparticles for solar cells, 185–187; silicon nanopyramid solar cells, 183–185; silicon nanowire flexible solar cells, 180–183; organic polymer-based, 174–175; silicon-wafer-based, 172
Silicon thin films, microcrystalline, 49
Silk fibroin, 94
Silver (Ag) nanowires, 275
Silver-zinc (Ag-Zn) batteries, 35, *36*
Simotwo, S. K., 62
Singh, R., 172
Single-walled CNTs (SWNTs), 54
Slot-die roll coating, 270–271
Snaith, H. J., 236
SnSe$_2$ nanodisks, 445–446, *446*
SnSe nanosheets, 445–446, *446*

Söderström, T., 177
Sodium dodecyl sulfate (SDS), 388, 390
Sodium-ion batteries (SIBs), 345; configurations, 345–346, *347*; belt-shaped, 346, *347*; 2D flexible Na-ion battery mode, 346, *347*; 1D tube-type flexible Na-ion battery, 346, *347*; fiber-shaped, 346, *347*; electrode materials for, 350–353, *351*, *352*; anode, 350, *352*, 353; cathode, 350, *351*; electrolytes in, 346, *348*, 349; hybrid electrolyte, 346, 349; ion conductive polymer/ceramic polymer electrolyte, 349; ionogel electrolyte, 349; organic liquid electrolytes, 346; plastic-crystal electrolytes, 349; solid ceramic-polymer composites, 346; solid-state electrolytes, 346; separators for, 353; glass fiber paper, 353; nonwoven organic polymers, 353; polyolefin microporous separators, 353
Softness of flexible energy devices, 28, 28–29
Solar cells, flexible, 14, 45, 61, 121, 211, 228, 262, 308; active materials, types of, 121, *124*; architecture for, 45–50; classification based on energetic material, *309*; dye-sensitized, 14–16, *17*, 45–47; electrode materials, *123*; inorganic materials based, 49, 211; *See also* (Inorganic-based flexible solar cells); inorganic-organic semiconductor for, 121; metal oxides used in, 198; a-Si:H thin-film solar cells, 205; CuIGSe solar cells, 207; dye-sensitized solar cells, 198–200, *201*; gallium arsenide (GaAs) type solar cells, 207; organic solar cells, 201–203, *202*, **203**; perovskite solar cells, 203–205, *206*; modern photovoltaic (PV) cells, 14, 48–50; organic materials based, 50, *50*; perovskite-based, 16–17, *18*; power output of lightweight solar cells, **287**; structure with functional materials, *122*, 198; substrates for, *123*, 198
Sol-gel-based nonprinting method, 178
Solid-electrolyte interface (SEI), 324, 334, 361, 363–364; K-ions, 361–363
Solid oxide fuel cells (SOFC), 134, 142; *See also* Fuel cells
Solid polymer electrolytes (SPEs), 335, 378–379
Solid-state electrolytes (SSE), 60, 104, 110, 125, 334
Solution dip coating, 404–406, *405*
Solvent processing, 129
Solvothermal method, 485
Son, D. I., 93
Soni, E., 59–79
Sonkawade, R. G., 539–549
Sonker, R. K., 539–549
Sontheimer, T., 189
Spin coating, 265, 267–268
Spinel oxides as catalysts, 146
Spine-type configuration, ESDs, *37*, 38
Spray coating, 270–271
Spray painting, *405*, 406
Spring-type structures, ESDs, *37*, 38
Strain-capacitance relationship, 3
Stretchable energy storage devices, 29

Strong metal-support interaction (SMSI), 146
Strontium-doped lanthanum manganite oxide (LSM), 145
Sua, L. S., 307–320
Subramani, K., 485
Substrate material toxicity, 128
Successive ionic-layer adsorption and reaction (SILAR), 233, 294, 298, 485
Sugiawati, V. A., 324
p-Sulfonated poly(allylphenylether) (SPAPE), 324
Sundriyal, P., 545
Sun, G., 106
Sun, H., 46, 548
Sun, L., 216
Suntivich, J., 145
Sun, Z., 139
Supercapacitors (SC), flexible, 1–2, 35, *37*, 53–54, 60, 90, 124–125, 418–419, 442, 461, 481–483, 501–502, 540, 551–552; *See also* Textile-based supercapacitors (TSCs); advantages of, 418–419; all-solid-state systems, 455; architectures for, 53–56; based on metal oxides and sulfides, 3–5, *6*, 35; batteries and, differences between, 418; capacitance determination, equations for, 463–464; carbon-based materials for, 2–3, *4*, 87–89, *88*, 419–423, *421*; bio-based carbon, 423, **425**; carbon nanotubes, 423, *424*; graphene, 420, 422, *422*; synthesis of, 423, 425, *426*; characteristics of, 463, **463**; classification of, 520–521, *521*; components of, 521–522; and conventional supercapacitor, *482*; current collectors, 522; demand and growth in development, 419; electrical double-layer capacitors, 2, 35, 462, 520; *See also* (Electrical double-layer capacitors (EDLCs)); electrode materials, 419, *420*, 521; electrolytes in, 465, **465**, 521–522; energy storage mechanisms for, 425; electrochemical double-layer capacitors, 426–428, *427*; hybrid capacitors, 429; pseudo-capacitors, 428–429; flexible materials for, 54, 56; hybrid supercapacitors (HCs), 2–3, 462, 521; metal oxide based, 464–476; nanocomposite-based, 5, *7*, 551–570; pseudocapacitors (PCs), 2, 3, 35, 462, 521; separators for, 522; solid-state supercapacitors, 124, *124*; structure design and basic concept, 54, *55*; substrates for, 483–484; 2D materials for, 441–455
Swarnkar, A., 312

Taghavi, M., 135
Takasu, Y., 150
Tang, Q., 31
Tavakoli, M. M., 272, 274
Te-based 2D TMDs, metallic, 447
Tebyetekerwa, M., 397–413
Tellurium toxicity, 127–128
Teoleken, A. C., 218
Textile-based nanogenerators, 577; classification of, 580; core-shell structure, *580*; hybrid

nanogenerators, 580–582; piezoelectric nanogenerators, 577–580, *580*; pyroelectric nanogenerators, 580–582, *581*, *582*; triboelectric nanogenerators, 582–584, *585*

Textile-based supercapacitors (TSCs), 520, 522, 540; *See also* Fabric-based supercapacitors; Fiber-based supercapacitors; Textile fabrics; applications, 529–533; all-in-one textile system, 529, 531, *532*; fiber-shaped electrodes and asymmetric supercapacitor, 532–533, *534*; fiber-shaped photocapacitor, 531, *533*; design strategies for, 546–548; development and challenges in, 539–548; fabrication of, 544–545, *546*; fabric-based, 520, 524; fiber-based, 520, 523, *523*; coaxial fiber structure, 523–524; parallel fiber structure, 523; twisted fiber structure, 523; future perspective, 533, 535, 548–549

Textile fabrics; *See also* Textile-based supercapacitors (TSCs); classification of, 540–543, *541*, *542*; man-made fiber, 541; natural fibers, 541; animal fibers, 541, 543; mineral-based fibers, 543; vegetable fibers, 543; properties of, for supercapacitor applications, 543–544

Theerthagiri, J., 495

Thermal evaporation method, 268; dual-source, 268; single-source, 268

Thin-film solar cells (TFSCs), 231, 242; *See also* Solar cells, flexible

Thirugnanam, L., 327

3D graphene aerogels, 106, *107*

Three-dimensional structural FSCs, 564–566, *565*

3D printed device, 112–113

3D printed electrodes, 105; carbon-based electrodes, 106–108, *107*, *108*; others, 109–110; polymer-based electrodes, 108–109, *109*

3D printed electrolytes, 110–112

3D printing, 408; challenges in, 113; for 3D MSCs, 448; for flexible supercapacitors, 442–443; integration with multi-materials printing and interface, 113; precision and resolution of, 113; technologies (*See* 3D-printing technologies)

3D-printing technologies, 99–100; binder jetting, 100; direct energy deposition, *101*, 102; direct ink writing, 100; fused deposition modelling, 100; material jetting, 100

Titanium dioxide (TiO$_2$), 273

Titanium (Ti) foils, 274

Titanium oxide-based supercapacitors, 467, 470–471, *472*

Titanium oxide nanotube (TNT), 199

Tomanika, S., 137

Transfer printing technology (TPT), 176

Transition-metal chalcogenides (TMCs), 429; for flexible supercapacitors, 443–448

Transition-metal dichalcogenides (TMDs), 26; for flexible supercapacitors, 443–448

Transition metal oxide-based FSCs electrodes, 502; binary metal oxides: Ce-oxides materials, *505*, 505–506; co-oxides materials, 503, *504*; Fe-oxides materials, 503–505; MOFs-derived metal oxide/composite, *513*, 513–514; ternary or mixed-metal oxides, 506; MnCo$_2$O$_4$ materials, *508*, 508–510, *509*; Ni$_x$Co$_{3-x}$O$_4$ materials, 510–513, *511*, *512*; ZnCo$_2$O$_4$ materials, 506–508, *507*

Transition metal oxides (TMOs), 228; 5d-orbitals, *230*

Transmission electron microscopy (TEM), 65–68; lithiation process for Si anode, *67*; open solid cell and liquid cell setup, *67*; set-up for, *66*

Transparent conducting layer (TCL), 275

Transparent conducting oxides (TCOs), 255

Transparent supercapacitors, 474–476, *475*

Triboelectric nanogenerators (TENGs), 84, 582–584, *585*; in-plane scrolling mode, 583; materials for triboelectrification, 583; with silver-coated glass microspheres, *585*; single-electrode, 583; working mechanisms of, *584*

Tripathi, S. K., 227–238

Tsakalakos, L., 180

Twisting method, 409–411, *410*

Two-dimensional film-shaped FSCs, 562–564, *563*

2D materials, for flexible supercapacitors, 441–455; challenges and prospects, 455; 2D conjugated MOFs (2D c-MOFs), 449–450, *451*; 2D structured metals, 450–451, *452*; 2D TMCs AND TMDs, 443–448, *445*, *446*; graphene and carbonaceous analogs, 443, *444*; hybrid 2D nanostructures, 451–454, *453*, *454*; MXene, 448, *449*, *450*; planar micro-supercapacitors (MSCs), 454–455; and structure design of FSCs, 442, *442*

Ultrafast charge transfer, 291–292

Ultrathin glass (UTG) substrates, 248, 274; CdTe solar cells, 250; CIGS solar cells, 248–249; CZTS/CZTSSe solar cells, 251

Unary-metal chalcogenides (UMCs), 485–487, *486*, **488**

Vanadium disulfide (VS$_2$) ultrathin nanosheets, 444, *445*

Vanadium oxide thin films, 149

Vapor-phase deposition method, 187

Vapor-phase polymerization method, 5, *8*

Vertical-aligned carbon nanotube (VACNT), 487

Vitamin K, as anode material for potassium ion batteries, 368

Wafer-based photovoltaics, 308

Waikar, M. R., 539–549

Wang, C., 333

Wang, D., 382

Wang, H., 135

Wang, J., 556

Wang, K., 70

Wang, L., 545

Wang, P.-C., 271, 272

Wang, Y., 137, 150, 544

Wang, Z., 269, 531, 566

Wastewater treatment, 167

Water-in-salt, 335
Wearable batteries, 398, 401; fabrication techniques,
     401–408; fiber batteries parts, 398; integration into
     textile fabrics, 411; knitting, 411, *412*; weaving,
     411, *412*; research goals of, 398, *399*; substrate-
     enabled fabrication techniques, 401, *401*, 402;
     biscrolling, 406; chemical vapor deposition, 402,
     402; electrochemical deposition, *403*, 404;
     electrospinning, 404, *405*; electrospraying, 404;
     hydrothermal deposition technique, 402–404, *403*;
     solution dip coating, 404–406, *405*; spray painting,
     405, 406; substrateless fabrication techniques, 401,
     401, 406; 3D printing, 408; electrospinning, 406,
     407; melt-spinning technique, *407*, 408; wet-
     spinning technique, *407*, 408; unification of
     electrodes, 408–409, *409*; coaxial assembly, 411;
     twisting method, 409–411, *410*; winding
     technique, 409, *410*
Wearable electronic devices, FSCs for, 566–568
Weaving process, 411, *412*
Weber, D., 235
Wee, S. H., 233
Wei, H., 232
Wei, L., 110
Wen, X., 219, 252
Wet-spinning technique, *407*, 408
Willow glass, 274
Winding technique, 409, *410*
Wronski, C. R., 178
Wu, C., 271
Wu, H., 138

Xia, B. Y., 142
Xia, H., 333
Xiaobing, X., 180
Xiaolin, S., 181
Xie, Y., 444
X-ray diffraction (XRD), *68*, 68–69, *69*
Xu, D., 350
Xu, J., 544
Xu, M., 268

Yadav, G. G., 390
Yakobson, B. I., 25
Yan, C., 397–413
Yang, F., 278
Yang, S., 397–413
Yang, W. S., 312
Yang, X., 551–570
Yang, Z., 269
Yao, B., 106, 107
Yao, Y., 350
Yarn electrospinning, 406, *407*
Yasin, G., 501–514
Yoon, S., 147
Yttria-stabilized zirconia (YSZ) electrolyte, 110
Yu, G., 434
Yu, H., 330

Yu, H. C., 145
Yuk, H., 108
Yuksel, R., 548
Yun, J., 568
Yu, X., 216, 349

Zang, J., 553
Zang, X., 286, 287
Zekoll, S., 110
Zeng, J., 556
Zeng, Y., 381
Zhai, T., 235
Zhang, B., 135, 165
Zhang, C., 178, 292
Zhang, F., 503
Zhang, H., 546
Zhang, L. C., 92
Zhang, Q., 560
Zhang, S., 181
Zhang, X., 297
Zhang, Y., 233
Zhao, B., 143
Zhao, C., 164
Zhao, X., 523
Zhao, Y., 338
Zheng, J., 381, 490
Zhi, C., 381, 382, 388
Zhi, M., 434
Zhong, Q. Z., 32
Zhou, G., 495
Zhou, H., 312
Zhou, Q., 544
Zhou, R., 143
Zhou, Y., 335
Zhou, Z., 564
ZIF-67, 513
Zinc-air battery, 12–13, *13*; kirigami design-inspired, 13
Zinc cobaltate ($ZnCo_2O_4$) materials, 506–508, *507*
Zinc-ion batteries (ZIBs), 376, *376*; cathode materials
     for, 377; challenges and perspectives:
     multifunctionalities, 392; structural
     enhancement, *391*, 391–392; voltage issue,
     388–391, *389*; charge storage mechanisms, 377;
     coaxial-fiber ZIB, 53; development of, 376;
     flexible device constructions, 385–388; cable-
     type structure, 385, *387*, 388; sandwich
     structure, 385, *387*; flexible materials for, 53;
     functionalities, 381–385; mechanisms for,
     377–378; polymer electrolytes, 378–379; PAM
     and derivatives, 381; PEO and derivatives, 379;
     PVA and derivatives, 379–381; as rocking chair
     batteries, 377; structure design and basic
     concept, 52, *52*
Zinc-manganese oxide ($Zn-MnO_2$) batteries, 35, *36*
Zinc oxide (ZnO), 273, 467; nanowires, 579; for
     supercapacitors applications, 467, *470*
Zn-ion supercapacitors (ZIC), 450–451, *452*
Zn nanosheets, 450